D1568606

S E M S M E S

EMS Textbooks in Mathematics

EMS Textbooks in Mathematics is a series of books aimed at students or professional mathematicians seeking an introduction into a particular field. The individual volumes are intended not only to provide relevant techniques, results, and applications, but also to afford insight into the motivations and ideas behind the theory. Suitably designed exercises help to master the subject and prepare the reader for the study of more advanced and specialized literature.

Jørn Justesen and Tom Høholdt, *A Course In Error-Correcting Codes*
Markus Stroppel, *Locally Compact Groups*
Peter Kunkel and Volker Mehrmann, *Differential-Algebraic Equations*
Dorothee D. Haroske and Hans Triebel, *Distributions, Sobolev Spaces, Elliptic Equations*
Thomas Timmermann, *An Invitation to Quantum Groups and Duality*
Oleg Bogopolski, *Introduction to Group Theory*
Marek Jarnicki and Peter Pflug, *First Steps in Several Complex Variables: Reinhardt Domains*
Tammo tom Dieck, *Algebraic Topology*
Mauro C. Beltrametti et al., *Lectures on Curves, Surfaces and Projective Varieties*
Wolfgang Woess, *Denumerable Markov Chains*
Eduard Zehnder, *Lectures on Dynamical Systems. Hamiltonian Vector Fields and Symplectic Capacities*
Andrzej Skowroński and Kunio Yamagata, *Frobenius Algebras I. Basic Representation Theory*
Piotr W. Nowak and Guoliang Yu, *Large Scale Geometry*
Joaquim Bruna and Juliá Cufí, *Complex Analysis*
Fabrice Baudoin, *Diffusion Processes and Stochastic Calculus*

Eduardo Casas-Alvero

Analytic Projective Geometry

European Mathematical Society

Author:

Eduardo Casas-Alvero
Departament d'Àlgebra i Geometria
Universitat de Barcelona
Gran Via 585
08007 Barcelona
Spain

E-mail: casasalvero@ub.edu

2010 Mathematics Subject Classification: 51-01, 51N15; 51N10, 51N20

Key words: Projective geometry, affine geometry, Euclidean geometry, linear varieties, cross ratio, projectivities, quadrics, pencils of quadrics, correlations

ISBN 978-3-03719-138-5

The Swiss National Library lists this publication in The Swiss Book, the Swiss national bibliography, and the detailed bibliographic data are available on the Internet at http://www.helveticat.ch.

Contact address:

European Mathematical Society Publishing House
Seminar for Applied Mathematics
ETH-Zentrum SEW A27
CH-8092 Zürich
Switzerland

Phone: +41 (0)44 632 34 36
Email: info@ems-ph.org
Homepage: www.ems-ph.org

Typeset using the author's TEX files: I. Zimmermann, Freiburg
Printing and binding: Beltz Bad Langensalza GmbH, Bad Langensalza, Germany
∞ Printed on acid free paper
9 8 7 6 5 4 3 2 1

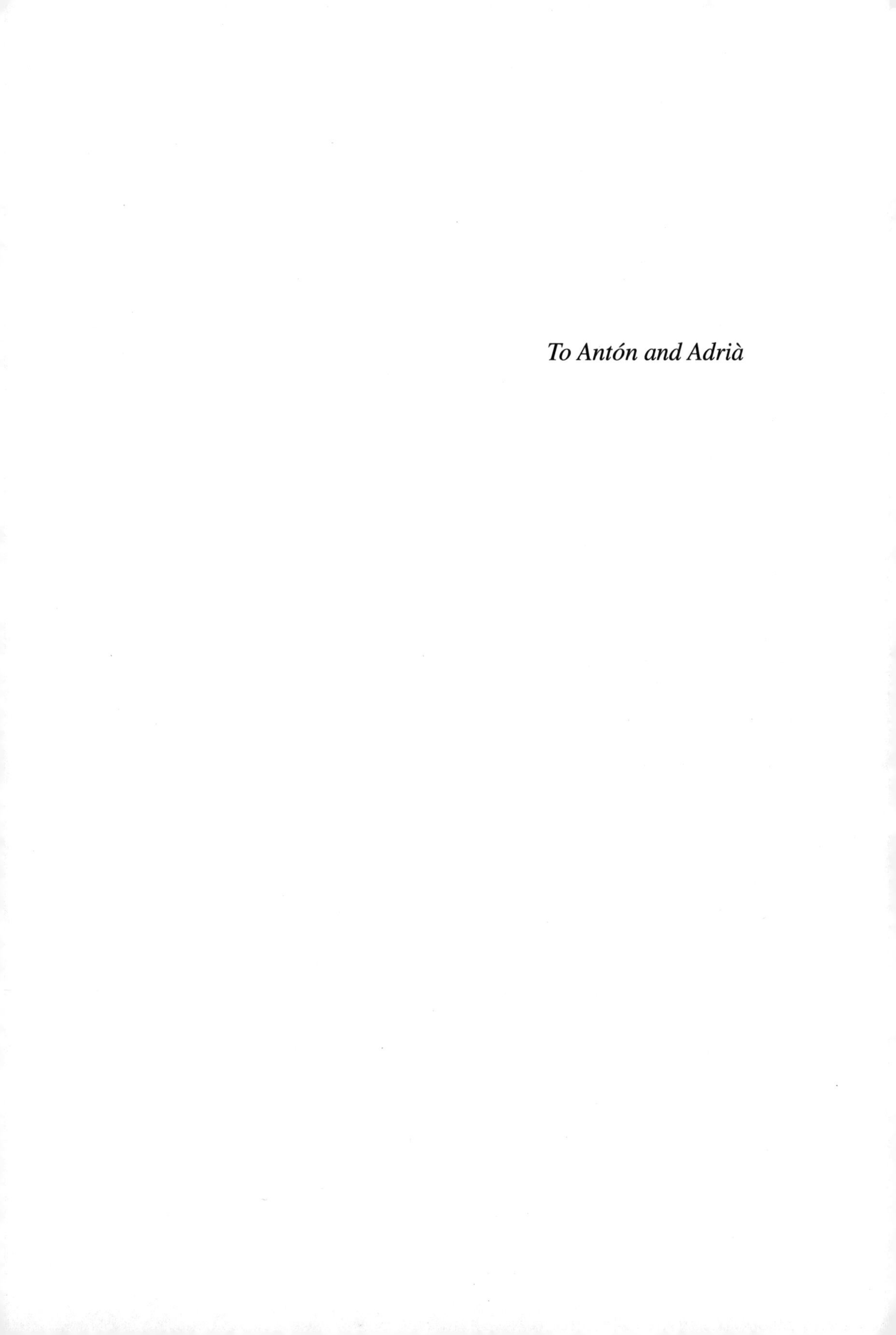

To Antón and Adrià

Contents

Introduction

Two of the most influential advances in the history of geometry occurred in France in the short period 1636–1639. One was the introduction and systematic use of coordinates conceived, independently, by R. Descartes (*La Géometrie*, 1637) and P. Fermat (*Ad locos planos and solidos isagoge*, around 1636). Coordinates allowed the use of resources from algebra and analysis in the foundations and development of geometry, which is usually called *analytic geometry*. Analytic geometry is thus not a part of geometry, but rather a method that applies to all parts of geometry; probably calling it *analytic method in geometry* would have been better. The other was the first study of the properties of figures invariant by perspectivities, due to G. Desargues (*Brouillon project d'une atteinte aux évenements des rencontres d'un cone avec un plan*, 1639). A perspectivity between two different planes of ordinary three-space consists in fixing a point O outside the planes and taking two points, one on each plane, as correspondent if and only if they are collinear with O. Perspectivities were already used in the studies of perspective originated by Renaissance painting and in the construction of sundials (Desargues himself wrote on both subjects). Also known to Desargues was the work of Apollonius of Perga (262–190 BC), who implicitly used perspectivities to relate arbitrary sections of a circular cone to one of its circular sections. The properties invariant by perspectivities are called *projective*; they are satisfied by large classes of figures. The study of projective properties, and of the classes of figures satisfying them, is the subject of *projective geometry*, which thus leaves aside notions such as parallelism and distance, which are not invariant by perspectivities.

Analytic and projective geometry provided two quite different ways to overcome the lack of generality inherent in the methods of the ancient Greek geometers. They both boosted geometry to a long period of continued progress in which the whole field of geometry was widely enlarged and far better understood. Analytic geometry had an immediate success: since the publication of *La Géometrie* till the end of the eighteenth century, there was a constant flourishment almost entirely based in the use of coordinates. Meanwhile, shadowed by analytic geometry, projective geometry made little progress. It was not until the beginning of the nineteenth century that interest in properties invariant by perspectivities was renewed: then the pioneering work of Desargues and his few followers was largely extended to subjects such as duality, cross-ratio and polarity. Cartesian coordinates being not well suited to it, in a first stage projective geometry was developed without using coordinates, which was called *pure* or *synthetic* (method in) geometry. After a while, the introduction of homogeneous coordinates allowed the fruitful application of the analytic method to projective geometry. All together, projective geometry emerged in its own right as an important – and very nice – part of geometry. However, the most important fact was that while projective geometry was growing, it was gradually realized that each of its notions and theorems had as specializations a number of already known non-projective notions or theorems; for instance, both the ratio of distances between three aligned points (*affine ratio*) and the angle between two lines could be expressed in terms of a more general projective invariant called *cross ratio*. Eventually it became clear that this situation was quite general: all the geometry known at the time could be seen as a specialization of projective geometry. This allowed a new presentation in which projective geometry was taken as the fundamentals of geometry, and provided a better and deeper understanding of the entire geometrical field, by unveiling the projective common roots of apparently unrelated parts of geometry.

During the second half of the nineteenth century, projective geometry gradually merged with the theory of algebraic functions giving rise to a specialized branch of geometry called *algebraic geometry*. An important part of it, named *projective algebraic geometry*, is devoted to the study of *projective algebraic varieties*, which are figures of projective spaces defined by polynomial equations. Very active at the research level today, projective algebraic geometry may be understood as the natural continuation of projective geometry.

The development of computer vision in recent years has brought a renewed interest in projective geometry and, especially, in its metric applications. Computer vision starts from the same objects that were at the basis of Desargues's work: the perspectivities, renamed *pinhole cameras*. Its basic goal is the analysis and reconstruction of three-dimensional scenes from perspective images of them. So it is not surprising to find many notions and results of classical projective geometry playing today a central role in computer vision.

This book is devoted to giving an analytic presentation, strongly based on linear algebra, of what may be considered to be the whole of n-dimensional projective geometry over the real and complex fields, together with their affine and metric specializations. When relevant, the specifics of the low-dimensional cases $n = 1, 2, 3$ are also dealt with in detail. According to the usual conventions, but for a few exceptions, we will limit ourselves to considering linear and quadratic geometric objects, the study of the higher degree ones belonging rather to algebraic geometry. The core of the projective part is the study of linear varieties, cross ratio, projective transformations and quadric hypersurfaces, including the projective classification of the latter. Special attention is paid to the projective structures on sets of certain geometric objects, such as hyperplanes or quadrics: considering these structures multiplies the applications of the abstract projective theorems and makes one of the main differences between old and modern geometry. In real projective geometry, imaginary points need to be considered as soon as non-linear equations do appear; they are introduced by a formal construction of the complex extension of a real projective space. The basic objects allowing the application of projective geometry to the affine and metric geometries – projective closure of an affine space, improper hyperplane, absolute quadric – are introduced and used to reformulate the basic elements of the affine and metric geometries in projective terms. In particular the affine and metric classifications of quadric hypersurfaces are presented as successive specializations of their projective classification. Since, besides their intrinsic interest, the affine and metric applications are very good illustrations of the projective results they come from, they are presented as soon as there is enough projective material to support them. The more technical projective classifications of collineations, pencils of quadrics and correlations, together with the algebraic background they require, are the contents of the last chapter. Two less usual applications are presented in two appendices: one goes back to the origins of projective geometry by showing the projective foundations of the practical rules of perspective; the other explains a parametric form of Klein's model of Euclidean and non-Euclidean plane geometries. A number of exercises are proposed at the end of each chapter; some of them are nice classical results, not central enough to have a place in the text. Results proved in the exercises are used in other exercises, but not in the text.

As already said, both the foundations and the general development of this book are analytic. When suitable, short synthetic arguments have been used at some points. The more elementary parts of projective geometry may be given purely synthetic presentations of which there are nice examples in the literature. These presentations proceed by direct and very clever proofs, and are the best for discussing foundations, as they start from a short number of axioms. However, it is quite unrealistic – and certainly non-practical – to try to cover the whole contents of this book by the exclusive use of the synthetic method: the more advanced parts of projective geometry cannot be developed without algebraic support and many applications need to use coordinates; this is notably the case for all the applications to computer vision, which run by numerical computation. Further, and more decisive, it makes no sense to hide the powerful underlying algebraic structures that are the basis of the analytical treatment, as they are so important as points and lines. The fecund relationship between algebraic and geometric structures alone boosted the continued progress which during the last one-hundred and fifty years and going far beyond the scope of this book, caused algebra, arithmetic and algebraic geometry to appear today as simply different views of a unique, rich and deep, field of knowledge.

Hopefully this book will be useful to undergraduate students taking a course on projective geometry, to graduate students and researchers working in fields that make use of projective geometry – such as algebraic geometry or computer vision – and also to anyone wishing to gain an advanced view on the whole of the geometrical field. All of them will find here a fairly complete account of what has been classically considered to be projective geometry and applications, written in a modern language and accordingly to today's standards of rigour. There are included a number of topics for which there is a lack of suitable references. Many of them, such as singular projectivities, correlations, Plücker coordinates, line-complexes and twisted cubics, are of use in computer vision.

Obviously, the contents of this book largely exceeds what is reasonable to teach in a single course, and many different courses may be given using parts of it. I have given introductory courses based on Chapters 1 and 2, selected parts of Chapters 3 to 6 and the projective and affine classifications of quadrics from Chapter 7. I have also given more advanced courses including most of Chapter 8, part of Chapter 9, the first four sections of Chapter 10 and one or both of the Appendices A, B.

Requirements for reading this book are the contents of a standard course on linear algebra, including vector spaces, linear maps, matrices, diagonalization, linear forms, dual space and bilinear forms, as well as some knowledge of the usual vectorial presentations of the linear affine and metric geometries, their basics being, however, quickly recalled in the text. Although not strictly needed, some background on the classic Euclidean presentation of the elementary geometry of lines, polygons, circles, etc., is advisable for a better understanding of the projective presentations of some of these subjects shown here.

I took my first – and very inspiring – course on projective geometry from Professors J. B. Sancho Guimerá and J. M. Ortega Aramburu, back in 1967: most of the spirit of that course is still present in this book. Other important sources have been classical Italian books, especially E. Bertini's *Introduzione alla Geometria Proiettiva degli Iperspazi* ([2]) and G. Castellnuovo's *Lezioni di Geometria Analitica* ([4]), the excellent *Algebraic Projective Geometry* by J. P. Semple and G. T. Kneebone ([25]) and, finally, O. Schreier and E. Sperner's *Projective Geometry of n dimensions* ([24]), which systematically deals with the n-dimensional case and makes intensive use of linear algebra.

After teaching courses on projective geometry for about thirty-five years, I am very grateful to all colleagues – too many to be named – who either taught parallel courses or collaborated in my own courses: I learned from all of them and hence they all have positively contributed to this book.

I am also very grateful to M. Casanellas, P. Cassou-Nogues, I. Giné, J. Ma. Giral, J. C. Naranjo, J. Roé, G. Solanes and G. Welters, who read parts of the manuscript and made valuable suggestions, and to M. Karbe, I. Zimmermann and the staff of the EMS Publishing House for their most helpful and efficient editorial work. The mistakes, of course, remain exclusively mine.

Barcelona, March 2014 Eduardo Casas-Alvero

General conventions

As said in the Introduction, this book is devoted to real and complex geometry; throughout it, k will denote a field (the *base field*) assumed to be either the real or the complex one. Unless otherwise said, k will remain fixed and all projective spaces, affine spaces and vector spaces will be assumed to be over k. The elements of the base field k will be called *scalars*, and sometimes also *constants*.

It is worth noting though, that with the only exceptions of Corollaries 2.4.3 and 2.4.4 (which require an infinite base field), the contents of Chapters 1 to 4 hold without changes if k is assumed to be a field with characteristic different from 2 (that is, with $1 + 1 \neq 0$) and other than the field $\mathbb{F}_3$ with three elements. Thus, for the first four chapters and with the above quoted exceptions, the reader may adopt this more general hypothesis as well.

As usual, $\mathbb{R}$ and $\mathbb{C}$ will denote the fields of the real and complex numbers, respectively, and $\mathbb{R}^+$ the set of all positive real numbers. The imaginary unit is denoted by $\boldsymbol{i}$, as the ordinary i is often used for other purposes. The identity map on a set X is written Id_X.

Usually, the entry in row i and column j of a matrix will be written in the form a^i_j, and also $a_{i,j}$ if the matrix is symmetric. The unit n-dimensional matrix will be denoted by $\mathbf{1}_n$, or just by $\mathbf{1}$ if no reference to the dimension is needed.

If A is any set, a subset $\{a, a'\} \subset A$ will be sometimes referred to as the *unordered pair* or just the *pair* composed by a, a'. This includes the case $a = a'$, in which we still consider $\{a\}$ as a pair: it will be called a *pair of coincident* or *repeated* elements of A, and often written $\{a, a\}$.

We will refer to the highest degree monomial (resp. coefficient) of a non-zero polynomial in one variable as its *leading monomial* (resp. *leading coefficient*). The polynomials which are non-zero and have leading coefficient equal to one are called *monic*. Greatest common divisors, minimal common multiples and irreducible factors of polynomials are always assumed to be monic. The polynomial 0 will be taken as a homogeneous polynomial of degree m for any non-negative integer m.

Unless otherwise stated, the roots of polynomials will be counted according to their multiplicities, that is, a root a of a polynomial $P(X)$ will be counted as many times as the number of factors $X - a$ appearing in the decomposition of $P(X)$ in irreducible factors.

Trigonometric functions will be taken as defined in function theory, for instance $\cos x = (e^{\boldsymbol{i}x} + e^{-\boldsymbol{i}x})/2$, and not by their relations to the elements of a right triangle, as the latter may or may not make sense or hold depending on the geometric context.

Many results of plane projective geometry give rise to graphic constructions which may be performed using a straight edge and, sometimes, a compass. We will make occasional references to some of these constructions, as they illustrate very well the related theory, but we do not intend to present them systematically. When dealing with graphic constructions the base field will be implicitly assumed to be the real one.

Chapter 1
Projective spaces and linear varieties

1.1 How projective spaces arise

Projective geometry is usually developed in spaces of a special type, called *projective spaces*, that are different from the usual affine or Euclidean spaces. In this section we give just an heuristic explanation of the reasons that lead to the abstract definition of projective space, as presented at the beginning of next Section 1.2. We will deal with the plane case only.

Early development of projective geometry took place at a time in which all geometry was done in the single three-dimensional Euclidean space intended to be our ambient physical space, and mostly on its planes. Then, as said in the introduction, projective geometry consisted in studying the notions and properties invariant by perspectivities between planes, making no distinction between a plane figure and any of its images by perspectivities. Later on, once other spaces were considered, perspectivities between them no longer made sense and more general transformations, called *projectivities*, were considered instead, but this is another story. The reader may see forthcoming Sections 1.5, 1.10 for the definition and role of projectivities. Perspectivities are introduced in Section 1.9 and related to projectivities in Section 5.9.

Let us first pay attention to the way in which perspectivities behave. Assume we have fixed two different and non-parallel planes α, α' of a Euclidean three-dimensional space, and a point O lying not on α or α'. According to its definition, the perspectivity with centre O, between α and α', takes points $p \in \alpha$ and $p' \in \alpha'$ as corresponding if and only if they are aligned with O, see Figure 1.1. Let s be

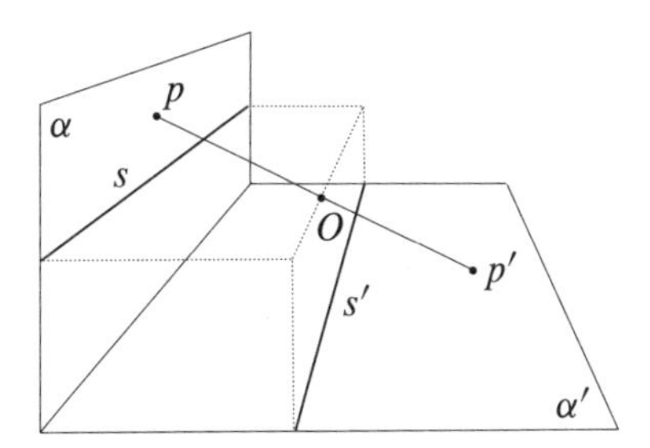

Figure 1.1. Perspectivity with centre O, between planes α and α': p and p' are corresponding points. The points on s have no corresponding point on α' and the points on s' have no corresponding point on α.

the line intersection of α and the plane parallel to α' through O, and s' the line intersection of α' and the plane parallel to α through O. If a point p on α does not belong to s, then the line Op is not parallel to α' and therefore intersects α' at a well-determined point p' which is the point corresponding to p. By contrast, if $p \in s$, then the line Op is parallel to α', has no intersection with it and therefore p has no corresponding point on α'. Similarly, all points on α' have a well-determined corresponding point on α except those lying on s', which have not.

Example 1.1.1 (See Figure 1.2). Fix a line $r \neq s$ on α and assume that it intersects s at a point a. While a point p varies on r, the line Op lies on the plane Or and therefore the point p' corresponding to p varies on the line $r' = (Or) \cap \alpha'$, which is taken as the line corresponding to r by the perspectivity. The reader may easily check that r' is parallel to Oa. If, similarly, one takes t' to be the line corresponding to a second line $t \neq r$, still on α and going through a, then also t' is parallel to Oa and so r', t' are a pair of distinct parallel (and hence disjoint) lines of α'. We see thus that a perspectivity may transform a pair of concurrent lines into a pair of parallel lines.

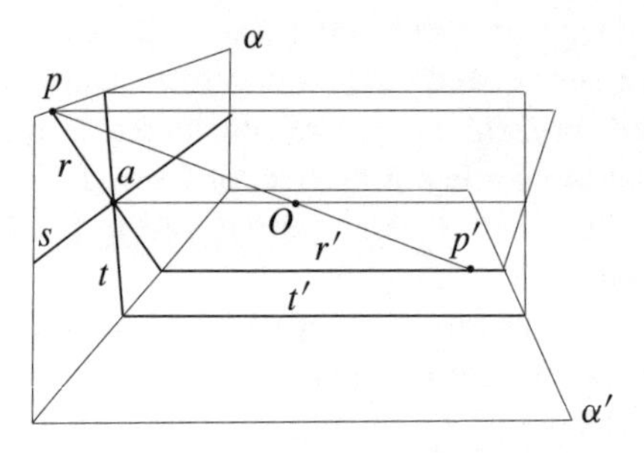

Figure 1.2. The perspectivity with centre O transforms the concurrent lines r, t into the parallel lines r', t'.

Example 1.1.2 (See Figure 1.3). Assume we have a circle C on the plane α intersecting s in two points x, y. While two points p, q describe each one of the open arcs the points x, y determine on C, their corresponding points p', q' describe two disjoint curves on α which actually are the two branches of the hyperbola intersection of α' and the cone projecting C from O.

The above examples show connected figures – a pair of concurrent lines and a circle – that are turned into unconnected ones by the action of a perspectivity, while from a projective viewpoint there should be no difference between a figure and its transform by a perspectivity. The reason for this apparent paradox is the existence of points – a, x, y in the above examples – that have no corresponding point by the

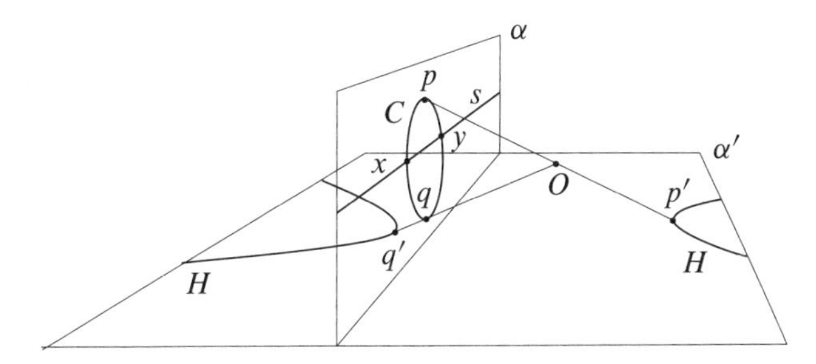

Figure 1.3. While p and q describe the arcs in which x, y divide C, their corresponding points p', q' describe the two branches of the hyperbola H.

perspectivity, or, in other words, the fact that not all points we are doing projective geometry with are shown on all planes, each plane offering only a partial view of the whole of points we have to consider. This situation was well understood by the early projective geometers, who used to move from one to another plane through perspectivities in order to get better views of the figures they have in hand. For instance, they would have made no distiction between the pairs of lines r, t and r', t', but would have taken them as projectively identical pairs of concurrent lines whose common point is shown on α but not on α'. Similarly, they would have taken C and H as just two views of the same projective curve, all of whose points are shown on α, but not on α'.

Jumping from plane to plane is not the best way of doing any kind of geometry, and so it is convenient to enlarge the planes in order to have all points we need lying on the same plane. One way to do this is to formally add new points to each plane, a new point for each direction parallel to the plane. These new points are called *improper points* or *points at infinity*, the old points being then called *proper points*. The improper point associated to a given direction is taken as belonging to all lines parallel to that direction and so, for instance, in the situation of Example 1.1.1, the point associated to the direction of the line aO is taken as the point corresponding to a by the perspectivity, and also as a point belonging to both the lines r' and t', which in such a way gain a common point. Addition of improper points to an affine space, in order to enlarge it to a projective space, will be explained in detail in Section 3.2.

The addition of improper points provides a sort of enlarged plane containing all points we want. Nevertheless, such an enlarged plane is not yet a satisfactory site to do projective geometry on. Indeed, the improper points being just formal objects, handling them requires in practice to move to another plane through a perspectivity in order to turn them into proper points, and so some wandering from plane to plane is still needed. Incidentally, note that distinguishing between proper and improper points has no projective sense, as improper points are turned into proper points by perspectivities, and conversely.

Let us continue fixing our attention on the pair of planes α, α' and the centre of perspectivity O. Actually, all points we want may be seen at once as the elements of a single set, placed midway between α and α', which is much better for developing projective geometry on: it is the set of all lines of the space which pass through the point O, denoted in the sequel by $\mathbb{P}$. Admittedly, the elements of $\mathbb{P}$ are lines, not points, but calling the elements of a given set points is just a convention. Therefore we will agree on calling the set of all lines through O a projective plane, and its elements – up to now called lines – points.

From a purely set-theoretic viewpoint, we have injective maps

$$\begin{aligned} \varphi\colon \alpha &\longrightarrow \mathbb{P}, \\ p &\longmapsto Op, \end{aligned}$$

and

$$\begin{aligned} \varphi'\colon \alpha' &\longrightarrow \mathbb{P}, \\ p' &\longmapsto Op, \end{aligned}$$

so that points $p \in \alpha$ and $p' \in \alpha'$ are mapped to the same point of $\mathbb{P}$ if and only if they are already corresponding by the perspectivity with centre O. Thus, after identifying the points of α and α' to their images by φ and φ', α and α' appear patched together in $\mathbb{P}$ in such a way that points $p \in \alpha$ and $p' \in \alpha'$ become identified if and only if they correspond by the perspectivity with centre O. Truth to tell, the line through O parallel to $\alpha \cap \alpha'$ is not covered by the images of α or α', but this is not a problem. Indeed, any plane missing O may be identified with a subset of $\mathbb{P}$ by a similar map, in such a way that still points of different planes become identified with the same point of $\mathbb{P}$ if and only if they correspond by the perspectivity with centre O. Then $\mathbb{P}$ is covered by the images of all these planes. In fact, to cover the whole of $\mathbb{P}$ it is enough to add to the images of α and α' above, the image of a third plane α'', missing O and non-parallel to $\alpha \cap \alpha'$.

Furthermore, $\mathbb{P}$ is far from being just an unstructured set, as we have, closely related to it, the vector space E of all space-vectors with origin at O: each non-zero vector $v \in E$ determines a point of $\mathbb{P}$, namely the line through O in the direction of v. This defines an exhaustive map $\pi: E - \{0\} \to \mathbb{P}$ through which two vectors have the same image if and only if they are proportional. The fact is that all the geometric structure that $\mathbb{P}$ could inherit from the different planes covering it may be obtained in a much easier way from the algebraic structure of the vector space E through the map π, as we will see in forthcoming sections.

To close, discarding all the accessory objects that appeared in the above discussion to keep just its final output, one has to retain:

- the set $\mathbb{P}$, taken as the sought-after projective plane, its elements being called points,

– the three-dimensional vector space E, which provides the algebraic background, and

– the map $\pi : E - \{0\} \to \mathbb{P}$, relating them.

The definition of n-dimensional projective space at the beginning of the next section is modeled on the above triple.

1.2 Projective spaces

From now on, unless otherwise said, k is either the real or the complex field and we refer to it as the *base field*. A *projective space* over the field k is a triple $(\mathbb{P}, E, \pi)$ where $\mathbb{P}$ is a set, E is a finite-dimensional vector space over k, $\dim E \geq 2$, and π is a map,

$$\pi : E - \{0\} \longrightarrow \mathbb{P}$$

satisfying:

(a) π is onto, and

(b) for any $v, w \in E - \{0\}$, $\pi(v) = \pi(w)$ if and only if there is a (necessarily non-zero) $\lambda \in k$ such that $v = \lambda w$.

Equivalently, in the above situation it is said that E and π (or just π) define a structure of projective space on $\mathbb{P}$. The elements of $\mathbb{P}$ are called the *points*, and in some cases also the *elements*, of the projective space. The set $\mathbb{P}$ is assumed to be the place where the figures lie and the geometrical constructions take place, while E and π provide an external structure allowing us to build up the geometry in the otherwise structureless set $\mathbb{P}$. E is called the *vector space* and π the *structural map of* (or *associated to*) the projective space.

A projective space over the field k is also called a k-projective space. Projective spaces over the fields $\mathbb{R}$ and $\mathbb{C}$, of the real and complex numbers, are called real and complex projective spaces, respectively.

Unless some confusion may arise, the same notation is used for the projective space and its set of points, thus denoting the above projective space $(\mathbb{P}, E, \pi)$ just $\mathbb{P}$. The reader may note that a similar practice is common in algebra when dealing with groups, rings, etc.

If $p = \pi(v)$, we will say that the point p is *represented* by the vector v and also that v is a *representative* of p. Usually, if no explicit reference to π is needed, we will write $\pi(v) = [v]$. The definition of projective space assures that every non-zero vector represents a point, that every point has a non-zero representative, and that two non-zero vectors represent the same point if and only if they are proportional. It is worth noting that the vector 0 represents no point and therefore $[v]$ has no meaning if $v = 0$.

The *dimension* of a projective space is defined as being one less than the dimension of its vector space. Notations being as above, by definition $\dim \mathbb{P} = \dim E - 1 \geq 1$. *Projective line*, *projective plane* and *projective three-space* mean projective space of dimension one, two and three, respectively. We will usually write the dimension of a projective space as a subscript, thus writing $\mathbb{P}_n$ to denote an arbitrary projective space of dimension n. If an explicit reference to the base field k is needed, we will write $\mathbb{P}_{n,k}$.

Next we give two examples of projective space, the first one is a projective structure on a set already familiar to the reader.

Example 1.2.1. Assume that $\mathbb{A}_2$ is an affine plane over a field k, and we have fixed affine coordinates x, y on $\mathbb{A}_2$. Let $\mathbb{P}$ be the set of all lines through the origin of $\mathbb{A}_2$. Take k^2 as vector space and π to map the couple (a, b) to the line with equation $ax + by = 0$. Then $(\mathbb{P}, k^2, \pi)$ is a one-dimensional projective space over k.

Example 1.2.2. Let E be a finite-dimensional vector space over a field k. Take $\mathbb{P}$ to be the set of all one-dimensional linear subspaces of E and define the map π by the rule $v \mapsto \langle v \rangle$, $v \neq 0$, $\langle\ \rangle$ meaning *subspace generated by*. Then $(\mathbb{P}, E, \pi)$ is a projective space over k, usually denoted by $\mathbb{P}(E)$ and called the *projectivization* of E. In particular $\mathbb{P}(k^{n+1})$ guarantees the existence of a projective space over k of dimension n, for each $n > 0$.

Assume that we have two projective structures on the same set $\mathbb{P}$, the first one defined by E, π as above, and the second one by a vector space E' and a map π'. We say that these projective structures are *equivalent* if and only if there is an isomorphism of vector spaces $\Phi \colon E \to E'$ so that $\pi = \pi' \circ \Phi$, that is, making the diagram

$$\begin{array}{ccc} E - \{0\} & \xrightarrow{\Phi} & E' - \{0\} \\ {\scriptstyle \pi}\downarrow & & \downarrow{\scriptstyle \pi'} \\ \mathbb{P} & = & \mathbb{P} \end{array}$$

commutative. We will leave it to the reader to check that every definition we shall make from now on remains unchanged if one substitutes equivalent projective structures for the original ones on the involved projective spaces. So, to all effects concerning the geometry we will develop in projective spaces, the choice between equivalent projective structures will be irrelevant. For instance, the reader may note that the already defined dimension of a projective space does not change if one takes an equivalent projective structure instead of the original one, just because isomorphic vector spaces have the same dimension.

1.3 Linear varieties

Let $\mathbb{P}_n$ be a projective space of dimension n, with associated vector space E and structural map π. A subset L of $\mathbb{P}_n$ is called a *linear variety* (or *projective subspace*) of $\mathbb{P}_n$ if and only if $L = \pi(F - \{0\})$ for a linear subspace F of E. In this situation, we will say that the subspace F *defines* or *represents* L, and write $L = [F]$. Linear varieties will be called *projective linear varieties* if some confusion with the affine linear varieties (see the forthcoming Section 3.4) may arise.

Clearly, a linear variety $L = [F]$ is determined by its defining subspace F. The next lemma shows how the variety L determines in turn F.

Lemma 1.3.1. *If a linear variety L is defined by a subspace F, then $F - \{0\} = \pi^{-1}(L)$.*

Proof. The inclusion $F - \{0\} \subset \pi^{-1}(L)$ is clear from the definition of linear variety. Conversely, if $v \neq 0$ has $\pi(v) \in L = \pi(F - \{0\})$, then there is $w \in F - \{0\}$ such that $\pi(w) = \pi(v)$. By the second condition in the definition of projective space, it is $v = \lambda w$ for a certain $\lambda \in k$ and therefore $v \in F$ as wanted. □

Remark 1.3.2. It follows in particular from Lemma 1.3.1 that all representatives of any point $p \in L = [F]$ belong to F.

The subspace F being determined by the linear variety L, the *dimension* of the linear variety $L = [F]$ is defined as $\dim L = \dim F - 1$. The *codimension* of a linear variety L of $\mathbb{P}_n$ is taken to be $n - \dim L$.

The empty set $\emptyset = [\{0\}]$ is thus a linear variety of dimension -1, while the whole $\mathbb{P}_n = [E]$ is a linear variety of dimension n. It is also clear that the subsets that consist of a single point $p = [v]$ are linear varieties of dimension zero, as $\{p\} = [\langle v\rangle]$. Conversely, any linear variety of dimension zero has a single point just because any non-zero vectors in a one-dimensional subspace are proportional. Although they are different objects, it is usual to denote a point p and the linear variety $\{p\}$, whose only point is p, by the same letter p, and refer to the zero-dimensional linear varieties just as *points*, rather than as *single-point sets*. We will follow this convention unless some confusion may result. Linear varieties of a projective space $\mathbb{P}_n$ of dimensions 1, 2, and $n-1$ are called *lines*, *planes* and *hyperplanes* of $\mathbb{P}_n$, respectively.

A direct consequence of the definition of linear variety and 1.3.1 above is:

Corollary 1.3.3. *The map*

$$\begin{aligned} \mathcal{S}_{d+1}(E) &\to \mathcal{LV}_d(\mathbb{P}_n),\\ F &\mapsto [F], \end{aligned}$$

between the set $\mathcal{S}_{d+1}(E)$, of the linear subspaces of E of dimension $d+1$, and the set $\mathcal{LV}_d(\mathbb{P}_n)$, of the linear varieties of $\mathbb{P}_n$ of dimension d, is bijective.

Remark 1.3.4. If $L = [F]$ is a linear variety of dimension $d > 0$, then it is easy to check that the restricted map

$$\pi_{|F-\{0\}}\colon F - \{0\} \longrightarrow L$$

defines on L a structure of projective space of dimension d, usually named the structure *induced* or *subordinated* on L by its ambient space $\mathbb{P}_n$. It is worth noting that the representatives of a point $p \in L$ are the same if p is taken as a point of L, with the induced structure, or as a point of $\mathbb{P}_n$. In the sequel, when needed, linear varieties of dimension not less than one will be considered as projective spaces with the induced structure without further mention. In particular, the double use of the words *line* and *plane* to name both the projective spaces and the linear varieties of dimensions 1 and 2 causes no confusion.

One may, of course, patch together the bijections of Corollary 1.3.3 to get a bijection

$$\begin{aligned} \mathcal{S}(E) &\to \mathcal{LV}(\mathbb{P}_n), \\ F &\mapsto [F], \end{aligned}$$

between the set $\mathcal{S}(E)$, of the linear subspaces of E and the set $\mathcal{LV}(\mathbb{P}_n)$, of the linear varieties of $\mathbb{P}_n$. The next lemma shows that this bijection is an isomorphism of ordered sets if both $\mathcal{S}(E)$ and $\mathcal{LV}(\mathbb{P}_n)$ are ordered by inclusion.

Lemma 1.3.5. *If $L_1 = [F_1]$ and $L_2 = [F_2]$ are linear varieties of a projective space, then $L_1 \subset L_2$ if and only if $F_1 \subset F_2$.*

Proof. If $F_1 \subset F_2$, then, clearly, $L_1 = \pi(F_1 - \{0\}) \subset \pi(F_2 - \{0\}) = L_2$. Conversely, if $L_1 \subset L_2$, then by 1.3.1, $F_1 - \{0\} = \pi^{-1}(L_1) \subset \pi^{-1}(L_2) = F_2 - \{0\}$ and the claim follows. □

It is now clear that if L_1, L_2 are linear varieties of $\mathbb{P}_n$, $\dim L_1 > 0$ and $L_1 \subset L_2$, then L_1 is also a linear variety of L_2, the latter taken as a projective space with the induced structure. Conversely, all linear varieties of L_2 are linear varieties of $\mathbb{P}_n$ contained in L_2. It is also clear that the dimension of L_1 is the same, whether it is taken as a linear variety of L_2 or $\mathbb{P}_n$. This is of course not true for its codimensions.

The next proposition relates inclusion and dimension of linear varieties. It will be used very often, mostly with no explicit reference.

Proposition 1.3.6. *For any linear varieties L_1, L_2, of a projective space,*

(a) $L_1 \subset L_2$ *implies* $\dim L_1 \leq \dim L_2$, *while*

(b) $L_1 \subset L_2$ *and* $\dim L_1 = \dim L_2$ *imply* $L_1 = L_2$.

Proof. Both claims follow from Lemma 1.3.5 and the similar claims for linear subspaces of a vector space. Indeed, assume $L_1 = [F_1]$ and $L_2 = [F_2]$:

$L_1 \subset L_2 \Rightarrow F_1 \subset F_2 \Rightarrow \dim F_1 \leq \dim F_2 \Rightarrow \dim L_1 = \dim F_1 - 1 \leq \dim F_2 - 1 = \dim L_2$, which proves claim (a).

Regarding claim (b), again $L_1 \subset L_2 \Rightarrow F_1 \subset F_2$, while $\dim L_1 = \dim L_2 \Rightarrow \dim F_1 = \dim F_2$. Both together give $F_1 = F_2$ and hence $L_1 = L_2$. □

Remark 1.3.7. For any linear variety L of $\mathbb{P}_n$, $\emptyset \subset L \subset \mathbb{P}_n$ and therefore, by Proposition 1.3.6 (a), $-1 \leq \dim L \leq n$. Furthermore, by Proposition 1.3.6 (b), $\dim L = -1$ if and only if $L = \emptyset$ and $\dim L = n$ if and only if $L = \mathbb{P}_n$.

1.4 Incidence of linear varieties

The intersection of two linear varieties is a linear variety too. More precisely:

Proposition 1.4.1. *If $L_1 = [F_1]$ and $L_2 = [F_2]$ are linear varieties of a projective space, then $L_1 \cap L_2$ is the linear variety defined by $F_1 \cap F_2$: $L_1 \cap L_2 = [F_1 \cap F_2]$.*

Proof. If p has a representative in $F_1 \cap F_2$, then p obviously belongs to $L_1 \cap L_2$. Conversely, if $p \in L_1 \cap L_2$, by Lemma 1.3.2, any of its representatives belongs to both F_1 and F_2, hence $p \in [F_1 \cap F_2]$. □

Once it is known that the intersection of any two linear varieties of a projective space $\mathbb{P}_n$ is a linear variety, the following proposition states an easy yet important property, namely that the intersection of L_1 and L_2 is the largest linear variety contained in both L_1 and L_2.

Proposition 1.4.2. *If L_1 and L_2 are linear varieties of a projective space, then $L_1 \cap L_2$ is a linear variety contained in L_1 and L_2, and any linear variety contained in L_1 and L_2 is contained in turn in $L_1 \cap L_2$. Furthermore, $L_1 \cap L_2$ is the only linear variety satisfying these properties.*

Proof. The first part of the claim is clear. If a linear variety T has the same property, then, just by that reason, $L_1 \cap L_2 \subset T$, while $T \subset L_1 \cap L_2$ by the first part of the claim. □

The reader familiar with ordered sets may have noticed that Proposition 1.4.2 just claims $L_1 \cap L_2 = \inf\{L_1, L_2\}$ if the set of all linear varieties of $\mathbb{P}_n$ is ordered by inclusion.

The definition of a linear variety playing the role of the smallest linear variety containing L_1 and L_2 will require using the linear structure of E, and not just elementary set theory as above. If still L_1 and L_2 are linear varieties of a projective space $\mathbb{P}_n$, $L_1 = [F_1]$, $L_2 = [F_2]$, we define the *join* $L_1 \vee L_2$, of L_1 and L_2, as $L_1 \vee L_2 = [F_1 + F_2]$. Then we have:

Proposition 1.4.3. *If L_1 and L_2 are linear varieties of a projective space, then $L_1 \vee L_2$ is a linear variety containing L_1 and L_2, and any linear variety containing L_1 and L_2 contains in turn $L_1 \vee L_2$. Furthermore, $L_1 \vee L_2$ is the only linear variety satisfying these properties.*

Proof. The first inclusions follow from 1.3.5 and the inclusions $F_1 \subset F_1 + F_2$ and $F_2 \subset F_1 + F_2$. If $L = [F]$ contains L_1 and L_2, then 1.3.5 gives $F_1 \subset F$ and $F_2 \subset F$, from which follows that $F_1 + F_2 \subset F$ and hence $L_1 \vee L_2 \subset L$. For the uniqueness the reader may argue as in the proof of 1.4.2. □

The join $L_1 \vee L_2$ is also called the linear variety *spanned* or *determined* by L_1 and L_2. A different convention, which will not be used here, writes $L_1 + L_2$ for $L_1 \vee L_2$ and calls it the *sum* of L_1 and L_2. Still ordering linear varieties by inclusion, we have proved that $L_1 \vee L_2 = \sup\{L_1, L_2\}$.

The next easy consequence of 1.4.3 is of frequent use, often with no explicit reference. A similar claim with intersections in the place of joins is obvious.

Corollary 1.4.4. *If $L_1, L_2, L'_1, L'_2 \subset \mathbb{P}_n$ are linear varieties and $L_1 \subset L'_1$, $L_2 \subset L'_2$, then $L_1 \vee L_2 \subset L'_1 \vee L'_2$.*

Proof. By 1.4.3, $L_i \subset L'_i \subset L'_1 \vee L'_2$ for i=1,2, after which, again by 1.4.3, $L_1 \vee L_2 \subset L'_1 \vee L'_2$. □

Assume that $L_i = [F_i], i = 1, \dots, r$, $r > 2$ are linear varieties of $\mathbb{P}_n$. Elementary set theory shows that the intersection $L_1 \cap \dots \cap L_r$ equals the intersection of any of the linear varieties and the intersection of the remaining ones. The reader may easily prove that

$$L_1 \cap \dots \cap L_r = [F_1 \cap \dots \cap F_r]$$

by either using 1.4.1 and induction on r or just arguing directly as in the proof of 1.4.1. Taking the join of two linear varieties is associative and commutative because so is the sum of linear subspaces of a vector space. In the sequel we will take the join $L_1 \vee \dots \vee L_r$ of $L_1, \dots, L_r$, as being, inductively, the join of any of the varieties, say L_i, and the join of the remaining ones. The result is independent of the choice of L_i just by the associativity and commutativity of the join as binary operation. Still, $L_1 \vee \dots \vee L_r$ is called the linear variety *spanned* or *determined* by $L_1, \dots, L_r$. Checking that

$$L_1 \vee \dots \vee L_r = [F_1 + \dots + F_r]$$

by induction on r is also left to the reader.

As usual, we will take $L_1 \cap \dots \cap L_r = L_1 \vee \dots \vee L_r = L_1$ for $r = 1$ and even $L_1 \cap \dots \cap L_r = \mathbb{P}_n$, $L_1 \vee \dots \vee L_r = \emptyset$ for $r = 0$, to cover the cases in which the set of linear varieties has a single element or is empty.

The next proposition easily follows from 1.4.2 and 1.4.3 using induction on r, the cases $r = 0, 1$ being obvious:

Proposition 1.4.5. *If $L_1, \dots, L_r$ are linear varieties of $\mathbb{P}_n$, then the following hold:*

(a) $L_1 \cap \dots \cap L_r \subset L_i$ *for* $i = 1, \dots, r$. *For any linear variety T of $\mathbb{P}_n$, $T \subset L_i$, $i = 1, \dots, r$, force $T \subset L_1 \cap \dots \cap L_r$.*

(b) $L_1 \vee \dots \vee L_r \supset L_i$ *for* $i = 1, \dots, r$. *For any linear variety T of $\mathbb{P}_n$, $T \supset L_i$, $i = 1, \dots, r$, force $T \supset L_1 \vee \dots \vee L_r$.*

The main relationship between the dimensions of two linear varieties, their intersection and their join, is as follows:

Theorem 1.4.6 (Grassmann formula). *For any two linear varieties L_1, L_2 of a projective space,*

$$\dim L_1 + \dim L_2 = \dim L_1 \cap L_2 + \dim L_1 \vee L_2.$$

Proof. If $L_1 = [F_1]$, $L_2 = [F_2]$ the claim directly follows from the similar formula for linear subspaces of a vector space, namely

$$\dim F_1 + \dim F_2 = \dim F_1 \cap F_2 + \dim(F_1 + F_2). \qquad \square$$

Next are two corollaries of the Grassmann formula which will be very useful in the sequel:

Corollary 1.4.7. *If L is a linear variety and p a point of $\mathbb{P}_n$, $p \notin L$, then*

$$\dim L \vee p = \dim L + 1.$$

Proof. Since $p \notin L$, $p \cap L = \emptyset$ and by 1.4.6 above

$$\dim L \vee p = \dim L + 0 - (-1) = \dim L + 1. \qquad \square$$

Corollary 1.4.8. *If L is a linear variety and H a hyperplane of $\mathbb{P}_n$, $H \not\supset L$, then*

$$\dim L \cap H = \dim L - 1.$$

Proof. We have $L \vee H \neq H$, otherwise $H = L \vee H \supset L$ against the hypothesis. Then, by 1.3.6 $\dim L \vee H > \dim H = n - 1$, which forces $\dim L \vee H = n$. After this, 1.4.6 gives

$$\dim L \cap H = \dim L + n - 1 - n = \dim L - 1. \qquad \square$$

Corollary 1.4.9. (a) *Any two different points of $\mathbb{P}_n$ span a line.*

(b) *A point and a line not containing it, both in $\mathbb{P}_n$, span a plane.*

(c) *The intersection of any two different lines of $\mathbb{P}_2$ is a point.*

(d) *The intersection of a line and a plane not containing it, both in $\mathbb{P}_3$, is a point.*

(e) *The intersection of any two different planes of* $\mathbb{P}_3$ *is a line.*

(f) *There are two possibilities for two different lines of* $\mathbb{P}_3$*, namely, either they span the whole space, which occurs if and only if they are disjoint* (***skew lines***)*, or they span a plane, which occurs if and only if their intersection is a point* (***coplanar lines***).

Proof. Claims (a) and (b) follow from 1.4.7, while claims (c), (d) and (e) are direct consequences of 1.4.8. Regarding claim (f), notice first that two different lines ℓ_1, ℓ_2 have $\dim \ell_1 \vee \ell_2 > 1$, otherwise by 1.3.6 both lines would be equal to their join and hence equal. On the other hand, the lines being in $\mathbb{P}_3$, $\dim \ell_1 \vee \ell_2 \leq 3$. This leaves the two possibilities of the claim, namely either $\dim \ell_1 \vee \ell_2 = 3$, which is equivalent to $\dim \ell_1 \cap \ell_2 = -1$, or $\dim \ell_1 \vee \ell_2 = 2$ which is equivalent to $\dim \ell_1 \cap \ell_2 = 0$, both by 1.4.6. □

The sign $\vee$ is often omitted when writing joins of points, for instance the line $p \vee q$ spanned by two different points p, q, is also denoted by pq.

Each of the situations described in 1.4.9 does actually occur in any projective space of the specified dimension. The reader may easily prove this by exhibiting an example of each situation by either choosing suitable subspaces of the associated vector space, or defining the linear varieties by their equations after forthcoming Chapter 2. Exercise 1.31 provides a more general result.

The reader may have noticed at this point that projective linear varieties behave quite differently from the affine ones, mainly due to the Grassmann formula 1.4.6, that does not hold for affine varieties. Indeed 1.4.8 implies that the intersection of a hyperplane and a positive-dimensional linear variety is always non-empty, and so, in particular, there are no disjoint lines in a projective plane or disjoint planes in a projective three-space. The same fact is illustrated by the next direct consequence of 1.4.6, which would be clearly false if claimed for affine varieties.

Corollary 1.4.10. *If* L_1, L_2 *are linear varieties of* P_n *and* $\dim L_1 + \dim L_2 > n-1$*, then* $L_1 \cap L_2 \neq \emptyset$.

The proof of the following corollary of 1.4.6 is also left to the reader:

Corollary 1.4.11. *If* L_1, L_2 *are linear varieties of* P_n*, any two of the following conditions imply the third one:*

(i) $L_1 \cap L_2 = \emptyset$,

(ii) $L_1 \vee L_2 = \mathbb{P}_n$,

(iii) $\dim L_1 + \dim L_2 = n - 1$.

A pair of linear varieties satisfying two (and hence all) of the conditions of 1.4.11 above are called *supplementary*, either linear variety being then said to be *supplementary to*, or *a supplementary linear variety of*, the other.

Remark 1.4.12. If the linear varieties of 1.4.11 are $L_i = [F_i]$, $i = 1, 2$, then the conditions (i) and (ii) are equivalent to $F_1 \cap F_2 = \{0\}$ and $F_1 + F_2 = E$ respectively. Thus L_1 and L_2 are supplementary if and only if $E = F_1 \oplus F_2$.

Corollary 1.4.13. *If $L_1, \dots, L_m$, $m \geq 2$, are linear varieties of $\mathbb{P}_n$, then*

$$n - \dim(L_1 \cap \cdots \cap L_m) \leq \sum_{i=1}^{m}(n - \dim L_i)$$

and the equality holds if and only if $(L_1 \cap \cdots \cap L_{i-1}) \vee L_i = \mathbb{P}_n$ for each $i = 2, \dots, m$.

Proof. If $L = L_1 \cap \cdots \cap L_{m-1}$, then, directly from 1.4.6,

$$n - \dim L \cap L_m = n - \dim L - \dim L_m + \dim L \vee L_m \leq 2n - \dim L - \dim L_m,$$

and the equality holds if and only if $L \vee L_m = \mathbb{P}_n$. This proves the case $m = 2$ and then also the general one using induction on m. □

A series of different and sometimes redundant words are used to describe incidence situations in order to make the discourse lighter and less boring. Although they are self-explanatory enough, we list some of them next. A point p in a linear variety L is also said to *belong to* or to *lie on* L, and L is said to *go* or *pass through* p. The intersection of two linear varieties is sometimes called the *section* of either of them by the other, especially when the latter is a hyperplane. Points belonging to the same line are called *collinear*. Points and/or lines contained in the same plane are called *coplanar*. Linear varieties *meet* or are *concurrent* when they have a non-empty intersection; if p belongs to their intersection, they are said to be *concurrent* or to *meet at* p. When a point p does not belong to a linear variety L it is often said that L *misses* p.

1.5 Linear independence of points

Proposition 1.5.1. *Let $q_0, \dots, q_m$ be points of $\mathbb{P}_n$. Then*

(a) $\dim q_0 \vee \cdots \vee q_m \leq m$*, and*

(b) $\dim q_0 \vee \cdots \vee q_m = m$ *if and only if, for each* $i = 1, \dots, m$, $q_i \notin q_0 \vee \cdots \vee q_{i-1}$.

Proof. For any point p and any linear variety L, $\dim L \vee q = \dim L$ if $q \in L$ and $\dim L \vee q = \dim L + 1$ otherwise (1.4.7). Thus, for each $i = 1, \dots, m$,

$$\dim q_0 \vee \cdots \vee q_i \leq \dim q_0 \vee \cdots \vee q_{i-1} + 1 \tag{1.1}$$

and the equality holds if and only if $q_i \notin q_0 \vee \cdots \vee q_{i-1}$. Since $\dim q_0 = 0$, the claim (a) follows by just adding up these inequalities. Furthermore the addition is an equality if and only if all inequalities (1.1) are equalities, which proves claim (b). □

Remark 1.5.2. Since the condition $\dim q_0 \vee \cdots \vee q_m = m$ in 1.5.1 is obviously independent of the ordering on the points, so is the equivalent condition $q_i \notin q_0 \vee \cdots \vee q_{i-1}$ for $i = 1, \ldots, m$.

Points $q_0, \ldots, q_m$ of a projective space $\mathbb{P}_n$ are called *linearly independent* if and only if $\dim q_0 \vee \cdots \vee q_m = m$. Otherwise, by 1.5.1, $\dim q_0 \vee \cdots \vee q_m < m$ and the points are called *linearly dependent*. In the sequel we will often drop the word *linearly* and say just independent or dependent points, this causing no confusion.

The definition says thus that points are independent if and only if they span a linear variety of the maximum expected dimension: two points are independent if and only if they span a line (and hence if and only if they are different), three points are independent if and only if they span a plane, and so on.

A direct consequence of 1.5.1 and 1.5.2 is

Proposition 1.5.3. *Points $q_0, \ldots, q_m$ are independent if and only if, for every i, $0 < i \leq m$, it holds that $q_i \notin q_0 \vee \cdots \vee q_{i-1}$. The latter condition is therefore independent of the ordering on the points.*

Corollary 1.5.4. *The points of any subset of a set of independent points are in turn independent.*

Proof. If the independent points are $q_0, \ldots, q_m$, after a suitable reordering it is not restrictive to assume that the given subset is $\{q_0, \ldots, q_{m'}\}$, $m' \leq m$, after which the claim follows from 1.5.3 above. □

Remark 1.5.5. It follows from the definition of independence that $d + 1$ points spanning a linear variety of dimension d are independent. Conversely, any $d + 1$ independent points of a linear variety L of dimension d span it, by 1.3.6. Thus, if $d + 1$ independent points of a linear variety L, of dimension d, belong to a linear variety L', then $L \subset L'$, by1.4.5 (b).

Theorem 1.5.6. *Assume that points $q_0, \ldots, q_m$ are independent and belong to a linear variety L of dimension d. Then*

(a) *$m \leq d$, and*

(b) *there exist points $q_{m+1}, \ldots, q_d$ in L such that $q_0, \ldots, q_d$ are independent.*

Proof. By 1.4.5, $q_0 \vee \cdots \vee q_m \subset L$ and thus, by the independence of the points, $m = \dim q_0 \vee \cdots \vee q_m \leq \dim L = d$, which proves (a). Claim (b) will be proved by induction on $d - m$. If $d - m = 0$ there is nothing to prove. Assume thus $d - m > 0$. Then, by 1.3.6, $q_0 \vee \cdots \vee q_m \neq L$ and we may choose $q_{m+1} \in L - q_0 \vee \cdots \vee q_m$. By 1.4.7, $\dim q_0 \vee \cdots \vee q_m \vee q_{m+1} = m + 1$ and hence the points $q_0, \ldots, q_m, q_{m+1}$ are independent. Then the existence of the remaining points $q_{m+2}, \ldots, q_d$ is assured by the induction hypothesis applied to $q_0, \ldots, q_{m+1}$. □

Remark 1.5.7. If $L \neq \emptyset$, the claim (b) of 1.5.6 and its proof still make sense for $m = -1$, that is, when there are no points q_i, $i \leq m$, by just taking $\emptyset$ as the variety spanned by an empty set of points. The existence of a set of $d + 1$ independent points in any linear variety of dimension $d > -1$ is thus guaranteed.

Remark 1.5.8. By 1.5.5, the points $q_0, \dots, q_d$ of 1.5.6 span L.

Corollary 1.5.9. *Any linear variety L of $\mathbb{P}_n$ has a supplementary.*

Proof. By 1.5.7 and 1.5.5, we may take independent points $q_0, \dots, q_d$ spanning L, $d = \dim L$. Theorem 1.5.6 assures the existence of $n-d$ points $q_{d+1}, \dots, q_n$ such that $q_0, \dots, q_n$ are independent. Then take $T = q_{d+1} \vee \dots \vee q_n$. Since $q_{d+1}, \dots, q_n$ are independent (by 1.5.4), $\dim T = n - d - 1$ and so $\dim L + \dim T = n - 1$. On the other hand $L \vee T = q_0 \vee \dots \vee q_n$ which in turn equals $\mathbb{P}_n$ by the independence of $q_0, \dots, q_n$. □

The existence of a chain of intermediate varieties, of all possible dimensions, between two given linear varieties included one into another, follows also from 1.5.6:

Corollary 1.5.10. *If L' and L are linear varieties of $\mathbb{P}_n$, $L' \subset L$, $m = \dim L'$ and $d = \dim L$, then there are linear varieties L_i, $i = m + 1, \dots, d - 1$ such that $\dim L_i = i$ and*

$$L' \subset L_{m+1} \subset \dots \subset L_{d-1} \subset L.$$

Proof. By 1.5.7 and 1.5.6, there are independent points $q_0, \dots, q_m$ spanning L' and then further points $q_{m+1}, \dots, q_d \in L$ such that $q_0, \dots, q_m$ are independent. Then it is enough to take $L_i = q_0 \vee \dots \vee q_i$. □

Proposition 1.5.11. *Any maximal subset of independent points of a finite set of points $\{q_0, \dots, q_r\}$ spans $q_0 \vee \dots \vee q_r$.*

Proof. Up to reordering $q_0, \dots, q_r$, we may assume that the maximal subset of independent points is $\{q_0, \dots, q_m\}$, $m \leq r$. By 1.4.5, $q_0 \vee \dots \vee q_m \subset q_0 \vee \dots \vee q_r$. If $q_i \in q_0 \vee \dots \vee q_m$ for $i = m + 1, \dots, r$, then $q_0 \vee \dots \vee q_m = q_0 \vee \dots \vee q_r$ as wanted. Otherwise, for some i, $q_i \notin q_0 \vee \dots \vee q_m$; then, by 1.4.7,

$$\dim q_0 \vee \dots \vee q_m \vee q_i = \dim q_0 \vee \dots \vee q_m + 1 = m + 1$$

and $q_0, \dots, q_m, q_i$ are independent, against the hypothesis of maximality. □

So far, we have dealt with independence of points without using representatives. Nevertheless, it is useful to have a reformulation of the independence of points in terms of their representatives, as presented next. After it the reader may easily re-prove most of the above results using linear algebra. It is worth recalling here that linear independence of vectors and linear independence of points are different notions. A close relationship between them is stated in the next lemma:

Lemma 1.5.12. *Assume that points $q_0, \dots, q_m$ have representatives $v_0, \dots, v_m$, respectively. Then the points $q_0, \dots, q_m$ are linearly independent if and only if the vectors $v_0, \dots, v_m$ are linearly independent.*

Proof. If $q_i = [v_i]$, $i = 0, \dots, m$, then

$$q_0 \vee \cdots \vee q_m = [\langle v_0 \rangle + \cdots + \langle v_m \rangle] = [\langle v_0, \dots, v_m \rangle],$$

$\langle\ \rangle$ still meaning *subspace generated by*. It follows that the points are independent if and only if $\dim\langle v_0, \dots, v_m \rangle = m + 1$ which in turn is equivalent to the linear independence of the $m + 1$ generators $v_0, \dots, v_m$. □

1.6 Projectivities

Let $(\mathbb{P}_n, E, \pi)$ and $(\mathbb{P}'_m, E', \pi')$ be projective spaces over the same field k and assume that φ is an isomorphism $\varphi \colon E \to E'$. Then, for any non-zero $v \in E$, the vector $\varphi(v)$ is not zero and therefore represents a point $p' \in \mathbb{P}'_m$. Furthermore, the image of a non-zero multiple λv of v is $\varphi(\lambda v) = \lambda\varphi(v)$, still a representative of the point p'. Thus $[\varphi(v)]$ is a point of $\mathbb{P}'_m$ which depends only on $p = [v]$, and not on its representative v, and we have a well-defined map

$$f \colon \mathbb{P}_n \longrightarrow \mathbb{P}'_m$$

by the rule

$$p = [v] \longmapsto [\varphi(v)].$$

The above map f is called the *projectivity induced* or *represented* by φ, written $f = [\varphi]$. The isomorphism φ is in turn called a *representative of* f.

We say that a map $f \colon \mathbb{P}_n \to \mathbb{P}'_m$ is a *projectivity* if and only if f is the projectivity induced by an isomorphism φ, that is, $f = [\varphi]$ for an isomorphism φ between the associated vector spaces. The projectivities are also called *homographies*. Since isomorphic vector spaces have the same base field and the same dimension, it is clear that the existence of a projectivity between projective spaces $\mathbb{P}_n$ and $\mathbb{P}'_m$ implies that both have the same base field and the same dimension $m = n$. A pair of *homographic projective spaces* is a triple consisting of two projective spaces and a projectivity between them.

Historically, there have been two different ways of defining projectivities, which are equivalent if the base field is the real one. In the synthetic way, projectivities are defined as bijective maps between projective spaces, preserving independence and dependence of points (with a further condition in the one-dimensional case); see for instance [23]. According to the analytic definition, a map between projective spaces is a projectivity if and only if, after fixing coordinates, there is a system of linear, homogeneous and invertible equations relating the coordinates of each point and those of its image. Our definition here is just an intrinsic version of the latter, avoiding the use of coordinates. After introducing projective coordinates, we will see that, indeed, the projectivities are the maps given by linear, homogeneous and invertible equations (see 2.8.4 and 2.8.5).

First we control the different representatives of a projectivity:

Lemma 1.6.1. *Two vector space isomorphisms $\varphi, \psi \colon E \to E'$ define the same projectivity if and only if there is a (necessarily non-zero) $\lambda \in k$ such that $\varphi = \lambda\psi$.*

Proof. The *if* part is obvious, as $\varphi(v) = \lambda\psi(v)$ and $\psi(v)$ represent the same point for any non-zero vector v. Conversely, if $[\varphi] = [\psi]$, then for any non-zero vector v, $[\varphi(v)] = [\psi(v)]$ and so there is a non-zero $\lambda_v \in k$ (which in principle may depend on v) such that $\varphi(v) = \lambda_v\psi(v)$. If v and w are independent vectors, then

$$\begin{aligned}\lambda_{v+w}(\psi(v) + \psi(w)) &= \lambda_{v+w}\psi(v + w) = \varphi(v + w)\\ &= \varphi(v) + \varphi(w) = \lambda_v\psi(v) + \lambda_w\psi(w)\end{aligned}$$

and thus

$$(\lambda_v - \lambda_{v+w})\varphi(v) + (\lambda_w - \lambda_{v+w})\varphi(w) = 0.$$

Since φ is an isomorphism, $\varphi(v)$ and $\varphi(w)$ are independent and so

$$\lambda_v = \lambda_{v+w} = \lambda_w.$$

The reader may check that the equality $\lambda_v = \lambda_w$ is still true in the easier case of v, w dependent. All together we have that λ_v does not depend on v, say $\lambda_v = \lambda$ for all non-zero v, from which $\varphi = \lambda\psi$ as wanted. □

Formal properties of projectivities are established next.

Proposition 1.6.2. (a) $\mathrm{Id}_{\mathbb{P}_n} = [\mathrm{Id}_E]$.

(b) *If $f = [\varphi] \colon \mathbb{P}_n \to \mathbb{P}'_n$ and $g = [\psi] \colon \mathbb{P}'_n \to \mathbb{P}''_n$ are projectivities, then $g \circ f = [\psi \circ \varphi]$; in particular the composition of f and g is a projectivity.*

(c) *If $f = [\varphi]$ is a projectivity, then it is bijective and its inverse map is the projectivity induced by φ^{-1}, $f^{-1} = [\varphi^{-1}]$.*

Proof. Claim (a) is obvious. For claim (b) just note that for any $p = [v] \in \mathbb{P}_n$,

$$(g \circ f)(p) = g(f(p)) = g(f([v])) = g([\varphi(v)]) = [\psi(\varphi(v))] = [(\psi \circ \varphi)(v)].$$

To close, claims (a) and (b) together give

$$[\varphi^{-1}] \circ [\varphi] = \mathrm{Id}_{\mathbb{P}_n} \quad \text{and} \quad [\varphi] \circ [\varphi^{-1}] = \mathrm{Id}_{\mathbb{P}'_n},$$

and hence claim (c). □

Projectivities from a projective space $\mathbb{P}_n$ into itself are called *collineations*, and also projectivities (or homographies) of $\mathbb{P}_n$. It is clear from 1.6.2 that the set of all projectivities of a fixed projective space $\mathbb{P}_n$, with the composition as operation, is a group, which is usually called the *projective group* of $\mathbb{P}_n$. We will denote it $\mathrm{PG}(\mathbb{P}_n)$.

Projectivities allow us to reformulate the equivalence of projective structures, namely:

Proposition 1.6.3. *Assume that the pairs E, π and E', π' define projective structures on the same set $\mathbb{P}$:*

(a) *The structures defined by E, π and E', π' are equivalent if and only if the identical map of $\mathbb{P}$ is a projectivity*

$$\mathrm{Id}_{\mathbb{P}} \colon (\mathbb{P}, E, \pi) \longrightarrow (\mathbb{P}, E', \pi').$$

(b) *If a map $f : \mathbb{P} \to \overline{\mathbb{P}}$ into a third projective space $(\overline{\mathbb{P}}, \overline{E}, \bar{\pi})$ is a projectivity for both projective structures on $\mathbb{P}$, then these structures are equivalent.*

Proof. The claim (a) directly follows from the definition of equivalent projective structures in Section 1.2. Regarding part (b), just note that if f is a projectivity, so is f^{-1}. Therefore the identical map of $\mathbb{P}$ being the composition of

$$(\mathbb{P}, E, \pi) \xrightarrow{f} (\overline{\mathbb{P}}, \overline{E}, \bar{\pi}) \xrightarrow{f^{-1}} (\mathbb{P}, E', \pi'),$$

it is a projectivity too, and part (a) applies. □

1.7 Projective invariance

Proposition 1.7.1. *If $f = [\varphi] \colon \mathbb{P}_n \to \overline{\mathbb{P}}_n$ is a projectivity and $L = [F]$ is a linear variety of $\mathbb{P}_n$, then:*

(a) *$f(L)$ is the linear variety of $\overline{\mathbb{P}}_n$ defined by $\varphi(F)$: $f(L) = [\varphi(F)]$. In particular the image of a linear variety by a projectivity is a linear variety of the same dimension.*

(b) *If* $\dim L > 0$, *then the restriction of f,*

$$f_{|L} \colon L \longrightarrow f(L),$$

is the projectivity induced by the restriction of φ,

$$\varphi_{|F} \colon F \longrightarrow \varphi(F).$$

Proof. A point q belongs to $f(L)$ if and only if $q = f(p)$ for some $p \in L$, that is, if and only if $q = f([v])$ for some non-zero $v \in F$. Since $f([v]) = [\varphi(v)]$, the first claim follows. The second one is clear from the definition of projectivity. □

Remark 1.7.2. According to 1.7.1 (a), the projectivity f induces a dimension-preserving map,

$$\begin{aligned} \mathcal{LV}(f)\colon \mathcal{LV}(\mathbb{P}_n) &\longrightarrow \mathcal{LV}(\overline{\mathbb{P}}_n), \\ L &\longmapsto f(L), \end{aligned}$$

between the sets of linear varieties of $\mathbb{P}_n$ and $\overline{\mathbb{P}}_n$. It is a bijection, as its inverse clearly is $\mathcal{LV}(f^{-1})$. The next corollary shows that $\mathcal{LV}(f)$ preserves inclusion, intersection and join.

Corollary 1.7.3. *If $f = [\varphi]\colon \mathbb{P}_n \to \overline{\mathbb{P}}_n$ is a projectivity and L_1, L_2 are linear varieties of $\mathbb{P}_n$, then*

(a) $L_1 \subset L_2$ *if and only if* $f(L_1) \subset f(L_2)$,

(b) $f(L_1 \cap L_2) = f(L_1) \cap f(L_2)$,

(c) $f(L_1 \vee L_2) = f(L_1) \vee f(L_2)$.

Proof. Any projectivity being a bijection, (a) and (b) may be deduced from elementary set theory, while (c) follows from 1.7.1 (a). □

Remark 1.7.4. Claim (c) of 1.7.3 assures that if points $q_0, \dots, q_m$ span L, then $f(q_0), \dots, f(q_m)$ span $f(L)$.

Corollary 1.7.5. *If $f = [\varphi]\colon \mathbb{P}_n \to \overline{\mathbb{P}}_n$ is a projectivity and $q_0, \dots, q_m$ are points of $\mathbb{P}_n$, then $q_0, \dots, q_m$ are independent if and only if $f(q_0), \dots, f(q_m)$ are independent.*

Proof. Just note that, by 1.7.1 (a) and 1.7.3 (c), $\dim q_0 \vee \dots \vee q_m = \dim f(q_0) \vee \dots \vee f(q_m)$. □

Example 1.7.6. If $p_0, p_1, p_2 \in \mathbb{P}_n$ are three independent points ($n > 1$ thus), by 1.5.4 and the definition of independence, the joins $p_0 \vee p_1$, $p_1 \vee p_2$ and $p_2 \vee p_0$ are three different lines, while $p_0 \vee p_1 \vee p_2$ is a plane. A *triangle* of $\mathbb{P}_n$ is the figure consisting of three independent points $p_0, p_1, p_2 \in \mathbb{P}_n$, its *vertices*, the three lines $p_0 \vee p_1$, $p_1 \vee p_2$, $p_2 \vee p_0$, its *sides*, and the plane $p_0 \vee p_1 \vee p_2$, the *plane* of the triangle. Clearly, the vertices determine the triangle, and sometimes one uses the term *the triangle $p_0 p_1 p_2$* to mean *the triangle of vertices p_0, p_1, p_2*. Each vertex and the side spanned by the other two are said to be *opposite*. As the reader may easily check, no vertex belongs to its opposite side and the sides of a triangle are coplanar and not concurrent. Conversely, any three coplanar and non-concurrent lines ℓ_1, ℓ_2, ℓ_3 have three non-collinear points as their pairwise intersections, these points being the vertices of the only triangle with sides ℓ_1, ℓ_2, ℓ_3. In case $n = 2$ the plane of the triangle is usually not considered, as it equals $\mathbb{P}_2$ for all triangles and so it carries no relevant information.

Example 1.7.7. If $p_0, \dots, p_m \in \mathbb{P}_n$ are independent points ($m \leq n$ thus), for each $d = 0, \dots, m$, the linear varieties $p_{i_0} \vee \dots \vee p_{i_d}$, $0 \leq i_0 < \dots < i_d \leq m$, spanned by each set of $d + 1$ points chosen among the p_i, are pairwise different and all have dimension d, again by 1.5.4 and the definition of independence. The figure consisting of all linear varieties $p_{i_0} \vee \dots \vee p_{i_d}$, $0 \leq i_0 < \dots < i_d \leq m$, $d = 0, \dots, m$, is called the *m-dimensional simplex* of *vertices* $p_0, \dots, p_m$. Obviously the vertices determine the simplex. Each $p_{i_0} \vee \dots \vee p_{i_d}$ is called a d-dimensional *face* (vertex if $d = 0$, *side* if $d = 1$, $m = 2$, *edge* if $d = 1$, $m > 2$, *face* if $d = 2$, $m = 3$) of the simplex. The vertices other than $p_{i_0}, \dots, p_{i_d}$ span an $(m-d-1)$-dimensional face which is called the face *opposite* $p_{i_0} \vee \dots \vee p_{i_d}$. Clearly *being opposite* is a symmetrical relation and opposite faces are supplementary varieties. In particular each vertex p_i is opposite the $(m-1)$-dimensional face $p_0 \vee \dots \vee p_{i-1} \vee p_{i+1} \vee \dots \vee p_m$, and conversely. A two-dimensional simplex is a triangle, while a three-dimensional simplex is called a *tetrahedron*; as the reader may easily check, a tetrahedron has four vertices, six edges and four faces.

A *linear figure* is a finite set of linear varieties (the *elements* of the linear figure) of a projective space $\mathbb{P}_n$, classified in a number of subsets. Usually the elements of each subset are given the same name, such as vertex, edge, face, etc., and all have the same dimension. For instance, a triangle is a linear figure whose elements are three points (its vertices), three lines (its sides) and one plane. More generally, any simplex is a linear figure: its elements are its faces, classified by dimension.

Projectivities act on linear figures in an obvious way: a projectivity $f : \mathbb{P}_n \to \mathbb{P}'_n$ transforms the linear figure composed of certain varieties into the linear figure composed of their images, classified accordingly; that is, the linear figure $\bigcup_{i=1}^r \{L_{i,1}, \dots, L_{i,s_i}\}$ of $\mathbb{P}_n$ into the linear figure $\bigcup_{i=1}^r \{f(L_{i,1}), \dots, f(L_{i,s_i})\}$ of $\mathbb{P}'_n$. A class of linear figures is called *projective* if and only if it is invariant by projectivities, that is, if and only if the image of any linear figure in the class still belongs to the class. Classes of linear figures defined by conditions formulated in terms of dimension, inclusion, intersection, join, and independence of points, are projective by 1.7.1, 1.7.3 and 1.7.5. Also, clearly, intersections and unions of projective classes are projective classes. For instance, Corollary 1.7.5 says that the class of all sets of $m + 1$ independent points is projective. The class of all triangles is projective. More generally, for each $m > 0$ the class of all m-dimensional simplexes is projective too.

In current language, the word *notion* is often used for *class*, by saying for instance that the notion of triangle is projective to mean that the class of all triangles is projective.

In the next two examples two projective classes of plane linear figures are defined. The details are left to the reader.

Example 1.7.8. See Figure 1.4. A *complete quadrilateral* (or just *quadrilateral* if no confusion with quadrilaterals of elementary geometry may occur) is the linear figure of $\mathbb{P}_2$ consisting of four lines, no three of which are concurrent, called the *sides* of the quadrilateral, together with the six different points which are the intersections of the pairs of different sides, named the *vertices* of the quadrilateral. Two vertices are called *opposite* if and only if no side contains both. Since there are exactly three vertices on each side, each vertex has a single opposite. This makes three pairs of opposite vertices and each join of two opposite vertices is called a *diagonal* of the quadrilateral, which gives three different diagonals. Clearly a quadrilateral is determined by its sides.

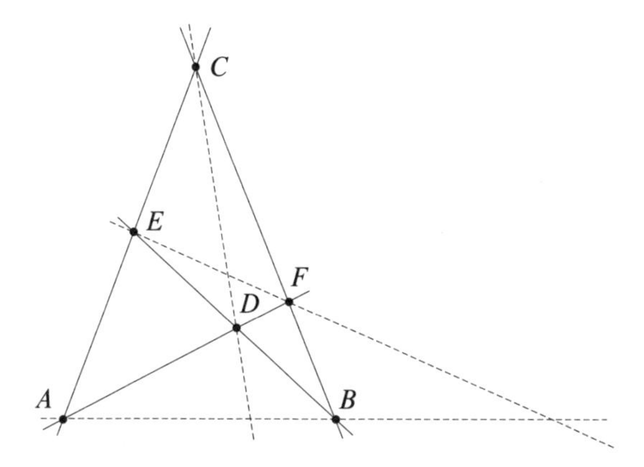

Figure 1.4. A complete quadrilateral and its diagonals.

Example 1.7.9. A *quadrivertex* (or *quadrangle*) is the linear figure of $\mathbb{P}_2$ consisting of four points, no three of which are aligned (its *vertices*), together with the six different lines spanned by the pairs of different vertices (the *sides* of the quadrivertex). Two sides are called *opposite* if and only if they share no vertex. Since there are exactly three sides through each vertex, each side has a single opposite. This makes three pairs of opposite sides and each intersection of two opposite sides is called a *diagonal point* of the quadrivertex, which gives three different diagonal points. A quadrivertex is determined by its vertices.

1.8 Pappus' and Desargues' theorems

In this section we present some incidence propositions and theorems belonging to plane projective geometry. Their proofs will all be achieved with the same analytic technique, namely taking suitable representatives of some points and then deriving representatives for the remaining points using the incidence relations and easy linear algebra computations.

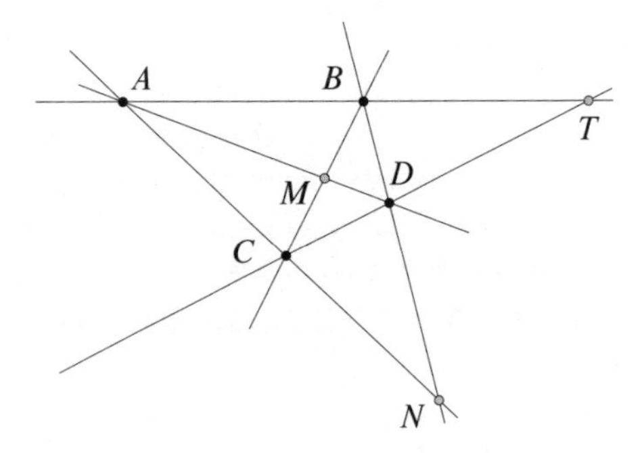

Figure 1.5. A quadrivertex, of vertices A, B, C, D and diagonal points M, N, T.

Proposition 1.8.1. *The three diagonals of a complete quadrilateral are not concurrent.*

Proof. Assume that the vertices of the quadrilateral are A, B, C, D, E, F, its sides being AEC, ADF, BDE and BFC, see Figure 1.6. Recall that, by the definition of complete quadrilateral, each vertex belongs to exactly two sides and therefore each side contains exactly three vertices, which in particular implies that the diagonals are three different lines. Take $A = [v']$, $C = [u]$. The point E being aligned with A and B, and different from any of them, $E = [\lambda u + \mu v']$, $\lambda, \mu \neq 0$. One may

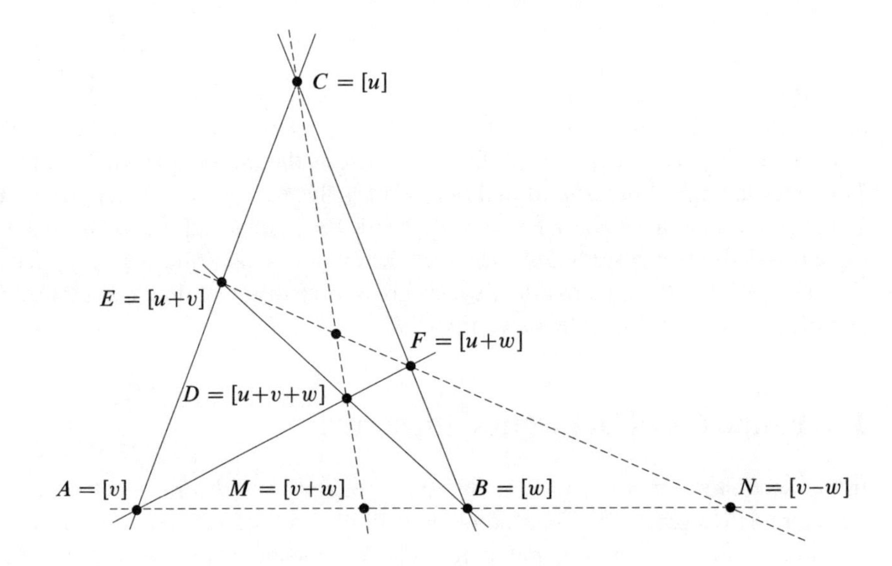

Figure 1.6. The diagonals of a quadrilateral are not concurrent.

take $v = (\mu/\lambda)v' \neq 0$ to have $A = [v]$, $C = [u]$ and $E = [u + v]$. Similarly, one may choose the representative w of B to have $B = [w]$ and $F = [u + w]$. Note that u, v, w are independent vectors, as they represent non-aligned points.

Since the vector $u+v+w$ may be written $u+v+w = v+(u+w) = w+(u+v)$, it belongs to $\langle v, u + w\rangle \cap \langle w, u + v\rangle$ and so represents the only point shared by the sides AF and BE, namely $D = [u + v + w]$,

Similarly, the equality $v + w = -u + (u + v + w)$ shows that the diagonals AB and CD intersect at $M = [v + w]$, while $N = AB \cap EF$ is represented by $v - w = (u + v) - (u + w)$. Now $M = N$ would contradict the independence of v, w. □

Proposition 1.8.2. *The three diagonal points of a quadrivertex are not aligned.*

The proof of Proposition 1.8.2 is similar to the one of 1.8.1 and left to the reader, who may also see Exercise 2.4. Anyway, Proposition 1.8.2 will be a direct consequence of Proposition 1.8.1 once duality arguments become available in Chapter 4 (see Example 4.3.9).

After 1.8.1 (resp. 1.8.2), the three diagonals (resp. diagonal points) of a quadrilateral (resp. quadrivertex) are the sides (resp. vertices) of a triangle, which is currently referred to as the *diagonal triangle* of the quadrilateral (resp. quadrivertex).

Theorem 1.8.3 (Desargues). *Assume we have triangles ABC and $A'B'C'$ in a projective plane $\mathbb{P}_2$, let a, b and c be the sides of the first triangle opposite A, B and C, respectively, and, similarly, a', b', and c' the sides of the second triangle opposite A', B' and C'. Assume furthermore that $A \neq A'$, $B \neq B'$, $C \neq C'$, $a \neq a'$, $b \neq b'$ and $c \neq c'$. If the lines AA', BB', CC' are concurrent, then the points $a \cap a'$, $b \cap b'$, $c \cap c'$ are aligned.*

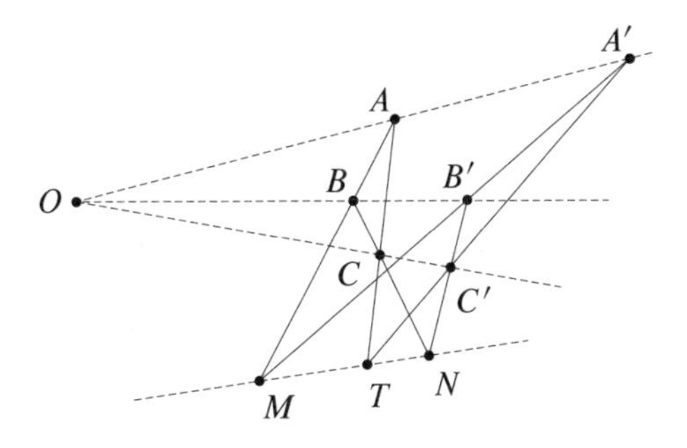

Figure 1.7. The theorem of Desargues.

Proof. Take representatives $A = [u]$, $B = [v]$, $C = [w]$, $A' = [u']$, $B' = [v']$, $C' = [w']$. If O is a point simultaneously lying on AA', BB' and CC', a representative of it may be written

$$\alpha u + \alpha' u' = \beta v + \beta' v' = \lambda w + \lambda' w'$$

for suitable $\alpha, \alpha', \beta, \beta', \lambda, \lambda' \in k$. Then

$$\alpha u - \beta v = \beta' v' - \alpha' u'$$

and since this vector is obviously non-zero, by the above equality it represents a point M that lies on both AB and $A'B'$, hence $M = c \cap c'$. Similarly, the vector

$$\beta v - \lambda w = \lambda' w' - \beta' v'$$

represents $N = a \cap a'$ and

$$\lambda w - \alpha u = \alpha' u' - \lambda' w'$$

represents $T = b \cap b'$. Since, clearly,

$$(\alpha u - \beta v) + (\beta v - \lambda w) + (\lambda w - \alpha u) = 0,$$

the claim follows. □

Remark 1.8.4. The claim of Desargues' theorem may be equivalently stated by saying that if there is a point O which is aligned with each of the pairs of points A, A', B, B' and C, C', then there is a line ℓ which is concurrent with each of the pairs of lines a, a', b, b' and c, c'. In this form the claim is still true in the cases in which there is a coincidence among corresponding vertices or sides, say, for instance, $A = A'$ or $a = a'$. These cases are very easy and may be given a separate proof by the reader.

Theorem 1.8.5 (Pappus). *Let r, s be two different lines of $\mathbb{P}_2$ and take $O = r \cap s$. If A, B, C are three different points of r, all different from O, and A', B', C' are three different points of s, also different from O, then the intersections $M = AB' \cap A'B$, $N = AC' \cap A'C$ and $T = BC' \cap B'C$ are three aligned points.*

Proof. If $AB' = A'B$, then A, B, A', B' would be aligned and therefore $r = AB = A'B' = s$ against the hypothesis. This proves that M is a point, and similarly for N and T. As in the proof of 1.8.1, we choose representatives $O = [u]$, $A = [v]$, $A' = [w]$ such that $B = [u + v]$ and $B' = [u + w]$. Since $C \in r = OA$, $C \neq O, A, B$, it may be written $C = [u + \lambda v]$, $\lambda \neq 0, 1$. Similarly $C' = [u + \mu w]$, $\mu \neq 0, 1$. Clearly, the vectors u, v, w are linearly independent by the hypothesis. Then $u + v + w \neq 0$ and the point

$$[u + v + w] = [(u + v) + w] = [v + (u + w)] = M,$$

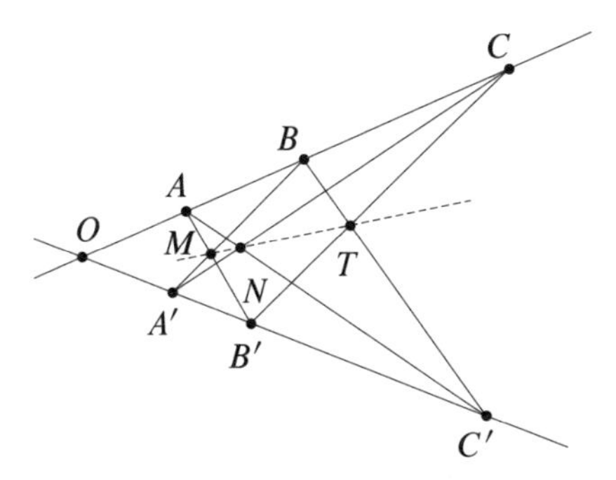

Figure 1.8. The theorem of Pappus.

as it belongs to both $A'B$ and AB'. In the same way,

$$[u + \lambda v + \mu w] = [(u + \lambda v) + \mu w] = [\lambda v + (u + \mu w)] = N$$

and

$$\begin{aligned}[(\lambda\mu - 1)u + \lambda(\mu - 1)v + \mu(\lambda - 1)w] &= [\lambda(\mu - 1)(u + v) + (\lambda - 1)(u + \mu w)] \\ &= [\mu(\lambda - 1)(u + w) + (\mu - 1)(u + \lambda v)] \\ &= T.\end{aligned}$$

Since

$$\lambda\mu(u + v + w) - (u + \lambda v + \mu w) - (\lambda\mu - 1)u - \lambda(\mu - 1)v - \mu(\lambda - 1)w = 0,$$

the claim is proved. □

1.9 Projection, section and perspectivity

If L is a d-dimensional linear variety of $\mathbb{P}_n$, $0 \leq d \leq n - 2$, we define the *bunch (of linear varieties)* with *centre* L, denoted by L^∇, as being the set of all linear varieties of dimension $d + 1$ containing L. For each point $p \notin L$, $p \vee L \in L^\nabla$, by 1.4.7. This gives a map

$$\begin{aligned}\delta_L \colon \mathbb{P}_n - L &\longrightarrow L^\nabla, \\ p &\longmapsto p \vee L.\end{aligned}$$

The linear variety $p \vee L$ is called the *projection of p from L*, or the linear variety *projecting p from L*, and the map δ_L the *projection* (or *projection map*) *from* (or *with centre*) L.

If T is a linear variety supplementary to L, then for any $L' \in L^\nabla$, $L' \vee T \supset L \vee T = \mathbb{P}_n$ (by 1.4.4) and therefore $L' \vee T = \mathbb{P}_n$. By 1.4.6, $\dim L' \cap T = 0$, after which mapping each $L' \in L^\nabla$ to the only point in $L' \cap T$ defines a map

$$\sigma_{L,T} \colon L^\nabla \to T$$

which is called a *section map*, and also the *section by* T (of the elements of L^∇).

Since $L \cap T = \emptyset$, the projection with centre L may be restricted to give a map $\delta_{L|T} \colon T \to L^\nabla$, still named *projection from* L (of the elements of T), and we have:

Proposition 1.9.1. *If L and T are supplementary linear varieties and $0 \leq \dim L \leq n-2$, then the section by T of the elements of L^∇, $\sigma_{L,T}$, and the projection from L of the elements of T, $\delta_{L|T}$, are reciprocal bijections.*

Proof. If $L' \in L^\nabla$ and $p = \sigma_{L,T}(L'))$, then p is the only point in $L' \cap T$. Thus, on one hand $p \in L'$ and therefore $p \vee L \subset L'$. On the other $p \in T \subset \mathbb{P}_n - L$, which gives, by 1.4.7, $\dim p \vee L = \dim L + 1 = \dim L'$. Both together prove $p \vee L = L'$, that is, that $\delta_{L|T}(\sigma_{L,T}(L')) = L'$.

If $p \in T$, clearly $p \in (p \vee L) \cap T$, which proves that $p = \sigma_{L,T}(\delta_{L|T}(p))$. □

We may also compose the projection map $\mathbb{P}_n - L \to L^\nabla$ and the section map $L^\nabla \to T$ to get a map

$$\begin{aligned} \delta_L^T \colon \mathbb{P}_n - L &\longrightarrow T, \\ p &\longmapsto (p \vee L) \cap T, \end{aligned}$$

usually called the *projection from L onto T*. Note that, by 1.9.1 above, δ_L^T restricts to the identity map of T and so it is, in particular, exhaustive.

Still assuming that $0 \leq d \leq n-2$, next we will show that the bunch L^∇ has a projective structure, in some sense inherited from $\mathbb{P}_n$, with which the projection and section maps $\delta_{L|T}$ and $\sigma_{L,T}$, of 1.9.1, are projectivities.

Proposition 1.9.2. *Let a projective space $\mathbb{P}_n$ have associated vector space E. If $L = [F]$ is a non-empty linear variety of $\mathbb{P}_n$ and has codimension at least two, then the map*

$$\begin{aligned} \pi_L \colon E/F - \{\bar{0}\} &\longrightarrow L^\nabla, \\ \bar{v} = v + F &\longmapsto [v] \vee L, \end{aligned}$$

is well defined and the triple $(L^\nabla, E/F, \pi_L)$ is a projective space of dimension $n - \dim L - 1$.

Proof. In order to check that π_L is well defined, note first that if $\bar{v} \neq \bar{0}$ in the quotient space E/F, then $v \notin F$ and hence $[v] \notin L$; then, by 1.4.7, $\dim [v] \vee L = \dim L + 1$ and so $[v] \vee L \in L^\nabla$, as required. Assume that v' is another representative of the class $\bar{v}$: then $v' = v + w$ for some $w \in F$ and so, in particular, $v' \in \langle v \rangle + F$; this gives $[v'] \in [v] \vee L$, from which (by 1.4.4) $[v'] \vee L \subset [v] \vee L$. Since $\bar{v}' = \bar{v} \neq \bar{0}$, arguing as above for v, $\dim [v'] \vee L = \dim L + 1$. It follows that $\dim [v'] \vee L = \dim [v] \vee L$ and so $[v'] \vee L = [v] \vee L$, showing that the image of $\bar{v}$ does not depend of the choice of the representative v.

We check now that π_L defines a projective structure on L^∇. For the exhaustivity, assume $L' \in L^\nabla$ and take $p = [v] \in L' - L$; then $[v] \vee L \subset L'$ and $\dim[v] \vee L = \dim L + 1 = \dim L'$, which force $L' = [v] \vee L$ and so $L' = \pi_L(\bar{v})$. Assume now that $[u] \vee L = [v] \vee L$ for $u, v \in E - F$. Then $[u] \in [v] \vee L$, which gives $u = \lambda v + w$ for some $\lambda \in k$ and $w \in F$. Hence $\bar{u} = \lambda \bar{v}$ as required.

To close, $\dim L^\nabla = \dim(E/F) - 1 = n - \dim L - 1$. □

Assume now that, as above, $T = [G]$ is supplementary to L and put $d = \dim L$. Then it is $L \cap T = \emptyset$ and $\dim T = n - d - 1$, or, equivalently, $F \cap G = \{0\}$ and $\dim G = n - d$. The first equality shows that the natural morphism $\rho\colon E \to E/F$ restricts to a monomorphism

$$\begin{aligned} \rho_{|G}\colon G &\longrightarrow E/F, \\ v &\longmapsto \bar{v}, \end{aligned}$$

which, by the second equality, is an isomorphism.

According to the definition of the projective structure in L^∇, the induced projectivity

$$[\rho_{|G}]\colon T \longrightarrow L^\nabla$$

is $\delta_{L|T}$. Indeed if $p = [v] \in T$, then $[\rho_{|G}](p) = [\rho_{|G}(v)] = [\bar{v}] = p \vee L$. In particular $\delta_{L|T}$ is a projectivity, and so is its inverse $\sigma_{L,T}$.

Assume now that T_1 and T_2 are two linear varieties, each supplementary to L. One may then compose the restriction to T_1 of the projection from L, $\delta_{L|T_1}$, and the section with T_2, σ_{L,T_2}, to obtain a projectivity $f\colon T_1 \to T_2$, which is called the *perspectivity from T_1 to T_2 with centre L*. Note that, by the definition of f, for any $p \in T_1$, $f(p)$ is the only point in $(p \vee L) \cap T_2$.

Summarizing, we have seen:

Theorem 1.9.3. *If L and T are supplementary linear varieties of $\mathbb{P}_n$, $0 \leq \dim L \leq n-2$, then the projection $\delta_{L|T}\colon T \to L^\nabla$ and the section $\sigma_{L,T}\colon L^\nabla \to T$ are reciprocal projectivities.*

If T_1 and T_2 are linear varieties of $\mathbb{P}_n$, both supplementaries to L, then the perspectivity with centre L, $f\colon T_1 \to T_2$, is a projectivity.

For a direct proof of the second claim of 1.9.3 using coordinates, see Exercise 2.9.

A pair of *perspective linear varieties* is a triple composed of two linear varieties T_1, T_2 and a perspectivity $f\colon T_1 \to T_2$.

Next we make use of Theorem 1.9.3 to describe the linear varieties of L^∇. The reader may provide a different proof using the well-known relationship between subspaces of E/F and subspaces of E containing F.

Corollary 1.9.4. *If L is a linear variety of $\mathbb{P}_n$, $0 \leq d = \dim L \leq n-2$, for any linear variety $S \supset L$ of dimension r, the set*

$$\check{S} = \{L' \in L^\nabla \mid L' \subset S\}$$

is a linear variety of the bunch L^∇ and has dimension $r-d-1$. Furthermore, each linear variety of L^∇ of dimension $r-d-1$ is $\check{S}$ for a uniquely determined $S \supset L$ of dimension r.

Proof. Take a linear variety T supplementary to L. We will check first that $\check{S} = \delta_{L|T}(S \cap T)$, which in particular will prove that $\check{S}$ is a linear variety. For, if $L' \in \check{S}$, then $L' \subset S$; its section $L' \cap T$ is thus a point of $S \cap T$, the projection of which, by 1.9.3, is L'. Conversely, the projection $L \vee p$ of any point $p \in S \cap T$ is contained in $L \vee (S \cap T) = \delta_{L|T}(\sigma_{L,T}(S)) = S$, and hence belongs to $\check{S}$.

In particular, since $\delta_{L|T}$ is a projectivity, it is $\dim \check{S} = \dim S \cap T$, and an easy count shows the latter to be $r-d-1$.

Regarding the last claim, note first that mapping $S \mapsto S \cap T$ is a bijection between the set of linear varieties of dimension r containing L, and the set of linear varieties of dimension $r-d-1$ of T, its inverse being $\bar{S} \mapsto \bar{S} \vee L$. On the other hand, since $\delta_{L|T}$ is a projectivity, $\bar{S} \mapsto \delta_{L|T}(\bar{S})$ is a bijection between the sets of linear varieties of dimension $r-d-1$ of T and L^∇. Since we have seen at the beginning of the proof that the composition of these bijections is the correspondence $S \to \check{S}$ of the claim, the latter is also a bijection, as claimed. □

1.10 What is projective geometry about?

It has been mentioned in the introduction and at the beginning of Section 1.1 that, at the early stages, perspectivities were taken as the fundamental transformations defining the subject of projective geometry. Nevertheless, perspectivities may only be defined between linear varieties of the same projective space, and therefore they are not suited to the consideration of abstract, unrelated, projective spaces: by this reason, projectivities are taken instead. When considering only the linear varieties of a single projective space – as the early projective geometers did – this makes no change, because perspectivities are projectivities (1.9.3), and we will see in 5.9.1 that any projectivity between linear varieties of a projective space can be resolved into a chain of perspectivities. The basic idea is thus that *projective geometry* deals with classes and properties of objects, and relations between them, that are invariant by projectivities; these classes, properties and relations are called *projective*.

The use of the term *invariant* presupposes that there is an action of the projectivities on the objects we are considering, and this is indeed the case when dealing with linear figures, as explained in Section 1.7. Let us name, for short, any arbitrary collection of subsets of a projective space $\mathbb{P}_n$ a *figure* of $\mathbb{P}_n$; bunches of liner varieties are good examples of figures. There is also an obvious action of projectivities on figures – so obvious that it is seldom mentioned: if I is an arbitrary set, $\mathcal{T} = \{T_i\}_{i\in I}$, $T_i \subset \mathbb{P}_n$, a figure of $\mathbb{P}_n$ and $g\colon \mathbb{P}_n \to \mathbb{P}'_n$ a projectivity, take $g(\mathcal{T}) = \{g(T_i)\}_{i\in I}$, a figure of $\mathbb{P}'_n$. Then it still makes sense to consider projective classes, properties and relations when dealing with non-necessarily linear figures. For instance, the reader may easily check that:

- The bunches of linear varieties compose a projective class.
- Having the centre of an already fixed dimension is a projective property of bunches.
- Being the centre of a bunch is a projective binary relation between linear varieties and bunches of the same projective space.

However, there are objects other than figures, each associated to a projective space, on which there is not a so obvious action of projectivities. For these objects to compose a *projective class*, such an action needs then to be separately defined. This means that for each object $\mathcal{T}$, associated to a projective space $\mathbb{P}_n$, and each projectivity $g\colon \mathbb{P}_n \to \mathbb{P}'_n$, we define an object $g(\mathcal{T})$, of the same class and associated to $\mathbb{P}'_n$. Usually $g(\mathcal{T})$ is called the image of $\mathcal{T}$ by g. In addition we require the definition to fulfill the two following conditions, which in the case of figures is obvious:

(a) $\mathrm{Id}_{\mathbb{P}_n}(\mathcal{T}) = \mathcal{T}$ for any object $\mathcal{T}$ associated to $\mathbb{P}_n$,

(b) $g'(g(\mathcal{T})) = (g' \circ g)(\mathcal{T})$ for any $\mathcal{T}$ associated to $\mathbb{P}_n$ and any pair of projectivities $g\colon \mathbb{P}_n \to \mathbb{P}'_n$ and $g'\colon \mathbb{P}'_n \to \mathbb{P}''_n$.

If these conditions are satisfied, then also

(c) $g^{-1}(g(\mathcal{T})) = \mathcal{T}$ for any object $\mathcal{T}$ associated to $\mathbb{P}_n$ and any projectivity $g\colon \mathbb{P}_n \to \mathbb{P}'_n$

is satisfied. Indeed, using first (b) and then (a), $g^{-1}(g(\mathcal{T})) = (g^{-1} \circ g)(\mathcal{T}) = \mathrm{Id}_{\mathbb{P}_n}(\mathcal{T}) = \mathcal{T}$.

In many cases most of the above formalism is omitted, because the action of projectivities is clear enough. This is notably the case for parts of projective classes that are invariant by the action of projectivities: they are turned into projective classes by just restricting to them the action of projectivities, the conditions (a) and (b) being then automatically satisfied.

In all cases, once an action of projectivities has been (explicitly or implicitly) defined on a certain class of objects so that the class becomes projective, a property is called *projective* if and only if it is satisfied by all the images by projectivities of all objects satisfying it. Similarly, a relation between objects of projective classes is *projective* if and only if the images of any two related objects by any projectivity are related too.

Example 1.10.1. The collineations $f: \mathbb{P}_n \to \mathbb{P}_n$ are objects associated to projective spaces which are not figures. Assume that $g: \mathbb{P}_n \to \mathbb{P}'_n$ is a projectivity. For any collineation f of $\mathbb{P}_n$ define $g(f) = g \circ f \circ g^{-1}$: $g(f)$ is a collineation of $\mathbb{P}'_n$ and it is direct to check that conditions (a) and (b) above are satisfied, after which the collineations of projective spaces compose a projective class.

A collineation f of $\mathbb{P}_n$ is called *cyclic* if and only if $f^m = \mathrm{Id}_{\mathbb{P}_n}$ for some $m > 0$; to be cyclic is an example of projective property. Indeed, for any $g: \mathbb{P}_n \to \mathbb{P}'_n$,

$$(g(f))^m = (g \circ f \circ g^{-1}) \circ \cdots \circ (g \circ f \circ g^{-1}) = g \circ f^m \circ g^{-1} = \mathrm{Id}_{\mathbb{P}'_n}$$

and $g(f)$ is cyclic too. Also, as the reader may check, the relation *being fixed by*, defined between points p and collineations f of the same projective space by the condition $f(p) = p$, is a projective one.

1.11 Exercises

Exercise 1.1. The definition of projective space being extended without changes to the case of k being an arbitrary field, prove that a projective space of dimension n over a finite field with q elements has $(q^n - 1)/(q - 1)$ points. *Hint*: Compute first the number of non-zero vectors in the associated vector space.

Exercise 1.2. Prove that a subset L of a projective space is a linear variety if and only if for each two different points $p, q \in L$ the line $p \vee q$ is contained in L.

Exercise 1.3. Let L_1, L_2 be different linear varieties of a projective space $\mathbb{P}_n$, both non-empty. Prove that $p \in L_1 \vee L_2$ if and only if there are $p_1 \in L_1$ and $p_2 \in L_2$, $p_1 \neq p_2$, for which $p \in p_1 \vee p_2$.

Exercise 1.4. Prove that the equality

$$L \cap (L_1 \vee L_2) = (L \cap L_1) \vee (L \cap L_2),$$

where $L, L_1, L_2 \subset \mathbb{P}_n$ are linear varieties, is false in general, but holds if either $L \supset L_1$ or $L \supset L_2$.

Exercise 1.5. Prove that three different and pairwise coplanary lines of $\mathbb{P}_3$ either lie on the same plane or go through the same point.

Exercise 1.6. Assume that ℓ_1, ℓ_2, ℓ_3 are three lines of a projective space, each disjoint with the span of the other two. Take points $p_i \in \ell_i$, $i = 1, 2, 3$, and $\pi = p_1 \vee p_2 \vee p_3$. Compute the dimensions of $\ell_1 \vee \ell_2 \vee \ell_3$ and $\ell_1 \vee \ell_2 \vee \pi$.

Exercise 1.7. Prove that if L_1, L_2, L_3, L_4 are pairwise disjoint linear varieties of a projective space, all of the same dimension, then

$$\dim(L_1 \vee L_2) \cap (L_3 \vee L_4) = \dim(L_1 \vee L_3) \cap (L_2 \vee L_4) = \dim(L_1 \vee L_4) \cap (L_2 \vee L_3).$$

Exercise 1.8. Let ℓ, ℓ' be two non-coplanar lines of $\mathbb{P}_3$ and s, s' two distinct lines, each with non-empty intersection with both ℓ and ℓ'. Prove that if s and s' meet, then their intersection point belongs to $\ell \cup \ell'$.

Exercise 1.9. Assume that L_1, L_2, L_3 are linear varieties of dimension 3 of a projective space $\mathbb{P}_5$, such that the joins $L_i \vee L_j$, for $i, j = 1, 2, 3$, $i \neq j$, are three different hyperplanes. Compute the dimension of $L_1 \cap L_2 \cap L_3$.

Exercise 1.10. Assume that $H_0, \dots, H_n$ are hyperplanes of $\mathbb{P}_n$ for which $H_0 \cap \dots \cap H_n = \emptyset$. Compute $\dim H_0 \cap \dots \cap H_i$ for $i = 1, \dots, n-1$.

Exercise 1.11. Assume that $L_1, \dots, L_m$ are linear varieties of dimension $n-2$ of a projective space $\mathbb{P}_n$ for which $\dim L_i \cap L_j = n-3$ and $\dim L_i \cap L_j \cap L_r = n-4$ for $i, j, r = 1, \dots, m$, $i < j < r$. Prove that there is a hyperplane H of $\mathbb{P}_n$ for which $L_i \subset H$, $i = 1, \dots, m$.

Exercise 1.12. Let ℓ_1 and ℓ_2 be two skew lines of a projective space $\mathbb{P}_3$. Prove that for any $p \in \mathbb{P}_3 - \ell_1 \cup \ell_2$ there is a unique line s through p meeting both ℓ_1 and ℓ_2. *Hint*: Use the planes $p \vee \ell_i$.

Exercise 1.13. Same as for Exercise 1.12, but for taking two supplementary linear varieties of a projective space $\mathbb{P}_n$ instead of ℓ_1, ℓ_2.

Exercise 1.14. Assume that L_1, L_2 are $(n-2)$-dimensional linear varieties of a projective space $\mathbb{P}_n$, $n \geq 3$, spanning $\mathbb{P}_n$. Prove that for any $p \in \mathbb{P}_n$, $p \notin L_i$, $i = 1, 2$, there is a unique linear variety L, of dimension $n-2$, going through p and such that $\dim L \cap L_1 = \dim L \cap L_2 = n-3$.

Exercise 1.15. Given two lines ℓ_1, ℓ_2 and a plane π of a projective space, assume that $\dim \ell_1 \cap \ell_2 = \dim \ell_2 \cap \pi = 0$ and $\ell_1 \cap \pi = \emptyset$ and compute $\dim \ell_1 \vee \ell_2 \vee \pi$.

Exercise 1.16. Assume that π_1, π_2, π_3 are three planes of a projective space for which $\dim \pi_1 \vee \pi_2 \vee \pi_3 = 6$ and there is a point $p \in \pi_1 \cap \pi_2 \cap \pi_3$. Prove that $\pi_1 \cap \pi_2 = \pi_2 \cap \pi_3 = \pi_3 \cap \pi_1 = \{p\}$.

Exercise 1.17. Let π_1, π_2, π_3 be planes of a projective space $\mathbb{P}_4$, all going through a certain point p. Assuming $\dim \pi_i \vee \pi_j = 3$ for $i, j = 1, 2, 3$, $i \neq j$, prove that either $\dim \pi_1 \cap \pi_2 \cap \pi_3 = 1$ or $\dim \pi_1 \vee \pi_2 \vee \pi_3 = 3$.

Exercise 1.18. Assume that a projective space $\mathbb{P}_n$ is spanned by three planes π_1, π_2, π_3, and also that no line of $\mathbb{P}_n$ has non-empty intersection with each of the planes π_1, π_2, π_3. Compute n.

Exercise 1.19. Let ℓ_1, ℓ_2, ℓ_3 be three pairwise disjoint lines of a projective space $\mathbb{P}_n$. Prove that $\dim \ell_1 \vee \ell_2 \vee \ell_3 = 4$ if and only if there exists a unique line ℓ of $\mathbb{P}_n$ for which $\ell \cap \ell_i \neq \emptyset$ for $i = 1, 2, 3$.

Exercise 1.20. Given four planes $\pi_1, \pi_2, \pi_3, \pi_4$ of a projective space $\mathbb{P}_{10}$, prove that there is a plane π of $\mathbb{P}_{10}$ which meets all the π_i, $i = 1, 2, 3, 4$.

Exercise 1.21. Prove that points $p_1, \dots, p_m \in \mathbb{P}_n$ are independent if and only if for any $i, j, i \neq j$, the join $p_i \vee p_j$ is a line and does not meet the linear variety $\bigvee_{s \neq i,j} p_s$, spanned by the remaining points.

Exercise 1.22. Let $p_1, \dots, p_n$ be points of a projective space $\mathbb{P}_n$ and assume that $p_1, \dots, p_{n-1}$ are independent. Prove that $p_1, \dots, p_n$ are independent if and only if there is a hyperplane H of $\mathbb{P}_n$ containing $p_1, \dots, p_{n-1}$ and containing not p_n.

Exercise 1.23. Prove that for any linear varieties $L_0, \dots, L_m$ of $\mathbb{P}_n$, the following conditions are equivalent:

(i) $L_i \cap \left(\bigvee_{j \neq i} L_j\right) = \emptyset$ for $i = 0, \dots, m$.

(ii) $\dim L_0 \vee \dots \vee L_m = \dim L_0 + \dots + \dim L_m + m$.

Linear varieties satisfying either (and so both) of the above conditions are called *independent*. What is independence of zero-dimensional linear varieties?

Exercise 1.24. Prove that the number of d-dimensional faces of an m-dimensional simplex is $\binom{m+1}{d+1}$.

Exercise 1.25. Prove that if a line ℓ intersects two faces of a tetrahedron τ of $\mathbb{P}_n$ in two different points, then ℓ meets all faces of τ.

Exercise 1.26. Prove that no vertex of the diagonal triangle of a quadrilateral lies on a side, and also that no side of the diagonal triangle of a quadrivertex contains a vertex.

Exercise 1.27. Assume that A, B, C, D are the vertices of a quadrivertex and P, Q, M, N are points other than the vertices and lying on the sides AB, BC, CD and DA, respectively. Prove that if PQ, MN and AC are concurrent, then PN, QM and BD are concurrent too. *Hint*: Consider the triangles with vertices D, M, N and B, P, Q.

Exercise 1.28. Assume that the vertices p_1, p_2, p_3, p_4 of a quadrivertex V belong, in this order, to the sides $\ell_1, \ell_2, \ell_3, \ell_4$ of a quadrilateral S, and that the intersection of the opposite sides p_1p_2 and p_3p_4 of V belongs to the diagonal $(\ell_1 \cap \ell_4) \vee (\ell_2 \cap \ell_3)$ of S. Prove that the intersections of the remaining two pairs of opposite sides of V belong each to one of the two remaining diagonals of S.

Exercise 1.29. The preceding definitions being readily extended to the case of an arbitrary base field k, prove that 1.8.1 is false if the characteristic of k is $\mathrm{char}(k) = 2$, the diagonals being then always concurrent. At which point the proof of 1.8.1 becomes wrong when $\mathrm{char}(k) = 2$?

Exercise 1.30. Two triangles T, T', in different planes π, π' of $\mathbb{P}_3$, are called *perspective* if and only if there is a perspectivity $f\colon \pi \to \pi'$ such that $f(T) = T'$. Prove that T and T' are perspective if and only if there is a one-to-one correspondence between the set of sides of T and the set of sides of T' such that each pair of corresponding sides have non-empty intersection.

Exercise 1.31. Assume we are given integers n, d_0, d_1, d_2 for which $n \geq 1$, $-1 \leq d_0 \leq d_i$ for $i = 1, 2$, and $d_1 + d_2 - d_0 \leq n$. Prove that any n-dimensional projective space contains linear varieties L_0, L_1, L_2 of respective dimensions d_0, d_1, d_2 and such that $L_0 = L_1 \cap L_2$. *Hint*: Deal first with the case $d_0 = -1$ and then use 1.9.4.

Exercise 1.32. Assume that $\pi \subset \mathbb{P}_3$ is a plane and T, T' two triangles in π, with vertices p_1, p_2, p_3 and p'_1, p'_2, p'_3, respectively. Assume furthermore that $p_i \neq p'_i$, $i = 1, 2, 3$ and $p_i \vee p_j \neq p'_i \vee p'_j$, $1 \leq i < j \leq 3$. Consider the following claims:

(i) The lines $p_i \vee p'_i$, $i = 1, 2, 3$, are concurrent.

(ii) The points $p_i p_j \cap p'_i p'_j$, $1 \leq i < j \leq 3$, are aligned.

(iii) There are a triangle $\bar{T}$, in a plane $\bar{\pi} \subset \mathbb{P}_3$, $\bar{\pi} \neq \pi$, and perspectivities $f, f'\colon \bar{\pi} \to \pi$ such that $f(\bar{T}) = T$ and $f'(\bar{T}) = T'$.

Prove that (i) $\iff$ (iii) and (ii) $\iff$ (iii). Re-prove Desargues' theorem and its converse from these equivalences.

Exercise 1.33. Given two tetrahedra of $\mathbb{P}_3$, with vertices $p_1, \dots, p_6$ and $p'_1, \dots, p'_6$, sharing no vertex, edge or face, assume that the lines $p_i p'_i$, $i = 1, \dots, 6$, are concurrent. Prove that the intersections of corresponding edges, $p_i p_j \cap p'_i p'_j, i, j = 1, \dots, 6$, $i \neq j$, and the intersections of corresponding faces, $p_i p_j p_s \cap p'_i p'_j p'_s$, $i, j, s = 1, \dots, 6$, all distinct, belong all to the same plane.

Chapter 2
Projective coordinates and cross ratio

2.1 Projective references

In order to lighten a bit the notations, from now on a hat $\widehat{}$ over an element of a list indicates that such element is dropped from the list. A *projective reference* (or *coordinate frame*) of a projective space $\mathbb{P}_n$ is an ordered set of $n+2$ points $(p_0, \ldots, p_n, A)$ that satisfy the following two conditions:

(1) $p_0, \ldots, p_n$ are independent, and

(2) for any $i = 0, \ldots, n$, $A \notin p_0 \vee \cdots \vee \widehat{p_i} \vee \cdots \vee p_n$.

The first $n+1$ points of the reference, $p_0, \ldots, p_n$, are called *vertices* of the reference, while the last point A is called the *unit point*. The simplex with vertices $p_0, \ldots, p_n$ is called the *simplex of the reference*, and sometimes the *fundamental simplex* (*triangle* if $n = 2$, *tetrahedron* if $n = 3$); its faces are called the *faces of the reference*. Then condition (2) above just says that the unit point does not belong to any of the $(n-1)$-dimensional faces of the reference. Usually the $(n-1)$-dimensional faces of the reference are numbered as the vertex they are opposite, $p_0 \vee \cdots \vee \widehat{p_i} \vee \cdots \vee p_n$ being thus called the i-th $(n-1)$-dimensional face of the reference.

Once the vertices $p_0, \ldots, p_n$ have been assumed to be independent, so are $p_0, \ldots, \widehat{p_i}, \ldots, p_n$ and so, by 1.4.7, $A \notin p_0 \vee \cdots \vee \widehat{p_i} \vee \cdots \vee p_n$ if and only if $p_0, \ldots, \widehat{p_i}, \ldots, p_n, A$ are independent points. It follows that the above conditions (1) and (2) together are equivalent to the single one:

(3) any $n+1$ points extracted from $\{p_0, \ldots, p_n, A\}$ are independent.

A projective reference allows us to select certain bases of the associated vector space E which will be used in turn to define the coordinates of the points: an ordered set of $n+1$ vectors $(e_0, \ldots, e_n)$ is said to be a *basis adapted to* the reference $\Delta = (p_0, \ldots, p_n, A)$ if and only if the vectors represent the corresponding vertices and their sum represents the unit point, that is: $p_i = [e_i]$, $i = 0, \ldots, n$, and $A = [e_0 + \cdots + e_n]$. The vectors $e_0, \ldots, e_n$ of an adapted basis compose, indeed, a basis of E, as they are independent, due to the independence of $p_0, \ldots, p_n$ (1.5.12), and in number equal to $\dim E$. An adapted basis obviously determines the reference, but the converse is not true, as we will see in 2.1.2 below.

Our first job is to prove the existence of adapted bases:

Lemma 2.1.1. *For any projective reference Δ, there is a basis adapted to it.*

Proof. If $\Delta = (p_0, \dots, p_n, A)$, select any representatives of the vertices and the unit point, say $p_i = [v_i], i = 0, \dots, n$, and $A = [v]$. As above, $v_0, \dots, v_n$ represent independent points and are in number equal to dim E, so they compose a basis of E. Thus, one may write $v = \lambda_0 v_0 + \dots + \lambda_n v_n$, where $\lambda_i \in k$ and, as we will see next, $\lambda_i \neq 0, i = 0, \dots, n$. Indeed, if $\lambda_i = 0$, then $v \in \langle v_0 \rangle + \dots + \widehat{\langle v_i \rangle} + \dots + \langle v_n \rangle$ and so $A \in p_0 \vee \dots \vee \widehat{p_i} \vee \dots \vee p_n$ against condition (2) of the definition of reference. Now, since $\lambda_i \neq 0$, we may take the vectors $e_i = \lambda_i v_i$ as new representatives of the vertices: $[e_i] = p_i, i = 0, \dots, n$. Furthermore $e_0 + \dots + e_n = v$ represents the unit point and so $e_0, \dots, e_n$ is a basis adapted to Δ. □

Corresponding vectors of two bases adapted to the same reference are proportional, all with the same ratio of proportionality, namely:

Lemma 2.1.2. *Assume that $v_0, \dots, v_n$ is a basis adapted to a reference Δ. Then $e_0, \dots, e_n$ is also a basis adapted to Δ if and only if there is a (necessarily non-zero) $\lambda \in k$ such that $e_i = \lambda v_i$, $i = 0, \dots, n$.*

Proof. Assume $\Delta = (p_0, \dots, p_n, A)$. The *if* part is clear: from $e_i = \lambda v_i, i = 0, \dots, n$, it follows that $[e_i] = [v_i] = p_i, i = 0, \dots, n$, and also

$$\Big[\sum_{i=0}^{n} e_i\Big] = \Big[\lambda \sum_{i=0}^{n} v_i\Big] = A.$$

For the converse, note first that for each $i = 0, \dots, n$, both e_i and v_i represent the i-th vertex of the reference, and thus there is $\lambda_i \in k$ such that $e_i = \lambda_i v_i$. In addition, both $\sum_0^n e_i$ and $\sum_0^n v_i$ represent the unit point, and therefore there is $\lambda \in k$ such that $\sum_0^n e_i = \lambda \sum_0^n v_i$. We get

$$\sum_0^n \lambda_i v_i = \lambda \sum_0^n v_i$$

and so, by the independence of the v_i, $\lambda_i = \lambda$, $i = 0, \dots, n$, as wanted. □

Sometimes it is useful to have a reference to which an already selected basis is adapted:

Lemma 2.1.3. *If $e_0, \dots, e_n$ is a basis of E, $\Delta = ([e_0], \dots, [e_n], [e_0 + \dots + e_n])$ is a projective reference, the only one with adapted basis $e_0, \dots, e_n$.*

Proof. The vertices $[e_0], \dots, [e_n]$ of Δ are independent by 1.5.12. If for some i, $[e_0 + \dots + e_n] \in [e_0] \vee \dots \vee \widehat{[e_i]} \vee \dots \vee [e_n]$, then

$$e_0 + \dots + e_n \in \langle e_0, \dots, \widehat{e_i}, \dots, e_n \rangle$$

and so

$$e_i \in \langle e_0, \dots, \widehat{e_i}, \dots, e_n \rangle$$

against the independence of $e_0, \dots, e_n$. So Δ is a reference. The remainder of the claim is clear from the definition of adapted basis. □

Remark 2.1.4. Since any non-zero vector space has a basis, Lemma 2.1.3 shows that any projective space $\mathbb{P}_n$ has a reference. It shows also also that any set of independent points $p_0, \dots, p_n \in \mathbb{P}_n$ may be taken as the set of vertices of a reference of $\mathbb{P}_n$. Indeed, if e_i is a representative of p_i, $i = 0, \dots, n$, the vectors $e_0, \dots, e_n$ are independent, due to the independence of $p_0, \dots, p_n$, and therefore a basis of E. Then it is enough to take the reference with adapted basis $e_0, \dots, e_n$.

2.2 Projective coordinates

Assume that Δ is a projective reference of $\mathbb{P}_n$. If $p \in \mathbb{P}_n$, we say that elements $x_0, \dots, x_n$ of k are *projective coordinates* of p relative to Δ if and only if there is a basis $e_0, \dots, e_n$ adapted to Δ for which $[x_0 e_0 + \cdots + x_n e_n] = p$. In other words, the components of any representative of p relative to any basis adapted to Δ are taken as projective coordinates of p relative to Δ.

Projective coordinates are also called *homogeneous coordinates* and we will often say just coordinates if no confusion may result. Once the reference is clear, we will drop any mention to it. Notations being as in the definition, we will say that the vector $(x_0, \dots, x_n) \in k^{n+1}$ is a coordinate vector of p. Note that the number of coordinates is the dimension of the space plus one.

For the remainder of this section we assume that an arbitrary projective reference $\Delta = (p_0, \dots, p_n, A)$ has been fixed and all projective coordinates will be relative to Δ. Since the existence of a basis adapted to Δ has been guaranteed in 2.1.1, it is clear that any point has coordinates. It is also clear from the definition that no point has coordinate vector $(0, \dots, 0)$, as the vector 0 represents no point. However, according to the definition of coordinates, a point p may have different coordinates relative to a reference Δ, depending on the choices of the representative of p and the basis adapted to Δ. We will control this after proving the following lemma:

Lemma 2.2.1. *If $p = [x_0 v_0 + \cdots + x_n v_n]$ for a basis $v_0, \dots, v_n$ adapted to Δ, then $p = [x_0 e_0 + \cdots + x_n e_n]$ for any basis $e_0, \dots, e_n$ adapted to Δ.*

Proof. Follows from 2.1.2 as, using the notations therein,

$$x_0 e_0 + \cdots + x_n e_n = \lambda (x_0 v_0 + \cdots + x_n v_n).$$

□

Proposition 2.2.2. *$x_0, \dots, x_n$ and $y_0, \dots, y_n$ are coordinates of the same point p if and only if there is $\lambda \in k - \{0\}$ such that $x_i = \lambda y_i$, $i = 0, \dots, n$.*

Proof. Fix a basis $e_0, \dots, e_n$ adapted to Δ. By 2.2.1, $x_0, \dots, x_n$ and $y_0, \dots, y_n$ are coordinates of the same point p if and only if

$$p = [x_0 e_0 + \cdots + x_n e_n] = [y_0 e_0 + \cdots + y_n e_n]$$

which in turn is equivalent to

$$x_0e_0 + \cdots + x_ne_n = \lambda(y_0e_0 + \cdots + y_ne_n),$$

and hence to $x_i = \lambda y_i$, $i = 0, \ldots, n$, for a certain non-zero $\lambda \in k$. □

So far, we have seen that any point p has coordinates $x_0, \ldots, x_n$ relative to Δ, that $x_i \neq 0$ for some i, and that $x_0, \ldots, x_n$ are determined by p up to a non-zero common factor. It is time to care about the converse, that is, the existence and uniqueness of a point with given coordinates:

Proposition 2.2.3. *Given $x_0, \ldots, x_n \in k$, not all equal to zero, there is one and only one point $p \in \mathbb{P}_n$ with projective coordinates $x_0, \ldots, x_n$.*

Proof. If $e_0, \ldots, e_n$ is a basis adapted to Δ, then, by the hypothesis, the vector $x_0e_0 + \cdots + x_ne_n$ is not zero and, clearly, the point it represents, $p = [x_0e_0 + \cdots + x_ne_n]$, has coordinates $x_0, \ldots, x_n$. If a point q has the same coordinates, then by the definition of coordinates and 2.2.1, $q = [x_0e_0 + \cdots + x_ne_n]$ and therefore $p = q$. □

In the sequel we will write $[x_0, \ldots, x_n]_\Delta$ (or just $[x_0, \ldots, x_n]$, if Δ is clear) to denote the point with projective coordinates $x_0, \ldots, x_n$ relative to Δ. Note that this makes sense if and only if at least one of the x_i is not zero. The following example follows directly from the definition of coordinates:

Example 2.2.4. The elements of Δ have coordinates as follows:

$$\begin{aligned} p_0 &= [1, 0, \ldots, 0]_\Delta, \\ &\ \vdots \\ p_n &= [0, \ldots, 0, 1]_\Delta, \\ A &= [1, 1, \ldots, 1]_\Delta. \end{aligned}$$

Remark 2.2.5. The projective coordinates of a point being determined up to a common factor, the reader may notice that the equality of points is equivalent to the proportionality, not the equality, of their coordinate vectors. Namely, by 2.2.2, $[x_0, \ldots, x_n] = [y_0, \ldots, y_n]$ if and only if there is $\lambda \in k$ such that $(x_0, \ldots, x_n) = \lambda(y_0, \ldots, y_n)$. The latter condition is often equivalently stated as

$$\operatorname{rk}\begin{pmatrix} x_0 & \ldots & x_n \\ y_0 & \ldots & y_n \end{pmatrix} < 2$$

or also as

$$\begin{vmatrix} x_i & x_j \\ y_i & y_j \end{vmatrix} = 0$$

for $0 \leq i < j \leq n$.

There is a choice of a coordinate frame on each face of the reference giving coordinates on the face easily related to the coordinates of $\mathbb{P}_n$, namely:

Proposition 2.2.6. *Let $\Delta = (p_0, \dots, p_n, A)$ be a projective reference. Assume that L is a d-dimensional face of Δ, $d > 0$, say $L = p_{i_0} \vee \dots \vee p_{i_d}$, $i_0 < \dots < i_d$, and denote by T the face opposite L. If A' is the image of A by projecting from T on L, then we have:*

(a) *$\Delta' = (p_{i_0} \dots p_{i_d}, A')$ is a reference of L.*

(b) *A point $p = [x_0, \dots, x_n]_\Delta \in \mathbb{P}_n$ belongs to L if and only if $x_i = 0$ for all $i \in \{0, \dots, n\} - \{i_0, \dots, i_d\}$ and in such a case $x_{i_0}, \dots, x_{i_d}$ are coordinates of p relative to Δ'.*

Proof. After reordering the vertices of Δ, we may assume without restriction that $L = p_0 \vee \dots \vee p_d$ and hence $T = p_{d+1} \vee \dots \vee p_n$. Let $e_0, \dots, e_n$ be a basis adapted to Δ. The equality $e_0 + \dots + e_d = (e_0 + \dots + e_n) - (e_{d+1} + \dots + e_n)$ shows that $[e_0 + \dots + e_d] \in L \cap (A \vee T)$ and thus that $A' = [e_0 + \dots + e_d]$. Then Δ' is the reference of L with adapted basis $e_0, \dots, e_d$, by 2.1.3. Any point $p = [x_0, \dots, x_d]_{\Delta'} \in L$ may be written $p = [x_0 e_0 + \dots + x_d e_d] = [x_0 e_0 + \dots + x_d e_d + 0 e_{d+1} + \dots + 0 e_n]$ and so $p = [x_0, \dots, x_d, 0, \dots, 0]_\Delta$. Conversely any point $p = [x_0, \dots, x_d, 0, \dots, 0]_\Delta$ is $p = [x_0 e_0 + \dots + x_d e_d]$ and so it belongs to L. □

The reference Δ' and the coordinates relative to it will be called the reference and coordinates of L *subordinated* or *induced* by Δ.

The fact that coordinates of a point are the components of one of its representatives, and conversely, makes it very easy to switch between points and their representatives: scalars $x_0, \dots, x_n$, not all zero, may be interpreted either as projective coordinates of a point $p \in \mathbb{P}_n$ (and then read up to a common non-zero factor) or as the components of a representative of p. The next proposition makes use of this fact. It is our first translation of a projective notion in terms of coordinates. Many others will be presented in forthcoming Sections 2.5 to 2.8.

Proposition 2.2.7. *Points $p_j = [b_j^0, \dots, b_j^n]$, $j = 0, \dots, m$, are linearly independent if and only if their coordinate vectors $(b_j^0, \dots, b_j^n)$, $j = 0, \dots, m$, are independent, or, equivalently, if and only if*

$$\operatorname{rk} \begin{pmatrix} b_0^0 & \dots & b_m^0 \\ \vdots & & \vdots \\ b_0^n & \dots & b_m^n \end{pmatrix} = m + 1.$$

Proof. Direct from 1.5.12 and the fact, remarked above, that coordinates of a point are the components of one of its representatives. □

2.3 Change of coordinates

In this section we will describe how coordinates of a point relative to two different references are related. Assume we have two projective references of $\mathbb{P}_n$, $\Delta = (p_0, \dots, p_n, A)$ and $\Omega = (q_0, \dots, q_n, B)$, and coordinates of the elements of one reference relative to the other, say,

$$\begin{aligned} q_0 &= [\bar{a}_0^0, \dots, \bar{a}_0^n]_\Delta, \\ &\vdots \\ q_n &= [\bar{a}_n^0, \dots, \bar{a}_n^n]_\Delta, \\ B &= [a^0, \dots, a^n]_\Delta. \end{aligned}$$

First we need to modify the coordinate vectors of the q_i, using that each is determined up to a non-zero factor:

Lemma 2.3.1. *If Δ and Ω are projective references of $\mathbb{P}_n$, one may choose coordinate vectors of the vertices of Ω which add up to a coordinate vector of the unit point of Ω, all coordinate vectors being relative to Δ.*

Proof. Take the notations as above. Our argument repeats the one we already used to prove the existence of adapted basis (2.1.1), this time using coordinate vectors. We make it explicit in order to show how to proceed in practice:

Since the points q_i, $i = 0, \dots, n$, are independent, by 2.2.7 so are their coordinate vectors and the system of linear equations in $\lambda_0, \dots, \lambda_n$,

$$\lambda_0 \begin{pmatrix} \bar{a}_0^0 \\ \vdots \\ \bar{a}_0^n \end{pmatrix} + \dots + \lambda_n \begin{pmatrix} \bar{a}_n^0 \\ \vdots \\ \bar{a}_n^n \end{pmatrix} = \begin{pmatrix} a^0 \\ \vdots \\ a^n \end{pmatrix},$$

has a unique solution; let us write it $\lambda_0, \dots, \lambda_n$. For each i, $\lambda_i \neq 0$, as otherwise, again by 2.2.7, the points $q_0, \dots, \widehat{q_i}, \dots, q_n, B$ would be dependent. Then just take new coordinate vectors

$$(a_i^0, \dots, a_i^n) = \lambda_i (\bar{a}_i^0, \dots, \bar{a}_i^n)$$

and the condition of the claim is fulfilled. □

Now the rule for changing coordinates is contained in the next proposition; it makes sense due to 2.3.1 above:

Proposition 2.3.2. *Take M to be a matrix whose columns are, in this order, coordinate vectors of $q_0, \dots, q_n$ whose sum is a coordinate vector of B, all coordinate* *vectors being relative to* Δ. *Then for any point* $p = [x_0, \dots, x_n]_\Delta = [y_0, \dots, y_n]_\Omega$,

$$\begin{pmatrix} x_0 \\ \vdots \\ x_n \end{pmatrix} = \rho M \begin{pmatrix} y_0 \\ \vdots \\ y_n \end{pmatrix} \tag{2.1}$$

for some non-zero $\rho \in k$.

Proof. Take a basis $e_0, \dots, e_n$ adapted to Δ. For $i = 0, \dots, n$, let $(a_i^0, \dots, a_i^n)$ be the i-th column vector of M; according to its definition,

$$v_i = a_i^0 e_0 + \dots + a_i^n e_n$$

is a representative of q_i and furthermore $B = [v_0 + \dots + v_n]$. Hence, $v_0, \dots, v_n$ is a basis adapted to Ω and a representative of p may be written

$$\sum_{i=0}^{n} y_i v_i = \sum_{i=0}^{n} y_i \Big(\sum_{j=0}^{n} a_i^j e_j\Big) = \sum_{j=0}^{n} \Big(\sum_{i=0}^{n} a_i^j y_i\Big) e_j.$$

It follows that

$$\sum_{i=0}^{n} a_i^0 y_i, \dots, \sum_{i=0}^{n} a_i^n y_i$$

are coordinates of p relative to Δ, and hence the claim. □

In the sequel we will refer to the matrix M of 2.3.2 as a *matrix changing coordinates relative to* Ω *into coordinates relative to* Δ. Its existence has been already proved in 2.3.1. By 2.2.7, it is an $(n+1) \times (n+1)$ invertible matrix and, obviously, its inverse changes coordinates relative to Δ into coordinates relative to Ω.

Remark 2.3.3. Assume we have fixed a reference Δ of a projective space $\mathbb{P}_n$ and denote by $x_0, \dots, x_n$ the coordinates relative to it. Given a regular $(n+1) \times (n+1)$ matrix M, there is a uniquely determined reference Ω of $\mathbb{P}_n$ such that M changes coordinates relative to Ω into coordinates relative to Δ. Indeed, for the existence, if $e_0, \dots, e_n$ is a basis adapted to Δ and $(a_i^0, \dots, a_i^n)$ is the i-th column vector of M, then, by 2.3.2, it is enough to take Ω to be the reference which has as adapted basis

$$v_i = a_i^0 e_0 + \dots + a_i^n e_n, \quad i = 0, \dots, n.$$

The uniqueness of Ω is clear because once M changes coordinates relative to Ω into coordinates relative to Δ,

$$M \begin{pmatrix} 1 \\ \vdots \\ 0 \end{pmatrix}, \dots, M \begin{pmatrix} 0 \\ \vdots \\ 1 \end{pmatrix}, M \begin{pmatrix} 1 \\ \vdots \\ 1 \end{pmatrix},$$

are coordinate vectors relative to Δ of the vertices and unit point of Ω, by 2.2.4.

Thus, a new reference Ω, and the coordinates $y_0, \dots, y_n$ relative to it, may be defined by just giving the regular matrix M changing coordinates relative to Ω into the actual ones. It is usual to do this by giving the equations that relate old and new coordinates, namely either

$$\begin{pmatrix} x_0 \\ \vdots \\ x_n \end{pmatrix} = \rho M \begin{pmatrix} y_0 \\ \vdots \\ y_n \end{pmatrix}, \qquad \det M \neq 0,$$

or, equivalently,

$$\begin{pmatrix} y_0 \\ \vdots \\ y_n \end{pmatrix} = \rho N \begin{pmatrix} x_0 \\ \vdots \\ x_n \end{pmatrix}, \qquad \det N \neq 0,$$

the matrix M being then N^{-1}.

2.4 Absolute coordinate

Assume we have fixed a projective reference $\Delta = (p_0, p_1, A)$ in a projective line $\mathbb{P}_1$. In Section 2.2, we have assigned to each point $p \in \mathbb{P}_1$ a pair of homogeneous coordinates (x_0, x_1) that determine p and are in turn determined by p up to a common factor. In order to remove this indeterminacy, if $x_1 \neq 0$, one may take the ratio x_0/x_1 instead of the pair (x_0, x_1). The exceptional case $x_1 = 0$ occurs only for the single point $[x_0, 0] = [1, 0] = p_0$. Then, we take ∞ (named infinity) to be a symbol, not an element of k, and define the *absolute coordinate* of the point $p = [x_0, x_1]$ relative to Δ as being x_0/x_1 if $x_1 \neq 0$, and ∞ otherwise. This provides a new type of coordinate with no indeterminacy:

Proposition 2.4.1. *Once a projective reference Δ of $\mathbb{P}_1$ has been fixed, the rule*

$$\mathbb{P}_1 \longrightarrow k \cup \{\infty\},$$

$$p = [x_0, x_1]_\Delta \longmapsto \begin{cases} x_0/x_1 & \text{if } x_1 \neq 0, \\ \infty & \text{otherwise}, \end{cases}$$

defines a bijective map that assigns to each point its absolute coordinate relative to Δ.

Proof. The map is well defined by 2.2.2. If points $[x_0, x_1]$ and $[y_0, y_1]$ have the same image in k, then both second coordinates are non-zero and $x_0/x_1 = y_0/y_1$, which in turn implies $[x_0, x_1] = [y_0, y_1]$ (2.2.5). If the same points are both mapped to ∞, then $x_1 = y_1 = 0$ and, obviously, $[x_0, 0] = [y_0, 0]$. To close, any $\theta \in k$ is the image of $[\theta, 1]$, while $[1, 0]$ is mapped to ∞. □

One says that an absolute coordinate is finite to mean that it is not ∞. If the usual rule $a/0 = \infty$, for any non-zero $a \in k$, is accepted, then the absolute coordinate of $[x_0, x_1]$ is x_0/x_1 in all cases. Conversely, as seen in the above proof, the point with finite absolute coordinate θ is $[\theta, 1]$, while the point with absolute coordinate ∞ is $[1, 0]$.

The next example should be familiar to the reader:

Example 2.4.2. In the one-dimensional projective space of Example 1.2.1, take as reference the lines $x = 0$, $y = 0$, $x - y = 0$. Then the line $ax + by = 0$ has projective coordinates $a, -b$ and hence its absolute coordinate is its slope $-a/b$.

Dividing by the last (or any other) coordinate in order to eliminate the indeterminacy of the projective coordinates is not useful if $n > 1$, as then there are too many points with last coordinate equal to zero which, therefore, would need a separate definition. If $n = 1$, the use of absolute coordinates usually leads to more compact computations, but it requires some skill, as the cases in which some coordinate takes value ∞ require a separate checking, usually done by switching back to projective coordinates. As an example, next we explain in detail the change of absolute coordinates.

Assume we have two projective references of $\mathbb{P}_1$, $\Delta = (p_0, p_1, A)$ and $\Omega = (q_0, q_1, B)$. In Section 2.3 we have seen that there is a regular matrix

$$M = \begin{pmatrix} a & b \\ c & d \end{pmatrix}$$

such that for any $p = [y_0, y_1]_\Omega \in \mathbb{P}_1$, it is $p = [x_0, x_1]_\Delta$ if and only if

$$\begin{aligned} x_0 &= \rho(ay_0 + by_1), \\ x_1 &= \rho(cy_0 + dy_1) \end{aligned} \tag{2.2}$$

for some non-zero $\rho \in k - \{0\}$. If $cy_0 + dy_1 \neq 0$, one may divide to obtain the absolute coordinate θ of p relative to Δ:

$$\theta = \frac{ay_0 + by_1}{cy_0 + dy_1}.$$

If the absolute coordinate ξ of p relative to Ω is finite, then $y_0 \neq 0$ and the last equality gives

$$\theta = \frac{a\xi + b}{c\xi + d}. \tag{2.3}$$

Now, for the cases we have excepted:

(a) If $\xi = \infty$, $p = [1, 0]_\Omega$ and so the equalities (2.2) give $\theta = a/c$ for $c \neq 0$ and $\theta = \infty$ if $c = 0$. In the latter case $a \neq 0$ because M is regular.

(b) If $\xi \neq \infty$ and $c\xi + d = 0$, again by (2.2), $\theta = \infty$. In such a case $a\xi + b \neq 0$, again by the regularity of M.

Thus the equation (2.3) covers all cases provided one uses the rule $\lambda/0 = \infty$, for any $\lambda \in k - \{0\}$, and sets the value of $(a\xi + b)/(c\xi + d)$ to be a/c for $\xi = \infty$ (the value given by the rules of elementary calculus for $\xi \to \infty$).

We close this section by presenting two consequences of 2.4.1. They still hold if the base field k is an arbitrary infinite field, but, obviously, both fail to be true if k is a finite field, by 2.4.1.

Corollary 2.4.3. *Any projective line contains infinitely many points.*

Proof. Direct from 2.4.1 and 1.3.4. □

Corollary 2.4.4. *If $L_1, \dots, L_m$ are linear varieties of a projective space $\mathbb{P}_n$ and* $\dim L_i < n$, *$i = 1, \dots, m$, then $\mathbb{P}_n \neq L_1 \cup \dots \cup L_m$.*

Proof. The case $m = 1$ being clear due to the hypothesis $\dim L_1 < n$, we proceed by induction on m: we assume thus $m > 1$ and the claim true for sets of $m' < m$ linear varieties. Assume we have $\mathbb{P}_n = L_1 \cup \dots \cup L_m$. Having $L_1 \subset L_2 \cup \dots \cup L_m$ would give $\mathbb{P}_n = L_2 \cup \dots \cup L_m$ against the induction hypothesis. We may thus pick a point $p \in L_1 - L_2 \cup \dots \cup L_m$. Since $\dim L_1 < n$, we may take a second point $q \in \mathbb{P}_n - L_1$. Then $p \neq q$ and $\ell = p \vee q$ is a line of $\mathbb{P}_n$. On one hand $\ell \not\subset L_1$ because $q \in \ell$, while, on the other, $\ell \not\subset L_i$, $i = 2, \dots, m$, because $p \in \ell$. As a consequence each of the intersections $\ell \cap L_i$ contains at most a single point and so $\ell = \ell \cap (L_1 \cup \dots \cup L_m)$ contains at most m points, against 2.4.3. □

2.5 Parametric equations

Let $L = [F]$ be a linear variety of dimension d of $\mathbb{P}_n$ and assume we are given $d + 1$ independent points $q_0, \dots, q_d \in L$, or, equivalently by 1.5.5, $d + 1$ points $q_0, \dots, q_d$ spanning L, $L = q_0 \vee \dots \vee q_d$. Any set $v_0, \dots, v_n$ of representatives of $q_0, \dots, q_d$, $q_i = [v_i]$, $i = 0, \dots, d$, is a basis of F, as they are independent vectors (by 1.5.12) in number equal to $\dim F$. Thus a point p belongs to L if and only if there exist $\lambda_0, \dots, \lambda_d \in k$, not all equal to zero, such that

$$p = [\lambda_0 v_0 + \dots + \lambda_d v_d]. \tag{2.4}$$

The above expression is in fact a condensed – vectorial – form of what is called a *system of parametric equations* of L. Note that while $(\lambda_0, \dots, \lambda_d)$ describes $k^{d+1} - (0, \dots, 0)$, the point p describes L. We will just turn (2.4) into scalar equations by using coordinates.

Assume that we fix a reference Δ of $\mathbb{P}_n$ and that $q_i = [b_i^0, \dots, b_i^n]_\Delta$, $i = 0, \dots, d$. Then the representatives v_i may be taken to be

$$v_i = \sum_{j=0}^{n} b_i^j e_j,$$

$e_0, \dots, e_n$ a basis adapted to Δ. The equality (2.4) may then be rewritten as

$$p = \Big[\lambda_0 \sum_{j=0}^{n} b_0^j e_j + \dots + \lambda_d \sum_{j=0}^{n} b_d^j e_j\Big] = \Big[\Big(\sum_{i=0}^{d} \lambda_i b_i^0\Big) e_0 + \dots + \Big(\sum_{i=0}^{d} \lambda_i b_i^n\Big) e_n\Big],$$

thus showing the coordinates of the points in L. We have proved:

Proposition 2.5.1. *A coordinate frame being chosen in P_n, assume that $d+1$ independent points $[b_i^0, \dots, b_i^n]$, $i = 0, \dots, d$, belong to a d-dimensional linear variety L. Then a point $p = [x_0, \dots, x_n]$ belongs to L if and only if there are $\lambda_0, \dots, \lambda_d \in k$, not all equal to zero, such that*

$$\begin{pmatrix} x_0 \\ \vdots \\ x_n \end{pmatrix} = \lambda_0 \begin{pmatrix} b_0^0 \\ \vdots \\ b_0^n \end{pmatrix} + \dots + \lambda_d \begin{pmatrix} b_d^0 \\ \vdots \\ b_d^n \end{pmatrix}. \tag{2.5}$$

The set of $n+1$ scalar equations in which the equation (2.5) splits (and often also (2.5) itself) is called a system of *parametric equations* of L, $\lambda_0, \dots, \lambda_n$ being called the *parameters* of p. The reader may notice that the column vectors

$$\begin{pmatrix} b_0^0 \\ \vdots \\ b_0^n \end{pmatrix}, \dots, \begin{pmatrix} b_d^0 \\ \vdots \\ b_d^n \end{pmatrix} \tag{2.6}$$

are linearly independent, as they are coordinate vectors of independent points. It is also worth noting that the parametric equations depend not only on the reference, but also on the choice of the independent points in L and the further choice of their coordinate vectors.

There is a converse of 2.5.1 that assures that any equations of the form of (2.5) are parametric equations of a linear variety of dimension d. To be precise,

Proposition 2.5.2. *Let Δ be a projective reference of a projective space $\mathbb{P}_n$. Assume that linearly independent vectors $(b_i^0, \dots, b_i^n) \in k^{n+1}$, $i = 0, \dots, d$, are given. Then the set of points $p = [x_0, \dots, x_n]_\Delta$ whose coordinates have the form*

$$\begin{pmatrix} x_0 \\ \vdots \\ x_n \end{pmatrix} = \lambda_0 \begin{pmatrix} b_0^0 \\ \vdots \\ b_0^n \end{pmatrix} + \dots + \lambda_d \begin{pmatrix} b_d^0 \\ \vdots \\ b_d^n \end{pmatrix}. \tag{2.7}$$

for some $\lambda_0, \dots, \lambda_d \in k$*, not all equal to zero, is a linear variety of dimension* d*, the one spanned by the points* $[b_i^0, \dots, b_i^n]_\Delta$, $i = 0, \dots, d$.

Proof. Take $e_0, \dots, e_n$ to be a basis adapted to Δ and $v_i = \sum_0^0 b_i^j e_j$. Then the vectors v_i are independent by the hypothesis, and equation (2.7) just says that $p \in [\langle v_0, \dots, v_n \rangle]$, hence the claim. □

Remark 2.5.3. The parameters of a system of parametric equations of a d-dimensional linear variety L may be understood as a set of $d+1$ variables whose free variation provides all coordinate vectors of all points in L. They have a deeper meaning though: take the notations as in Proposition 2.5.1, and consider in L the projective reference which has adapted basis $v_i = \sum_{j=0}^n b_i^j e_j$, $i = 1, \dots, d$, (2.1.3); the equation (2.5) says then that the parameters $\lambda_0, \dots, \lambda_d$ of the point $p \in L$ are just coordinates of p relative to Ω. The arbitrary point $p \in L$ has thus coordinates $\lambda_0, \dots, \lambda_d$ in L and coordinates $x_0, \dots, x_n$ in $\mathbb{P}_n$, the parametric equations relating them.

In particular the reader may notice that the parameters of a point $p \in L$ are determined by p up to a common factor, which is also a direct consequence of the linear independence of the columns of the parametric equations.

2.6 Implicit equations

Let us first recall that if F is a subspace of dimension $d+1$ of a vector space E of dimension $n+1$, then the linear forms ω vanishing on all vectors $v \in F$ are the elements of an $(n-d)$-dimensional subspace $F^\perp$ of the dual space $E^\vee$ of E, usually called the *incident* or the *orthogonal* of F. In other words,

$$F^\perp = \{\omega \in E^\vee \mid \omega(v) = 0 \text{ for all } v \in F\}.$$

Furthermore F may be recovered from $F^\perp$ as the set of all vectors on which all forms in $F^\perp$ do vanish, that is,

$$F = \{v \in E \mid \omega(v) = 0 \text{ for all } \omega \in F^\perp\}.$$

Assume again that $L = [F]$ is a linear variety of P_n. A point $p = [v]$ belongs to L if and only if $v \in F$, which, as recalled above, is equivalent to

$$\omega(v) = 0 \quad \text{for all } \omega \in F^\perp. \tag{2.8}$$

If $\tau_1, \dots, \tau_{n-d}$ is a basis of $F^\perp$, then any form $\omega \in F^\perp$ is a linear combination of $\tau_1, \dots, \tau_{n-d}$ and therefore condition (2.8) above may be equivalently stated as

$$\tau_1(v) = 0, \dots, \tau_{n-d}(v) = 0. \tag{2.9}$$

Again these equalities are condensed forms of the equations we want. Next they will be turned into scalar equations by taking coordinates.

Let Δ be a projective reference and $e_0, \dots, e_n$ a basis adapted to it. If $\omega_1, \dots, \omega_n$ is the basis of $E^\vee$ dual of $e_0, \dots, e_n$, each τ_i may be written

$$\tau_i = a_i^0 \omega_0 + \dots + a_i^n \omega_n, \quad i = 1, \dots, n-d.$$

Using these expressions and that, as well known, $\omega_i(e_i) = 1$ and $\omega_i(e_j) = 0$ for $i \neq j, i, j = 0, \dots, n$, the equalities (2.9) may be rewritten as

$$\begin{gathered} a_1^0 x_0 + \dots + a_1^n x_n = 0, \\ \vdots \\ a_{n-d}^0 x_0 + \dots + a_{n-d}^n x_n = 0, \end{gathered}$$

which are linearly independent linear equations because the forms $\tau_1, \dots, \tau_{n-d}$ are linearly independent. The above is thus a set of $n-d$ independent homogeneous linear equations which are satisfied by the coordinates of a point p if and only if $p \in L$. Such a set is called a *system of implicit* (or *homogeneous*) *equations of* L. It is often called just a *system of equations of* L if no confusion may result, and its elements are referred to as *implicit* or *homogeneous equations* (or just *equations*) *of* L. The adjective *homogeneous* is used to distinguish the present equations from the equations of the affine linear varieties. We already exhibited a system of implicit equations of each face of the reference in Proposition 2.2.6. The reader may notice that the homogeneity of the equations makes irrelevant the arbitrary factor involved in the coordinates of a point in order for these to be a solution of the system. It cannot be otherwise, as coordinates $x_0, \dots, x_n$ of a point p are a solution if and only if $p \in L$, and the latter condition is obviously independent of the arbitrary factor.

We have thus proved:

Proposition 2.6.1. *Once a projective reference is fixed in $\mathbb{P}_n$, for each linear variety L of dimension d there is a system Σ, of $n-d$ independent homogeneous linear equations in $n+1$ variables, such that a point $[x_0, \dots, x_n]$ belongs to L if and only if $x_0, \dots, x_n$ is a solution of Σ.*

The reference being fixed, the system Σ obviously determines L. If $\Sigma = \{h_1 = 0, \dots, h_{n-d} = 0\}$, L is referred to as the linear variety *defined by the* (or *with*) *equations* $h_1 = 0, \dots, h_{n-d} = 0$ or, often, just as the linear variety $h_1 = 0, \dots, h_{n-d} = 0$. If H is a hyperplane, the notation $H : h = 0$ is used to mean that it is the hyperplane with equation $h = 0$.

On the other hand, for a given L and relative to the same reference, there may be many systems of implicit equations of L, as any $n-d$ independent linear combinations of the equations of the system give rise to a system with the same solutions. The converse of 2.6.1 does hold:

Proposition 2.6.2. *Assume that a projective reference Δ is fixed in $\mathbb{P}_n$ and a system Σ of $n-d$ independent homogeneous linear equations in $x_0, \dots, x_n$,*

$$a_1^0 x_0 + \dots + a_1^n x_n = 0,$$
$$\vdots$$
$$a_{n-d}^0 x_0 + \dots + a_{n-d}^n x_n = 0,$$

is given. Then the set of points $p = [x_0, \dots, x_n]$, for which $x_0, \dots, x_n$ is a solution of Σ, is a linear variety L of $\mathbb{P}_n$ of dimension d.

Proof. Let $e_0, \dots, e_n$ be a basis adapted to Δ. It is a well-known fact from the theory of linear equations, that the vectors $x_0e_0 + \dots + x_ne_n$ for which $(x_0, \dots, x_n)$ is a solution of Σ describe a linear subspace F of dimension $n+1-(n-d) = d+1$ of E. Then, clearly, $L = [F]$. □

Obviously the linear variety L of 2.6.2 is the one defined by Σ.

Remark 2.6.3. If L is a linear variety of dimension d in $\mathbb{P}_n$, we have defined a system of equations of L as any system of $n-d$ linearly independent homogeneous linear equations whose non-zero solutions are the coordinate vectors of the points of L. After 2.6.2, the condition on the number of equations is redundant, as the non-zero solutions of any system of $r \neq n-d$ linearly independent homogeneous linear equations are the coordinate vectors of the points of a linear variety L' of dimension $n-r \neq d$, and hence different from L.

Remark 2.6.4. It is an easy fact from linear algebra that a system of linear equations Σ' and any maximal subsystem $\Sigma \subset \Sigma'$ composed of linearly independent equations have the same solutions. Furthermore, the number of equations of Σ is the rank $\operatorname{rk}\Sigma'$ of Σ'. Then it follows from 2.6.2 that the set of points of $\mathbb{P}_n$ whose coordinate vectors are solutions of a system Σ' of homogeneous, linear and non-necessarily independent equations is a linear variety L of dimension $n - \operatorname{rk}\Sigma'$, any maximal subsystem of linearly independent equations $\Sigma \subset \Sigma'$ being a system of equations of L.

2.7 Incidence with coordinates

let L be a linear variety of $\mathbb{P}_n$ of dimension d and assume that a reference in $\mathbb{P}_n$ has been fixed. Since any coordinate vectors of $d+1$ independent points of L may be taken as the column vectors of a system of parametric equations of L, one may get parametric equations of L from a system Σ of implicit equations of L by just computing $d+1$ independent solutions of Σ. This is usually done as follows:

Assume that the implicit equations of L are

$$\begin{aligned} a_1^0 x_0 + \cdots + a_1^n x_n &= 0, \\ &\vdots \\ a_{n-d}^0 x_0 + \cdots + a_{n-d}^n x_n &= 0. \end{aligned} \tag{2.10}$$

Since these equations are independent, the matrix (a_i^j) of the system has rank $n-d$ and therefore a square $(n-d) \times (n-d)$ submatrix S with non-zero determinant. For simplicity the reader may assume that this submatrix is the one composed by first $n-d$ columns of (a_i^j), namely

$$S = \begin{pmatrix} a_1^0 & \dots & a_1^{n-d-1} \\ \vdots & & \vdots \\ a_{n-d}^0 & \dots & a_{n-d}^{n-d-1} \end{pmatrix}.$$

In order to have a coordinate vector of a point $q \in L$, one may give arbitrary values (not all zero) to the coordinates corresponding to columns not in S ($x_{n-d}, \dots, x_n$ if S is taken as above) and compute the remaining ones as the unique solution of the system resulting from (2.10), which has matrix S. Then it is enough to take the values of the former variables first equal to $1, 0, \dots, 0$, then equal to $0, 1, \dots, 0$, and so on, until taking them equal to $0, 0, \dots, 1$. Each choice giving a coordinate vector, we get a set of $d+1$ coordinate vectors of points of L, which clearly are independent and therefore may be taken as the columns of a system of parametric equations of L.

Before explaining how to get implicit equations from the parametric ones, we prove an easy algebraic lemma:

Lemma 2.7.1. *Assume that E is a vector space and $v_0, \dots, v_d \in E$ are linearly independent vectors. Take a further $v \in E$. Then $v, v_0, \dots, v_d$ are linearly dependent if and only if v is a linear combination of $v_0, \dots, v_d$.*

Proof. The *if* part is obvious. For the converse, assume we have

$$\lambda v + \lambda_0 v_0 + \cdots + \lambda_d v_d = 0$$

with either $\lambda \neq 0$ or $\lambda_i \neq 0$ for some i. If $\lambda \neq 0$, then v is a linear combination of $v_0, \dots, v_d$, namely

$$v = -\lambda^{-1}(\lambda_0 v_0 + \cdots + \lambda_d v_d).$$

Otherwise it is

$$\lambda_0 v_0 + \cdots + \lambda_d v_d = 0,$$

with some $\lambda_i \neq 0$, against the independence of $v_0, \dots, v_d$. □

Now assume that $(b_i^0, \dots, b_i^n)$, $i = 0, \dots, d$, are the columns of a system of parametric equations of L (or, equivalently, coordinate vectors of $d+1$ independent points in L) and proceed as follows:

(1) Take the $(b_i^0, \dots, b_i^n)$, $i = 0, \dots, d$, as the rows of a matrix B' and add to it a row of indeterminates to get the $(d+2) \times (n+1)$ matrix

$$B = \begin{pmatrix} x_0 & \dots & x_n \\ b_0^0 & \dots & b_0^n \\ \vdots & & \vdots \\ b_d^0 & \dots & b_d^n \end{pmatrix}.$$

(2) Since $\operatorname{rk} B' = d + 1$, choose a regular $(d+1) \times (d+1)$ submatrix D of B'. The other cases being similar, in the sequel we assume for simplicity that D is composed of the first $d + 1$ columns of B', namely,

$$D = \begin{pmatrix} b_0^0 & \dots & b_0^d \\ \vdots & & \vdots \\ b_d^0 & \dots & b_d^d \end{pmatrix}.$$

(3) Collect the determinants δ_i of the $(d+2) \times (d+2)$ submatrices D_i of B whose columns are the columns of B that meet D, plus a further column of B meeting not D, that is

$$\delta_i = \begin{vmatrix} x_0 & \dots & x_d & x_i \\ b_0^0 & \dots & b_0^d & b_0^i \\ \vdots & & \vdots & \vdots \\ b_d^0 & \dots & b_d^d & b_d^i \end{vmatrix},$$

for $i = d + 1, \dots, n$.

Then we have:

Proposition 2.7.2. *A system of implicit equations of L is $\delta_i = 0$, $i = d+1, \dots, n$.*

Proof. After developing the determinants by its first row, it is clear that the $\delta_i = 0$, $i = d + 1, \dots, n$, are $n - d$ homogeneous linear equations in $x_0, \dots, x_n$. Their independence follows from the fact that, for $i = d + 1, \dots, n$, the indeterminate x_i appears, with coefficient $\pm \det D \neq 0$, in the i-th equation, and does not appear in the others.

Now it remains to see that $p = [x_0, \dots, x_n]$ belongs to L if and only if $x_0, \dots, x_n$ satisfy $\delta_i = 0$, $i = d + 1, \dots, n$. By 2.5.1, $p \in L$ if and only if $(x_0, \dots, x_n)$ is a linear combination of the $(b_i^0, \dots, b_i^n)$, $i = 0, \dots, d$, which, by 2.7.1, is in turn equivalent to having $\operatorname{rk} B < d + 2$, where, as above,

$$B = \begin{pmatrix} x_0 & \dots & x_d & \dots & x_n \\ b_0^0 & \dots & b_0^d & \dots & b_0^n \\ \vdots & & \vdots & & \vdots \\ b_d^0 & \dots & b_d^d & \dots & b_d^n \end{pmatrix}.$$

If this condition is satisfied, clearly $\delta_i = 0$, $i = d + 1, \dots, n$. For the converse, note that if $\delta_i = 0$, then the columns of D_i are linearly dependent and so, again by 2.7.1, the last column of D_i is a linear combination of the preceding ones. This, for $i = d + 1, \dots, n$, assures that each of the remaining columns of B is a linear combination of the $d + 1$ columns that meet D, and hence that $\operatorname{rk} B < d + 2$. □

The reader may note that we have in fact re-proved 2.6.1.

Let L_1, L_2 be linear varieties of a projective space $\mathbb{P}_n$. We will explain next how to get equations of $L_1 \cap L_2$ and $L_1 \vee L_2$ from equations of L_1 and L_2, and also how to check an inclusion or equality between L_1 and L_2. Assume that

$$f_1 = 0, \ \dots, \ f_{n-d} = 0$$

is a system of implicit equations of L_1 and $v_0, \dots, v_d$ are coordinate vectors of independent points spanning L_1, or, equivalently, the columns of a system of parametric equations of L_1. In particular $d = \dim L_1$. Similarly, let

$$g_1 = 0, \ \dots, \ g_{n-s} = 0$$

and $u_0, \dots, u_s$ be implicit equations of L_2 and coordinate vectors of independent points spanning L_2, $s = \dim L_2$.

2.7.3. Computing equations of $L_1 \cap L_2$. Putting together the systems of implicit equations of L_1 and L_2 results in the system of homogeneous linear equations

$$f_1 = 0, \ \dots, \ f_{n-d} = 0, \quad g_1 = 0, \ \dots, \ g_{n-s} = 0,$$

whose non-zero solutions are, clearly, the coordinate vectors of the points of $L_1 \cap L_2$. Nevertheless, these equations need not be linearly independent, and so, as explained in 2.6.4, it is necessary to extract from them a maximal system of independent equations (those involved in a maximal non-zero minor of the corresponding matrix) to get a system of implicit equations of $L_1 \cap L_2$.

2.7.4. Computing equations of $L_1 \vee L_2$. The vectors $v_0, \dots, v_d, u_0, \dots, u_s$ are coordinate vectors of points spanning $L_1 \vee L_2$. They (and hence their corresponding points) need not be independent. Extracting from them a maximal system of independent vectors (again those involved in a non-zero minor of maximal order) gives coordinate vectors of a system of independent points still spanning $L_1 \vee L_2$ (by 1.5.11): they may be taken as the columns of a system of parametric equations of $L_1 \vee L_2$.

2.7.5. Checking $L_1 \subset L_2$. The condition $L_1 \subset L_2$ is equivalent to $L_1 \cap L_2 = L_1$ and so, by 1.3.6, also to $\dim L_1 = \dim L_1 \cap L_2$, which in turn may be written $n - \dim L_1 = n - \dim L_1 \cap L_2$, that is,

$$n - d = \operatorname{rk}\{f_1 = 0, \dots, f_{n-d} = 0,\ g_1 = 0, \dots, g_{n-s} = 0\}.$$

In other words, $L_1 \subset L_2$ if and only if adding the equations of a system of implicit equations of L_2 to a system of implicit equations of L_1 does not increase the rank.

Alternatively, $L_1 \subset L_2$ is equivalent to $L_1 \vee L_2 = L_2$ and so, as above, also to $\dim L_2 = \dim L_1 \vee L_2$. The latter condition is satisfied if and only if

$$s = \operatorname{rk}\{v_0, \dots, v_d,\ u_0, \dots, u_s\},$$

that is, adding the columns of a system of parametric equations of L_1 to the columns of a system of parametric equations of L_2 does not increase the rank.

2.7.6. Checking $L_1 = L_2$. By 1.3.6, either of the conditions of 2.7.5 above characterizes the equality $L_1 = L_2$ in case when $d = s$. In particular, if L_1 and L_2 are hyperplanes, each has a single equation and then they agree if and only if their equations are proportional.

Example 2.7.7. Assume that $\Delta = (p_0, \dots, p_n, A)$ is a reference. Let L and T be opposite faces of Δ, $-1 < d = \dim L < n - 1$. For simplicity assume that $L = p_0 \vee \dots \vee p_d$ and $T = p_{d+1} \vee \dots \vee p_n$. As seen in 2.2.6, L has equations $x_{d+1} = 0, \dots, x_n = 0$ and T has equations $x_0 = 0, \dots, x_d = 0$.

Assume now that $p = [a_0, \dots, a_n] \notin L$; then $a_i \neq 0$ for some $i = d+1, \dots, n$. It results from a direct computation using 2.7.2 that a system of equations of the projection $p \vee L$ is

$$a_i x_j - a_j x_i = 0, \quad j = d+1, \dots, n,\ j \neq i.$$

Anyway, the computation may be avoided using the following argument: the above equations being independent, they define a linear variety V of dimension $d + 1$, and it is clear from their coordinates that $p, p_0, \dots, p_d \in V$; then $p \vee L \subset V$ and so $p \vee L = V$, because both have dimension $d + 1$.

Then, solving the system

$$\begin{aligned} x_j &= 0, \quad j = 0, \dots, d, \\ a_i x_j - a_j x_i &= 0, \quad j = d+1, \dots, n,\ j \neq i, \end{aligned} \tag{2.11}$$

one sees that the section $(p \vee L) \cap T$ is the point $\bar{p} = [0, \dots, 0, a_{d+1}, \dots, a_n]$. In fact, it is enough to see that the above coordinates of $\bar{p}$ satisfy the equations (2.11), as we have already seen in Section 1.9 that $(p \vee L) \cap T$ is a point.

Using the reference Δ' subordinated by Δ on T (2.2.6), $\bar{p} = [a_{d+1}, \dots, a_n]_{\Delta'}$. Thus, if L is a face of the reference, $-1 < \dim L < n-1$, dropping from the coordinates of any point $p \notin L$ those corresponding to the vertices spanning L, gives coordinates of the image of p by the projection from L onto the face opposite L.

Remark 2.7.8. Any two supplementary linear varieties L, T of $\mathbb{P}_n$ are opposite faces of a reference of $\mathbb{P}_n$, and therefore are in the situation of Example 2.7.7. Indeed, if $d = \dim L$, take $p_0, \dots, p_d$ spanning L and $p_{d+1}, \dots, p_n$ spanning T. Then $p_0, \dots, p_n$ span $\mathbb{P}_n$ and therefore are independent. Any reference with vertices $p_0, \dots, p_n$ (2.1.4) has L and T as opposite faces.

2.8 Determination and matrices of a projectivity

Let $f: \mathbb{P}_n \to \mathbb{P}'_n$ be a projectivity and $\Delta = (p_0, \dots, p_n, A)$ a reference of $\mathbb{P}_n$. Since f preserves the independence of points (1.7.5), the image of Δ, $f(\Delta) = (f(p_0), \dots, f(p_n), f(A))$ is a reference of $\mathbb{P}'_n$. The theorem we present next assures that a projectivity $f: \mathbb{P}_n \to \mathbb{P}'_n$ is uniquely determined by an arbitrary choice of the images of the elements of Δ, provided they compose a reference too. Some authors call this result the *main theorem of projective geometry*, see for instance [26].

Theorem 2.8.1. *Assume that $\mathbb{P}_n$ and $\mathbb{P}'_n$ are projective spaces over k and that $\Delta = (p_0, \dots, p_n, A)$ and $\Delta' = (p'_0, \dots, p'_n, A')$ are projective references of $\mathbb{P}_n$ and $\mathbb{P}'_n$, respectively. Then:*

(a) *There is a unique projectivity*

$$f: \mathbb{P}_n \longrightarrow \mathbb{P}'_n$$

such that $f(\Delta) = \Delta'$.

(b) *If f is the above projectivity, then for any point $p = [x_0, \dots, x_n]_\Delta \in \mathbb{P}_n$, $f(p) = [x_0, \dots, x_n]_{\Delta'}$.*

Proof. Let E and E' be the vector spaces associated to $\mathbb{P}_n$ and $\mathbb{P}'_n$, respectively. Take $e_0, \dots, e_n$ to be a basis adapted to Δ and $e'_0, \dots, e'_n$ to be a basis adapted to Δ'. If φ is the (only) linear map

$$\varphi: E \longrightarrow E'$$

such that $\varphi(e_i) = e'_i$, $i = 0, \dots, n$, then φ is an isomorphism because it maps a basis to a basis. By the linearity of φ, $\varphi(x_0 e_0 + \dots + x_n e_n) = x_0 e'_0 + \dots + x_n e'_n$ for any $(x_0, \dots, x_n) \in k^{n+1}$. Thus, if we take $f = [\varphi]$, then

$$f([x_0, \dots, x_n]_\Delta) = [\varphi(x_0 e_0 + \dots + x_n e_n)] = [x_0 e'_0 + \dots + x_n e'_n] = [x_0, \dots, x_n]_{\Delta'}$$

for any $(x_0, \dots, x_n) \in k^{n+1} - \{(0, \dots, 0)\}$, as claimed in (b). Also $f(\Delta) = \Delta'$, as the equality displayed above in particular gives $f([1, 0, \dots, 0]_\Delta) = [1, 0, \dots, 0]_{\Delta'}$, ..., $f([0, \dots, 0, 1]_\Delta) = [0, \dots, 0, 1]_{\Delta'}$, $f([1, 1, \dots, 1]_\Delta) = [1, 1, \dots, 1]_{\Delta'}$.

It remains to prove the uniqueness claimed in part (a). To this end, assume that for a second projectivity

$$g = [\psi]\colon \mathbb{P}_n \longrightarrow \mathbb{P}'_n$$

it is $g(\Delta) = \Delta'$. Then

$$\begin{aligned} p'_0 &= g(p_0) = [\psi(e_0)], \\ &\vdots \\ p'_n &= g(e_n) = [\psi(e_n)], \\ A' &= g(A) = [\psi(e_0 + \dots + e_n)] = [\psi(e_0) + \dots + \psi(e_n)], \end{aligned}$$

which proves that $\psi(e_0), \dots, \psi(e_n)$ is a basis adapted to Δ'. By 2.1.2, there is $\lambda \neq 0$ such that $\psi(e_i) = \lambda e'_i = \lambda\varphi(e_i)$, $i = 0, \dots, n$. It follows that $\psi(v) = \lambda\varphi(v)$ for all $v \in E$, hence $\psi = \lambda\varphi$ and so $g = f$, as wanted. □

Corollary 2.8.2. *If a projectivity $f\colon \mathbb{P}_n \to \mathbb{P}_n$ leaves fixed all elements of a reference Δ of $\mathbb{P}_n$, then f is the identical map.*

Proof. Follows from the uniqueness in 2.8.1, as both f and the identical map are mapping Δ to Δ. □

Remark 2.8.3. It follows from 2.8.1 and 2.1.4 that if $\mathbb{P}_n$ and $\mathbb{P}'_n$ are projective spaces over k of the same dimension, then there is a projectivity $f\colon \mathbb{P}_n \to \mathbb{P}'_n$.

Again assume that $f\colon \mathbb{P}_n \to \mathbb{P}'_n$ is a projectivity and Δ a reference of $\mathbb{P}_n$. By part (a) of Theorem 2.8.1, f is the only projectivity mapping Δ to $f(\Delta)$. Then part (b) of 2.8.1 provides coordinates of $f(p)$ for any $p \in \mathbb{P}_n$. Nevertheless, these coordinates are relative to $f(\Delta)$, while usually coordinates of $f(p)$ relative to a different, already fixed, reference of $\mathbb{P}'_n$ are required. Next we will show how to get them.

Assume that we choose references $\Delta = (p_0, \dots, p_n, A)$ in $\mathbb{P}_n$ and Ω in $\mathbb{P}'_n$. First of all note that, $f(\Delta)$ being a reference of $\mathbb{P}'_n$, by 2.3.1 there are coordinate vectors of $f(p_0), \dots, f(p_n)$ whose sum is a coordinate vector of $f(A)$, all coordinate vectors being relative to Ω. Then we may take M to be a matrix whose columns are coordinate vectors of $f(p_0), \dots, f(p_n)$ chosen such that their sum is a coordinate vector of $f(A)$: such an M is called a *matrix of f relative to the references Δ, Ω*. Note that, from its own definition, M is a regular $(n+1) \times (n+1)$ matrix, as $f(p_0), \dots, f(p_n)$ are independent points.

In case $\mathbb{P}'_n = \mathbb{P}_n$, that is, f is a collineation of $\mathbb{P}_n$, it is usual to take $\Omega = \Delta$ and call M a *matrix of f relative to Δ*, or just a *matrix of f* if no mention of Δ is needed.

A matrix of a projectivity allows us to compute the image of any point, namely:

Proposition 2.8.4. *Assume that M is a matrix of the projectivity $f\colon \mathbb{P}_n \to \mathbb{P}'_n$ relative to references Δ, Ω. Then, for any $p = [x_0, \dots, x_n]_\Delta$, $f(p) = [y_0, \dots, y_n]_\Omega$ if and only if*

$$\begin{pmatrix} y_0 \\ \vdots \\ y_n \end{pmatrix} = \rho M \begin{pmatrix} x_0 \\ \vdots \\ x_n \end{pmatrix} \tag{2.12}$$

for some non-zero $\rho \in k$. In particular M determines f.

Proof. By 2.8.1, $f(p) = [x_0, \dots, x_n]_{f(\Delta)}$, after which the claim follows from 2.3.2, as, by its own definition, M is a matrix that changes coordinates relative to $f(\Delta)$ into coordinates relative to Ω. □

The $n + 1$ scalar equations in which the matricial equality (2.12) splits, are called the *equations* relative to Δ, Ω. Clearly, they are linear and homogeneous. Furthermore, since M is regular, these equations can be inverted to give $x_0, \dots, x_n$ as linear homogeneous functions of $y_0, \dots, y_n$. As already said at the beginning of Section 1.6, these are the conditions classically imposed on the equations of a map between projective spaces for it to be a projectivity. Next is the converse of 2.8.4. Its proof is left to the reader.

Proposition 2.8.5. *Assume that $\mathbb{P}_n$ and $\mathbb{P}'_n$ are projective spaces over k, with fixed references Δ and Ω respectively. If M is any regular $(n+1) \times (n+1)$ matrix, for any $(x_0, \dots, x_n) \in k^{n+1}$ write*

$$\begin{pmatrix} y_0 \\ \vdots \\ y_n \end{pmatrix} = M \begin{pmatrix} x_0 \\ \vdots \\ x_n \end{pmatrix}.$$

Then mapping any $p = [x_0, \dots, x_n]_\Delta \in P_n$ to $[y_0, \dots, y_n]_\Omega$ is a projectivity $f\colon \mathbb{P}_n \to \mathbb{P}'_n$, namely the projectivity which has matrix M relative to Δ, Ω.

Once the references are fixed, different matrices of a projectivity are proportional:

Lemma 2.8.6. *Assume that M is a matrix of a projectivity $f\colon \mathbb{P}_n \to \mathbb{P}'_n$ relative to references Δ, Ω. Then M' is also a matrix of f relative to Δ, Ω if and only if $M' = \lambda M$ for some non-zero $\lambda \in k$.*

Proof. The *if* part is obvious. By the definition of matrix of f, the columns of M are the components of the vectors of a basis adapted to $f(\Delta)$ relative to a basis adapted to Ω. The same being true for M', it is enough to apply 2.1.2. □

It will be useful to prove that any matrix that works like a matrix of f is actually a matrix of f:

Lemma 2.8.7. *Let M be an $(n+1)\times(n+1)$ matrix. If for any $p = [x_0, \dots, x_n]_\Delta \in \mathbb{P}_n$,*

$$M\begin{pmatrix} x_0 \\ \vdots \\ x_n \end{pmatrix}$$

is a coordinate vector of $f(p)$ relative to Ω, then M is a matrix of f relative to Δ, Ω.

Proof. By successively taking $(x_0, \dots, x_n) = (1, 0, \dots, 0), \dots, (0, \dots, 0, 1)$, it follows from the hypothesis that the columns of M are coordinate vectors of the images of the vertices of Δ. Taking $(x_0, \dots, x_n) = (1, \dots, 1)$ shows that the sum of the columns of M is a coordinate vector of the image of the unit point. □

Proposition 2.8.8. *Denote by Δ, Δ' and Δ'' projective references in projective spaces $\mathbb{P}_n$, $\mathbb{P}'_n$ and $\mathbb{P}''_n$, respectively, by f, g projectivities, $f\colon \mathbb{P}_n \to \mathbb{P}'_n$ and $g\colon \mathbb{P}'_n \to \mathbb{P}''_n$, by M any matrix of f relative to Δ, Δ' and by N any matrix of g relative to Δ, Δ''. Then:*

(a) *The unit matrix is a matrix of the identical map of $\mathbb{P}_n$ relative to Δ.*

(b) *NM is a matrix of $g \circ f$ relative to Δ, Δ''.*

(c) *M^{-1} is a matrix of f^{-1} relative to Δ', Δ.*

Proof. Part (a) is obvious, while parts (b) and (c) directly follow from 2.8.7. □

Changing coordinates affects the matrix of a projectivity in the following way:

Proposition 2.8.9. *Assume that a projectivity $f\colon \mathbb{P}_n \to \mathbb{P}'_n$ has matrix M relative to references Δ and Ω. If Δ' and Ω' are references of $\mathbb{P}_n$ and $\mathbb{P}'_n$ and $P_{\Delta',\Delta}$ (resp. $Q_{\Omega',\Omega}$) is a matrix changing coordinates relative to Δ' (resp. Ω') into coordinates relative to Δ (resp. Ω), then*

$$Q^{-1}_{\Omega',\Omega} M P_{\Delta',\Delta}$$

is a matrix of f relative to references Δ' and Ω'.

Proof. Direct from 2.3.2, 2.8.4 and 2.8.7. □

Remark 2.8.10. If the projectivity f of 2.8.9 is a collineation, that is, $\mathbb{P}'_n = \mathbb{P}_n$, and, as usual, one takes $\Omega = \Delta$ and $\Omega' = \Delta'$, then the matrix of f relative to Δ' is just

$$P^{-1}_{\Delta',\Delta} M P_{\Delta',\Delta},$$

M the matrix of f relative to Δ.

The matrices of a projectivity are those of its representatives:

Proposition 2.8.11. *M is a matrix of a projectivity f relative to references Δ, Ω if and only if M is a matrix of a representative of f relative to bases adapted to Δ and Ω.*

Proof. Assume that $f = [\varphi]$ and M is the matrix of φ relative to adapted bases $\{e_i\}$ and $\{u_i\}$. As it is well known from linear algebra, any $v = \sum_{i=0}^{n} x_i e_i$ has $\varphi(v) = \sum_{i=0}^{n} y_i u_i$ if and only if

$$\begin{pmatrix} y_0 \\ \vdots \\ y_n \end{pmatrix} = M \begin{pmatrix} x_0 \\ \vdots \\ x_n \end{pmatrix}.$$

Then 2.8.7 and the definition of coordinates prove that M is a matrix of f. If M' is a matrix of f, then $M' = \lambda M$, $\lambda \in k - \{0\}$ by 2.8.6 and so M' is the matrix of $\lambda\varphi$. □

Bunches and projections may be easily handled using the coordinates defined below:

Proposition 2.8.12. *Let L be a linear variety, $0 \leq \dim L \leq n-2$, and assume that $\Delta = (p_0, \dots, p_n, A)$ is a reference whose first $d+1$ vertices span L. Then $\Omega = (p_{d+1} \vee L, \dots, p_n \vee L, A \vee L)$ is a reference of the bunch L^∇ relative to which the projection $p \vee L$ of any point $p = [a_0, \dots, a_n] \notin L$ has coordinates $a_{d+1}, \dots, a_n$.*

Proof. Take $T = p_{d+1} \vee \dots \vee p_n$ the face of Δ opposite L, and take in T the reference $\Delta' = (p_{d+1}, \dots, p_n, A')$ subordinated by Δ. Consider the pair of reciprocal projectivities section and projection $L^\nabla \leftrightarrow T$, according to 1.9.3. By the definition of Δ' in 2.2.6, A' is the section of $A \vee L$, hence $A \vee L$ is the projection of A'. Therefore Ω is the projection of Δ' and so it is, in particular, a reference of L^∇. Furthermore, by 2.8.1, the matrix of the projection map relative to Δ', Ω is the unit matrix. Since we have seen in 2.7.7 that the section of $p \vee L$ has coordinates $a_{d+1}, \dots, a_n$, the second claim follows. □

2.9 Cross ratio

Let q_1, q_2, q_3, q_4 be points of a one-dimensional projective space $\mathbb{P}_1$, at least three of them different. Assume that a projective reference $\Delta = (p_0, p_1, A)$ of $\mathbb{P}_1$ has been fixed and that $q_i = [x_i, y_i]_\Delta$, $i = 1, \dots, 4$. Since there are at least three different points among q_1, q_2, q_3, q_4, at most one of the determinants

$$\begin{vmatrix} x_i & y_i \\ x_j & y_j \end{vmatrix}, \quad i \neq j,$$

is zero, (by 2.2.5). Thus, using again the symbol ∞ of Section 2.4 and the convention $a/0 = \infty$ for $a \in k$, $a \neq 0$, take

$$\rho = \frac{\begin{vmatrix} x_3 & y_3 \\ x_1 & y_1 \end{vmatrix}}{\begin{vmatrix} x_3 & y_3 \\ x_2 & y_2 \end{vmatrix}} : \frac{\begin{vmatrix} x_4 & y_4 \\ x_1 & y_1 \end{vmatrix}}{\begin{vmatrix} x_4 & y_4 \\ x_2 & y_2 \end{vmatrix}} \in k \cup \{\infty\}, \tag{2.13}$$

the colon : meaning division. The important point here is:

Lemma 2.9.1. *ρ, given by the equality* (2.13) *above, does not depend on the choice of the reference Δ, nor on the particular choice of coordinates for each point q_i.*

Proof. Assume that $\bar{\Delta}$ is any reference of $\mathbb{P}_1$ and $q_i = [\bar{x}_i, \bar{y}_i]_{\bar{\Delta}}$, $i = 1, \dots, 4$, and write

$$\bar{\rho} = \frac{\begin{vmatrix} \bar{x}_3 & \bar{y}_3 \\ \bar{x}_1 & \bar{y}_1 \end{vmatrix}}{\begin{vmatrix} \bar{x}_3 & \bar{y}_3 \\ \bar{x}_2 & \bar{y}_2 \end{vmatrix}} : \frac{\begin{vmatrix} \bar{x}_4 & \bar{y}_4 \\ \bar{x}_1 & \bar{y}_1 \end{vmatrix}}{\begin{vmatrix} \bar{x}_4 & \bar{y}_4 \\ \bar{x}_2 & \bar{y}_2 \end{vmatrix}}. \tag{2.14}$$

According to 2.3.2, there is a regular 2×2 matrix M and non-zero scalars r_i such that, for each $i = 1, \dots, 4$,

$$\begin{pmatrix} \bar{x}_i \\ \bar{y}_i \end{pmatrix} = r_i M \begin{pmatrix} x_i \\ y_i \end{pmatrix}.$$

Thus, for each i, j,

$$\begin{pmatrix} \bar{x}_i & \bar{x}_j \\ \bar{y}_i & \bar{y}_i \end{pmatrix} = M \begin{pmatrix} r_i x_i & r_j x_j \\ r_i y_i & r_j y_j \end{pmatrix}$$

and hence

$$\begin{vmatrix} \bar{x}_i & \bar{x}_j \\ \bar{y}_i & \bar{y}_i \end{vmatrix} = r_i r_j \det M \begin{vmatrix} x_i & x_j \\ y_i & y_j \end{vmatrix}.$$

After substituting in (2.14) we obtain

$$\bar{\rho} = \frac{r_1 r_3 \det M \begin{vmatrix} x_3 & y_3 \\ x_1 & y_1 \end{vmatrix}}{r_2 r_3 \det M \begin{vmatrix} x_3 & y_3 \\ x_2 & y_2 \end{vmatrix}} : \frac{r_1 r_4 \det M \begin{vmatrix} \bar{x}_4 & \bar{y}_4 \\ \bar{x}_1 & \bar{y}_1 \end{vmatrix}}{r_2 r_4 \det M \begin{vmatrix} \bar{x}_4 & \bar{y}_4 \\ \bar{x}_2 & \bar{y}_2 \end{vmatrix}} = \rho,$$

and hence the claim □

Once we know that ρ, given by the equation (2.14) above, is independent of the choices of the reference and the coordinates of the points, we define it to be the *cross ratio* of q_1, q_2, q_3, q_4, written in the sequel $\rho = (q_1, q_2, q_3, q_4)$. The reader may note that it is defined if and only if the points q_1, q_2, q_3, q_4 belong to the same

$\mathbb{P}_1$ and least three of them are different. Sometimes these conditions are implicitly assumed to be satisfied once the cross ratio is mentioned.

Further expressions for the cross ratio are presented next:

Proposition 2.9.2. *If, after fixing a reference, each point q_i has finite absolute coordinate θ_i, $i = 1, \dots, 4$, then*

$$(q_1, q_2, q_3, q_4) = \frac{\theta_3 - \theta_1}{\theta_3 - \theta_2} : \frac{\theta_4 - \theta_1}{\theta_4 - \theta_2}. \tag{2.15}$$

Proof. Just take $q_i = [\theta_i, 1]$ and use 2.9.1 and the equality (2.13). □

Remark 2.9.3. Since a projective line $\mathbb{P}_1$ contains more than four points (due to 2.4.1), for any $q_1, q_2, q_3, q_4 \in \mathbb{P}_1$ there is a reference (p_0, p_1, A) of $\mathbb{P}_1$ with $p_0 \neq q_i$, $i = 1, \dots, 4$, allowing thus to compute (q_1, q_2, q_3, q_4) via the formula (2.15).

Remark 2.9.4. As the reader may check case by case going back to (2.13), the formula (2.15) still holds if one of the absolute coordinates θ_i is ∞, provided the expression on the right is computed as a limit for $\theta_i \to \infty$, using the formal rules of elementary calculus. For instance, for $\theta_4 = \infty$ it gives

$$(q_1, q_2, q_3, q_4) = \frac{\theta_3 - \theta_1}{\theta_3 - \theta_2}.$$

Proposition 2.9.5. *If q_3 and q_4 have finite and non-zero absolute coordinates θ_3 and θ_4 relative to a reference whose first and second vertices are q_1 and q_2, then*

$$(q_1, q_2, q_3, q_4) = \frac{\theta_4}{\theta_3}.$$

Proof. Apply the equality (2.13) to $q_1 = [1, 0]$, $q_2 = [0, 1]$, $q_3 = [\theta_3, 1]$ and $q_4 = [\theta_4, 1]$, or use 2.9.4. □

Corollary 2.9.6 (Multiplicativity of the cross ratio). *If $q_1, q_2, q_3, q_4, q_5 \in \mathbb{P}_1$, $q_1 \neq q_2$ and $q_i \neq q_1, q_2$ for $i = 3, 4, 5$, then*

$$(q_1, q_2, q_3, q_4)(q_1, q_2, q_4, q_5) = (q_1, q_2, q_3, q_5).$$

Proof. Take a reference with vertices q_1, q_2 and use 2.9.5 to compute the cross ratios. □

The next result is very important. Although slightly less general than the definition, it provides a far deeper understanding of the cross ratio as a measure of the position of the fourth point in relation to the first three ones.

Theorem 2.9.7. *If $q_1, q_2, q_3 \in \mathbb{P}_1$ are three different points, then, for any $q_4 \in \mathbb{P}_1$, the cross ratio (q_1, q_2, q_3, q_4) is the absolute coordinate of q_4 relative to the reference (q_1, q_2, q_3)*

Proof. As above, compute from $q_1 = [1,0]$, $q_2 = [0,1]$, $q_3 = [1,1]$ and $q_4 = [x_4, y_4]$ using equality (2.13). □

Corollary 2.9.8. *Given three different points $q_1, q_2, q_3 \in \mathbb{P}_1$, for each $\rho \in k \cup \{\infty\}$ there is a unique point $q^\rho \in \mathbb{P}_1$ such that $(q_1, q_2, q_3, q^\rho) = \rho$.*

Proof. Follows from 2.9.7 and 2.4.1 □

Theorem 2.9.7 may be reformulated in the following way:

Corollary 2.9.9. *Still assume $q_1, q_2, q_3, q_4 \in \mathbb{P}_1$ and q_1, q_2, q_3 pairwise different. Then $(q_1, q_2, q_3, q_4) = \rho \neq \infty$ if and only if there exist vectors v, w such that*

$$q_1 = [v], \quad q_2 = [w], \quad q_3 = [v + w], \quad q_4 = [\rho v + w].$$

Proof. Direct from 2.9.7 and the definition of coordinates. □

Often the one-dimensional projective space $\mathbb{P}_1$, the points q_i are assumed to belong to, is a line of a projective space $\mathbb{P}_n, n > 1$, and one has projective coordinates of the points q_i relative to a reference of $\mathbb{P}_n$. Then, computing the cross ratio (q_1, q_2, q_3, q_4) by one of the above formulas (2.13) or (2.15) would require having coordinates of the points q_i relative to a reference of $\mathbb{P}_1$ and it is easier to compute from the coordinates in $\mathbb{P}_n$ using 2.9.9 as follows: since coordinate vectors of points are component vectors of their representatives, one chooses coordinate vectors of q_1 and q_2 whose sum is a coordinate vector of q_3 and then writes a coordinate vector of q_4 as a linear combination of them; the ratio of the coefficients is then $\rho = (q_1, q_2, q_3, q_4)$.

The cases in which two of the points coincide are easily characterized in terms of cross ratio, namely:

Proposition 2.9.10. *If, $q_1, q_2, q_3, q_4 \in \mathbb{P}_1$ and at least three of them are different, then:*

$$\begin{aligned} q_1 = q_3 \text{ or } q_2 = q_4 \quad &\text{if and only if} \quad (q_1, q_2, q_3, q_4) = 0,\\ q_2 = q_3 \text{ or } q_1 = q_4 \quad &\text{if and only if} \quad (q_1, q_2, q_3, q_4) = \infty,\\ q_1 = q_2 \text{ or } q_3 = q_4 \quad &\text{if and only if} \quad (q_1, q_2, q_3, q_4) = 1. \end{aligned}$$

Proof. Choose a reference $\Delta = (p_0, p_1, A)$ of $\mathbb{P}_1$ such that $q_i \neq p_0$ for all i. Then 2.9.2 applies and all three claims are direct consequences of the equality (2.15) and Proposition 2.4.1. Indeed, (2.15) shows that $(q_1, q_2, q_3, q_4) = 0$ is equivalent to $(\theta_3 - \theta_1)(\theta_4 - \theta_2) = 0$, which in turn, by 2.4.1, is equivalent to being either

$q_1 = q_3$ or $q_2 = q_4$, proving the first claim. A similar argument proves the second one, while for the third, first note that a direct computation allows us to write

$$\frac{\theta_3 - \theta_1}{\theta_3 - \theta_2} : \frac{\theta_4 - \theta_1}{\theta_4 - \theta_2} = 1$$

in the form

$$(\theta_1 - \theta_2)(\theta_3 - \theta_4) = 0,$$

and then proceed as in the former cases. □

The next theorem establishes the invariance of the cross ratio by projectivities:

Theorem 2.9.11 (Projective invariance of the cross ratio). *If $f : \mathbb{P}_1 \to \mathbb{P}'_1$ is a projectivity between one-dimensional projective spaces, $q_1, q_2, q_3, q_4 \in \mathbb{P}_1$ and at least three of them are different, then at least three of their images are different and*

$$(q_1, q_2, q_3, q_4) = (f(q_1), f(q_2), f(q_3), f(q_4)).$$

Proof. Take a reference Δ in $\mathbb{P}_1$ and $\Delta' = f(\Delta)$ in $\mathbb{P}'_1$. By 2.8.1, if $q_i = [x_i, y_i]_\Delta$, then $f(q_i) = [x_i, y_i]_{\Delta'}$, $i = 1, \dots, 4$. Since by 2.9.1 we are allowed to compute both cross ratios using the formula (2.13) and these coordinates, the claim follows. □

Remark 2.9.12. The equality of cross ratios of 2.9.11 still holds if q_1, q_2, q_3, q_4 are aligned points of an n-dimensional projective space $\mathbb{P}_n$, at least three of them different, and f is a projectivity defined in $\mathbb{P}_n$, as then the restriction of f to the line spanned by the q_i, $i = 1, \dots, 4$, is a projectivity (by 1.7.1 (b)) and 2.9.11 applies to it.

Conversely, having equal cross ratios is a sufficient condition for two ordered sets of four aligned points to be mapped one to the other by a projectivity, namely:

Proposition 2.9.13. *Assume that $\mathbb{P}_1$ and $\mathbb{P}'_1$ are one-dimensional projective spaces over k. If $q_1, q_2, q_3, q_4 \in \mathbb{P}_1$ and $q'_1, q'_2, q'_3, q'_4 \in \mathbb{P}'_1$, all points in each set being different, and $(q_1, q_2, q_3, q_4) = (q'_1, q'_2, q'_3, q'_4)$, then there is a projectivity $f : \mathbb{P}_1 \to \mathbb{P}'_1$ such that $f(q_i) = q'_i$, $i = 1, \dots, 4$.*

Proof. Using 2.8.1, there is a projectivity $f : \mathbb{P}_1 \to \mathbb{P}'_1$ such that $f(q_i) = q'_i$ for $i = 1, 2, 3$. Furthermore, by 2.9.7 and the equality of cross ratios, the absolute coordinates of q_4 and q'_4, relative to the references (q_1, q_2, q_3) and (q'_1, q'_2, q'_3), respectively, are equal. Then $f(q_4) = q'_4$ by 2.8.1 (b). □

Projective equivalence, projective classification and projective invariants will be explained in forthcoming Section 2.11; next is an advance example:

Example 2.9.14. Let us call, for short, *collinear quadruples* the ordered sets of four different points belonging to the same one-dimensional projective space. We take as equivalent two collinear quadruples if and only if there is a projectivity transforming one in the other, which is an equivalence relation, as the reader may easily check. This equivalence relation is called the *projective equivalence* of collinear quadruples, and the partition in equivalence classes it induces, their *projective classification*. Theorem 2.9.11 assures that the cross ratio is constant on any equivalence class: for this reason, the cross ratio is said to be a *projective invariant* of the collinear quadruples. On the other hand, Proposition 2.9.13 asserts that two collinear quadruples with the same cross ratio are projectively equivalent. Therefore the cross ratio completely describes the projective classification of collinear quadruples: two collinear quadruples are projectively equivalent if and only if they have the same cross ratio. Due to this property, it is said that the cross ratio composes a *complete system of projective invariants* for collinear quadruples.

The next proposition, together with 2.9.11, provides a characterization of the projectivities between one-dimensional projective spaces in terms of cross ratio:

Proposition 2.9.15. *Let $f : \mathbb{P}_1 \to \mathbb{P}'_1$ be a map between one-dimensional projective spaces over k. Assume that for any four points $q_1, q_2, q_3, q_4 \in \mathbb{P}_1$, three at least different, three at least of their images $f(q_1)$, $f(q_2)$, $f(q_3)$, $f(q_4)$ are also different and*

$$(f(q_1), f(q_2), f(q_3), f(q_4)) = (q_1, q_2, q_3, q_4).$$

Then f is a projectivity.

Proof. Take a reference $\Delta = (p_0, p_1, A)$ of $\mathbb{P}_1$. Its image

$$\Omega = (f(p_0), f(p_1), f(A))$$

is a reference of $\mathbb{P}'_1$, as otherwise there would be at most two different points among $f(p_0), f(p_1), f(A), f(A)$, against the hypothesis. Then for any $p = [x_0, x_1]_\Delta \in \mathbb{P}_1$,

$$(p_0, p_1, A, p) = (f(p_0), f(p_1), f(A), f(p))$$

which, by 2.9.7, gives $f(p) = [x_0, x_1]_\Omega$ and hence the claim, using either 2.8.1 or 2.8.5. □

Remark 2.9.16. In 2.9.15 it is just assumed that if the cross ratio of any four points is defined, then the cross ratio of their images is also defined and both cross ratios agree, which is often shortened by saying that f *preserves cross ratios*. Then 2.9.15 says that any map between projective lines preserving cross ratios is a projectivity.

2.10 Harmonic sets

Assume that $q_1, q_2, q_3, q_4 \in \mathbb{P}_1$ are any four different points. First of all we will describe how reordering them does affect their cross ratio; in other words, we will consider the action of the group $\mathbf{S}_4$, of the permutations of four elements, on $\{q_1, q_2, q_3, q_4\}$, and describe the effect of this action on the values of the cross ratio (q_1, q_2, q_3, q_4).

Proposition 2.10.1. *If $q_1, q_2, q_3, q_4 \in \mathbb{P}_1$ are four different points and $\rho = (q_1, q_2, q_3, q_4)$, then*

$$\begin{aligned}
\rho &= (q_1, q_2, q_3, q_4) = (q_2, q_1, q_4, q_3) = (q_3, q_4, q_1, q_2) = (q_4, q_3, q_2, q_1),\\
\frac{1}{\rho} &= (q_2, q_1, q_3, q_4) = (q_1, q_2, q_4, q_3) = (q_3, q_4, q_2, q_1) = (q_4, q_3, q_1, q_2),\\
1-\rho &= (q_1, q_3, q_2, q_4) = (q_3, q_1, q_4, q_2) = (q_2, q_4, q_1, q_3) = (q_4, q_2, q_3, q_1),\\
\frac{\rho-1}{\rho} &= (q_2, q_3, q_1, q_4) = (q_3, q_2, q_4, q_1) = (q_1, q_4, q_2, q_3) = (q_4, q_1, q_3, q_2),\\
\frac{1}{1-\rho} &= (q_3, q_1, q_2, q_4) = (q_1, q_3, q_4, q_2) = (q_2, q_4, q_3, q_1) = (q_4, q_2, q_1, q_3),\\
\frac{\rho}{\rho-1} &= (q_3, q_2, q_1, q_4) = (q_2, q_3, q_4, q_1) = (q_1, q_4, q_3, q_2) = (q_4, q_1, q_2, q_3).
\end{aligned}$$

Proof. First note that $\rho \neq 0, 1, \infty$ by 2.9.10, which assures that all expressions in ρ appearing in the claim are well-defined elements of k. We choose a reference such that all the absolute coordinates θ_i of q_i, $i = 1, \dots, 4$ are finite, and we may thus use the formula

$$(q_1, q_2, q_3, q_4) = \frac{\theta_3 - \theta_1}{\theta_3 - \theta_2} : \frac{\theta_4 - \theta_1}{\theta_4 - \theta_2}. \tag{2.16}$$

Since checking one by one the twenty-three claimed equalities would be too long and a little boring, we will proceed differently. We denote by (i, j, s, ℓ) the permutation that changes q_1, q_2, q_3, q_4 into q_i, q_j, q_s, q_ℓ. The first equality in the first row just defines ρ, while the second and third equalities may be easily checked using (2.16). These equalities mean that the permutations $(2, 1, 4, 3)$ and $(3, 4, 1, 2)$ leave invariant the cross ratio; hence so does the permutation $(4, 3, 2, 1) = (3, 4, 1, 2)(2, 1, 4, 3)$, which proves the last equality in the first row.

Since the permutations $(2, 1, 4, 3)$, $(3, 4, 1, 2)$ and $(4, 3, 2, 1)$ leave invariant the cross ratio, any permutation σ and the permutations $(2, 1, 4, 3)\sigma$, $(3, 4, 1, 2)\sigma$ and $(4, 3, 2, 1)\sigma$ have the same effect on the cross ratio. In the sequel we will make repeated use of this fact.

The first equality in the second row is evident from (2.16). The remaining equalities in the same row follow from it using the fact noted above, due to being

$$\begin{aligned}(1,2,4,3) &= (2,1,4,3)(2,1,3,4),\\ (3,4,2,1) &= (3,4,1,2)(2,1,3,4),\\ (4,3,1,2) &= (4,3,2,1)(2,1,3,4).\end{aligned}$$

The reader may check that

$$\frac{\theta_3-\theta_1}{\theta_3-\theta_2}:\frac{\theta_4-\theta_1}{\theta_4-\theta_2}+\frac{\theta_2-\theta_1}{\theta_2-\theta_3}:\frac{\theta_4-\theta_1}{\theta_4-\theta_3}=1,$$

which proves the first equality in the third row. From it, the remaining ones follow as in the preceding case.

Once the effects of $(1,3,2,4)$ and $(2,1,3,4)$ on the cross ratio are known, the equality $(2,3,1,4) = (1,3,2,4)(2,1,3,4)$ shows that $(q_2,q_3,q_1,q_4) = 1-1/\rho = (\rho-1)/\rho$, which is the first equality in the fourth row; the other follow from it as above.

The proofs of the equalities in the two remaining rows are achieved similarly from $(3,1,2,4) = (2,1,3,4)(1,3,2,4)$ and $(3,2,1,4) = (2,1,3,4)(2,3,1,4)$. □

Ordered sets of four points $q_1,q_2,q_3,q_4 \in \mathbb{P}_1$, three at least different, for which $(q_1,q_2,q_3,q_4) = -1$ are called *harmonic sets*. As we shall see, harmonic sets play an important role in projective geometry. They are characterized in the following way:

Proposition 2.10.2. *Points $q_1,q_2,q_3,q_4 \in \mathbb{P}_1$, in this order, compose a harmonic set if and only if they are pairwise different and have*

$$(q_1,q_2,q_3,q_4) = (q_2,q_1,q_3,q_4).$$

Proof. Take $\rho = (q_1,q_2,q_3,q_4)$. If $\rho = -1$ the points are pairwise different by 2.9.10 and the equality $(q_1,q_2,q_3,q_4) = (q_2,q_1,q_3,q_4)$ follows from 2.10.1. Conversely, if the points are pairwise different and this equality is satisfied, then 2.10.1 gives $\rho = 1/\rho$. It follows that either $\rho = -1$, as claimed, or $\rho = 1$, against the assumption of being all points different. □

One sees from 2.10.1 that there are eight permutations leaving invariant the cross ratio of an harmonic set, and hence preserving its condition of harmonic set: these are the permutations in the first two rows of equalities of 2.10.1. One may in particular swap over the points q_1, q_2, or the points q_3, q_4, or the pairs $\{q_1,q_2\}$, $\{q_3,q_4\}$, and the resulting set still is harmonic. When giving a harmonic set it is thus not necessary to specify the ordering of the points, but just how they are paired off, say $\{q_1,q_2\}$, $\{q_3,q_4\}$, the ordering of the points in each pair as well as the ordering of the pairs themselves being irrelevant. Because of this, when q_1, q_2, q_3, q_4 is a harmonic set, it is often equivalently said that the pairs $\{q_1, q_2\}$, $\{q_3, q_4\}$ *harmonically divide* each other. It is clear from 2.10.1 that the other values of the cross ratio that may be obtained by permutation of the points of an harmonic set are 2 and $1/2$.

Theorem 2.10.3 (Complete quadrilateral theorem). *On any diagonal of a complete quadrilateral, the pair of vertices and the pair of intersection points with the other diagonals harmonically divide each other.*

Proof. The reader may give a direct proof by means of the representatives already used in the proof of 1.8.1. For a different proof, let A and B be the vertices on the diagonal we are considering. Assume that C, D and E, F are the other pairs of opposite vertices, chosen in such a way that $E \in AC$ and therefore $F \in BC$. Take $M = AB \cap CD$, $N = AB \cap EF$ and $T = CD \cap EF$ (see Figure 2.1). The perspectivity with centre E from CD to AB maps C, D, M, T to A, B, M, N, in this order. Similarly, the perspectivity with centre F between the same lines maps C, D, M, T to B, A, M, N. Since cross ratios are preserved by perspectivities (1.9.3 and 2.9.11), one gets

$$(A, B, M, N) = (C, D, M, T) = (B, A, M, N).$$

On the other hand, it is easy to check from the definition of quadrilateral that A, B, M, N are four different points, after which the claim follows from 2.10.2. □

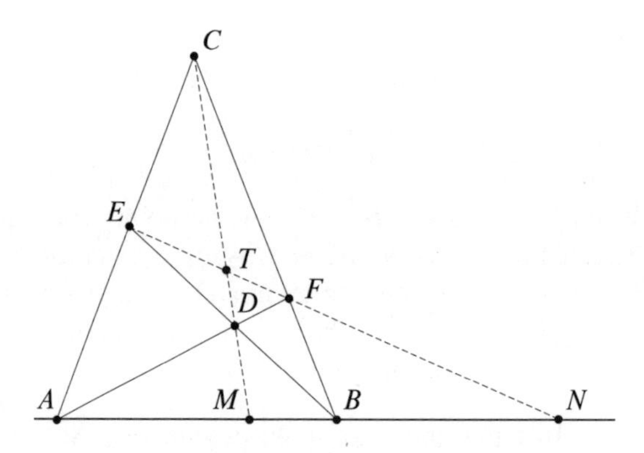

Figure 2.1. Theorem of the complete quadrilateral.

It is a particular case of 2.9.8 that given any three different points $q_1, q_2, q_3 \in \mathbb{P}_1$ there is one and only point q_4 making q_1, q_2, q_3, q_4 a harmonic set: such a point is called the *fourth harmonic* of q_1, q_2, q_3, and also the fourth harmonic of q_3 relative to q_1, q_2. Theorem 2.10.3 provides a construction (using a straight edge) of the fourth harmonic point N of three aligned points A, B, M drawn on a line of a projective plane. It is called the *harmonic construction* and consists in drawing a complete quadrilateral with two opposite vertices at A and B and one diagonal other than AB going through M, which can be done in infinitely many different ways. Then the intersection of the remaining diagonal and AB is the fourth harmonic N, regardless of the particular choice of the complete quadrilateral.

It is worth investigating in which cases, besides that of the harmonic sets, there are more than four permutations leaving invariant the cross ratio ρ of four different points q_1, q_2, q_3, q_4. According to 2.10.1, $\rho \neq 0, 1, \infty$ should equal one of the values $1/\rho, 1-\rho, (\rho-1)/\rho, 1/(1-\rho), \rho/(\rho-1)$. If $\rho = 1/\rho$ we have the harmonic sets. Nothing essentially new comes from the cases $\rho = 1-\rho$ and $\rho = \rho/(\rho-1)$, as they give $\rho = 1/2, 2$ and hence correspond to points that suitably reordered give a harmonic set. Both cases $\rho = (\rho-1)/\rho$ and $\rho = 1/(1-\rho)$ lead to the equation $\rho^2 - \rho + 1 = 0$ which has no solution if $k = \mathbb{R}$, and in case $k = \mathbb{C}$ gives $\rho = \frac{1}{2}(1 \pm \boldsymbol{i}\sqrt{3})$. A set of four points q_1, q_2, q_3, q_4 of a complex projective line is called an *equinharmonic set* if and only if

$$(q_1, q_2, q_3, q_4) = \frac{1 \pm \boldsymbol{i}\sqrt{3}}{2}.$$

The reader may note that this condition does not depend on the ordering of the points, as twelve of the permutations of the four points of an equinharmonic set leave invariant the cross ratio, while the remaining twelve turn the cross ratio into its complex-conjugate.

We close this section by presenting a relation, defined in terms of cross ratio, which causes a real projective line to split in disjoint sets. Two disjoint pairs of different points p_1, p_2 and q_1, q_2 of $\mathbb{P}_{1,\mathbb{R}}$ are said to *separate* each other if and only if $(p_1, p_2, q_1, q_2) < 0$, this relation being clearly symmetric and independent of the ordering on each pair by 2.10.1. For instance, the two pairs of an harmonic set separate each other.

Proposition 2.10.4. *Fix two different points $p_1, p_2 \in \mathbb{P}_{1,\mathbb{R}}$. There is a unique decomposition*

$$\mathbb{P}_{1,\mathbb{R}} - \{p_1, p_2\} = S \cup S', \quad S \cap S' = \emptyset$$

such that points $q_1, q_2 \in \mathbb{P}_{1,\mathbb{R}} - \{p_1, p_2\}$ have $(p_1, p_2, q_1, q_2) > 0$ if and only if either $q_1, q_2 \in S$ or $q_1, q_2 \in S'$

Proof. Define an equivalence relation on $\mathbb{P}_{1,\mathbb{R}} - \{p_1, p_2\}$ by taking as equivalent points q_1, q_2 if and only if $(p_1, p_2, q_1, q_2) > 0$: it is reflexive by 2.9.10, symmetric by 2.10.1 and transitive by 2.9.6. Take any $q \in \mathbb{P}_{1,\mathbb{R}} - \{p_1, p_2\}$: there are points non-equivalent to q by 2.9.8: any two such points q_1, q_2 have $(p_1, p_2, q_1, q) < 0$ and $(p_1, p_2, q, q_2) < 0$. Hence, again by 2.9.6,

$$(p_1, p_2, q_1, q_2) = (p_1, p_2, q_1, q)(p_1, p_2, q, q_2) > 0$$

and q_1, q_2 are equivalent. The above equivalence relation has thus just two equivalence classes, namely the set of all points equivalent to q and the set of all points non-equivalent to q. After this, clearly, sets S, S' satisfy the claim if and only if they are the above equivalence classes. $\square$

The sets S, S' of 2.10.4 are called the *projective segments* (*of* $\mathbb{P}_{1,\mathbb{R}}$) *with ends* p_1, p_2; we will say just *segments* if no confusion with the affine segments may occur. Points $q_1, q_2 \in \mathbb{P}_{1,\mathbb{R}} - \{p_1, p_2\}$ belong thus to different segments with ends p_1, p_2 if and only if they are different and separate p_1, p_2.

Remark 2.10.5. If $f : \mathbb{P}_{1,\mathbb{R}} \to \mathbb{P}'_{1,\mathbb{R}}$ is any projectivity, then it follows from the invariance of the cross ratio 2.9.11 that $\{p_1, p_2\}$ and $\{q_1, q_2\}$ separate each other if and only if $\{f(p_1), f(p_2)\}$ and $\{f(q_1), f(q_2)\}$ separate each other. As a consequence, the images by f of the segments with ends $\{p_1, p_2\}$ are the segments with ends $\{f(p_1), f(p_2)\}$.

2.11 Projective classification

It is time to add some further methodological comments. Assume that we have fixed a projective class $\mathbf{F}$ (in the sense of Section 1.10): two objects $\mathcal{T}, \mathcal{T}' \in \mathbf{F}$, associated to projective spaces $\mathbb{P}_n$ and $\mathbb{P}'_m$ respectively, are called *projectively equivalent* if and only if there is a projectivity $g : \mathbb{P}_n \to \mathbb{P}'_m$ ($m = n$ thus) such that $g(\mathcal{T}) = \mathcal{T}'$. This defines an equivalence relation due to the properties (a), (b), and (c) of Section 1.10: the corresponding partition of $\mathbf{F}$ in equivalence classes is called the *projective classification* of the objects of $\mathbf{F}$. We have already dealt with the projective classification of collinear quadruples in Example 2.9.14. The equivalence class of an object is called its *projective class* and also its *projective type*. Usually, a projective classification theorem lists all projective types and provides some procedure to determine the projective type of a given object. This is often done by using one or more maps, from $\mathbf{F}$ into certain sets, that assign the same image to any two projectively equivalent objects. Each such map is called a *projective invariant* of the objects of $\mathbf{F}$.

Often the name of the images is used for the invariant itself: for instance, in Example 2.9.14, the cross ratio has been said to be a projective invariant of the collinear quadruples, rather than saying that mapping each collinear quadruple to its cross ratio is a projective invariant. Assume to have a set (it is customary to say *system*) of projective invariants $\{\mathcal{I}_j\}_{j\in J}$. By definition, if objects $\mathcal{T}, \mathcal{T}'$ are projectively equivalent, then $\mathcal{I}_j(\mathcal{T}) = \mathcal{I}_j(\mathcal{T}')$ for any $j \in J$. When the converse holds true, that is, when for any pair $\mathcal{T}, \mathcal{T}'$, the equalities $\mathcal{I}_j(\mathcal{T}) = \mathcal{I}_j(\mathcal{T}')$, $j \in J$, imply that $\mathcal{T}$ and $\mathcal{T}'$ are projectively equivalent, it is said that the system of projective invariants $\{\mathcal{I}_j\}_{j\in J}$ is *complete*. The cross ratio alone composes a complete system of projective invariants for collinear quadruples by 2.9.11 and 2.9.13, as already explained in Example 2.9.14.

2.12 Exercises

Exercise 2.1. Assume that $\Delta = (p_0, p_1, p_2, p_3, A)$ is a reference of $\mathbb{P}_3$ and that $p = [a_0, a_1, a_2, a_3] \in \mathbb{P}_3$ belongs to no two-dimensional face of Δ and is $p \neq A$. Proving on the way that all the implicit assertions are true, compute:

(1) a system of equations of the line pp_3,

(2) an equation of the plane spanned by p, p_2, p_3,

(3) a system of equations of the image of the line pA by projection from p_3 onto the face of Δ opposite p_3,

(4) parametric and implicit equations of the only line through p that meets $p_0 p_1$ and $p_2 p_3$ and

(5) coordinates of the intersection point of the line pA and the plane $x_0 + x_1 + x_2 + x_3 = 0$.

Exercise 2.2. Give necessary and sufficient conditions on the coefficients of their implicit equations for four planes of $\mathbb{P}_3$ to have non-empty intersection. Same for three planes to share a line.

Exercise 2.3. Choose suitable references and re-prove Desargues' and Pappus' theorems using coordinates.

Exercise 2.4. Taking the vertices of a quadrivertex as the points of a reference, use coordinates to prove that the diagonal points of the quadrivertex are non-aligned.

Exercise 2.5. Assume that $\Delta = (p_0, \dots, p_n, A)$ is a reference of $\mathbb{P}_n$ and take $A' = [a_0, \dots, a_n]_\Delta$ with $a_i \neq 0$, $i = 0, \dots, n$. Prove that $\Delta' = (p_0, \dots, p_n, A')$ is also a reference and write down equations relating coordinates of an arbitrary point relative to Δ and Δ'.

Exercise 2.6. Let Δ and Ω be projective references of a projective space $\mathbb{P}_n$. Assume that the unit point of Ω is $[b_0, \dots, b_n]_\Delta$ and that, for $i = 0, \dots, n$, using coordinates relative to Δ,

$$a_0^i x_0 + \dots + a_n^i x_n = 0$$

is an equation of the i-th $(n-1)$-dimensional face of Ω chosen in such a way that $a_0^i b_0 + \dots + a_n^i b_n = 1$. Prove that for any $p = [x_0, \dots, x_n]_\Delta$,

$$\begin{aligned} y_0 &= a_0^0 x_0 + \dots + a_n^0 x_n, \\ &\vdots \\ y_n &= a_0^n x_0 + \dots + a_n^n x_n \end{aligned}$$

are coordinates of p relative to Ω. (This is mostly the way in which projective coordinates were first introduced by Plücker.) *Hint*: Prove that the matrix $(a^i_j)_{i,j=0,\dots,n}$ is regular and use 2.3.3 to prove that $y_0, \dots, y_1$ are homogeneous coordinates of p relative to a certain reference Ω'; then prove that $\Omega' = \Omega$.

Exercise 2.7. Let p_0, p_1, p_2 be the vertices of a triangle T of $\mathbb{P}_2$ and p'_0, p'_1, p'_2 points lying on the sides of T opposite p_0, p_1 and p_2, respectively, $p'_i \neq p_j$ for $i, j = 1, 2, 3, i \neq j$. Prove that the points $p_i p_j \cap p'_i p'_j$, $i, j = 1, 2, 3, i \neq j$, are aligned if and only if either the lines $p_i p'_i$, $i = 1, 2, 3$, are concurrent, or the points p'_0, p'_1, p'_2 are aligned.

Exercise 2.8. Let $\mathbb{P}_n$ and $\mathbb{P}'_n$ be projective spaces of the same dimension over the same field k. Prove that the projective groups $\mathrm{PG}(\mathbb{P}_n)$ and $\mathrm{PG}(\mathbb{P}'_n)$ are isomorphic. That is the reason why often no distinction between two such groups is made, any of them being denoted by $\mathrm{PG}(n, k)$ (or $\mathrm{PGL}(n+1, k)$, in algebraic contexts). *Hint*: Use 2.8.3.

Exercise 2.9. Let $L, T_1, T_2 \subset \mathbb{P}_n$ be linear varieties. Assume that L has dimension $d, 0 \leq d \leq n-2$, and that T_1, T_2 are both supplementaries to L. Let $f : T_1 \to T_2$ be the perspectivity with centre L. Take a reference Δ with $d+1$ vertices in L and the remaining ones in T_2, and assume given parametric equations of T_1. Express the coordinates of $f(p)$ as functions of the parameters of an arbitrary $p \in T_1$, thus re-proving in particular the second half of Theorem 1.9.3 and computing a matrix of f (by 2.5.3 and 2.8.5).

Exercise 2.10. A reference in $\mathbb{P}_3$ being fixed, check that the points

$$[1, 1, 2, 1], \quad [3, -1, 1, 2], \quad [6, 2, 7, 5], \quad [2, -2, -1, 1]$$

are aligned. Compute their cross ratio.

Exercise 2.11. Assume that four points of $\mathbb{P}_1$ have finite absolute coordinates $\theta_1, \theta_2, \varepsilon_1, \varepsilon_2$. Prove that they compose a harmonic set if and only if at least three of them are different and

$$\theta_1\theta_2 + \varepsilon_1\varepsilon_2 - \frac{1}{2}(\theta_1 + \theta_2)(\varepsilon_1 + \varepsilon_2) = 0.$$

Exercise 2.12. Assume that $q_s = [\alpha^0_s, \dots, \alpha^n_s]$, $s = 1, 2, 3, 4$ are four aligned points of $\mathbb{P}_n$, at least three of them different. Fix $i, j, 0 \leq i < j \leq n$, and take

$$\delta_{s,t} = \begin{vmatrix} \alpha^i_s & \alpha^j_s \\ \alpha^i_t & \alpha^j_t \end{vmatrix}.$$

Prove that either $\delta_{3,1} = \delta_{3,2} = \delta_{4,1} = \delta_{4,2} = 0$ or, otherwise,

$$(q_1, q_2, q_3, q_4) = \frac{\delta_{3,1}}{\delta_{3,2}} : \frac{\delta_{4,1}}{\delta_{4,2}}.$$

Exercise 2.13. Let $T = \{q_1, q_2, q_3, q_4\}$ be a set of four different points of a projective line $\mathbb{P}_1$. Let $\rho_1, \dots, \rho_6$ be the cross ratios of q_1, q_2, q_3, q_4 taken in all possible orderings, and take $J(T) = \sum_{i<j} \rho_i \rho_j$. Prove that

$$J(T) = 6 - \frac{(\rho^2 - \rho + 1)^3}{\rho^2(\rho - 1)^2}$$

for $\rho = \rho_1, \dots, \rho_6$.

Assume that T' is a set of four different points of a projective line $\mathbb{P}'_1$, over the same field as $\mathbb{P}_1$. Prove that there is a projectivity $f\colon \mathbb{P}_1 \to \mathbb{P}'_1$ such that $f(T) = T'$ if and only if $J(T) = J(T')$. *Hint*: For the first part take $\rho_i = \rho$ and the remaining values of the cross ratio as given by 2.10.1. For the second one, prove that $\{\rho_1, \dots, \rho_s\}$ is the set of roots of the polynomial

$$(X^2 - X + 1)^3 + (J - 6)(X^2(X - 1)^2).$$

The cases in which T, suitably ordered, is a harmonic or equinharmonic set require separate attention.

Exercise 2.14. Assume to have in $\mathbb{P}_2$ a triangle with vertices A, B, C and two different points M, N, no one lying on a side of the triangle. Let A', B', C' be, respectively, the intersections of the line MN with the sides of the triangle opposite A, B, C. Denote by $\bar{A}$, $\bar{B}$, $\bar{C}$ the fourth harmonics of A', B', C' with respect to M, N. Prove that the lines $A\bar{A}$, $B\bar{B}$, $C\bar{C}$ are concurrent.

Exercise 2.15. Let ℓ_1, ℓ_2, ℓ_3 be three pairwise skew lines of a k-projective space $\mathbb{P}_3$. Prove that there is a reference of $\mathbb{P}_3$ relative to which these lines have equations

$$\begin{aligned} \ell_1 &: x_0 = x_1 = 0, \\ \ell_2 &: x_2 = x_3 = 0, \\ \ell_3 &: x_0 - x_2 = x_1 - x_3 = 0. \end{aligned}$$

Assume that ℓ'_1, ℓ'_2, ℓ'_3 are three pairwise skew lines of a k-projective space $\mathbb{P}'_3$. Prove that there is a projectivity $f\colon \mathbb{P}_3 \to \mathbb{P}'_3$ such that $f(\ell_i) = \ell'_i$. Is such a projectivity unique?

Exercise 2.16. Let Σ be an n-dimensional simplex of $\mathbb{P}_n$ with vertices $p_0, \dots, p_n$. For each point p belonging to no $(n-1)$-dimensional face of Σ and any i, j, $0 \leq i < j \leq n$, take $p_{i,j}$ to be the projection of p from the face of Σ opposite $p_i p_j$ onto $p_i p_j$. Let $\bar{p}_{i,j}$ be the fourth harmonic of p_i, p_j, $p_{i,j}$ and prove that:

(a) The points $\bar{p}_{i,j}$, $0 \leq i < j \leq n$, span a hyperplane $H(p)$, which is called the *harmonic polar* of p relative to Σ.

(b) Mapping p to $H(p)$ is a bijection between the set of all points lying on no $(n-1)$-dimensional face of Σ and the set of all hyperplanes going through no vertex of Σ. The point p is called the *harmonic pole* of the hyperplane $H(p)$ relative to Σ.

Hint: Use coordinates relative to a reference with vertices $p_0, \dots, p_n$ and prove that the harmonic polar of $[a_0, \dots, a_n]$ has equation $a_0^{-1}x_0 + \dots + a_n^{-1}x_n = 0$.

Exercise 2.17. Notations and hypothesis being those of Exercise 2.16, for each $i = 0, \dots, n$, let H_i be the face of Σ opposite p_i, denote by Σ_i the simplex of H_i with vertices p_j, $j = 0, \dots, n$, $j \neq i$, and by p_i' the harmonic pole of $H(p) \cap H_i$ relative to Σ_i. Prove that the lines $p_i p_i'$ intersect at p.

Exercise 2.18. Assume given in a projective plane $\mathbb{P}_2$ a triangle T and a point p lying on no side of T. Take T as the triangle of the reference and p as the unit point and compute the coordinates of the remaining vertices and the sides of a quadrivertex of $\mathbb{P}_2$ with one vertex at p and diagonal triangle T. This will in particular prove the existence and uniqueness of a quadrivertex with given diagonal triangle T and vertex p, p a point lying on no side of T. *Hint*: use the theorem of the complete quadrilateral 2.10.3.

Exercise 2.19. Describe a graphic construction of the remaining vertices of the quadrivertex of Exercise 2.18, once T and p are given on the drawing plane.

Exercise 2.20. Fix two different points $p_1, p_2 \in \mathbb{P}_{1,\mathbb{R}}$. Prove that there is a projectivity of $\mathbb{P}_{1,\mathbb{R}}$ leaving p_1 and p_2 invariant and mapping to each other the segments with ends p_1, p_2, no projective distinction between these segments being thus possible.

Exercise 2.21. Assume given, in a projective plane, a triangle with vertices A_1, A_2, A_3 and for each $i = 1, 2, 3$, let a_i be the side opposite A_i. Assume also given a line ℓ going through no vertex and, for each $i = 1, 2, 3$, a point $C_i \in a_i$ different from any vertex. If $B_i = \ell \cap a_i$, $i = 1, 2, 3$, then prove that

$$(A_1, A_2, B_3, C_3)(A_2, A_3, B_1, C_1)(A_3, A_1, B_2, C_2) = -1$$

if and only if A_1C_1, A_2C_2, A_3C_3 are concurrent.

Exercise 2.22. Notations and hypothesis being as in Exercise 2.21, prove that

$$(A_1, A_2, B_3, C_3)(A_2, A_3, B_1, C_1)(A_3, A_1, B_2, C_2) = 1$$

if and only if C_1, C_2, C_3 are aligned.

Exercise 2.23. Let ℓ_i, $i = 1, 2, 3, 4$, be the sides of a quadrilateral of $\mathbb{P}_2$, $A_{i,j} = \ell_i \cap \ell_j$, $1 \leq i < j \leq 4$, its vertices and $B_i = \ell_i \cap s$, $i = 1, 2, 3, 4$, the intersections of the sides with a line s containing no vertex. Prove that

$$\frac{(A_{1,4}, A_{1,3}, A_{1,2}, B_1)}{(A_{2,4}, A_{2,3}, A_{2,1}, B_2)} = (B_1, B_2, B_3, B_4).$$

Exercise 2.24. Assume to have three pairwise skew lines $r, s, t \subset \mathbb{P}_3$ and four different lines $\ell_1, \ell_2, \ell_3, \ell_4 \subset \mathbb{P}_3$, each ℓ_i intersecting r, s and t in points $A_i = r \cap \ell_i$, $B_i = s \cap \ell_i$, $C_i = t \cap \ell_i$, $i = 1, 2, 3, 4$. Prove that

$$(A_1, A_2, A_3, A_4) = (B_1, B_2, B_3, B_4) = (C_1, C_2, C_3, C_4).$$

Chapter 3
Affine geometry

Now that the basic elements of linear projective geometry have been presented in the preceding chapters, we devote the present one to relate affine and projective geometries. Development of projective geometry will be continued in Chapter 4.

3.1 Recalling basic facts of affine geometry

In this section we briefly recall, without proofs, the most basic facts of linear affine geometry, mainly to fix the notations and the nomenclature.

An *affine space* of dimension n, $n \geq 1$, over the field k is a triple $(\mathbb{A}_n, F, +)$ where $\mathbb{A}_n$ is a set whose elements are called *points*, F is an n-dimensional vector space over k whose elements are called *free vectors* or just *vectors*, and $+$ is an operation of vectors on points, that is, a map

$$\begin{aligned} F \times \mathbb{A}_n &\longrightarrow \mathbb{A}_n, \\ (v, p) &\longmapsto p + v, \end{aligned}$$

that satisfies

(1) $p + (v + w) = (p + v) + w$ for any $p \in \mathbb{A}_n$, $v, w \in F$,

(2) $p + 0 = p$ for any $p \in \mathbb{A}_n$, and

(3) for any $p, q \in \mathbb{A}_n$ there is a unique $v \in F$ such that $p = q + v$. We will write $v = p - q$ (also $v = \vec{qp}$ is a usual notation).

The same notation will be used for both the affine space and its set of points, this causing no confusion. The vector space F is called the vector space *associated* to $\mathbb{A}_n$. In the sequel $\mathbb{A}_n$ (or $\mathbb{A}_{n,k}$, if mentioning the base field k is relevant) will denote an arbitrary n-dimensional affine space over k.

An (affine) *reference* Δ of $\mathbb{A}_n$ consists of a point $O \in \mathbb{A}_n$ and a basis $e_1, \dots, e_n$ of F. The point O is called the *origin* of Δ, each vector e_i a *vector* of Δ, and the ordered set $e_1, \dots, e_n$ the *basis* of Δ. The (*affine*) *coordinates* of a point p, relative to such a reference, are the components of $p - O$ relative to the basis $e_1, \dots, e_n$; once the reference is fixed, the coordinates are uniquely determined by the point and in turn determine it; we will use the customary notation $(X_1, \dots, X_n)$ to denote the point whose affine coordinates are $X_1, \dots, X_n$.

An (affine) *linear variety* L of $\mathbb{A}_n$ is any set of the form

$$L = p + G = \{p + v \mid v \in G\}$$

where $p \in \mathbb{A}_n$ and G is a subspace of F. Note that $p \in L$ and so, in particular, $L \neq \emptyset$. The same linear variety L may be written $L = p' + G$ for any $p' \in L$. The vector subspace G may be recovered form L as $G = \{v = p - q \mid p, q \in L\}$. It is called the *director subspace* of L and its vectors are called the *vectors on* L. The dimension of L is taken as $\dim L = \dim G$. Linear varieties are *parallel* if and only if their director subspaces are included one into the other in either sense.

Once a reference of $\mathbb{A}_n$ is fixed, the points whose coordinates $X_1, \dots, X_n$ satisfy a compatible system of independent linear equations

$$\begin{aligned} a_1 + a_1^1 X_1 + \cdots + a_1^n X_n &= 0, \\ &\vdots \\ a_{n-d} + a_{n-d}^1 X_1 + \cdots + a_{n-d}^n X_n &= 0 \end{aligned} \tag{3.1}$$

describe a linear variety of dimension d and, conversely, any linear variety L of $\mathbb{A}_n$ may be obtained in this form. The system (3.1) is called a system of equations of L. In the sequel, their equations will be called *affine*, in order to distinguish them from the homogeneous equations of projective varieties. A vector belongs to the director subspace of the variety L above if and only if its components (relative to the basis of the reference) satisfy the system of (independent, linear and homogeneous) equations

$$\begin{aligned} a_1^1 X_1 + \cdots + a_1^n X_n &= 0, \\ &\vdots \\ a_{n-d}^1 X_1 + \cdots + a_{n-d}^n X_n &= 0 \end{aligned}$$

obtained by dropping the independent terms in (3.1).

If $L_1, \dots, L_r$ are linear varieties of $\mathbb{A}_n$, its intersection $L_1 \cap \cdots \cap L_r$ is either the empty set or a linear variety of $\mathbb{A}_n$. In the latter case the director subspace of $L_1 \cap \cdots \cap L_r$ is $G_1 \cap \cdots \cap G_r$, where each G_i is the director subspace of L_i, $i = 1, \dots, r$.

On the other hand, there is a well-determined linear variety which is minimal among the linear varieties of $\mathbb{A}_n$ containing $L_1, \dots, L_r$: it is called the *join of*, or the linear variety *spanned by*, $L_1, \dots, L_r$, and denoted by $L_1 \vee \cdots \vee L_r$. Points $p_1, \dots, p_s$ of $\mathbb{A}_n$ are called independent if and only if the linear variety they span has dimension $s - 1$.

An *affine transformation*, or *affinity*, f between affine spaces $(\mathbb{A}_n, F, +)$ and $(\mathbb{B}_n, G, +)$ (necessarily of the same dimension) is any map of the form

$$\begin{aligned} f \colon \mathbb{A}_n &\longrightarrow \mathbb{B}_n, \\ p &\longmapsto q_2 + \varphi(p - q_1), \end{aligned}$$

for certain fixed points $q_1 \in A_n$, $q_2 \in \mathbb{B}_n$ and an isomorphism of vector spaces $\varphi \colon F \to G$. It easily turns out that $q_2 = f(q_1)$ and that the same map is defined if one substitutes any pair $q, f(q)$ for q_1, q_2 in the above rule. The isomorphism φ may be recovered from f through the equality $\varphi(p-q) = f(p) - f(q)$ for any $p, q \in \mathbb{A}_n$; it is called the *isomorphism* (and often just the *morphism* o the *linear map) associated* to f.

Once an affine reference in each space has been chosen, for each $p \in \mathbb{A}_n$, the coordinates $Y_1, \dots, Y_n$ of $f(p)$ may be computed from the coordinates $X_1, \dots, X_n$ of p by linear equations

$$\begin{aligned} Y_1 &= b_1 + a_1^1 X_1 + \cdots + a_n^1 X_n, \\ &\vdots \\ Y_n &= b_n + a_1^n X_1 + \cdots + a_n^n X_n, \end{aligned} \tag{3.2}$$

where

$$N = \begin{pmatrix} a_1^1 & \dots & a_n^1 \\ \vdots & & \vdots \\ a_1^n & \dots & a_n^n \end{pmatrix}$$

is the matrix of the associated morphism φ relative to the bases of the references, and is therefore a regular matrix. Conversely, any equations as (3.2) above, with $\det N \neq 0$, give rise to an affinity $f : \mathbb{A}_n \to \mathbb{B}_n$ if the point with coordinates $Y_1, \dots, Y_n$ is taken as the image of the point with coordinates $X_1, \dots, X_n$, for arbitrary values of $X_1, \dots, X_n$. The morphism associated to such an f has matrix N.

The *affine ratio* of three aligned points q_1, q_2, q_3, two at least distinct, is defined as being ∞ if $q_2 = q_3$ or the scalar ρ for which $q_1 = q_3 + \rho(q_2 - q_3)$ otherwise. As usual, we will denote it by (q_1, q_2, q_3). The *midpoint* of two different points q_1, q_2 is defined as the only point q_3 which is aligned with q_1, q_2 and has $(q_1, q_2, q_3) = -1$; equivalently,

$$q_3 = q_1 + \frac{1}{2}(q_2 - q_1) = q_2 + \frac{1}{2}(q_1 - q_2).$$

In the sequel, when dealing with lines ℓ which have two distinguished points q_1, q_2, the midpoint of q_1, q_2 will be sometimes referred to as the *midpoint* of ℓ: for instance the midpoint of a side of a triangle will be the midpoint of the two vertices on it. This will keep the usual nomenclature of elementary geometry, in which segments are taken in place of these lines.

The affine space $\mathbb{A}_n$ being not a vector space, an expression of the form

$$\lambda_1 q_1 + \cdots + \lambda_m q_m,$$

where $\lambda_i \in k$ and $q_i \in \mathbb{A}_n$, $i = 1, \dots, m$, makes in general no sense. However, if $\sum_{i=1}^m \lambda_i = 1$, then one defines

$$\lambda_1 q_1 + \cdots + \lambda_m q_m = O + \lambda_1(q_1 - O) + \cdots + \lambda_m(q_m - O)$$

for an arbitrary choice of $O \in \mathbb{A}_n$, because the member on the right does not depend on O. Indeed, for any $O' \in \mathbb{A}_n$,

$$\begin{aligned} O &+ \lambda_1(q_1 - O) + \cdots + \lambda_m(q_m - O) \\ &= O + \lambda_1(q_1 - O) + \cdots + \lambda_m(q_m - O) + (O' - O) + \sum_{i=0}^{m} \lambda_i(O - O') \\ &= O' + \lambda_1(q_1 - O') + \cdots + \lambda_m(q_m - O'). \end{aligned}$$

3.2 The projective closure of an affine space

Assume that we have fixed an n-dimensional affine space $\mathbb{A}_n$ over the field k, with associated vector space F. Roughly speaking, to construct the projective closure of $\mathbb{A}_n$ we will first add new points to $\mathbb{A}_n$, one for each direction, and then give to the whole set of the old and new points a structure of projective space.

If, as usual, the *direction* of a non-zero vector of F is taken to be its class modulo proportionality, the directions of all non-zero vectors are the points of the projectivization $\mathbb{P}(F)$ of the vector space F (see Section 1.2). We shall write $H_\infty = \mathbb{P}(F)$ and denote by $[v]$ the direction of $v \in F - \{0\}$, v being, by definition, a representative of its direction.

We take the disjoint union $\overline{\mathbb{A}}_n = \mathbb{A}_n \cup H_\infty$ to be the set of points of a projective space. An element p of $\overline{\mathbb{A}}_n$ is thus either a point of $\mathbb{A}_n$ or a direction: in the first case p is called a *proper* or *actual point*; in the second case it is called an *improper point* or a *point at infinity*. The reader may note that the names *direction* and *improper point* denote the same mathematical object; in the sequel they will be taken as synonymous, the use of one or the other depending on the context. $\mathbb{A}_n$ is often called the *proper* (or *finite* or *actual*) *part* of $\overline{\mathbb{A}}_n$. After defining the projective structure on $\overline{\mathbb{A}}_n$, it will turn out that H_∞ is a hyperplane of $\overline{\mathbb{A}}_n$: it is called the *improper hyperplane* or the *hyperplane at infinity* of $\overline{\mathbb{A}}_n$, and also the *improper hyperplane* or the *hyperplane at infinity* of $\mathbb{A}_n$. Of course, if $n = 1, 2, 3$ we say *point*, *line* or *plane* instead of *hyperplane*.

In order to get a suitable vector space we will consider vector fields on the affine space $\mathbb{A}_n$: a *vector field* θ on $\mathbb{A}_n$ consists in assigning to each point $p \in \mathbb{A}_n$ a vector $\theta(p) \in F$. In other words a vector field on $\mathbb{A}_n$ is a map

$$\begin{aligned} \theta\colon \mathbb{A}_n &\longrightarrow F, \\ p &\longmapsto \theta(p). \end{aligned}$$

We will denote by Θ the set of all vector fields on $\mathbb{A}_n$. Once equipped with the operations defined by the rules

$$\begin{aligned} (\theta_1 + \theta_2)(p) &= \theta_1(p) + \theta_2(p), \\ (\lambda\theta)(p) &= \lambda(\theta(p)), \end{aligned}$$

for any $\theta, \theta_1, \theta_2 \in \Theta$, $p \in \mathbb{A}_n$ and $\lambda \in k$, Θ obviously becomes a vector space whose zero element is the field assigning the zero vector to all points.

The vector field that assigns the same vector $v \in F$ to all $p \in \mathbb{A}_n$ will be called the *constant vector field* with vector v, denoted by $\hat{v}$. The reader may easily prove:

Lemma 3.2.1. *The set $\widehat{F}$ of all constant vector fields is a subspace of Θ, and the map*

$$\begin{aligned} \kappa\colon F &\longrightarrow \widehat{F}, \\ v &\longmapsto \hat{v}, \end{aligned}$$

is an isomorphism of vector spaces. In particular $\widehat{F}$ has dimension n.

Fix $q \in \mathbb{A}_n$ and $\delta \in k - \{0\}$. The *radial vector field* with *centre* q and *ratio* δ is the vector field $\theta_{q,\delta}$ defined by the rule

$$\begin{aligned} \theta_{q,\delta}\colon \mathbb{A}_n &\longrightarrow F, \\ p &\longmapsto \delta(q-p). \end{aligned}$$

Note that no radial field is constant and also that $\theta_{q_1,\delta_1} = \theta_{q_2,\delta_2}$ if and only if $q_1 = q_2$ and $\delta_1 = \delta_2$. The next lemma provides the rules for operating with constant and radial vector fields.

Lemma 3.2.2. *For any $v, v_1, v_2 \in F$, $q, q_1, q_2 \in \mathbb{A}_n$, $\lambda \in k$ and $\delta, \delta_1, \delta_2 \in k - \{0\}$,*

(a) $\hat{v}_1 + \hat{v}_2 = \widehat{v_1 + v_2}$.

(b) $\lambda\hat{v} = \widehat{\lambda v}$.

(c) *If $\delta_1 + \delta_2 \neq 0$, then $\theta_{q_1,\delta_1} + \theta_{q_2,\delta_2} = \theta_{q',\delta_1+\delta_2}$, where*

$$q' = \frac{\delta_1}{\delta_1+\delta_2}q_1 + \frac{\delta_2}{\delta_1+\delta_2}q_2.$$

(d) *If $\delta_1 + \delta_2 = 0$, then $\theta_{q_1,\delta_1} + \theta_{q_2,\delta_2} = \delta_1(\widehat{q_1 - q_2})$.*

(e) *If $\lambda \neq 0$, then $\lambda\theta_{q,\delta} = \theta_{q,\lambda\delta}$.*

(f) $\hat{v} + \theta_{q,\delta} = \theta_{q+\delta^{-1}v,\delta}$.

Proof. The claims (a) and (b) are part of 3.2.1. For claim (c), take any $p \in \mathbb{A}_n$; then

$$\begin{aligned} (\theta_{q_1,\delta_1} + \theta_{q_2,\delta_2})(p) &= \delta_1(q_1 - p) + \delta_2(q_2 - p) \\ &= (\delta_1+\delta_2)\Big(\frac{\delta_1}{\delta_1+\delta_2}q_1 + \frac{\delta_2}{\delta_1+\delta_2}q_2 - p\Big) \\ &= \theta_{q',\delta_1+\delta_2}(p). \end{aligned}$$

Regarding (d), also for any p,

$$(\theta_{q_1,\delta_1} + \theta_{q_2,\delta_2})(p) = \delta_1(q_1 - p) + \delta_2(q_2 - p) = \delta_1(q_1 - q_2).$$

Proving the remaining claims is similar and left to the reader. $\square$

Take E to be the set of all vector fields that are either constant or radial. It is clear from 3.2.2 above that E is a subspace of Θ, and obviously $\widehat{F} \subset E$. Our choice of vector space for the projective closure of $\mathbb{A}_n$ will be E. The next lemma proves in particular that it has the right dimension.

Lemma 3.2.3. *Assume that $e_1, \dots, e_n$ is a basis of F, $q \in \mathbb{A}_n$ and $\delta \in k - \{0\}$. Then $\theta_{q,\delta}, \hat{e}_1, \dots, \hat{e}_n$ is a basis of E and therefore* $\dim E = n + 1$.

Proof. Since by 3.2.1, the vector fields $\hat{e}_1, \dots, \hat{e}_n$ compose a basis of $\widehat{F}$, they are in particular linearly independent. Since $\theta_{q,\delta} \notin \widehat{F}$, $\theta_{q,\delta}, \hat{e}_1, \dots, \hat{e}_n$ are linearly independent too. Using again that $\hat{e}_1, \dots, \hat{e}_n$ is a basis of $\widehat{F}$, it is clear that any constant vector field is a linear combination of $\theta_{q,\delta}, \hat{e}_1, \dots, \hat{e}_n$. Take thus an arbitrary radial vector field $\theta_{q',\delta'}$. By 3.2.2, the vector field

$$\theta_{q',\delta'} - \frac{\delta'}{\delta}\theta_{q,\delta}$$

is constant; it may thus be written

$$\theta_{q',\delta'} - \frac{\delta'}{\delta}\theta_{q,\delta} = \sum_{i=1}^{n} \lambda_i \hat{e}_i$$

with $\lambda_1, \dots, \lambda_n \in k$, after which $\theta_{q',\delta'}$ is a linear combination of $\theta_{q,\delta}, \hat{e}_1, \dots, \hat{e}_n$. $\square$

Defining the map π, which will be the third piece of the projective closure, is easy: we just map non-zero constant vector fields to the directions of their corresponding vectors and radial vector fields to their centres. That is, we take

$$\pi : E - \{0\} \longrightarrow \overline{\mathbb{A}}_n$$

defined by the rules:

– $\pi(\hat{v}) = [v]$ if $\hat{v}$ is a non-zero constant vector field, and

– $\pi(\theta_{q,\delta}) = q$ if $\theta_{q,\delta}$ is a radial vector field.

Then we have:

Theorem 3.2.4. (a) *The triple* $(\overline{\mathbb{A}}_n, E, \pi)$ *defined above is an* n*-dimensional projective space.*

(b) H_∞ *is a hyperplane of* $\overline{\mathbb{A}}_n$*. The projective structure induced by* $\overline{\mathbb{A}}_n$ *on* H_∞ *is equivalent to its original structure* $H_\infty = \mathbb{P}(F)$*.*

Proof. Claim (a) is direct using 3.2.2 (b), 3.2.2 (e) and 3.2.3. The equivalence of projective structures on H_∞ is induced by the isomorphism κ of 3.2.1. □

Remark 3.2.5. Any proper point q has a unique representative $\theta_{q,1}$ with ratio 1. All representatives of an improper point are non-zero constant vector fields.

Remark 3.2.6. Due to part (b) of 3.2.4, it is often said that the projective structure of $\overline{\mathbb{A}}_n$ extends the one we already have on H_∞. Note that one may write $[v] = [\hat{v}]$ for any non-zero $v \in F$.

Remark 3.2.7. Due to 3.2.2 (f), the translated of a proper point $q = [\theta_{q,1}]$ by a vector v is $q + v = [\theta_{q+v,1}] = [\theta_{q,1} + \hat{v}]$

We have enlarged the affine space $\mathbb{A}_n$ to the projective space $(\overline{\mathbb{A}}_n, E, \pi)$, which in the sequel will be denoted by just $\overline{\mathbb{A}}_n$ and called the *projective closure* of $\mathbb{A}_n$. $\overline{\mathbb{A}}_n$ is a projective space with a distinguished hyperplane, the improper one, that will play a central role in the reinterpretation of the affine geometry of $\mathbb{A}_n$ in the frame of the projective geometry of $\overline{\mathbb{A}}_n$.

3.3 Affine and projective coordinates

In this section we will associate to each affine reference Δ of $\mathbb{A}_n$ a projective reference $\overline{\Delta}$ of $\overline{\mathbb{A}}_n$ in such a way that affine and homogeneous coordinates of an arbitrary point of $\mathbb{A}_n$ are easily related.

Assume we have fixed an affine reference $\Delta = (O; e_1, \dots, e_n)$ in $\mathbb{A}_n$. Take $p_0 = O$, the origin of Δ, $p_i = [e_i]$, $i = 1, \dots, n$, the directions of the axes of Δ, and $A = (1, \dots, 1)_\Delta$, the proper point whose affine coordinates are all equal to 1. Then we have:

Lemma 3.3.1. $\overline{\Delta} = (p_0, \dots, p_n, A)$ *is a projective reference of* $\overline{\mathbb{A}}_n$*. A basis adapted to* $\overline{\Delta}$ *is* $\theta_{O,1}, \hat{e}_1, \dots, \hat{e}_n$*.*

Proof. We know from 3.2.3 that $\theta_{O,1}, \hat{e}_1, \dots, \hat{e}_n$ is a basis of E. After the definition of the structure map π, it is clear that $O = [\theta_{O,1}]$ and $p_i = [e_i] = [\hat{e}_i]$ for $i = 1, \dots, n$. Furthermore, A having coordinates $1, \dots, 1$, according to 3.2.7,

$$A = O + e_1 + \dots + e_n = [\theta_{O,1} + \hat{e}_1 + \dots + \hat{e}_n].$$

Then, $\overline{\Delta}$ is the reference with adapted basis $\theta_{O,1}, \hat{e}_1, \dots, \hat{e}_n$ (see 2.1.3). □

The reference $\overline{\Delta}$ is called the projective reference *associated* to Δ. It is worth noting that its vertices $p_1, \dots, p_n$ are improper points, or, equivalently, the 0-th face of $\overline{\Delta}$ is H_∞.

The next proposition relates affine and projective coordinates of proper points, and also components of vectors and projective coordinates of their directions. In the sequel the projective structure of $\overline{\mathbb{A}}_n$ will be usually handled through it, rather than using the construction of Section 3.2.

Proposition 3.3.2. *Fix an affine reference Δ of $\mathbb{A}_n$.*

(a) *If a proper point p has affine coordinates $X_1, \dots, X_n$ relative to Δ, then it has projective coordinates $1, X_1, \dots, X_n$ relative to the projective reference associated to Δ.*

(b) *If a non-zero vector $v \in F$ has components $\alpha_1, \dots, \alpha_n$ relative to the basis of Δ, then its direction $[v]$ has projective coordinates $0, \alpha_1, \dots, \alpha_n$ relative to the projective reference associated to Δ.*

Proof. By the definition of affine coordinates and 3.2.7,

$$p = O + X_1 e_1 + \dots + X_n e_n = [\theta_{O+X_1e_1+\dots+X_ne_n,1}] = [\theta_{O,1} + X_1\hat{e}_1 + \dots + X_n\hat{e}_n]$$

after which claim (a) follows from 3.3.1. If $v = \alpha_1 e_1 + \dots + \alpha_n e_n$, then, by 3.2.6 and 3.2.1,

$$[v] = [\alpha_1 e_1 + \dots + \alpha_n e_n] = [\alpha_1 \hat{e}_1 + \dots + \alpha_n \hat{e}_n],$$

which proves claim (b). □

Remark 3.3.3. It has been already noticed that the associated reference $\overline{\Delta}$ has H_∞ as its 0-th face. Hence, the proper points are those with 0-th homogeneous coordinate different from zero. Any proper point $p = [x_0, x_1, \dots, x_n]_{\overline{\Delta}}$ has $x_0 \neq 0$, it may thus be written $p = [1, x_1/x_0, \dots, x_n/x_0]$ and so it has affine coordinates $x_1/x_0, \dots, x_n/x_0$.

Remark 3.3.4. If one takes in H_∞ the reference subordinated by $\overline{\Delta}$, 3.3.2 (b) says that the components of a non-zero vector are homogeneous coordinates of its direction and, conversely, any homogeneous coordinates of a direction are the components of one of its representatives. Equivalently, the reader may note that the basis $e_1, \dots, e_n$ is adapted to the reference subordinated by $\overline{\Delta}$ in H_∞.

Remark 3.3.5. If $n = 1$, the affine coordinate of any proper point is its absolute coordinate relative to the reference (p_1, p_0, A), obtained from $\overline{\Delta}$ by swapping over its vertices.

It is worth characterizing which projective references are associated to affine references:

Proposition 3.3.6. *Mapping each affine reference Δ to its associate projective reference $\overline{\Delta}$ is a bijection from the set of all affine references of $\mathbb{A}_n$ onto the set of all projective references of $\overline{\mathbb{A}}_n$ with 0-th face H_∞.*

Proof. That all associated references have 0-th face H_∞ has been already noticed. Assume that $\Omega = (p_0, \dots, p_n, A)$ is a projective reference of $\overline{\mathbb{A}}_n$ and $p_1 \vee \dots \vee p_n = H_\infty$. Then, clearly, p_0 is a proper point and so it has $\theta_{p_0,1}$ as representative. By multiplying the elements of an adapted basis by a suitable scalar, we may get a basis still adapted to Ω and with 0-th element $\theta_{p_0,1}$. The remaining vectors of this basis are of course independent and represent improper points, so they will be $\hat{e}_1, \dots, \hat{e}_n$, where $e_1, \dots, e_n$ are independent vectors, and therefore a basis, of F (By 3.2.1). We take $\Delta = (p_0; e_1, \dots, e_n)$ as affine reference. Next we will check that $\overline{\Delta} = \Omega$. Indeed, p_0 is the origin of Δ and the remaining vertices of Ω are the directions of the vectors of Δ, taken in the same order, so it remains only to see that A is the point with affine coordinates $(1, \dots, 1)$, and this is clear because, the former basis being adapted to Ω,

$$A = [\theta_{p_0,1} + \hat{e}_1 + \dots + \hat{e}_n] = [\theta_{p_0+e_1+\dots+e_n}] = p_0 + e_1 + \dots + e_n$$

by 3.2.7. To prove the injectivity, assume that another affine reference $\Delta_1 = (q, v_1, \dots, v_n)$ has $\overline{\Delta}_1 = \Omega$. Then, first, $q = p_0$. Also $[v_i] = p_i = [e_i]$ which gives $v_i = \lambda_i e_i$, $\lambda_i \in k - \{0\}$, $i = 1, \dots, n$. Furthermore,

$$q + v_1 + \dots + v_n = A = p_0 + e_1 + \dots + e_n.$$

Since $q = p_0$,

$$v_1 + \dots + v_n = e_1 + \dots + e_n,$$

or, equivalently,

$$\lambda_1 e_1 + \dots + \lambda_n e_n = e_1 + \dots + e_n.$$

The independence of the e_i forces $\lambda_i = 1$, $i = 1, \dots, n$, and so $\Delta' = \Delta$ as wanted. □

Remark 3.3.7. The projective references of $\overline{\mathbb{A}}_n$ with 0-th face H_∞ have been seen to be those associated to the affine references of $\mathbb{A}_n$: in the sequel they will be called *affine references of* $\overline{\mathbb{A}}_n$, in order to distinguish them from the other projective references of $\overline{\mathbb{A}}_n$. In fact, since an affine reference Δ of $\mathbb{A}_n$ and its associated projective reference $\overline{\Delta}$ determine each other, in the sequel we will make no distinction between Δ and $\overline{\Delta}$, and take either of them (or both together) as being an affine reference relative to which the points have both affine coordinates (those relative to Δ) and homogeneous coordinates (those relative to $\overline{\Delta}$). Under this convention, Proposition 3.3.2 and Remark 3.3.3 just relate affine and homogeneous coordinates relative to the same affine reference.

The construction of the projective closure starts from an affine space $\mathbb{A}_n$ and extends it to an n-dimensional projective space in which $\mathbb{A}_n$ appears as the complementary of a hyperplane. There is a sort of inverse construction that starting from a projective space $\mathbb{P}_n$ and one of its hyperplanes H, intrinsically defines on the set $\mathbb{P}_n - H$ a structure of n-dimensional affine space whose projective closure and improper hyperplane may be identified with $\mathbb{P}_n$ and H, respectively. This construction making use of certain collineations, it is outlined in forthcoming Exercise 5.29.

3.4 Affine linear varieties

In this section we will explain the relationship between the affine linear varieties of an affine space $\mathbb{A}_n$ and the projective linear varieties of its projective closure $\overline{\mathbb{A}}_n$. The linear varieties of $\overline{\mathbb{A}}_n$ contained in the improper hyperplane H_∞ are called *improper*, while those containing some proper point are called *proper*. We will see that the linear varieties of a fixed dimension d of $\mathbb{A}_n$ are in one-to-one correspondence with the proper linear varieties of dimension d of $\overline{\mathbb{A}}_n$, each affine linear variety being the set of proper points of its corresponding proper linear variety.

Let L be a proper linear variety of $\overline{\mathbb{A}}_n$, of dimension d. After fixing an affine reference Δ, assume that L has equations

$$\begin{aligned} a_1^0 x_0 + a_1^1 x_1 + \cdots + a_1^n x_n &= 0,\\ &\vdots \\ a_{n-d}^0 x_0 + a_{n-d}^1 x_1 + \cdots + a_{n-d}^n x_n &= 0, \end{aligned} \tag{3.3}$$

where $x_0, x_1, \ldots, x_n$ are homogeneous coordinates relative to Δ. A proper point p, with affine coordinates $X_1, \ldots, X_n$, belongs to L if and only if its homogeneous coordinates $1, X_1, \ldots, X_n$ (see 3.3.2 and 3.3.7) are a solution of the equations (3.3) above, that is, if and only if

$$\begin{aligned} a_1^0 + a_1^1 X_1 + \cdots + a_1^n X_n &= 0,\\ &\vdots \\ a_{n-d}^0 + a_{n-d}^1 X_1 + \cdots + a_{n-d}^n X_n &= 0. \end{aligned} \tag{3.4}$$

Now the linear equations (3.4) are independent, because so are the equations (3.3), and compatible, because L has a proper point. As recalled in Section 3.1, the set of all points of $\mathbb{A}_n$ whose affine coordinates satisfy a given system of $n-d$ independent and compatible equations is an affine linear variety of dimension d. We have thus proved:

Proposition 3.4.1. *If L is a proper linear variety of $\overline{\mathbb{A}}_n$ of dimension d, then $\widetilde{L} = L \cap \mathbb{A}_n$ is a linear variety of $\mathbb{A}_n$ of dimension d.*

The affine linear variety $\widetilde{L}$ is called the *proper* (or *actual*, or *finite*) *part* of L. $L - \widetilde{L} = L \cap H_\infty$ is called the *improper part* of L; it is an improper linear variety of dimension $\dim L - 1$ (by 1.4.8).

Remark 3.4.2. Note that the equations (3.4) of $\widetilde{L}$ are obtained by just substituting $1, X_1, \dots, X_n$ for $x_0, x_1, \dots, x_n$ in the equations (3.3) of L, which is called *dehomogenizing* the equations (3.3). Conversely, the equations (3.3) may be recovered by substituting $x_1/x_0, \dots, x_n/x_0$ for $X_1, \dots, X_n$ in (3.4) and then multiplying each equation by x_0; this is called *homogenizing* the equations (3.4). In particular it is clear that any affine linear variety T of $\mathbb{A}_n$ is the proper part of the projective linear variety L of $\overline{\mathbb{A}}_n$, whose equations are obtained by homogenizing a system of equations of T.

Notations being as above, let us write $T = \widetilde{L}$ and pay some attention to the points in the improper part of L, $L - T = L \cap H_\infty$. On one hand, the coordinates of the points of $L \cap H_\infty$ are the non-trivial solutions of the system

$$\begin{aligned} x_0 &= 0, \\ a_1^0 x_0 + a_1^1 x_1 + \dots + a_1^n x_n &= 0, \\ &\vdots \\ a_{n-d}^0 x_0 + a_{n-d}^1 x_1 + \dots + a_{n-d}^n x_n &= 0. \end{aligned} \tag{3.5}$$

Thus, the points of $L \cap H_\infty$ are those whose 0-th coordinate is zero and whose remaining coordinates are a non-zero solution of the equations

$$\begin{aligned} a_1^1 x_1 + \dots + a_1^n x_n &= 0, \\ &\vdots \\ a_{n-d}^1 x_1 + \dots + a_{n-d}^n x_n &= 0. \end{aligned} \tag{3.6}$$

On the other hand (see Section 3.1), dropping the independent terms of the equations of an affine linear variety T gives rise to a system of equations whose solutions are the components of the vectors in the director subspace of T. If T has the equations (3.4), this system is

$$\begin{aligned} a_1^1 X_1 + \dots + a_1^n X_n &= 0, \\ &\vdots \\ a_{n-d}^1 X_1 + \dots + a_{n-d}^n X_n &= 0. \end{aligned} \tag{3.7}$$

Since the systems (3.6) and (3.7) differ only in the the name of the variables, we get:

Proposition 3.4.3. *An improper point belongs to a proper linear variety L of $\bar{\mathbb{A}}_n$ if and only if it is the direction of a non-zero vector on the proper part $\tilde{L} = L \cap \mathbb{A}_n$ of L.*

Proof. If $p = [0, x_1, \dots, x_n] \in L \cap H_\infty$, then $p = [x_1 e_1 + \dots + x_n e_n]$ (by 3.3.2) and $x_1, \dots, x_n$ is a solution of (3.6), and so also of (3.7). It follows that the vector $x_1 e_1 + \dots + x_n e_n$ belongs to the director subspace of $\tilde{L}$. The converse follows by just reversing the argument. □

Theorem 3.4.4. *Taking proper parts gives a bijection*

$$L \longmapsto \tilde{L} = L \cap \mathbb{A}_n$$

between the set of all proper (projective) linear varieties of $\bar{\mathbb{A}}_n$ of dimension d and the set of all (affine) linear varieties of $\mathbb{A}_n$ of dimension d. Its inverse consists in adding to each affine linear variety the directions of the non-zero vectors on it,

$$T \longmapsto \bar{T} = T \cup [F_T],$$

F_T the director subspace of T. In particular, the improper part of $\bar{T}$ is $[F_T]$.

Proof. That $L \cap \mathbb{A}_n$ is an affine linear variety of dimension d has been seen in 3.4.1, and the exhaustivity of the map comes from 3.4.2. Since we have seen in 3.4.3 that L may be recovered as $L = T \cup [F_T]$, $T = L \cap \mathbb{A}_n$, the injectivity and the description of the inverse map follow. □

The reader may note that Theorem 3.4.4 assures that any affine linear variety T may be extended to a uniquely determined proper linear variety $\bar{T}$ of $\bar{\mathbb{A}}_n$ whose proper part is T, and also that such an extension $\bar{T}$ is obtained by just adding to T the directions of the non-zero vectors of the director subspace F_T of T. $\bar{T}$ is called the *projective extension* or the *projective closure* of T. The improper part of $\bar{T}$, $\bar{T} \cap H_\infty$, is also called the *improper part* of T, and its points the *improper points of T*.

Remark 3.4.5. If $\dim T > 0$, T is itself an affine space, with vector space F_T, and therefore its projective closure is defined: we provisionally denote it by T'. Since the directions of the non-zero vectors of F_T are the elements of $\mathbb{P}(F_T)$, we have just said that, as sets, $\bar{T} = T'$. According to the construction of Section 3.2, the structure on T' is defined by the space E_T, of the constant and radial vector fields on T, and the map π_T, mapping radial vector fields to their centres and non-zero constant vector fields to the directions of their corresponding vectors. On the other hand, the representatives of the points of $\bar{T}$ are the radial vector fields $\theta_{q,\delta}$, $q \in T$, and the constant vector fields $\hat{v}$, $v \in F_T$: they, together with the zero vector field, are the elements of the subspace $G \subset E$ defining $\bar{T}$ as projective linear variety. Then the projective structure induced by $\bar{\mathbb{A}}_n$ on $\bar{T}$ (see 1.3.4) is defined by G and the restriction of π to $G - \{0\}$. It is direct to check that restricting the vector fields on $\mathbb{A}_n$ to T defines an isomorphism $r: G \to E_T$ that makes commutative the diagram

$$\begin{array}{ccc} G - \{0\} & \xrightarrow{r} & E_T - \{0\} \\ {\scriptstyle \pi_{|G-\{0\}}}\downarrow & & \downarrow{\scriptstyle \pi_T} \\ \overline{T} & = & T' \end{array}$$

and is thus an equivalence between the projective structures of $\overline{T}$ and T'. This shows that the linear variety $\overline{T}$, with the projective structure induced by $\overline{\mathbb{A}}_n$, may be identified with the projective closure of T as affine space. In particular, neither the notation $\overline{T}$ nor naming $\overline{T}$ the projective closure of T are misleading.

The second claim of the next proposition shows that $\overline{T}$ is the smallest projective linear variety containing T:

Proposition 3.4.6. (a) *If T_1, T_2 are linear varieties of $\mathbb{A}_n$, then*

$$T_1 \subset T_2 \iff \overline{T}_1 \subset \overline{T}_2.$$

(b) *If a linear variety L of $\overline{\mathbb{A}}_n$ contains the affine linear variety T, then $L \supset \overline{T}$.*

Proof. For claim (a) note that $T_1 \subset T_2$ forces the inclusion of their director subspaces, $F_{T_1} \subset F_{T_2}$, after which $\overline{T}_1 \subset \overline{T}_2$ by 3.4.3. The converse is clear because, by 3.4.4, $T_i = \overline{T}_i \cap \mathbb{A}_n$, $i = 1, 2$.

Regarding claim (b), from $L \supset T$ we get $\widetilde{L} \supset T$ and then, by claim (a),

$$L = \overline{\widetilde{L}} \supset \overline{T}. \qquad \square$$

Joins and intersections of affine and projective linear varieties are related next:

Proposition 3.4.7. *If T_1, T_2 are linear varieties of $\mathbb{A}_n$ and L_1, L_2 are proper linear varieties of $\overline{\mathbb{A}}_n$, then:*

(a) $\overline{T}_1 \vee \overline{T}_2 = \overline{T_1 \vee T_2}$.

(b) *If* $T_1 \cap T_2 \neq \emptyset$, $\overline{T}_1 \cap \overline{T}_2 = \overline{T_1 \cap T_2}$.

(c) $\widetilde{L}_1 \vee \widetilde{L}_2 = \widetilde{L_1 \vee L_2}$.

(d) *Either* $\widetilde{L}_1 \cap \widetilde{L}_2 = \emptyset$ *or* $\widetilde{L}_1 \cap \widetilde{L}_2 = \widetilde{L_1 \cap L_2}$.

Proof. We will make repeated use of 3.4.6 without further mention. To prove claim (a), we will see that $\overline{T_1 \vee T_2}$ is the smallest projective linear variety containing $\overline{T}_1$ and $\overline{T}_2$. Indeed, $T_1 \vee T_2 \supset T_i$ gives $\overline{T_1 \vee T_2} \supset \overline{T}_i$, $i = 1, 2$. If $L \supset \overline{T}_i$, $i = 1, 2$, then L contains some proper point and $\widetilde{L} \supset \widetilde{\overline{T}}_i = T_i$. It follows that $\widetilde{L} \supset T_1 \vee T_2$ and so $L = \overline{\widetilde{L}} \supset \overline{T_1 \vee T_2}$.

Denote by G_i the director subspace of T_i, $i = 1, 2$. As recalled in Section 3.1, if $T_1 \cap T_2$ is non-empty, then $T_1 \cap T_2$ is an affine linear variety with director subspace $G_1 \cap G_2$. Claim (b) directly follows from this fact.

To prove (c) note first that from $L_1 \vee L_2 \supset L_i$ it follows that $\widetilde{L_1 \vee L_2} \supset \widetilde{L}_i$, $i = 1, 2$. If T is affine and contains $\widetilde{L}_i$, $i = 1, 2$, then $\overline{T} \supset \overline{\widetilde{L}}_i = L_i$, $i = 1, 2$, hence $\overline{T} \supset L_1 \vee L_2$ and $T = \widetilde{\overline{T}} \supset \widetilde{L_1 \vee L_2}$. This proves that $\widetilde{L_1 \vee L_2}$ is the smallest affine linear variety containing $\widetilde{L}_1$ and $\widetilde{L}_2$, hence it equals $\widetilde{L}_1 \vee \widetilde{L}_2$.

Claim (d) is obvious, as in any case $\widetilde{L}_1 \cap \widetilde{L}_2 = L_1 \cap L_2 \cap \mathbb{A}_n$. □

Corollary 3.4.8. *Points $q_0, \ldots, q_m \in \mathbb{A}_n$ are independent as points of the affine space if and only if they are independent as points of $\overline{\mathbb{A}}_n$.*

Proof. Since in both cases the points are independent if and only if the linear variety they span has dimension m, just note that, for each i, $\overline{\{q_i\}} = \{q_i\}$ and so, by 3.4.7 (a), the closure of the affine linear variety spanned by the points in $\mathbb{A}_n$ is the projective linear variety spanned by the same points in $\overline{\mathbb{A}}_n$. □

The projective extensions of parallel affine linear varieties are also said to be *parallel*. To close this section we characterize parallelism of linear varieties in terms of their improper parts:

Proposition 3.4.9. *Affine linear varieties T_1, T_2 are parallel if and only if their improper parts $\overline{T}_1 \cap H_\infty$, $\overline{T}_2 \cap H_\infty$ are included in either sense.*

Proof. By 3.4.4 and using the notations therein, the improper parts are $\overline{T}_i \cap H_\infty = [F_{T_i}]$. Thus $\overline{T}_i \cap H_\infty \subset \overline{T}_j \cap H_\infty$ if and only if $F_{T_i} \subset F_{T_j}$, $\{i, j\} = \{1, 2\}$, the definitory condition of parallelism. □

Remark 3.4.10. It follows from 3.4.4 that two affine varieties of equal dimension are parallel if and only if their improper parts agree. Due to this, the improper part of an affine linear variety is also called its *direction*.

3.5 Affine transformations, affine ratio

Let $\mathbb{A}_n$, $\mathbb{A}'_n$ be affine spaces of the same dimension, $\overline{\mathbb{A}}_n$, $\overline{\mathbb{A}}'_n$ their projective closures and H_∞, H'_∞ the corresponding improper hyperplanes. We will consider projectivities

$$f : \overline{\mathbb{A}}_n \longrightarrow \overline{\mathbb{A}}'_n$$

that satisfy the condition $f(H_\infty) \subset H'_\infty$, which is obviously equivalent to $f(H_\infty) = H'_\infty$ and hence also to $f(\mathbb{A}_n) = \mathbb{A}'_n$. These are thus the projectivities mapping proper points to proper points and improper points to improper points.

Assume that we fix affine references Δ in $\mathbb{A}_n$ and Δ' in $\mathbb{A}'_n$ and that, relative to these references, f has matrix

$$M = \begin{pmatrix} a_0^0 & a_1^0 & \dots & a_n^0 \\ a_0^1 & a_1^1 & \dots & a_n^1 \\ \vdots & \vdots & & \vdots \\ a_0^n & a_1^n & \dots & a_n^n \end{pmatrix}.$$

First of all we have:

Lemma 3.5.1. *The condition $f(H_\infty) \subset H'_\infty$ is equivalent to $a_1^0 = \dots = a_n^0 = 0$.*

Proof. Since in both spaces the improper points are those with 0-th coordinate equal to zero, the condition $f(H_\infty) \subset H'_\infty$ is fulfilled if and only if all images $f([0, x_1, \dots, x_n])$ have 0-th coordinate equal to zero. Computing the images using the above matrix, the latter condition is in turn equivalent to $a_1^0 x_1 + \dots + a_n^0 x_n = 0$ for all values of $x_1, \dots, x_n$, and hence to $a_1^0 = \dots = a_n^0 = 0$. □

Remark 3.5.2. The matrix M being regular, $a_1^0 = \dots = a_n^0 = 0$ forces $a_0^0 \neq 0$ and so, up to dividing M by a_0^0, in the sequel we will assume that $a_0^0 = 1$. Note that, by the regularity of M,

$$\begin{vmatrix} a_1^1 & \dots & a_n^1 \\ \vdots & & \vdots \\ a_1^n & \dots & a_n^n \end{vmatrix} \neq 0.$$

By the hypothesis, f maps proper points to proper points. We are now able to compute equations of the restricted map

$$\tilde{f} = f_{|\mathbb{A}_n} \colon \mathbb{A}_n \longrightarrow \mathbb{A}'_n.$$

Indeed, by 3.3.2, a proper point of $\mathbb{A}_n$, with affine coordinates $X_1, \dots, X_n$, has homogeneous coordinates $1, X_1, \dots, X_n$; therefore its image has homogeneous coordinate vector

$$\begin{pmatrix} 1 & 0 & \dots & 0 \\ a_0^1 & a_1^1 & \dots & a_n^1 \\ \vdots & \vdots & & \vdots \\ a_0^n & a_1^n & \dots & a_n^n \end{pmatrix} \begin{pmatrix} 1 \\ X_1 \\ \vdots \\ X_n \end{pmatrix}$$

and hence affine coordinates

$$\begin{aligned} Y_1 &= a_0^1 + a_1^1 X_1 + \dots + a_n^1 X_n, \\ &\vdots \\ Y_n &= a_0^n + a_1^n X_1 + \dots + a_n^n X_n. \end{aligned} \tag{3.8}$$

Since the determinant of

$$\widetilde{M} = \begin{pmatrix} a_1^1 & \dots & a_n^1 \\ \vdots & & \vdots \\ a_1^n & \dots & a_n^n \end{pmatrix}$$

is non-zero, as noticed in 3.5.2, we see that $\tilde{f}$ is an affinity.

Conversely, any affinity $g\colon \mathbb{A}_n \to \mathbb{A}'_n$ has equations as (3.8) above, with $\det \widetilde{M} \neq 0$. Then the matrix

$$\begin{pmatrix} 1 & 0 & \dots & 0 \\ a_0^1 & a_1^1 & \dots & a_n^1 \\ \vdots & \vdots & & \vdots \\ a_0^n & a_1^n & \dots & a_n^n \end{pmatrix}$$

is regular, and so it is the matrix of a projectivity $f\colon \overline{\mathbb{A}}_n \to \overline{\mathbb{A}}'_n$ that clearly satisfies $f(H_\infty) \subset H'_\infty$ and $g = \tilde{f}$.

Now, once we know that the restriction of f to proper points, $\tilde{f}$, is an affinity, we will see how the restriction of f to the improper points,

$$f_{|H_\infty}\colon H_\infty \longrightarrow H'_\infty,$$

is determined by $\tilde{f}$.

On one hand, it is clear that taking in the improper hyperplanes the projective references subordinated by Δ and Δ', the image of the improper point with coordinates $x_1, \dots, x_n$ has coordinate vector

$$\begin{pmatrix} a_1^1 & \dots & a_n^1 \\ \vdots & & \vdots \\ a_1^n & \dots & a_n^n \end{pmatrix} \begin{pmatrix} x_1 \\ \vdots \\ x_n \end{pmatrix},$$

and so the restriction $f_{|H_\infty}$ has matrix $\widetilde{M}$.

On the other, it is well known (and very easy to check) that the morphism ψ associated to the affinity $\tilde{f}$ of equations (3.8), has matrix $\widetilde{M}$ relative to the bases of the affine references. Since these bases are adapted to the projective references subordinated in the improper hyperplanes (3.3.4), it turns out that $f_{|H_\infty} = [\psi]$. In other words, f maps the direction of a non-zero vector v to the direction of the image of v by the morphism associated to the affinity $\tilde{f}$. In particular, $\tilde{f}$ determines f.

All together, we have proved:

Theorem 3.5.3. *Restricting to the proper points defines a bijection between the set of all projectivities $f\colon \overline{\mathbb{A}}_n \to \overline{\mathbb{A}}'_n$ such that $f(H_\infty) \subset H'_\infty$ and the set of* *all affinities* $g\colon \mathbb{A}_n \to \mathbb{A}'_n$. *The inverse map associates to each affinity* g, *with associated linear map* ψ, *the projectivity defined by the rule*

$$\bar{g}(p) = \begin{cases} g(p) & \text{if } p \text{ is proper,} \\ [\psi(v)] & \text{if } p \text{ is the direction of } v. \end{cases}$$

Corollary 3.5.4. *The composition of two projectivities that map improper points to improper points,* $f : \overline{\mathbb{A}}_n \to \overline{\mathbb{A}}'_n$ *and* $f' : \overline{\mathbb{A}}'_n \to \overline{\mathbb{A}}''_n$, *still maps improper points to improper points and*

$$\widetilde{f' \circ f} = \tilde{f}' \circ \tilde{f}.$$

Conversely, if $g\colon \mathbb{A}_n \to \mathbb{A}'_n$ *and* $g'\colon \mathbb{A}'_n \to \mathbb{A}''_n$ *are affinities, then*

$$\overline{g' \circ g} = \overline{g'} \circ \bar{g}.$$

Also,

$$\widetilde{\mathrm{Id}}_{\overline{\mathbb{A}}_n} = \mathrm{Id}_{\mathbb{A}_n}, \quad \overline{\mathrm{Id}}_{\mathbb{A}_n} = \mathrm{Id}_{\overline{\mathbb{A}}_n}$$

and

$$\widetilde{f^{-1}} = (\tilde{f})^{-1}, \quad \overline{g^{-1}} = (\bar{g})^{-1}.$$

Proof. The first claim is obvious, while the second one follows from it just using that the map of 3.5.3 is bijective. The remaining claims may be proved similarly. □

Theorem 3.5.3 says in particular that each affinity $g\colon \mathbb{A}_n \to \mathbb{A}'_n$ may be uniquely extended to projectivity $\bar{g}\colon \overline{\mathbb{A}}_n \to \overline{\mathbb{A}}'_n$, which is usually called the *extension*, or the *projective extension*, of g. In 3.5.4 we have seen that the extension of a composition of affinities is the composition of their extensions, that the inverse of an affinity extends to the inverse of its extension and also that $\mathrm{Id}_{\mathbb{A}_n}$ extends to $\mathrm{Id}_{\overline{\mathbb{A}}_n}$. In particular extending affinities $g\colon \mathbb{A}_n \to \mathbb{A}_n$ is an isomorphism between the affine group of $\mathbb{A}_n$ and the subgroup of $\mathrm{PG}(\overline{\mathbb{A}}_n)$ of the homographies that leave invariant the improper hyperplane.

Because of the above, usually no distinction is made between an affinity g and its extension $\bar{g}$: both are seen as a unique transformation which one may think of as acting either on the proper points only (g) or on the whole of the proper and improper points ($\bar{g}$). For instance, it is usual to say that translations and homotheties leave all the improper points invariant, which, strictly speaking, means that their extensions do. Incidentally, this fact may be proved to be true by the reader. Thus, in the sequel, the projectivities $f : \overline{\mathbb{A}}_n \to \overline{\mathbb{A}}'_n$ that map improper points to improper points, which by 3.5.3 are the extensions of the affinities $g\colon \mathbb{A}_n \to \mathbb{A}'_n$, will be also called *affinities* (between projective closures), and the word *affinity* will be used in this sense unless otherwise said.

Remark 3.5.5. The affinities compose a subclass of the class of all projectivities between projective closures of affine spaces, which contains all identical maps and is closed by composition and inversion of maps. In particular, as already noticed, the affinities $f: \overline{\mathbb{A}}_n \to \overline{\mathbb{A}}_n$ compose a subgroup of $\mathrm{PG}(\overline{\mathbb{A}}_n)$ isomorphic to the affine group of $\mathbb{A}_n$. Often this subgroup is also named *affine group*.

The affine ratio of three aligned points equals the cross ratio of these points together with the improper point of the line containing them:

Proposition 3.5.6. *Let $q_1, q_2, q_3 \in \mathbb{A}_n$ be aligned points, at least two of them different. If L is the line spanned by them and q_∞ denotes the improper point of L, then the affine ratio of q_1, q_2, q_3 equals the cross ratio of q_1, q_2, q_3, q_∞, that is,*

$$(q_1, q_2, q_3) = (q_1, q_2, q_3, q_\infty).$$

Proof. If $q_2 = q_3$, then by the definition of affine ratio and 2.9.10, both sides of the claimed equality are ∞. We assume thus in the sequel $q_2 \neq q_3$ and write $\rho = (q_1, q_2, q_3) \neq \infty$. Fix an affine reference and assume that q_3 has coordinates $(a_1, \dots, a_n)$, while the free vector $v = q_2 - q_3 \neq 0$ has components $(\alpha_1, \dots, \alpha_n)$. Then q_2 has coordinates $(a_1 + \alpha_1, \dots, a_n + \alpha_n)$ and, by the definition of affine ratio, q_1 has coordinates $(a_1 + \rho\alpha_1, \dots, a_n + \rho\alpha_n)$. Using homogeneous coordinates, by 3.3.2,

$$\begin{aligned} q_1 &= [1, a_1 + \rho\alpha_1, \dots, a_n + \rho\alpha_n], \\ q_2 &= [1, a_1 + \alpha_1, \dots, a_n + \alpha_n], \\ q_3 &= [1, a_1, \dots, a_n], \\ q_\infty &= [0, \alpha_1, \dots, \alpha_n], \end{aligned}$$

the last equality due to the fact that q_∞ is the direction of v, by 3.4.3. The claim follows from the above equalities by direct computation (or just by reversing the ordering of the points and using 2.9.9). □

The next corollary characterizes midpoints in projective terms; it provides the usual way of handling midpoints when using projective techniques.

Corollary 3.5.7. *The midpoint of two different proper points q_1, q_2 is the fourth harmonic of q_1, q_2, q_∞, q_∞ being the improper point of $q_1 \vee q_2$.*

Proof. Direct from 3.5.6 and the definition of midpoint. □

3.6 Affine geometry in the projective frame

In the preceding sections we have reformulated the basic notions of affine geometry in terms of objects of the projective closures and their relation to the improper hyperplanes. We have seen that the linear varieties of $\mathbb{A}_n$ are the finite parts of the proper linear varieties of $\overline{\mathbb{A}}_n$, and also that the director subspaces of the former define the improper parts of the latter; in particular, the parallelism of two linear varieties turned out to be equivalent to the inclusion of their improper parts. The affinities have been seen to be (the restrictions of) the projectivities between projective closures preserving the improperness (and therefore also the properness) of points. Lastly, the affine ratio of three aligned points appeared as the cross ratio of the same points together with the improper point of the line they span. This being done, subsequent definitions of affine geometry may be reformulated in terms of the projective geometry of the projective closures and the improper hyperplanes (the improper hyperplanes themselves making not a projective class). Then the affine results appear as direct consequences, often just particular cases, of projective theorems. Next we will not attempt to make a systematic development of linear affine geometry following this way, but just show some relevant examples to illustrate the method. First of all, we prove a useful lemma showing how a proper linear variety may be recovered from one of its proper points and its improper part. The reader may find a similarity to the equality $L = p + G$, L an affine linear variety, G its director subspace and $p \in L$.

Lemma 3.6.1. *For any proper linear variety $T \subset \overline{\mathbb{A}}_n$ and any proper point $p \in T$, $T = p \vee (T \cap H_\infty)$.*

Proof. By 1.4.8, $\dim T \cap H_\infty = \dim T - 1$ and so by, 1.4.7, $\dim p \vee (T \cap H_\infty) = \dim T$. Since, clearly, $T \supset p \vee (T \cap H_\infty)$ the claim follows. □

Proposition 3.6.2. *Parallel affine linear varieties are either included or disjoint.*

Proof. If L_1 and L_2 are parallel affine linear varieties, up to swapping them over, assume $\overline{L}_1 \cap H_\infty \subset \overline{L}_2 \cap H_\infty$. If $p \in L_1 \cap L_2$, then, by 3.6.1,

$$\overline{L}_1 = p \vee (\overline{L}_1 \cap H_\infty) \subset p \vee (\overline{L}_2 \cap H_\infty) = \overline{L}_2,$$

which yields $L_1 \subset L_2$. □

The incidence of lines and planes in a three-dimensional affine space may be easily described from 1.4.9 taking into account the positions relative to the improper hyperplane. Note that included linear varieties are parallel. We have:

Proposition 3.6.3 (Incidence of lines and planes). *All lines and planes belonging to a fixed three-dimensional affine space,*

(a) *two lines are coplanar if and only if they are parallel or have as intersection a single point,*

(b) *a plane and a line either intersect at a point or are parallel, and*

(c) *two planes either intersect at a line or are parallel.*

Proof. Let r and r' be two lines of $\mathbb{A}_3$ both contained in a plane π. Their projective extensions, $\bar{r}$ and $\bar{r}'$, are coplanar lines of $\bar{\mathbb{A}}_3$, because both lie in the projective extension of π. If they have the same improper point, then r and r' are parallel. Otherwise $\bar{r}$, $\bar{r}'$ are coplanar and distinct. Then, by 1.4.9, their intersection is a point p which needs to be proper because $\bar{r}$ and $\bar{r}'$ have different improper points. It follows $\{p\} = r \cap r'$. Conversely, if r and r' are parallel or meet at a point, then $\bar{r} \cap \bar{r}' \neq \emptyset$ and by 1.4.9 there is a plane H of $\bar{\mathbb{A}}_3$ containing $\bar{r}$ and $\bar{r}'$. The plane H is proper because $\bar{r}$ has proper points and so the proper part of H is a plane of $\mathbb{A}_3$ containing r and r'.

Regarding claim (b), let r and π be a line and a plane. If their projective extensions are included, then so are their improper parts and therefore r and p are parallel. Otherwise, by 1.4.9, the projective extensions intersect at a point p. If p is an improper point, then r and π are parallel. If p is proper, $r \cap \pi = p$. The proof of claim (c) is similar and left to the reader. □

Next is the affine specialization of 1.4.8:

Proposition 3.6.4. *Assume that L and H are, respectively, a linear variety of dimension d and a hyperplane of $\mathbb{A}_n$. Then either L and H are parallel or $L \cap H$ is a linear variety of dimension equal to $d - 1$.*

Proof. If $L \subset H$, then they are parallel and there is nothing to prove. We assume thus $L \not\subset H$, after which $\bar{L} \not\subset \bar{H}$ by 3.4.6 and so $\dim \bar{L} \cap \bar{H} = d - 1$ by 1.4.8. If $\bar{L} \cap \bar{H}$ is proper, then, by 3.4.7 (d), its proper part is $L \cap H$ which therefore is an affine linear variety of dimension $d - 1$. Otherwise $\bar{L} \cap \bar{H} \subset H_\infty$ which gives

$$\bar{L} \cap \bar{H} = \bar{L} \cap \bar{H} \cap H_\infty \subset \bar{L} \cap H_\infty.$$

Since both ends have the same dimension, they agree. In particular $\bar{L} \cap H_\infty \subset \bar{H} \cap H_\infty$ and so L and H are parallel, as wanted. □

Proposition 3.6.5. *If p is a point and L a linear variety of an affine space $\mathbb{A}_n$, then there is a unique linear variety of $\mathbb{A}_n$ that has the same dimension as L, goes through p and is parallel to L.*

Proof. Let S be the improper part of L and put $d = \dim L$. The proper part of $p \vee S$ obviously satisfies the conditions of the claim. If a linear variety L' has dimension d and is parallel to L, then it has improper part S (3.4.10). If in addition L' contains p, then, by 3.6.1, its projective extension is $\bar{L}' = p \vee S$ and so L' equals the proper part of $p \vee S$. □

Theorem 3.6.6 (Affine invariance of the affine ratio). *If points q_1, q_2, q_3 of an affine space $\mathbb{A}_n$ are aligned and two at least distinct, and $f: \mathbb{A}_n \to \mathbb{A}'_n$ is an affinity, then the affine ratio $(f(q_1), f(q_2), f(q_3))$ is defined and*

$$(f(q_1), f(q_2), f(q_3)) = (q_1, q_2, q_3).$$

Proof. Let H_∞ and H'_∞ be the improper hyperplanes of $\mathbb{A}_n$ and $\mathbb{A}'_n$, respectively, assume that r is the line spanned by q_1, q_2, q_3 in $\bar{\mathbb{A}}_n$ and denote by $\bar{f}$ the extension of f to the projective closures. Then clearly $f(q_1), f(q_2), f(q_3)$ are proper points, two at least distinct, and they are aligned because they all belong to the proper part of $\bar{f}(r)$: their affine ratio is thus defined. Take $q_\infty = r \cap H_\infty$, the improper point of r. Since $\bar{f}$ is (the extension of) an affinity, $\bar{f}(H_\infty) = H'_\infty$ and therefore

$$\bar{f}(q_\infty) = \bar{f}(r \cap H_\infty) = \bar{f}(r) \cap H'_\infty$$

is the improper point of $\bar{f}(r)$, which in turn is the line spanned by $f(q_1)$, $f(q_2)$ and $f(q_3)$ in $\bar{\mathbb{A}}_n$. Then the invariance of the cross ratio (2.9.11) and 3.5.6 give

$$\begin{aligned}(q_1, q_2, q_3) &= (q_1, q_2, q_3, q_\infty)\\ &= (\bar{f}(q_1), \bar{f}(q_2), \bar{f}(q_3), \bar{f}(q_\infty))\\ &= (f(q_1), f(q_2), f(q_3))\end{aligned}$$

as claimed. □

Projecting onto a proper linear variety T from a supplementary S is called *central projection* when S is proper, and *parallel projection* when S is improper. Assume that the centre S is improper, after which assuming T proper is redundant, as no supplementary of an improper linear variety may be improper. Usually an affine linear variety L with improper part S is chosen and the projection with centre S is called *projection parallel to* L. Its centre being improper, a parallel projection is defined on all proper points. Furthermore, since for any proper point p, $(p \vee S) \cap H_\infty = S$ (otherwise p would be improper), a parallel projection maps proper points to proper points and therefore restricts to a map $\mathbb{A}_n \to T \cap \mathbb{A}_n$ which still is called parallel projection. As seen in the proof of 3.6.5, for any proper point p, the proper part L_p of $p \vee S$ is the only affine linear variety that has the same dimension as L, goes through p and is parallel to L; then $f(p) = T \cap L_p$, which is the usual description of a parallel projection in a purely affine setting.

Perspectivities $f: T' \to T$ with improper centre are restrictions of parallel projections. Therefore they map proper points to proper points and are thus, by 1.9.3 and 3.5.3, affinities. This, together with 3.6.6, yields one of the oldest and more important theorems of affine geometry, namely:

Theorem 3.6.7 (Thales). *The affine ratio is preserved by parallel projection.*

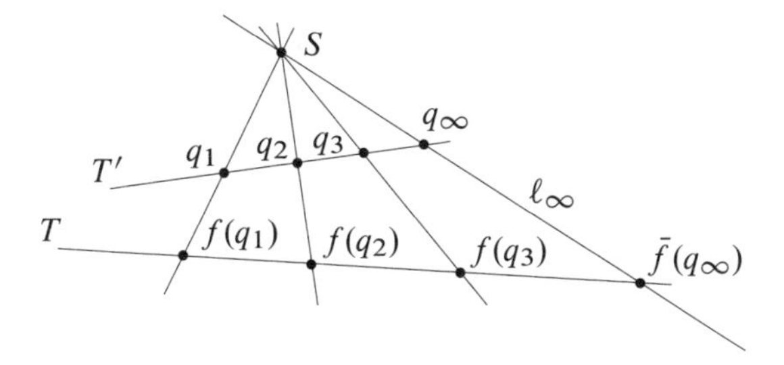

Figure 3.1. Parallel projection in the plane and Thales theorem. ℓ_∞ is the improper line.

Once the basic affine notions have been reinterpreted in terms of projective relations with the improper hyperplanes, affine geometry is developed systematically using the linear varieties of the projective closures, the proper ones being taken instead of their finite parts. Linear figures and their elements are often renamed taking into account their position relative to the improper hyperplane. For instance, quadrilaterals of $\overline{\mathbb{A}}_2$ with proper sides are called *quadrilaterals of* $\mathbb{A}_2$ or *affine quadrilaterals*. A *parallelogram* is a quadrilateral of $\mathbb{A}_2$ with two (necessarily opposite) vertices improper: its four sides compose two pairs of parallel lines. The proper vertices of the quadrilateral are then called the *vertices* of the parallelogram. One diagonal of the quadrilateral being the improper line, the other two are proper lines, each joining two opposite vertices of the parallelogram: they are called the *diagonals* of the parallelogram. Similarly, a simplex (in particular a triangle or a tetrahedron) of $\overline{\mathbb{A}}_n$ with proper vertices is called a *simplex* (or *triangle, or tetrahedron*) *of* $\mathbb{A}_n$ or an *affine simplex*; all their faces are proper linear varieties.

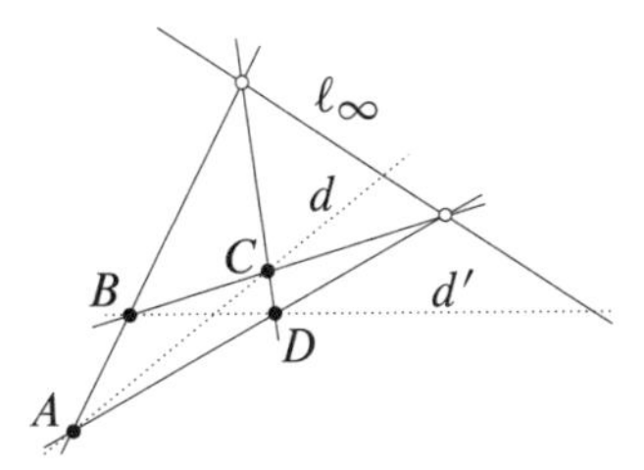

Figure 3.2. A parallelogram with vertices A, B, C, D and diagonals d, d'. ℓ_∞ is the improper line.

We close this section with a further example of a well-known affine result obtained as a particular case of a projective theorem:

Proposition 3.6.8. *Both pairs of opposite vertices of a parallelogram have as midpoint the intersection point of the diagonals of the parallelogram.*

Proof. Just apply the complete quadrilateral theorem 2.10.3 to each diagonal of the parallelogram and use 3.5.7. □

3.7 The Erlangen Program

Developing affine geometry in the projective closures of affine spaces, rather than in the affine spaces themselves, makes it easier to compare it to projective geometry. In this section we present a very clarifying explanation of the relationship between projective, affine and Euclidean geometries, originally given by F. Klein in 1872 and known as the *Erlangen Program* (See [20], [21]). Up to the end of the section, we restrict ourselves to considering only projective spaces that are projective closures of affine spaces, projectivities between them and objects of projective families belonging to them. Each of our projective spaces has thus its improper hyperplane as a distinguished hyperplane and besides the class of all projectivities, we have the smaller class of the affinities, which, as said above, are the projectivities mapping improper points to improper points. We already explained, in Section 1.10, that projective geometry deals with the families, properties and relations that are invariant by projectivities; the object of *affine geometry* is set now as being the study of the families, properties and relations that are invariant by affinities.

An *affine class* is any part of a projective class which is invariant by affinities. For instance, parallelograms, defined as the quadrilaterals which have two opposite vertices improper, compose an affine class (or, equivalently, the notion of parallelogram is affine), because if two opposite vertices of a quadrilateral T are improper, then their images by an affinity f are a pair of opposite and improper vertices of the quadrilateral $f(T)$. Note that this is no longer true if f is taken an arbitrary projectivity.

If the objects satisfying a given property compose an affine class, then the property itself is called *affine*. For instance being proper is an affine property of linear varieties. Relations between objects belonging to the same space are called *affine* if and only any affine images – by the same projectivity – of related objects are also related. Being parallel is an affine relation between pairs of proper linear varieties of the same space due to 3.4.9.

Two objects in an affine class are said to be *affinely equivalent* if and only if one is the image of the other by an affinity. The affine equivalence is indeed an equivalence relation because identical maps, compositions of affinities and inverses of affinities are all affinities; the classification it induces is called an *affine classification*. The affine equivalence class of an object is also called its *affine type*; it is the smallest affine class containing the object.

Maps defined on affine families assigning the same image to any two affinely equivalent objects are called *affine invariants*. Theorem 3.6.6 states that the affine ratio is an affine invariant of the ordered triples of aligned proper points, two at least distinct. As an example, the reader may easily check using 3.5.6 and 2.9.13

that ordered triples of distinct proper and aligned points are affinely equivalent if and only if they have the same affine ratio. As in the projective case, systems of invariants that suffice to characterize the affine equivalence are called *complete*. The affine ratio alone composes thus a complete system of affine invariants for the ordered triples of distinct, proper and aligned points.

Since all affinities are projectivities, any projective class (or notion) is in particular affine. Likewise, all projective properties, relations and invariants are also affine. The converses are not true: we have noticed above that the class of all parallelograms is not projective, and the reader may easily see that properness and improperness of points, parallelism or affine ratio are not invariant by projectivities and therefore do not belong to projective geometry.

The projective families being larger than the affine ones, projective notions, relations and properties are more general than the affine ones in the sense that they apply to or are satisfied by larger classes of objects. By contrast, affine geometry deals with notions, relations and properties that are not projective, such as midpoint and parallelism, and therefore goes into more detail: affine geometry is less general than projective geometry, but is finer.

The picture becomes complete after considering the even smaller class of the affinities preserving distance (called *isometries*): these transformations lead to a third geometry, the *metric* or *Euclidean geometry*: it deals with the notions, relations and properties that are invariant by isometries. (The basic points of metric geometry should be already familiar to the reader, they will be related to the projective context in forthcoming Section 6.9.) As it happens with affine geometry with respect to projective geometry, its transformations being scarcer, metric geometry is less general and finer than affine geometry: its results apply to narrower classes of objects and involve notions, relations and properties, such as angle and distance, that do not belong to affine geometry.

Since the ancient Greeks, geometry grew in a purely metric context and, but for the scarcely influential work of Desargues in the 17th century, it was not until the 19th century that affine and projective geometries gradually emerged as distinct geometries. The definitive clarification came from the Erlangen Program, which proposed a hierarchy of three geometries, projective, affine and metric, each characterized by its own class of transformations as described above. Somewhat reversing the historical path, the Erlangen Program takes projective geometry as the fundamentals of geometry: projective notions, relations and properties being the most general, they are presented first, and then the affine and metric ones come as specializations of them. This procedure unveils the common projective origin of apparently unrelated parts of affine and metric geometry, providing a far clearer insight into geometry. As a sample, the reader may compare Proposition 3.6.8 and Exercise 3.9, which would appear as unrelated if independently proved in a purely affine setting.

3.8 Exercises

Exercise 3.1 (*Affine Grassmann formulas*). Assume that L_1 and L_2 are linear varieties of $\mathbb{A}_n$ which have director subspaces F_1 and F_2, respectively. Use 1.4.6 to prove that

$$\dim L_1 \vee L_2 + \dim L_1 \cap L_2 = \dim L_1 + \dim L_2$$

if $L_1 \cap L_2 \neq \emptyset$ and, otherwise,

$$\dim L_1 \vee L_2 + \dim F_1 \cap F_2 = \dim L_1 + \dim L_2 + 1.$$

Exercise 3.2. Prove that the intersection with the improper hyperplane of the faces and edges of a tetrahedron of $\mathbb{A}_3$ (with proper vertices) are the sides and vertices of a complete quadrilateral of the improper hyperplane.

Exercise 3.3. Determine which proper planes intersect the faces of a tetrahedron of $\mathbb{A}_3$ in the sides of

(a) a quadrilateral,

(b) an affine quadrilateral,

(c) a parallelogram.

Exercise 3.4. Hypothesis and notations being as in Exercise 1.12, assume $\mathbb{P}_3 = \overline{\mathbb{A}}_3$ and the lines ℓ_1, ℓ_2 proper. Determine for which points p the line s is proper and intersects ℓ_1 and ℓ_2 in proper points.

Exercise 3.5. In Exercise 1.19, assume $\mathbb{P}_n = \overline{\mathbb{A}}_n$ and the lines ℓ_1, ℓ_2, ℓ_3 to be proper. Reformulate the claim in terms of the finite parts of ℓ_1, ℓ_2, ℓ_3 and linear varieties of $\mathbb{A}_n$.

Exercise 3.6. Use 3.3.5 and 3.5.6 to obtain a formula for the affine ratio of three points given by their affine coordinates on the line they belong to.

Exercise 3.7. Let $q_1, q_2, q_3, q_4 \in \mathbb{A}_n$ be aligned and at least three distinct. Prove that their cross ratio (as aligned points of $\overline{\mathbb{A}}_n$) is a ratio of affine ratios, according to the equality

$$(q_1, q_2, q_3, q_4) = \frac{(q_1, q_2, q_3)}{(q_1, q_2, q_4)}.$$

Exercise 3.8. Fix two distinct points $p_1, p_2 \in \mathbb{A}_{1,\mathbb{R}}$. According to the usual definitions in affine geometry, the points *lying between* p_1, p_2 are those $q \in \mathbb{A}_{1,\mathbb{R}}$ for which $(p_1, p_2, q) < 0$; they are called the *interior points* – and the set they compose the *interior* – of the *affine segment* with ends p_1, p_2; the *affine segment* itself is in turn composed of its interior points together with its *ends* p_1, p_2. Terminology and notations being as at the end of Section 2.10, prove that a point $q \in \mathbb{A}_{1,\mathbb{R}}$ lies between p_1, p_2 if and only if q and the improper point separate p_1, p_2. Prove also that the interior of the affine segment with ends p_1, p_2 is the projective segment with ends p_1, p_2 that does not contain the improper point.

Exercise 3.9. Use the complete quadrilateral theorem (2.10.3) to prove that if A, B, C are the vertices of a triangle of an affine plane, then the lines parallel to AB and AC through C and B, respectively, intersect the line parallel to BC through A in two points whose midpoint is A.

Exercise 3.10. Let ℓ, ℓ' be proper lines of $\overline{\mathbb{A}}_2$, $O \notin \ell$ a proper point and $\delta\colon \ell \to \ell'$ the restriction of the central projection onto ℓ' with centre O. Prove that:

(a) If there are three distinct proper points $A, B, C \in \ell$ whose images $\delta(A), \delta(B), \delta(C)$ are proper and have $(\delta(A), \delta(B), \delta(C)) = (A, B, C)$, then ℓ and ℓ' are parallel.

(b) If ℓ and ℓ' are parallel, then δ is an affinity, and so, in particular, preserves affine ratios.

Exercise 3.11. Prove that the line joining the midpoints of two sides of a triangle is parallel to the third side.

Exercise 3.12. Let ℓ, ℓ', s be three pairwise skew proper lines of $\overline{\mathbb{A}}_3$ and denote by

$$\delta\colon \overline{\mathbb{A}}_3 - s \longrightarrow \ell'$$

the projection onto ℓ' with centre s. Prove that:

(a) If for three distinct proper points $A, B, C \in \ell$, $\delta(A), \delta(B), \delta(C)$ are proper and have $(\delta(A), \delta(B), \delta(C)) = (A, B, C)$, then there is a plane parallel to ℓ, ℓ' and s.

(b) If there is a plane parallel to ℓ, ℓ' and s, then

$$\delta_{|\ell}\colon \ell \longrightarrow \ell'$$

is an affinity, and so, in particular, preserves affine ratios.

Exercise 3.13. Let $f : \overline{\mathbb{A}}_n \to \overline{\mathbb{A}}'_n$ be a projectivity and $\Delta = (p_0, \dots, p_n, A)$ a reference of $\overline{\mathbb{A}}_n$, all of whose elements are proper. For $0 \le i < j \le n$, denote by $p_{i,j}$ the image of A by projecting onto $p_i p_j$ from the opposite face of Δ. Prove that f is an affinity if and only if for any $i, j, 0 \le i < j \le n$, the points $f(p_i), f(p_j), f(p_{i,j})$ are proper and have $(f(p_i), f(p_j), f(p_{i,j})) = (p_i, p_j, p_{i,j})$.

Exercise 3.14. Let a triangle T of an affine plane have (proper) vertices A, B, C. Assume to have proper points A', B', C' on the sides of T opposite A, B, C respectively, all different from the vertices. Use Exercises 2.21 and 2.22 to prove:

(a) (*Menelaos' theorem*) A', B', C' are aligned if and only if

$$(A, B, C')(B, C, A')(C, A, B') = -1.$$

(b) (*Ceva's theorem*) AA', BB', CC' are concurrent if and only if

$$(A, B, C')(B, C, A')(C, A, B') = 1.$$

Exercise 3.15. The *barycentre* of points $p_0, \dots, p_m \in \mathbb{A}_n$ is defined as being the point

$$\sum_{i=0}^{m} \frac{1}{m+1} p_i.$$

Assume that $m = n$ and the points $p_0, \dots, p_n$ are independent, and prove that their barycentre is the harmonic pole of the improper hyperplane relative to the simplex with vertices $p_0, \dots, p_n$. Note that this proves in particular that the lines joining the vertices of a triangle to the midpoints of the opposite sides (the *medians* of the triangle) intersect at the barycentre of the vertices of the triangle (usually called the *barycentre* or the *centroid* of the triangle).

Exercise 3.16 (*Barycentric coordinates*). Assume that $p_0, \dots, p_n \in \mathbb{A}_n$ are independent points and let A be their barycentre. Prove that:

(a) $\Delta = (p_o, \dots, p_n, A)$ is a reference of $\overline{\mathbb{A}}_n$ relative to which the improper hyperplane has equation $x_0 + \dots + x_n = 0$.

(b) The rule

$$p = [x_0, \dots, x_n]_\Delta \longmapsto (y_0, \dots, y_n) = \frac{1}{\sum_{i=0}^n x_i}(x_0, \dots, x_n)$$

defines a bijective map

$$\mathbb{A}_n \longrightarrow \{(y_0, \dots, y_n) \in k^{n+1} | y_0 + \dots + y_n = 1\}.$$

The above $y_0, \dots, y_n$ are called the *barycentric coordinates* of p relative to the simplex with vertices $p_0, \dots, p_n$. Note that they are homogeneous coordinates of p relative to Δ normalized by the condition $\sum_{i=0}^n y_i = 1$. Sometimes this normalization condition is dropped and the name *barycentric coordinates* refers to any homogeneous coordinates relative to a reference of $\overline{\mathbb{A}}_n$ which, as Δ above, has proper vertices and their barycentre as unit point.

Exercise 3.17. Assume to have a simplex Δ of $\mathbb{A}_n$, $n > 1$, with (proper) vertices $p_0, \dots, p_n$. Denote by A the barycentre of all the vertices (usually called the *barycentre* of Δ) and by A_i the barycentre of the vertices lying on the face opposite p_i. Use barycentric coordinates to prove that the lines $p_i A_i$, $i = 0, \dots, n$, intersect at A, and that

$$(A_i, p_i, A) = -\frac{1}{n},$$

for $i = 0, \dots, n$.

Exercise 3.18. With the same hypothesis and notations, generalize Exercise 3.17 by proving that:

(1) For a fixed r, $1 \leq r \leq n-1$, the lines joining the barycentre of each r vertices $p_{i_1}, \dots, p_{i_r}$, $i_1 < \dots < i_r$, of Δ to the barycentre of the remaining ones are concurrent at A.

(2) It holds that

$$(B, B', A) = 1 - \frac{n+1}{r},$$

where B is the barycentre of $p_{i_1}, \dots, p_{i_r}$ and B' is the barycentre of the remaining vertices.

Chapter 4
Duality

One of the more interesting features of projective geometry is that certain sets of objects belonging to a projective space may in turn receive a structure of projective space. This provides a new reading of projective notions and theorems already set for an arbitrary projective space in terms of these objects, thus leading to new notions and theorems. In this chapter we present an important occurrence of this situation: the set of all hyperplanes of a projective space will be given a structure of projective space, after which, as we will see, points and hyperplanes play in some sense symmetric – the proper word is *dual* – roles in the geometry of $\mathbb{P}_n$.

4.1 The space of hyperplanes

Assume that $\mathbb{P}_n$ is an n-dimensional projective space, with associated vector space E and structural map π, and consider the set, denoted by $\mathbb{P}_n^\vee$, of all the hyperplanes of $\mathbb{P}_n$. Our first task is to define a projective structure on $\mathbb{P}_n^\vee$. To this end, take $E^\vee$ to be the dual space of E. Any non-zero linear form $\omega \in E^\vee$ is a rank-one linear map $\omega\colon E \to k$. Therefore $\ker \omega$ is an n-dimensional subspace of E and represents the hyperplane $[\ker \omega]$ of $\mathbb{P}_n$. It thus makes sense to define

$$\pi^\vee\colon E^\vee - \{0\} \longrightarrow \mathbb{P}_n^\vee$$

by the rule

$$\omega \longmapsto [\ker \omega].$$

We have

Lemma 4.1.1. *The map $\pi^\vee$ defines on $\mathbb{P}_n^\vee$ a structure of n-dimensional projective space.*

Proof. Assume that $H = [F]$ is a hyperplane of $\mathbb{P}_n$ and therefore $\dim F = n$. Then, it is a standard result of linear algebra that $F^\perp = \{\omega \in E^\vee \mid \omega(F) = \{0\}\}$ is a subspace of $E^\vee$ of dimension $n + 1 - n = 1$. For any non-zero $\omega \in F^\perp$, $\ker \omega \supset F$ and, as noted above, $\dim \ker \omega = n$. It follows that $\ker \omega = F$ and hence $\pi^\vee(\omega) = H$.

Clearly $\ker \omega = \ker \lambda\omega$ and thus $\pi^\vee(\omega) = \pi^\vee(\lambda\omega)$ for any non-zero $\lambda \in k$. Conversely, if $\pi^\vee(\omega) = \pi^\vee(\omega') = H = [F]$ for $\omega, \omega' \in E^\vee$, $\omega, \omega' \neq 0$, $\ker \omega = \ker \omega' = F$ and so $\omega, \omega' \in F^\perp$. Since, as already seen, $\dim F^\perp = 1$, the forms ω, ω' are proportional. That $\dim \mathbb{P}_n^\vee = n$ is an obvious consequence of being $\dim E = \dim E^\vee$. $\square$

Remark 4.1.2. We have seen above that the representatives of a hyperplane $H = [F]$ are the non-zero forms in $F^{\perp}$. Each of them is thus a basis of $F^{\perp}$ and gives rise to an equation of H, as explained in Section 2.6.

Remark 4.1.3. The definition of $\pi^{\vee}$ may be equivalently stated by saying that a point $p = [v]$ belongs to the hyperplane H represented by ω, $H = [\omega]$, if and only if $\omega(v) = 0$. Conversely, given a form $\omega \neq 0$, $H = [\omega]$ if and only if $\omega(v) = 0$ for all v for which $[v] \in H$. In this way we relate the description of H as a subset of $\mathbb{P}_n$, by its points, and its description as an element of the projective space $\mathbb{P}_n^{\vee}$, by a representative.

The projective space $\mathbb{P}_n^{\vee}$ above is named the *dual space*, or the *space of hyperplanes*, of $\mathbb{P}_n$ (*plane of lines* if $n = 2$, *space of planes* if $n = 3$). The use of the word *points* for the elements of $\mathbb{P}_n^{\vee}$ is sometimes avoided by calling them *elements of* $\mathbb{P}_n^{\vee}$, or just *hyperplanes*, in order to prevent any confusion with the points of $\mathbb{P}_n$.

4.2 Bundles of hyperplanes

Our first task is to characterize the linear varieties of the dual space $\mathbb{P}_n^{\vee}$ among all sets (often it is equivalently said *families*) of hyperplanes. We begin by showing some linear varieties of the space of hyperplanes. For any linear variety L of $\mathbb{P}_n$, we denote by L^* the set of all hyperplanes containing L. The sets L^*, for L a linear variety of $\mathbb{P}_n$, are called *bundles of hyperplanes*, and sometimes just *bundles* if no confusion may arise. We will refer to L^* as the bundle of the hyperplanes through L.

Lemma 4.2.1. *If $L = [F]$ is a linear variety of $\mathbb{P}_n$ of dimension d, then $L^* = [F^{\perp}]$, in particular L^* is a linear variety of $\mathbb{P}_n^{\vee}$ of dimension $n - d - 1$.*

Proof. Let $H = [\omega]$ be a hyperplane. By 4.1.3, $H \supset L$ if and only if $\omega(v) = 0$ for any representative v of any point $p \in L$, that is, if and only if $\omega \in F^{\perp}$, hence the claim. □

Note that the dimensions of L and L^* are symmetrically related, that is, $\dim L = n - \dim L^* - 1$.

We have seen that all bundles of hyperplanes are linear varieties of $\mathbb{P}_n^{\vee}$. In fact they are the only ones, as seen next:

Proposition 4.2.2. *Any linear variety T of $\mathbb{P}_n^{\vee}$ is a bundle of hyperplanes, $T = L^*$, for a uniquely determined linear variety L of $\mathbb{P}_n$.*

Proof. As in Section 1.3, $\mathcal{S}(\,)$ and $\mathcal{LV}(\,)$ mean *set of subspaces* and *set of linear varieties*, respectively. By 4.2.1, the map

$$\begin{aligned} * : \mathcal{LV}(\mathbb{P}_n) &\longrightarrow \mathcal{LV}(\mathbb{P}_n^{\vee}), \\ L &\longmapsto L^*, \end{aligned}$$

equals the composition of

$$\mathcal{L}\mathcal{V}(\mathbb{P}_n) \longrightarrow \mathcal{S}(E) \xrightarrow{\perp} \mathcal{S}(E^\vee) \longrightarrow \mathcal{L}\mathcal{V}(\mathbb{P}_n^\vee),$$

$$L = [F] \longmapsto F \longmapsto F^\perp \longmapsto [F^\perp] = L^*.$$

That $\perp$ is a bijection is a standard result of linear algebra. Since the other two maps have been seen to be bijections in Section 1.3, the map $*$ is also a bijection as claimed. □

The linear variety L is called the *kernel* or the *base variety* (or base point, base line, etc.) of L^*.

Remark 4.2.3. We have seen in 4.2.1 that $\dim L^* = n - \dim L - 1$, so the above bijection $*$ restricts to bijections between sets of linear varieties of fixed dimension,

$$\begin{aligned} *_d \colon \mathcal{L}\mathcal{V}_d(\mathbb{P}_n) &\longrightarrow \mathcal{L}\mathcal{V}_{n-d-1}(\mathbb{P}_n^\vee), \\ L &\longmapsto L^*, \end{aligned}$$

for $-1 \le d \le n$.

The sets consisting of a single hyperplane are the zero-dimensional bundles of hyperplanes, they have the form $\{H\} = H^*$, for H a hyperplane. The one-dimensional linear varieties of $\mathbb{P}_n^\vee$ (the "lines" of hyperplanes) are the bundles of hyperplanes with $(n-2)$-dimensional kernel: they are called *pencils of hyperplanes* (*pencils of lines* if $n = 2$, *pencils of planes* if $n = 3$). The one-codimensional linear varieties of $\mathbb{P}_n^\vee$ (the "hyperplanes" of hyperplanes) are the bundles of the hyperplanes through a fixed point, they are called *hyperbundles* (just *bundles* if $n = 3$).

Thus, if $n = 2$, the bundles of lines are the empty one, $\emptyset = \mathbb{P}_2^*$, the one composed of all lines, $\mathbb{P}_2^\vee = \emptyset^*$, those composed of a single line $\ell \subset \mathbb{P}_2$, $\{\ell\} = \ell^*$, and the pencils of lines p^* for $p \in \mathbb{P}_2$; they have dimensions $-1, 2, 0$ and 1, respectively. If $n = 3$, besides the extremal cases $\emptyset$ and $\mathbb{P}_3^\vee$, and the bundles composed of a single plane, we have the pencils of planes ℓ^*, for ℓ a line of $\mathbb{P}_3$, which have dimension 1, and the bundles of planes p^*, $p \in \mathbb{P}$, which have dimension 2.

Once the linear varieties of hyperplanes have been shown to be the bundles of hyperplanes, we describe the basic incidence relations between them in terms of incidence relations between their kernels:

Proposition 4.2.4. *If L, L_1 and L_2 denote linear varieties of $\mathbb{P}_n$, then*

(a) $L^* = \emptyset$ *if and only if* $L = \mathbb{P}_n$,

(b) $L^* = \mathbb{P}_n^\vee$ *if and only if* $L = \emptyset$,

(c) $L_1^* \subset L_2^*$ *if and only if* $L_1 \supset L_2$,

(d) $L_1^* \vee L_2^* = (L_1 \cap L_2)^*$,

(e) $L_1^* \cap L_2^* = (L_1 \vee L_2)^*$.

Proof. Some of the claims are obvious, such as (a), (b) and the *if* part of (c), while (e) follows from 1.4.2. Anyway, all claims easily follow from 4.2.1 and the properties of the map $\perp$, $\perp(F) = F^\perp$. For instance, if $L_1 = [F_1]$, $L_2 = [F_2]$, for claim (c)

$$L_1 \subset L_2 \iff F_1 \subset F_2 \iff F_1^\perp \supset F_2^\perp \iff L_1^* \supset L_2^*,$$

while for claim (d)

$$L_1^* \vee L_2^* = [F_1^\perp + F_2^\perp] = [(F_1 \cap F_2)^\perp] = (L_1 \cap L_2)^*.$$

Claim (e) may be proved similarly. □

The reader may pay a special attention to claim (d). It describes the join of two bundles as the bundle of hyperplanes through the intersection of their kernels. For instance the join of two different pencils of planes in $\mathbb{P}_3$ is either the bundle of planes through the intersection point of their base lines if these meet, or the whole set of all planes if the base lines are disjoint. Claim (e) is easier just because taking the intersection is a set-theoretical operation, while taking the join is not.

Remark 4.2.5. Claims (d) and (e) of 4.2.4 may be read as saying that the kernel of a join of bundles is the intersection of their kernels and that the kernel of an intersection of bundles is the join of their kernels.

The notion of independence being defined for the points of any projective space is in particular defined for hyperplanes. Next we characterize the independence of hyperplanes in terms of the same hyperplanes viewed as linear varieties of $\mathbb{P}_n$, rather than as elements of $\mathbb{P}_n^\vee$.

Proposition 4.2.6. *Hyperplanes $H_1, \dots, H_s$ of $\mathbb{P}_n$ are independent if and only if* $\dim H_1 \cap \dots \cap H_s = n - s$.

Proof. By the definition, $H_1, \dots, H_s$ are independent if and only if

$$\dim\{H_1\} \vee \dots \vee \{H_s\} = s - 1.$$

(The reader may note that this time we are correctly writing $\{H_i\}$ and not H_i for the variety of $\mathbb{P}_n^\vee$ whose only element is H_i, as writing $H_1 \vee \dots \vee H_s$ would have a quite different meaning.) Since $\{H_i\} = H_i^*$ the above condition may be equivalently written

$$s - 1 = \dim H_1^* \vee \dots \vee H_s^* = \dim(H_1 \cap \dots \cap H_s)^*,$$

using 4.2.4. By 4.2.1, the last condition is equivalent to

$$\dim H_1 \cap \dots \cap H_s = n - (s - 1) - 1 = n - s$$

as claimed. □

Corollary 4.2.7. *Hyperplanes $H_1, \dots, H_s$ are independent if and only if for each $i = 2, \dots, s$ there is a point $p_i \in H_1 \cap \dots \cap H_{i-1}$ that does not belong to H_i.*

Proof. By 1.4.8, if p_i exists, then

$$\dim(H_1 \cap \dots \cap H_i) = \dim(H_1 \cap \dots \cap H_{i-1}) - 1.$$

Otherwise

$$\dim(H_1 \cap \dots \cap H_i) = \dim(H_1 \cap \dots \cap H_{i-1}).$$

Then it is enough to add up the suitable of the above equalities, for $i = 2, \dots, s$, to get

$$\dim(H_1 \cap \dots \cap H_s) = n - 1 - (s - 1) = n - s$$

if all points p_i do exist, and

$$\dim(H_1 \cap \dots \cap H_s) > n - s$$

if one of the p_i fails to exist, hence the claim. □

Corollary 4.2.8. *For any linear variety $L \subset \mathbb{P}_n$, $L \neq \mathbb{P}_n$, of dimension d, there exist $n - d$ independent hyperplanes containing L. Furthermore, for any $n - d$ independent hyperplanes $H_1, \dots, H_{n-d}$ containing L,*

$$L = H_1 \cap \dots \cap H_{n-d}.$$

Proof. Since $\dim L^* = n - d - 1$, the existence of the hyperplanes follows from 1.5.7. The inclusion $L \subset H_1 \cap \dots \cap H_{n-d}$ is clear, and both sides have the same dimension by 4.2.6. □

We already know (from 4.2.2) that a bundle of hyperplanes determines its kernel. The next corollary makes explicit this determination.

Corollary 4.2.9. *For any linear variety $L \neq \mathbb{P}_n$ of $\mathbb{P}_n$,*

$$L = \bigcap_{H \in L^*} H.$$

Proof. Obviously $L \subset H$ for all $H \in L^*$ and so the inclusion $\subset$ is clear. Put $\dim L = d < n$. By 4.2.8, we are allowed to choose independent hyperplanes $H_1, \dots, H_{n-d} \in L^*$, after which it holds that

$$L = H_1 \cap \dots \cap H_{n-d} \supset \bigcap_{H \in L^*} H,$$

which completes the proof. □

Corollary 4.2.10. *For any set of independent points* $\{p_0, \dots, p_m\} \subset \mathbb{P}_n$ *there is a hyperplane of* $\mathbb{P}_n$ *containing* $p_0, \dots, p_{m-1}$ *and missing* p_m.

Proof. The hyperplanes containing $p_0, \dots, p_{m-1}$ are those containing $L = p_0 \vee \dots \vee p_{m-1}$. If no one is missing p_m, then, by 4.2.9, $p_m \in L$, against the independence of the points. □

Since the definition of cross ratio of Section 2.9 applies to points of an arbitrary one-dimensional projective space, the cross ratio of four hyperplanes in a pencil, at least three of which are different, is defined. The next proposition relates it to the cross ratio of four points:

Proposition 4.2.11. *Assume that* H_1, H_2, H_3, H_4 *are hyperplanes, at least three of them different, belonging to a pencil of hyperplanes* L^*, $\dim L = n-2$. *If* ℓ *is any line supplementary to* L, *then each intersection* $H_i \cap \ell$ *is a point* p_i, $i = 1, \dots, 4$, *at least three of the points* p_1, p_2, p_3, p_4 *are different and*

$$(H_1, H_2, H_3, H_4) = (p_1, p_2, p_3, p_4).$$

Proof. If ℓ is supplementary to L, then $\ell \not\subset H_i$; therefore, by 1.4.8, $\ell \cap H_i$ is a point, say $\ell \cap H_i = \{p_i\}$, and by 1.4.7, $H_i = L \vee p_i$, for $i = 1, \dots, 4$. After this, $p_i = p_j$ if and only if $H_i = H_j$. This proves the first half of the claim. Regarding the equality of cross ratios, if for some $i \neq j$, $H_i = H_j$, then, as seen above, $p_i = p_j$ and the claim follows from 2.9.10. Otherwise, H_1, H_2, H_3, H_4 are four different hyperplanes and therefore p_1, p_2, p_3, p_4 are four different points. Write $\rho = (H_1, H_2, H_3, H_4)$ and $\rho' = (p_1, p_2, p_3, p_4)$. By 2.9.9 there are forms ω_1, ω_2 such that

$$H_1 = [\omega_1], \quad H_2 = [\omega_2], \quad H_3 = [\omega_1 + \omega_2], \quad H_4 = [\rho\omega_1 + \omega_2].$$

Similarly, there are vectors v_1, v_2 such that

$$p_1 = [v_1], \quad p_2 = [v_2], \quad p_3 = [v_1 + v_2], \quad p_4 = [\rho' v_1 + v_2].$$

Now the conditions $p_1 \in H_1$ and $p_2 \in H_2$ imply, by 4.1.3,

$$\omega_1(v_1) = 0, \quad \omega_2(v_2) = 0,$$

after which the conditions $p_3 \in H_3$ and $p_4 \in H_4$ imply

$$\begin{aligned} 0 &= (\omega_1 + \omega_2)(v_1 + v_2) = \omega_1(v_2) + \omega_2(v_1), \\ 0 &= (\rho\omega_1 + \omega_2)(\rho' v_1 + v_2) = \rho\omega_1(v_2) + \rho'\omega_2(v_1) \end{aligned}$$

and so

$$0 = (\rho - \rho')\omega_1(v_2).$$

Now $\omega_1(v_2) = 0$ would give $p_2 \in H_1$ and so $H_1 = H_2$, as then both hyperplanes would contain the point $p_2 \notin L$. Thus $\omega_1(v_2) \neq 0$ and $\rho = \rho'$, as claimed. □

Remark 4.2.12. Assume $n = 1$. Then the hyperplanes are just the single-point subsets $\{p\}$, $p \in \mathbb{P}_1$, Proposition 4.2.11 applies with $L = \emptyset$, $\ell = \mathbb{P}_1$ and $H_i = \{p_i\}$, and shows that the map

$$\begin{aligned} \iota\colon \mathbb{P}_1 &\longrightarrow \mathbb{P}_1^\vee, \\ p &\longmapsto \{p\}, \end{aligned}$$

preserves cross ratios and therefore (by 2.9.15) is a projectivity. Making, as usual, no distinction between p and $\{p\}$ identifies $\mathbb{P}_1$ and $\mathbb{P}_1^\vee$ through ι. Since ι is a projectivity there is no essential difference between considering hyperplanes $\{p\} \in \mathbb{P}_1^\vee$ or their corresponding points $p \in \mathbb{P}_1$. Usually one deals just with $\mathbb{P}_1$ and $\mathbb{P}_1^\vee$ is seldom mentioned. Duality for one-dimensional projective spaces has thus little interest.

Let L be any $n-2$ linear variety of $\mathbb{P}_n$. Obviously any hyperplane $H \in L^*$ is $H = L \vee p$ for a point $p \notin L$ and conversely, which proves that, as sets of hyperplanes, the pencil of kernel L equals the bunch of centre L. We have thus two projective structures on the set of hyperplanes through L, namely the bunch structure, defined in Section 1.9, and its structure as a bundle of hyperplanes. The fact is that these two structures are equivalent:

Proposition 4.2.13. *The bunch and bundle structures on the set of hyperplanes of $\mathbb{P}_n$ through a fixed $(n-2)$-dimensional linear variety L are equivalent.*

Proof. Let T be a line supplementary to L. On one hand, it follows from 2.9.15 and 4.2.11 that the section map

$$L^* \longrightarrow T,$$

which maps each hyperplane H to the only point in $H \cap T$, is a projectivity. On the other, we have seen in 1.9.3 that the projection map

$$\begin{aligned} T &\longrightarrow L^\nabla, \\ p &\longmapsto L \vee p, \end{aligned}$$

is also a projectivity. Since the composition of these projectivities is the identical map

$$\mathrm{Id}\colon L^* \longrightarrow L^\nabla,$$

the claim follows from 1.6.3. □

4.3 The principle of duality

The word *principle* has not a well-defined meaning in mathematics. Sometimes it refers to a fundamental fact which receives no justification or just a very vague one. An example is the *principle of continuity* which was rather obscurely used at the beginnings of the systematic development of projective geometry in the 19th century (see [12], 2.2 and 2.3) and has no use today. This is not the case of the *principle of duality*, which is a useful general trick that essentially consists in getting new true statements by claiming for hyperplanes statements already proved for points. Indeed, in the preceding sections we have seen that for each projective space $\mathbb{P}_n$ we have two n-dimensional projective structures: the structure of $\mathbb{P}_n$ itself, on its set of points, and the structure of $\mathbb{P}_n^\vee$, on the set of hyperplanes of $\mathbb{P}_n$. Assume that we fix our attention on a statement (theorem, proposition, etc.) $\mathcal{S}$ that holds for the linear varieties of any n-dimensional projective space, for a fixed n, and whose claim involves only dimensions, inclusions, intersections and joins of linear varieties. Choose an arbitrary projective space $\mathbb{P}_n$: on one hand our statement $\mathcal{S}$ holds for the linear varieties of $\mathbb{P}_n$; on the other, it also holds for the linear varieties of $\mathbb{P}_n^\vee$, the bundles of $\mathbb{P}_n$. Then 4.2.1, 4.2.2 and 4.2.4 may be used to turn the bundle-version of $\mathcal{S}$ into an equivalent statement in terms of the kernels of the bundles. This new statement, named the *dual* of $\mathcal{S}$, needs no proof, because it is just an equivalent reformulation of $\mathcal{S}$ read in $\mathbb{P}_n^\vee$ and therefore, by the arbitrariness of $\mathbb{P}_n$, does hold for the linear varieties of any n-dimensional projective space.

Before making this idea precise we show an easy example. After 1.4.9, we know to be true the statement:

$\mathcal{S}$: *The intersection of any two different planes in a three-dimensional projective space is a line.*

The dual space $\mathbb{P}_3^\vee$ of any three-dimensional projective space $\mathbb{P}_3$ being also three-dimensional, $\mathcal{S}$ does hold in $\mathbb{P}_3^\vee$ as well, and there it reads:

$\mathcal{S}'$: *The intersection of any two different two-dimensional bundles of planes of a three-dimensional projective space is a pencil of planes.*

Now, assume we have any pair of different points p_1, p_2 in a three-dimensional projective space. By 4.2.3, p_1^*, p_2^* are two different two-dimensional bundles which, by $\mathcal{S}'$, have $\dim p_1^* \cap p_2^* = 1$. Then, since by 4.2.4 it is $p_1^* \cap p_2^* = (p_1 \vee p_2)^*$, using 4.2.1 we obtain $\dim p_1 \vee p_2 = 3 - \dim p_1^* \cap p_2^* - 1 = 1$ and hence it holds that:

$\mathcal{S}^\vee$: *The join of any two different points p_1, p_2 of a three-dimensional projective space is a line.*

Of course, we already know $\mathcal{S}^\vee$ to be true, as it is a particular case of 1.4.7, but the important point here is that, once $\mathcal{S}$ is proved, $\mathcal{S}^\vee$ does not require any proof, because it is equivalent to $\mathcal{S}$ claimed in the dual space.

Back to the general situation, let $\mathcal{S}$ be a statement of the form

$$\mathcal{H} \Longrightarrow \mathcal{T}$$

in terms of the linear varieties of an arbitrary projective space $\mathbb{P}_n$, their dimensions, inclusions, and finite intersections and joins. Then, make in the statement $\mathcal{S}$ formal replacements according to the following rules:

4.3.1. Rules for dualizing a statement about linear varieties of an n-dimensional projective space.

(a) replace each dimension d of a linear variety with $n - d - 1$,

(b) replace each inclusion between linear varieties with the reversed one,

(c) replace each intersection of linear varieties with a join, and

(d) replace each join of linear varieties with an intersection.

Then we get a new statement $\mathcal{S}^\vee$:

$$\mathcal{H}^\vee \Longrightarrow \mathcal{T}^\vee$$

which is, by definition, the *dual statement* of $\mathcal{S}$. Its hypothesis $H^\vee$ and thesis $\mathcal{T}^\vee$ are called the duals of $\mathcal{H}$ and $\mathcal{T}$, respectively. What the *principle of duality* says is that once the statement S is known to be true, its dual $S^\vee$ is automatically true, with no need of proof. In other words, $S \Rightarrow S^\vee$. The argument for this is as follows. Assume we have in an arbitrary n-dimensional projective space $\mathbb{P}_n$ linear varieties $L_1, \dots, L_r$ in the situation described by $\mathcal{H}^\vee$. By 4.2.1 and 4.2.4 the bundles $L_1^*, \dots, L_r^*$ are in the situation described by $\mathcal{H}$. Since $\mathcal{S}$ holds in $\mathbb{P}_n^\vee$, $L_1^*, \dots, L_r^*$ satisfy $\mathcal{T}$ which in turn, again by 4.2.1 and 4.2.4, guarantees that their kernels $L_1, \dots, L_r$ satisfy $\mathcal{T}^\vee$. This proves that the statement $\mathcal{S}^\vee$ holds true, as claimed.

Thus, the principle of duality is not a theorem of projective geometry, but rather a procedure for getting new true projective statements out of already proved ones. One may think of it as a *metatheorem* (a theorem on theorems) and give it a claim such as:

4.3.2 (Principle of duality). *If a statement about linear varieties of an arbitrary n-dimensional projective space, involving only dimensions, inclusions and finite intersections and joins, holds true, then the statement obtained from it by turning each dimension d of a linear variety into $n - d - 1$, each inclusion into its reverse, each intersection into a join and each join into an intersection, holds true as well.*

Anyway, there is no need of entering the domains of logic, trying to make a proof of the principle of duality. Instead, it is easier to understand the principle of duality just as a rule whose applications could be checked case by case, by a specific argument following the pattern of the general argument explained above.

The remainder of this section contains some remarks, comments and examples that are intended to make easier the application of the principle of duality. The first one says that dualizing statements is involutive, which is clear from the definition of dual statement.

Remark 4.3.3. The dual of the statement $\mathcal{S}^\vee$ is $\mathcal{S}$. Thus the statements we are considering are grouped in unordered pairs $\mathcal{S}$, $\mathcal{S}^\vee$ of mutually dual statements.

Remark 4.3.4. Sometimes the dual statement is not new, but coincides with the original one. An easy example is given by the Grassmann formula 1.4.6. Indeed, once dualized it reads:

For any two linear varieties L_1, L_2 of a projective space $\mathbb{P}_n$,

$$(n-\dim L_1-1)+(n-\dim L_2-1) = (n-\dim L_1 \vee L_2-1)+(n-\dim L_1 \cap L_2-1),$$

which is obviously equivalent to the original statement.

So far, dualizing a statement requires to have it expressed in the basic terms of dimension, inclusion, intersection and join of linear varieties. There is a number of other terms or phrases, such as *concurrent lines* or *triangle*, each defined in terms of dimension, inclusion, intersection and join, that appear in the statements. They would need to be replaced with their definitions before dualizing. For instance, one should replace *concurrent lines* with *lines containing the same point.* Making all these replacements is rather boring and often gives rise to almost unreadable statements. Therefore it is preferred to proceed as follows: when a new term (or phrase) $\mathcal{U}$ is defined using the basic ones, we perform in its definition the substitutions prescribed by the rules 4.3.1; the result is taken as the definition of what is called the *dual term* $\mathcal{U}^\vee$ of $\mathcal{U}$. Then when dualizing a statement that contains the term $\mathcal{U}$, one just replaces it with $\mathcal{U}^\vee$. If the definition of a new term $\mathcal{V}$ involves $\mathcal{U}$, the same substitution is done to get the definition of the dual term $\mathcal{V}^\vee$. As for statements, the dual of $\mathcal{U}^\vee$ is $\mathcal{U}$. Collecting pairs of mutually dual terms enlarges the basic substitution rules of 4.3.1 and builds a sort of symmetric dictionary for dualization: the larger it is, the easier dualizing becomes.

Remark 4.3.5. Dualizing strongly depends on the dimension n of the ambient space, because of its occurrence in the rule 4.3.1 (a). For instance *line* and *point* are dual terms for $n = 2$, while *line* is a self-dual term if $n = 3$.

Remark 4.3.6. The cross ratio of four hyperplanes in a pencil being defined as their cross ratio as points of $\mathbb{P}_n^\vee$, the term *hyperplanes in a pencil with cross ratio* ρ dualizes into *points in a line with cross ratio* ρ, and conversely.

We list below some examples, details are mostly left to the reader.

Example 4.3.7. In dimension two, the dual of *aligned points* is *concurrent lines*, and conversely.

Example 4.3.8. A triangle of $\mathbb{P}_2$ consists of three non-aligned points p_0, p_1, p_2, its vertices, and the three lines $p_i \vee p_j$, $0 \leq i < j \leq 2$, its sides. Dualizing we have to consider figures consisting of three non-concurrent lines ℓ_0, ℓ_1, ℓ_2 and the three points $\ell_i \cap \ell_j$, $0 \leq i < j \leq 2$ which is an equivalent definition of triangle (see 1.7.6). The term *triangle* is thus, for $n = 2$, self-dual. However, *vertex of a triangle* dualizes to *side of a triangle* and conversely.

Example 4.3.9. According to 1.7.8 and 1.7.9, *quadrilateral* and *quadrivertex* (of $\mathbb{P}_2$) are dual terms. *Side*, *vertex*, *pair of opposite vertices* and *diagonal* of a quadrilateral have as duals, respectively, *vertex*, *side*, *pair of opposite sides* and *diagonal point* of a quadrivertex, and conversely. As a consequence, 1.8.1 and 1.8.2 are dual of each other and hence the proof of 1.8.2 that was omitted in Section 1.8 is in fact not needed.

Example 4.3.10. A *trihedron* of $\mathbb{P}_3$ consists of three independent planes of $\mathbb{P}_3$ (its *faces*), the three lines intersections of the pairs of faces (its *edges*) and the point intersection of the three faces (its *vertex*). In $\mathbb{P}_3$, *trihedron* and *triangle* are dual terms; *face*, *edge* and *vertex of a trihedron* are dual of *vertex*, *side* and *plane of a triangle*, and conversely.

Example 4.3.11. In $\mathbb{P}_n$, *projection of a point from a d-dimensional variety* and *section of an $(n-d-1)$-dimensional variety by a hyperplane* are dual of each other.

Example 4.3.12. Like *triangle of $\mathbb{P}_2$*, *n-dimensional simplex of $\mathbb{P}_n$* is self-dual, while *d-dimensional face* and *$(n-d-1)$-dimensional face* are mutually dual. *Opposite faces* is self-dual.

We close this series of examples by dualizing Desargues' theorem. As stated in 1.8.3, it reads:

Theorem (Desargues). *Assume triangles ABC and $A'B'C'$ in a projective plane $\mathbb{P}_2$, let a, b and c be the sides of the first triangle opposite A, B and C, respectively, and, similarly, let a', b', and c' be the sides of the second triangle opposite A', B' and C'. Assume furthermore that $A \neq A'$, $B \neq B'$, $C \neq C'$, $a \neq a'$, $b \neq b'$ and $c \neq c'$. If the lines AA', BB', CC' are concurrent, then the points $a \cap a'$, $b \cap b'$, $c \cap c'$ are aligned.*

Using the above rules, its dual, which in this case is also its converse, is just:

Theorem 4.3.13 (Converse of Desargues'). *Assume triangles with sides a, b, c and a', b', c' in a projective plane $\mathbb{P}_2$, let A, B and C be the vertices of the first triangle opposite a, b and c, respectively, and, similarly, let A', B', and C' be the vertices of the second triangle opposite a', b' and c'. Assume furthermore that $a \neq a'$, $b \neq b'$, $c \neq c'$, $A \neq A'$, $B \neq B'$ and $C \neq C'$. If the points $a \cap a'$, $b \cap b'$, $c \cap c'$ are aligned, then the lines AA', BB', CC' are concurrent.*

The geometers of the 19th century considered the principle of duality as one of the most relevant facts of projective geometry. Today the relevance is rather given to the whole phenomena of duality, namely to the existence of two projective structures, on points and hyperplanes, together with duality relationships between objects of the same kind belonging to one and another structure, for instance the relationship between linear varieties and bundles explained in Section 4.2. Further relationships of this kind will appear in the sequel.

We close this section with a heuristic comment about the reason why there is no extension of the duality principle to affine geometry. The hyperplanes of an affine space $\mathbb{A}_n$ being identified with their projective closures, they are the points of the dual space $\overline{\mathbb{A}}_n^\vee$ other than the improper hyperplane H_∞, which appears as a distinguished point of $\overline{\mathbb{A}}_n^\vee$. As the reader may easily check, there is no point of $\overline{\mathbb{A}}_n$ invariant by all affinities, after which no hyperplane of $\overline{\mathbb{A}}_n^\vee$ is invariant by all affinities. The lack of a distinguished hyperplane in $\overline{\mathbb{A}}_n^\vee$ prevents us from defining the duals of the affine (non-projective) notions, which are formulated in $\overline{\mathbb{A}}_n$ using H_∞. For instance, parallelism of lines $\ell, \ell' \subset \overline{\mathbb{A}}_n$ is characterized by the condition $\ell \cap \ell' \cap H_\infty \neq \emptyset$. Having no distinguished hyperplane in $\overline{\mathbb{A}}_n^\vee$ playing the role of H_∞, there is no similar condition for lines of $\overline{\mathbb{A}}_n^\vee$, and so no notion of *parallel pencils of hyperplanes* dual to that of *parallel lines*. In other words, the point is that the further structures that affine geometry determines in $\overline{\mathbb{A}}_n$ and $\overline{\mathbb{A}}_n^\vee$ (a distinguished hyperplane and a distinguished point, respectively) are quite different, and so definitions involving them cannot be readily translated from one to another space, as we did in the projective case. However, when doing affine geometry, the dual space $\overline{\mathbb{A}}_n^\vee$, with the distinguished point H_∞, may still be considered and is quite useful.

4.4 Hyperplane coordinates

Corresponding to their double interpretation, as linear varieties of $\mathbb{P}_n$ and as points of $\mathbb{P}_n^\vee$, hyperplanes may be determined by their equations in $\mathbb{P}_n$ and also by their coordinates in $\mathbb{P}_n^\vee$: in this section we will relate both.

Lemma 4.4.1. *Assume that Δ is a reference of $\mathbb{P}_n$, that $e_0, \dots, e_n$ is a basis adapted to Δ and that $\omega_0, \dots, \omega_n$ is the basis dual of $e_0, \dots, e_n$. Then, a hyperplane H has equation*

$$u_0x_0 + \cdots + u_nx_n = 0 \tag{4.1}$$

relative to Δ if and only if it has representative

$$\omega = u_0\omega_0 + \cdots + u_n\omega_n.$$

Proof. Assume that $H = [F]$ has equation (4.1). A non-zero vector $v = \sum_0^n x_i e_i$ belongs to F if and only if the point it represents, $[v] = [x_0, \dots, x_n]$ belongs to H, and so, if and only if

$$u_0x_0 + \cdots + u_nx_n = 0.$$

Recalling that, by the definition of dual basis, $\omega_j(e_i) = 1$ if $i = j$ and $\omega_j(e_i) = 0$ otherwise, this equality may be written

$$\omega(v) = \Big(\sum_0^n u_j\omega_j\Big)\Big(\sum_0^n x_ie_i\Big) = \sum_0^n u_ix_i = 0.$$

This proves that $F = \ker\omega$ and so $H = [\omega]$.

Conversely, if $H = [\omega]$, then the same computation shows that $[v] = [x_0, \dots, x_n]$ belongs to H if and only if

$$\sum_0^n u_ix_i = \omega(v) = 0,$$

which proves that (4.1) is an equation of H. □

Theorem 4.4.2. *For each reference Δ of $\mathbb{P}_n$ there is a reference $\Delta^\vee$ of $\mathbb{P}_n^\vee$ such that a hyperplane H has equation $u_0x_0 + \cdots + u_nx_n = 0$ relative to Δ if and only if it has coordinates $u_0, \dots, u_n$ relative to $\Delta^\vee$. Furthermore the reference $\Delta^\vee$ is uniquely determined by Δ and this property.*

Proof. Let us prove the uniqueness first. As for any reference, the vertices and unit point of $\Delta^\vee$ are

$$[1, 0, \dots, 0]_{\Delta^\vee}, \quad \dots, \quad [0, 0, \dots, 1]_{\Delta^\vee}, \quad [1, 1, \dots, 1]_{\Delta^\vee}.$$

Then, according to the property claimed in 4.4.2, the vertices of $\Delta^\vee$ are the hyperplanes with equations

$$x_0 = 0, \ x_1 = 0, \ \dots, \ x_n = 0,$$

and its unit element is the hyperplane with equation

$$x_0 + x_1 + \cdots + x_n = 0.$$

For the existence, as suggested by the above, define $\Delta^\vee$ by taking the $(n-1)$-dimensional faces of Δ, $x_i = 0$, $i = 0, \dots, n$, as vertices and the hyperplane $\sum_0^n x_i = 0$ as unit point. If $e_0, \dots, e_n$ is a basis adapted to Δ and $\omega_0, \dots, \omega_n$ is the basis dual of $e_0, \dots, e_n$, then, by 4.4.1, each ω_i is a representative of the hyperplane $x_i = 0$, $i = 0, \dots, n$ and $\sum_0^n \omega_i$ represents the unit hyperplane $\sum_0^n x_i = 0$. Then 2.1.3 proves that $\Delta^\vee$ is a projective reference and $\omega_0, \dots, \omega_n$ is a basis adapted to it.

Now, again using 4.4.1, a hyperplane H has equation $\sum_0^n u_ix_i = 0$ if and only if $H = [\sum_0^n u_i\omega_i]$ and thus, by the above, if and only if $H = [u_0, \dots, u_n]_{\Delta^\vee}$. □

The reference $\Delta^\vee$ is called the *dual reference* of Δ; its unit point is often called the *unit hyperplane*.

Remark 4.4.3. The elements of the dual reference have been explicitly determined in the proof of 4.4.2, namely:

- The vertices of $\Delta^\vee$ are the hyperplanes $x_i = 0$, $i = 0, \dots, n$, that is, the $(n-1)$-dimensional faces of Δ numbered as their opposite vertices.
- The unit hyperplane of $\Delta^\vee$ is $\sum_{i=0}^n x_i = 0$; see Exercise 4.3 for another description.

Remark 4.4.4. We have seen in the course of the proof of 4.4.2 that if a basis is adapted to Δ, then its dual basis is adapted to $\Delta^\vee$.

The reader may note that the fact that both the equation of a hyperplane (as seen in Section 2.7) and its homogeneous coordinates are determined up to a non-zero constant factor is consistent with Theorem 4.4.2.

We have seen that Δ determines $\Delta^\vee$, which is the reason why $\Delta^\vee$ is seldom mentioned in practice. The coordinates of a hyperplane H relative to $\Delta^\vee$ are called *hyperplane coordinates* (*line coordinates*, *plane coordinates* if $n = 2, 3$) or *dual coordinates* of H relative to Δ. To avoid confusion, the homogeneous coordinates, as defined in Section 2.2 are sometimes called *point coordinates*. Using such terms, the hyperplane coordinates of H relative to Δ are, by definition, its point coordinates relative to $\Delta^\vee$. We shall write $[u_0, \dots, u_n]_\Delta$ to denote the hyperplane with coordinates $u_0, \dots, u_n$ relative to Δ. Note that, according to the formerly introduced notation for point coordinates, $[u_0, \dots, u_n]_\Delta = [u_0, \dots, u_n]_{\Delta^\vee}$. As for point coordinates, the mention of Δ is often omitted.

Example 4.4.5. Assume we have fixed a reference Δ in a one-dimensional projective space $\mathbb{P}_1$. If a point $p \in \mathbb{P}_1$ has (point) coordinates x_0, x_1, then the hyperplane $\{p\}$ has hyperplane coordinates

$$u_0 = x_1, \quad u_1 = -x_0,$$

just because $x_1 x_0 - x_0 x_1 = 0$ and so an equation of $\{p\}$ has coefficients $x_1, -x_0$. The equations displayed above are equations of the projectivity ι of 4.2.12: They re-prove that ι is a projectivity.

If, as usual, the point coordinate vectors are taken as column vectors, it is convenient to take the hyperplane coordinate vectors as row vectors, as this allows us to write the condition for the point $[x_0, \dots, x_n]_\Delta$ to belong to the hyperplane $[u_0, \dots, u_n]_\Delta$ in matricial form

$$\begin{pmatrix} u_0 & \dots & u_n \end{pmatrix} \begin{pmatrix} x_0 \\ \vdots \\ x_n \end{pmatrix} = 0.$$

Roughly speaking, Theorem 4.4.2 says that the coefficients of an equation of a hyperplane may be taken as projective coordinates of the hyperplane in the dual projective space. This provides a very easy way of handling duality: hyperplanes of $\mathbb{P}_n$ are dealt with as points of $\mathbb{P}_n^\vee$ by just taking the coefficients of an equation of H as their coordinates. Consider for instance the relation

$$u_0x_0 + \cdots + u_nx_n = 0,$$

expressing that the point $p = [a_0, \ldots, a_n]$ belongs to the hyperplane

$$H = [u_0, \ldots, u_n].$$

If H is taken fixed and p variable, it is just an equation of H, but if p is taken fixed and H variable, then it is just an equation of the hyperbundle p^*. Next we explain a procedure, based in this fact, allowing us to compute a linear variety L and its associated bundle L^* from each other.

Remark 4.4.6. Assume that L is a d-dimensional linear variety of $\mathbb{P}_n$ spanned by $d+1$ (necessarily independent) points $q_i = [a_i^0, \ldots, a_i^n]$, $i = 0, \ldots, d$. Then a hyperplane H with equation $u_0x_0 + \cdots + u_nx_n = 0$ contains L if and only if it contains all points q_i, $i = 0, \ldots, d$, and so, if and only if,

$$u_0a_i^0 + \cdots + u_na_i^n = 0, \quad i = 0, \ldots, d. \tag{4.2}$$

Now, again because the coefficients u_i are also coordinates, the equations (4.2) are a system of implicit equations (in $u_0, \ldots, u_n$) of the bundle L^*.

Conversely, if the above equations (4.2) are a system of equations of L^*, then the equations themselves show that the points

$$[a_i^0, \ldots, a_i^n], \quad i = 0, \ldots, d,$$

whose coordinates are the coefficients of the equations, belong to all the hyperplanes in L^*, and hence to L. These points being independent, because so are the equations, and $d+1$ in number, they span L by 4.2.9.

Note that not only the first member of the equation $u_0x_0 + \cdots + u_nx_n = 0$ of a hyperplane and its coordinate vector $(u_0, \ldots, u_n)$ carry the same information, but also linear operations may be equivalently performed using coordinate vectors or equations. Due to this, often just equations instead of coordinate vectors are used to handle hyperplanes as points of $\mathbb{P}_n^\vee$. Next are three examples:

Proposition 4.4.7. *Hyperplanes H_i, with equations*

$$\sum_{i=0}^{n} u_i^j x_i = 0, \quad j = 1, \ldots, s,$$

are independent if and only if their equations are linearly independent.

Proof. Since the linear independence of the equations $\sum_{i=0}^{n} u_i^j x_i = 0$, $j = 1, \dots, s$, is equivalent to the linear independence of the coordinate vectors $(u_0^j, \dots, u_n^j)$, $j = 1, \dots, s$, the claim follows from 2.2.7. □

The reader may give a different proof of 4.4.7 using 4.2.6.

Proposition 4.4.8. *Let $L \subset \mathbb{P}_n$ be a linear variety of dimension d and assume that independent hyperplanes H_i, with equations*

$$\sum_{i=0}^{n} u_i^j x_i = 0, \quad j = 1, \dots, n-d,$$

are going through L. Then a hyperplane H goes through L if and only if it has an equation of the form

$$\lambda_1\Big(\sum_{i=0}^{n} u_i^1 x_i\Big) + \dots + \lambda_{n-d}\Big(\sum_{i=0}^{n} u_i^{n-d} x_i\Big) = 0 \tag{4.3}$$

for some $\lambda_1, \dots, \lambda_{n-d} \in k$, not all zero.

Proof. Assume that $H = [u_0, \dots, u_n]$. By 2.5.1, $H \in L^*$ if and only if

$$\begin{pmatrix} u_0 \\ \vdots \\ u_n \end{pmatrix} = \lambda_1 \begin{pmatrix} u_0^1 \\ \vdots \\ u_n^1 \end{pmatrix} + \dots + \lambda_{n-d} \begin{pmatrix} u_0^{n-d} \\ \vdots \\ u_n^{n-d} \end{pmatrix}$$

for some $\lambda_1, \dots, \lambda_{n-d} \in k$, not all zero, which is obviously equivalent to the condition of the claim. □

Remark 4.4.9. By Remark 2.5.3, the parameters $\lambda_1, \dots, \lambda_{n-d}$ in (4.3) are homogeneous coordinates of H relative to the reference of L^* which has the hyperplanes $H_1, \dots, H_{n-d}$ as vertices and

$$\sum_{j=1}^{n-d}\Big(\sum_{i=0}^{n} u_i^j x_i\Big) = 0$$

as unit hyperplane.

By 4.2.6 and 4.4.7, one may equivalently assume in 4.4.8 that

$$\sum_{i=0}^{n} u_i^j x_i = 0, \quad j = 1, \dots, n-d,$$

is a system of implicit equations of L. In this form, the reader may provide a direct proof of 4.4.8 using 2.7.5.

Proposition 4.4.10. *Four different hyperplanes H_1, H_2, H_3, H_4 belong to the same pencil and have cross ratio ρ if and only if there are equations $f = 0$ of H_1 and $g = 0$ of H_2 such that $f + g = 0$ is an equation of H_3 and $\rho f + g = 0$ is an equation of H_4*

Proof. H_1 and H_2 being different, they are independent and therefore span a pencil. By 4.4.8, if H_3 and H_4 have equations as claimed, then they belong to the pencil spanned by H_1 and H_2, and the value of the cross ratio is ρ by 2.9.9. The converse is direct from 2.9.9. □

We close this section by showing how a change of reference in $\mathbb{P}_n$ does affect hyperplane coordinates: we will see that the matrices used to change point coordinates in Section 2.3 may be used to change hyperplane coordinates in the opposite sense, provided they act by right multiplication on row hyperplane coordinate vectors. That is:

Proposition 4.4.11. *Let Δ and Ω be references of $\mathbb{P}_n$. If M is a matrix that changes coordinates relative to Ω into coordinates relative to Δ, then for any hyperplane $H = [u_0, \dots, u_n]_\Delta = [v_0, \dots, v_n]_\Omega$,*

$$\begin{pmatrix} v_0 & \dots & v_n \end{pmatrix} = \delta \begin{pmatrix} u_0 & \dots & u_n \end{pmatrix} M \tag{4.4}$$

for some non-zero $\delta \in k$.

Proof. If $p = [y_0, \dots, y_n]_\Omega$ is a point and $H = [u_0, \dots, u_n]_\Delta$ is a hyperplane, by the hypothesis,

$$M \begin{pmatrix} y_0 \\ \vdots \\ y_n \end{pmatrix}$$

are coordinates of p relative to Δ and so a necessary and sufficient condition for $p \in H$ is

$$\begin{pmatrix} u_0 & \dots & u_n \end{pmatrix} \left(M \begin{pmatrix} y_0 \\ \vdots \\ y_n \end{pmatrix} \right) = 0.$$

If this condition is rewritten

$$\left(\begin{pmatrix} u_0 & \dots & u_n \end{pmatrix} M \right) \begin{pmatrix} y_0 \\ \vdots \\ y_n \end{pmatrix} = 0,$$

it clearly is an equation of H relative to Ω, its coefficients are thus hyperplane coordinates of H relative to Ω and the claim follows. □

In order to change hyperplane coordinates relative to Ω into hyperplane coordinates relative to Δ using column vectors, the equality (4.4) may be rewritten

$$\delta \begin{pmatrix} u_0 \\ \vdots \\ u_n \end{pmatrix} = (M^t)^{-1} \begin{pmatrix} v_0 \\ \vdots \\ v_n \end{pmatrix}.$$

Then we have:

Corollary 4.4.12. *If M is a matrix that changes coordinates relative to Ω into coordinates relative to Δ, then $(M^t)^{-1}$ changes coordinates relative to $\Omega^\vee$ into coordinates relative to $\Delta^\vee$.*

4.5 The dual of a projectivity

Assume that $f : \mathbb{P}_n \to \overline{\mathbb{P}}_n$ is a projectivity between projective spaces $\mathbb{P}_n$, $\overline{\mathbb{P}}_n$. Since a projectivity transforms hyperplanes into hyperplanes (1.7.1), f induces a pair of mutually reciprocal maps,

$$\begin{aligned} \mathbb{P}_n^\vee &\longrightarrow \overline{\mathbb{P}}_n^\vee, \\ H &\longmapsto f(H), \end{aligned}$$

and

$$\begin{aligned} \overline{\mathbb{P}}_n^\vee &\longrightarrow \mathbb{P}_n^\vee, \\ \bar{H} &\longmapsto f^{-1}(\bar{H}). \end{aligned}$$

The main result in this section is that these maps are projectivities. We will fix our attention on the second map, as it is easier to describe. We have:

Theorem 4.5.1. *If $f : \mathbb{P}_n \to \overline{\mathbb{P}}_n$ is a projectivity, represented by a linear map φ, then the map*

$$\begin{aligned} f^\vee : \overline{\mathbb{P}}_n^\vee &\longrightarrow \mathbb{P}_n^\vee, \\ \bar{H} &\longmapsto f^{-1}(\bar{H}), \end{aligned}$$

is the projectivity represented by the dual map of φ.

Proof. Denote by E and $\bar{E}$ the vector spaces associated to $\mathbb{P}_n$ and $\overline{\mathbb{P}}_n$, respectively, and by $\varphi^\vee : \bar{E}^\vee \to E^\vee$ the dual map of φ. We claim $f^\vee = [\varphi^\vee]$. Assume that $\bar{H} = [\omega]$ is a hyperplane of $\overline{\mathbb{P}}_n^\vee$, and so $\omega \in \bar{E}^\vee - \{0\}$. A point $p = [e] \in \mathbb{P}_n$ belongs to $f^\vee(\bar{H}) = f^{-1}(\bar{H})$ if and only if $f(p) = [\varphi(e)] \in \bar{H}$, which is equivalent to

$$0 = \omega(\varphi(e)) = (\varphi^\vee(\omega))(e).$$

By 4.1.3, this shows that $f^{-1}(\bar{H})$ is the hyperplane of $\mathbb{P}_n$ represented by $\varphi^\vee(\omega)$, as wanted. $\square$

The projectivity $f^\vee$ is called the *dual* of f. Obviously, the first of the maps displayed above, $H \mapsto f(H)$, is the projectivity represented by $(\varphi^\vee)^{-1}$.

Proposition 4.5.2. *If a projectivity $f : \mathbb{P}_n \to \bar{\mathbb{P}}_n$ has matrix M relative to references Δ of $\mathbb{P}_n$ and $\bar{\Delta}$ of $\bar{\mathbb{P}}_n$, then its dual $f^\vee$ has matrix M^t relative to the references $\bar{\Delta}^\vee$ and $\Delta^\vee$.*

Note that in 4.5.2 we stick to the general conventions of Section 2.8 about matrices of projectivities. Thus, the above claim means that if (u) is a column coordinate vector of any hyperplane $\bar{H} \in \bar{\mathbb{P}}_n$, then $(v) = M^t(u)$ is a column coordinate vector of $\varphi^\vee(\bar{H})$. If computing with row coordinate vectors is preferred, then the equivalent equality $(v)^t = (u)^t M$ may be used.

Proof of 4.5.2. By 2.8.11, M is the matrix of φ relative to certain bases adapted to Δ and $\bar{\Delta}$. Then $\varphi^\vee$ has matrix M^t relative to the dual bases which, by 4.4.4 and again 2.8.11, shows that M^t is a matrix of $f^\vee$ relative to $\bar{\Delta}^\vee$, $\Delta^\vee$, as claimed.

Alternatively, the reader may find it interesting to repeat the argument of the proof of 4.5.1 using coordinate vectors instead of representatives: if (u) is a column coordinate vector of a hyperplane $\bar{H} \in \bar{\mathbb{P}}_n^\vee$, and (x) a column coordinate vector of a point $p \in \mathbb{P}_n$, then $M(x)$ is a column coordinate vector of $f(p)$ and thus p belongs to $f^\vee(\bar{H}) = f^{-1}(\bar{H})$ if and only if

$$0 = (u)^t(M(x)) = ((u)^t M)(x).$$

We see that the above is an equation of $f^\vee(\bar{H})$, $(u)^t M$ is thus one of the row coordinate vectors of $f^\vee(\bar{H})$ and therefore $M^t(u)$ is a column coordinate vector of $f^\vee(\bar{H})$. Then using 2.8.7 ends the proof. □

Proposition 4.5.3. *If $f : \mathbb{P}_n \to \bar{\mathbb{P}}_n$ is a projectivity and L a linear variety of $\mathbb{P}_n$, then $f(L)^* = (f^\vee)^{-1}(L^*)$.*

Proof. For any $\bar{H} \in \bar{\mathbb{P}}_n$,

$$\bar{H} \supset f(L) \Longleftrightarrow f^\vee(\bar{H}) = f^{-1}(\bar{H}) \supset L \Longleftrightarrow f^\vee(\bar{H}) \in L^* \Longleftrightarrow \bar{H} \in (f^\vee)^{-1}(L^*).$$

□

The following formal properties directly follow from the definition of the dual projectivity. The reader may also prove them by either using 4.5.1 and standard facts from linear algebra, or just 4.5.2:

Proposition 4.5.4. (a) *The dual of the identical map of $\mathbb{P}_n$ is the identical map of $\mathbb{P}_n^\vee$.*

(b) *If $f : \mathbb{P}_n \to \bar{\mathbb{P}}_n$ and $g : \bar{\mathbb{P}}_n \to \bar{\bar{\mathbb{P}}}_n$ are projectivities, then $(g \circ f)^\vee = f^\vee \circ g^\vee$.*

(c) *If f is a projectivity, then $(f^{-1})^\vee = (f^\vee)^{-1}$.*

4.6 Biduality

Fix a projective space $\mathbb{P}_n$. The dual projective space of any projective space being defined, we may in particular consider the dual space of the dual space of $\mathbb{P}_n$, $\mathbb{P}_n^{\vee\vee} = (\mathbb{P}_n^\vee)^\vee$, named the *bidual space* of $\mathbb{P}_n$. Its elements are the hyperplanes of $\mathbb{P}_n^\vee$, which we have seen to be in turn the bundles of hyperplanes of $\mathbb{P}_n$ with zero-dimensional kernel, that is, the bundles p^* for $p \in \mathbb{P}_n$. We have:

Theorem 4.6.1. *The map*

$$\begin{aligned}\Phi_{\mathbb{P}_n}\colon \mathbb{P}_n &\longrightarrow \mathbb{P}_n^{\vee\vee},\\ p &\longmapsto p^*,\end{aligned}$$

is a projectivity.

In the sequel, while dealing with a single $\mathbb{P}_n$, we will just write $\Phi_{\mathbb{P}_n} = \Phi$.

Proof of 4.6.1. Choose a projective reference Δ in $\mathbb{P}_n$, let $\Delta^\vee$ be the reference of $\mathbb{P}_n^\vee$ dual of Δ and $\Delta^{\vee\vee}$ the reference of $\mathbb{P}_n^{\vee\vee}$ dual of $\Delta^\vee$. Take any point $p = [x_0, \dots, x_n]_\Delta \in \mathbb{P}_n$. By 4.4.2, the hyperplane $H = [u_0, \dots, u_n]_{\Delta^\vee}$, of $\mathbb{P}_n$, goes through p if and only if

$$u_0x_0 + \dots + u_nx_n = 0 \tag{4.5}$$

and so, the u_i being taken as variables, this relation is an equation of the bundle p^* relative to the reference $\Delta^\vee$. Again by 4.4.2, this time used in $\mathbb{P}_n^\vee$, the coefficients $x_0, \dots, x_n$ of the equation (4.5) are the coordinates of p^* relative to $\Delta^{\vee\vee}$. Thus p and $p^* = \Phi(p)$ have the same coordinates, after which the claim follows from 2.8.1 or 2.8.5. □

Corollary 4.6.2 (of the proof of 4.6.1). *For any reference Δ of $\mathbb{P}_n$, $\Phi(\Delta) = \Delta^{\vee\vee}$.*

Proof. In the proof of 4.6.1 we have seen that $\Phi([x_0, \dots, x_n]_\Delta) = [x_0, \dots, x_n]_{\Delta^{\vee\vee}}$, for any $p = [x_0, \dots, x_n]_\Delta \in \mathbb{P}_n$, and we are just claiming the same equality for p ranging on the elements of Δ. □

Remark 4.6.3. It is worth noting that the above projectivity Φ is independent of any choice. The projective space $\mathbb{P}_n$ is usually identified with its bidual space $\mathbb{P}_n^{\vee\vee}$ through Φ. Then, as seen in the proof of 4.6.1, no matter which reference Δ is fixed in $\mathbb{P}_n$, the point $[x_0, \dots, x_n]_\Delta$ is identified with the hyperbundle of hyperplanes whose hyperplane coordinates relative to $\Delta^\vee$ are $x_0, \dots, x_n$.

After the above identification, the one-to-one map $*$ of Section 4.2 between linear varieties of $\mathbb{P}^\vee$ and $\mathbb{P}_n^{\vee\vee}$, maps each bundle to its kernel; in other words, it is just the inverse of the ordinary

$$*\colon \mathcal{LV}(\mathbb{P}_n) \longrightarrow \mathcal{LV}(\mathbb{P}_n^\vee),$$

as shown next:

Proposition 4.6.4. *For any linear variety L of $\mathbb{P}_n$, $(L^*)^* = \Phi(L)$.*

Proof. Take $p \in \mathbb{P}_n$. By 4.2.4, $p \in L$ if and only if $p^* \supset L^*$, but by the definition of $(L^*)^*$ this is just $\Phi(p) = p^* \in (L^*)^*$. □

The next proposition shows that when the projective spaces are identified with their respective biduals, any projectivity f becomes the dual of its dual $f^\vee$:

Proposition 4.6.5. *If $f : \mathbb{P}_n \to \overline{\mathbb{P}}_n$ is a projectivity, then for any $p \in \mathbb{P}_n$,*

$$f^{\vee\vee}(p^*) = f(p)^*.$$

Proof. By the definition of the dual projectivity, $f^{\vee\vee}(p^*) = (f^\vee)^{-1}(p^*)$, which, by 4.5.3, equals $f(p)^*$. □

Remark 4.6.6. The claim of 4.6.5 above just says that the diagram

$$\begin{array}{ccc} \mathbb{P}_n & \xrightarrow{\ f\ } & \overline{\mathbb{P}}_n \\ {\scriptstyle \Phi_{\mathbb{P}_n}}\downarrow & & \downarrow{\scriptstyle \Phi_{\overline{\mathbb{P}}_n}} \\ \mathbb{P}_n^{\vee\vee} & \xrightarrow{\ f^{\vee\vee}\ } & \overline{\mathbb{P}}_n^{\vee\vee} \end{array}$$

commutes.

The last result in this section allows us to select a reference Δ of $\mathbb{P}_n$ by the choice of the elements of its dual $\Delta^\vee$. In other words, a reference Δ of $\mathbb{P}_n$ is well determined by giving its $(n-1)$-dimensional faces $H_0, \dots, H_n$ and the hyperplane H_{n+1} with equation $\sum_{i=0}^n x_i = 0$ relative to Δ, provided of course that the independence conditions

$$\bigcap_{\substack{0 \le i \le n+1 \\ i \ne j}} H_i = \emptyset \quad \text{for } j = 0, \dots, n+1$$

are satisfied.

Proposition 4.6.7. *The map $\Delta \mapsto \Delta^\vee$ is a bijection between the sets of references of $\mathbb{P}_n$ and $\mathbb{P}_n^\vee$, respectively.*

Proof. By 4.6.2, $\Phi(\Delta) = \Delta^{\vee\vee}$, which determines Δ from $\Delta^\vee$ by the rule $\Delta = \Phi^{-1}((\Delta^\vee)^\vee)$ and therefore proves the injectivity. For the exhaustivity, if Ω is any reference of $\mathbb{P}_n^\vee$, take $\Delta = \Phi^{-1}(\Omega^\vee)$. Then $\Omega^\vee = \Phi(\Delta) = \Delta^{\vee\vee}$ by 4.6.2. Since we have already proved that the references of an arbitrary projective space are injectively mapped to their duals, we use this fact for the references of $\mathbb{P}_n^\vee$ getting $\Omega = \Delta^\vee$, which completes the proof. □

4.7 Duals of linear varieties and bunches

Fix a linear variety $L \subset \mathbb{P}_n$ of dimension $d > 0$. Since L itself is a projective space, it has a dual $L^\vee$ whose elements are the $(d-1)$-dimensional linear varieties of $\mathbb{P}_n$ contained in L. The reader may note that $L^\vee$ is quite different from the bundle of hyperplanes L^*. We will show next that the dual $L^\vee$ of L has a nice interpretation in terms of hyperplanes of $\mathbb{P}_n$, as it may be identified with the bunch $(L^*)^\nabla$ of the dual space $\mathbb{P}_n^\vee$.

Proposition 4.7.1. *If $L \subset \mathbb{P}_n$ is a linear variety of dimension $d > 0$, then mapping each linear variety T to the bundle T^* gives a projectivity*

$$\sigma : L^\vee \longrightarrow (L^*)^\nabla.$$

Proof. First we see that defining σ as above makes sense. Indeed if $T \in L^\vee$, then $\dim T = d - 1$ and so $\dim T^* = n - d = \dim L^* + 1$. Furthermore, $T \subset L$ assures that $T^* \supset L^*$. Once we know that σ is well defined, we use a little of linear algebra. Assume $L = [F]$. Then restricting linear forms to F is an exhaustive linear map

$$E^\vee \longrightarrow F^\vee$$

which has kernel $F^\perp$ and therefore induces an isomorphism

$$\begin{aligned} E^\vee / F^\perp &\longrightarrow F^\vee, \\ \bar{\omega} &\longmapsto \omega_{|F}, \end{aligned}$$

$\bar{\omega}$ the class of ω modulo $F^\perp$. We will see that this isomorphism is a representative of σ^{-1}. Assume that $\omega \in E^\vee - F^\perp$. Then ω represents a hyperplane $H = [G] \not\supset L$ and so $H \cap L$ is a hyperplane of L. Also the restriction $\omega_{|F} \in F^\vee$ is a non-zero form that takes value zero on the representatives of all points in $H \cap L$. Thus $\omega_{|F}$ represents the hyperplane $H \cap L$ of L. On the other hand, by the definition of the projective structure in a bunch (see Section 1.9) the non-zero class $\bar{\omega}$ represents the bundle

$$[\langle\omega\rangle + F^\perp] = [G^\perp + F^\perp] = [(G \cap F)^\perp] = [(G \cap F)]^* = (H \cap L)^*$$

as wanted. $\square$

In 1.9.4 we have obtained bijections mapping the linear varieties of a bunch L^∇ of $\mathbb{P}_n$ to linear varieties of $\mathbb{P}_n$ of the same codimension. We can see now that from these bijections, the one acting on hyperplanes is a projectivity:

Proposition 4.7.2. *If $L \subset \mathbb{P}_n$ is a linear variety of dimension $d < n-1$, then the map*

$$\begin{aligned}\delta\colon L^* &\longrightarrow (L^\nabla)^\vee,\\ S &\longmapsto \check{S} = \{L' \in L^\nabla \mid L' \subset S\},\end{aligned}$$

is a projectivity.

Proof. Assume that $L = [F]$. It is an easy fact from linear algebra that composing the forms $\omega \in (E/F)^\vee$ with the natural morphism

$$\begin{aligned}\rho\colon E &\longrightarrow E/F,\\ v &\longmapsto \bar{v},\end{aligned}$$

gives an isomorphism

$$\begin{aligned}\rho^\vee\colon (E/F)^\vee &\to F^\perp,\\ \omega &\longmapsto \rho \circ \omega.\end{aligned}$$

We will prove that $\delta = [(\rho^\vee)^{-1}]$. Take hyperplanes represented by forms that correspond by $\rho^\vee$, namely $S = [\omega \circ \rho]$ and $[\omega]$, $\omega \in (E/F)^\vee - \{0\}$. Then $p = [v] \in S - L$ if and only if $\omega(\bar{v}) = 0$ and $\bar{v} \neq 0$, which in turn is equivalent to having $p \notin L$ and $p \vee L = [\bar{v}] \in [\omega]$. This shows that $[\omega] = \check{S}$, as wanted. □

4.8 Exercises

Exercise 4.1. Write the dual of 4.2.7 and prove 4.2.10 after it.

Exercise 4.2. Prove that the four planes of $\mathbb{P}_3$ with equations $x_0 + 2x_1 + 2x_4 = 0$, $x_0 - x_1 - x_3 + x_4 = 0$, $x_0 + 5x_1 + x_3 + x_4 = 0$, $3x_0 - 2x_3 + 3x_4 = 0$ belong to a pencil. Compute their cross ratio.

Exercise 4.3. Prove that the unit hyperplane of the dual reference $\Delta^\vee$ of a projective reference Δ is the harmonic polar (see Exercise 2.16) of the unit point of Δ relative to the simplex of Δ.

Exercise 4.4. Describe a construction, using a straight edge, allowing us to draw the fourth harmonic of three concurrent lines on a plane

Exercise 4.5. Let H, H' be two distinct hyperplanes of $\mathbb{P}_n$ and p a point not in H or H'. Take $L = H \cap H'$, $T = p \vee L$ and T' the fourth harmonic (in L^*) of T with respect to H, H'. Prove that p and a point $q \neq p$ harmonically divide the intersections of pq with H and H' if and only if $q \in T' - L$.

Exercise 4.6. Assume given two parallel and distinct lines ℓ, ℓ' of an affine plane. Prove that the midpoints of all pairs p, p', $p \in \ell$, $p' \in \ell'$, lie on a line which is the fourth harmonic of the improper line with respect to ℓ, ℓ'.

Exercise 4.7. Assume we have in $\mathbb{P}_2$ a triangle T with vertices A, B, C, a fourth point D lying on no side of T and a line ℓ through D missing all vertices of T. Denote by A', B' and C' the intersections of ℓ and the sides opposite A, B and C, respectively, and prove that

$$(AD, BD, CD, \ell) = (A', B', C', D).$$

Exercise 4.8. Let T be a tetrahedron of $\mathbb{P}_3$ and ℓ a line containing no vertex of T. Prove that the cross ratio of the four points in which ℓ intersects the faces of T equals the cross ratio of the four planes projecting ℓ from the vertices of T, taken in a suitable order.

Exercise 4.9. Write down the duals of Exercises 1.7, 1.11, 1.14, 1.17 and 1.31.

Chapter 5
Projective transformations

In this chapter we will study projectivities $f: \mathbb{P}_n \to \mathbb{P}'_n$ in cases in which there is some relationship between the source and target spaces: we will consider the cases $\mathbb{P}'_n = \mathbb{P}_n$ (collineations of $\mathbb{P}_n$), $\mathbb{P}'_n = \mathbb{P}_n^\vee$ (correlations of $\mathbb{P}_n$), and also cases in which both the source and the target are linear varieties of the same projective space. The one-dimensional cases will be given a separate treatment by elementary methods going further at some points than in the general case. The more technical projective classifications of collineations and correlations will be delayed to Chapter 11. A final section deals with singular projectivities: they are correspondences given by non-invertible linear equations and include the projections from linear varieties, already introduced in Section 1.9.

Since the study of the projectivities of a projective space $\mathbb{P}_{n,k}$ involves algebraic non-linear equations, assuming that the base field k is $\mathbb{R}$ or $\mathbb{C}$ becomes now essential. In the case $k = \mathbb{R}$, it is useful to handle points with imaginary coordinates: the first section of this chapter is devoted to support this.

5.1 Complex extension of a real projective space

In this section we will explain how any real projective space may be intrinsically enlarged to a complex projective space, after which we will be allowed to consider new points with complex, non-real, coordinates. These points, called *imaginary*, are a useful tool in the development of the real projective geometry, as they provide solutions to real algebraic equations that otherwise would have none. It is maybe worth noting that the contents of this section strictly belongs to real projective geometry, a similar procedure for complex projective spaces having of course no sense. In the sequel we will denote $\bar{z} = a - bi$ the complex-conjugate of a complex number $z = a + bi$ and make frequent use, without further mention, of the basic properties of the complex conjugation map $z \mapsto \bar{z}$, namely that it is an automorphism of the field $\mathbb{C}$, that $\bar{\bar{z}} = z$ for any $z \in \mathbb{C}$ and that $z = \bar{z}$ if and only if $z \in \mathbb{R}$. The absolute value of the complex number z will be written $|z| = \sqrt{z\bar{z}}$.

We first enlarge real vector spaces to complex vector spaces. Let E be a real vector space of (finite) dimension $n+1$. Consider the product $E \times E$ with its usual structure of real vector space, its addition and external product being given by the rules

$$(v_1, w_1) + (v_2, w_2) = (v_1 + v_2, w_1 + w_2),$$
$$a(v, w) = (av, aw),$$

$v, v_1, v_2, w, w_1, w_2 \in E$, $a \in \mathbb{R}$.

Define an external complex product by the rule

$$(a + b\boldsymbol{i})(v, w) = a(v, w) + b(-w, v),$$

for any $(v, w) \in E \times E$, $a + b\boldsymbol{i} \in \mathbb{C}$. The reader may easily check that with this product, the additive group of $E \times E$ is turned into a complex vector space that we will denote by $\mathbb{C}E$ and name the *complexification* (or *complexified space*) of E. The reader may also remark that for a real scalar a, both external products give the same result: $(a + 0\boldsymbol{i})(v, w) = a(v, w)$. In other words, the complex external product extends the real external product we already have.

In the sequel $\mathbb{C}E$ will be taken as a complex vector space unless otherwise is said. Nevertheless, since (as any complex vector space) it is also a real vector space by just restricting the external product to the real numbers, we will use the prefixes $\mathbb{R}$- and $\mathbb{C}$- to indicate which structure a notion is relative to.

Obviously $E \times \{0\}$ is an $\mathbb{R}$-subspace of $\mathbb{C}E$ and the map

$$\begin{aligned} \tau\colon E &\longrightarrow E \times \{0\}, \\ v &\longmapsto (v, 0), \end{aligned}$$

is an isomorphism (of real vector spaces). In the sequel we will identify E and $E \times \{0\}$ by means of this isomorphism and so we will write $v = (v, 0)$ for all $v \in E$. This allows us to rewrite any element of $\mathbb{C}E$ as

$$(v, w) = (v, 0) + (0, w) = (v, 0) + \boldsymbol{i}(w, 0) = v + \boldsymbol{i}w$$

and so the operating rules in the far more pleasant form

$$\begin{aligned} (v_1 + \boldsymbol{i}w_1) + (v_2 + \boldsymbol{i}w_2) &= (v_1 + v_2) + \boldsymbol{i}(w_1 + w_2), \\ (a + b\boldsymbol{i})(v + \boldsymbol{i}w) &= av - bw + \boldsymbol{i}(aw + bv). \end{aligned}$$

It is also worth recalling that $u = v + \boldsymbol{i}w = (v, w)$ uniquely determines v and w, which are called, respectively, the real and imaginary part of u. In particular $v + \boldsymbol{i}w = 0$ if and only if $v = w = 0$. The vectors of $\mathbb{C}E$ with 0 as imaginary part are those already identified with the vectors of E: they are called the *real* vectors of $\mathbb{C}E$, while the remaining vectors are called *imaginary*.

We will make use of the following easy lemmas; the proof of the second one is left to the reader:

Lemma 5.1.1. (a) *If vectors $e_0, \dots, e_m \in E$ are independent, then they are $\mathbb{C}$-independent in $\mathbb{C}E$.*

(b) *If vectors $e_0, \dots, e_m$ generate E, then they are $\mathbb{C}$-generators of $\mathbb{C}E$.*

(c) *If $e_0, \dots, e_n$ is a basis of E, then it is also a $\mathbb{C}$-basis of $\mathbb{C}E$ and the components of any $v \in E$ relative to $e_0, \dots, e_n$ are the same, no matter if v is considered as a vector of E or as a vector of $\mathbb{C}E$. In particular the $\mathbb{C}$-dimension of $\mathbb{C}E$ is $n + 1 = \dim E$.*

Proof. For claim (a) just note that an equality

$$(a_0 + b_0 \boldsymbol{i})e_0 + \cdots + (a_m + b_m \boldsymbol{i})e_m = 0$$

may be written

$$a_0 e_0 + \cdots + a_m e_m + \boldsymbol{i}(b_0 e_0 + \cdots + b_m e_m) = 0,$$

thus giving

$$\begin{aligned} a_0 e_0 + \cdots + a_m e_m &= 0, \\ b_0 e_0 + \cdots + b_m e_m &= 0, \end{aligned}$$

and so, by the independence of $e_0, \dots, e_m$,

$$a_0 = \cdots = a_m = b_0 = \cdots = b_m = 0.$$

For claim (b) take any $v + \boldsymbol{i}w \in \mathbb{C}E$ and, by the hypothesis, write

$$\begin{aligned} v &= a_0 e_0 + \cdots + a_m e_m, \\ w &= b_0 e_0 + \cdots + b_m e_m \end{aligned}$$

to get

$$v + \boldsymbol{i}w = (a_0 + b_0 \boldsymbol{i})e_0 + \cdots + (a_m + b_m \boldsymbol{i})e_m.$$

The last claim is now clear. □

Lemma 5.1.2. (a) *If*

$$\varphi\colon E \to E'$$

is a linear map between real vector spaces, then it has a unique extension to a $\mathbb{C}$*-linear map*

$$\varphi_{\mathbb{C}}\colon \mathbb{C}E \longrightarrow \mathbb{C}E',$$

and such an extension $\varphi_{\mathbb{C}}$ *satisfies*

$$\varphi_{\mathbb{C}}(v + \boldsymbol{i}w) = \varphi(v) + \boldsymbol{i}\varphi(w)$$

for any $v + \boldsymbol{i}w \in \mathbb{C}E$.

(b) $(\mathrm{Id}_E)_{\mathbb{C}} = \mathrm{Id}_{\mathbb{C}E}$.

(c) *If* $\psi\colon E' \to E''$ *is a second linear map between real vector spaces, then* $(\psi \circ \varphi)_{\mathbb{C}} = \psi_{\mathbb{C}} \circ \varphi_{\mathbb{C}}$.

(d) *If* φ *is an isomorphism, then* $\varphi_{\mathbb{C}}$ *is a* $\mathbb{C}$*-isomorphism and* $(\varphi_{\mathbb{C}})^{-1} = (\varphi^{-1})_{\mathbb{C}}$.

The vector $\overline{v + \boldsymbol{i}\, w} = v - \boldsymbol{i}\, w$ is called the *conjugate* of $v + \boldsymbol{i}\, w$. Mapping each vector to its conjugate defines a map

$$\begin{aligned} \mathbb{C}E &\longrightarrow \mathbb{C}E, \\ e = v + \boldsymbol{i}\, w &\longmapsto \bar{e} = v - \boldsymbol{i}\, w, \end{aligned}$$

called the *conjugation* of vectors. Its main properties are summarized in the lemma below, whose easy proof is also left to the reader:

Lemma 5.1.3. *The following hold for any $e, e' \in \mathbb{C}E$ and any $z \in \mathbb{C}$:*

(a) $\overline{e + e'} = \bar{e} + \bar{e}'$,

(b) $\overline{ze} = \bar{z}\bar{e}$,

(c) *e is real if and only if $\bar{e} = e$,*

(d) *$\bar{\bar{e}} = e$, after which*

(e) *$e = 0$ if and only if $\bar{e} = 0$, and*

(f) *the conjugation of vectors is bijective.*

The reader may note that the conjugation of vectors is not a $\mathbb{C}$-linear map, due to the property (b) above.

Until the end of this section, $\mathbb{P}_n$ will be a real projective space of dimension n with associated vector space E. Define the *complex extension* (or *complexification*) of $\mathbb{P}_n$ as $\mathbb{C}\mathbb{P}_n = \mathbb{P}(\mathbb{C}E)$, the projectivization of the complexification of E. Since proportional non-zero vectors $v, w \in E$ are $\mathbb{C}$-proportional in $\mathbb{C}E$, mapping the point $[v] \in \mathbb{P}_n$ to the point of $\mathbb{C}\mathbb{P}_n$ represented by the same vector v (this time viewed as an element of $\mathbb{C}E$) is a well-defined map

$$\begin{aligned} \iota\colon \mathbb{P}_n &\longrightarrow \mathbb{C}\mathbb{P}_n, \\ [v] &\longmapsto [v], \end{aligned} \tag{5.1}$$

that is injective by 5.1.1.a. From now on we will identify p and $\iota(p)$ for all $p \in \mathbb{P}_n$, which in particular unifies the two different meanings of $[v]$ in (5.1). This turns the real projective space $\mathbb{P}_n$ into a subset of the complex projective space $\mathbb{C}\mathbb{P}_n$: we will often refer to the points of $\mathbb{P}_n$ as *real points*, and call those of $\mathbb{C}\mathbb{P}_n - \mathbb{P}_n$ the *imaginary points* of $\mathbb{P}_n$ (even if they do not belong to $\mathbb{P}_n$). The real points are thus the points of $\mathbb{C}\mathbb{P}_n$ that have a real representative, but the reader may note that a real point also has imaginary representatives: for instance $[v] = [\boldsymbol{i}\, v]$ for any non-zero real vector v. Thus, the imaginary points are not those with an imaginary representative. By 1.6.1 (c) both $\mathbb{P}_n$ and $\mathbb{C}\mathbb{P}_n$ have dimension n; $\mathbb{C}\mathbb{P}_n$ is larger than $\mathbb{P}_n$ because it is a projective space over a larger field, not because it has higher dimension.

The properties (b) and (e) in 5.1.3 assure that the conjugation of vectors induces a well-defined map

$$\begin{aligned}\sigma\colon \mathbb{CP}_n &\longrightarrow \mathbb{CP}_n,\\ [e] &\longmapsto [\bar{e}],\end{aligned}$$

called *complex conjugation*, or just *conjugation* when no confusion with the conjugation relative to a quadric (see Chapter 6) may arise. The point $\sigma(p)$ will be called the *complex-conjugate* (or just *conjugate*) of p. Clearly from 5.1.3 (d) $\sigma^2 = \mathrm{Id}_{\mathbb{CP}_n}$, which in particular implies that σ is bijective. Real points are easily characterized using conjugation:

Proposition 5.1.4. *A point $p \in \mathbb{CP}_n$ is real if and only if $\sigma(p) = p$.*

Proof. The *only if* part is clear from 5.1.3 (c). For the converse, assume that $p = [e] = \sigma(p) = [\bar{e}]$. If e is real we are done. Otherwise $e - \bar{e} \neq 0$ and since both vectors represent p, also $p = [\boldsymbol{i}\,(e - \bar{e})]$. Now

$$\overline{\boldsymbol{i}\,(e-\bar{e})} = (-\boldsymbol{i}\,)(\bar{e} - e) = \boldsymbol{i}\,(e - \bar{e}).$$

This proves that $\boldsymbol{i}\,(e - \bar{e})$ is real, and therefore so is $p = [\boldsymbol{i}\,(e - \bar{e})]$. □

The next proposition describes the effect of the complex conjugation on the cross ratio. We agree in taking ∞ as the complex-conjugate of itself: $\overline{\infty} = \infty$.

Proposition 5.1.5. *If $q_1, q_2, q_3, q_4 \in \mathbb{CP}_n$ are aligned and three at least different, then also $\sigma(q_1)$, $\sigma(q_2)$, $\sigma(q_3)$, $\sigma(q_4)$ are aligned and three at least different, and their cross ratio is*

$$(\sigma(q_1), \sigma(q_2), \sigma(q_3), \sigma(q_4)) = \overline{(q_1, q_2, q_3, q_4)}.$$

Proof. If two points are coincident, then the claim follows from 2.9.10. Otherwise, take

$$\rho = (q_1, q_2, q_3, q_4) \in \mathbb{C} - \{0, 1\}.$$

Then, by 2.9.9 there are vectors e, e' such that

$$q_1 = [e], \quad q_2 = [e'], \quad q_3 = [e + e'], \quad q_4 = [\rho e + e'].$$

By taking conjugates,

$$\sigma(q_1) = [\bar{e}], \quad \sigma(q_2) = [\bar{e}'], \quad \sigma(q_3) = [\bar{e}+\bar{e}'], \quad \sigma(q_4) = [\overline{\rho e + e'}] = [\bar{\rho}\bar{e}+\bar{e}'],$$

where $\bar{\rho} \neq 0, 1$. Then it is clear that $\sigma(q_1)$, $\sigma(q_2)$, $\sigma(q_3)$, $\sigma(q_4)$ are four aligned and different points, and furthermore, using again 2.9.9, also that

$$\bar{\rho} = (\sigma(q_1), \sigma(q_2), \sigma(q_3), \sigma(q_4)),$$

as claimed. □

Proposition 5.1.6. (a) *If $(p_0, \dots, p_n, A)$ is a reference of $\mathbb{P}_n$, then the same points, taken as points of $\mathbb{C}\mathbb{P}_n$, compose a reference Δ' of $\mathbb{C}\mathbb{P}_n$.*

(b) *If a point $p \in \mathbb{P}_n$ has coordinates $x_0, \dots, x_n$ relative to Δ, then the same point p has, in $\mathbb{C}\mathbb{P}_n$, coordinates $x_0, \dots, x_n$ relative to Δ'.*

Proof. Part (a) is a direct consequence of 5.1.1 (a) and 1.5.12. To prove part (b), just note that if $e_0, \dots, e_n$ is a basis adapted to Δ, then it is also a basis adapted to Δ'. If $p = [v]$ and $v = x_0 e_0 + \dots + x_n e_n$ in E, then the same equalities hold in $\mathbb{C}\mathbb{P}_n$ and $\mathbb{C}E$ respectively. □

Remark 5.1.7. In the sequel we will make no distinction between Δ and Δ' above and any projective reference of $\mathbb{P}_n$ will be taken as being a reference of $\mathbb{C}\mathbb{P}_n$ too. These references will be called *real references* (of $\mathbb{C}\mathbb{P}_n$), to distinguish them from the remaining references of $\mathbb{C}\mathbb{P}_n$. Note that a real reference always has a real adapted basis.

Remark 5.1.8. Since, by 2.8.1 (a), a projectivity is determined by the images of the elements of a reference, the existence of real references assures that a projectivity defined on $\mathbb{C}\mathbb{P}_n$ is determined by the images of the real points.

As the reader may see, beyond the algebraic formalism needed to intrinsically define the complexification, the situation is quite simple: once a reference of $\mathbb{P}_n$ has been fixed, we are allowed to consider points with arbitrary, not all zero, complex coordinates: these points describe the larger complex projective space $\mathbb{C}\mathbb{P}_n$, still of dimension n, and of course include the points $\mathbb{P}_n$, which have real coordinates. It is worth noting that, relative to a real reference, the real points have imaginary coordinates too, as the arbitrary constant up to which the homogeneous coordinates in $\mathbb{C}\mathbb{P}_n$ are defined, may be given imaginary values: if $p = [a_0, \dots, a_n]$, $a_i \in \mathbb{R}$, $i = 1, \dots, n$, then also $p = [\boldsymbol{i}\, a_0, \dots, \boldsymbol{i}\, a_n]$. Proposition 5.1.10 below characterizes the real points in terms of their coordinates.

Proposition 5.1.9. *If Δ is a real reference of $\mathbb{C}\mathbb{P}_n$ and $p = [z_0, \dots, z_n]_\Delta \in \mathbb{C}\mathbb{P}_n$, then $\sigma(p) = [\bar{z}_0, \dots, \bar{z}_n]_\Delta$.*

Proof. If $e_0, \dots, e_n$ is a real basis adapted to Δ and we write $z_j = x_j + \boldsymbol{i}\, y_j$, $j = 0, \dots, n$, a representative of p is

$$\sum_{j=0}^{n} z_j e_j = \sum_{j=0}^{n} (x_j + \boldsymbol{i}\, y_j) e_j = \sum_{j=0}^{n} x_j e_j + \boldsymbol{i} \Big(\sum_{j=0}^{n} y_j \Big) e_j$$

and therefore

$$\sum_{j=0}^{n} x_j e_j - \boldsymbol{i} \Big(\sum_{j=0}^{n} y_j \Big) e_j = \sum_{j=0}^{n} (x_j - \boldsymbol{i}\, y_j) e_j = \sum_{j=0}^{n} \bar{z}_j e_j$$

represents $\sigma(p)$. □

Proposition 5.1.10. *Assume that a real reference Δ of $\mathbb{C}\mathbb{P}_n$ has been fixed, and that $p = [z_0, \dots, z_n]_\Delta \in \mathbb{C}\mathbb{P}_n$ has $z_j \neq 0$ for a certain j. Then p is real if and only if all ratios z_r/z_j, $r \neq j$, are real.*

Proof. For simplicity of notation we make the proof for $j = 0$, the other cases being dealt with similarly. Since $p = [1, z_1/z_0, \dots, z_n/z_0]$, one has $\sigma(p) = [1, \bar{z}_1/\bar{z}_0, \dots, \bar{z}_n/\bar{z}_0]$ by 5.1.9. Then 5.1.4 assures that p is real if and only if $z_r/z_0 = \bar{z}_r/\bar{z}_0$, $r = 1, \dots, n$, from which the claim. □

Proposition 5.1.11. (a) *Points $q_0, \dots, q_r \in \mathbb{P}_n$ span in $\mathbb{P}_n$ and $\mathbb{C}\mathbb{P}_n$ linear varieties of the same dimension, in particular they are independent (or dependent) in one space if and only if they are so in the other.*

(b) *By claim* (a)*, four points of $\mathbb{P}_n$, three at least different, are aligned in $\mathbb{P}_n$ if and only if they are aligned in $\mathbb{C}\mathbb{P}_n$. Then, if this is the case, their cross ratio is the same in both spaces.*

Proof. After fixing a real reference, by 5.1.6, the coordinates of the points may be taken to be the same in both spaces. Selecting from $\{q_0, \dots, q_r\}$ a maximal subset of independent points, such as those involved in a maximal non-zero minor of the matrix of the coordinates of the points, may be made at once in both spaces with the same result. Then the first claim follows, as by 1.5.11 the dimension of the variety spanned by $q_0, \dots, q_r$ in either space is one less than the number of independent points selected. The second claim follows similarly: the case of two points being coincident is obvious and the procedure for computing the cross ratio of four different points after 2.9.9 uses only the coordinates of the points and may be performed equally in both spaces. □

The complex-conjugate of a linear equation

$$\ell = (a_0 + \boldsymbol{i}\, b_0) z_0 + \cdots + (a_n + \boldsymbol{i}\, b_n) z_n = 0,$$

with variables $z_0, \dots, z_n$, will be taken to be

$$\bar{\ell} = (a_0 - \boldsymbol{i}\, b_0) z_0 + \cdots + (a_n - \boldsymbol{i}\, b_n) z_n = 0.$$

Clearly $\bar{\bar{\ell}} = \ell$ and $\ell = \bar{\ell}$ if and only if ℓ is real (that is, has real coefficients).

The rank of a matrix is unaffected by an extension of the base field, because it may be computed using determinants. In particular, real linear equations are linearly independent over $\mathbb{R}$ if and only if, viewed as complex equations, they are linearly independent over $\mathbb{C}$. In the sequel we will make use of this fact without further mention, referring only to independent equations and making no reference to the base field.

Proposition 5.1.12. *Assume that a real reference of $\mathbb{CP}_n$ has been fixed. If T is a linear variety of dimension d of $\mathbb{CP}_n$ and has equations*

$$\ell_1 = 0, \ \ldots, \ \ell_{n-d} = 0,$$

then $\sigma(T)$ is also a linear variety of dimension d of $\mathbb{CP}_n$ and it has equations

$$\bar{\ell}_1 = 0, \ \ldots, \ \bar{\ell}_{n-d} = 0.$$

Proof. By 5.1.9 and the properties of complex conjugation, the coordinates of a point p satisfy an equation $\ell_j = 0$ if and only if the coordinates of $\sigma(p)$ satisfy $\bar{\ell}_j = 0$. On the other hand, the equations $\bar{\ell}_j = 0$, $j = 1, \ldots, n-d$ are linearly independent, as otherwise a non-trivial dependence relation $\sum_{j=1}^{n-d} a_j \bar{\ell}_j = 0$ would give rise by conjugation to $\sum_{j=1}^{n-d} \bar{a}_j \ell_j = 0$ against the linear independence of the $\ell_j = 0$. This proves that $\sigma(T)$ is the linear variety with equations $\bar{\ell}_j = 0$, $j = 1, \ldots, n-d$, and hence the claim. □

The linear variety $\sigma(T)$ is called the *complex-conjugate* (or just the *conjugate*) of T.

Proposition 5.1.13. *If T is a linear variety of $\mathbb{CP}_n$, then the set of real points of T, $\mathbb{P}_n \cap T$, is a linear variety of $\mathbb{P}_n$ and*

$$2 \dim T - n \leq \dim \mathbb{P}_n \cap T \leq \dim T.$$

Proof. Fix a real reference. The real points being those having real coordinates, the real points of T are those which have real coordinates satisfying a system of equations of T, say

$$\ell_1 = 0, \ \ldots, \ \ell_{n-d} = 0, \tag{5.2}$$

$d = \dim T$. The coordinates being real, they also satisfy the conjugate equations, and hence $p \in \mathbb{P}_n \cap T$ if and only if p has real coordinates and these coordinates satisfy

$$\begin{aligned} \ell_1 = 0, \ \ldots, \ \ell_{n-d} = 0, \\ \bar{\ell}_1 = 0, \ \ldots, \ \bar{\ell}_{n-d} = 0. \end{aligned} \tag{5.3}$$

If r is the rank of this system of equations, then $n - d \leq r \leq 2(n-d)$, due to the independence of the equations of T. Obviously the equations

$$\begin{aligned} \bar{\ell}_1 + \ell_1 = 0, \ \ldots, \ \bar{\ell}_{n-d} + \ell_{n-d} = 0, \\ i(\bar{\ell}_1 - \ell_1) = 0, \ \ldots, \ i(\bar{\ell}_{n-d} - \ell_{n-d}) = 0, \end{aligned} \tag{5.4}$$

which are the real and imaginary parts of the equations (5.2), have real coefficients and compose a system of equations equivalent to (5.3); in particular it also has rank

r. It is enough to extract from the equations (5.4) a maximal subset of independent equations to get an equivalent system of real independent equations

$$g_1 = 0, \ \dots, \ g_r = 0$$

whose real solutions are the coordinates in $\mathbb{P}_n$ of the points of $\mathbb{P}_n \cap T$. Thus $P_n \cap T$ is a linear variety of $\mathbb{P}_n$ of dimension $n-r$. To close, the above inequalities $n - d \leq r \leq 2(n-d)$ yield the claimed ones. □

Remark 5.1.14. The proof of 5.1.13 shows that a maximal set of independent equations extracted from the real and imaginary parts of the equations of T is a system of equations of $\mathbb{P}_n \cap T$.

We will pay now some attention to the linear varieties T of $\mathbb{C}\mathbb{P}_n$ for which $\dim \mathbb{P}_n \cap T$ reaches its maximum $\dim T$.

Proposition 5.1.15. *For a linear variety T of $\mathbb{C}\mathbb{P}_n$, the following conditions are equivalent:*

(i) $T = \sigma(T)$,

(ii) *T has a system of real equations,*

(iii) $\dim T = \dim \mathbb{P}_n \cap T$.

Proof. (i) ⇒ (ii): Let

$$\ell_1 = 0, \ \dots, \ \ell_{n-d} = 0$$

be a system of equations of T, $d = \dim T$. By 5.1.12, their conjugates compose a system of equations of $\sigma(T)$ and therefore a maximal set of independent equations extracted from

$$\begin{aligned} \ell_1 = 0, \ \dots, \ \ell_{n-d} = 0, \\ \bar{\ell}_1 = 0, \ \dots, \ \bar{\ell}_{n-d} = 0 \end{aligned} \tag{5.5}$$

is a system of equations of $T \cap \sigma(T) = T$. As in the proof of 5.1.13, the equations

$$\begin{aligned} \ell_1 + \ell_1 = 0, \ \dots, \ell_{n-d} + \ell_{n-d} = 0, \\ i(\bar{\ell}_1 - \ell_1) = 0, \ \dots, \ i(\bar{\ell}_{n-d} - \ell_{n-d}) = 0 \end{aligned} \tag{5.6}$$

are real and have the same solutions as the system (5.5) above. Thus also a maximal set of independent equations extracted from (5.6) is a system of equations of T, and these are real.

(ii) ⇒ (iii): By 5.1.14, a system of real equations of T is also a system of equations of $\mathbb{P}_n \cap T$, hence the equality of dimensions.

(iii) ⇒ (i): The real points being invariant by σ, $\mathbb{P}_n \cap T = \mathbb{P}_n \cap (T \cap \sigma(T))$. Therefore, by 5.1.13,

$$\dim T \cap \sigma(T) \geq \dim \mathbb{P}_n \cap (T \cap \sigma(T)) = \dim \mathbb{P}_n \cap T = \dim T$$

and so $T = \sigma(T)$. □

The linear varieties of $\mathbb{C}\mathbb{P}_n$ satisfying the equivalent conditions of 5.1.15 are called *real*.

Corollary 5.1.16. *For any linear variety T of $\mathbb{C}\mathbb{P}_n$, $T \cap \sigma(T)$ is real and its real points are the real points of T.*

Proof. The first claim is clear after the condition (i) of 5.1.15, while the equality $\mathbb{P}_n \cap T = \mathbb{P}_n \cap (T \cap \sigma(T))$ follows from the invariance of the real points by σ, as already said in the proof of 5.1.15. □

The following proposition shows in particular that taking traces on $\mathbb{P}_n$ defines a dimension-preserving one-to-one map between the set of the real linear varieties of $\mathbb{C}\mathbb{P}_n$ and the set of the linear varieties of $\mathbb{P}_n$.

Proposition 5.1.17. *For any linear variety L of $\mathbb{P}_n$ there is a unique linear variety $L_{\mathbb{C}}$ of $\mathbb{C}\mathbb{P}_n$ such that $L \subset L_{\mathbb{C}}$ and* $\dim L = \dim L_{\mathbb{C}}$. *Furthermore:*

(a) *$L_{\mathbb{C}}$ is real.*

(b) $\mathbb{P}_n \cap L_{\mathbb{C}} = L$.

(c) *$L_{\mathbb{C}}$ is spanned (in $\mathbb{C}\mathbb{P}_n$) by any set of points spanning L.*

(d) *Any system of equations of L may be taken as a system of equations of $L_{\mathbb{C}}$ relative to the same (real) reference.*

(e) *For any real linear variety T of $\mathbb{C}\mathbb{P}_n$, $T = (\mathbb{P}_n \cap T)_{\mathbb{C}}$.*

(f) *Mapping $L \mapsto L_{\mathbb{C}}$ is a bijection between the set of all linear varieties of $\mathbb{P}_n$ and the set of all real linear varieties of $\mathbb{C}\mathbb{P}_n$. Its inverse map is $T \mapsto \mathbb{P}_n \cap T$.*

Proof. For the existence, fix a reference of $\mathbb{P}_n$ and take as $L_{\mathbb{C}}$ the linear variety defined in $\mathbb{C}\mathbb{P}_n$ by a system of equations of L: being defined by the same equations, $L_{\mathbb{C}}$ obviously contains L and has the same dimension. Now, assume that there is a linear variety L' of $\mathbb{C}\mathbb{P}_n$ such that $L \subset L'$ and $\dim L = \dim L'$. Then $L \subset \mathbb{P}_n \cap (L_{\mathbb{C}} \cap L')$ and by 5.1.13,

$$\dim L_{\mathbb{C}} = \dim L' = \dim L \leq \dim \mathbb{P}_n \cap (L_{\mathbb{C}} \cap L') \leq \dim L_{\mathbb{C}} \cap L',$$

which yields $L_{\mathbb{C}} = L'$ and proves the uniqueness. As said above, any system of equations of L defines in $\mathbb{C}\mathbb{P}_n$ a linear variety L' that contains L and has the same dimension: by the uniqueness, $L' = L_{\mathbb{C}}$, which proves claim (d). Claim (a) directly follows from claim (d). For claim (b), just note that $L \subset \mathbb{P}_n \cap L_{\mathbb{C}}$ and both have the same dimension because $L_{\mathbb{C}}$ is real. Regarding claim (c), points spanning L span in $\mathbb{C}\mathbb{P}_n$ a linear variety L' that contains L and has the same dimension by 5.1.11: again $L' = L_{\mathbb{C}}$ by the uniqueness. If T is real, then $\dim T = \dim \mathbb{P}_n \cap T$ and obviously $T \supset (\mathbb{P}_n \cap T)$, so T satisfies the conditions defining $(\mathbb{P}_n \cap T)_{\mathbb{C}}$ and hence equals it; this proves claim (e). To close, claim (f) is direct from claims (b) and (e). □

The linear variety $L_{\mathbb{C}}$ is called the *complex extension* of L. The points of $L_{\mathbb{C}} - L$ are called the *imaginary points* of L. Usually the name *real linear varieties* applies to both the linear varieties of $\mathbb{P}_n$ and the real linear varieties of $\mathbb{C}\mathbb{P}_n$ as defined above. This causes no confusion because no essential distinction between L and $L_{\mathbb{C}}$ is made in practice: they are thought of as a unique linear variety of $\mathbb{P}_n$, of which one may consider either just the real points (L) or both the real and the imaginary points ($L_{\mathbb{C}}$). The linear varieties of $\mathbb{C}\mathbb{P}_n$ that are not real are called *imaginary*.

Proposition 5.1.18. *Assume that*

$$f : \mathbb{P}_n \longrightarrow \mathbb{P}'_n$$

is a projectivity between real projective spaces. Then:

(a) f *has a unique extension to a projectivity*

$$f_{\mathbb{C}} : \mathbb{C}\mathbb{P}_n \longrightarrow \mathbb{C}\mathbb{P}'_n.$$

(b) *Any matrix of f relative to references Δ of $\mathbb{P}_n$ and Ω of $\mathbb{P}'_n$ is a matrix of $f_{\mathbb{C}}$ relative to the same references taken as references of the complex extensions.*

(c) $(\mathrm{Id}_{\mathbb{P}_n})_{\mathbb{C}} = \mathrm{Id}_{\mathbb{C}\mathbb{P}_n}$.

(d) *If $g : \mathbb{P}'_n \longrightarrow \mathbb{P}''_n$ is a second projectivity between real projective spaces, then* $(g \circ f)_{\mathbb{C}} = g_{\mathbb{C}} \circ f_{\mathbb{C}}$.

(e) $(f_{\mathbb{C}})^{-1} = (f^{-1})_{\mathbb{C}}$

(f) *For any linear variety L of $\mathbb{P}_n$, $f_{\mathbb{C}}(L_{\mathbb{C}}) = f(L)_{\mathbb{C}}$.*

Proof. The existence follows from 5.1.2, while the uniqueness is clear after 5.1.8, because the images by $f_{\mathbb{C}}$ of the real elements of $\mathbb{C}\mathbb{P}_n$ are already determined. Claims (b) to (e) are a direct consequence of the uniqueness, as, for instance, for claim (b) it is enough to note that the projectivity between the complexified spaces defined by the matrix of f is an extension of f. For claim (f) take $q_0, \dots, q_d \in \mathbb{P}_n$ spanning L. On one hand they span $L_{\mathbb{C}}$ by 5.1.17 (c) and so $f_{\mathbb{C}}(q_0), \dots, f_{\mathbb{C}}(q_d)$ span $f_{\mathbb{C}}(L_{\mathbb{C}})$. On the other, $f(q_0), \dots, f(q_d)$ span $f(L)$ and hence also $f(L)_{\mathbb{C}}$, again by 5.1.17 (c). Since $f_{\mathbb{C}}$ is an extension of f, $f(q_i) = f_{\mathbb{C}}(q_i)$, $i = 0, \dots, d$, and so $f_{\mathbb{C}}(L_{\mathbb{C}}) = f(L)_{\mathbb{C}}$ because they are spanned by the same points. □

The projectivity $f_{\mathbb{C}}$ is called the *complex extension* of f. As for linear varieties, usually no distinction is made between f and $f_{\mathbb{C}}$, which are viewed as a single transformation acting on the real points and, if needed, also on the imaginary ones.

If $\mathbb{P}_n$ is a real projective space, one may consider the dual of its complex extension $(\mathbb{C}\mathbb{P}_n)^\vee$. The bijection of 5.1.17 induces an injective map

$$\begin{aligned} \mathbb{P}_n^\vee &\longrightarrow (\mathbb{C}\mathbb{P}_n)^\vee, \\ H &\longmapsto H_C, \end{aligned}$$

whose image is the set of all real hyperplanes of $\mathbb{C}\mathbb{P}_n$. If Δ is any reference of $\mathbb{P}_n$ and it is taken also as reference of $\mathbb{C}\mathbb{P}_n$, any hyperplane H of $\mathbb{P}_n$ and its complex extension $H_{\mathbb{C}}$ have the same hyperplane coordinates relative to Δ, by 5.1.2 (d). Identifying each hyperplane H of $\mathbb{P}_n$ to its complex extension, turns $\mathbb{P}_n^\vee$ into a subset of $(\mathbb{C}\mathbb{P}_n)^\vee$ which, once a reference of $\mathbb{P}_n$ is fixed, is the set of the hyperplanes of $\mathbb{C}\mathbb{P}_n$ that have real coordinates. The dual space $(\mathbb{C}\mathbb{P}_n)^\vee$ and the complex extension of $\mathbb{P}_n^\vee$ play thus very similar roles with respect to $\mathbb{P}_n^\vee$. The next proposition shows that $(\mathbb{C}\mathbb{P}_n)^\vee$ and the complex extension of $\mathbb{P}_n^\vee$ may be identified, as extensions of $\mathbb{P}_n^\vee$, through a projectivity depending on no choice and compatible with the induced projectivities (the reader familiar with theory of categories may recognize g as a functorial isomorphism).

Proposition 5.1.19. (a) *If $\mathbb{P}_n$ is a projective space over $\mathbb{R}$, then there is a unique projectivity*

$$g\colon \mathbb{C}(\mathbb{P}_n^\vee) \longrightarrow (\mathbb{C}\mathbb{P}_n)^\vee$$

satisfying $g(H) = H_{\mathbb{C}}$ for any hyperplane H of $\mathbb{P}_n$.

(b) *If $f\colon \mathbb{P}_n \to \overline{\mathbb{P}}_n$ is a projectivity between real projective spaces and g and $\bar{g}$ denote the projectivities of part* (a) *relative to $\mathbb{P}_n$ and $\overline{\mathbb{P}}_n$, respectively, then the diagram*

$$\begin{array}{ccc} \mathbb{C}(\mathbb{P}_n^\vee) & \xleftarrow{(f^\vee)_{\mathbb{C}}} & \mathbb{C}(\overline{\mathbb{P}}_n^\vee) \\ {\scriptstyle g}\big\downarrow & & \big\downarrow{\scriptstyle \bar{g}} \\ (\mathbb{C}\mathbb{P}_n)^\vee & \xleftarrow[(f_{\mathbb{C}})^\vee]{} & (\mathbb{C}\overline{\mathbb{P}}_n)^\vee \end{array}$$

is commutative, that is, $g \circ (f^\vee)_{\mathbb{C}} = (f_{\mathbb{C}})^\vee \circ \bar{g}$.

Proof. For claim (a), fix a reference Δ of $\mathbb{P}_n$. On one hand, its dual $\Delta^\vee$ is a reference of $\mathbb{P}_n^\vee$ which may be taken as a reference of $\mathbb{C}(\mathbb{P}_n^\vee)$. On the other, Δ may be taken as reference of $\mathbb{C}\mathbb{P}_n$, and, as such, it has a dual which is a reference of $(\mathbb{C}\mathbb{P}_n)^\vee$: we denote it by Δ^* to avoid any confusion. For the existence, we take g to be the projectivity mapping $\Delta^\vee$ to Δ^* (2.8.1): next we will see that it maps any hyperplane H of $\mathbb{P}_n$ to $H_{\mathbb{C}}$. Indeed, as in any complex extension, any coordinates of H in $\mathbb{P}_n^\vee$ are also coordinates of H in $\mathbb{C}(\mathbb{P}_n^\vee)$, provided the same reference ($\Delta^\vee$ in this case) is used in both spaces (5.1.6). On the other hand, we have noted above that, due to 5.1.2 (d), any coordinates of H relative to $\Delta^\vee$ are also coordinates of $H_{\mathbb{C}} \in (\mathbb{C}\mathbb{P}_n)^\vee$ relative to Δ^*. Thus, H and $H_{\mathbb{C}}$ have equal coordinates, after which, by 2.8.1, $g(H) = H_{\mathbb{C}}$. The uniqueness follows from 5.1.8, because the condition of the claim determines the images of the real elements of $\mathbb{C}(\mathbb{P}_n^\vee)$.

For claim (b), by 5.1.8, it is enough to see that $(f_{\mathbb{C}})^\vee(\bar{g}(H)) = g((f^\vee)_{\mathbb{C}}(H))$ for any $H \in \overline{\mathbb{P}}_n^\vee$, and indeed, using 5.1.18,

$$(f_{\mathbb{C}})^\vee(\bar{g}(H)) = f_{\mathbb{C}}^{-1}(H_{\mathbb{C}}) = (f^{-1}(H))_{\mathbb{C}} = g(f^\vee(H)) = g((f^\vee)_{\mathbb{C}}(H)). \quad \square$$

Remark 5.1.20. It follows from 5.1.19 and its proof that, no matter which reference Δ of $\mathbb{P}_n$ is chosen, the projectivity g maps the reference $\Delta^\vee$ of $\mathbb{C}(\mathbb{P}_n^\vee)$ to the dual of Δ taken as reference of $\mathbb{C}\mathbb{P}_n$. Therefore, g maps the element of $\mathbb{C}(\mathbb{P}_n^\vee)$ which has coordinates $a_0, \dots, a_n$ relative to $\Delta^\vee$, to the hyperplane of $\mathbb{C}\mathbb{P}_n$ which has hyperplane coordinates $a_0, \dots, a_n$ relative to Δ.

When needed, we will make use of the complex extension $\mathbb{C}\bar{\mathbb{A}}_n$ of the projective closure of any real affine space $\mathbb{A}_n$. The imaginary points of $\mathbb{C}\bar{\mathbb{A}}_n$ will be called *imaginary improper points* if they belong to $(H_\infty)_\mathbb{C}$, and *imaginary proper points* otherwise.

5.2 Equations of projectivities between lines

In this section we will present different forms of the equations of a projectivity that apply to the one-dimensional case only. They will be used in the forthcoming Section 5.5, and may also be used to provide alternative proofs of the results of the next Section 5.3.

Assume to have fixed projective references Δ and Ω in one-dimensional projective spaces $\mathbb{P}_1$ and $\mathbb{P}'_1$, respectively. By 2.8.4, for each projectivity $f : \mathbb{P}_1 \to \mathbb{P}'_1$ there are $a, b, c, d \in k$, determined by f up to a common factor and such that $ad - bc \neq 0$, in such a way that $[y_0, y_1]_\Omega = f([x_0, x_1]_\Delta)$ if and only if

$$\begin{aligned} \rho y_0 &= a x_0 + b x_1, \\ \rho y_1 &= c x_0 + d x_1 \end{aligned} \tag{5.7}$$

for some non-zero $\rho \in k$. Furthermore, by 2.8.5, $a, b, c, d \in k$, fulfilling the condition $ad - bc \neq 0$, may be arbitrarily prescribed to determine a projectivity $f : \mathbb{P}_1 \to \mathbb{P}'_1$ given by the equations (5.7) as described above.

The equations (5.7) are the equations of f; using absolute coordinates $x = x_0/x_1$ and $y = y_0/y_1$, they are turned into the single equation

$$y = \frac{ax + b}{cx + d}, \tag{5.8}$$

called *the equation of f* relative to Δ, Ω. Obviously (5.7) and (5.8) carry the same information. The conventions for $x = \infty$ and $y = \infty$ in (5.8) are as set for the equation of change of coordinates at the end of Section 2.4.

The existence of $\rho \in k$ satisfying equations (5.7) may be equivalently written

$$\begin{vmatrix} y_0 & a x_0 + b x_1 \\ y_1 & c x_0 + d x_1 \end{vmatrix} = 0 \tag{5.9}$$

as this is the condition of compatibility of (5.7) as a system of equations in ρ; note that $\rho = 0$ is never a solution, due to the fact that $ad - bc \neq 0$ and $(x_0, x_1) \neq (0, 0)$.

The equality (5.9) is said to be the result of *eliminating* ρ from (5.7); once expanded, it takes the form

$$Ax_0y_0 + Bx_0y_1 + B'x_1y_0 + Cx_1y_1 = 0, \tag{5.10}$$

where

$$A = c, \quad B = -a, \quad B' = d, \quad C = -b$$

and thus $BB' - AC \neq 0$. Hence equation (5.10) provides a new condition equivalent to $f([x_0, x_1]_\Delta) = [y_0, y_1]_\Omega$. It is called the *symmetric equation* of f, because by swapping the roles of x_0, x_1 and y_0, y_1 it may be used to represent f^{-1} as well. The reader may note that, in spite of this name, the first member of equation (5.10) is not symmetric in the two pairs of variables x_0, x_1 and y_0, y_1, unless $B = B'$. Clearly, a, b, c, d and A, B, B', C are equivalent data, also the latter are determined by f up to a common factor and, provided $BB' - AC \neq 0$, they may be arbitrarily fixed to define a projectivity through (5.10).

Using absolute coordinates as above, equation (5.10) may be rewritten in a shorter form as

$$Axy + Bx + B'y + C = 0, \tag{5.11}$$

which still is a necessary and sufficient condition for two points, this time with absolute coordinates x, y, to correspond through f. As in former cases the conventions to cover the cases $x = \infty$ and $y = \infty$ can be established going back to (5.10): the equations $Ay + B = 0$ and $Ax + B' = 0$ appear as resulting from (5.11) for $x = \infty$ and $y = \infty$, respectively. Heuristically, for $x = \infty$ the same equation results from dividing (5.11) by x and taking limit for $x \to \infty$, and similarly for $y = \infty$.

5.3 Projectivities between distinct lines

Throughout this section r and s will be two different lines of a projective space $\mathbb{P}_n$ and we will study the projectivities $f : r \to s$. We begin by assuming $n = 2$, the easier case $n > 2$ will be considered at the end of the section.

Assume thus that r and s are distinct lines of $\mathbb{P}_2$. For each point $T \in \mathbb{P}_2$, lying not on r or s, we have the perspectivity with centre T, which is the map

$$\begin{aligned} g_T : r &\longrightarrow s, \\ p &\longmapsto (p \vee T) \cap s, \end{aligned}$$

and has been seen to be a projectivity in 1.9.3. Equivalently, points $p \in r$ and $q \in s$ correspond by g_T if and only if p, q and T are aligned. The next proposition gives a useful characterization of perspectivities and in particular, combined with 2.8.1, shows that not every projectivity $f : r \to s$ is a perspectivity:

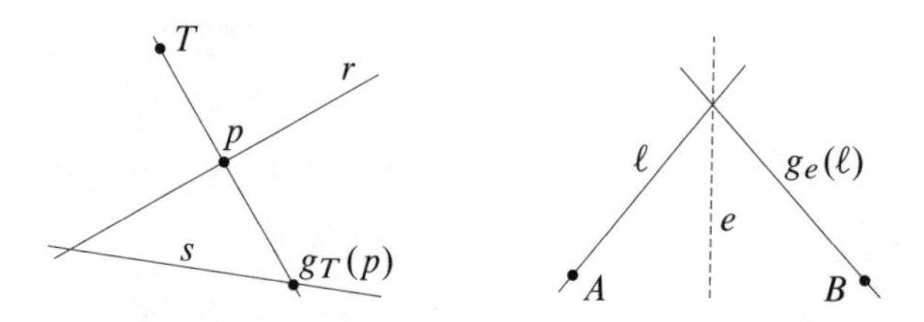

Figure 5.1. Perspectivities: on the left, between lines r, s with centre T; on the right, between pencils A^*, B^* with axis e.

Proposition 5.3.1. *Assume that r, s are distinct lines of $\mathbb{P}_2$ and write $O = r \cap s$ their intersection point. Then a projectivity $f: r \to s$ is a perspectivity if and only if $f(O) = O$.*

Proof. The *only if* part is clear. For the *if* part, assume that $f(O) = O$ and choose two distinct points $A, B \in r, A \neq O, B \neq O$. Then, on one hand, $A \notin s$ and $B \notin s$. On the other, $O = f(O)$, $f(A)$ and $f(B)$ are three different points and in particular neither $f(A)$ nor $f(B)$ belong to r. It follows that $A \neq f(A)$ and therefore $Af(A)$ is a line, $Af(A) \cap r = A$ because $f(A) \notin r$ and $Af(A) \cap s = f(A)$ because $A \notin s$. Similarly $Bf(B)$ is a line, $Bf(B) \cap r = B$ and $Bf(B) \cap s = f(B)$. Since the lines $Af(A)$ and $Bf(B)$ intersect both r and s at different points, they are different lines and their intersection point $T = Af(A) \cap Bf(B)$ does not belong to r or s. We may thus consider the perspectivity $g_T : r \to s$ with centre T: by the definition of T, $g_T(A) = f(A)$ and $g_T(B) = f(B)$, while obviously $g_T(O) = O$. We see thus that $f = g_T$, because both give the same image to the three distinct points A, B, O (2.8.1). In particular f is a perspectivity as wanted. □

Dually, we have the perspectivities between pencils of lines: if A, B are distinct points of $\mathbb{P}_2$ and e is a line not going through A or B, the perspectivity from A^* onto B^* with *axis* e is the projectivity

$$\begin{aligned} g_e : A^* &\longrightarrow B^*, \\ \ell &\longmapsto (\ell \cap e) \vee B. \end{aligned}$$

Equivalently, lines $\ell \in A^*$ and $t \in B^*$ correspond by g_e if and only if ℓ, t and e are concurrent.

The dual of 2.8.1 needs, of course, no proof:

Proposition 5.3.2. *If A, B are distinct points of $\mathbb{P}_2$, then a projectivity $f: A^* \to B^*$ is a perspectivity if and only if $f(AB) = AB$.*

The definition of perspectivities describes them in terms of incidence, and makes obvious how to construct the image of an arbitrary point by a perspectivity, once the centre or two different pairs of distinct corresponding points are given. We are interested in having an incidence description of the projectivities that are not perspectivities. It is provided by Theorem 5.3.5 below and applies to perspectivities as well. As above assume that r, s are two distinct lines of $\mathbb{P}_2$, take $O = r \cap s$ and let $f : r \to s$ be a projectivity.

Lemma 5.3.3. *Fix any point $A \in r$, $A \neq O, f^{-1}(O)$. There is a unique line e_A, $e_A \neq r, s$, such that $f(A)B$, $Af(B)$ and e_A are concurrent lines for any $B \in r$.*

Proof. By the hypothesis $A \neq O$, $f(A) \neq O$, and so $f(A) \notin r$ and $A \notin s$. We may thus take g to be the composition of

$$f(A)^* \xrightarrow{\sigma_r} r \xrightarrow{f} s \xrightarrow{\delta_A} A^*,$$

where σ_r and δ_A are the section and projection maps. Then g is a projectivity. After an easy checking, $g(f(A)A) = Af(A)$ and hence g is a perspectivity, by 5.3.2. Take e_A to be the axis of g: it is different from r and s because $A \in r$ and $f(A) \in s$. Furthermore, for any $B \in r$, $Bf(A) = \sigma_r^{-1}(B)$ and $Af(B) = \delta_A(f(B))$ are lines that correspond by g and therefore are concurrent with e_A. For the uniqueness, just note that any line satisfying the claimed properties is spanned by the points $Af(B) \cap f(A)B$ and $Af(B') \cap f(A)B'$ for any two different points $B, B' \in r$, $B, B' \neq A$. □

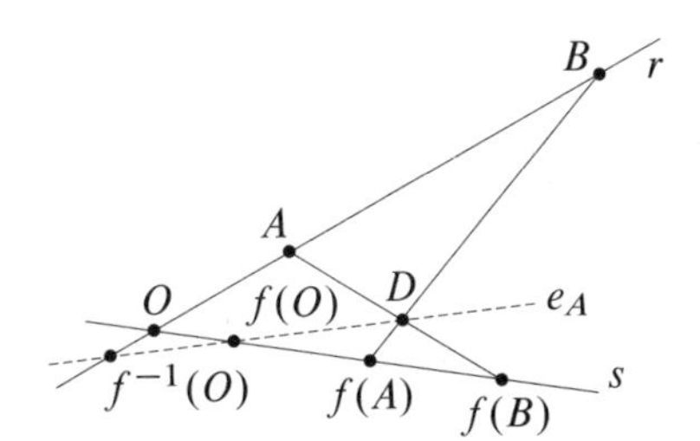

Figure 5.2. Lemmas 5.3.3 and 5.3.4: while B varies on r, the point D lies on the line e_A, whose intersections with r and s are $f^{-1}(O)$ and $f(O)$, respectively.

Lemma 5.3.4. *If e_A is as in* 5.3.3 *above, then $e_A \cap r = f^{-1}(O)$, $e_A \cap s = f(O)$ and e_A is independent of the choice of the point A.*

Proof. Take $U = f^{-1}(O)$: it is a point of $r = AO = Af(U)$ and obviously also a point of $Uf(A)$. These lines being different because $f(A) \neq O$ and hence $f(A) \notin r$, $U = Af(U) \cap f(A)U \in e_A$ by 5.3.3, and hence $U = r \cap e_A$. Proving that $e_A \cap s = f(O)$ by a similar argument is left to the reader.

Now, if $U = f^{-1}(O) \neq f(O) = V$, $e_A = UV$ is obviously independent of the choice of A. Otherwise $O = f^{-1}(O) = f(O) \in e_A$. If $C \neq A$ is a point in the same conditions as A, then also $O \in e_C$. Furthermore $D = Af(C) \cap f(A)C$ clearly is a point different from O which belongs to both e_A and e_C by 5.3.3 applied to A and C respectively. It follows that $e_A = OD = e_C$ as wanted. □

Lemmas 5.3.3 and 5.3.4 together yield:

Theorem 5.3.5 (Cross-axis theorem for lines). *If $f : r \to s$ is a projectivity between distinct lines r, s of $\mathbb{P}_2$, $r \cap s = O$, then there exists a unique line e, called the* ***cross-axis*** *of f, satisfying*

(1) *$e \cap r = f^{-1}(O)$, $e \cap s = f(O)$, and*

(2) *for any two points $A, B \in r$, $\{A, B\} \neq \{O, f^{-1}(O)\}$, $Af(B)$ and $Bf(A)$ and e are concurrent lines.*

Proof. By 5.3.4, we may take $e = e_A$ for all $A \in r$, $A \neq O, f^{-1}(O)$. Then claim (1) is part of 5.3.4. For claim (b), the cases $A = B = O$ and $A = B = f^{-1}(O)$ are obvious. Otherwise, either $A \neq O, f^{-1}(O)$ or $B \neq O, f^{-1}(O)$, after which 5.3.3 applies. For the uniqueness, pick any $A \in r$, $A \neq O, f^{-1}(O)$ and note that any line e' satisfying claim (2) is $e' = e_A = e$. □

The reader may easily check that any point $p \in e$ is $p = Af(B) \cap Bf(A)$ for some A, B as in 5.3.5 (2), and note that 5.3.5 (2) would fail for $\{A, B\} = \{O, f^{-1}(O)\}$.

Remark 5.3.6. The cross-axis e of 5.3.5 may be easily constructed from three different points, $A, B, C \in r$ and their images. Indeed, the points being different, one of them, say A, is $A \neq O, f^{-1}(O)$, and then e is the line spanned by the obviously different points $Af(B) \cap Bf(A)$ and $Af(C) \cap Cf(A)$.

Once the cross-axis e is determined, constructing the image of an arbitrary point $p \neq A$ is direct from the condition $Af(p) \cap pf(A) \in e$, as

$$f(p) = ((pf(A) \cap e) \vee A) \cap s,$$

see Figure 5.3. In particular f is determined by its cross-axis and a pair of corresponding points, both different from O. Note also that the above re-proves that f is determined by A, B, C and their images.

The cross-axis theorem and the Pappus theorem 1.8.5 are essentially equivalent. The latter may be re-proved from the former using the projectivity that maps A, B, C to A', B', C' respectively (notations as in 1.8.5). Conversely the reader may give an easy alternative proof of the cross-axis theorem using Pappus.

The next proposition shows that, but for some mild incidence requirements, a pair of corresponding points and the cross-axis of a projectivity $f : r \to s$ may be chosen arbitrarily. Note that they determine f, as already shown in 5.3.6:

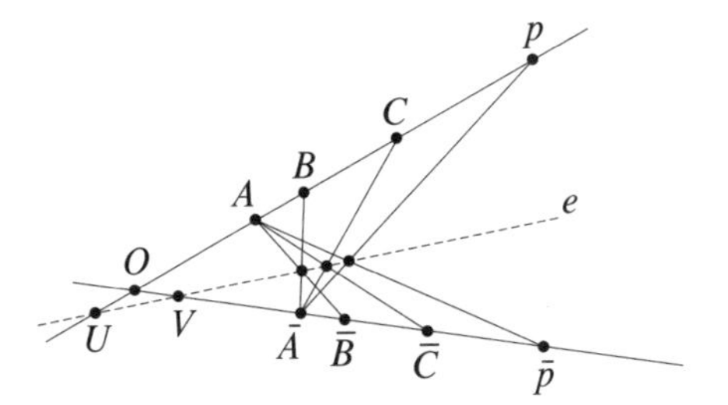

Figure 5.3. The cross-axis e constructed from points A, B, C and their respective images $\bar{A}$, $\bar{B}$, $\bar{C}$. U and V are the inverse and direct images of O, while the image $\bar{p}$ of a fourth point p is constructed from e and the pair $(A, \bar{A})$.

Proposition 5.3.7. *Let r, s be two different lines of $\mathbb{P}_2$, $O = r \cap s$. Assume that there are given points $A \in r$, $\bar{A} \in s$, both different from O, and a line e, different from r and s and not going through A or $\bar{A}$. Then taking as correspondent two points $p \in r$ and $\bar{p} \in s$ if and only if the lines $p\bar{A}$, $\bar{p}A$ and e are concurrent, defines a projectivity $f : r \to s$ that maps A to $\bar{A}$ and has cross-axis e.*

Proof. Note first that, indeed, both $p\bar{A}$, $\bar{p}A$ are lines because $A \neq O$, $\bar{A} \neq O$ force $A \notin s$, $\bar{A} \notin r$ and hence $p \neq \bar{A}$, $\bar{p} \neq A$. According to the claim p, $\bar{p}$ correspond if and only if the lines $p\bar{A}$, $\bar{p}A$ correspond by the perspectivity $g \colon \bar{A}^* \to A^*$ with axis e. Then f is the composition of

$$r \xrightarrow{\delta_{\bar{A}}} \bar{A}^* \xrightarrow{g} A^* \xrightarrow{\sigma_s} s,$$

$\delta_{\bar{A}}$ and σ_s still being the projection and section maps. After this, the claim readily follows. □

We will close this section by showing that the situation is much simpler when the lines r, s are not coplanar, as then any projectivity $f : r \to s$ is a perspectivity. We will need a very elementary incidence lemma, already proposed as Exercise 1.12:

Lemma 5.3.8. *If $\ell_1, \ell_2 \subset \mathbb{P}_n$ are non-coplanar lines and $p \in \ell_1 \vee \ell_2$ a point not belonging to ℓ_1 or ℓ_2, then there is a unique line t through p meeting both ℓ_1 and ℓ_2.*

Proof. The planes $\pi_i = p \vee \ell_i, i = 1, 2$, are distinct, as otherwise ℓ_1 and ℓ_2 would be coplanar; they both are contained in $\ell_1 \vee \ell_2$, which in turn clearly has dimension three. It follows that $t = \pi_1 \cap \pi_2$ is a line, obviously coplanar with ℓ_1 and ℓ_2, and hence meeting both. Now, for the uniqueness, if t' is any line through p and meeting ℓ_i, then, since $p \notin \ell_i$, $q_i = t' \cap \ell_i$ is a point different from p. Since both p and q_i lie on π_i we obtain $t' = p \vee q_i \subset \pi_i$ for $i = 1, 2$ and therefore $t' = t$. □

Remark 5.3.9. Since $p \notin \ell_i, t \neq \ell_i$ and necessarily $t \cap \ell_i$ is a point, for $i = 1, 2$.

Now, as already announced:

Theorem 5.3.10. *If $r, s \subset \mathbb{P}_n$ are non-coplanar lines, then any projectivity $f : r \to s$ is a perspectivity.*

Proof. Take $L = r \vee s$; obviously $\dim L = 3$. Choose three different points $q_0, q_1, q_2 \in r$ and write $\bar{q}_i = f(q_i)$, $i = 0, 1, 2$. The lines r and s being disjoint, $q_i \neq \bar{q}_i$ and we may consider the lines $\ell_i = q_i\bar{q}_i$, $i = 0, 1, 2$, all in L. A plane π containing two of them, say $\pi \supset \ell_i, \ell_j, i \neq j$, would contain $q_i, \bar{q}_i, q_j$ and $\bar{q}_j$ and therefore also $r = q_i \vee q_j$ and $s = \bar{q}_i \vee \bar{q}_j$, against the hypothesis. The lines ℓ_i are thus pairwise non-coplanar, and hence any two span L. Now we choose a point $p_0 \in \ell_0$, $p_0 \neq q_0, \bar{q}_0$. Since $p_0 \notin \ell_i$, $i = 1, 2$, by 5.3.8, we take t to be the line through p_0 meeting ℓ_1 and ℓ_2 at, say, points p_1 and p_2, respectively. Now let us check that $p_1 \neq q_1, \bar{q}_1$: if for instance $p_1 = q_1$, both t and r go through such a point and meet ℓ_0 and ℓ_2; 5.3.8 forces $t = r$ and hence $p_0 = t \cap \ell_0 = r \cap \ell_0 = q_0$ against the choice of p_0; the other case is similar. Arguing as above, if a plane contains t and r, then it contains p_0, p_1, q_0 and q_1 and therefore also $\ell_0 = p_0q_0$ and $\ell_1 = p_1q_1$, against what we have already proved. This proves that t and r are not coplanar, and a similar argument shows that t and s are not coplanar either.

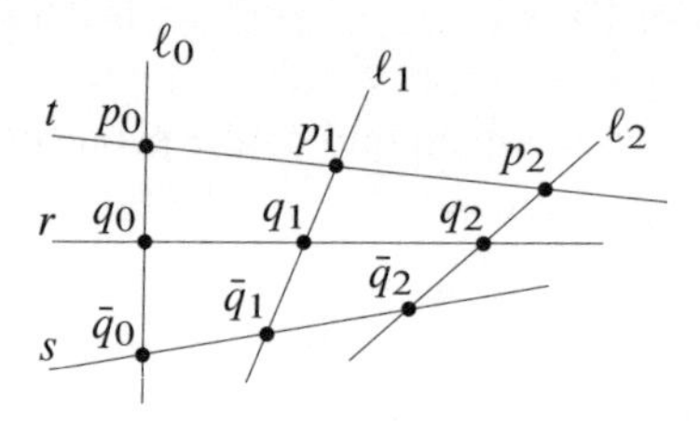

Figure 5.4. The six lines of the proof of 5.3.10; their only intersections are those pictured.

If $n = 3$, then the line t may be taken as the centre of the perspectivity we are looking for. For $n > 3$ we need a higher-dimensional centre, so we will enlarge t. Take T' to be a linear variety supplementary to L ($T' = \emptyset$ if $n = 3$) and $T = t \vee T'$. We have $\dim T' = n - 4$ and therefore, t and T' being disjoint, $\dim T = n - 2$ as required. Furthermore, since clearly $L \vee T = \mathbb{P}_n$, $\dim L \cap T = n - 3 - n + 2 = 1$, which proves that $L \cap T = t$. After this it is clear that T and r are disjoint and hence, by their dimensions, supplementary. The same may be said of T and s and therefore it makes sense to consider the perspectivity of centre T, $g : r \to s$. Now it is clear that

$$\bar{q}_i \in \ell_i \vee T = q_i \vee p_i \vee T = q_i \vee T$$

which proves that $f(q_i) = \bar{q}_i = g(q_i)$ for $i = 0, 1, 2$. Therefore, by 2.8.1, $f = g$ is a perspectivity. □

Remark 5.3.11. The above proof shows that, even in case $n = 3$, the centre of f, as a perspectivity, is not unique.

5.4 Pairs of points on projective lines

In this section, we present very elementary facts regarding pairs of points of $\mathbb{P}_1$ and the equations defining them, as they will appear related to the projectivities, and especially to the involutions, of $\mathbb{P}_1$. We will come back to pairs of points of $\mathbb{P}_1$ in the forthcoming Chapter 6, as they are the quadric hypersurfaces of $\mathbb{P}_1$ and play a crucial role in the study of the quadric hypersurfaces of higher-dimensional spaces.

Assume to have fixed a one-dimensional projective space $\mathbb{P}_1$ over the field k, still $k = \mathbb{R}$ or $\mathbb{C}$, and a reference of it. We will consider homogeneous equations of the second degree in two variables, namely of the form

$$F = F(x_0, x_1) = ax_0^2 + 2bx_0x_1 + cx_1^2 = 0, \tag{5.12}$$

$a, b, c \in k$, one of them at least being non-zero. (Taking the middle term in the form $2b$ makes some expressions shorter.) A point $p = [\alpha_0, \alpha_1]$ is said to *satisfy*, or to be a *solution* of, the above equation (5.12) if and only if $F(\alpha_0, \alpha_1) = 0$. Note that this condition depends only on the point and not on the choice of its coordinates, due to the homogeneity of the equation, as $F(\lambda\alpha_0, \lambda\alpha_1) = \lambda^2 F(\alpha_0, \alpha_1)$. The *discriminant* of the equation (5.12) will be taken to be $D = ac - b^2$.

Assume $k = \mathbb{R}$. Then one may consider solutions in $\mathbb{C}\mathbb{P}_1$ as well: the solutions of $F = 0$ in $\mathbb{P}_1$ are called *real*, while those in $\mathbb{C}\mathbb{P}_1 - \mathbb{P}_1$ are called *imaginary*. The coefficients of F being real, for any complex α_0, α_1, $F(\bar{\alpha}_0, \bar{\alpha}_1)$ is the complex-conjugate of $F(\alpha_0, \alpha_1)$. It follows that if p is an imaginary solution of $F = 0$, then so is its conjugate point.

Proposition 5.4.1. *Let $F = 0$ be the equation* (5.12) *above and D its discriminant. If $k = \mathbb{R}$, then the* (*real or imaginary*) *solutions of $F = 0$ are either:*

(r1) *two, both real, if $D < 0$, or*

(r2) *a single one, real, if $D = 0$, or*

(r3) *two, imaginary and conjugate, if $D > 0$.*

If $k = \mathbb{C}$, the solutions of $F = 0$ are either

(c1) *two if $D \neq 0$, or*

(c2) *a single one if $D = 0$.*

In cases (r2) and (c2) above it is often said that the only solution is a *double solution*, or even that the equation has *two coincident solutions*. To avoid confusion, in the remaining cases it is usual to say that there are two *distinct* solutions.

Proof of 5.4.1. Assume first that $a \neq 0$. Then the point $[1, 0]$ is not a solution and so the solutions are the points $[\alpha_0, \alpha_1]$ whose absolute coordinate $\alpha = \alpha_0/\alpha_1$ is finite and satisfies

$$a\alpha^2 + 2b\alpha + c = 0.$$

Then the claim follows because $4b^2 - 4ac = -4D$.

In case $a = 0$, the equation may be written

$$x_1(2bx_0 + cx_1) = 0.$$

Its solutions have thus either $\alpha_1 = 0$ or $2b\alpha_0 + c\alpha_1 = 0$, and so they are $[1, 0]$ and $[c, -2b]$. These points are different if and only if $b \neq 0$. Since in this case $D = -b^2$, the claim is satisfied too. □

Conversely, any pair of points, conjugate of each other in case of being imaginary, is the set of solutions of a homogeneous equation of the second degree, determined by the points up to a constant factor:

Proposition 5.4.2. *Let $q_1 = [\alpha_0, \alpha_1]$ and $q_2 = [\beta_0, \beta_1]$ be either points of $\mathbb{P}_1$ or, in case $k = \mathbb{R}$, a pair of conjugate imaginary points of $\mathbb{P}_1$. In the latter case, assume, as allowed by* (5.1.9), *that their coordinates have been chosen such that $\beta_0 = \bar{\alpha}_0$ and $\beta_1 = \bar{\alpha}_1$. Then the equation*

$$(\alpha_1 x_0 - \alpha_0 x_1)(\beta_1 x_0 - \beta_0 x_1) = 0 \tag{5.13}$$

is a degree-two homogeneous equation with coefficients in k, whose set of (*maybe imaginary*) *solutions is $\{q_1, q_2\}$. Any equation satisfying the same properties is the product of the above by a factor $\lambda \in k - \{0\}$.*

Proof. Clearly equation (5.13) is homogeneous and has degree two. In case $q_1, q_2 \in \mathbb{P}_1$, it is also clear that it has its coefficients in the base field k. Since it may be written

$$\alpha_1\beta_1 x_0^2 - (\alpha_1\beta_0 + \alpha_0\beta_1)x_0x_1 + \alpha_0\beta_0 x_1^2 = 0,$$

for $k = \mathbb{R}$ and $q_1, q_2 \in \mathbb{C}\mathbb{P}_1 - \mathbb{P}_1$, all its coefficients are invariant by complex conjugation and hence real. The coefficients of the equation (5.13) belong thus to k in any case. Once (5.13) is rewritten

$$\begin{vmatrix} x_0 & x_1 \\ \alpha_0 & \alpha_1 \end{vmatrix} \begin{vmatrix} x_0 & x_1 \\ \beta_0 & \beta_1 \end{vmatrix} = 0,$$

one sees that its solutions are just q_1, q_2, by 2.2.5.

For the uniqueness, let $ax_0^2 + 2bx_0x_1 + cx_1^2 = 0$, $a, b, c \in k$, be an equation whose solutions are q_1, q_2. Assume first that $a \neq 0$. Then, as in the proof of 6.2.1, $\alpha_1, \beta_1 \neq 0$ and the absolute coordinates α_0/α_1 and β_0/β_1 are the roots of the polynomial $ax^2 + 2bx + c$. It follows that

$$ax^2 + 2bx + c = a(x - \alpha_0/\alpha_1)(x - \beta_0/\beta_1)$$

and therefore

$$\begin{aligned} ax_0^2 + 2bx_0x_1 + cx_1^2 &= x_1^2\left(a\left(\frac{x_0}{x_1}\right)^2 + b\frac{x_0}{x_1} + c\right) \\ &= x_1^2 a\left(\frac{x_0}{x_1} - \frac{\alpha_0}{\alpha_1}\right)\left(\frac{x_0}{x_1} - \frac{\beta_0}{\beta_1}\right) \\ &= \frac{a}{\alpha_1\beta_1}(\alpha_1x_0 - \alpha_0x_1)(\beta_1x_0 - \beta_0x_1). \end{aligned}$$

Since in the case of imaginary points we have taken $\beta_1 = \bar{\alpha}_1$, in all cases $a/\alpha_1\beta_1 \in k - \{0\}$.

If $a = 0$, we have seen in the proof of 5.4.1 that the solutions are $[1,0]$ and $[-c, 2b]$. Thus $\alpha_0 \neq 0$, $\alpha_1 = 0$ and $\beta_0 = -\rho c$, $\beta_1 = 2\rho b$ for some $\rho \neq 0$. It follows that

$$ax_0^2 + 2bx_0x_1 + cx_1^2 = 2bx_0x_1 + cx_1^2 = \frac{-1}{\rho\alpha_0}(\alpha_1x_0 - \alpha_0x_1)(\beta_1x_0 - \beta_0x_1). \quad \square$$

Remark 5.4.3. The second-degree homogeneous equations whose solutions are the points $q_1, q_2 \in \mathbb{P}_1$, described in 5.4.2, are called *equations* of the pair q_1, q_2, which in turn is said to be *defined* by any of its equations. It is also worth noting that the first member of any equation of q_1, q_2 is the product of two linear factors, which are equations of q_1 and q_2 as zero-dimensional linear varieties of $\mathbb{P}_1$. In the case of a single solution, that is, $q_1 = q_2$, there is a single factor with multiplicity two, which somewhat justifies calling double the only solution.

5.5 Projectivities of a line

In this section we will study collineations $f : \mathbb{P}_1 \to \mathbb{P}_1$ using matrices and the different equations introduced in Section 5.2. Since all these forms of representing a projectivity carry the same information, and switching between them is easy, we will use in each case the one best suited to our computations. We will use $*$ as a superscript to denote the absolute or homogeneous coordinate(s) of the image.

The next proposition describes the *fixed points* of f, that is, the points $p \in \mathbb{P}_1$ for which $f(p) = p$:

Proposition 5.5.1. *Let f be a non-identical projectivity of $\mathbb{P}_1$. In case $k = \mathbb{R}$, f has either two fixed points, or a single fixed point, or no fixed point. If $k = \mathbb{C}$, then f has either two fixed points or a single fixed point.*

Proof. Take a reference of $\mathbb{P}_1$ and assume that, relative to it, f has symmetric equation (see Section 5.2)

$$Ax_0x_0^* + Bx_0x_1^* + B'x_1x_0^* + Cx_1x_1^* = 0, \tag{5.14}$$

$BB' - AC \neq 0$. The point $[x_0, x_1]$ is thus fixed if and only if

$$Ax_0^2 + (B + B')x_0x_1 + Cx_1^2 = 0. \tag{5.15}$$

Note that in case $A = B + B' = C = 0$ it is $B \neq 0$; then equation (5.14) becomes

$$B(x_0x_1^* - x_1x_0^*) = 0,$$

that is,

$$\begin{vmatrix} x_0 & x_1 \\ x_0^* & x_1^* \end{vmatrix} = 0,$$

and therefore (by 2.2.5) $f = \mathrm{Id}_{\mathbb{P}_1}$, which has been already excluded. Thus (5.15) is a non-trivial homogeneous equation of the second degree whose solutions are the fixed points of f, which turns the claim into a direct consequence of 5.4.1. □

Projectivities with two fixed points are called *hyperbolic*, those with a unique fixed point are called *parabolic* and those with no fixed points are called *elliptic*. After 5.5.1 there are no elliptic projectivities if $k = \mathbb{C}$.

The above equation (5.15) is called the *equation of the fixed points* of f. The proof of 5.5.1 provides further information, namely:

Corollary 5.5.2 (of the proof of 5.5.1). *Assume that f has equation*

$$Ax_0x_0^* + Bx_0x_1^* + B'x_1x_0^* + Cx_1x_1^* = 0.$$

Then, for $k = \mathbb{R}$,

- *f is elliptic if and only if $(B + B')^2 - 4AC < 0$,*
- *f is parabolic if and only if $(B + B')^2 - 4AC = 0$, and*
- *f is hyperbolic if and only if $(B + B')^2 - 4AC > 0$.*

For $k = \mathbb{C}$,

- *f is parabolic if and only if $(B + B')^2 - 4AC = 0$, and*
- *f is hyperbolic if and only if $(B + B')^2 - 4AC \neq 0$.*

Proof. Just note that $(B+B')^2 - 4AC$ is four times the opposite of the discriminant of the equation of the fixed points and again apply 5.4.1. □

Remark 5.5.3. All cases appearing in 5.5.1 actually occur: just check the projectivities with symmetric equations

$$x_0x_0^* + x_1x_1^* = 0, \quad x_0x_0^* + x_0x_1^* - x_1x_0^* = 0 \quad \text{and} \quad x_0x_0^* - x_1x_1^* = 0.$$

Remark 5.5.4. The imaginary solutions of the equation of fixed points of an elliptic projectivity f are a pair of imaginary and conjugate (and hence distinct) points. They are the fixed points of the complex extension $f_{\mathbb{C}}$ of f, as f and $f_{\mathbb{C}}$ have the same equation (by 5.1.18) and therefore the same equation of fixed points. Usually these points are called the *fixed points* of f, or the *imaginary fixed points* of f if the fact that they do not belong to $\mathbb{P}_1$ needs to be recalled.

The next proposition proves in particular that being elliptic, parabolic or hyperbolic are projective properties:

Proposition 5.5.5. *Given any projectivity $g: \mathbb{P}_1 \to \mathbb{P}'_1$, a point p is fixed by a projectivity f of $\mathbb{P}_1$ if and only if $g(p)$ is fixed by the image $g(f)$ of f by g. The projectivity f is elliptic, parabolic or hyperbolic if and only if so is $g(f)$.*

Proof. By definition (see Section 1.10), $g(f) = g \circ f \circ g^{-1}$. If $f(p) = p$, then $g \circ f \circ g^{-1}(g(p)) = g(p)$. Conversely, if the last equality is true, applying g^{-1} to both sides gives $f(p) = p$. The second half of the claim is direct from the first one. □

We will make separate studies of the projectivities of $\mathbb{P}_1$ according to their configuration of fixed points.

Hyperbolic projectivities. We begin by showing a nice form of the equation of a hyperbolic projectivity:

Proposition 5.5.6. *If f is a hyperbolic projectivity of $\mathbb{P}_1$, then its equation relative to any reference whose vertices are the fixed points of f has the form*

$$x^* = \lambda x, \quad \lambda \in k - \{0, 1\}. \tag{5.16}$$

Proof. Assume that f has matrix

$$\begin{pmatrix} a & b \\ c & d \end{pmatrix}.$$

A direct computation using equations (5.7) or (5.8) shows that $[1, 0]$ is fixed if and only if $c = 0$, while $[0, 1]$ is fixed if and only if $b = 0$. Then, the matrix being regular, $a, d \neq 0$ and it is enough to take $\lambda = a/d$. □

The converse of 5.5.6 is clearly true: any projectivity with an equation of the form of (5.16) is hyperbolic and has the vertices of the reference as fixed points. The simple form of equation (5.16) leads us to an interesting fact:

Theorem 5.5.7. *If f is hyperbolic and p_0 and p_1 are its fixed points, then the cross ratio*

$$\delta = (p_0, p_1, p, f(p))$$

is the same for all points $p \neq p_0, p_1$.

Proof. Take a reference with vertices p_0, p_1. According to 5.5.6, if p has absolute coordinate x ($x \neq 0, \infty$), then the absolute coordinate of $f(p)$ is λx. By 2.9.5, $\delta = (p_0, p_1, p, f(p)) = \lambda$, which is independent of p. □

Remark 5.5.8. By its definition, the common value δ of all cross ratios

$$(p_0, p_1, p, f(p)),$$

$p \in \mathbb{P}_1 - \{p_0, p_1\}$, depends only on f and the order in which the fixed points p_0, p_1 are taken. It is called the *invariant* or the *modulus* of f. Swapping over the fixed points turns δ into δ^{-1} and so δ is taken up to this indeterminacy, but for the case in which the order of the fixed points is specified.

Remark 5.5.9. Obviously the modulus δ is $\delta \neq 0, 1$. Sometimes 1 is taken as the modulus of the identity map, as 5.5.7 is obviously true for $f = \mathrm{Id}$, $\delta = 1$ and any choice of different points p_0, p_1.

Remark 5.5.10. A hyperbolic projectivity of $\mathbb{P}_{1,\mathbb{R}}$ with fixed points p_0, p_1 and positive modulus leaves invariant the segments with ends p_0, p_1, while it maps them to each other in case of having negative modulus.

Remark 5.5.11. As seen in the proof of 5.5.7, any equation of f of the form of (5.16) above has $\lambda = \delta$, provided the fixed points are taken as vertices in the same order as they are taken to define the invariant. Otherwise $\lambda = \delta^{-1}$.

A hyperbolic projectivity may be defined by its fixed points and modulus:

Proposition 5.5.12. *Given an ordered pair (p_0, p_1), of different points of $\mathbb{P}_1$, and $\delta \in k - \{0, 1\}$, there is a unique projectivity f of $\mathbb{P}_1$ with fixed points p_0, p_1 and modulus (relative to the specified ordering of the fixed points) δ.*

Proof. For the existence, just take a reference with vertices p_0, p_1, in this order, and the projectivity with equation $x^* = \delta x$. The uniqueness is obvious, as being p_0, p_1 fixed and $(p_0, p_1, p, f(p)) = \delta$ for any $p \neq p_0, p_1$, determines $f(p)$ for all $p \in \mathbb{P}_1$, by 2.9.8. □

In the sequel we will refer to the projectivity of 5.5.12 as the *projectivity with fixed points p_0, p_1 and invariant* (or *modulus*) δ.

The next proposition shows that the values of the invariant classify the hyperbolic projectivities, and in particular justifies the use of the name *invariant*:

Proposition 5.5.13. *Hyperbolic projectivities f, f', of one-dimensional projective spaces $\mathbb{P}_1$ and $\mathbb{P}'_1$ over the same field, are projectively equivalent if and only if they have equal (or reciprocal) invariants.*

Proof. If $g\colon \mathbb{P}_1 \to \mathbb{P}'_1$ is a projectivity and f has fixed points p_0, p_1, then $g(f)$ has fixed points $g(p_0)$, $g(p_1)$, by 5.5.5. The invariance of the cross ratio gives

$$(p_0, p_1, p, f(p)) = (g(p_0), g(p_1), g(p), g(f(p))),$$

for any $p \neq p_0, p_1$. This proves that the invariants of f and $g(f)$ are equal, as, by definition, $g(f)(g(p))) = g(f(p))$. Conversely, take g to be any projectivity mapping the fixed points p_0, p_1 of f to the fixed points p'_0, p'_1 of f', the pairs of fixed points being ordered to give equal moduli. Then, for any $p \neq p_0, p_1$, $g(p) \neq p'_0, p'_1$ and by the equality of moduli

$$(p_0, p_1, p, f(p)) = (p'_0, p'_1, g(p), f'(g(p))).$$

Since, on the other hand, again by the invariance of the cross ratio,

$$(p_0, p_1, p, f(p)) = (p'_0, p'_1, g(p), g(f(p)),$$

we get

$$f'(g(p)) = g(f(p)) = g \circ f \circ g^{-1}(g(p))$$

which proves

$$f'(p') = (g \circ f \circ g^{-1})(p') = g(f)(p')$$

for any $p' \neq p'_0, p'_1$. The same equality being obviously true for $p' = p'_0, p'_1$, $f' = g(f)$ as wanted. □

Parabolic projectivities. Assume now that $f : \mathbb{P}_1 \to \mathbb{P}_1$ is parabolic. Take its only fixed point q, any $q' \neq q$ and $f(q')$ to compose a reference of $\mathbb{P}_1$, any two of these points being obviously different. If relative to this reference f has equation

$$x^* = \frac{ax + b}{cx + d}, \quad ad - bc \neq 0,$$

having the point q, with coordinate ∞, fixed causes $c = 0$. Then $d \neq 0$ and taking $a' = a/d$, $b' = b/d$ the former equation may be rewritten

$$x^* = a'x + b', \quad a' \neq 0.$$

Since f is parabolic and has fixed point q, no point with finite coordinate may be fixed, which forces $a' = 1$. Lastly, the point with coordinate 1 being the image of the point with coordinate 0 yields $b' = 1$ and we have proved:

Proposition 5.5.14. *If $f : \mathbb{P}_1 \to \mathbb{P}_1$ is parabolic, then its fixed point q, any $q' \neq q$ and $f(q')$ compose a reference relative to which f has equation*

$$x^* = x + 1. \tag{5.17}$$

Also in this case the converse is clearly true: any projectivity with an equation of the form of (5.17) is parabolic, has the first vertex of the reference as fixed point and maps the second vertex to the unit point. The absence of parameters in equation (5.17) causes all parabolic projectivities to be projectively equivalent:

Proposition 5.5.15. *Any two parabolic projectivities f, f', of one-dimensional projective spaces $\mathbb{P}_1$ and $\mathbb{P}'_1$ over the same field, are projectively equivalent.*

Proof. Let Δ and Δ' be references relative to which f and f', respectively, have equations of the form of (5.17) and take g the projectivity mapping Δ to Δ'. Then, as the reader may easily check using 2.8.1(b), $g \circ f = f' \circ g$ and so $g(f) = g \circ f \circ g^{-1} = f'$ as claimed. □

Elliptic projectivities. We are in case $k = \mathbb{R}$. As it is clear from 5.1.18 and 5.5.2, the complex extension of any elliptic projectivity is hyperbolic. This causes the elliptic projectivities to behave much as the hyperbolic ones, but for having their fixed points imaginary, and hence non-suitable to be taken as vertices of a real reference. In fact we have:

Proposition 5.5.16. *If f is an elliptic projectivity of $\mathbb{P}_1$, with imaginary fixed points $q, \bar{q}$, then $q, \bar{q}$ are complex-conjugate, the cross ratio*

$$\delta = (q, \bar{q}, p, f(p)) \in \mathbb{C}$$

is independent of p, $\delta \neq 1$ and $|\delta| = 1$.

Conversely, for any two complex-conjugate imaginary points $q, \bar{q}$ and any complex number $\delta \neq 1$ with $|\delta| = 1$, mapping each point $p \in \mathbb{P}_1$ to the only point $p' \in \mathbb{C}\mathbb{P}_1$ for which

$$(q, \bar{q}, p, p') = \delta$$

defines an elliptic projectivity f of $\mathbb{P}_1$ with fixed points $q, \bar{q}$.

Proof. The fixed points of f are conjugate by 5.5.4. The constancy of the cross ratio δ follows from 5.5.7 applied to the complex extension $f_{\mathbb{C}}$ of f. Note that the cross ratio is always defined as, p being real, $p \neq q, \bar{q}$. Clearly, $\delta \neq 1$ because $p \neq f(p)$ for any $p \in \mathbb{P}_1$. By 5.1.5,

$$\bar{\delta} = (\sigma(q), \sigma(\bar{q}), \sigma(p), \sigma(f(p))) = (\bar{q}, q, p, f(p)) = \delta^{-1},$$

and so $|\delta| = 1$.

For the converse, by 5.5.12, take $\tilde{f}$ to be the projectivity of $\mathbb{C}\mathbb{P}_1$ with ordered pair of fixed points $q, \bar{q}$ and invariant δ. Write $p' = \tilde{f}(p)$. Next we will show that for any $p \in \mathbb{P}_1$, $p' \in \mathbb{P}_1$. Again by 5.1.5,

$$(\sigma(q), \sigma(\bar{q}), \sigma(p), \sigma(p')) = (\bar{q}, q, p, \sigma(p')) = \bar{\delta} = \delta^{-1},$$

the last equality due to being $|\delta| = 1$. Thus

$$(q, \bar{q}, p, \sigma(p')) = \delta,$$

which, together with the condition defining p',

$$(q, \bar{q}, p, p') = \delta,$$

yields $p' = \sigma(p')$ and p' is real. Now the map defined in the claim is

$$f = \tilde{f}_{|\mathbb{P}_1} : \mathbb{P}_1 \longrightarrow \mathbb{P}_1.$$

Since $\tilde{f}$ preserves cross ratios, so does f and therefore, by 2.9.15, f itself is a projectivity. Lastly, the uniqueness of the complex extension of f (5.1.18) shows that $\tilde{f} = f_{\mathbb{C}}$ and then, by 5.5.4, the fixed points of f are $q, \bar{q}$. □

As for the hyperbolic projectivities, the complex number δ of the first half of 5.5.16 is called the *modulus* or the *invariant* of f. Again f determines only the pair of values $\delta^{\pm 1}$ and not δ itself, unless an ordering on the fixed points has been specified. Note that, as shown in 5.5.16, the invariant δ of f need not be real and always has $|\delta| = 1$.

Next we will do some work in order to get a nice form of the matrix of an elliptic projectivity f. First we swap over $q, \bar{q}$, if needed, in order to have the argument β of δ satisfying $0 < \beta \leq \pi$. Then we pick any point $p \in \mathbb{P}_1$ and take a reference Ω of $\mathbb{C}\mathbb{P}_1$ with vertices the fixed points $q, \bar{q}$ of f and unit point p, the three points being obviously distinct. According to 5.5.6, relative to this reference $f_{\mathbb{C}}$ has matrix

$$\begin{pmatrix} \delta & 0 \\ 0 & 1 \end{pmatrix},$$

where δ is the invariant of f. Let μ be the square root of δ which has argument $\alpha = \beta/2$: then $0 < \alpha \leq \pi/2$. Still $|\mu| = 1$ and so $\bar{\mu} = \mu^{-1}$. After dividing the above matrix by μ, we take

$$\begin{pmatrix} \mu & 0 \\ 0 & \bar{\mu} \end{pmatrix} \tag{5.18}$$

as matrix of $f_{\mathbb{C}}$ relative to Ω.

Of course we need a real reference: we take its first vertex $p_0 = p$; its second vertex p_1 is taken to be the fourth harmonic of q, $\bar{q}$, p. Then p_0, p_1 and q are three different points; this allows us to take as unit point the only point A in $\mathbb{C}\mathbb{P}_1$ for which

$$(p_0, p_1, q, A) = \boldsymbol{i},$$

which in particular assures that p_0, p_1, A are three different points. The point $p = p_0$ has been taken real, so to see that $\Delta = (p_0, p_1, A)$ is a reference of $\mathbb{P}_1$ we need to see that both p_1 and A are real too, which we will do next.

Taking complex-conjugates, the equality

$$-1 = (q, \bar{q}, p, p_1), \tag{5.19}$$

defining p_1, yields

$$-1 = (\bar{q}, q, p, \sigma(p_1)) = (q, \bar{q}, p, \sigma(p_1)). \tag{5.20}$$

Then the equalities (5.19) and (5.20) together force $p_1 = \sigma(p_1)$, and so p_1 is real.

Similarly, from

$$(p_0, p_1, q, A) = \boldsymbol{i}$$

we get

$$(p_0, p_1, \bar{q}, \sigma(A)) = -\boldsymbol{i}\,. \tag{5.21}$$

On the other hand, by 2.9.6, we get

$$(p_0, p_1, q, \sigma(A))(p_0, p_1, \sigma(A), \bar{q}) = (p_0, p_1, q, \bar{q}) = -1$$

and so we have

$$(p_0, p_1, q, \sigma(A)) = -(p_0, p_1, \sigma(A), \bar{q})^{-1} = -(p_0, p_1, \bar{q}, \sigma(A)) = \boldsymbol{i}\,. \tag{5.22}$$

As above, the equalities (5.21) and (5.22) give $\sigma(A) = A$, and so also the point A is real.

From their definitions,

$$p_0 = [1, 1]_\Omega, \quad p_1 = [1, -1]_\Omega \quad \text{and} \quad A = [\boldsymbol{i} + 1, \boldsymbol{i} - 1]_\Omega.$$

One may thus compute a matrix M of $f_{\mathbb{C}}$ relative to Δ from the matrix (5.18) above, using 2.8.10. This, up to a scalar factor, gives

$$M = \begin{pmatrix} \mu + \bar{\mu} & -\boldsymbol{i}\,\mu + \boldsymbol{i}\,\bar{\mu} \\ \boldsymbol{i}\,\mu - \boldsymbol{i}\,\bar{\mu} & \mu + \bar{\mu} \end{pmatrix},$$

which is a real matrix. Since Δ is a real reference M is a matrix of f relative to Δ, by 5.1.18 and 2.8.6. Just writing $\mu = \cos\alpha + \boldsymbol{i}\sin\alpha$ gives:

Proposition 5.5.17. *Assume that f is an elliptic projectivity of $\mathbb{P}_1$, with imaginary fixed points q, $\bar{q}$. For any choice of a point $p_0 \in \mathbb{P}_1$, p_0 itself, its fourth harmonic p_1 with respect to q, $\bar{q}$ and the point A for which $(p_0, p_1, q, A) = \boldsymbol{i}$ compose a reference of $\mathbb{P}_1$. Relative to it, f has matrix*

$$\begin{pmatrix} \cos\alpha & -\sin\alpha \\ \sin\alpha & \cos\alpha \end{pmatrix}, \tag{5.23}$$

where $\alpha \in (0, \pi/2]$ and 2α is the argument of (the suitable determination of) the invariant of f.

Remark 5.5.18. An easy computation shows that any projectivity of a real $\mathbb{P}_1$, with matrix (5.23) and $\alpha \in (0, \pi/2]$, is elliptic and has invariant $e^{2i\alpha}$.

Like the hyperbolic ones, the elliptic projectivities are classified by their moduli:

Proposition 5.5.19. *Elliptic projectivities f and f', of one-dimensional projective spaces $\mathbb{P}_1$ and $\mathbb{P}'_1$, are projectively equivalent if and only if they have equal (or reciprocal) invariants.*

Proof. If $f' = g(f)$, then, clearly, $f'_{\mathbb{C}} = g_{\mathbb{C}}(f_{\mathbb{C}})$ and so, an elliptic projectivity and its complex extension having equal invariants, the *only if* part follows from 5.5.13. For the converse just note that two elliptic projectivities with same or reciprocal invariants may be represented by the same matrix (5.23) above and then proceed as in the proof of 5.5.15. □

Of course, the argument $\alpha \in (0, \pi/2]$ may be used in Proposition 5.5.19 instead of the invariant $\delta = e^{2i\alpha}$.

5.6 Involutions

The projectivities τ of a one-dimensional projective space $\mathbb{P}_1$ that satisfy $\tau^2 = \mathrm{Id}_{\mathbb{P}_1}$ and $\tau \neq \mathrm{Id}_{\mathbb{P}_1}$ are called *involutions* of $\mathbb{P}_1$. As we will see, involutions do appear quite often in different geometric contexts and may be thought of in many different manners (see the forthcoming Remark 9.7.3). Their higher-dimensional analogues, namely the non-identical collineations f of $\mathbb{P}_n$, $n > 1$, for which $f^2 = \mathrm{Id}_{\mathbb{P}_n}$ and $f \neq \mathrm{Id}_{\mathbb{P}_n}$, are less interesting: we will call them *involutive collineations* (or projectivities) , thus reserving the name *involutions* for the one-dimensional case.

Remark 5.6.1. If $k = \mathbb{R}$, it directly follows from 5.1.18 that a projectivity f of $\mathbb{P}_1$ is an involution if and only if so is its complex extension.

The condition $\tau^2 = \mathrm{Id}_{\mathbb{P}_1}$ may be equivalently written $\tau = \tau^{-1}$ and so the involutions are the non-identical projectivities τ of $\mathbb{P}_1$ such that for any $p, q \in \mathbb{P}_1$, $q = \tau(p)$ if and only if $p = \tau(q)$. In other words, when two points correspond by an involution, then they correspond in both senses, making it unnecessary to specify which point is mapped to which. The unordered pairs of corresponding points $\{p, \tau(p)\}$, $p \in \mathbb{P}_1$, are said to be the *pairs of* (or *belonging to*) the involution.

First of all we characterize the involutions among all the projectivities of $\mathbb{P}_1$:

Proposition 5.6.2. *A projectivity f of $\mathbb{P}_1$ with symmetric equation*

$$Ax_0x_0^* + Bx_0x_1^* + B'x_1x_0^* + Cx_1x_1^* = 0, \quad BB' - AC \neq 0, \tag{5.24}$$

is an involution if and only if $B = B'$.

Proof. Assume $B = B'$. Since a symmetric equation of $\mathrm{Id}_{\mathbb{P}_1}$ is

$$x_0 x_1^* - x_0^* x_1 = 0,$$

clearly $f \neq \mathrm{Id}_{\mathbb{P}_1}$. Furthermore, if $B = B'$ the first member of the equation (5.24) is symmetric in the two pairs of variables (x_0, x_1) and (x_0^*, x_1^*); this means that the condition for having $q = f(p)$ is the same as for having $p = f(q)$ for any $p, q \in \mathbb{P}_1$, as wanted.

Conversely, assume that f is an involution. Take $p = [\alpha_0, \alpha_1]$ to be a non-fixed point of f and write $q = f(p) = [\alpha_0^*, \alpha_1^*]$. Then

$$A\alpha_0\alpha_0^* + B\alpha_0\alpha_1^* + B'\alpha_1\alpha_0^* + C\alpha_1\alpha_1^* = 0,$$

and since $f(q) = p$, also

$$A\alpha_0^*\alpha_0 + B\alpha_0^*\alpha_1 + B'\alpha_1^*\alpha_0 + C\alpha_1^*\alpha_1 = 0.$$

Subtracting these equalities gives

$$(B - B')\begin{vmatrix} \alpha_0 & \alpha_1 \\ \alpha_0^* & \alpha_1^* \end{vmatrix} = 0$$

and so $B - B' = 0$, the determinant being not zero due to the hypothesis $p \neq q$. □

Remark 5.6.3. The above condition $B = B'$, for f to be an involution, written in terms of a matrix

$$\begin{pmatrix} a & b \\ c & d \end{pmatrix}$$

of f, reads just $d = -a$ (see Section 5.2).

In the proof of 5.6.2 we have proved $B = B'$ by just using the existence of a single pair of different points p, q such that $q = f(p)$ and $p = f(q)$. Since in turn $B = B'$ assures that f is an involution, we have in fact proved a condition for f to be an involution that is apparently far weaker than the definition itself:

Corollary 5.6.4 (of the proof of 5.6.2). *If f is a projectivity of $\mathbb{P}_1$ and there are two distinct points $p, q \in \mathbb{P}_1$ for which $q = f(p)$ and $p = f(q)$, then f is an involution.*

If g is a transformation of any kind, a pair of points p, q such that $q = g(p)$ and $p = g(q)$ is called an *involutive pair* of g. Corollary 5.6.4 says thus that if a projectivity f has an involutive pair of different points, then it is an involution, and therefore all pairs p, $f(p)$ are involutive.

Note that, in the hypothesis of 5.6.4, it is essential to assume that the points of the involutive pair are distinct. In fact any projectivity with fixed point p has the involutive pair p, p.

Remark 5.6.5. In case $k = \mathbb{R}$, the claim of 5.6.4 still holds if p and q are assumed to be distinct imaginary points corresponding to each other by the complex extension of f, as then 5.6.4 applies to $f_{\mathbb{C}}$ showing that $f_{\mathbb{C}}$ is an involution, and therefore so is f (5.6.1).

According to 5.6.2, in the sequel the symmetric equations of the involutions will be written in the form

$$Ax_0x_0^* + B(x_0x_1^* + x_1x_0^*) + Cx_1x_1^* = 0, \quad B^2 - AC \neq 0, \tag{5.25}$$

any such equation representing in turn an involution. The corresponding equation of fixed points has then the form

$$Ax_0^2 + 2Bx_0x_1 + Cx_1^2 = 0$$

and discriminant $AC - B^2 \neq 0$ due to the initial requirement on the coefficients of the equation. There are thus no parabolic involutions. In fact we can say more, namely:

Proposition 5.6.6. *A projectivity f of $\mathbb{P}_1$ is an involution if and only if it is hyperbolic or elliptic and has invariant -1.*

Proof. If f is an involution, then $f \neq \mathrm{Id}_{\mathbb{P}_n}$ and we have just seen that it is not parabolic: it is thus hyperbolic or elliptic. In either case, let $q, \bar{q}$ be its fixed points and pick any non-fixed point p of f. According to 5.5.7 or 5.5.16, the invariant δ of f is $\delta \neq 1$ and

$$\delta = (q, \bar{q}, p, f(p)).$$

Also, p being the image of $f(p)$,

$$\delta = (q, \bar{q}, f(p), p) = (q, \bar{q}, p, f(p))^{-1}.$$

Then it follows that $\delta = \delta^{-1}$ and so $\delta = -1$. By 5.5.6 and 5.5.11 any hyperbolic projectivity with invariant -1 has equation

$$x^* = -x$$

and so it clearly is an involution. The same occurs in the elliptic case, as then 5.5.17 provides the equation

$$x^* = -\frac{1}{x}. \qquad \square$$

Remark 5.6.7. We have seen in the above proof that, relative to a suitable reference, any hyperbolic involution has equation $x^* = -x$, while any elliptic involution has equation $x^* = -1/x$. Details about the reference are in 5.5.6 and 5.5.17.

Any involution consists thus in mapping each point to its fourth harmonic with respect to a pair of already selected points. To be precise:

Corollary 5.6.8. *Assume that q, $\bar{q}$ is either a pair of distinct points of $\mathbb{P}_1$ or, in case $k = \mathbb{R}$, a pair of conjugate imaginary points. Leaving q and $\bar{q}$ fixed, in case $q, \bar{q} \in \mathbb{P}_1$, and mapping any $p \neq q, \bar{q}$ to its fourth harmonic relative to q, $\bar{q}$, is an involution of $\mathbb{P}_1$. Any involution τ of $\mathbb{P}_1$ may be obtained in this way, q and $\bar{q}$ being then the fixed points of τ.*

Proof. Proposition 5.6.6 turns all affirmations in the claim into particular cases of 5.5.7, 5.5.12 and 5.5.16. □

We state part of 5.6.8 separately, for future reference:

Corollary 5.6.9. *An involution is determined by its fixed points. Points p, p', both different from q_1 and q_2, correspond by the involution with fixed points q_1, q_2 if and only if $(q_1, q_2, p, p') = -1$.*

Two pairs of corresponding points may be assigned to determine an involution:

Proposition 5.6.10. *Take any $p, p', q, q' \in \mathbb{P}_1$ such that $\{p, p'\} \cap \{q, q'\} = \emptyset$. Then there is a unique involution τ of $\mathbb{P}_1$ such that $\tau(p) = p'$ and $\tau(q) = q'$.*

Proof. If $p = p'$ and $q = q'$, then the claim follows from 5.6.8. Otherwise, up to renaming the points, we may assume that $p \neq p'$. Then $\Delta = (p, q, p')$ and $\Omega = (p', q', p)$ are references of $\mathbb{P}_1$ and, by 2.8.1, we take τ to be the projectivity for which $\tau(p) = p'$, $\tau(q) = q'$ and $\tau(p') = p$. Then τ satisfies the conditions of the claim and, since it has the involutive pair of distinct points p, p', τ is an involution by 5.6.4. For the uniqueness, note that any involution mapping p to p' and q to q', maps also p' to p and so Δ to Ω: by 2.8.1 there is a unique projectivity, and hence at most one involution, doing so. □

In the sequel we will refer to the only involution mapping p to p' and q to q' as the involution *determined by* $\{p, p'\}$ and $\{q, q'\}$, the ordering on each pair being of course irrelevant.

Clearly, taking $\{p, p'\} = \{q, q'\}$ in 5.6.10 would not determine τ, while taking $\{p, p'\}$ and $\{q, q'\}$ with a single common point would give no solution, as then the common point would have two different images by τ. Hence the need of the hypothesis $\{p, p'\} \cap \{q, q'\} = \emptyset$.

The cross ratio of two different pairs of corresponding points allows us to distinguish between elliptic and hyperbolic involutions, namely,

Proposition 5.6.11. *Assume that p, p' and q, q' are two different pairs of distinct points corresponding by an involution τ of $\mathbb{P}_{1,\mathbb{R}}$: if τ is hyperbolic, then $(p, p', q, q') > 0$, while for τ elliptic, $(p, p', q, q') < 0$.*

Proof. In the hyperbolic case, taking the reference of 5.6.7, the pairs of corresponding points have absolute coordinates $x, -x$ and $y, -y$, all finite and non-zero because neither of the points is fixed. A direct computation gives

$$(p, p', q, q') = \frac{(x-y)^2}{(x+y)^2} > 0.$$

In the elliptic case, again using the reference of 5.6.7, the points have absolute coordinates $x, -1/x$ and $y, -1/y$. If all these coordinates are finite we get

$$(p, p', q, q') = -\frac{(x-y)^2}{(xy+1)^2} < 0,$$

because $xy + 1 = 0$ would force the two pairs to coincide. Otherwise, up to renaming the points, assume $x = \infty$: then $1/x = 0$, the first pair is the pair of vertices of the reference and, the pairs being not coincident, $y \neq 0, \infty$. By 2.9.5,

$$(p, p', q, q') = -1/y^2 < 0,$$

as wanted. □

Next we will make some computations that re-prove 5.6.10 and have further and interesting consequences. Fix a reference in $\mathbb{P}_1$, still let $\{p, p'\}$ and $\{q, q'\}$ be two disjoint pairs of points and assume now that

$$p = [y_0, y_1], \quad p' = [y_0^*, y_1^*], \quad q = [z_0, z_1], \quad q' = [z_0^*, z_1^*].$$

An involution τ, with symmetric equation

$$Ax_0x_0^* + B(x_0x_1^* + x_1x_0^*) + Cx_1x_1^* = 0,$$

satisfies the conditions of 5.6.10 if and only if the coefficients A, B, C satisfy

$$\begin{aligned} Ay_0y_0^* + B(y_0y_1^* + y_1y_0^*) + Cy_1y_1^* &= 0, \\ Az_0z_0^* + B(z_0z_1^* + z_1z_0^*) + Cz_1z_1^* &= 0, \end{aligned}$$

which is a system of homogeneous linear equations in A, B, C. We have:

Lemma 5.6.12. *The system of equations*

$$\begin{aligned} Ay_0y_0^* + B(y_0y_1^* + y_1y_0^*) + Cy_1y_1^* &= 0, \\ Az_0z_0^* + B(z_0z_1^* + z_1z_0^*) + Cz_1z_1^* &= 0 \end{aligned} \tag{5.26}$$

in A, B, C, has rank two. Furthermore A, B, C is a non-trivial solution of it if and only if the pairs $\{[y_0, y_1], [y_0^*, y_1^*]\}$ and $\{[z_0, z_1], [z_0^*, z_1^*]\}$ correspond by the involution of symmetric equation

$$Ax_0x_0^* + B(x_0x_1^* + x_1x_0^*) + Cx_1x_1^* = 0.$$

Proof. We need only to prove that the system has rank two, as the remainder of the claim has been seen above. To this end, recall from Section 5.2 that

$$y_0 y_0^*, \quad -(y_0 y_1^* + y_1 y_0^*), \quad y_1 y_1^*$$

are the coefficients of an equation $F = 0$ of the pair p, p'. Similarly

$$z_0 z_0^*, \quad -(z_0 z_1^* + z_1 z_0^*), \quad z_1 z_1$$

are the coefficients of an equation $G = 0$ of q, q'. Now, the equations $F = 0$ and $G = 0$ are not proportional because $\{p, p'\} \neq \{q, q'\}$ and therefore their rows of coefficients are linearly independent. Turning into their opposites the central terms of these rows gives the rows of coefficients of the equations (5.26); thus the latter are also linearly independent and the system of equations (5.26) has rank two. □

Lemma 5.6.12 gives a new proof of 5.6.10:

Proof. Second proof of 5.6.10: Since the system of equations (5.26) has rank two, it has a non-trivial solution, unique up to a scalar factor, which proves the uniqueness of τ. For the existence we need to check that this solution satisfies $B^2 - AC \neq 0$. As it is well known,

$$A = \begin{vmatrix} y_0 y_1^* + y_1 y_0^* & y_1 y_1^* \\ z_0 z_1^* + z_1 z_0^* & z_1 z_1^* \end{vmatrix}, \qquad B = -\begin{vmatrix} y_0 y_0^* & y_1 y_1^* \\ z_0 z_0^* & z_1 z_1^* \end{vmatrix},$$
$$C = \begin{vmatrix} y_0 y_0^* & y_0 y_1^* + y_1 y_0^* \\ z_0 z_0^* & z_0 z_1^* + z_1 z_0^* \end{vmatrix}$$

may be taken as a non-trivial solution of (5.26). Then it is straightforward to check that

$$B^2 - AC = \begin{vmatrix} y_0 & y_1 \\ z_0 & z_1 \end{vmatrix} \cdot \begin{vmatrix} y_0 & y_1 \\ z_0^* & z_1^* \end{vmatrix} \cdot \begin{vmatrix} y_0^* & y_1^* \\ z_0 & z_1 \end{vmatrix} \cdot \begin{vmatrix} y_0^* & y_1^* \\ z_0^* & z_1^*, \end{vmatrix}$$

which is non-zero due to the hypothesis $\{p, p'\} \cap \{q, q'\} = \emptyset$. □

It is now easy to give a direct representation of the involution determined by two pairs of points:

Proposition 5.6.13. *The points $[x_0, x_1]$, $[x_0^*, x_1^*]$ correspond by the involution τ, determined by the disjoint pairs $\{[y_0, y_1], [y_0^*, y_1^*]\}$ and $\{[z_0, z_1], [z_0^*, z_1^*]\}$, if and only if*

$$\begin{vmatrix} x_0 x_0^* & x_0 x_1^* + x_1 x_0^* & x_1 x_1^* \\ y_0 y_0^* & y_0 y_1^* + y_1 y_0^* & y_1 y_1^* \\ z_0 z_0^* & z_0 z_1^* + z_1 z_0^* & z_1 z_1^* \end{vmatrix} = 0. \tag{5.27}$$

Proof. By 5.6.12, the points $[x_0, x_1]$, $[x_0^*, x_1^*]$ correspond by τ if and only if

$$Ax_0x_0^* + B(x_0x_1^* + x_1x_0^*) + Cx_1x_1^* = 0,$$

for A, B, C a non-trivial solution of (5.26). In turn, this is equivalent to the existence of a non-trivial solution of the system of equations in A, B, C,

$$\begin{aligned} Ax_0x_0^* + B(x_0x_1^* + x_1x_0^*) + Cx_1x_1^* &= 0,\\ Ay_0y_0^* + B(y_0y_1^* + y_1y_0^*) + Cy_1y_1^* &= 0,\\ Az_0z_0^* + B(z_0z_1^* + z_1z_0^*) + Cz_1z_1^* &= 0, \end{aligned} \tag{5.28}$$

which in turn is equivalent to the equality (5.27). □

Note that the equality (5.27) is, as an equation in x_0, x_1, x_0^*, x_1^*, a form of the symmetric equation of an involution written directly from two pairs of points determining it. We close this section with a presentation of a form of representing involutions which is quite different from the preceding ones. It is called the *parametric representation* of the involution and describes it as a set of unordered pairs of points, rather than as a map.

Proposition 5.6.14. *Let $\{p, p'\}$ and $\{q, q'\}$ be disjoint pairs of points of $\mathbb{P}_1$ that, after fixing coordinates, have equations $F = 0$ and $G = 0$ (see Section 5.2). A pair of points $\{t, t'\}$, with equation $H = 0$, belongs to the involution determined by $\{p, p'\}$ and $\{q, q'\}$ if and only if*

$$H = \lambda F + \mu G$$

for some $\lambda, \mu \in k$, not both zero.

Proof. The claim is independent of the choice of the equations of the pairs, as each equation is determined by the pair up to a constant factor. Assume that $t = [x_0, x_1]$, $t' = [x_0^*, x_1^*]$ and, as already, $p = [y_0, y_1]$, $p' = [y_0^*, y_1^*]$, $q = [z_0, z_1]$ and $q' = [z_0^*, z_1^*]$. Then, using X_0, X_1 as variables, we may take

$$\begin{aligned} H &= x_0x_0^*X_1^2 - (x_0x_1^* + x_1x_0^*)X_1X_0 + x_1x_1^*X_0^2,\\ F &= y_0y_0^*X_1^2 - (y_0y_1^* + y_1y_0^*)X_1X_0 + y_1y_1^*X_0^2,\\ G &= z_0z_0^*X_1^2 - (z_0z_1^* + z_1z_0^*)X_1X_0 + z_1z_1^*X_0^2. \end{aligned}$$

As already noted, F and G are linearly independent polynomials because $\{p, p'\} \neq \{q, q'\}$. Then, by 2.7.1, the equality of the claim is equivalent to the linear dependence of H, F, G, that is to having

$$\begin{vmatrix} x_0x_0^* & -(x_0x_1^* + x_1x_0^*) & x_1x_1^*\\ y_0y_0^* & -(y_0y_1^* + y_1y_0^*) & y_1y_1^*\\ z_0z_0^* & -(z_0z_1^* + z_1z_0^*) & z_1z_1^* \end{vmatrix} = 0.$$

Since the above determinant is the opposite of the one in (5.27), the claim follows from 5.6.13. □

Remark 5.6.15. Hypothesis and notations being as in 5.6.14, we name τ the involution determined by $\{p, p'\}$ and $\{q, q'\}$. If $k = \mathbb{C}$, then any equation $\lambda F + \mu G = 0$ defines a pair of points of $\mathbb{P}_1$, which correspond by τ by 5.6.14. Thus, while $(\lambda, \mu) \in \mathbb{C}^2 - \{0, 0\}$, the equations $\lambda F + \mu G = 0$ define all the pairs of points of $\mathbb{P}_1$ that are corresponding by τ. If $k = \mathbb{R}$ this is no longer true. Note first that the complex extension $\tau_{\mathbb{C}}$ clearly is the involution of $\mathbb{C}\mathbb{P}_1$ determined by $\{p, p'\}$ and $\{q, q'\}$. Then, $\lambda F + \mu G = 0$ defines either a pair of real points which, by 5.6.14, are corresponding by τ, or a pair of imaginary and conjugate points which, using 5.6.14 in $\mathbb{C}\mathbb{P}_1$, are corresponding by $\tau_{\mathbb{C}}$. Conversely, 5.6.14, assures that any pair of corresponding real points has an equation $\lambda F + \mu G = 0$, $(\lambda, \mu) \in \mathbb{R}^2 - \{0, 0\}$. A corresponding pair of imaginary and conjugate points has a real equation $H = 0$, and, again using 5.6.14 in $\mathbb{C}\mathbb{P}_1$, $H = \lambda F + \mu G$, for some $(\lambda, \mu) \in \mathbb{C}^2 - \{0, 0\}$. Now, H, F and G being real polynomials, necessarily $\lambda, \mu \in \mathbb{R}$, because they may be computed as the only solution of a system of real linear equations. So, also in this case $H = \lambda F + \mu G$, for some $(\lambda, \mu) \in \mathbb{R}^2 - \{0, 0\}$.

Summarizing, for $k = \mathbb{C}$, $\lambda F + \mu G = 0$, $(\lambda, \mu) \in k^2 - \{0, 0\}$, are the equations of all pairs of points of $\mathbb{P}_1$ corresponding by τ, while for $k = \mathbb{R}$ the same equations define all pairs of corresponding real points together with all pairs of corresponding (by $\tau_{\mathbb{C}}$) imaginary and conjugate points.

The reader may rewrite all the above using absolute coordinates: the expressions will become lighter, but suitable conventions for the cases in which some coordinate equals ∞ should be set.

5.7 Fixed points of collineations

In this section we will consider collineations of an n-dimensional projective space, $f : \mathbb{P}_n \to \mathbb{P}_n$, and study the points and hyperplanes they leave invariant. Let us begin by giving some definitions: as in the one-dimensional case, a point $p \in \mathbb{P}_n$ is said to be a *fixed point* of a collineation $f : \mathbb{P}_n \to \mathbb{P}_n$ if and only if $f(p) = p$. In such a case it is equivalently said that p is *invariant* or *fixed* by f.

More generally, a linear variety $L \subset \mathbb{P}_n$ said to be *fixed* or *invariant* by f if and only if $f(L) = L$, or, equivalently by 1.3.6 (b), just $f(L) \subset L$. If, in particular, $L = H$ is a hyperplane, then it is fixed by f if and only if $f^{\vee}(H) = f^{-1}(H) = H$, that is, if and only if H, as a point of $\mathbb{P}_n^{\vee}$, is fixed by the dual $f^{\vee}$ of f (see Section 4.5).

The linear varieties all whose points are fixed by f are called *varieties of fixed points*. Obviously, they are invariant by f. The converse is not true if $\dim L > 0$, as having $f(p) \in L$ for all $p \in L$ does not force $f(p) = p$ for all $p \in L$. For an easy example any non-identical collineation of $\mathbb{P}_n$ has $\mathbb{P}_n$ as an invariant variety which is not a variety of fixed points. It is also clear (from 1.7.3) that intersections and joins of invariant linear varieties are invariant too, which will be used in the sequel without further reference.

The next lemma is the key to all results in this section:

Lemma 5.7.1. *Assume that f is a collineation of $\mathbb{P}_n$, represented by an automorphism φ of the vector space E associated to $\mathbb{P}_n$. Then a vector $v \in E - \{0\}$ represents a fixed point of f if and only if v is an eigenvector of φ.*

Proof. Since $f([v]) = [\varphi(v)]$, the point $[v]$ is fixed if and only if $[\varphi(v)] = [v]$, that is, if and only if there is $\lambda \in k$ such that $\varphi(v) = \lambda v$, which is just the definition of an eigenvector. □

Next we will prove the main result about fixed points of collineations. It asserts that the fixed points of a collineation are those of a finite set of linear varieties none of which intersects the join of the others. Furthermore it gives a precise determination of these linear varieties in terms of a representative of the collineation.

Theorem 5.7.2. *Let f be a collineation of $\mathbb{P}_n$ and choose a representative φ of f. Denote by $\lambda_1, \dots, \lambda_m$, $0 \leq m \leq n$, the different roots of the characteristic polynomial of φ, $c_\varphi(X) = \det(\varphi - X\,\mathrm{Id})$. For each $i = 1, \dots, m$ define the linear variety V_i of $\mathbb{P}_n$ as*

$$V_i = [\ker(\varphi - \lambda_i \mathrm{Id})].$$

Then:

(1) *The linear varieties V_i, $i = 1, \dots, m$, are non-empty and do not depend on the choice of the representative φ of f.*

(2) *A point $p \in \mathbb{P}_n$ is fixed by f if and only if $p \in V_i$ for some $i = 1, \dots, m$.*

(3) *(Segre's theorem) If $m \geq 2$, for each $i = 1, \dots, m$,*

$$\emptyset = V_i \cap \Big(\bigvee_{j \neq i} V_j\Big).$$

Proof. First, by the definition of V_i,

$$\dim V_i = \dim \ker(\varphi - \lambda_i \mathrm{Id}) - 1 = n - \mathrm{rk}(\varphi - \lambda_i \mathrm{Id}) \geq n - n = 0,$$

the last inequality due to the hypothesis $\det(\varphi - \lambda_i \mathrm{Id}) = 0$. Any other representative of f being $\alpha\varphi$, $\alpha \in k - \{0\}$, it has characteristic polynomial $c_{\alpha\varphi}(X) = \det(\alpha\varphi - X\,\mathrm{Id}) = \alpha^{n+1}\det(\varphi - \frac{X}{\alpha}\mathrm{Id}) = \alpha^{n+1} c_\varphi(X/\alpha)$, with roots $\alpha\lambda_i$, $i = 1, \dots, m$. Then

$$\ker(\alpha\varphi - \alpha\lambda_i \mathrm{Id}) = \ker(\alpha(\varphi - \lambda_i \mathrm{Id})) = \ker(\varphi - \lambda_i \mathrm{Id}).$$

This proves claim (1).

Using Lemma 5.7.1, claim (2) is equivalently stated by saying that the eigenvectors of φ are just the non-zero vectors belonging to one of the subspaces $\ker(\varphi - \lambda_i \mathrm{Id})$, $i = 1, \dots, m$, which is in turn a well-known fact of linear algebra. Anyway we sketch its easy proof for the convenience of the reader: it is clear that any non-zero v with $(\varphi - \lambda_i \mathrm{Id})(v) = 0$ is an eigenvector (with eigenvalue λ_i). Conversely, if $v \neq 0$ and $\varphi(v) = \lambda v$, then $v \in \ker(\varphi - \lambda \mathrm{Id})$; since this forces $\varphi - \lambda \mathrm{Id}$ to be non-injective, $\det(\varphi - \lambda \mathrm{Id}) = 0$ and λ is one of the roots of c_φ.

Likewise, claim (3) follows via 5.7.1 from the theorem of linear algebra asserting that eigenvectors with different eigenvalues are always independent. Indeed if

$$p \in V_i \cap \Big(\bigvee_{j \neq i} V_j\Big),$$

then any representative v of p is an eigenvector with eigenvalue λ_i and may be written $v = \sum_{j \neq i} \alpha_j v_j$, where each v_j is an eigenvector with eigenvalue λ_j, against the theorem quoted above. □

Remark 5.7.3. We have seen in the proof of 5.7.2 that

$$\dim V_i = n - \mathrm{rk}(\varphi - \lambda_i \mathrm{Id}).$$

The above linear varieties V_i are called the *fundamental varieties* of f. By 5.7.2 (2), the set of fixed points of f is $\bigcup_{i=1}^m V_i$ and the equalities of 5.7.2 (3) imply the much weaker ones $V_i \cap V_j = \emptyset$ for $i \neq j$. Zero-dimensional fundamental varieties are also called *isolated fixed points*. The fundamental varieties are maximal among the linear varieties all of whose points are fixed. In other words:

Corollary 5.7.4. *All points of a linear variety L are fixed by a collineation f of $\mathbb{P}_n$ if and only if L is contained in one of the fundamental varieties of f.*

Proof. The *if* part is clear after 5.7.2. Conversely, there is nothing to prove if $L = \emptyset$. Otherwise, pick $p = [v] \in L$: the point p being fixed, by 5.7.2, $p \in V_i$ for some $i = 1, \dots, m$. If $L \subset V_i$ we are done. If not, there is $q = [w] \in L - V_i$ and obviously $q \neq p$. The point q being fixed too, it should be $q \in V_j$, $j \neq i$. Thus v and w are linearly independent and $\varphi(v) = \lambda_i v$, $\varphi(w) = \lambda_j w$ with $\lambda_i \neq \lambda_j$. It easily follows that $v + w$ and $\varphi(v + w) = \lambda_i v + \lambda_j w$ are linearly independent vectors, after which, by 5.7.1, $[v+w] \in L$ is not a fixed point, against the hypothesis. □

Corollary 5.7.5. *If two linear varieties of fixed points have a non-empty intersection, then they both are contained in the same fundamental variety.*

Proof. By 5.7.4, linear varieties of fixed points L, L' are contained in fundamental varieties V_i, V_j. Having $i \neq j$ and $V_i \cap V_j \supset L \cap L' \neq \emptyset$ would contradict 5.7.2 (3). □

Remark 5.7.6. Fix a reference of $\mathbb{P}_n$ and assume that M is a matrix of f. Then M is the matrix of a representative of f relative to an adapted basis (2.8.11) and therefore the λ_i of 5.7.2 may be taken to be the roots of the characteristic polynomial of M, $\det(M - X\mathbf{1}_{n+1})$.

Remark 5.7.7. By the definition of V_i, a point $[x_0, \dots, x_n]$ belongs to V_i if and only if

$$(M - \lambda_i \mathbf{1}_{n+1}) \begin{pmatrix} x_0 \\ \vdots \\ x_n \end{pmatrix} = \begin{pmatrix} 0 \\ \vdots \\ 0 \end{pmatrix}. \tag{5.29}$$

Thus, if the matricial equality (5.29) is equivalently written as a system of $n+1$ scalar equations, linear and homogeneous in $x_0, \dots, x_n$, then any maximal set of independent equations extracted from it is a system of implicit equations of V_i, and the dimension of V_i is $\dim V_i = n - \operatorname{rk}(M - \lambda_i \mathbf{1}_{n+1})$ (see 2.6.4), in accordance with 5.7.3.

Remark 5.7.8. As it is well known (and we have in fact seen while proving 5.7.2 (2)) $\lambda_1, \dots, \lambda_m$ are the eigenvalues of the representative φ of f. Since φ is an automorphism, $\lambda_i \neq 0$, $i = 1, \dots, m$. We have seen in the proof of 5.7.2 (1) that the set $\{\lambda_1, \dots, \lambda_m\}$ is determined by f up to a common non-zero scalar factor. If $m \geq 2$, then the set of all ratios λ_i/λ_j, $i, j = 1, \dots, n$, $i \neq j$, is non-empty and intrinsically related to f. Each of the above ratios λ_i/λ_j is called an *invariant* of f (see Exercise 5.11). These invariants were already considered for projectivities of $\mathbb{P}_1$ in Section 5.5. Of course considering all of them is redundant: usually one index j is fixed by some convention and then, with no loss of information, only the invariants λ_i/λ_j, $i = 1, \dots, n$, $i \neq j$, are kept. We will come back to the invariants of f from a more geometrical viewpoint later in this section, and also in Sections 11.4 and 11.5. There we will see that, in spite of being usually named *the* invariants of f, the ratios λ_i/λ_j do not compose a complete system of projective invariants.

We have noticed at the beginning of this section that a hyperplane H of $\mathbb{P}_n$ is fixed by f if and only if, as a point of $\mathbb{P}_n^\vee$, it is fixed by $f^\vee$. Thus, all the above may be applied to $f^\vee$ to get finitely many non-empty linear bundles, the fundamental varieties of $f^\vee$, whose elements are the hyperplanes fixed by f. A closer look will reveal a nice symmetry between the structures of fixed points and fixed hyperplanes of f.

As usual, denote by E the vector space associated to $\mathbb{P}_n$. Choose a representative φ of f and take its dual $\varphi^\vee$ as representative of $f^\vee$ (4.5.1). Since φ and $\varphi^\vee$ have the same characteristic polynomial, namely

$$\det(\varphi^\vee - X\mathrm{Id}_{E^\vee}) = \det((\varphi - X\mathrm{Id}_E)^\vee) = \det(\varphi - X\mathrm{Id}_E),$$

Theorem 5.7.2 may be applied to $f^\vee$ using the same $\lambda_1, \dots, \lambda_m$ to obtain the fundamental varieties of $f^\vee$,

$$W_i^* = [\ker(\varphi^* - \lambda_i \mathrm{Id}_{E^\vee})], \quad i = 1, \dots, m,$$

which are called the *fundamental bundles* of f. Note that, as usual for linear bundles, we have named the fundamental bundles W_i^* by their kernels W_i. Each W_i is thus a linear variety of $\mathbb{P}_n$ and $p \in W_i$ if and only if $p \in H$ for all H in the i-th fundamental bundle W_i^* (4.2.9). Furthermore, since each kernel W_i is an intersection of fixed hyperplanes (4.2.8), it is a linear variety invariant by f. The points of the W_i need not be fixed points though, see for instance Example 5.7.10 below.

Theorem 5.7.2 asserts that a hyperplane is fixed by f if and only if it belongs to one of the fundamental bundles W_i^*, $i = 1, \dots, m$, and also that

$$\emptyset = W_i^* \cap \Big(\bigvee_{j \neq i} W_j^*\Big),$$

which may be equivalently written in terms of the kernels of the fundamental bundles as

$$\mathbb{P}_n = W_i \vee \Big(\bigcap_{j \neq i} W_j\Big).$$

If M is a matrix of f relative to a reference Δ, then M^t is a matrix of $f^\vee$ relative to $\Delta^\vee$ (by 4.5.2) and both M and M^t have the same eigenvalues $\lambda_1, \dots, \lambda_m$. As in 5.7.6, maximal systems of independent equations extracted from

$$(M^t - \lambda_i \mathbf{1}_{n+1}) \begin{pmatrix} u_0 \\ \vdots \\ u_n \end{pmatrix} = \begin{pmatrix} 0 \\ \vdots \\ 0 \end{pmatrix},$$

$i = 1, \dots, m$, are systems of equations of the fundamental bundles W_i^*, or, if preferred, maximal sets of independent columns of $M - \lambda_i \mathbf{1}_{n+1}$ (rows of $M^t - \lambda_i \mathbf{1}_{n+1}$) are coordinate vectors of independent points spanning the kernels W_i (see 4.4.6).

Regarding the already announced symmetry between fundamental varieties and fundamental bundles, let us recall first that after fixing a representative φ of f, we have associated to each eigenvector λ_i of φ, $i = 1, \dots, m$, on one hand, the fundamental variety

$$V_i = [\ker(\varphi - \lambda_i \mathrm{Id}_E)],$$

and, on the other, the fundamental bundle

$$W_i^* = [\ker(\varphi^\vee - \lambda_i \mathrm{Id}_{E^\vee})].$$

Mapping V_i to W_i^* clearly is a bijection between the sets of fundamental varieties and fundamental bundles. Furthermore it is clear that such a bijection does not depend on the choice of the representative φ of f, because, as seen while proving 5.7.2 (1), both V_i and W_i^* are also associated the same eigenvalue $\alpha\lambda_i$ of any other representative $\alpha\varphi$ of f. In addition we have

$$\mathrm{rk}(\varphi^\vee - \lambda_i \mathrm{Id}_{E^\vee}) = \mathrm{rk}((\varphi - \lambda_i \mathrm{Id}_E)^\vee) = \mathrm{rk}(\varphi - \lambda_i \mathrm{Id}_E)$$

and thus (5.7.3)

$$\dim V_i = n - \mathrm{rk}(\varphi - \lambda_i \mathrm{Id}_E) = n - \mathrm{rk}(\varphi^\vee - \lambda_i \mathrm{Id}_{E^\vee}) = \dim W_i^*.$$

For proving this equality, the reader may also compare the already described systems of equations of V_i and W_i^*, which is not an essentially different argument. All together we have proved:

Theorem 5.7.9. *Assume that $f = [\varphi]$ is a collineation of $\mathbb{P}_n$ and write $\lambda_1, \dots, \lambda_m$ the eigenvalues of φ. Mapping each fundamental variety to the fundamental bundle associated to the same eigenvalue,*

$$V_i = [\ker(\varphi - \lambda_i \mathrm{Id}_E)] \longmapsto W_i^* = [\ker(\varphi^* - \lambda_i \mathrm{Id}_{E^\vee})],$$

for $i = 1, \dots, m$, is a bijection between the sets of fundamental varieties and fundamental bundles of f, which is independent of the choice of the representative φ of f. Furthermore, $\dim V_i = \dim W_i^*$ *for $i = 1, \dots, m$.*

In the sequel we will say that the fundamental variety and the fundamental bundle associated to the same eigenvalue are *corresponding*.

Clearly, each equality $\dim W_i^* = \dim V_i$ may be equivalently written $\dim W_i = n - \dim V_i - 1$. Theorem 5.7.9 says in particular that there are the same number of fundamental varieties and fundamental bundles of each dimension, which strongly restricts the possible sets of fundamental varieties and fundamental bundles of a collineation.

Example 5.7.10 (Collineations with $n+1$ isolated fixed points). If the linear automorphism φ representing a collineation $f = [\varphi]$ of $\mathbb{P}_n$ has $n+1$ different eigenvalues $\lambda_0, \dots, \lambda_n$, then each eigenvalue gives rise to a one-dimensional space of associated eigenvectors and therefore f has $n+1$ isolated fixed points as fundamental varieties, say $V_i = \ker(\varphi - \lambda_i \mathrm{Id}) = \{p_i\}$. The points $p_0, \dots, p_n$ are thus the only fixed points, and they are independent by 5.7.2 (3). By 5.7.9, f has $n+1$ fundamental bundles, each of dimension zero and therefore consisting of a single fixed hyperplane. Since, on the other hand, it is clear that the hyperplanes $H_i = p_0 \vee \dots \vee \widehat{p_i} \vee \dots \vee p_n$, $i = 0, \dots, n$, are fixed, by 5.7.4, the fundamental bundles of f are $\{H_i\} = H_i^*$, $i = 0, \dots, n$. By a direct computation using a reference with vertices $p_0, \dots, p_n$, the reader may easily confirm all the above and furthermore check that H_i^* is the fundamental bundle corresponding to $\{p_i\}$. Therefore, with the notations already introduced, $W_i = H_i$, $i = 0, \dots, n$.

Example 5.7.11 (Plane rotations). A rotation $\mathcal{R}$ of a Euclidean plane $\mathbb{A}_2$ has equations

$$\begin{aligned} Y_1 &= \cos\alpha X_1 - \sin\alpha X_2, \\ Y_2 &= \sin\alpha X_1 + \cos\alpha X_2 \end{aligned}$$

relative to any orthonormal reference with origin at the centre O of $\mathcal{R}$, and so its projective extension $\bar{\mathcal{R}}$ has matrix

$$\begin{pmatrix} 1 & 0 & 0 \\ 0 & \cos\alpha & -\sin\alpha \\ 0 & \sin\alpha & \cos\alpha \end{pmatrix}.$$

Assume that α, the angle of the rotation, is $\alpha \neq 0, \pi$. A direct computation shows that $\{O\}$ is the only fundamental variety of $\bar{\mathcal{R}}$. The complex extension $\bar{\mathcal{R}}_{\mathbb{C}}$ has two further isolated fixed points, namely the improper imaginary points $I = [0, 1, \boldsymbol{i}]$ and $J = [0, 1, -\boldsymbol{i}]$, whose relevant metric significance will be shown in Section 6.9. The eigenvalues of a representative of $\bar{\mathcal{R}}_{\mathbb{C}}$ are $1, e^{\alpha \boldsymbol{i}}, e^{-\alpha \boldsymbol{i}}$, which shows the relationship between the angle of $\mathcal{R}$ and the invariants of $\bar{\mathcal{R}}_{\mathbb{C}}$.

Example 5.7.12 (Three-space rotations). A rotation $\mathcal{R}$ of a Euclidean three-space $\mathbb{A}_3$ has equations

$$\begin{aligned} Y_1 &= \cos\alpha X_1 - \sin\alpha X_2, \\ Y_2 &= \sin\alpha X_1 + \cos\alpha X_2, \\ Y_3 &= X_3 \end{aligned}$$

relative to any orthonormal reference whose third axis is the axis e of $\mathcal{R}$, and so its projective extension $\bar{\mathcal{R}}$ has matrix

$$\begin{pmatrix} 1 & 0 & 0 & 0 \\ 0 & \cos\alpha & -\sin\alpha & 0 \\ 0 & \sin\alpha & \cos\alpha & 0 \\ 0 & 0 & 0 & 1 \end{pmatrix}.$$

Assume as above that the angle α of the rotation is $\alpha \neq 0, \pi$. It turns out, after direct computation, that e is the only fundamental variety of $\mathcal{R}$. If $\bar{e}$ denotes the base line of the pencil of planes orthogonal to e, the only fundamental bundle of $\bar{\mathcal{R}}$ is $\bar{e}^*$. As in the plane case, the complex extension has two further isolated fixed points; their joins with e are isolated fixed planes, each corresponding to the other isolated fixed point. The eigenvalues of a representative still being $1, e^{\alpha \boldsymbol{i}}, e^{-\alpha \boldsymbol{i}}$, the angle of $\mathcal{R}$ and the invariants of $\bar{\mathcal{R}}_{\mathbb{C}}$ are related as in the plane case.

Corollary 5.7.13. *If a non-identical collineation of $\mathbb{P}_n$ has a hyperplane of fixed points, then it has an $(n-1)$-dimensional bundle of fixed hyperplanes, and conversely.*

Proof. Not all points being fixed by f, by 5.7.4 the hyperplane of fixed points is a fundamental variety, after which, by 5.7.9, there is an $(n-1)$-dimensional fundamental bundle. The converse is the dual. □

The non-identical collineations of $\mathbb{P}_n, n \geq 2$, satisfying the equivalent conditions of 5.7.13 are called *homologies*. A homology has thus a hyperplane of fixed points and an $(n-1)$-dimensional bundle of fixed hyperplanes, both unique due to their dimensions and 5.7.5. The hyperplane of fixed points is called the *hyperplane of homology* (*axis of homology* if $n = 2$, *plane of homology* if $n = 3$). The kernel of the $(n-1)$-dimensional bundle of fixed hyperplanes is a point, which is called the *centre of homology*. The kernel of any fundamental bundle being invariant, the centre of homology is a fixed point, and so it may or may not belong to the hyperplane of homology: the homology is called *special* if and only if its center belongs to the hyperplane of homology; otherwise the homology is called *general*. As the reader may easily check, the collineations with matrices

$$\begin{pmatrix} -1 & 0 & 0 & \dots & 0 \\ 0 & 1 & 0 & \dots & 0 \\ 0 & 0 & 1 & \dots & 0 \\ \vdots & \vdots & \vdots & \ddots & \vdots \\ 0 & 0 & 0 & \dots & 1 \end{pmatrix} \quad \text{and} \quad \begin{pmatrix} 1 & 0 & 0 & \dots & 0 \\ 1 & 1 & 0 & \dots & 0 \\ 0 & 0 & 1 & \dots & 0 \\ \vdots & \vdots & \vdots & \ddots & \vdots \\ 0 & 0 & 0 & \dots & 1 \end{pmatrix}$$

are examples of general and special homologies, respectively. The following example may be more familiar to the reader:

Example 5.7.14. Both the non-identical translations and the homotheties of $\mathbb{A}_n$ transform any free vector into a proportional one. Therefore, by 3.5.3, their extensions to $\overline{\mathbb{A}}_n$ leave all the improper points invariant and are thus homologies, with the improper hyperplane as hyperplane of homology. The extensions of the homotheties clearly have the centre of homothety as centre of homology, and hence are non-special. A translation having no fixed (proper) point, its extension must be special, and indeed it is: its centre is the direction of the vector of translation, as the reader may easily check.

Next we describe the fundamental varieties and fundamental bundles of the general and the special homologies:

Proposition 5.7.15. *Assume that f is a homology of $\mathbb{P}_n$ with hyperplane and centre of homology H and O, respectively. Then:*

(a) *If f is general, then its fundamental varieties are O and H, and its fundamental bundles are $H^* = \{H\}$ and O^*.*

(b) *If f is special, then it has H as its only fundamental variety and O^* as its only fundamental bundle.*

Proof. If f is general, O is fixed and does not belong to H: by 5.7.4 it needs to belong to a fundamental variety other than H. Such a fundamental variety being disjoint with H by 5.7.2 (3), it must equal O. Since $\mathbb{P}_n = H \vee O$, again by 5.7.2 (3), there is no other fundamental variety. Both O^* and H^* are bundles of fixed hyperplanes, and they are the only fundamental bundles by 5.7.4 and 5.7.9.

Assume now that f is special. If it has a fixed point $p \notin H$, in particular $p \neq O$. Let H' be any hyperplane through p. An easy count of dimensions shows that is $H = p \vee (H' \cap H)$. Both p and $H' \cap H$ being fixed, so is H'. Thus all hyperplanes in p^* are fixed, against the uniqueness of the $(n-1)$-dimensional bundle of fixed hyperplanes. This proves that f has no fixed point other than those in H, and hence no fundamental variety other than H. By 5.7.9, there is no fundamental bundle other than O^* either. □

Assume that f is a general homology with centre O and hyperplane of homology H. Let us take H as the first fundamental variety, say $H = V_1$. By 5.7.9, corresponding fundamental varieties and fundamental bundles have equal dimensions. Therefore the fundamental bundle corresponding to H is $W_1^* = O^*$, while the one corresponding to $O = V_2$ is $W_2^* = H^*$. It follows that $V_1 = W_2 = H$ and $V_2 = W_1 = \{O\}$ and so, for a general homology the kernels of the fundamental bundles are fundamental varieties. Nevertheless, this does not hold true for arbitrary collineations, as shown by the collineations with $n+1$ isolated fixed points (5.7.10) and the special homologies. Next we will prove a weaker relationship:

Proposition 5.7.16. *If V_i, W_i^*, $i = 1, \dots, m$ are the pairs of corresponding fundamental varieties and fundamental bundles of a collineation f, then $V_i \subset W_j$ for each $i, j = 1, \dots, m$, $i \neq j$.*

Proof. As previously, fix a representative φ of f and denote by λ_i the eigenvalue of φ corresponding to V_i. If $p \in V_i$, it is $p = [v]$ with $\varphi(v) = \lambda_i(v)$. We need to see that $p \in W_j$, that is, that $p \in H$ for all $H \in W_j^*$. Since, by its definition, the hyperplanes of W_j^* are those represented by eigenforms of $\varphi^\vee$ with eigenvalue λ_j, we need to see (by 4.1.3) that $\omega(v) = 0$ for all forms ω for which $\varphi^\vee(\omega) = \lambda_j\omega$, and this will follow from an easy computation. We have

$$\lambda_i\omega(v) = \omega(\lambda_i v) = \omega(\varphi(v)) = \varphi^\vee(\omega)(v) = (\lambda_j\omega)(v) = \lambda_j\omega(v),$$

which together with $\lambda_i \neq \lambda_j$ gives the wanted equality $\omega(v) = 0$. □

For the relative position of V_i and W_i, see Exercise 11.13.

Remark 5.7.17. Proposition 5.7.16 may be used to recover the correspondence of 5.7.9, between fundamental varieties and fundamental bundles, with no use of eigenvalues, which is useful in many cases. For instance, 5.7.16 directly shows that the pairs of corresponding fundamental varieties and fundamental bundles in Example 5.7.10 are $\{p_i\}$, H_i^*, $i = 0, \dots, n$.

Proposition 5.7.18. *Let $f = [\varphi]$ be a collineation of $\mathbb{P}_n$ and, as before, denote by V_i and W_i^* the fundamental variety and the fundamental bundle associated to the eigenvalue λ_i of φ, for $i = 1, \dots, m$. Assume that $p \notin V_i$, $p \notin W_i$ for $i = 1, \dots, m$. Then,*

(1) *The point p and its image $f(p)$ span a line that intersects the kernel of each fundamental bundle at a single point, say $q_i = pf(p) \cap W_i$.*

(2) $(p, f(p), q_i, q_j) = \lambda_j/\lambda_i$ *for any $i, j = 1, \dots, m$, $i \neq j$.*

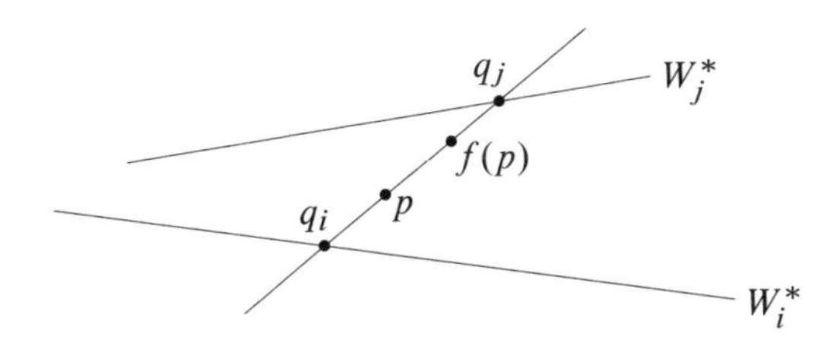

Figure 5.5. Proposition 5.7.18: while p varies in $\mathbb{P}_n - \left(\bigcup_r V_r\right) \cup \left(\bigcup_r W_r^*\right)$, the points p, $f(p)$ span a line that meets W_i and W_j at points q_i, q_j, and the cross ratio $(p, f(p), q_i, q_j)$ remains constant.

Proof. By the hypothesis, p is not fixed and so $pf(p)$ is a line. Since $p \notin W_i$, $pf(p) \not\subset W_i$ and therefore we need only to exhibit a point $q_i \in pf(p) \cap W_i$ to see that $pf(p) \cap W_i = q_i$. Take $p = [v]$; then $f(p) = [\varphi(v)]$, v and $\varphi(v)$ are independent and we choose $q_i = [\lambda_i v - \varphi(v)]$, obviously a point of $pf(p)$. For any eigenform ω of $\varphi^\vee$ with eigenvalue λ_i,

$$\begin{aligned}\omega(\lambda_i v - \varphi(v)) &= \lambda_i\omega(v) - \omega(\varphi(v)) = \lambda_i\omega(v) - \varphi^\vee(\omega)(v)\\ &= \lambda_i\omega(v) - (\lambda_i\omega)(v) = 0.\end{aligned}$$

As explained while proving 5.7.16, this proves that $q_i \in W_i$ and thus ends the proof of claim (1). Claim (2) follows from a direct computation using 2.9.5 and the representatives v, $\varphi(v)$, $\lambda_i v - \varphi(v)$ and $\lambda_j v - \varphi(v)$. □

Next are some remarks and two examples concerning Proposition 5.7.18:

Remark 5.7.19. Obviously the claim of 5.7.18 is empty in case $f = \mathrm{Id}_{\mathbb{P}_n}$. Otherwise $\dim V_i < n$ and $\dim W_i < n$ for $i = 1, \dots, m$, the latter because, being a fundamental bundle, $W_i^* \neq \emptyset$. Hence, 2.4.4 assures that if $f \neq \mathrm{Id}_{\mathbb{P}_n}$, then there are points $p \in \mathbb{P}_n$ for which the hypothesis of 5.7.18 are satisfied. Note also that the hypothesis $p \notin V_i$ can be dropped if $m \geq 2$, by 5.7.16.

Remark 5.7.20. Claim (1) of 5.7.18 imposes strong restrictions on $f(p)$ but for the cases in which W_i is a hyperplane, as it forces $f(p) \in p \vee W_i$, $i = 1, \dots, m$. For instance, if f is a homology, then $f(p)$ needs to be aligned with p and the centre of homology.

Remark 5.7.21. Claim (2) of 5.7.18 is empty if $m = 1$. Otherwise it provides the already announced geometric interpretation of the ratios between eigenvalues of any fixed representative of f. It re-proves the fact that the set $\{\lambda_1, \dots, \lambda_m\}$ is determined by f but for an arbitrary non-zero common scalar factor. Claim (2) also implies that for each $i \neq j$ the cross ratio $(p, f(p), q_i, q_j)$ is the same for all points p in the conditions of 5.7.18, which is a quite remarkable fact.

Example 5.7.22. By 5.7.15 (b), special homologies have no invariants. For them, Proposition 5.7.18 just assures that any line $pf(p)$, for p a non-fixed point, goes through the center of homology, which follows also from the definition of center of homology.

Example 5.7.23. Assume that f is a general homology with centre of homology O and hyperplane of homology H. Let λ_1 and λ_2 be the eigenvalues of a representative of f corresponding to the fundamental varieties H and O, respectively, and hence also to the fundamental bundles O^* and H^*, in this order. Besides assuring that for any non-fixed point p, the line $pf(p)$ goes through O (as already noted in 5.7.20), Proposition 5.7.18 also assures that $p' = pf(p) \cap H$ is a point and $(p, f(p), O, p') = \lambda_2/\lambda_1$, independently of the choice of p. From the pair of reciprocal invariants $\lambda_2/\lambda_1, \lambda_1/\lambda_2$, usually the first one is taken as the invariant of f (see Remark 5.7.8). Note that if p is not fixed, then $f(p)$ is determined as being the only point on the line Op which satisfies $(p, f(p), O, p') = \lambda_2/\lambda_1$. General homologies with invariant -1 are called *harmonic homologies*; they are involutive collineations due to 2.10.2.

The next two lemmas are often used to define homologies:

Lemma 5.7.24 (Determination of general homologies). *Assume given a point O and a hyperplane H of a projective space $\mathbb{P}_n$, $O \notin H$, a scalar $\lambda \in k - \{0, 1\}$ and two different points $p, q \in \mathbb{P}_n$, both different from O, aligned with it and not belonging to H. Then:*

(a) *There is a unique homology f of $\mathbb{P}_n$ with centre O, hyperplane of homology H and invariant λ.*

(b) *There is a unique homology f' of $\mathbb{P}_n$ with centre O, hyperplane of homology H and mapping p to q.*

Clearly, the homologies f, f' above are general.

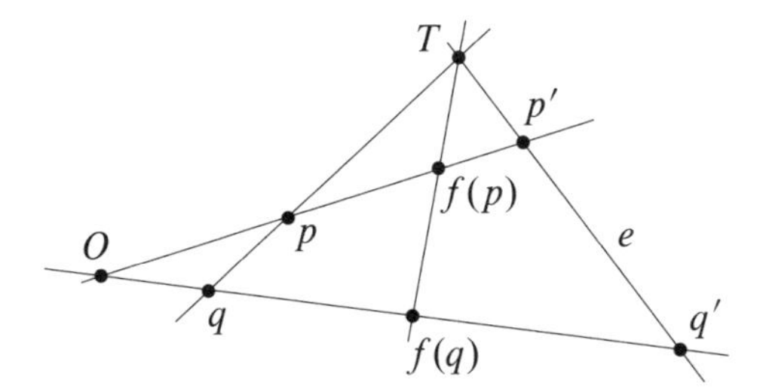

Figure 5.6. Images of points p, q, assumed non-fixed and non-aligned with O, by a plane general homology f with centre O and axis e. The pairs p, $f(p)$ and q, $f(q)$ are aligned with O by 5.7.20. The point $T = pq \cap e$ being fixed, the line $f(p)f(q) = f(pq)$ goes also through it. This determines $f(q) = Oq \cap Tf(p)$ and provides a way of constructing it once the centre, the axis and the corresponding pair p, $f(p)$ are given. For the image of a point $\tilde{q} \in Op$, proceed similarly, using the already constructed pair q, $f(q)$ instead of p, $f(p)$. Also the equality $(p, f(p), O, p') = (q, f(q), O, q')$ is clear from the figure.

Proof of 5.7.24. For the existence in (a) take any reference $\Delta = (p_0, \dots, p_n, A)$ of $\mathbb{P}_n$ with $p_0 = O$ and $p_1, \dots, p_n \in H$, and then f to be the collineation with matrix

$$\begin{pmatrix} \lambda & & & \\ & 1 & & \\ & & \ddots & \\ & & & 1 \end{pmatrix}$$

which clearly fulfils the conditions. The uniqueness is clear, as the image of an arbitrary point p' is $f(p') = p'$ if either $p' = O$ or $p' \in H$, and otherwise it is determined as explained in 5.7.23.

The existence in (b) follows by taking f' as f above, but for $\lambda = (p, q, O, q')$, $q' = Op \cap H$. For the uniqueness just note that any homology fulfilling the conditions of (b) has invariant (p, q, O, q'), $q' = Op \cap H$, after which claim (a) applies. □

Lemma 5.7.25 (Determination of special homologies). *Assume given, in a projective space $\mathbb{P}_n$, a hyperplane H and two different points p, q not belonging to H. Then there is a unique special homology f of $\mathbb{P}_n$ with hyperplane of homology H and mapping p to q.*

Proof. Take any reference $\Delta = (p_0, \dots, p_n, A)$ of $\mathbb{P}_n$ with $p_0 = p$, $p_1 = pq \cap H$ and $p_2, \dots, p_n \in H$. Then $q = [1, \delta, \dots, 0]$, $\delta \in k - \{0\}$ and taking f to be the collineation with matrix

$$\begin{pmatrix} 1 & 0 & 0 & \dots & 0 \\ \delta & 1 & 0 & \dots & 0 \\ 0 & 0 & 1 & \dots & 0 \\ \vdots & \vdots & \vdots & \ddots & \vdots \\ 0 & 0 & 0 & \dots & 1 \end{pmatrix}$$

proves the existence. For the uniqueness, note first that any collineation g leaving all points of H invariant has a matrix relative to Δ of the form

$$\begin{pmatrix} a_0 & 0 & 0 & \dots & 0 \\ a_1 & 1 & 0 & \dots & 0 \\ a_2 & 0 & 1 & \dots & 0 \\ \vdots & \vdots & \vdots & \ddots & \vdots \\ a_n & 0 & 0 & \dots & 1. \end{pmatrix}$$

If g is a special homology, then $a_0 = 1$, as otherwise a_0 would be an eigenvalue different from 1 giving rise to a fundamental variety other than H. Then the condition of mapping p to q forces $a_1 = \delta$ and $a_2 = \dots = a_n = 0$, after which $g = f$. □

Example 5.7.26. If a homothety of $\mathbb{A}_n$ is seen as the restriction of a general homology of $\bar{\mathbb{A}}_n$ (see 5.7.14), then the ratio of the homothety is the inverse of the invariant of the homology. In particular the *point reflections* of $\mathbb{A}_n$ (homotheties of ratio -1) are the restrictions of the harmonic homologies of $\mathbb{A}_n$ with hyperplane of homology H_∞.

Example 5.7.27. The *reflection in a hyperplane* H of a Euclidean space $\mathbb{A}_n$ is the restriction of the harmonic homology of $\bar{\mathbb{A}}_n$ with hyperplane of homology the projective extension of H, and centre the direction orthogonal to H.

Example 5.7.28. Assume that f is a non-identical collineation of $\mathbb{P}_3$ that has two skew lines e_1, e_2 of fixed points. The reader may argue as in the proof of 5.7.15 to see that e_1 and e_2 are the only fundamental varieties of f. They are called the *axes* of f, and f itself a *biaxial collineation.* If λ_1, λ_2 are the eigenvalues of a representative of f, the invariants of f are λ_1/λ_2 and λ_2/λ_1. Either of them is taken as defined up to reciprocation and called the invariant of f. Biaxial collineations with invariant -1 are called *harmonic*; as the harmonic homologies, they are involutive. Since each plane through one of the axes meets the other at a point which is fixed by f, it is clear that both e_1^* and e_2^* are pencils of fixed planes, and so, by 5.7.9, they are the fundamental bundles of f. Furthermore, by 5.7.16, each fundamental bundle corresponds with the kernel of the other, and so if, using the notations of 5.7.18, $V_1 = e_1$ and $V_2 = e_2$, then the kernels of the fundamental bundles are $W_1 = e_2$ and $W_2 = e_1$. Now, by 5.7.18 (1), for any point $p \notin e_i$, $i = 1, 2$, the line $\ell_p = pf(p)$

is the only line that goes through p and meets both axes (see 5.3.8). If $\ell_p \cap e_i = p_i$, $i = 1, 2$, then $f(p)$ is the point of ℓ_p that satisfies $(p, f(p), p_1, p_2) = \lambda_1/\lambda_2$.

5.8 Correlations

A *correlation* g of a projective space $\mathbb{P}_n$ is a projectivity from $\mathbb{P}_n$ onto its dual space,

$$g\colon \mathbb{P}_n \longrightarrow \mathbb{P}_n^\vee.$$

The correlations are thus the projectivities mapping points to hyperplanes of the same projective space; they map linear varieties of points to bundles of hyperplanes of the same dimension (due to 1.7.1 and 4.2.2). If L is a linear variety of $\mathbb{P}_n$, its image by the correlation g is often written $g(L) = (\bar{L})^*$ and $\bar{L}$, the kernel of $g(L)$, is called the *transform* of L. While a point p describes L, its image $g(p)$ describes the bundle of hyperplanes through $\bar{L}$. Again by 1.7.1 and 4.2.2, $\dim \bar{L} = n - \dim L$. As a direct consequence of 1.7.3 and 4.2.4 we have:

Proposition 5.8.1. *If L_1 and L_2 are linear varieties and g a correlation of $\mathbb{P}_n$, and we write $g(L_i) = (\bar{L}_i)^*$, $i = 1, 2$, then*

(a) $L_1 \subset L_2$ *if and only if* $\bar{L}_1 \supset \bar{L}_2$,

(b) $\overline{L_1 \cap L_2} = \bar{L}_1 \vee \bar{L}_2$,

(c) $\overline{L_1 \vee L_2} = \bar{L}_1 \cap \bar{L}_2$.

The action of projectivities on correlations is defined by taking the image of the correlation g of $\mathbb{P}_n$ by the projectivity $f\colon \mathbb{P}_n \to \mathbb{P}'_n$ as being

$$f(g) = (f^{-1})^\vee \circ g \circ f^{-1}.$$

The reader may easily check that, indeed, $f(g)$ thus defined is a correlation of $\mathbb{P}'_n$, and also that the properties of Section 1.10 are satisfied.

Unless otherwise stated, matrices of correlations are always taken relative to a reference of $\mathbb{P}_n$ and its dual. In this way, a matrix of a correlation g of $\mathbb{P}_n$, relative to a reference Δ of $\mathbb{P}_n$ and its dual $\Delta^\vee$, is called a *matrix of g relative to Δ*, or just a matrix of g if there is no need of mentioning Δ.

As a particular case of 2.8.4, the equations of a correlation provide the hyperplane coordinates of the image at any point:

Proposition 5.8.2. *Assume to have fixed a reference of $\mathbb{P}_n$ relative to which the correlation g has matrix M. Then g maps the point with coordinates $y_0, \dots, y_n$ to the hyperplane with equation $u_0x_0 + \cdots + u_nx_n = 0$ if and only if, for some non-zero $\rho \in k$,*

$$\begin{pmatrix} u_0 \\ \vdots \\ u_n \end{pmatrix} = \rho M \begin{pmatrix} y_0 \\ \vdots \\ y_n \end{pmatrix}.$$

Proposition 5.8.3. *If M is the matrix of a correlation g relative to a reference Δ and the matrix P changes coordinates relative to a second reference Ω into coordinates relative to Δ, then*

$$P^t M P$$

is a matrix of g relative to Ω.

Proof. Just recall that, by 4.4.12, P^t changes hyperplane coordinates relative to Δ into hyperplane coordinates relative to Ω and use 2.8.9. □

Since a correlation maps points to hyperplanes, one may ask for the points that lie in their own images: they describe a set which is called the *incidence set* of the correlation. After fixing a reference of $\mathbb{P}_n$, assume that the correlation g has matrix

$$M = \begin{pmatrix} b_0^0 & \dots & b_n^0 \\ \vdots & & \vdots \\ b_0^n & \dots & b_n^n \end{pmatrix}.$$

Then the point $p = [x_0, \dots, x_n]$ belongs to $g(p)$ if and only if

$$(x_0 \ \dots \ x_n)\, M \begin{pmatrix} x_0 \\ \vdots \\ x_n \end{pmatrix} = \sum_{i,j=0}^{n} b_j^i x_i x_j = 0. \tag{5.30}$$

Remark 5.8.4. As the reader may easily check, the image of the incidence set of a correlation g of $\mathbb{P}_n$ by any projectivity $f : \mathbb{P}_n \to \mathbb{P}'_n$ is the incidence set of $f(g)$.

We will first pay some attention to the correlations with incidence set equal to $\mathbb{P}_n$: they are called *null-systems* and the next proposition characterizes them:

Proposition 5.8.5. *If a correlation has a skew-symmetric matrix, then it is a null-system. Conversely, any matrix of a null-system is skew-symmetric.*

Proof. After rewriting the equation (5.30) in the form

$$\sum_{i=0}^{n} b_i^i x_i^2 + \sum_{0 \le i < j \le n} (b_j^i + b_i^j) x_i x_j = 0, \tag{5.31}$$

it clearly becomes trivial when the matrix M is skew-symmetric, hence the first claim. For the second one, assume that the above M is a non-skew-symmetric matrix of g. Then either one of the coefficients b_i^i is non-zero, or $b_i^i = 0$ for $i = 0, \dots, n$ and $b_i^j + b_j^i \neq 0$ for some $i \neq j$. In the first case $x_i = 1$ and $x_j = 0$ for $j \neq i$ is not a solution of the equation (5.31) and therefore the corresponding point does not belong to the incidence set. The same occurs in the second case if we take $x_i = x_j = 1$ and $x_r = 0$ for $r \neq i, j$, and so g is not a null-system. □

Remark 5.8.6. The even-dimensional projective spaces have no null-systems, because there are no regular skew-symmetric matrices of odd dimension: indeed, by definition, a skew-symmetric $(n+1)$-dimensional matrix M has $M^t = -M$ and therefore $\det M = \det M^t = (-1)^{n+1} \det M$, which for n even forces $\det M = 0$. By contrast, the even-dimensional matrices

$$\begin{pmatrix} & & & & & 1 \\ & & & & \cdot^{\cdot^{\cdot}} & \\ & & & 1 & & \\ & & -1 & & & \\ & \cdot^{\cdot^{\cdot}} & & & & \\ -1 & & & & & \end{pmatrix}, \tag{5.32}$$

with all entries but those on the secondary diagonal equal to zero, are skew-symmetric and regular: therefore all odd-dimensional projective spaces have null-systems.

If a correlation is not a null-system, then the corresponding equation (5.30) is non-trivial. This equation is then homogeneous of the second degree in the coordinates x_i: we will see in the forthcoming Section 6.1 that the set of points whose coordinates are solutions of such an equation, and so, in particular, the incidence set of any correlation other than a null-system, is the set of points of a quadric hypersurface. The reader may see also Exercise 6.7.

According to the general definition of Section 4.5, the dual of a correlation $g\colon \mathbb{P}_n \to \mathbb{P}_n^\vee$ is the projectivity

$$\begin{aligned} g^\vee\colon \mathbb{P}_n^{\vee\vee} &\longrightarrow \mathbb{P}_n^\vee, \\ p^* &\longmapsto g^{-1}(p^*) = \{q \mid p \in g(q)\}. \end{aligned}$$

As seen in Section 4.6, mapping each point $p \in \mathbb{P}_n$ to the bundle p^* is a projectivity $\Phi_{\mathbb{P}_n}$ through which $\mathbb{P}_n$ and its bidual $\mathbb{P}_n^{\vee\vee}$ are identified. It is usual to consider $g^\bullet = g^\vee \circ \Phi_{\mathbb{P}_n}$ instead of $g^\vee$ and still call it the *dual* of g, this causing no confusion. The dual $g^\bullet$ of g is thus the correlation

$$\begin{aligned} g^\bullet\colon \mathbb{P}_n &\longrightarrow \mathbb{P}_n^\vee, \\ p &\longmapsto \{q \mid p \in g(q)\}. \end{aligned}$$

The next proposition provides two ways of handling it:

Proposition 5.8.7. (a) *For any $p, q \in \mathbb{P}_n$, $q \in g^\bullet(p)$ if and only if $p \in g(q)$.*

(b) *If M is a matrix of g relative to a reference Δ, then M^t is a matrix of $g^\bullet$ relative to Δ.*

Proof. The first claim is direct from the above description of $g^\bullet$, while the second one follows from its definition and 4.6.3. □

Corollary 5.8.8. *A correlation and its dual have the same incidence set.*

Proof. Just note that the equation (5.30), of the incidence set, remains unchanged after replacing $b_{i,j}$ with $b_{j,i}$ for all $i, j = 0, \dots, n$. □

As one may expect, g and $g^\bullet$ are projectively related, namely:

Proposition 5.8.9. *For any projectivity $f : \mathbb{P}_n \to \overline{\mathbb{P}}_n$, $f(g^\bullet) = (f(g))^\bullet$.*

Proof. The reader may provide an alternative proof using matrices. Write $\Phi = \Phi_{\mathbb{P}_n}$ and $\overline{\Phi} = \Phi_{\overline{\mathbb{P}}_n}$. Using 4.6.6, it is

$$\begin{aligned} f(g^\bullet) &= (f^{-1})^\vee \circ g^\vee \circ \Phi \circ f^{-1} \\ &= (f^{-1})^\vee \circ g^\vee \circ (f^{-1})^{\vee\vee} \circ \overline{\Phi} \\ &= \big((f^{-1})^\vee \circ g \circ f^{-1}\big)^\vee \circ \overline{\Phi} \\ &= \big((f^{-1})^\vee \circ g \circ f^{-1}\big)^\bullet \\ &= (f(g))^\bullet. \end{aligned}$$

□

Both g and its dual $g^\bullet$ being correlations of P_n, they may be compared; in the remainder of this section we will pay attention to the case in which g is its own dual, that is $g = g^\bullet$. The study of arbitrary correlations will be continued, for $k = \mathbb{C}$, in the forthcoming Section 11.7. By 5.8.7, the equality $g = g^\bullet$ is equivalent to having

$$p \in g(q) \iff q \in g(p) \quad \text{for any } p, q \in P_n,$$

which is called the *reciprocity law* or just *reciprocity*. So, when a correlation is its own dual, it is equivalently said that it *satisfies reciprocity*.

Null-systems are a first example of correlations satisfying reciprocity: indeed, any matrix M of a null-system g being skew-symmetric by 5.8.5, it satisfies $M^t = -M$, which gives $g^\bullet = g$.

The correlations that satisfy reciprocity and are not null-systems are called *polarities*. They are characterized in terms of their matrices as follows:

Proposition 5.8.10. *If a correlation has a symmetric matrix, then it is a polarity. Any matrix of a polarity is symmetric.*

Proof. Assume that a correlation g has a symmetric matrix M. Then $M^t = M$, which gives $g^\bullet = g$ and so g satisfies reciprocity. In addition, since no non-zero matrix is simultaneously symmetric and skew-symmetric, g is not a null-system by 5.8.5.

Assume now that a correlation g of $\mathbb{P}_n$ satisfies reciprocity and is not a null-system. Fix any reference Δ of $\mathbb{P}_n$ and assume that g has matrix M. The condition $g^\bullet = g$ is equivalent to $M^t = \rho M$ for some $\rho \in k - \{0\}$; after transposing both sides we get $M = \rho M^t$ and so $M = \rho^2 M$. At least one entry of M being non-zero, it results that $\rho = \pm 1$. Since in case $M^t = -M$ the correlation g would be a null-system (again by 5.8.5), it is $M^t = M$ and M is symmetric as claimed. □

After 5.8.10, it is clear that any projective space has polarities, as there are regular symmetric matrices of any dimension.

The next proposition summarizes what we have seen about correlations satisfying reciprocity.

Proposition 5.8.11. *A correlation satisfies reciprocity if and only if it is either a null-system or a polarity. No polarity is a null-system.*

The notions of null-system and polarity are projective, namely:

Corollary 5.8.12. *A correlation projectively equivalent to a null-system* (*resp. polarity*) *is a null-system* (*resp. polarity*) *too.*

Proof. The null-systems being the correlations whose incidence set is the whole space, the claim for them follows from 5.8.4. If a correlation g' is projectively equivalent to a polarity g, then it satisfies reciprocity by 5.8.9. Furthermore g' cannot be a null-system, as if this were the case, by the part of the claim already proved, so would be g, against 5.8.11 or the definition of polarity. □

The study of the polarities being equivalent to the study of the non-degenerate quadric hypersurfaces, it will not be pursued here, but delayed to Chapters 6 and 7, see in particular Section 6.5 and Exercise 7.10. We close this section by proving that any null-system has a matrix of the form of (5.32) above and, as a consequence, that any two null-systems of projective spaces of the same dimension, over the same field, are projectively equivalent.

Theorem 5.8.13. *If g is a null-system of a projective space $\mathbb{P}_n$, then $m = (n-1)/2$ is a positive integer and there is a reference of $\mathbb{P}_n$ relative to which g has equations*

$$\begin{aligned} u_0 &= \rho x_n, \\ &\ \vdots \\ u_m &= \rho x_{m+1}, \\ u_{m+1} &= -\rho x_m, \\ &\ \vdots \\ u_n &= -\rho x_0. \end{aligned}$$

Proof. By 5.8.6, the existence of g assures that n is odd and hence m is a positive integer as claimed. We begin by choosing the vertices $p_0, \dots, p_n$ of the reference. Once each p_i is defined, we write $\pi_i = g(p_i)$ and have thus $p_i \in \pi_i$, $i = 0, \dots, n$.

Take p_0 to be any point: as noted above, $p_0 \in \pi_0$. Inductively, assume we have independent points $p_0, \dots, p_i \in \pi_0 \cap \dots \cap \pi_i$; then it holds that $\dim \pi_0 \cap \dots \cap \pi_i = n - i - 1$ due to the independence of $p_0, \dots, p_i$. As far as

$$\dim p_0 \vee \dots \vee p_i = i + 1 < n - i - 1 = \dim \pi_0 \cap \dots \cap \pi_i,$$

take

$$p_{i+1} \in \pi_0 \cap \cdots \cap \pi_i - p_0 \vee \cdots \vee p_i$$

after which $p_0, \dots, p_i, p_{i+1}$ are independent. Furthermore $p_{i+1} \in \pi_{i+1}$ and by reciprocity $p_0, \dots, p_i \in \pi_{i+1}$; it follows that $p_0, \dots, p_{i+1} \in \pi_0 \cap \cdots \cap \pi_{i+1}$. Since n is odd, this may be done for $i = 0, \dots, m-1$ and we obtain independent points $p_0, \dots, p_m$ belonging to $\pi_0 \cap \cdots \cap \pi_m$.

We continue with another induction. Assume we have chosen further points $p_{m+1}, \dots, p_{m+i}$, $0 \le i \le m$, such that

$$p_0, \dots, p_m, p_{m+1}, \dots, p_{m+i}$$

are independent and furthermore

$$p_s \in \pi_j \quad \text{for } 0 \le s, j \le m+i \text{ and } s+j \ne 2m+1,$$

which is clearly true for $i = 0$. Then the hyperplanes $\pi_0, \dots, \pi_{m+i}$ are independent, as they are the images of $p_0, \dots, p_{m+i}$. Also $m - i \ge 0$ and

$$p_0, \dots, p_{m+i} \in \pi_0 \cap \cdots \cap \pi_{m-i}.$$

By comparing dimensions,

$$p_0 \vee \cdots \vee p_{m+i} = \pi_0 \cap \cdots \cap \pi_{m-i} \tag{5.33}$$

because both the points and the hyperplanes involved are independent. Now we choose

$$p_{m+i+1} \in \pi_0 \cap \cdots \cap \pi_{m-i-1} \cap \pi_{m-i+1} \cap \cdots \cap \pi_{m+i} \tag{5.34}$$

and such that

$$p_{m+i+1} \notin \pi_0 \cap \cdots \cap \pi_{m-i}. \tag{5.35}$$

This is possible because

$$\pi_0 \cap \cdots \cap \pi_{m-i-1} \cap \pi_{m-i+1} \cap \cdots \cap \pi_{m+i} \not\subset \pi_0 \cap \cdots \cap \pi_{m-i},$$

as otherwise we would have

$$\pi_0 \cap \cdots \cap \pi_{m-i-1} \cap \pi_{m-i+1} \cap \cdots \cap \pi_{m+i} \subset \pi_{m-i},$$

and so

$$\pi_0 \cap \cdots \cap \pi_{m-i-1} \cap \pi_{m-i+1} \cap \cdots \cap \pi_{m+i} = \pi_0 \cap \cdots \cap \pi_{m+i},$$

against the independence of $\pi_0 \dots \pi_{m+i}$. The condition (5.35) and the equality (5.33) together guarantee that $p_0, \dots, p_{m+i+1}$ are independent. On the other hand by adding $p_{m+i+1} \in \pi_{m+i+1}$ to (5.34) we get

$$p_{m+i+1} \in \pi_j \quad \text{for } 0 \le j \le m+i+1 \text{ and } (m+i+1)+j \ne 2m+1,$$

and, by reciprocity, also

$$p_s \in \pi_{m+i+1} \quad \text{for } 0 \le s \le m+i+1 \text{ and } s+(m+i+1) \ne 2m+1.$$

All together, $p_0, \dots, p_{m+i+1}$ are independent and such that

$$p_s \in \pi_j \quad \text{for } 0 \le s, j \le m+i+1 \text{ and } s+j \ne 2m+1,$$

as required for the induction.

Since $2m+1 = n$, after the last step, made for $i = m$, we have independent points $p_0, \dots, p_n$ such that

$$p_s \in \pi_j \quad \text{for } 0 \le s, j \le n \text{ and } s+j \ne n,$$

or, equivalently,

$$g(p_j) = \pi_j = p_0 \vee \cdots \vee p_{n-j-1} \vee p_{n-j+1} \vee \cdots \vee p_n$$

for $j = 0, \dots, n$. After this, the matrix of g relative to any reference $\Delta = \{p_0, \dots, p_n, B\}$, with vertices $p_0, \dots, p_n$, has the form

$$\begin{pmatrix} & & & & & a_0 \\ & & & & \iddots & \\ & & & a_m & & \\ & & -a_m & & & \\ & \iddots & & & & \\ -a_0 & & & & & \end{pmatrix},$$

with all entries outside the secondary diagonal equal to zero and $a_i \ne 0$ for $i = 0, \dots, m$.

Now take $A = [a_0^{-1}, \dots, a_m^{-1}, 1, \dots, 1]_\Delta$ as a new unit point, which is possible because all its coordinates are non-zero. The matrix changing coordinates relative to $\Omega = \{p_0, \dots, p_n, A\}$ into coordinates relative to Δ is the diagonal matrix

$$\begin{pmatrix} a_0^{-1} & & & & & \\ & \ddots & & & & \\ & & a_m^{-1} & & & \\ & & & 1 & & \\ & & & & \ddots & \\ & & & & & 1 \end{pmatrix},$$

after which using 5.8.3 ends the proof. □

Corollary 5.8.14. *Any two null-systems of projective spaces of the same dimension, over the same field k, are projectively equivalent.*

Proof. If g and g' are null-systems of spaces $\mathbb{P}_n$ and $\mathbb{P}'_n$, respectively, take Δ and Δ' to be references relative to which g and g' have the equations of 5.8.13 and hence the same matrix M. If $f: \mathbb{P}_n \to \mathbb{P}'_n$ is the projectivity mapping Δ to Δ', then its matrix relative to these references is the unit one (by 2.8.1). Therefore it is clear that both g' and $f(g) = (f^{-1})^\vee \circ g \circ f^{-1}$ have matrix relative to Δ' equal to M, and hence they agree. □

5.9 Projectivities and perspectivities

We will prove in this section that any projectivity between linear varieties of the same projective space is a composition of perspectivities, that is:

Theorem 5.9.1. *Let $f: L \to L'$ be a projectivity between linear varieties L, L' of a projective space $\mathbb{P}_n$, $\dim L = \dim L' < n$. There exist linear varieties $L_0, \dots, L_s$ of $\mathbb{P}_n$, $L_0 = L$, $L_s = L'$ and perspectivities $g_i : L_{i-1} \to L_i$ such that $f = g_s \circ \dots \circ g_1$.*

Proof. In case $L' \neq L$, fix any perspectivity $g: L' \to L$. Then g^{-1} is also a perspectivity and once the theorem is proved for the projectivity $g \circ f: L \to L$, it obviously holds for $f = g^{-1} \circ (g \circ f)$. Thus, in the sequel, we will consider the case $L' = L$ only.

We want to reduce ourselves to the case in which L is a hyperplane of $\mathbb{P}_n$. To this end, if $\dim L < n - 1$, take a linear variety T of $\mathbb{P}_n$ such that $T \supset L$ and $\dim T = \dim L + 1$, and a supplementary linear variety T' of T. The reader may easily check that if L_1 and L_2 are linear varieties contained in T and $g: L_1 \to L_2$ is a perspectivity of T (as a projective space) with centre F, then g is also a perspectivity of $\mathbb{P}_n$, the perspectivity with centre $F \vee T'$ in fact. Thus if f is a composition of perspectivities of T, then it is also a composition of perspectivities of $\mathbb{P}_n$ and so, by taking the above T instead of $\mathbb{P}_n$, we are allowed to assume in the sequel that $\dim L = n - 1$.

Assume thus that we have $f: L \to L$, L a hyperplane of $\mathbb{P}_n$. Take V to be either a fundamental variety of f with maximal dimension, or $V = \emptyset$ if f has no fixed point. We will make the proof by reverse induction on $d = \dim V$. In case $d = \dim L$, it is $V = L$, $f = \mathrm{Id}_L$ and the claim is obviously true. Assume now that $d < \dim L$ and also that, by induction, any projectivity of L with a fundamental variety of dimension greater than d is a composition of perspectivities.

If f has no fixed point other than those in V, take any point $p \in L - V$. Obviously, $f(p) \neq p$. Choose any hyperplane L' such that $V \subset L'$ and $p, f(p) \notin L'$. If q is any point in $L' - L \cap L'$, then neither of the lines $q \vee f(p)$ and $q \vee p$ is contained in L or L'. We may thus take points $A_1 \in q \vee f(p)$ and $A_2 \in q \vee p$ as the centres of perspectivities $h_1: L \to L'$ and $h_2: L' \to L$, respectively. By the above choices, p is a fixed point of $f_1 = h_2 \circ h_1 \circ f$. On the other hand, the condition $V \subset L'$ assures that all points of V are invariant by h_1 and h_2, and so they all are fixed points of f_1. Obviously if f_1 satisfies the claim, so does $f = h_1^{-1} \circ h_2^{-1} \circ f_1$. Now, if f_1 has a fundamental variety of dimension greater than d, we are done, because f_1 satisfies the claim by induction. Note that this is the case if $d = -1$ because f_1 has the fixed point p. Otherwise, by 5.7.4, V is a fundamental variety of f_1 of maximal dimension, $\dim V = d$ and f_1 has a further fixed point $p \notin V$.

If f has a fixed point p not in V take $f_1 = f$ and note that in this case $V \neq \emptyset$, that is, V actually is a fundamental variety, due to the existence of the fixed point p.

In this way, all remaining cases will be covered if the claim is proved for any projectivity f_1 with a fundamental variety V of maximal dimension, $\dim V = d$, and a fixed point $p \notin V$, which we do next.

Take any point $p'' \in V$ and $r = p \vee p''$, obviously a line invariant by f_1. We choose $p' \in r - \{p, p''\}$ and still $f(p') \in r - \{p, p''\}$. Take L' to be any hyperplane through V and missing p; clearly it misses also p' and $f(p')$. Choose any $q \in L' - L \cap L'$. Then $q \notin r \subset L$ and hence $\pi = r \vee q$ is a plane that contains the lines $s_1 = q \vee f(p')$ and $s_2 = q \vee p'$, neither of which is contained in L or L' because $q \notin L$ and $p, f_1(p) \notin L'$. Take s to be any line on π through p different from r and $p \vee q$, and then $B_1 = s_1 \cap s$, $B_2 = s_2 \cap s$. As it is clear from the construction, B_1 and B_2 may be taken as centres of perspectivities $g_1 : L \to L'$ and $g_2 : L' \to L$, respectively, whose composition $g_2 \circ g_1$ satisfies $(g_2 \circ g_1)(f(p')) = p'$ and $(g_2 \circ g_1)(p) = p$. Hence the projectivity $f_2 = g_2 \circ g_1 \circ f_1$, of L, leaves invariant p and p'. Furthermore all points in V are fixed by $g_2 \circ g_1$, and hence also by f_2, because $V \subset L \cap L'$. In particular p'' is fixed by f_2, after which the three distinct points p, p', p'' are fixed by f_2 and so all points in r are. Now, since $V \cap r = \{p''\}$, by 5.7.5, both V and r are contained in a fundamental variety V' of f_2. Clearly $\dim V' > \dim V = d$, as $p \notin V$, and so, by induction, f_2 satisfies the claim. As above, this assures that the claim is satisfied by f_1, and hence ends the proof. □

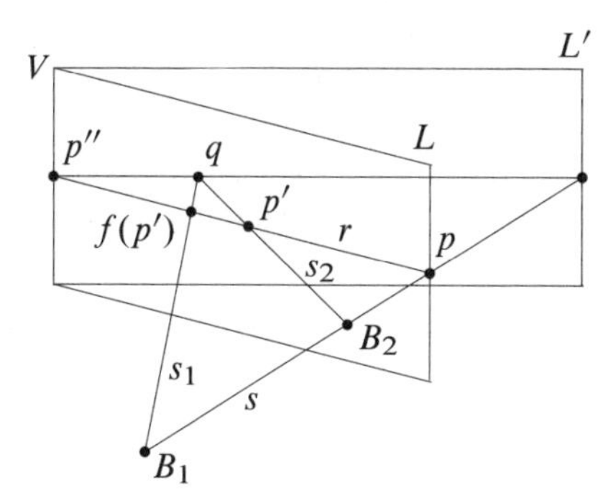

Figure 5.7. The argument ending the proof of Theorem 5.9.1.

5.10 Singular projectivities

In Section 1.6 we have defined the projectivities between two projective spaces $\mathbb{P}_n$, $\mathbb{P}'_m$ as being the maps induced by the isomorphisms between their associated vector spaces. A monomorphism between the associated vector spaces obviously gives rise to a projectivity from $\mathbb{P}_n$ onto a linear variety of $\mathbb{P}'_m$, which is not an essentially new object. In this section we will study the correspondences between projective spaces induced by non-injective linear maps between their associated vector spaces.

Let $(\mathbb{P}_n, E, \pi)$ and $(\mathbb{P}'_m, E', \pi')$ be projective spaces and $\varphi\colon E \to E'$ a non-zero linear map that we will suppose to be not injective. Then $[\ker(\varphi)]$ is a non-empty linear variety of $\mathbb{P}_n$ and if $p = [v] \in [\ker(\varphi)]$ then $\varphi(v) = 0$ represents no point. Thus, φ may not be used to define a map in the whole of $\mathbb{P}_n$ but just

$$\begin{aligned} f = [\varphi]\colon \mathbb{P}_n - [\ker(\varphi)] &\longrightarrow \mathbb{P}'_m, \\ p = [v] &\longmapsto [\varphi(v)]. \end{aligned}$$

Clearly, $[\varphi(v)]$ is well defined, because $v \notin \ker(\varphi)$, and does not depend on the choice of the representative v of p, by the linearity of φ. The map f is called the *singular projectivity* defined or represented by φ, while the linear map φ is called a *representative*, of f.

Example 5.10.1. We have seen in Section 1.9 that if $L = [F] \subset \mathbb{P}_n$ is a non-empty linear variety, then the projection map $\mathbb{P}_n - L \to L^\nabla$ is a singular projectivity, induced by the natural morphism $E \to E/F$.

The next lemma proves that the representative φ of f is determined by f up to a non-zero factor:

Lemma 5.10.2. *If a second linear map $\psi\colon E \to E'$ induces on $\mathbb{P}_n - [\ker(\varphi)]$ the same map as φ, that is, it has $\psi(v) \neq 0$ and $[\psi(v)] = [\varphi(v)]$ for all $v \notin \ker(\varphi)$, then $\psi = \lambda\varphi$ for some $\lambda \in k - \{0\}$, and conversely.*

Proof. The hypothesis says that $\ker\psi \subset \ker\varphi$ and that for any $v \notin \ker(\varphi)$ there is $\lambda_v \in k - \{0\}$ such that $\varphi(v) = \lambda_v\psi(v)$. We will see first that λ_v does not depend on v.

Assume $v, w \notin \ker(\varphi)$ and $\psi(v), \psi(w)$ independent. Then, if $v + w \notin \ker(\varphi)$,

$$\begin{aligned} \lambda_v\psi(v) + \lambda_w\psi(w) &= \varphi(v) + \varphi(w) = \varphi(v + w) \\ &= \lambda_{v+w}\psi(v + w) = \lambda_{v+w}\psi(v) + \lambda_{v+w}\psi(w), \end{aligned}$$

from which the independence of $\psi(v), \psi(w)$ gives $\lambda_v = \lambda_{v+w} = \lambda_w$ as wanted. If $v + w \in \ker(\varphi)$, then $v - w \notin \ker(\varphi)$ (otherwise $v, w \in \ker(\varphi)$) and a similar argument using $v - w$ instead of $v + w$ leads to the same conclusion.

If still $v, w \notin \ker(\varphi)$ and $\psi(v), \psi(w)$ are dependent, it is $\psi(v) = \mu\psi(w)$, $\mu \in k - \{0\}$. Then $\psi(v - \mu w) = 0$, hence, by the hypothesis, $\varphi(v - \mu w) = 0$ and so $\varphi(v) = \mu\varphi(w)$. It follows that

$$\mu\lambda_w\psi(w) = \mu\varphi(w) = \varphi(v) = \lambda_v\psi(v) = \mu\lambda_v\psi(w)$$

and so, again, $\lambda_v = \lambda_w$.

Once we know that there is $\lambda \neq 0$ such that $\varphi(v) = \lambda\psi(v)$ for all $v \notin \ker(\varphi)$, choose a vector $w \notin \ker(\varphi)$ and take any $v \in \ker(\varphi)$. Obviously $v + w \notin \ker(\varphi)$ and

$$\lambda\psi(w) = \varphi(w) = \varphi(v + w) = \lambda\psi(v + w) = \lambda\psi(v) + \lambda\psi(w).$$

It follows that $\psi(v) = 0$ and, all together, $\varphi(v) = \lambda\psi(v)$, $\lambda \neq 0$, for all $v \in E$. The converse being obvious, this ends the proof. □

Remark 5.10.3. It results from 5.10.2 above that $\ker(\psi) = \ker(\varphi)$ and therefore $[\ker(\psi)] = [\ker(\varphi)]$. Therefore, the singular projectivity $[\varphi]$ is not the restriction of a projectivity or a singular projectivity defined in a set strictly larger than $\mathbb{P}_n - [\ker(\varphi)]$.

The name *singular projectivity* is classical (see for instance [2]), although a bit confusing, as, according to our definitions, a singular projectivity is not a projectivity. As noted in 5.10.3, the linear variety $\mathrm{Sing}(f) = [\ker(\varphi)]$ is independent of the choice of the representative φ: it is called the *singular variety* of f, and its points the *singular points* of f. A singular projectivity $f : \mathbb{P}_n - \mathrm{Sing}(f) \to \mathbb{P}'_m$ is often viewed as a correspondence between the projective spaces and then called a singular projectivity from $\mathbb{P}_n$ into $\mathbb{P}'_m$, written $f : \mathbb{P}_n \to \mathbb{P}'_m$. Whether f is viewed as a map or as a correspondence, the points of its singular variety have undefined (or, as said in old texts, undetermined) image. Singular projectivities from a projective space into its dual space are called *singular correlations*.

As the reader may check, the image of a singular projectivity $f = [\varphi]$ is the linear variety $\mathrm{Im}(f) = [\mathrm{Im}(\varphi)]$. We define the *rank* of f as being $\mathrm{rk}(f) = 1 + \dim \mathrm{Im}(\mathrm{f}) = \mathrm{rk}(\varphi)$. The well-known relation between the dimensions of the kernel and the image of a linear map directly gives

Lemma 5.10.4. *If $f : \mathbb{P}_n \to \mathbb{P}'_m$ is a singular projectivity, then*

$$n = \dim \mathrm{Sing}(f) + \mathrm{rk}(f).$$

Note that the singular projectivities of rank one map all points to a single one. By its definition, it is $\mathrm{rk}(f) \leq m + 1$, and by the hypothesis $\mathrm{Sing}(f) \neq \emptyset$, $\mathrm{rk}(f) \leq n$.

It is easy to see that composing a projectivity and a singular projectivity, in either order, gives rise to a singular projectivity. Composing singular projectivities is not always possible. We have:

Proposition 5.10.5. *If* $f\colon \mathbb{P}_n - \mathrm{Sing}(f) \to \mathbb{P}'_m$ *and* $g\colon \mathbb{P}'_m - \mathrm{Sing}(g) \to \mathbb{P}''_r$ *are singular projectivities and* $\mathrm{Im}(f) \not\subset \mathrm{Sing}(g)$, *then*

$$g \circ f\colon \mathbb{P}_n - \mathrm{Sing}(f) \cup f^{-1}(\mathrm{Sing}(g)) \longrightarrow \mathbb{P}''_r$$

is a singular projectivity with singular variety $\mathrm{Sing}(f) \cup f^{-1}(\mathrm{Sing}(g))$.

Proof. Just note that if $f = [\varphi]$ and $g = [\psi]$, then $\psi \circ \varphi$ is not zero, has kernel $\varphi^{-1}(\ker \psi)$ and defines $g \circ f$. □

Restricting a singular projectivity to a positive-dimensional linear variety not contained in the singular locus gives rise to either a projectivity or a singular projectivity:

Proposition 5.10.6. *Let* $f\colon \mathbb{P}_n \to \mathbb{P}'_m$ *be a singular projectivity and* L *a linear variety of* $\mathbb{P}_n$, $\dim L > 0$*:*

(a) *If* $L \cap \mathrm{Sing}(f) = \emptyset$, *then* $f(L)$ *is a linear variety of* $\mathbb{P}'_m$ *and* $f_{|L}\colon L \to f(L)$ *is a projectivity.*

(b) *If* $L \cap \mathrm{Sing}(f) \neq \emptyset$ *and* $L \not\subset \mathrm{Sing}(f)$, *then*

$$f_{|L-\mathrm{Sing}(f)}\colon L - \mathrm{Sing}(f) \longrightarrow \mathbb{P}'_m$$

is a singular projectivity with singular variety $L \cap \mathrm{Sing}(f)$ *and* $f(L - \mathrm{Sing}(f))$ *is a linear variety of dimension* $\dim L - \dim L \cap \mathrm{Sing}(f) - 1$.

Proof. Assume $f = [\varphi]$ and $L = [F]$. In case (a), it is $\ker \varphi_{|F} = F \cap \ker \varphi = \{0\}$ and $\varphi_{|F}$ defines $f_{|L}\colon L \to f(L)$ as a projectivity. In case (b) write $f_L = f_{|L-\mathrm{Sing}(f)}$. By the hypothesis, we have $F \cap \ker \varphi \neq \{0\}$ and $F \not\subset \ker \varphi$. Then, since still $\ker \varphi_{|F} = F \cap \ker \varphi$, $\varphi_{|F}$ is a non-injective and non-zero linear map and defines a singular projectivity from L to $\mathbb{P}'_m$ which obviously has singular variety $L \cap \mathrm{Sing}(f)$ and agrees with f_L. Furthermore, $f(L - \mathrm{Sing}(f))$ being the image of f_L, it is a linear variety of dimension

$$\mathrm{rk}(f_F) - 1 = \dim L - \dim \mathrm{Sing}(f_L) - 1 = \dim L - \dim L \cap \mathrm{Sing}(f) - 1,$$

as claimed. □

As for projectivities, the above definition of singular projectivities is just an intrinsic way of saying that they are correspondences given by linear equations, this time with a matrix of rank at most equal to the dimension of the source:

Proposition 5.10.7. *Assume we have fixed references* Δ *and* Ω *in projective spaces* $\mathbb{P}_n$ *and* $\mathbb{P}'_m$, *respectively. Let* $f\colon \mathbb{P}_n - \mathrm{Sing}(f) \to \mathbb{P}'_m$ *be a singular projectivity.*

There exists a matrix M, with m rows and n columns, such that an arbitrary point $[x_0, \dots, x_n] \in \mathbb{P}_n - S(f)$ has image $[y_0, \dots, y_m]$ if and only if

$$\begin{pmatrix} y_0 \\ \vdots \\ y_m \end{pmatrix} = \rho M \begin{pmatrix} x_0 \\ \vdots \\ x_n \end{pmatrix} \tag{5.36}$$

for some $\rho \neq 0$. Furthermore M is determined by this property up to a non-zero constant factor and $\mathrm{rk}(M) = \mathrm{rk}(f)$

Proof. If $f = [\varphi]$, then any matrix of φ relative to bases $e_0, \dots, e_n$ and $e'_0, \dots, e'_m$, adapted to Δ and Ω respectively, clearly satisfies the property of the claim. If a second matrix M' satisfies the same property, then the linear map φ' whose matrix relative to $e_0, \dots, e_n$ and $e'_0, \dots, e'_m$ is M' also induces f. By 5.10.2, $\varphi' = \lambda\varphi$ for some $\lambda \in k - \{0\}$ and thus $M' = \lambda M$. That $\mathrm{rk}(f) = \mathrm{rk}(\varphi) = \mathrm{rk}(M)$ is clear. □

Notations being as in 5.10.7, M is said to be a matrix of f relative to Δ, Ω. Once the references have been fixed, the matrix M obviously determines f, which is called the singular projectivity defined by M. Furthermore one may choose the matrix in order to define f, namely:

Proposition 5.10.8. *Fix references Δ and Ω in projective spaces $\mathbb{P}_n$ and $\mathbb{P}'_m$ respectively. Assume that M is a matrix with m rows and n columns and* $0 < \mathrm{rk}(M) \leq n$. *For each $p = [x_0, \dots, x_n] \in \mathbb{P}_n$ take $y_0, \dots, y_m$ given by the equations* (5.36)*: mapping p to $f(p) = [y_0, \dots, y_m]$ if some $y_i \neq 0$, and leaving $f(p)$ undefined otherwise, defines a singular projectivity f from $\mathbb{P}_n$ into $\mathbb{P}_m$ that has matrix M relative to Δ and Ω.*

Proof. As in the proof of 5.10.7, just choose bases adapted to Δ and Ω, define φ as the linear map whose matrix relative to them is M: then $f = [\varphi]$. □

Remark 5.10.9. Notations being as above, a maximal set of independent equations extracted from

$$M \begin{pmatrix} x_0 \\ \vdots \\ x_n \end{pmatrix} = 0$$

is a system of equations of the singular variety of f.

The projection maps appeared as examples of singular projectivities in 5.10.1. The next theorem shows that, up to projectivities and inclusions of linear varieties, they are the only ones. We drop the rank-one case to avoid trivialities.

Theorem 5.10.10. *Let $f : \mathbb{P}_n \to \mathbb{P}'_m$ be a singular projectivity with* $\mathrm{rk}(f) > 1$. *Then:*

(a) *The map*

$$\tilde{f}\colon \mathrm{Sing}(f)^{\nabla} \longrightarrow \mathrm{Im}(f),$$
$$p \vee \mathrm{Sing}(f) \longmapsto f(p),$$

is a projectivity.

(b) *The diagram*

$$\begin{array}{ccc} \mathbb{P}_n - \mathrm{Sing}(f) & \xrightarrow{\;f\;} & \mathbb{P}'_m \\ \downarrow \pi & & \uparrow \iota \\ S(F)^{\nabla} & \xrightarrow[\tilde{f}]{} & \mathrm{Im}(f), \end{array}$$

where π and ι are, respectively, the projection from $\mathrm{Sing}(f)$ *and the inclusion in $\mathbb{P}'_m$, is commutative, that is, $f = \iota \circ \tilde{f} \circ \pi$.*

Proof. Let E be the vector space associated to $\mathbb{P}_n$ and assume that $f = [\varphi]$. As it is well known, the linear map φ gives rise to the isomorphism

$$\tilde{\varphi}\colon E/\ker(\varphi) \longrightarrow \mathrm{Im}(\varphi),$$
$$\bar{v} = v + \ker(\varphi) \longmapsto \varphi(v).$$

According to the way we have defined the projective structure on a bunch in Section 1.9, $[\tilde{\varphi}] = \tilde{f}$, after which the commutativity of the diagram is clear. □

Corollary 5.10.11. *For two different points p, q, both non-singular for a singular projectivity f, the following conditions are equivalent:*

(i) $f(p) = f(q)$,

(ii) $p \vee S(f) = q \vee S(f)$,

(iii) *p and q are collinear with a singular point of f.*

Proof. The equivalence (i) $\Longleftrightarrow$ (ii) follows from 5.10.10, while (ii) $\Longleftrightarrow$ (iii) is an easy exercise on incidence of linear varieties. □

In particular the restriction of a singular projectivity f to a line $\ell \not\subset \mathrm{Sing}(f)$ containing a singular point is constant. We have already seen that in case $\ell \cap \mathrm{Sing}(f) = \emptyset$ the restriction of f is a projectivity $\ell \to f(\ell)$. Therefore, by Theorem 2.9.11, the following holds:

Proposition 5.10.12. *If $f\colon \mathbb{P}_n - \mathrm{Sing}(f) \to \mathbb{P}'_m$ is a singular projectivity and q_1, q_2, q_3, q_4 are aligned points of $\mathbb{P}_n$, non-aligned with a singular point and three at least distinct, then $f(q_1)$, $f(q_2)$, $f(q_3)$, $f(q_4)$ are aligned, three at least distinct, and have*

$$(f(q_1), f(q_2), f(q_3), f(q_4)) = (q_1, q_2, q_3, q_4).$$

5.11 Exercises

Exercise 5.1. Let L be a linear variety of a real projective space $\mathbb{P}_n$, $n > 1$. Prove that if g is the projectivity of Proposition 5.1.19, then $g((L^*)_{\mathbb{C}}) = (L_{\mathbb{C}})^*$.

Exercise 5.2. Prove that for any projectivity f between different lines $r, s \subset \mathbb{P}_2$, there is a third line $t \subset \mathbb{P}_2$, $t \neq r, s$, and perspectivities $g_1 : r \to t$, $g_2 : t \to s$ such that $f = g_2 \circ g_1$.

Exercise 5.3. Prove that the cross-axis of the perspectivity with centre p between different lines $r, s \subset \mathbb{P}_2$ is the fourth harmonic of r, s, pO in O^*, $O = r \cap s$.

Exercise 5.4. Given three different and concurrent lines of $\mathbb{P}_2$, ℓ_1, ℓ_2, ℓ_3, and perspectivities $f_1 : \ell_1 \to \ell_2$ and $f_2 : \ell_2 \to \ell_3$, prove that $f_2 \circ f_1$ is a perspectivity whose centre is aligned with those of f_1 and f_2. Use this fact to re-prove Desargues' theorem for triangles sharing no vertex or side.

Exercise 5.5. Describe the correspondence between two projective lines defined by the equation (5.10) in the case where $BB' - AC = 0$. *Hint*: Factor the equation.

Exercise 5.6. Prove that no parabolic projectivity of a projective line is cyclic.

Exercise 5.7. Prove that no two imaginary and complex-conjugate points correspond by an elliptic involution of $\mathbb{P}_{1,\mathbb{R}}$. *Hint*: Use 5.6.7.

Exercise 5.8. Prove that three pairwise disjoint pairs of points of $\mathbb{P}_1$, $\{a_1, a_2\}$, $\{b_1, b_2\}$ and $\{c_1, c_2\}$, with $c_1 \neq c_2$, belong to the same involution if and only if

$$(a_1, b_1, c_1, c_2) = (b_2, a_2, c_1, c_2).$$

Hint: For the *if* part, consider the involution mapping a_1 to a_2 and c_1 to c_2.

Exercise 5.9. Prove that two different pairs of distinct points corresponding by an involution τ of $\mathbb{P}_{1,\mathbb{R}}$ separate each other if and only if the fixed points of τ are imaginary. *Hint*: Take three of the points to compose the reference.

Exercise 5.10. If τ_1, τ_2 are two different involutions of $\mathbb{P}_1$, prove that the following conditions are equivalent:

(i) $\tau_1 \circ \tau_2 = \tau_2 \circ \tau_1$.

(ii) $\tau_3 = \tau_1 \circ \tau_2$ is an involution.

(iii) The pairs of fixed points of τ_1 and τ_2 harmonically divide each other.

The above conditions being satisfied, prove that $(\{\tau_1, \tau_2, \tau_3, \mathrm{Id}_{\mathbb{P}_1}\}, \circ)$ is a group and write down its table.

Exercise 5.11. Prove that projectively equivalent collineations have the same invariants, thus justifying the name.

Exercise 5.12. Assume that $f\colon \mathbb{P}_3 \to \mathbb{P}_3$ is a collineation leaving invariant two skew lines ℓ_1, ℓ_2 and a point $p \notin \ell_1 \cup \ell_2$. Assume furthermore that neither ℓ_1 nor ℓ_2 have all points invariant by f and prove that:

(a) All points of the line ℓ through p that meets both ℓ_1 and ℓ_2 are invariant.

(b) There is a pencil of fixed planes whose kernel meets both ℓ_1 and ℓ_2.

Exercise 5.13. Assume given a tetrahedron T of $\mathbb{P}_3$. Prove that there is no collineation of $\mathbb{P}_3$ leaving all edges of T invariant and whose restrictions to them are all involutions.

Exercise 5.14. Describe the fundamental varieties and fundamental bundles of the projectivity of $\bar{\mathbb{A}}_3$ which, relative to a system of barycentric coordinates, has matrix

$$\begin{pmatrix} 0 & 1 & 1 & 1 \\ 1 & 0 & 1 & 1 \\ 1 & 1 & 0 & 1 \\ 1 & 1 & 1 & 0 \end{pmatrix}.$$

Prove in particular that it is an affinity and describe it.

Exercise 5.15. Assume given three concurrent non-coplanary lines of $\mathbb{P}_3$. Describe the fixed points and fixed planes of the homographies of $\mathbb{P}_3$ that leave invariant these lines and whose restriction to each of them is parabolic.

Exercise 5.16. Assume that a collineation f of $\mathbb{P}_n$ has a fundamental variety L which is also the kernel of its corresponding fundamental bundle. Prove that L is the only fundamental variety of f, n is odd and $\dim L = (n-1)/2$.

Exercise 5.17. Determine the points and planes invariant by the composition of three harmonic biaxial collineations of $\mathbb{P}_3$ with axes the pairs of opposite edges of a given tetrahedron.

Exercise 5.18. Assume given in $\mathbb{P}_2$ two distinct lines ℓ_1, ℓ_2 and a point $p \notin \ell_1 \cup \ell_2$. Describe the collineations of $\mathbb{P}_2$ that leave p fixed and whose restriction to ℓ_1 is the perspectivity $g\colon \ell_1 \to \ell_2$ with centre p.

Exercise 5.19. Prove that:

(a) The dual of a special homology is a special homology.

(b) The dual of a general homology is a general homology and both have the same invariant.

Exercise 5.20. Prove that for any general homology f of $\mathbb{P}_n$ with invariant λ, there is a reference relative to which f has matrix

$$\begin{pmatrix} \lambda & & & \\ & 1 & & \\ & & \ddots & \\ & & & 1 \end{pmatrix}.$$

Exercise 5.21. Prove that for any special homology f of $\mathbb{P}_n$ there is a reference relative to which f has matrix

$$\begin{pmatrix} 1 & 0 & 0 & \dots & 0 \\ 1 & 1 & 0 & \dots & 0 \\ 0 & 0 & 1 & \dots & 0 \\ \vdots & \vdots & \vdots & \ddots & \vdots \\ 0 & 0 & 0 & \dots & 1 \end{pmatrix}.$$

Exercise 5.22 (*Projective classification of homologies*). Prove that two homologies of n-dimensional projective spaces are projectively equivalent if and only if either they both are special or they both are general and have the same invariant. *Hint*: For the *if* parts take references according to Exercise 5.20 or Exercise 5.21, and the projectivity mapping one reference to the other.

Exercise 5.23. Assume that T and T' are triangles of $\mathbb{P}_2$ sharing no vertex or side and prove that the following conditions are equivalent:

(i) There is a homology of $\mathbb{P}_2$ mapping T to T'.

(ii) There is a one-to-one correspondence between the sets of vertices of T and T' and a point aligned with each pair of corresponding vertices.

(iii) There is a one-to-one correspondence between the sets of sides of T and T' and a line concurrent with each pair of corresponding sides.

Note that proving (i) $\Longleftrightarrow$ (ii) and (i) $\Longleftrightarrow$ (iii) re-proves Desargues' theorem for triangles sharing no vertex or side. Triangles satisfying the above conditions are called *homological* (due to condition (i)) and also *perspective* (due to condition (ii)).

Exercise 5.24. Assume we are given two different planes π, π' of $\mathbb{P}_3$ and two different points $p_1, p_2 \in \mathbb{P}_3 - \pi \cup \pi'$, and write κ_i the perspectivity

$$\kappa_i : \pi' \longrightarrow \pi$$

with centre $p_i, i = 1, 2$. Prove that $\kappa_2 \circ \kappa_1^{-1}$ is a homology of π. Determine its centre and axis. The pair of images of a figure F of π' by κ_1 and κ_2 is called a *stereoscopic projection* of F. The above proves that the components of a stereoscopic projection correspond by a homology. Relate to Exercises 1.32 and 5.23.

Exercise 5.25. Assume given a homology φ of a plane π of $\mathbb{P}_3$. Prove that there are a plane π' and two different points p_1, p_2 as in Exercise 5.24 such that, with the notations used there, $\varphi = \kappa_2 \circ \kappa_1^{-1}$.

Exercise 5.26. Let H be a hyperplane of a projective space $\mathbb{P}_n$, $n \geq 2$, and $\Delta = \{p_0, \dots, p_n, A\}$ a reference of $\mathbb{P}_n$ with $p_1, \dots, p_n \in H$.

(a) Prove that if a collineation f of $\mathbb{P}_n$ leaves fixed all points of H, then it has a unique matrix relative to Δ of the form

$$\begin{pmatrix} a_0 & 0 & \dots & 0 \\ a_1 & 1 & \dots & 0 \\ \vdots & & & \vdots \\ a_n & 0 & \dots & 1 \end{pmatrix},$$

with $a_0, \dots, a_n \in k$, $a_0 \neq 0$. Prove also that, conversely, any such matrix is the matrix of a collineation that leaves fixed all points of H.

(b) Prove that if $f \neq \mathrm{Id}$, then f is a homology with centre $[a_0 - 1, a_1, \dots, a_n]$ and plane of homology H.

(c) Prove that f is a general homology if and only if $a_0 \neq 1$, and hence that f is a special homology if and only if $a_0 = 1$ and $a_i \neq 0$ for some $i = 1, \dots, n$.

Exercise 5.27. Assume that f and g are special homologies of $\mathbb{P}_n$, $n \geq 2$, both with plane of homology H and prove that:

(a) $f \circ g$ is either the identity or a special homology with hyperplane of homology H.

(b) $f \circ g = g \circ f$.

(c) If $f \circ g$ is a special homology, then the centres of f, g and $f \circ g$ are aligned.

Hint: Use Exercise 5.26.

Exercise 5.28. Assume that f is a special homology of $\mathbb{P}_n$, $n \geq 2$, with centre O and hyperplane of homology H, and $\rho \in k$. Define a map $\rho f : \mathbb{P}_n \to \mathbb{P}_n$ by the rule

$$\rho f(p) = \begin{cases} p & \text{if } p \in H, \\ \text{the only point } p' \in pf(p) \text{ for which } (O, p, f(p), p') = \rho & \text{otherwise.} \end{cases}$$

Prove that ρf is the identity map if $\rho = 0$, and a special homology with centre O and hyperplane of homology H if $\rho \neq 0$.

Exercise 5.29 (*Affine structure on the complementary of a hyperplane*). Let H be a hyperplane of a projective space $\mathbb{P}_n$, $n \geq 2$, and put $U = \mathbb{P}_n - H$. Let F be the set consisting of the identical map of $\mathbb{P}_n$ and all special homologies of $\mathbb{P}_n$ with hyperplane of homology H. For any $\rho \in k$, take ρf as in Exercise 5.28 if $f \in F - \{\mathrm{Id}\}$ and $\rho \mathrm{Id} = \mathrm{Id}$. Use Exercises 5.26, 5.27 and 5.28 to prove that:

(a) The addition $f + g = f \circ g$ and the multiplication $\rho \cdot f = \rho f$ define on F a structure of n-dimensional vector space.

(b) The map

$$\begin{aligned} F \times U &\longrightarrow U, \\ (f, p) &\longmapsto f(p), \end{aligned}$$

defines on U a structure of n-dimensional affine space.

(c) Mapping each point of $\mathbb{P}(F)$ to the common centre of its representatives (see Exercise 5.28), extends the identical map of U to a projectivity $\Gamma \colon \bar{U} \to \mathbb{P}_n$ ($\bar{U}$ the projective closure of U as affine space). The projectivity Γ^{-1} allows us to identify $\mathbb{P}_n$ and H to the projective closure and the improper hyperplane of the affine space U, respectively. *Hint*: In order to define a linear map representing Γ take $\Delta = \{p_0, \dots, p_n, A\}$ to be a reference of $\mathbb{P}_n$ with $p_1, \dots, p_n \in H$ and fix a basis $e_0, \dots, e_n$ adapted to Δ. If $f \in F$ has matrix

$$\begin{pmatrix} 1 & 0 & \dots & 0 \\ a_1 & 1 & \dots & 0 \\ \vdots & & & \vdots \\ a_n & 0 & \dots & 1 \end{pmatrix}$$

(see Exercise 5.26), then map the constant vector field $\hat{f}$ to $a_1 e_1 + \dots + a_n e_n$. If $\theta_{q,\delta}$ is a radial vector field and $q = [1, b_1, \dots, b_n]_\Delta$, then map it to $\delta e_0 + \delta b_1 e_1 + \dots + \delta b_n e_n$.

Exercise 5.30. Prove that two biaxial collineations are projectively equivalent if and only if they have equal invariants.

Exercise 5.31. Claim and prove a result similar to 5.7.24 for biaxial collineations of $\mathbb{P}_3$.

Exercise 5.32. Prove that the restrictions of a biaxial collineation f of $\mathbb{P}_3$ to any two planes containing different axes are general homologies with reciprocal invariants. Prove also that the restrictions of f to two planes containing the same axis are general homologies with the same invariant.

Exercise 5.33. If g is a correlation of $\mathbb{P}_n$, $L \subset \mathbb{P}_n$ a linear variety and $\bar{L}$ the transform of L by g, prove that $g^\bullet(\bar{L}) = L^*$.

Exercise 5.34. Prove that the images of the points of a line $\ell \subset \mathbb{P}_n$ by a singular projectivity $f : \mathbb{P}_n \to \mathbb{P}_m$ either

(1) are all undefined, or

(2) are all equal to the same point of $\mathbb{P}_m$ but for a single one which is undefined, or

(3) are all defined and are the points of a line ℓ' of $\mathbb{P}_m$. Furthermore, in this case, $f_{|\ell} : \ell \to \ell'$ is a projectivity.

Exercise 5.35. Determine the fixed points of a collineation of a projective space $\mathbb{P}_{3,\mathbb{R}}$ that leaves invariant all the planes of a certain pencil and whose restrictions to two of them have three isolated fixed points.

Chapter 6
Quadric hypersurfaces

6.1 The notion of quadric

In an elementary context, quadric hypersurfaces (or quadrics for short) are usually defined as sets of points whose coordinates satisfy an equation of the second degree. This is a naive and rather inconvenient definition, because the equation of a quadric encloses essential information that not always can be read from the set of points it defines. For instance, after such a definition all quadrics without points would be equal, and no distinction could be made between a hyperplane such as $x_0 = 0$ and the quadric $x_0^2 = 0$. Defining quadrics as being equations or polynomials of the second degree, taken up to a scalar factor, is more satisfactory, but it makes the definition dependent on the previous choice of a system of coordinates, and removing such a dependence leads to a rather long and boring discussion.

We will define quadrics using symmetric bilinear forms, which will play the role of equations without any need of coordinates, just as linear forms did for linear varieties in Section 2.6. Second-degree equations will appear after a short while, depending on the choice of a projective reference (see equation (6.4) below). Before using symmetric bilinear forms to define quadrics, we recall some basic facts about them.

A symmetric bilinear form on a finite-dimensional vector space E over a field k is any map

$$\eta\colon E \times E \longrightarrow k$$

that satisfies

(a) $\eta(v + v', w) = \eta(v, w) + \eta(v', w)$,

(b) $\eta(\lambda v, w) = \lambda\eta(v, w)$,

(c) $\eta(v, w) = \eta(w, v)$

for any $v, v', w \in E$ and $\lambda \in k$. The reader may note that the properties analogous to (a) and (b) for the second variable w do hold by virtue of (c). The usual definitions of sum of maps into k and their product by elements of k,

$$(\eta_1 + \eta_2)(v, w) = \eta_1(v, w) + \eta_2(v, w),$$
$$(\lambda\eta)(v, w) = \lambda\eta(v, w),$$

turn the set of all bilinear symmetric forms on E into a finite-dimensional vector space over k: we will denote it $S^2(E)$.

If $e_0, \dots, e_n$ is a basis of E and

$$v = y_0 e_0 + \dots + y_n e_n, \quad w = x_0 e_0 + \dots + x_n e_n,$$

an easy computation gives

$$\eta(v, w) = \sum_{i,j=0}^{n} y_i x_j \eta(e_i, e_j)$$

which may be written in matricial form

$$\eta(v, w) = \begin{pmatrix} y_0 & \dots & y_n \end{pmatrix} \begin{pmatrix} \eta(e_0, e_0) & \dots & \eta(e_0, e_n) \\ \vdots & & \vdots \\ \eta(e_n, e_0) & \dots & \eta(e_n, e_n) \end{pmatrix} \begin{pmatrix} x_0 \\ \vdots \\ x_n \end{pmatrix}. \tag{6.1}$$

The matrix

$$N = \begin{pmatrix} \eta(e_0, e_0) & \dots & \eta(e_0, e_n) \\ \vdots & & \vdots \\ \eta(e_n, e_0) & \dots & \eta(e_n, e_n) \end{pmatrix}$$

is called the matrix of η relative to the basis $e_0, \dots, e_n$: it is symmetric, determines η and is in turn determined by η once the basis has been fixed. Furthermore, any $(n+1) \times (n+1)$ symmetrical matrix N defines a symmetric bilinear form η by the rule

$$\eta\Big(\sum_{i=0}^{n} y_i e_i, \sum_{j=0}^{n} x_j e_j\Big) = \begin{pmatrix} y_0 & \dots & y_n \end{pmatrix} N \begin{pmatrix} x_0 \\ \vdots \\ x_n \end{pmatrix}$$

and the matrix of such an η, relative to the basis $e_0, \dots, e_n$, is N.

Let $\mathbb{P}_n$ be a projective space with associated vector space E. A *quadric hypersurface* (or just a *quadric*) of $\mathbb{P}_n$ is a class of non-zero symmetric bilinear forms on E modulo proportionality. In other words, the quadrics of $\mathbb{P}_n$ are the points of the projective space $\mathbb{P}(S^2(E))$. The quadrics of a projective plane $\mathbb{P}_2$ are called *conics*. The fact that the quadrics of $\mathbb{P}_n$ describe in turn a projective space is very relevant regarding families of quadrics, as we will see in Chapter 9. For the moment we will focus on the fact that each individual quadric Q is, by definition, a class $Q = \{\lambda\eta\}_{\lambda \in k - \{0\}}$, η a non-zero symmetric bilinear form on E. As for the elements of any projective space, we will say that Q is *represented* by any of the $\lambda\eta$, $\lambda \neq 0$, and also that the $\lambda\eta$ are *representatives* of Q, and write $Q = [\eta]$ to indicate that Q is the quadric represented by η. Needless to say, the form η determines $Q = [\eta]$ and is in turn determined by Q up to a non-zero factor. The quadrics of real projective spaces are called *real quadrics*, while those of complex projective spaces are called *complex quadrics*.

A point $p = [v] \in \mathbb{P}_n$ is said to be a *point* (and sometimes an *element*) of a quadric $Q = [\eta]$ if and only if $\eta(v, v) = 0$. This condition is independent of the representatives η of Q and v of p, as for any $\lambda, \mu \in k - \{0\}$,

$$(\lambda\eta)(\mu v, \mu v) = \lambda\mu^2\eta(v, v) = 0$$

if and only if $\eta(v, v) = 0$. We will write $p \in Q$ to indicate that p is a point of the quadric Q. When p is a point of a quadric Q it is equivalently said that p *belongs to* or *lies on* Q, and also that Q *contains*, or *goes through*, p.

The set of points of a quadric Q will be denoted by $|Q|$. It is worth noting that a quadric and its set of points are different objects and in some cases the quadric is not determined by its set of points, see Section 7.3. Nevertheless, provided no confusion may result, we will follow the very usual convention of mentioning the quadric instead of its set of points in set-theoretical relations involving the latter. For instance when all points of a set S belong to a quadric Q it is said that Q contains (or goes through) S, and it is written $S \subset Q$ instead of $S \subset |Q|$. Similarly, if Q_1 and Q_2 are quadrics of the same space, $|Q_1| \cap |Q_2|$ is usually written $Q_1 \cap Q_2$ and called the intersection of Q_1 and Q_2.

Assume that $\Delta = (p_0, \dots, p_n, A)$ is a reference of $\mathbb{P}_n$. The matrix of any representative of Q relative to any basis adapted to Δ is called a *matrix* of Q relative to Δ. Matrices of a quadric relative to a fixed reference are determined up to a non-zero factor:

Lemma 6.1.1. *If N is a matrix of a quadric Q, a matrix N' is also a matrix of Q relative to the same reference if and only if $N' = \rho N$ for some non-zero $\rho \in k$.*

Proof. If $Q = [\eta]$ and $e_0, \dots, e_n$ is a basis adapted to Δ, then

$$N = \begin{pmatrix} a_{0,0} & \dots & a_{n,0} \\ \vdots & & \vdots \\ a_{0,n} & \dots & a_{n,n} \end{pmatrix},$$

with $a_{i,j} = \eta(e_i, e_j)$, $i, j = 0, \dots, n$, is a matrix of Q relative to Δ. Since any other representative of Q is $\lambda\eta$ for some $\lambda \neq 0$, and any basis adapted to Δ is $\mu e_0, \dots, \mu e_n$, for some $\mu \neq 0$, an arbitrary matrix of Q relative to Δ has coefficients

$$(\lambda\eta)(\mu e_i, \mu e_j) = \lambda\mu^2 a_{i,j}, \quad i, j = 0, \dots, n,$$

and so equals $\lambda\mu^2 N$. Conversely for any $\rho \in k - \{0\}$, ρN is the matrix of $\rho\eta$ relative to $e_0, \dots, e_n$, and therefore is a matrix of Q relative to Δ. □

The next lemma allows us to define a quadric by giving its matrix:

Lemma 6.1.2. *Once a projective reference Δ of $\mathbb{P}_n$ has been fixed, for each non-zero symmetric $(n + 1) \times (n + 1)$ matrix N there is one and only one quadric of $\mathbb{P}_n$ which has matrix relative to Δ equal to N.*

Proof. Fix a basis $e_0, \dots, e_n$ adapted to Δ. We have already noted that there is a symmetric bilinear form η whose matrix relative to $e_0, \dots, e_n$ is N; $N \neq 0$ forces $\eta \neq 0$ and then the quadric $Q = [\eta]$ has matrix N. If a quadric Q' also has matrix N relative to the same reference, then N is the matrix of a certain representative ζ of Q' relative to a basis adapted to Δ which, by 2.1.2, has the form $\rho e_0, \dots, \rho e_n$, $\rho \in k - \{0\}$. It follows that for any i, j it holds that

$$\rho^2 \zeta(e_i, e_j) = \zeta(\rho e_i, \rho e_j) = \eta(e_i, e_j),$$

which in turn, by the bilinearity of ζ and η, gives $\rho^2 \zeta = \eta$ and thus $Q' = Q$. □

Still assume that a reference Δ has been fixed and

$$N = \begin{pmatrix} a_{0,0} & \dots & a_{n,0} \\ \vdots & & \vdots \\ a_{0,n} & \dots & a_{n,n} \end{pmatrix}, \tag{6.2}$$

is a matrix of Q relative to Δ. According to the above equality (6.1), the condition for a point $p = [x_0, \dots, x_n]$ to belong to Q may be written as either

$$\begin{pmatrix} x_0 & \dots & x_n \end{pmatrix} \begin{pmatrix} a_{0,0} & \dots & a_{n,0} \\ \vdots & & \vdots \\ a_{0,n} & \dots & a_{n,n} \end{pmatrix} \begin{pmatrix} x_0 \\ \vdots \\ x_n \end{pmatrix} = 0 \tag{6.3}$$

or, equivalently,

$$\sum_{i,j=0}^{n} a_{i,j} x_i x_j = 0. \tag{6.4}$$

The equality (6.4) (and also its equivalent matricial form (6.3)) is called an *equation* of Q relative to Δ. Note that the first member of this equation is a non-zero homogeneous polynomial of the second degree that has been written in symmetric form, namely, for each i, j with $i \neq j$, it has the monomials in $x_i x_j$ and $x_j x_i$ written separately and with equal coefficients $a_{i,j} = a_{j,i}$. Of course any homogeneous polynomial of degree two

$$\sum_{n \geq i \geq j \geq 0} b_{i,j} x_i x_j$$

may be uniquely written in this form by taking $a_{i,i} = b_{i,i}$, for $0 \leq i \leq n$, and $a_{i,j} = a_{j,i} = b_{i,j}/2$ for $0 \leq j < i \leq n$. Clearly, the equation (6.4) and the matrix N enclose the same information. In particular, by 6.1.1 and 6.1.2, once a reference has been fixed, the equation (6.4) is determined by the quadric Q up to a non-zero factor and, conversely, any equation of the form of (6.4) with some $a_{i,j} \neq 0$ is an equation of a uniquely determined quadric. We will use the notation

$Q : \sum_{i,j} a_{i,j} x_i x_j = 0$ to mean that Q is the quadric defined by the equation $\sum_{i,j} a_{i,j} x_i x_j = 0$, sometimes referred to just as the quadric $\sum_{i,j} a_{i,j} x_i x_j = 0$.

Quadrics being not sets of points, the action of projectivities on them is not obvious and needs to be defined. Assume, as above, that $Q = [\eta]$ is a quadric of $\mathbb{P}_n$ and $f : \mathbb{P}_n \to \mathbb{P}'_n$ is a projectivity, represented by the isomorphism $\varphi : E \to E'$ between the corresponding associated vector spaces. The inverse φ^{-1} of φ being linear, the reader may easily see that the composition of the map

$$\begin{aligned} \varphi^{-1} \times \varphi^{-1} : E' \times E' &\longrightarrow E \times E, \\ (w_1, w_2) &\longmapsto (\varphi^{-1}(w_1), \varphi^{-1}(w_2)), \end{aligned}$$

and the form η is a non-zero symmetric bilinear form $\eta \circ (\varphi^{-1} \times \varphi^{-1})$ on E'. By definition the quadric of $\mathbb{P}'_n$ defined by this form is the *image* or the *transform* of Q by f, written

$$f(Q) = [\eta \circ (\varphi^{-1} \times \varphi^{-1})].$$

First of all we show that this definition is compatible with the action of f on points, namely:

Proposition 6.1.3. *For any quadric Q of $\mathbb{P}_n$ and any projectivity $f : \mathbb{P}_n \to \mathbb{P}'_n$,*

$$|f(Q)| = f(|Q|).$$

Proof. The notations being as above, by the definition of $f(Q)$, a point $q = [w]$ belongs to $f(Q)$ if and only if

$$\eta(\varphi^{-1}(w), \varphi^{-1}(w)) = 0,$$

which is also the condition for the point $f^{-1}(q) = [\varphi^{-1}(w)]$ to belong to Q, hence the claim. □

The action of projectivities on quadrics has the required properties:

Proposition 6.1.4. *For any quadric Q of a projective space $\mathbb{P}_n$,*

(a) $\mathrm{Id}_{\mathbb{P}_n}(Q) = Q$,

(b) *for any two projectivities $f : \mathbb{P}_n \to \mathbb{P}'_n$ and $g : \mathbb{P}'_n \to \mathbb{P}''_n$, $g(f(Q)) = (g \circ f)(Q)$,*

(c) *for any projectivity $f : \mathbb{P}_n \to \mathbb{P}'_n$, $f^{-1}(f(Q)) = Q$.*

Proof. Claim (a) is obvious from the definition. For claim (b) just use that if $f = [\varphi]$ and $g = [\psi]$, then $(\varphi^{-1} \times \varphi^{-1}) \circ (\psi^{-1} \times \psi^{-1}) = (\psi \circ \varphi)^{-1} \times (\psi \circ \varphi)^{-1}$. Claim (c) is a direct consequence of the preceding ones. □

Also the section of a quadric by a linear variety needs a definition. Assume that $Q = [\eta]$ and $L = [F]$ are, respectively, a quadric and a positive-dimensional linear variety of $\mathbb{P}_n$. We will distinguish two cases, namely:

1. *The restriction $\eta_{|F}$ of η to $F \times F$ is non-zero.* Then $\eta_{|F}$ represents a quadric of L which is called the *section* of Q by L, and also the *intersection* of Q and L. It is denoted by $Q \cap L$ and its points are the points of Q lying on L, that is, $|Q \cap L| = |Q| \cap L$. Indeed, for any point $p = [v] \in L$, $v \in F$ and then the conditions $\eta_{|F}(v, v) = 0$ and $\eta(v, v) = 0$ are obviously equivalent. Sections of quadrics by hyperplanes are called *hyperplane sections* (*line sections*, *plane sections* if $n = 2, 3$).

2. $\eta_{|F} = 0$. Then $\eta(v, v) = 0$ for any $v \in F$ and therefore all points of L belong to Q. Conversely, if all points of L belong to Q, $\eta(v, v) = 0$ for all $v \in F$. Then, using the bilinearity of η, for any $v, w \in F$ one may compute

$$\eta(v, w) = \frac{1}{2}(\eta(v + w, v + w) - \eta(v, v) - \eta(w, w)) = 0,$$

the last equality due to the fact that both v, w and $v + w$ belong to F. We have thus seen that $\eta_{|F} = 0$ is equivalent to $L \subset |Q|$. If any of these conditions is satisfied we will write $L \subset Q$ (or $Q \supset L$) and also $Q \cap L = L$, and say that L is *contained in* or *lies on* Q (and also that Q *contains* L). Also, following a classical use, we will say in this case that the section of Q by L is *undetermined*.

Remark 6.1.5. The above argument applies to the case $L = \mathbb{P}_n$ and shows that $|Q| \neq P_n$, as in this case $\eta_{|F} = \eta \neq 0$.

The reader may easily prove the next lemma from the above definitions:

Lemma 6.1.6. *If Q is a quadric and $L_1 \subset L_2$ a pair of embodied linear varieties of $\mathbb{P}_n$ and* $\dim L_1 > 0$, *then*

$$(Q \cap L_2) \cap L_1 = Q \cap L_1.$$

The next proposition shows that the image of a section is a section of the image. More precisely:

Proposition 6.1.7. *If Q is a quadric of $\mathbb{P}_n$, $L \subset \mathbb{P}_n$ is a linear variety of positive dimension and $f : \mathbb{P}_n \to \mathbb{P}'_n$ is a projectivity, then*

$$f_{|L}(Q \cap L) = f(Q) \cap f(L).$$

Proof. Assume that $Q = [\eta]$, $L = [F]$ and $f = [\varphi]$, after which $f(L) = [\varphi(F)]$. The reader may then easily check that the equality of bilinear forms on $\varphi(F)$

$$\eta_{|F} \circ ((\varphi_{|F})^{-1} \times (\varphi_{|F})^{-1}) = (\eta \circ (\varphi^{-1} \times \varphi^{-1}))_{|\varphi(F)}$$

does hold. If $Q \cap L = L$, then $\eta_{|F} = 0$, the above equality gives $(\eta \circ (\varphi^{-1} \times \varphi^{-1}))_{|\varphi(F)} = 0$ and hence $f(Q) \cap f(L) = f(L)$ as claimed. Otherwise both members of the equality are non-zero, the one on the left defines $f_{|L}(Q \cap L)$, while the other defines $f(Q) \cap f(L)$, from which the claim. □

Remark 6.1.8. It is easy to get a matrix of $Q \cap L$ from a matrix of Q and parametric equations of L. Assume that, after fixing a reference Δ of $\mathbb{P}_n$, Q has matrix N and L, of dimension d, has parametric equations

$$x_j = \sum_{i=0}^{d} b_i^j \lambda_i, \quad j = 0, \dots, n.$$

Write $(b_i) = (b_i^0, \dots, b_i^n)^t$, $i = 0, \dots, d$, the column vectors of the parametric equations. If $e_0, \dots, e_n$ is a basis adapted to Δ, take in L the reference Ω which has adapted basis $v_i = \sum_{j=0}^n b_i^j e_j$, $i = 0, \dots, d$, relative to which the parameters $\lambda_0, \dots, \lambda_d$ are coordinates of the point $[x_0, \dots, x_n]_\Delta$ (see 2.5.3). Then the coefficients of the matrix of $\eta_{|F}$ relative to the basis $v_0, \dots, v_d$ of F may be computed as

$$\eta_{|F}(v_i, v_j) = \eta(v_i, v_j) = (b_i)^t N(b_j).$$

If all of them are equal to zero, then $\eta_{|F} = 0$ and the section is undetermined. Otherwise they are the coefficients of a matrix (or equation) of $Q \cap L$ relative to Ω. We see thus that the equation

$$\sum_{i,j=0}^{d} \big((b_i)^t N(b_j)\big)\lambda_i \lambda_j = 0$$

is an identity $0 = 0$ if and only if $Q \cap L = L$; otherwise it is an equation of $Q \cap L$.

Since the above equation may be equivalently written

$$\Big(\sum_{i=0}^{d} \lambda_i (b_i)\Big)^t N \Big(\sum_{j=0}^{d} \lambda_j (b_j)\Big) = 0,$$

we see that substituting parametric equations of L in an equation of Q allows us to check if $Q \cap L$ is a quadric and, in the affirmative case, provides an equation of it, namely:

Lemma 6.1.9. *Assume that $f(x_0, \dots, x_n) = 0$ is an equation of a quadric Q of $\mathbb{P}_n$. Assume also that*

$$x_j = \sum_{i=0}^{d} b_i^j \lambda_i, \quad j = 0, \dots, n,$$

are parametric equations of a d-dimensional linear variety $L \subset \mathbb{P}_n$, $d > 0$, and take $\lambda_0, \dots, \lambda_d$ as projective coordinates on L (2.5.3). Put

$$g(\lambda_0, \dots, \lambda_d) = f\Big(\sum_{i=0}^{d} b_i^0 \lambda_i, \dots, \sum_{i=0}^{d} b_i^d \lambda_i\Big).$$

Then g is zero, as a polynomial in $\lambda_0, \dots, \lambda_d$, if and only if $Q \supset L$. Otherwise $g(\lambda_0, \dots, \lambda_d) = 0$ is an equation of $Q \cap L$ in the coordinates $\lambda_0, \dots, \lambda_d$.

If in particular L is a face of the reference Δ we have:

Lemma 6.1.10. *Assume that a quadric Q has matrix $N = (a_{i,j})$ relative to a reference $\Delta = (p_0, \dots, p_n, A)$ and let*

$$L = \bigvee_{i \in I} p_i, \quad I \subset \{0, \dots, n\},$$

be a positive-dimensional face of Δ. Then the matrix $N' = (a_{i,j})_{i,j \in I}$ is $N' = 0$ if and only if $L \subset Q$. Otherwise N' is a matrix of $Q \cap L$ relative to the reference of L subordinated by Δ.

Proof. Follows from the computation preceding 6.1.9, as, in the present case, the vectors v_i, $i = 0, \dots, d$, are the e_i, $i \in I$. □

Remark 6.1.11. Equivalently, in the situation of 6.1.10, the equation

$$\sum_{i,j \in I} a_{i,j} x_i x_j = 0$$

is an identity $0 = 0$ if and only if $L \subset Q$. Otherwise it is an equation of $Q \cap L$ relative to the reference subordinated by Δ.

Next we give the rule that relates matrices of a quadric relative to two different references of $\mathbb{P}_n$:

Proposition 6.1.12. *If a quadric Q has matrix N relative to a reference Δ and P is the matrix that changes coordinates relative to a second reference Δ' into coordinates relative to Δ, then $P^t N P$ is a matrix of Q relative to Δ'.*

Proof. Let $e_0, \dots, e_n$ be a basis adapted to Δ. As seen in the proof of 2.3.2, if $(c_i^0, \dots, c_i^n)^t$ is the i-th column of the matrix P, then the vectors

$$v_i = \sum_{s=0}^{n} c_i^s e_s, \quad i = 0, \dots, n,$$

compose a basis adapted to Δ'. The coefficients of a matrix of Q relative to Δ' are thus

$$\eta(v_i, v_j) = \sum_{s=0}^{n} \sum_{\ell=0}^{n} c_i^s c_j^\ell \eta(e_s, e_\ell),$$

$i, j = 0, \dots, n$, from which the claim follows. □

Remark 6.1.13. Denote by (x) and (y) column vectors of coordinates relative to Δ and Δ', respectively. According to 6.1.12, an equation of Q relative to Δ' is

$$(y)^t P^t N P(y) = 0.$$

Clearly, this equation results from substituting the value given by the rule of changing coordinates

$$(x) = P(y)$$

for (x) in the equation of Q relative to Δ

$$(x)^t N(x) = 0.$$

Proposition 6.1.12 provides also the matrices of the images of quadrics by projectivities:

Corollary 6.1.14. *Assume that Q is a quadric of $\mathbb{P}_n$, $f : \mathbb{P}_n \to \mathbb{P}'_n$ a projectivity and Δ, Ω references of $\mathbb{P}_n$ and $\mathbb{P}'_n$ respectively and N a matrix of Q relative to Δ. We have:*

(a) *N is a matrix of $f(Q)$ relative to $f(\Delta)$.*

(b) *If M is a matrix of f relative to Δ, Ω, then*

$$(M^{-1})^t N M^{-1}$$

is a matrix of $f(Q)$ relative to Ω.

Proof. Assume that $f = [\varphi]$. If N is the matrix of a representative η of Q relative to a basis $e_0, \dots, e_n$ adapted to Δ, then $\varphi(e_0), \dots, \varphi(e_n)$ is a basis adapted to $f(\Delta)$ and clearly

$$\eta(e_i, e_j) = (\eta \circ (\varphi^{-1} \times \varphi^{-1}))(\varphi(e_i), \varphi(e_j))$$

which proves (a). Since, by its definition, M is a matrix that changes coordinates relative to $f(\Delta)$ into coordinates relative to Ω, the claim (b) follows from (a) using 6.1.12. □

For the remainder of this section assume that the base field is $\mathbb{R}$. The reader may easily check that any bilinear form η on a vector space E extends to a bilinear form $\eta_{\mathbb{C}}$ on its complexification $\mathbb{C}E$ (see Section 5.1) by the rule

$$\eta_{\mathbb{C}}(v + iv', w + iw') = \eta(v, w) - \eta(v', w') + i(\eta(v, w) + \eta(v', w)),$$

and also that $\eta_{\mathbb{C}}$ is symmetric and non-zero if η is so. Furthermore, if $e_0, \dots, e_n$ is a basis of E, then it is also a $\mathbb{C}$-basis of $\mathbb{C}E$ (5.1.1) and it follows directly from the definition of $\eta_{\mathbb{C}}$ that both η and $\eta_{\mathbb{C}}$ have the same matrix relative to $e_0, \dots, e_n$.

If $Q = [\eta]$ is a quadric of a (real) projective space $\mathbb{P}_n$, the extended form $\eta_{\mathbb{C}}$ is not zero and therefore defines a quadric $Q_{\mathbb{C}} = [\eta_{\mathbb{C}}]$ of the complexified space $\mathbb{C}\mathbb{P}_n$ which obviously does not depend on the choice of the representative η of Q. The quadric $Q_{\mathbb{C}}$ is called the *complexification* or the *complex extension* of Q. Since the restriction of $\eta_{\mathbb{C}}$ to E is η, the conditions $\eta(v, v) = 0$ and $\eta_{\mathbb{C}}(v, v) = 0$ are equivalent for $v \in E$ and therefore the points of Q are the real points of $Q_{\mathbb{C}}$. The imaginary points of $Q_{\mathbb{C}}$ are called the *imaginary points* of Q and it is usually said that Q *contains* or *goes through* any of them. The points of Q, as already defined in this section, will be sometimes called the *real points* of Q to avoid confusion with the imaginary ones.

Lemma 6.1.15. (1) *If Q and Δ are, respectively, a quadric and a reference of $\mathbb{P}_{n,\mathbb{R}}$, then any matrix of Q relative to Δ is also a matrix of $Q_{\mathbb{C}}$ relative to Δ, this time taken as a reference of $\mathbb{C}\mathbb{P}_{n,\mathbb{R}}$.*

(2) *If real quadrics Q, Q' have $Q_{\mathbb{C}} = Q'_{\mathbb{C}}$, then $Q = Q'$.*

Proof. The first claim is a direct consequence of the already noticed equality of the matrices η and $\eta_{\mathbb{C}}$ relative to any basis of E. For claim (2), assume that Q and Q' have matrices N and N' relative to a reference Δ of $\mathbb{P}_{n,\mathbb{R}}$. Since the complex extensions of Q and Q' have the same matrices, $Q_{\mathbb{C}} = Q'_{\mathbb{C}}$ forces $N = \rho N'$ for some $\rho \in \mathbb{C} - \{0\}$. Both N and N' being real and non-zero, it follows that $\rho \in \mathbb{R}$ and so $Q = Q'$. □

We have seen that Q and $Q_{\mathbb{C}}$ have equal matrices, and hence equations, relative to any real reference. Thus the coordinates of the points of the complexification $Q_{\mathbb{C}}$ of Q are the solutions in $\mathbb{C}^{n+1} - \{0\}$ of any fixed equation of Q. This equation being real, it follows from 5.1.9 that the set of points of $Q_{\mathbb{C}}$ is invariant by complex conjugation, which we state for future reference as:

Lemma 6.1.16. *If p is an imaginary point of a quadric Q of a real projective space, then its complex-conjugate $\sigma(p)$ is also an imaginary point of Q.*

The easy proof of the following lemma is left to the reader.

Lemma 6.1.17. *If Q is a quadric of a real projective space $\mathbb{P}_n$, $L \subset \mathbb{P}_n$ is a positive-dimensional linear variety and $f : \mathbb{P}_n \to \mathbb{P}'_n$ a projectivity, then*

$$(Q \cap L)_{\mathbb{C}} = Q_{\mathbb{C}} \cap L_{\mathbb{C}}$$

and

$$(f(Q))_{\mathbb{C}} = f_{\mathbb{C}}(Q_{\mathbb{C}}).$$

6.2 Quadrics of $\mathbb{P}_1$

In Section 5.4 we have in fact dealt with the quadrics of $\mathbb{P}_1$ without mentioning them, but only their points and equations. In the present one we just translate the results of Section 5.4 in terms of quadrics.

Proposition 6.2.1. *Let Q be a quadric of $\mathbb{P}_1$ and N a matrix of Q. If the base field k is $\mathbb{R}$, then:*

(r1) $\det N < 0$ *if and only if Q has two different points.*

(r2) $\det N = 0$ *if and only if Q has a single point.*

(r3) $\det N > 0$ *if and only if Q has no point.*

If $k = \mathbb{C}$:

(c1) $\det N \neq 0$ *if and only if Q has two different points.*

(c2) $\det N = 0$ *if and only if Q has a single point.*

Proof. Just write

$$N = \begin{pmatrix} a & b \\ b & c \end{pmatrix}$$

and use 5.4.1. □

In case r3, the complexified $Q_{\mathbb{C}}$ of Q is in case c1 and so Q has two different imaginary points, which are complex-conjugate by 6.1.16. Since Q has no real points, we will refer to these imaginary points as *the points* of Q, its condition of imaginary being recalled if some confusion may arise.

Conversely, any pair of points, real or imaginary and complex-conjugate in case $k = \mathbb{R}$, is the pair of points of a well-determined quadric:

Proposition 6.2.2. *Let q_1, q_2 be either points of $\mathbb{P}_1$ or, in case $k = \mathbb{R}$, a pair of complex-conjugate imaginary points of $\mathbb{P}_1$. Then there is a unique quadric Q of $\mathbb{P}_1$ whose (maybe imaginary) points are q_1, q_2.*

Assume that, after fixing a reference, $q_1 = [\alpha_0, \alpha_1]$ and $q_2 = [\beta_0, \beta_1]$. Assume furthermore that in case of q_1, q_2 being imaginary, their coordinates have been chosen such that $\beta_0 = \bar{\alpha}_0$ and $\beta_1 = \bar{\alpha}_1$, as allowed by 5.1.9. Then Q has equation

$$(\alpha_1 x_0 - \alpha_0 x_1)(\beta_1 x_0 - \beta_0 x_1) = 0.$$

Proof. Direct from 5.4.2. □

Remark 6.2.3. The coefficients of the equation of Q in 6.2.2 are

$$a_{0,0} = \alpha_1\beta_1, \quad a_{0,1} = -\frac{1}{2}(\alpha_1\beta_0 + \alpha_1\beta_1), \quad a_{1,1} = \alpha_0\beta_0.$$

According to 6.2.2, a quadric of $\mathbb{P}_1$ and the unordered pair of its (maybe imaginary and complex-conjugate) points determine each other, hence the names *pair of points* and *point-pair* given to the quadrics of $\mathbb{P}_1$. In the sequel we will make no distinction between a quadric of $\mathbb{P}_1$ and the pair of its points, the name *pair of points* being used to denote either of them.

When the quadric Q has a single point q (that is, in cases (r2) and (c2) of 6.2.1), it is said that Q is a *pair of coincident points*, or a point *counted twice*, and it is often written $Q = 2q = q + q$. The word *twice* is intended to recall the fact, shown by 6.2.2, that the only point q of Q appears twice as a solution of its equation, or, equivalently, that the multiplicity of the corresponding factor in an equation of Q is two. Otherwise Q is called a *pair of distinct points*, which if $k = \mathbb{R}$ may be either real or imaginary and complex-conjugate. If these points are q_1, q_2 we will write $Q = q_1 + q_2$.

The following is an old fashioned and very expressive way of stating part of the contents of 6.2.1.

6.2.4. *A quadric of $\mathbb{P}_1$ consists of two points, distinct or coincident, which, for a real quadric, may be either real or imaginary and complex-conjugate.*

We close this section with an easy and useful consequence of 6.2.1.

Lemma 6.2.5. *If a line ℓ contains three (maybe imaginary) different points of a quadric Q of $\mathbb{P}_n$, $n \geq 2$, then $\ell \subset Q$.*

Proof. It follows from 6.2.1 that no quadric of ℓ has three different points: thus, $Q \cap \ell$ being not a quadric, $\ell \subset Q$. □

6.3 Tangent lines

Assume that Q and ℓ are, respectively, a quadric and a line of a projective space $\mathbb{P}_n$. The next proposition describes the intersection $Q \cap \ell$.

Proposition 6.3.1. *The intersection $Q \cap \ell$ is either*

(a) *the whole line ℓ, or*

(b) *a point counted twice, or*

(c) *a pair of distinct points, which in case $k = \mathbb{R}$ may be either real or imaginary and complex-conjugate.*

Proof. Follows from the results of the preceding section, as we already know from its definition that the intersection $Q \cap \ell$ is either ℓ or a quadric of ℓ. □

The line ℓ is said to be *tangent* to the quadric Q if and only if either $\ell \subset Q$ or $Q \cap \ell$ is a pair of coincident points. Then it is equivalently said that ℓ is a *tangent line* or just a *tangent* to C. If ℓ is tangent to Q, then for any point $p \in Q \cap \ell$ it is said that ℓ is *tangent to Q at p*, and also that p is a *contact point* of ℓ and Q.

The tangent lines not contained in the quadric are called *proper tangents* (or said to be *properly tangent*) to distinguish them from the lines contained in the quadric. A line is thus a proper tangent if and only if it meets the quadric at a single point, this point being then its only contact point. By contrast, a line contained in a quadric is tangent to it at each of its points.

The lines that are not tangent to Q are called *secant lines* or just *secants*, and also *chords*. By 6.2.1, they meet the quadric at two different points which, for real quadrics, may be either real or imaginary and complex-conjugate. These points are often called the *ends* of the chord, the word *end* being reminiscent of elementary Euclidean geometry, in which the chords are taken as line segments rather than as complete lines.

Since tangent lines have been defined in terms of their intersection with the quadric, the next proposition is a direct consequence of 6.1.7.

Proposition 6.3.2 (Invariance of tangency). *Let Q be a quadric of $\mathbb{P}_n$ and $f : \mathbb{P}_n \to \mathbb{P}'_n$ a projectivity. A line ℓ is tangent to Q at a point p if and only if $f(\ell)$ is tangent to $f(Q)$ at $f(p)$.*

In order to handle tangency, we first obtain an equation of the intersection of a line and a quadric:

Lemma 6.3.3. *Let $Q = [\eta]$ be a quadric and $\ell = [v] \vee [w]$ a line, both of a projective space $\mathbb{P}_n$. The equation in λ, μ,*

$$\eta(v, v)\lambda^2 + 2\eta v(v, w)\lambda\mu + \eta(w, w)\mu^2 = 0, \tag{6.5}$$

is an identity $0 = 0$ if and only if $\ell \subset Q$. Otherwise it is an equation of $\ell \cap Q$ relative to the reference of ℓ with adapted basis v, w.

Proof. Direct from the description of $Q \cap L$ given in 6.1.8. □

The next proposition gives a characterization of tangent lines.

Proposition 6.3.4. *A line ℓ spanned by points $p = [v]$ and $q = [w]$ is tangent to a quadric $Q = [\eta]$ if and only if*

$$\eta(v, v)\eta(w, w) - \eta(v, w)^2 = 0. \tag{6.6}$$

Proof. By the definition of tangency, if ℓ is tangent to Q, then either the equation (6.5) is identically zero, in which case the condition is obviously satisfied, or otherwise (6.5) defines a pair of coincident points, in which case the equality (6.6) follows from 6.2.1. Conversely, if (6.6) holds, then either the equation (6.5) is identically zero or, again by 6.2.1, defines a point counted twice. □

Remark 6.3.5. Obviously the line ℓ of 6.3.4 is a chord if and only if

$$\eta(v,v)\eta(w,w) - \eta(v,w)^2 \neq 0.$$

When $k = \mathbb{R}$, the ends of ℓ are real if

$$\eta(v,v)\eta(w,w) - \eta(v,w)^2 < 0$$

and imaginary if

$$\eta(v,v)\eta(w,w) - \eta(v,w)^2 > 0,$$

by 6.3.3.

After fixing a reference, assume that Q has matrix N and the points q and p have column coordinate vectors $(x) = (x_0, \dots, x_n)^t$ and $(y) = (y_0, \dots, y_n)^t$, respectively. Then the equality (6.6) reads

$$\big((x)^t N(x)\big)\big((y)^t N(y)\big) - ((x)^t N(y))^2 = 0. \tag{6.7}$$

If in particular one fixes the point p, the first member of the above equality is a (maybe zero) homogeneous polynomial of degree two in the coordinates $x_0, \dots, x_n$. By 6.3.4, the equation (6.7) is satisfied by the coordinates of the points $q \neq p$ spanning with p a line tangent to Q and, obviously, also by the coordinates of p itself. It follows that p and the points that together with p span a line tangent to Q describe either the whole space $\mathbb{P}_n$ (when the equation (6.7) is trivial) or, otherwise, the set of points of the quadric $T_p(Q)$ defined by the equation (6.7). $T_p(Q)$ is called the *tangent cone to Q with vertex* (or *from*) p. The meaning of the words *cone* and *vertex* will be explained in Section 6.7; we will prove in the forthcoming Corollary 6.7.5 that the tangent cone does not depend on the choice of the coordinates.

To close the present section we extend the definition of tangency to linear varieties of arbitrary dimension. A linear variety L is said to be *tangent* to Q at a point p if and only if $p \in Q$, $p \in L$ and every point $q \in L$, $q \neq p$, spans with p a line tangent to Q at p. The point p is then called a *contact point* of Q and L. Clearly, a linear variety included in Q is tangent to Q at each of its points. As for lines, *tangent* will mean *tangent at some point*. When L is tangent to Q, it is equivalently said that L and Q are tangent, and also that Q is tangent to L. *To touch* is sometimes used as a synonymous of *to be tangent*. The reader may note that the definition is tautological if $\dim L = 1$, as it should. Note also that each point $p \in Q$ is taken as being the only zero-dimensional linear variety tangent to Q at p, this being useful to make some claims simpler. The invariance of tangency (6.3.2) holds also for linear varieties of arbitrary dimension, as the above definition uses projectively invariant terms only.

6.4 Conjugation

The *conjugation relative to a quadric* of $\mathbb{P}_n$ is a binary relation between points of $\mathbb{P}_n$ which is the most relevant object associated to a quadric. Its definition is quite easy: points $p = [v]$ and $q = [w]$ are said to be *conjugate with respect to a quadric* $Q = [\eta]$ if and only if $\eta(v, w) = 0$. Obviously, the definition is independent of the choice of the representatives v, w and η. We will equivalently say that the quadric *Q has p, q as a conjugate pair*. When no confusion may arise, it is usual to say just *conjugate* instead of *conjugate with respect to Q*. Due to the symmetry of the form η, the conjugation clearly is a symmetric relation: p and q are conjugate with respect to a quadric Q if and only if so are q and p.

Example 6.4.1. If Q has equation $\sum_{i,j=0}^{n} a_{i,j} x_i x_j = 0$ relative to a reference $\Delta = \{p_0, \dots, p_n, A\}$ and $e_0, \dots, e_n$ is a basis adapted to Δ, we have seen in Section 6.1 that $a_{i,j} = \eta(e_i, e_j)$ for a suitable representative η of Q. It follows that the vertices p_i, p_j of Δ are conjugate with respect to Q if and only if $a_{i,j} = 0$.

The conjugation unifies a number of different situations that may arise between a quadric and two points, namely:

Theorem 6.4.2. *The conjugation being relative to a given quadric Q of $\mathbb{P}_n$:*

(a) *A point p is conjugate to itself if and only if $p \in Q$.*

(b) *Two different points p, q, $p \in Q$, are conjugate if and only if the line pq is tangent to Q.*

(c) *Two different points p, q, $p \notin Q$, $q \notin Q$, are conjugate if and only if pq is a chord of Q and the pair p, q harmonically divides the ends of pq.*

Proof. Assume that $p = [v]$, $q = [w]$ and $Q = [\eta]$. Claim (a) is obvious from the definitions. If $p \in Q$, then $\eta(v, v) = 0$, after which the condition of 6.3.4 for pq to be tangent to Q reads just $\eta(v, w)^2 = 0$, and this is of course equivalent to $\eta(v, w) = 0$. For claim (c), note first that since $p \notin Q$, $Q \cap pq$ is a point-pair and the claim makes sense. Assume $Q \cap pq = q_1 + q_2$. Take in $\ell = pq$ the reference which has adapted basis v, w. Since $p \notin Q, q \notin Q$, the points q_1, q_2 have finite and non-zero absolute coordinates θ_1, θ_2 which, by 6.3.3, are the roots of the equation

$$\eta(v, v)\theta^2 + 2\eta(v, w)\theta + \eta(w, w) = 0.$$

In it, $\eta(v, v) \neq 0$ because $p \notin Q$. It is thus

$$\theta_1 + \theta_2 = -\frac{2\eta(v, w)}{\eta(v, v)}. \tag{6.8}$$

Now, p, q are conjugate if and only if $\eta(v, w) = 0$, which, by (6.8) is equivalent to $\theta_1 + \theta_2 = 0$. This equality being written $\theta_2/\theta_1 = -1$, by 2.9.5 it is equivalent to $(p, q, q_1, q_2) = -1$, which ends the proof. □

Remark 6.4.3. If the equivalent conditions of claim (b) of 6.4.2 are satisfied, then p is a contact point of the tangent line pq. Therefore, in case $p, q \in Q$ and still $p \neq q$, p and q are conjugate if and only if $p \vee q$ is tangent at both p and q, that is, if and only if $p \vee q \subset Q$. See also 6.4.6 below.

Note that the points of a quadric are characterized in terms of conjugation as being self-conjugate points. Also, the tangent lines at a point $p \in Q$ are the lines through p all of whose points are conjugate to p. We see thus that both 6.1.3 and 6.3.2 are particular cases of the following, more general, proposition:

Proposition 6.4.4 (Invariance of conjugation). *If $f : \mathbb{P}_n \to \mathbb{P}'_n$ is a projectivity and Q a quadric of $\mathbb{P}_n$, then points $p, q \in P_n$ are conjugate with respect to Q if and only if $f(p), f(q)$ are conjugate with respect to $f(Q)$.*

Proof. If $f = [\varphi]$, $Q = [\eta]$, $p = [v]$ and $q = [w]$, just use that

$$\eta(v, w) = (\eta \circ (\varphi^{-1} \times \varphi^{-1}))(\varphi(v), \varphi(w)). \qquad \square$$

The reader may also prove 6.4.4 using Theorem 6.4.2 and then, case by case, 6.1.3, 6.3.2 and 2.9.15.

Proposition 6.4.5. *If Q and L are, respectively, a quadric and a positive-dimensional linear variety of $\mathbb{P}_n$, then*

(a) *$L \subset Q$ if and only if any pair of points $p, q \in L$ are conjugate with respect to Q.*

(b) *If, otherwise, $L \not\subset Q$, then points $p, q \in L$ are conjugate with respect to Q if and only if they are conjugate with respect to the quadric $Q \cap L$ of L.*

Proof. Assume that $Q = [\eta]$ and $L = [F]$. By definition, $L \subset Q$ is equivalent to $\eta(v, w) = 0$ for any $v, w \in F$. This equality being obviously satisfied if either $v = 0$ or $w = 0$, the latter condition is in turn equivalent to having p, q conjugate for any $p, q \in L$, as claimed in (a). If $L \not\subset Q$, then the restricted form $\eta_{|F}$ is not zero and defines $Q \cap L$. Then points $p = [v], q = [w] \in L$ are conjugate with respect to Q if and only if $\eta(v, w) = 0$. Since $v, w \in F$, this condition may be equivalently written $\eta_{|F}(v, w) = 0$, which is the condition for p, q to be conjugate with respect to $Q \cap L$. $\square$

The condition of part (a) of 6.4.5 may be easily weakened:

Corollary 6.4.6. *A positive-dimensional linear variety L, spanned by points $q_0, \dots, q_r$, is contained in a quadric Q if and only if q_i and q_j are conjugate with respect to Q for all $i, j = 0, \dots, r$.*

Proof. The condition is obviously necessary, by 6.4.5 (a). For the sufficiency, assume that $q_i = [v_i]$ and take the other notations as in the proof of 6.4.5. Then $F = \langle v_0, \dots, v_n \rangle$ and the conditions of conjugation give $\eta(v_i, v_j) = 0$ for $i, j = 0, \dots, r$; by the bilinearity of η, $\eta_{|F} = 0$ and so $L \subset Q$. □

A better understanding of the conjugation relative to a quadric Q will be achieved by considering for each fixed point p the set

$$H_p = \{q \in \mathbb{P}_n \mid p \text{ and } q \text{ are conjugate}\}$$

which will be written $H_{p,Q}$ if an explicit reference to the quadric Q is needed.

First we will have a look at H_p using coordinates. Fix a reference of $\mathbb{P}_n$ and assume that Q has matrix N, and that q and p have column coordinate vectors $(x) = (x_0, \dots, x_n)^t$ and $(y) = (y_0, \dots, y_n)^t$, respectively. The condition for q and p to be conjugate may be written

$$(x)^t N(y) = 0. \tag{6.9}$$

We will consider two cases. First the column vector $N(y)$ may be zero: then the equation (6.9) is satisfied by the coordinates of any point q and therefore $H_p = \mathbb{P}_n$. If, otherwise, $N(y) \neq 0$, then (6.9) is a non-trivial homogeneous linear equation in the coordinates $x_0, \dots, x_n$ of the variable point q and so H_p is the hyperplane with equation (6.9). In the latter case, H_p is called the *polar hyperplane* (*polar line* or just *polar* if $n = 2$, *polar plane* if $n = 3$) of the point p relative to the quadric Q. In case $H_p = \mathbb{P}_n$ it is said that the polar hyperplane is *undetermined* or *undefined*. We have thus that the vector $N(y)$ is zero if and only if $H_p = \mathbb{P}_n$, and otherwise its components are hyperplane coordinates of the polar hyperplane H_p. In other words, either the above equation (6.9), in the variables (x), is an identity $0 = 0$ and the polar hyperplane is undetermined, or, otherwise, it is an equation of the polar hyperplane. From the above we retain for further reference:

Proposition 6.4.7. *Let Q be a quadric and p a point of $\mathbb{P}_n$ which, relative to an arbitrarily fixed reference, have matrix N and column coordinate vector (y), respectively. Then either*

(a) *$H_{p,Q}$ is a hyperplane with coordinate vector $N(y)$, in particular $N(y) \neq 0$, or*

(b) *$H_{p,Q} = \mathbb{P}_n$ and $N(y) = 0$.*

Since in both cases H_p is a linear variety, regarding conjugation we have:

Corollary 6.4.8. *If points $p_1, \dots, p_m$ are all conjugate to a point p with respect to a quadric Q, then any point of $p_1 \vee \dots \vee p_m$ is conjugate of p with respect to Q.*

Proof. The hypothesis just says that $p_i \in H_p$ for $i = 1, \dots, m$, after which $p_1 \vee \dots \vee p_m \subset H_p$ and the claim follows. □

The assignation $(p, Q) \mapsto H_{p,Q}$ is projective:

Proposition 6.4.9 (Invariance of taking polar hyperplanes). *If $f : \mathbb{P}_n \to \mathbb{P}'_n$ is a projectivity and Q a quadric of $\mathbb{P}_n$, for any $p \in P_n$, $f(H_{p,Q}) = H_{f(p),f(Q)}$.*

Proof. Direct from the definition and 6.4.4 □

The next lemma is just a reformulation of the symmetry of the conjugation in terms of polar hyperplanes. Note that it holds even if one or both polar hyperplanes are undetermined.

Lemma 6.4.10 (Reciprocity of polar hyperplanes). *For any two points p and q, q belongs to H_p if and only if p belongs to H_q.*

Proof. Obvious. □

The following geometric description of the polar hyperplanes is just a reformulation of 6.4.2. It will be very important in the sequel.

Theorem 6.4.11. *Assume that Q is a quadric and p a point of $\mathbb{P}_n$:*

(a) *$p \in Q$ if and only if $p \in H_p$. If this is the case, the points of H_p are p itself and all points $q \neq p$ spanning with p a line tangent to Q at p.*

(b) *If $p \notin Q$, then $p \notin H_p$ and a point q belongs to H_p if and only if either*

  (b1) *$q \in Q$ and the line pq is a proper tangent to Q at q, or*

  (b2) *$q \notin Q$, pq is a chord and p, q harmonically divide the ends of pq.*

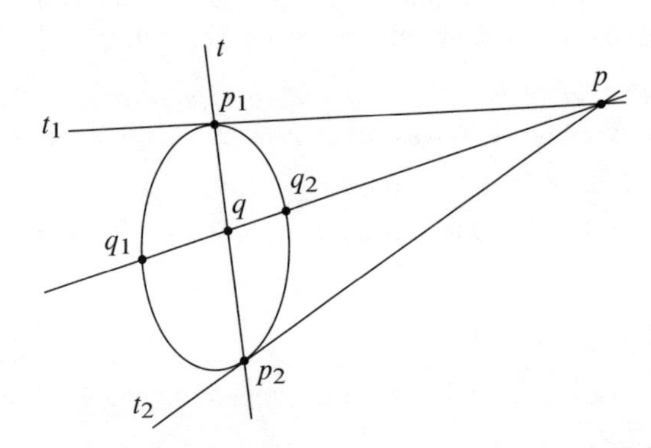

Figure 6.1. Many of the situations described in 6.4.11: by 6.4.11 (a) t_1, t_2 are the polars of p_1, p_2; by reciprocity, or by 6.4.11 (b1), t is the polar of p and then, by 6.4.11 (b2), $(p, q, q_1, q_2) = -1$.

Many consequences will follow from 6.4.11. Let us first pay some attention to the points p with undetermined polar hyperplane: a point p is called a *double point* of Q (or said to be double for Q) if and only if $H_p = \mathbb{P}_n$. Equivalently, p is a double point of Q if and only if it is conjugate to all points $q \in \mathbb{P}_n$. Sometimes the double points of a quadric are also called *singular points* of the quadric. By 6.4.11 (a), the double points of Q belong to Q and may be characterized as the points of Q all lines through which are tangent to Q. In particular, for $n = 1$, p is a double point of the point-pair $p_1 + p_2$ if and only if $p_1 = p_2 = p$.

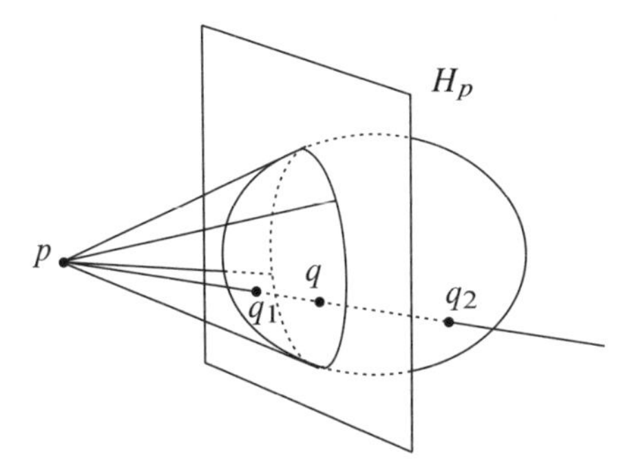

Figure 6.2. Theorem 6.4.11 (b) in three-space: there are shown four tangents and a chord through p. On the latter $(p, q, q_1, q_2) = -1$.

In the sequel the set of double points of a quadric Q will be denoted by $\mathcal{D}(Q)$; as shown next, it is a linear variety, so it is called the *linear variety* (or just the *variety*) *of double points* of Q.

Proposition 6.4.12. *Fix a system of coordinates and assume that a quadric Q has matrix N. Then:*

(a) $p = [y_0, \dots, y_n]$ *is a double point of Q if and only if*

$$N \begin{pmatrix} y_0 \\ \vdots \\ y_n \end{pmatrix} = \begin{pmatrix} 0 \\ \vdots \\ 0 \end{pmatrix}. \tag{6.10}$$

(b) $\mathcal{D}(Q)$ *is a linear variety of $\mathbb{P}_n$ of dimension $n - \operatorname{rk} N < n$.*

(c) *All matrices of Q, relative to different references, have the same rank.*

Proof. Claim (a) is direct from 6.4.7 and the definition of double point. Claim (b) follows from (a) and 2.6.4, as (6.10) is a system of homogeneous linear equations with rank equal to $\operatorname{rk} N$. In turn, claim (b) gives claim (c) because $\dim \mathcal{D}(Q)$ is independent of the choice of the reference. □

All matrices of Q have thus the same rank, which will be called the *rank* of Q, and denoted by $\mathbf{r}(Q)$ in the sequel. The next corollary shows in particular that the rank of a quadric is a projective invariant.

Corollary 6.4.13. *If Q is a quadric of $\mathbb{P}_n$, and $f\colon \mathbb{P}_n \to \mathbb{P}'_n$ a projectivity, then $\mathcal{D}(f(Q)) = f(\mathcal{D}(Q))$ and $\mathbf{r}(f(Q)) = \mathbf{r}(Q)$.*

Proof. The first equality follows from the invariance of conjugation (6.4.4) and gives in turn the second one using 6.4.12 (b). As an alternative, the reader may prove the second equality from 6.1.14 (a). □

As a consequence of 6.4.12, a quadric Q of $\mathbb{P}_n$ has no double points if and only if $\mathbf{r}(Q) = n + 1$. Quadrics without double points are thus those having one (and hence any) of its matrices N with $\det N \neq 0$ or, equivalently, invertible. These quadrics are called *non-degenerate*. Consequently, a quadric is called *degenerate* if and only if it has at least one double point. The degenerate quadrics are thus those whose matrices are non-invertible. The name *degenerate* is reminiscent of the early treatises on conics, in which degenerate conics were not considered as proper conics and called *degenerations of conics*.

As a direct consequence of 6.4.13 and the above definitions, we have:

Corollary 6.4.14. *If Q is a quadric of $\mathbb{P}_n$, and $f\colon \mathbb{P}_n \to \mathbb{P}'_n$ a projectivity, then Q is degenerate (resp. non-degenerate) if and only if $f(Q)$ is degenerate (resp. non-degenerate).*

Example 6.4.15. It follows directly from the definitions, and also from 6.2.1, that a point-pair Q is degenerate if and only if it is a point counted twice. Such a point being then the only point of Q, it is also its only double point.

Example 6.4.16 (Pairs of distinct hyperplanes). Assume that H_1, H_2 are two different hyperplanes of $\mathbb{P}_n$, $n \geq 2$, with equations $F_1(x_0, \dots, x_n) = 0$ and $F_2(x_0, \dots, x_n) = 0$ relative to a certain reference Δ. Both F_1, F_2 being linear forms, $F_1 F_2 = 0$ is an equation of a quadric Q whose set of points is $H_1 \cup H_2$. Clearly the lines disjoint with $H_1 \cap H_2$ are chords of Q, as they meet the hyperplanes at different points. On the contrary, all lines meeting $H_1 \cap H_2$ are tangent lines, as they are either contained in one of the hyperplanes or meet both at the same point. It follows that $H_1 \cap H_2$ is the variety of double points of Q and in particular Q has rank 2. The reader may confirm these facts by computing from the equation $F_1 F_2 = 0$; using 6.4.30 below will shorten the computation.

We will check next that there is a unique quadric whose set of points contains $H_1 \cup H_2$. Indeed, take points $p_0, p_3, \dots, p_n$ spanning $H_1 \cap H_2$, and $p_i \in H_i - H_1 \cap H_2, i = 1, 2$. Then the points $p_0, \dots, p_n$ are independent and so may be taken as the vertices of a reference Ω which has H_1 and H_2 as faces. Using coordinates relative to Ω, it follows from 6.1.10 that a quadric containing both H_1 and H_2 has equation $y_1 y_2 = 0$ relative to Δ, hence the uniqueness.

The only quadric containing $H_1 \cup H_2$ is thus the above $Q : F_1F_2 = 0$: this shows that Q is determined by the hyperplanes H_1, H_2 and is in particular independent of the reference Δ used in its definition. The quadric Q is referred to as *the quadric composed of* H_1 *and* H_2 and denoted by $Q = H_1 + H_2$. The quadrics $H_1 + H_2$, for any two different hyperplanes $H_1, H_2 \subset \mathbb{P}_n$, are called *pairs of distinct hyperplanes*,

The next claim has been already proved as part of Example 6.4.16; we state it separately for future reference.

Lemma 6.4.17. *If* H_1, H_2 *are distinct hyperplanes of* $\mathbb{P}_n$, *then* $H_1 + H_2$ *is the only quadric whose set of points contains* $H_1 \cup H_2$.

Example 6.4.18 (Pairs of imaginary hyperplanes). In this example assume $k = \mathbb{R}$. If $F = 0$ is an equation of an imaginary hyperplane H (of $\mathbb{C}\mathbb{P}_n$), its complex-conjugate $\bar{F} = 0$ is an equation of the complex-conjugate of H, $\sigma(H)$. Then $F\bar{F}$ is a real polynomial (because it is invariant by complex conjugation) and therefore defines a quadric $Q : F\bar{F} = 0$ of $\mathbb{P}_n$ with complex extension $Q_{\mathbb{C}} = H + \sigma(H)$. By 6.1.15 and 6.4.17, Q is the only quadric of $\mathbb{P}_n$ whose complex extension contains $H \cup \sigma(H)$, and so it is determined by H: it is called the *pair of imaginary hyperplanes* composed of H and $\sigma(H)$, and denoted by $Q = H + \sigma(H)$ if no confusion with $Q_{\mathbb{C}}$ may arise. A real point of Q needs to belong to one of the hyperplanes, and therefore also to the other, because they are complex conjugate. Thus, a real point of Q is a real point of $H \cap \sigma(H)$. The converse being clear, it follows (5.1.13) that the real points of Q describe a linear variety of $\mathbb{P}_n$ of dimension $n - 2$. All real points of Q are then double points, because any line through one of them is either contained in Q or has a single intersection with Q; in particular $\mathbf{r}(Q) = 2$.

Example 6.4.19 (Double hyperplanes). Take a hyperplane $H \subset \mathbb{P}_n$, with equation $F(x_0, \dots, x_n) = 0$, and consider the quadric Q with equation $F^2 = 0$. It is clear that Q and H have the same points, $|Q| = H$. Then, any line of $\mathbb{P}_n$ either is contained in Q or intersects Q at a single point, and therefore is tangent to Q. It follows that all points of Q are double points, $\mathcal{D}(Q) = H$, and so $\mathbf{r}(Q) = 1$. For any point $p \notin H$, the polar hyperplane H_p of p contains H by reciprocity (6.4.11), and therefore equals it: $H_p = H$. Again a direct computation may confirm these facts.

The pairs of coincident points, as defined in Section 6.2, are the double hyperplanes of $\mathbb{P}_1$; we will not call them *double points* to avoid any confusion with the double points of quadrics. The double hyperplanes of $\mathbb{P}_2$ and $\mathbb{P}_3$ will be called *double lines* and *double planes*. The next lemma gives some further information about double hyperplanes:

Lemma 6.4.20. *After fixing a reference of $\mathbb{P}_n$, assume that $F(x_0,\dots,x_n) = 0$ is the equation of a hyperplane H. Then, for a quadric Q of $\mathbb{P}_n$, the following conditions are equivalent:*

(i) *Q has equation $F^2 = 0$.*

(ii) *$|Q| = H$.*

(iii) *$\mathcal{D}(Q) = H$.*

Proof. That (i) $\Rightarrow$ (ii) and (ii) $\Rightarrow$ (iii) have been seen in Example 6.4.19 above. To complete the proof we will show that the condition $\mathcal{D}(Q) = H$ uniquely determines Q, after which, by the parts already proved, Q is the quadric $F^2 = 0$. Indeed, choose any reference $\Omega = (p_0,\dots,p_n, A)$ with 0-th face H. Let $N = (a_{i,j})$ be a matrix of Q relative to Ω; since p_i is conjugate to p_j for $i = 1,\dots,n$, $j = 0,\dots,n$, it is $a_{i,j} = 0$ for $(i,j) \neq (0,0)$ (see Example 6.4.1). This determines N up to a non-zero factor, and hence determines Q. □

Lemma 6.4.20 proves in particular that the hyperplane H determines the quadric Q, which is usually referred to as H *counted twice*, and also as *two times* H, and denoted by $Q = H + H = 2H$. The quadrics $2H$, for H a hyperplane of $\mathbb{P}_n$, are called *double hyperplanes*, and also *pairs of coincident hyperplanes*.

Remark 6.4.21. By the condition (iii) of 6.4.20, all rank-one quadrics are hyperplanes counted twice, and conversely.

In the sequel any quadric $H_1 + H_2$, where H_1, H_2 are hyperplanes, will be called a *pair of hyperplanes* or a *hyperplane-pair*. Summarizing, a pair of hyperplanes may be of distinct or coincident hyperplanes; in case $k = \mathbb{R}$, the hyperplanes may be either both real, or imaginary and complex-conjugate. In the latter case, they are, of course, distinct. The words *line* and *plane* are used instead of *hyperplane* when $n = 2, 3$.

A quadric containing a hyperplane H is a pair of hyperplanes:

Lemma 6.4.22. *If a quadric Q of $\mathbb{P}_n$ contains a hyperplane H, then $Q = H + H'$ for H' a hyperplane of $\mathbb{P}_n$.*

Proof. Take a coordinate frame having H as its 0-th face. Then a quadric Q : $\sum_{i,j=0}^{n} a_{i,j}x_ix_j = 0$ containing H has $a_{i,j} = 0$ for $i,j = 1,\dots,n$ (6.1.11) and hence equation

$$x_0(a_{0,0}x_0 + 2a_{0,1}x_0x_1 + \dots + 2a_{0,n}x_0x_n) = 0.$$

Then, just take H' to be the hyperplane defined by the second factor equated to zero. □

Back to considering polar hyperplanes, the following are corollaries of 6.4.11:

Corollary 6.4.23. *A linear variety L is tangent to a quadric Q at a point p if and only if $p \in L \subset H_p$.*

Proof. Follows from the definition of tangent linear variety and 6.4.11 (a). □

Since a double point p of Q has $H_p = \mathbb{P}_n$, all linear varieties through p are tangent to Q at p. The points $p \in Q$ that are not double points of Q are called *simple points* of Q (or said to be *simple* for Q). The polar hyperplane of a simple point p is, by 6.4.23, the only hyperplane tangent to Q at p. By this reason it is also called the *tangent hyperplane* to Q at p. Of course one says *tangent line* (sometimes just *tangent*) if $n = 2$, and *tangent plane* if $n = 3$. The reader may easily check that the equation (6.7) of Section 6.3, of the locus described by the tangent lines through p, identically vanishes if p is a double point of Q, while it is the square of an equation of the polar hyperplane H_p if p is a simple point of Q.

Two quadrics are said to be *tangent* at a common point p if and only if either p is a double point of one of them or, otherwise, the two quadrics have the same tangent hyperplane at p.

Tangency may also be characterized in terms of double points, namely:

Corollary 6.4.24. *A linear variety L is tangent to a quadric Q at a point p if and only if either $p \in L \subset Q$ or, otherwise, p is a double point of $Q \cap L$.*

Proof. By 6.4.23, L is tangent to Q at p if and only if $p \in L$ and any point of L is conjugate to p with respect to Q. If this is the case and $L \not\subset Q$, then, by 6.4.5, all points of L are conjugate to p with respect to $Q \cap L$ and therefore p is a double point of $Q \cap L$. The converse is clear if $p \in L \subset Q$. If $L \not\subset Q$ and p is a double point of $Q \cap L$, then p belongs to L and is conjugate to any point of L with respect to $Q \cap L$, and so (by 6.4.5) also with respect to Q. □

Remark 6.4.25. In case $L \not\subset Q$, 6.4.24 assures that the set of contact points of L and Q is the linear variety $\mathcal{D}(Q \cap L)$. In particular L is not tangent to Q if and only if $Q \cap L$ is non-degenerate.

Corollary 6.4.26. *The contact points of the tangent lines to Q going through a fixed point $p \notin Q$ are the points of $Q \cap H_p$.*

Proof. Direct from 6.4.11 (b). □

Note that in 6.4.26 H_p is a hyperplane (because $p \notin Q$) and the tangent lines through p are the projections from p of the points of $Q \cap H_p$.

The next corollary provides a nice way to draw polar lines

Corollary 6.4.27. *Assume that C is a conic, $p \notin C$ is a point and A, A' and B, B' are the ends of two different chords of C through p. Then A, A', B, B' are the vertices of a quadrivertex whose diagonal triangle has p as a vertex and the polar of p as its opposite side.*

Proof. (See Figure 6.3.) If three of the ends A, A', B, B' are collinear, then the chords need to share one end. Since they share also the point p, which is different from any end because $p \notin C$, they agree, against the hypothesis. The sides AA' and BB' share thus no vertex and therefore $p = AA' \cap BB'$ is a diagonal point. Write M, N the other vertices of the diagonal triangle, $T = AA' \cap MN$ and $S = BB' \cap MN$; T and S are points by 1.8.2. Now 2.10.3 applies to the diagonals AA' and BB' of the complete quadrilateral with sides AB, $A'B'$, AB', AB' to show that $-1 = (A, A', p, T) = (B, B', p, S)$, after which the claim follows from 6.4.11 (b2). □

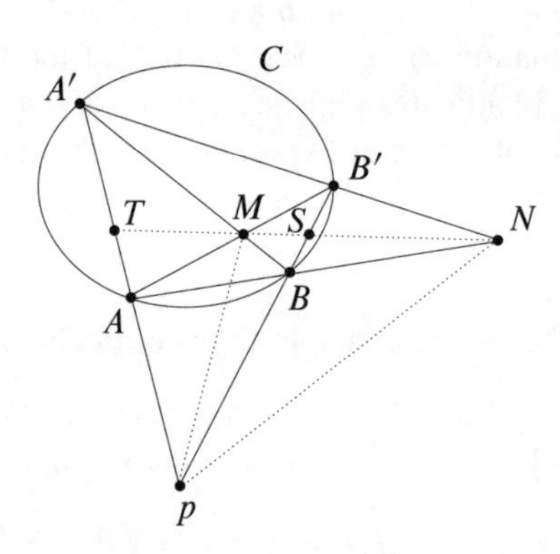

Figure 6.3. Proof of 6.4.27.

Remark 6.4.28. Note that the polar of p in 6.4.27 is the same for all conics through A, A', B, B' missing p.

Remark 6.4.29. Corollary 6.4.27 provides a construction of the polar of any point $p \notin C$, from just the ends of two different chords through p. If the whole of the points of C are already drawn, then the tangents to C through p may be drawn by joining p to the intersection points of its polar with C, by 6.4.26.

Remark 6.4.30. Write $F(x_0, \dots, x_n) = (x)^t N(x)$, where $(x) = (x_0, \dots, x_n)^t$ and N is a matrix of a quadric Q. A direct computation shows that the equation (6.9) of H_p, $p = [y_0, \dots, y_n]$, may be written

$$\sum_{i=0}^{n} \left(\frac{\partial F}{\partial x_i}\right)_{(y)} x_i = 0,$$

$(y) = (y_0, \dots, y_n)$. In particular the equations (6.10) characterizing the double points of Q read

$$\left(\frac{\partial F}{\partial x_i}\right)_{(y)} = 0, \quad i = 0, \dots, n.$$

The above are forms of the equation of the polar hyperplane and the equations of the double points that use derivatives instead of matrices. They are sometimes easier to compute. For instance, the reader may use the latter and the equation (6.7) of the Section 6.3 to easily show that p is a double point of the tangent cone $T_p(Q)$.

In case $k = \mathbb{R}$, there are in $\mathbb{P}_n$ quadrics of all possible ranks with no point other than their double points, namely, as the reader may easily check,

$$Q_r : x_0^2 + \cdots + x_{r-1}^2 = 0, \quad r = 1, \ldots, n+1.$$

In particular Q_{n+1} is non-degenerate and therefore has no points. The situation is quite different if $k = \mathbb{C}$, as then any quadric Q other than a double hyperplane (for them, see 6.4.20) has simple points. Indeed, if $\mathbf{r}(Q) > 1$, then $\dim \mathcal{D}(Q) < n-1$ and there is a line ℓ disjoint with $\mathcal{D}(Q)$. Since there is at least one point of Q on ℓ (by 6.3.1), such a point is a simple point of Q.

By the above, the next proposition applies in particular to all complex quadrics Q with $\mathbf{r}(Q) > 1$.

Proposition 6.4.31. *If a quadric Q of $\mathbb{P}_n$ has a simple point, then no hyperplane of $\mathbb{P}_n$ contains $|Q|$.*

Proof. Let p be a simple point of Q and assume that there is a hyperplane H for which $|Q| \subset H$. Take any point q lying not on H or on the tangent hyperplane to Q at p (2.4.4). Then $q \neq p$ and the line pq contains a point $q' \in Q$, $q' \neq p$, because it is not tangent to Q at p. By the hypothesis on H, $q' \in H$ and $pq = pq' \subset H$ against the choice of q. □

Remark 6.4.32. The claim of 6.4.31 may be equivalently stated by saying that if Q has a simple point, then there are points $p_0, \ldots, p_n \in Q$ spanning $\mathbb{P}_n$. For, it is clear that there is $p_0 \in Q$. Inductively, assume that there are $p_0, \ldots, p_i \in Q$ spanning an i-dimensional linear variety L_i. As far as $i < n$, there is a hyperplane H containing L_i (by 1.5.10). Using 6.4.31, there is a point $p_{i+1} \in Q$ with $p_{i+1} \notin H$ and therefore $p_{i+1} \notin L_i$. Then $p_0, \ldots, p_{i+1}$ belong to Q and span a linear variety of dimension $i+1$. This eventually provides $p_0, \ldots, p_n$ in Q spanning $\mathbb{P}_n$. The converse is clear.

Corollary 6.4.33. *If a linear variety of positive dimension L goes through a point p of a quadric Q and is not tangent to Q at p, then L is the minimal linear variety containing all points of $Q \cap L$.*

Proof. The linear variety L being not tangent to Q at p, by 6.4.24 $Q \cap L$ is a quadric of L which has p as a simple point, after which 6.4.31 applies. □

The last result of this section compares a real quadric and its complex extension regarding conjugation:

Proposition 6.4.34. *If Q is a quadric of a real projective space $\mathbb{P}_n$, then*

(a) *Points $p, q \in \mathbb{P}_n$ are conjugate with respect to Q if and only if they are conjugate with respect to $Q_{\mathbb{C}}$.*

(b) *For any $p \in \mathbb{P}_n$, $(H_{p,Q})_{\mathbb{C}} = H_{p,Q_{\mathbb{C}}}$.*

(c) *For any $q \in \mathbb{C}\mathbb{P}_n$, $\sigma(H_{q,Q_{\mathbb{C}}}) = H_{\sigma(q),Q_{\mathbb{C}}}$, σ being the complex-conjugation map.*

(d) $\mathcal{D}(Q_{\mathbb{C}}) = \mathcal{D}(Q)_{\mathbb{C}}$.

(e) $\mathbf{r}(Q_{\mathbb{C}}) = \mathbf{r}(Q)$.

(f) *A linear variety L is tangent to Q at p if and only if $L_{\mathbb{C}}$ is tangent to $Q_{\mathbb{C}}$ at p.*

Proof. All claims directly follow from the fact that a matrix of Q is also a matrix of $Q_{\mathbb{C}}$, except the last one which is a direct consequence of (b) and 6.4.23 □

According to 6.4.34 (a), the conjugation with respect to $Q_{\mathbb{C}}$ extends the conjugation with respect to C: often no distinction between them is made, and the imaginary points conjugate with respect to $Q_{\mathbb{C}}$ are said to be conjugate with respect to Q.

It follows in particular from 6.4.34 (b) that the complexified of the hyperplanes tangent to a quadric Q of $\mathbb{P}_n$ are the hyperplanes of $\mathbb{C}\mathbb{P}_n$ tangent to $Q_{\mathbb{C}}$ at its real points; they are, indeed, real hyperplanes of $\mathbb{C}\mathbb{P}_n$. We will see in the forthcoming 6.7.16 that the remaining tangent hyperplanes to $Q_{\mathbb{C}}$, that is, those which are not tangent at a real point, are all imaginary. They are often called the *imaginary tangent hyperplanes to Q*. The ordinary tangent hyperplanes to Q are called *real* if some confusion may arise.

6.5 Non-degenerate quadrics. Polarity

In this section we will assume that Q is a non-degenerate quadric. Then Q has no double points and therefore taking the polar hyperplane defines a map

$$\begin{aligned} \mathcal{P}_Q \colon \mathbb{P}_n &\longrightarrow \mathbb{P}_n^{\vee}, \\ p &\longmapsto H_p. \end{aligned}$$

Assume that a reference Δ of $\mathbb{P}_n$ has been fixed and that Q has matrix N relative to it. We have seen in Section 6.4 that a coordinate vector of the polar hyperplane of $p = [y_0, \dots, y_n]$ is

$$N \begin{pmatrix} y_0 \\ \vdots \\ y_n \end{pmatrix}.$$

The quadric Q being non-degenerate, we have $\det N \neq 0$ and so $\mathcal{P}_Q$ is the projectivity with matrix relative to $\Delta, \Delta^\vee$ equal to N (2.8.5). Using the terms introduced in Section 5.8, $\mathcal{P}_Q$ is the correlation of $\mathbb{P}_n$ with matrix relative to Δ equal to N. Actually, since N is a symmetric matrix, $\mathcal{P}_Q$ is a polarity: it is called the *polarity* relative (or associated) to Q. Both the matrices of a quadric and a projectivity relative to fixed references being determined up to a non-zero factor, we have proved:

Theorem 6.5.1. *If Q is a non-degenerate quadric of $\mathbb{P}_n$, then mapping each point $p \in \mathbb{P}_n$ to its polar hyperplane H_p is a polarity $\mathcal{P}_Q$ of $\mathbb{P}_n$. Furthermore Q and $\mathcal{P}_Q$ have the same matrices relative to any fixed reference.*

The polarity determines in turn the quadric, in fact we have:

Corollary 6.5.2. *Mapping $Q \mapsto \mathcal{P}_Q$ is a bijection between the set of the non-degenerate quadrics of $\mathbb{P}_n$ and the set of polarities of $\mathbb{P}_n$.*

Proof. Fix a reference of $\mathbb{P}_n$. If g is a polarity with matrix N, then N is regular and symmetric. By 6.5.1, $\mathcal{P}_Q = g$ if and only if the quadric Q has matrix N. On the other hand, by 6.1.2, there is a unique quadric Q with matrix N, which is non-degenerate because N is regular. Hence the claim. □

The next proposition directly follows from 6.4.9 and the definition of the action of projectivities on correlations. Its proof is left to the reader.

Proposition 6.5.3. *For any non-degenerate quadric Q of a projective space $\mathbb{P}_n$ and any projectivity $f : \mathbb{P}_n \to \mathbb{P}'_n$, $f(\mathcal{P}_Q) = \mathcal{P}_{f(Q)}$.*

Remark 6.5.4. Points p, p' are conjugate with respect to a pair of distinct points $q_1 + q_2$ of $\mathbb{P}_1$ if and only if they correspond by the involution of $\mathbb{P}_1$ with fixed points q_1, q_2, as both conditions are equivalent to having either $p = p' = q_1$, or $p = p' = q_2$, or, otherwise, $(p, p', q_1, q_2) = -1$, by 6.4.11 and 5.6.9. The conjugation with respect to $q_1 + q_2$ is thus the involution with fixed points q_1, q_2, which is called the *involution of conjugation*. If the conjugate to a point p is p', then the polarity $\mathcal{P}_{q_1+q_2}$ maps p to $\{p'\}$, and so there is little difference between the conjugation and the polarity relative to a pair of distinct points of $\mathbb{P}_1$; the polarity is seldom used, as the involution of conjugation usually takes its role.

If, after fixing coordinates, $q_1 + q_2$ has matrix

$$\begin{pmatrix} a & b \\ b & c \end{pmatrix},$$

then this is also the matrix of the polarity $\mathcal{P}_{q_1+q_2}$. Assume that the conjugate to a point $p = [x_0, x_1]$ is p': then $\{p'\}$ has hyperplane coordinates $ax_0 + bx_1$, $bx_0 + cx_1$ and, by 4.4.5, p' has point coordinates $bx_0 + cx_1$, $-ax_0 - bx_1$. The matrix of the involution of conjugation is thus

$$\begin{pmatrix} b & c \\ -a & -b \end{pmatrix}.$$

Back to the n-dimensional case, the fact that $\mathcal{P}_Q$ is a projectivity has many important consequences that will be presented next. The first and more obvious one is that the map $\mathcal{P}_Q$ is bijective: each hyperplane H of $\mathbb{P}_n$ is the polar hyperplane of a unique point p which is called the *pole* of H (with respect to Q). In particular the hyperplanes tangent to Q are tangent at a single point, its pole, which is called *the contact point* of the tangent hyperplane.

The pairing $p \leftrightarrow H_p$, between points and hyperplanes, may be extended to the linear varieties of intermediate dimensions. Assume that L is a linear variety of dimension d, $0 \leq d \leq n-1$. By 6.5.1, the image of L by $\mathcal{P}_Q$, $\mathcal{P}_Q(L) = \{H_p \mid p \in L\}$ is a d-dimensional bundle of hyperplanes. Its kernel $\bar{L}$ is thus a linear variety of dimension $n-d-1$: it is called the *polar variety of* L with respect (or relative) to Q, or just the polar variety of L if no confusion may occur. Of course it is called a *polar line* if $d = n-2$ and a *polar plane* if $d = n-3$.

The kernel of a bundle being the intersection of all hyperplanes in the bundle (4.2.9), the polar variety $\bar{L}$ may be viewed as

$$\bar{L} = \bigcap_{p \in L} H_p,$$

that is, as the set of points q conjugate to all $p \in L$. Take now the polar variety $\bar{\bar{L}}$ of $\bar{L}$: it has dimension $n-(n-d-1)-1 = d$ and furthermore, since any point of L is conjugate to all points in $\bar{L}$, $L \subset \bar{\bar{L}}$. This gives $L = \bar{\bar{L}}$, that is, the symmetry of the pairing $L \leftrightarrow \bar{L}$. Because of this symmetry, a pair $L, \bar{L}$ is often referred to as a pair of mutually polar varieties. Summarizing, we have proved:

Proposition 6.5.5. *Once a non-degenerate quadric of $\mathbb{P}_n$ has been fixed, for any linear variety $L \subset \mathbb{P}_n$ of dimension d, $0 \leq d \leq n-1$, the set*

$$\bar{L} = \{q \in \mathbb{P}_n \mid q \text{ is conjugate to } p \text{ for all } p \in L\}$$

is a linear variety of dimension $n-d-1$ called the polar variety of L relative to Q. For any L as above, $\bar{\bar{L}} = L$.

The extremal cases $d = 0, n-1$ give the already considered pairing pole-polar hyperplane, $p \leftrightarrow H_p$, so the easier interesting case of the above is that of the pairs of mutually polar lines with respect to a non-degenerate quadric of a three-space.

The polar variety may also be written as an intersection of finitely many hyperplanes:

Lemma 6.5.6. *For any linear variety L of dimension d and any $d+1$ independent points $p_0, \dots, p_d \in L$,*

$$\bar{L} = H_{p_0} \cap \dots \cap H_{p_d}.$$

Proof. By 6.5.1, the hyperplanes $H_{p_0}, \dots, H_{p_d}$ are independent and they obviously belong to $\bar{L}^*$. Then 4.2.8 applies. □

The basic properties of taking the polar variety are summarized next:

Proposition 6.5.7. *The map*

$$\begin{aligned}\mathcal{LV}_d(\mathbb{P}_n) &\longrightarrow \mathcal{LV}_{n-d-1}(\mathbb{P}_n),\\ L &\longmapsto \overline{L},\end{aligned}$$

is a bijection between the sets of linear varieties of $\mathbb{P}_n$ of dimensions d and $n-d-1$. Furthermore for any linear varieties $L, L' \subset \mathbb{P}_n$,

(a) $L \subset L'$ *if and only if* $\overline{L} \supset \overline{L}'$,

(b) $\overline{L \cap L'} = \overline{L} \vee \overline{L}'$,

(c) $\overline{L \vee L'} = \overline{L} \cap \overline{L}'$.

Proof. Since by definition $\overline{L}$ is the kernel of $\mathcal{P}_Q(L)$, the first claim follows from being bijections both $\mathcal{P}_Q$ (6.5.1) and taking the kernel (4.2.3), and also from the equality $\overline{\overline{L}} = L$ of 6.5.5. The remaining claims are direct from 1.7.3 and 4.2.4 (or 4.2.5). □

A system of equations of the polar variety $\overline{L}$ is easy to obtain. Assume we have fixed a reference, let $(a^i) = (a_i^0, \dots, a_i^n)^t$, $i = 0, \dots, d$, be column coordinate vectors of a set of $d+1$ independent points $p_0, \dots, p_d \in L$ and still write N a matrix of Q. Then, the hyperplanes H_{p_i}, $i = 0, \dots, d$, have coordinate vectors $N(a^i)$ and, by 6.5.6, $\overline{L} = H_{p_0} \cap \dots \cap H_{p_d}$. Thus

$$\begin{pmatrix} x_0 & \dots & x_n \end{pmatrix} N \begin{pmatrix} a_i^0 \\ \vdots \\ a_i^n \end{pmatrix} = 0, \quad i = 0, \dots d,$$

is a system of equations of $\overline{L}$. It is left to the reader to prove that multiplying by N^{-1} the vectors of coefficients of the equations of any system of equations of L gives coordinate vectors of a set of independent points spanning $\overline{L}$.

Next are some consequences of 6.5.5:

Corollary 6.5.8. *Assume that $q \in Q$. Then $q \in \overline{L}$ if and only if L is contained in the tangent hyperplane to Q at q.*

Proof. By 6.5.5, q belongs to $\overline{L}$ if and only if $H_q \supset L$. Just add that if $p \in Q$ then H_q is the tangent hyperplane to Q at q. □

Corollary 6.5.9. *The following conditions are equivalent:*

(i) *The point p belongs to $L \cap \overline{L}$.*

(ii) *$p \in Q$ and L is tangent to Q at p.*

(iii) *$p \in Q$ and $\overline{L}$ is tangent to Q at p.*

Proof. If $p \in L \cap \overline{L}$, then, $p \in L$ and by the definition of $\overline{L}$, p is conjugate to all points in L. This gives $p \in L \subset H_p$ and hence (ii), by 6.4.23. Conversely, again by 6.4.23, if L is tangent to Q at p, then $p \in L$ and $L \subset H_p$. The latter assures that p is conjugate to all points in L, and therefore $p \in \overline{L}$. Since $\overline{\overline{L}} = L$ the equivalence (i) $\iff$ (iii) follows from (i) $\iff$ (ii) by just taking $\overline{L}$ instead of L. □

Corollary 6.5.10. *$L \subset Q$ if and only if $L \subset \overline{L}$.*

Proof. If $L \subset Q$, then, for any $p \in L$, L is tangent to Q at p and therefore, by 6.5.9, $p \in \overline{L}$. Conversely, if $L \subset \overline{L}$, any $p \in L$ is conjugate to itself and hence $p \in Q$. □

Corollary 6.5.11. *For any linear variety L contained in a non-degenerate quadric of $\mathbb{P}_n$, $\dim L \leq (n-1)/2$.*

Proof. By 6.5.10, $L \subset \overline{L}$ and therefore

$$\dim L \leq \dim \overline{L} = n - \dim L - 1,$$

from which the claim. □

There are three possibilities regarding the hyperplanes of a pencil tangent to a non-degenerate quadric. They are described next in terms of the relative position of the kernel of the pencil and its polar line:

Corollary 6.5.12. *If L is an $(n-2)$-dimensional linear variety and Q a non-degenerate quadric of $\mathbb{P}_n$, $n > 1$, then the three conditions within each of the following three groups are equivalent to each other. Furthermore, each L satisfies the conditions in one and only one of the groups:*

(a1) $L \supset \overline{L}$.

(a2) $\overline{L} \subset Q$.

(a3) *Any hyperplane through L is tangent to Q.*

(b1) *$L \cap \overline{L}$ is a point.*

(b2) *$\overline{L}$ is properly tangent to Q.*

(b3) *There is a single hyperplane through L tangent to Q.*

(c1) $L \cap \overline{L} = \emptyset$.

(c2) *$\overline{L}$ is a chord of Q.*

(c3) *There are exactly two distinct hyperplanes through L tangent to Q or, only in case $k = \mathbb{R}$, none.*

Proof. By 6.5.5, $\overline{L}$ is a line and so it makes sense to call it a proper tangent or a chord. In each group, the equivalence between the first and second conditions is direct from 6.5.9, while the equivalence between the second and third conditions follows from 6.5.8, using that the polarity $\mathcal{P}_Q$ is a bijection. For the last claim, just note that (a1), (b1) and (c1) are pairwise incompatible and one of them is always satisfied because $\overline{L}$ is a line. □

Remark 6.5.13. Some additions to 6.5.12 follow. If the conditions (a) are satisfied, then the hyperplanes through L are the tangent hyperplanes at the points of $\overline{L}$, by 6.5.8. If the conditions (b) are satisfied, put $L \cap \overline{L} = \{q\}$; then, by 6.5.9, $\overline{L}$ is tangent to Q at q, the tangent hyperplane through L is H_q and $H_q = L \vee \overline{L}$, as the reader may easily check. Finally, the conditions (c) are satisfied, then the tangent planes through L are tangent at the ends of $\overline{L}$, again by 6.5.8. If $Q \cap \overline{L} = \{q_1, q_2\}$, then $q_i \notin L$ and therefore the tangent hyperplanes through L are $q_i \vee L$, $i = 1, 2$.

The smallest dimension for which Corollary 6.5.9 brings new information is $n = 3$: then both L and $\overline{L}$ are lines, which makes the situation more symmetric. The lines L and $\overline{L}$ can either:

(a) agree, in which case $L = \overline{L} \subset Q$ and all planes through L are tangent to Q, or

(b) meet at a single point p: then $p \in Q$, both L and $\overline{L}$ are proper tangents at p and together span the tangent plane H_p, which is the only tangent plane through L, and also the only tangent plane through $\overline{L}$, or

(c) be disjoint, in which case each line is a chord whose ends are the contact points of the tangent planes through the other.

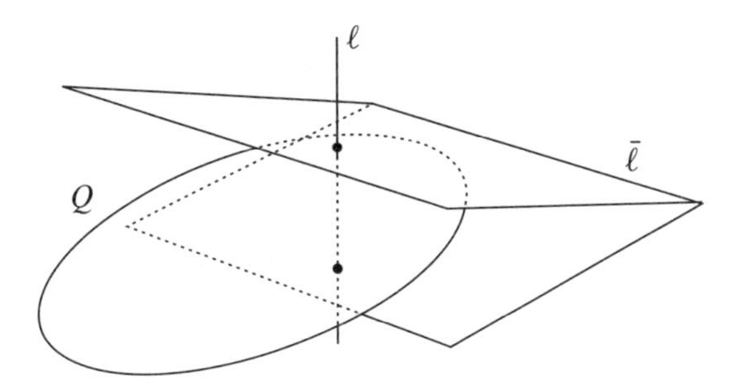

Figure 6.4. A pair of disjoint mutually polar lines ℓ, $\bar{\ell}$ in a three-space.

6.6 Non-degenerate quadric envelopes

Once a projective space $\mathbb{P}_n$ has been fixed, the quadrics of its dual space $\mathbb{P}_n^\vee$ are called *quadric envelopes* or *hyperplane-quadrics* of $\mathbb{P}_n$ (*conic envelopes* or *line-conics* if $n = 2$), the ordinary quadrics of $\mathbb{P}_n$ being then called *quadric locus* or *point-quadrics* if some confusion may arise. In this section we will describe the non-degenerate quadric envelopes.

First we will show some non-degenerate quadrics of $\mathbb{P}_n^\vee$. Fix a non-degenerate quadric locus Q and consider its image $Q^* = \mathcal{P}_Q(Q)$ by its own polarity. By 6.4.13, Q^* is a non-degenerate quadric envelope and, by 6.1.3, a hyperplane belongs to Q^* if and only if it is the polar hyperplane of a point of Q, that is, if and only if it is a tangent hyperplane to Q. Because of this the quadric Q^* is named the *envelope* of Q; its elements, which are the tangent hyperplanes to Q, are said *to envelop* Q.

Fix a reference Δ of $\mathbb{P}_n$ and assume that Q has matrix

$$N = \begin{pmatrix} a_{0,0} & \dots & a_{n,0} \\ \vdots & & \vdots \\ a_{n,0} & \dots & a_{n,n} \end{pmatrix}.$$

By 6.1.14 and 6.5.1, the envelope Q^* of Q has matrix $(N^{-1})^t N N^{-1} = N^{-1}$ relative to the dual reference $\Delta^\vee$. To avoid dividing by the determinant, it is often preferred to take $(\det N)N^{-1}$, the matrix of cofactors of N, as a matrix of Q^*. Using hyperplane coordinates relative to Δ, Q^* has thus equation

$$\sum_{i,j=0}^{n} A_{i,j} u_i u_j = 0,$$

each $A_{i,j}$ being the cofactor of $a_{i,j}$ in N. This equation may be viewed as a necessary and sufficient condition on the coordinates of a hyperplane for it to be tangent to Q.

Since proportional regular matrices have proportional inverses and conversely, mapping each non-degenerate quadric Q to its envelope Q^* is a bijection between the set of all non-degenerate point-quadrics and the set of all non-degenerate quadric envelopes, which means that:

(1) All non-degenerate quadric envelopes are envelopes of non-degenerate point-quadrics, which, incidentally, is the reason for their name.

(2) Each non-degenerate point-quadric is determined by its envelope; the point-quadric Q is referred to as the quadric *enveloped* by Q^*.

The next theorem summarizes the above and adds some further information:

Theorem 6.6.1. *Any non-degenerate quadric envelope is the envelope Q^* of a uniquely determined non-degenerate point-quadric Q, and the following hold:*

(a) *A hyperplane H belongs to Q^* if and only if H is tangent to Q.*

(b) *Two hyperplanes are conjugate with respect to Q^* if and only if their poles are conjugate with respect to Q.*

(c) *Two hyperplanes are conjugate with respect to Q^* if and only if one of them contains the pole of the other.*

(d) *If H is the polar hyperplane of p with respect to Q, then p^* is the polar hyperbundle of H with respect to Q^*, and conversely.*

(e) *N is a matrix of Q relative to a reference Δ if and only if N^{-1} is a matrix of Q^* relative to $\Delta^\vee$.*

Proof. Only the claims (b) (c) and (d) have not been already proved. The claim (b) is a particular case of 6.4.4, while the claims (c) and (d) directly follow from it, as the poles p, p' of hyperplanes H_p, $H_{p'}$ are conjugate with respect to Q if and only if one of them, say p, belongs to the hyperplane polar $H_{p'}$ of the other. □

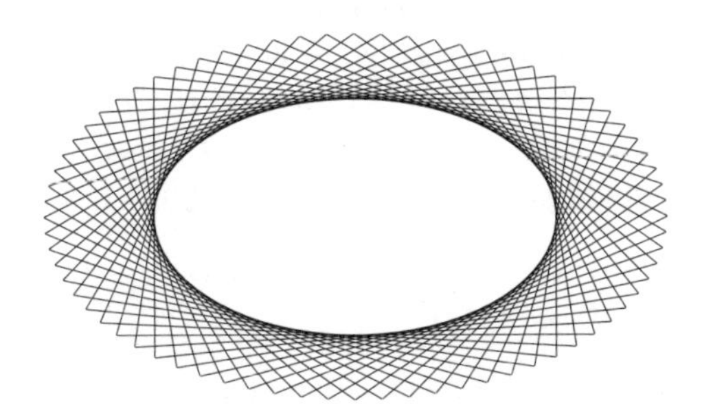

Figure 6.5. A conic envelope.

Note that claim (a) of 6.6.1 is a particular case of (d), and also that (b), (c) and (d) are just different versions of the same fact.

As usual with duality, nothing essentially new appears for $n = 1$:

Proposition 6.6.2. *If Q is a non-degenerate quadric of $\mathbb{P}_1$, then the projectivity ι of* 4.2.12 *maps Q into Q^*. In particular, points $p, q \in \mathbb{P}_1$ are conjugate with respect to Q if and only if the hyperplanes $\{p\}$, $\{q\}$ are conjugate with respect to Q^*.*

Proof. If Q has equation $ax_0^2 + 2bx_0x_1 + cx_2^2 = 0$, then Q^* has equation $cu_0^2 - 2bu_0u_1 + au_2^2 = 0$. Since we know from 4.4.5 that ι has matrix

$$\begin{pmatrix} 0 & 1 \\ -1 & 0 \end{pmatrix},$$

the claim follows from a direct checking which is left to the reader. □

Proposition 6.6.3. *If $f : \mathbb{P}_n \to \mathbb{P}'_n$ is a projectivity and Q a non-degenerate quadric of $\mathbb{P}_n$, $f^\vee((f(Q))^*) = Q^*$.*

Proof. Take references Δ and Δ' in $\mathbb{P}_n$ and $\mathbb{P}'_n$ respectively, and assume that f and Q have matrices M and N. Using 6.1.14, 6.6.1 and 4.5.2 the claim is implied by the matricial equality

$$M^{-1}((M^{-1})^t N M^{-1})^{-1}(M^t)^{-1} = N^{-1},$$

which in turn is clear. □

Identify $\mathbb{P}_n$ with its bidual space $\mathbb{P}_n^{\vee\vee}$ through the projectivity Φ of Theorem 4.6.1, and so each point p with the hyperbundle p^*. By 6.6.1 (d), the pole p of any hyperplane H is identified with the polar hyperbundle of H: it is said then that taking the quadric envelope Q^* instead of Q, swaps over the roles of pole and polar hyperplane, or, equivalently, that the polarity with respect to Q^* is the inverse of the polarity relative to Q. In particular the point identified with the hyperbundle tangent to Q^* at $H \in Q^*$ is the contact point of H. The fact that the points of Q are identified with the tangent hyperbundles to Q^* suggests that the envelope of the envelope, Q^{**}, should be identified with the original Q, which is indeed true:

Proposition 6.6.4. *If $\Phi\colon \mathbb{P}_n \to \mathbb{P}_n^{\vee\vee}$ is the projectivity of* 4.6.1, *then for any non-degenerate quadric Q of $\mathbb{P}_n$, $\Phi(Q) = Q^{**}$.*

Proof. Fix any reference Δ of $\mathbb{P}_n$ and assume that Q has matrix N. By 6.6.1 (e), Q^{**} has matrix $(N^{-1})^{-1} = N$ relative to $\Delta^{\vee\vee}$. Then the claim follows, as by 4.6.3 the matrix of Φ is the unit matrix. □

Regarding duality, it is convenient to take each non-degenerate quadric and its envelope together, and so to think of non-degenerate quadrics as being pairs (Q, Q^*), Q a non-degenerate point-quadric and Q^* its envelope (these pairs are called *complete quadrics*). This makes no essential difference because the quadric locus and its envelope determine each other, but turns the notion of non-degenerate quadric into a self-dual one. Then the duality principle may be enlarged by taking the term *non-degenerate quadric* as a self-dual one and those of *pole* and *polar hyperplane* as duals of each other. In particular, *point of* and *hyperplane tangent to* a quadric are mutually dual, and so are *contact point of a tangent hyperplane* and *tangent hyperplane at a point*. The term *pair of mutually polar varieties* is self-dual, as is *linear variety tangent to a non-degenerate quadric*, because a linear variety is tangent if and only if it meets its polar (6.5.9). For instance, the fact that lines non-tangent to a non-degenerate quadric Q contain exactly two distinct (maybe imaginary) points of Q provides by duality the next proposition which has been already proved as part of 6.5.12.

Proposition 6.6.5. *If an $(n-2)$-dimensional linear variety L is not tangent to a non-degenerate quadric Q, then there are exactly two distinct (maybe imaginary) hyperplanes through L tangent to Q.*

6.7 Degenerate quadrics. Cones

We start by proving a lemma that provides easy to handle matrices of degenerate quadrics. As above, $\mathcal{D}(Q)$ denotes the variety of double points of a quadric Q:

Lemma 6.7.1. *Assume that vertices $p_0, \dots, p_d$ of a reference $\Delta = (p_0, \dots, p_n, A)$ are double points of a quadric Q. Then:*

(a) *Any matrix of Q relative to Δ has the form, in blocks,*

$$N = \begin{pmatrix} 0 & 0 \\ 0 & N' \end{pmatrix} \tag{6.11}$$

where N' is an $(n-d) \times (n-d)$ square matrix.

(b) *If $d < n-1$ and $T = p_{d+1} \vee \dots \vee p_n$, then $Q' = Q \cap T$ is a quadric that has matrix N' relative to the reference Δ' subordinated by Δ in T.*

(c) *The vertices $p_0, \dots, p_d$ span $\mathcal{D}(Q)$ if and only if* $\det N' \neq 0$.

Proof. The entries of N are $f(e_i, e_j)$, $i, j = 0, \dots, n$, where $e_0, \dots, e_n$ is a basis adapted to Δ and f a suitable representative of Q. By the hypothesis, for $i = 0, \dots, d$, $p_i = [e_i]$ is conjugate to all points, hence $f(e_i, e_j) = 0$ if either $i \leq d$ or $j \leq d$, from which claim (a) follows.

Write $T = [F]$. As seen in the proof of 2.2.6, $e_{d+1}, \dots, e_n$ is a basis of F adapted to Δ'. Thus the entries of N' are

$$f(e_i, e_j) = f_{|F}(e_i, e_j), \quad i, j = d+1, \dots, n,$$

that is, either the entries of a matrix of $Q \cap T$ relative to Δ' if $Q \cap T$ is a quadric, or all equal to zero in case $T \subset Q$. The latter possibility would give $N = 0$ which is obviously not allowed.

Regarding claim (c) just note that

$$n - d - \operatorname{rk}(N') = n - d - \operatorname{rk}(N) = n - d - (n - \dim \mathcal{D}(Q)) = \dim \mathcal{D}(Q) - d.$$

□

Next we introduce a certain type of quadrics, called *cones*: they will help to describe the degenerate quadrics, as it will turn out in Corollary 6.7.7 that all degenerate quadrics but the double hyperplanes are cones over non-degenerate quadrics.

Proposition 6.7.2. *Assume that D and T are supplementary linear varieties of $\mathbb{P}_n$, $n > \dim T > 0$, and Q' a quadric of T.*

(a) *There is one and only one quadric Q of $\mathbb{P}_n$ such that $Q \cap T = Q'$ and $\mathcal{D}(Q) \supset D$.*

(b) *Denote by $\tau\colon \mathbb{P}_n - D \to T$ the projection from D onto T. Points $p, p' \notin D$ are conjugate with respect to Q if and only if their projections $\tau(p), \tau(p')$ are conjugate with respect to Q'. In particular a point $p \notin D$ belongs to Q if and only if $\tau(p)$ belongs to Q'.*

(c) $\mathbf{r}(Q) = \mathbf{r}(Q')$ *and* $\mathcal{D}(Q) = D \vee \mathcal{D}(Q')$.

Proof. Take a reference Δ with its first $d+1$ vertices spanning D and the remaining ones spanning T (by 2.7.8). Take in T the reference Δ' subordinated by Δ. By 6.7.1, a matrix of Q is

$$N = \begin{pmatrix} 0 & 0 \\ 0 & N' \end{pmatrix}$$

where N' is a matrix of Q' relative to Δ'. This proves the uniqueness of Q. For the existence just take Q the quadric having the above matrix N. Any point of D having coordinates $x_0, \dots, x_d, 0, \dots, 0$, it clearly is a double point of Q. This is in particular true for $p_0, \dots, p_d$, so 6.7.1 applies proving that $Q \cap T$ is a quadric and has matrix N' relative to Δ'. It follows that $Q \cap T = Q'$ as required.

Claim (b) is a direct consequence of the form of the matrix N and 2.7.7. Since $\operatorname{rk}(N) = \operatorname{rk}(N')$, it is clear that $\mathbf{r}(Q) = \mathbf{r}(Q')$. To close, $D \subset \mathcal{D}(Q)$ by the hypothesis, while $\mathcal{D}(Q') \subset \mathcal{D}(Q)$ by direct computation. This gives $D \vee \mathcal{D}(Q') \subset \mathcal{D}(Q)$. Furthermore $D \cap \mathcal{D}(Q') = \emptyset$ and so

$$\dim(D \vee \mathcal{D}(Q')) = d + (n-d-1) - \mathbf{r}(Q') + 1 = n - \mathbf{r}(Q) = \dim(\mathcal{D}(Q))$$

giving the equality. □

The quadric Q of 6.7.2 is called the *cone over Q' with vertex D*, and also the *cone projecting Q' from D*. The simple term *cone* or *quadric cone* is used if no reference to D or Q' is needed.

Remark 6.7.3. As it is clear from 6.7.2, if $|Q'| \neq \emptyset$, then

$$|Q| = \bigcup_{p \in Q'} D \vee p,$$

and otherwise $|Q| = D$. On the other hand (by 1.10.1) $(D \vee p) \cap (D \vee p') = D$ if $p, p' \in Q'$, $p \neq p'$.

Example 6.7.4. In $\mathbb{P}_3$, the cones projecting a non-degenerate conic with points from a point are called *ordinary cones*. Note that their points are the points of the lines projecting the points of the conic from the vertex, according to the usual definition of cone in elementary geometry.

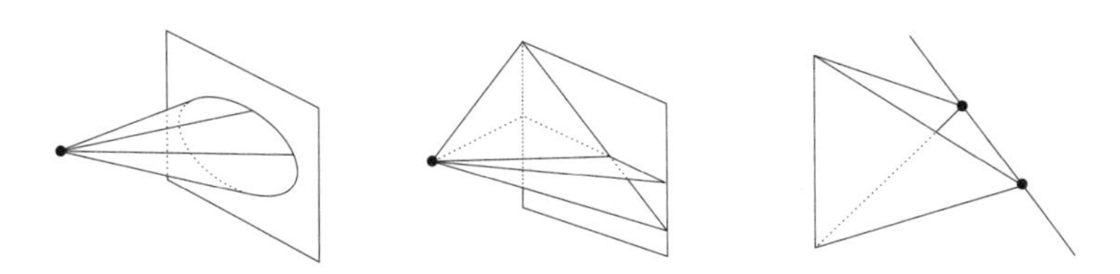

Figure 6.6. Three cones in three-space: from left to right, over a non-degenerate conic, over a line-pair and over a point-pair. Note that the last two are plane-pairs.

Assume that Q is a quadric that has matrix M relative to a certain projective reference. The tangent cone to Q with vertex a point $p = [y_0, \dots, y_n]$ has been defined in Section 6.3 as being the quadric $TC_p(Q)$ with equation

$$\big((x)^t N(x)\big)\big((y)^t N(y)\big) - ((x)^t N(y))^2 = 0 \tag{6.12}$$

provided this equation is not identically zero. We can now give a more intrinsic description of the tangent cone. It is clear that the equation (6.12) is an identity $0 = 0$ if p is a double point of Q, as then $N(y) = 0$. If p is a simple point of Q, then $(y)^t N(y) = 0$ and (6.12) is an equation of the tangent hyperplane H_p counted twice. If $p \notin Q$ and Q has rank one, then $\mathcal{D}(Q)$ is a hyperplane and therefore all lines through p are tangent to Q. Since the coordinates of all points on these lines are solutions of the equation (6.12), again it is $0 = 0$, this time by 6.1.5. Thus we may focus our attention on the case $p \notin Q$ and $\mathbf{r}(Q) > 1$:

Corollary 6.7.5. *If Q is a quadric with $\mathbf{r}(Q) > 1$ and p a point not belonging to Q, then p and H_p are supplementary, $H_p \cap Q$ is a quadric and the tangent cone $TC_p(Q)$ is the cone over $Q \cap H_p$ with vertex p.*

Proof. First note that $p \notin Q$ forces $p \notin H_p$ and hence p and H_p are supplementary. We will check now that $Q \cap H_p$ is a quadric. Indeed, otherwise $H_p \subset Q$ and any $q \in H_p$ would be conjugate to all points of H_p (by 6.4.11), and of course also to p; then $H_q = \mathbb{P}_n$ and so any $q \in H_p$ would be a double point of Q, against $\mathbf{r}(Q) > 1$.

The reader may easily check that $p = [y_0, \dots, y_n]$ is a double point of $TC_p(Q)$, see 6.4.30. It remains just to see that $TC_p(Q) \cap H_p = Q \cap H_p$. We have already seen, in 6.4.11, that $|TC_p(Q) \cap H_p| = |Q \cap H_p|$, which of course is not enough. We will compare equations of $TC_p(Q) \cap H_p$ and $Q \cap H_p$. To this end, assume that $(x) = B(\lambda)$, $(x) = (x_0, \dots, x_n)^t$, $(\lambda) = (\lambda_0, \dots, \lambda_{n-1})^t$

and B an $n \times (n-1)$-matrix, is a system of parametric equations of H_p. Then $(\lambda)^t B^t N(y) = 0$ identically in (λ), after which applying the substitution rule of 6.1.9 to the equation (6.12) of $TC_p(Q)$ and the equation $(x)^t N(x) = 0$ of Q gives equations of $TC_p(Q) \cap H_p$ and $Q \cap H_p$ that are proportional, as wanted. □

If Q is a non-degenerate quadric and H a hyperplane non-tangent to Q, the pole p of H is determined by H and does not belong to Q. Then H determines the tangent cone $TC_p(Q)$, which is also denoted by $TC_H(Q)$ and called the *cone tangent to Q along its section by H*. Let Q' be a second non-degenerate quadric Q' and assume that $Q \cap H = Q' \cap H$, which in particular, by 6.4.24, assures that H is not tangent to Q' either. The condition $Q \cap H = Q' \cap H$ being satisfied, it is said that Q and Q' are *tangent along their common section by H* if and only if $TC_H(Q) = TC_H(Q)$. By 6.7.5 above, this is equivalent to the equality of the poles of H relative to Q and Q'.

Next we will see that all degenerate quadrics but the pairs of coincident points of $\mathbb{P}_1$ are cones, namely:

Corollary 6.7.6. *If Q is a degenerate quadric, $D \subset \mathfrak{D}(Q)$ is a linear variety, $0 \leq \dim D \leq n-2$, and T a linear variety supplementary to D, then $Q \cap T$ is a quadric and Q is the cone over $Q \cap T$ with vertex D.*

Proof. Assume that $T \subset Q$. Then, by 6.4.2, any point $q \in T$ is conjugate to all points of T, and also to all points in D because the latter are double. T and D being supplementary, $H_q = \mathbb{P}_n$ and any $q \in T$ is double. This would give $\mathbb{P}_n = D \vee T \subset \mathfrak{D}(Q)$ which is not possible. After this, the claim follows because $n > \dim T > 0$ and Q obviously fulfills the conditions that determine the cone according to Proposition 6.7.2. □

If $\dim \mathfrak{D}(Q) > 0$, Q is a cone in many different ways, by different choices of $D \subset \mathfrak{D}(Q)$. If $\mathbf{r}(Q) > 1$, the most useful choice is taking $D = \mathfrak{D}(Q)$, because then Q appears as a cone over a non-degenerate quadric:

Corollary 6.7.7. *If Q is a degenerate quadric with $\mathbf{r}(Q) > 1$ and T a linear variety supplementary to $\mathfrak{D}(Q)$, then $Q \cap T$ is a non-degenerate quadric of T and Q is the cone over $Q \cap T$ with vertex $\mathfrak{D}(Q)$.*

Proof. We have already seen (in 6.7.6) that Q is the cone over $Q \cap T$ with vertex $\mathfrak{D}(Q)$. Then by 6.7.2, $\mathfrak{D}(Q) = \mathfrak{D}(Q) \vee \mathfrak{D}(Q \cap T)$. The linear varieties $\mathfrak{D}(Q)$ and T being disjoint, this forces $\mathfrak{D}(Q \cap T) = \emptyset$, as claimed. □

Rank-one quadrics are not covered by the above description, but they have been already described in 6.4.20.

Example 6.7.8. Any rank-two quadric is a cone over a pair of distinct points, hence it is a pair of distinct (real or imaginary if $k = \mathbb{R}$) hyperplanes, by 6.7.2 and 6.4.17.

Still assume that Q is degenerate and $\mathbf{r}(Q) > 1$. As usual with ordinary cones of a three-space, the linear varieties $\mathcal{D}(Q) \vee p$, for p a simple point of Q, will be called the *generators* of Q. As it follows from 6.7.2, all generators are contained in $|Q|$ and, once a supplementary variety T of $\mathcal{D}(Q)$ has been fixed, any generator of Q is $\mathcal{D}(Q) \vee p$ for a unique $p \in Q \cap T$. Clearly, the set of points of Q is the union of the generators of Q, but for the case in which Q has no simple point.

After 6.7.7, the bound of 6.5.11 may be easily extended to degenerate quadrics:

Corollary 6.7.9. *If L is a linear variety contained in a quadric Q of $\mathbb{P}_n$, then*

$$\dim L \leq n - \frac{\mathbf{r}(Q)}{2}.$$

Proof. The claim is obvious if $\mathbf{r}(Q) = 1$ and is just 6.5.11 if $\mathbf{r}(Q) = n + 1$. We assume thus $1 < \mathbf{r}(Q) < n + 1$. Take T a linear variety supplementary to $\mathcal{D}(\mathbf{Q})$. Then $\dim T = \mathbf{r}(Q) - 1$ and therefore

$$\dim L \leq n + \dim L \cap T - \dim T = n + \dim L \cap T - \mathbf{r}(Q) + 1.$$

Since $L \cap T \subset Q \cap T$ and the latter is a non-degenerate quadric of T by 6.7.7, according to 6.5.11,

$$\dim L \cap T \leq \frac{\mathbf{r}(Q) - 2}{2}$$

which, together with the former inequality, gives the claim. □

Taking the polar hyperplane with respect to a quadric no longer is a correlation if the quadric is degenerate:

Proposition 6.7.10. *Let Q be a degenerate quadric of $\mathbb{P}_n$. Mapping each point of $\mathbb{P}_n - \mathcal{D}(Q)$ to its polar hyperplane relative to Q is a singular correlation*

$$\begin{aligned} \mathbb{P}_n - \mathcal{D}(Q) &\longrightarrow \mathbb{P}_n^\vee, \\ p &\longmapsto H_{p,Q}, \end{aligned}$$

with singular variety $\mathcal{D}(Q)$ and image $\mathcal{D}(Q)^*$.

The singular correlation of 6.7.10 will be still called the polarity relative to Q (even if it is not a polarity in the sense of Section 5.8) and denoted by $\mathcal{P}_Q$.

Proof of 6.7.10. Fix a reference and assume that a degenerate quadric Q has matrix N. As for non-degenerate quadrics, coordinates $(u_0, \dots, u_n)$ of the polar hyperplane of $p = [x_0, \dots, x_n]$, if defined, are computed by the rule

$$\begin{pmatrix} u_0 \\ \vdots \\ u_n \end{pmatrix} = N \begin{pmatrix} x_0 \\ \vdots \\ x_n \end{pmatrix},$$

the matrix N being now non-regular. After this, taking the polar hyperplane is a singular correlation, by 5.10.7. It has singular variety $\mathcal{D}(Q)$ by 5.10.9. By reciprocity (6.4.10), for any double point q and any p, it is $p \in H_q = \mathbb{P}_n$ and hence $q \in H_p$. This proves that the image of the polarity is included in $\mathcal{D}(Q)^*$. The equality follows by comparing dimensions: by 5.10.7, the image has dimension $\mathrm{rk}(N) - 1 = \mathbf{r}(Q) - 1$, while by 6.4.12, $\dim \mathcal{D}(Q)^* = n - \dim \mathcal{D}(Q) - 1 = \mathbf{r}(Q) - 1$. □

As an immediate consequence of 6.7.10 and 5.10.11 we have the following corollary which may also be proved directly from 6.7.2 and 6.7.7:

Corollary 6.7.11. *For any two different simple points p, q of a degenerate quadric Q, the following conditions are equivalent:*

(i) *p and q have the same polar hyperplane with respect to Q,*

(ii) $p \vee \mathcal{D}(Q) = q \vee \mathcal{D}(Q)$,

(iii) *p and q are collinear with a double point of Q.*

It follows in particular from 6.7.11 that if Q is degenerate and has simple points (so in particular $\mathbf{r}(Q) > 1$), then the tangent hyperplane to Q at a simple point p is also tangent to Q at all simple points in the generator $\mathcal{D}(Q) \vee p$. It is sometimes called the tangent hyperplane to Q along $\mathcal{D}(Q) \vee p$.

Quadrics are determined by their polarities:

Theorem 6.7.12. *If quadrics Q and Q' have $\mathcal{P}_Q = \mathcal{P}_{Q'}$, then $Q = Q'$.*

Proof. Assume that $\mathcal{P}_Q = \mathcal{P}_{Q'}$: then

$$\mathcal{D}(Q) = \mathrm{Sing}(\mathcal{P}_Q) = \mathrm{Sing}(\mathcal{P}_{Q'}) = \mathcal{D}(Q')$$

and in particular $\mathbf{r}(Q) = \mathbf{r}(Q')$. If this common rank is one, then both quadrics have the same hyperplane as a linear variety of double points, in which case the claim is true by 6.4.20. If the common rank is $n+1$, then both quadrics are non-degenerate and then the claim has been already proved in 6.5.2. Thus we assume in the sequel that $n + 1 > \mathbf{r}(Q) = \mathbf{r}(Q') > 1$ and choose a linear variety T, supplementary to $\mathcal{D}(Q) = \mathcal{D}(Q')$. By 6.7.7, $Q \cap T$ is a non-degenerate quadric and, by 6.4.5 the polar hyperplane (in T) of any $p \in T$ with respect to $Q \cap T$ is $\mathcal{P}_Q(p) \cap T$. Since the same holds for Q', we see that $Q \cap T$ and $Q' \cap T$ are non-degenerate quadrics with the same polarity. By 6.5.2, $Q \cap T = Q' \cap T$ and then, by 6.7.6, $Q = Q'$. □

Corollary 6.7.13. *If the conjugations relative to two quadrics Q and Q' of a projective space $\mathbb{P}_n$ do agree (as binary relations on $\mathbb{P}_n$), then $Q = Q'$.*

Proof. By the definition of the polarity relative to a quadric Q (in both the degenerate and non-degenerate cases), $\mathcal{P}_Q(p)$ is defined if and only if there is some point q non-conjugate to p, and in such a case $\mathcal{P}_Q(p)$ is the set of all points conjugate to p. Thus, quadrics with equal conjugations have equal polarities and 6.7.12 above applies. □

The quadrics of a bunch L^∇ of $\mathbb{P}_n$ are closely related to the quadrics of $\mathbb{P}_n$ for which all points of L are double:

Proposition 6.7.14 (Quadrics of a bunch). *Let L be a linear variety of $\mathbb{P}_n$, $0 \leq \dim L \leq n-2$. There is a bijective map $Q \leftrightarrow \bar{Q}$ between the set of all quadrics Q of $\mathbb{P}_n$ with $\mathcal{D}(Q) \supset L$ and the set of all quadrics of L^∇, such that for any choice of a supplementary linear variety T of L, $\bar{Q}$ is the image of $Q \cap T$ by the projection map $T \to L^\nabla$.*

Proof. Assume that $L = [F]$ and $Q = [\eta]$, write E for the vector space associated to $\mathbb{P}_n$ and denote by $\bar{v} \in E/F$ the class modulo F of $v \in E$. Since, by the hypothesis, any point of L is double for Q, $\eta(v, w) = 0$ for any $v \in E$ and $w \in F$. Therefore setting

$$\bar{\eta}(\bar{v}_1, \bar{v}_2) = \eta(v_1, v_2)$$

makes $\bar{\eta}(\bar{v}_1, \bar{v}_2)$ independent of the choice of the representatives v_1, v_2 and defines a non-zero symmetric bilinear form $\bar{\eta}$ on E/F: we take $\bar{Q} = [\bar{\eta}]$. Conversely, if τ is any non-zero symmetric bilinear form on E/F, the rule

$$\tilde{\tau}(v_1, v_2) = \tau(\bar{v}_1, \bar{v}_2)$$

defines a non-zero symmetric bilinear form $\tilde{\tau}$ on E that satisfies $\tilde{\tau}(v, w) = 0$ for any $v \in E$ and $w \in F$, and hence defines a quadric $[\tilde{\tau}]$ of $\mathbb{P}_n$ with $\mathcal{D}([\tilde{\tau}]) \supset L$. It is direct from the definitions that $\tilde{\bar{\eta}} = \eta$ and $\bar{\tilde{\tau}} = \tau$. Thus the maps $\eta \mapsto \bar{\eta}$ and $\tau \mapsto \tilde{\tau}$ are reciprocal bijections between the set of all representatives of the quadrics Q of $\mathbb{P}_n$ with $\mathcal{D}(Q) \supset L$ and the set of all representatives of all quadrics of L^∇. Since both bijections map proportional forms to proportional forms, $Q \mapsto \bar{Q}$ is a bijection as claimed.

Now assume that $T = [G]$ is a supplementary of L, while still $Q = [\eta]$ is a quadric of $\mathbb{P}_n$ with $\mathcal{D}(Q) \supset L$. Then $Q \cap T$ is a quadric by 6.7.6. Recall from the proof of 1.9.3 that the projection map $T \to L^\nabla$ is the projectivity represented by the isomorphism of vector spaces

$$\begin{aligned} G &\longrightarrow E/F, \\ v &\longmapsto \bar{v}. \end{aligned}$$

Since $Q \cap T = [\eta_{|G}]$ and, by the definition of $\bar{\eta}$,

$$\bar{\eta}(\bar{v}_1, \bar{v}_2) = \eta(v_1, v_2) = \eta_{|G}(v_1, v_2),$$

if $v_1, v_2 \in G$, the second claim directly follows from the definition of the image of a quadric by a projectivity (see Section 6.1). □

Remark 6.7.15. It follows from 6.7.6 and 6.7.2 that Q and $\bar{Q}$ have the same rank. In particular $\bar{Q}$ is non-degenerate if and only if $\mathcal{D}(Q) = L$.

Proposition 6.7.16. *If Q is a quadric of a real projective space $\mathbb{P}_n$, then for any $q \in \mathbb{C}\mathbb{P}_n$, $H_{q,Q_\mathbb{C}}$ is real if and only if there is a real point p for which $H_{p,Q_\mathbb{C}} = H_{q,Q_\mathbb{C}}$. If Q is non-degenerate, then $H_{q,Q_\mathbb{C}}$ is real if and only if its pole q is real.*

Proof. The *if* part of the first claim is clear from 6.4.34 (b). For the converse, assume that q is imaginary, and hence different from its complex-conjugate $\sigma(q)$, as otherwise the claim is obvious. If q is a double point of $Q_\mathbb{C}$, then $Q_\mathbb{C}$ is degenerate, so is Q by 6.4.34 (e) and any double point of Q may be taken as p. If q is not a double point, by 6.4.34 (c), $H_{\sigma(q),Q_\mathbb{C}} = H_{q,Q_\mathbb{C}}$ and so, by 6.7.11, all non-double points on the line $\ell = q \vee \sigma(q)$ have the same polar hyperplane. Since ℓ is real and, by 6.4.34 (d), $\ell \not\subset \mathcal{D}(Q_\mathbb{C})$, there is a real non-double point $p \in \ell$: for such a point the first claim is satisfied. If in addition Q is non-degenerate, then so is $Q_\mathbb{C}$ and q is the only point in $\mathbb{C}\mathbb{P}_n$ with polar hyperplane $H_{q,Q_\mathbb{C}}$. Then, by the first part of the claim, $H_{q,Q_\mathbb{C}}$ is real if and only if q is real. □

Remark 6.7.17. As it is clear from the above proof, if Q is degenerate, then the polar hyperplane $H_{q,Q_\mathbb{C}}$ of an imaginary point q may be real: for an example it is enough to choose a real line containing a single double point of Q and then any imaginary point q on it.

6.8 Degenerate quadric envelopes

The elements of a quadric envelope K being hyperplanes, we will refer to those which are double for K as the *singular hyperplanes* of K. We will not call them *double hyperplanes* to avoid any confusion with the quadrics of 6.4.19.

In Section 6.5 we have seen that the non-degenerate quadric envelopes of a projective space $\mathbb{P}_n$, $n > 1$ (that is, the non-degenerate quadrics of $\mathbb{P}_n^\vee$) are the envelopes of the non-degenerate quadrics of $\mathbb{P}_n$. In this section we will deal with degenerate quadric envelopes; we will see that a quadric envelope K of $\mathbb{P}_n$ envelops a quadric of the kernel of the bundle of singular hyperplanes $\mathcal{D}(K)$ of K. Only in case of K being non-degenerate is it $\mathcal{D}(K) = \emptyset$ and the enveloped quadric belongs to $\mathbb{P}_n$.

Rank-one quadric envelopes will be given a separate description, as they have no simple hyperplanes and therefore do not fit in the general context. Directly from 6.4.20 and 6.4.19, read in terms of hyperplanes, we have:

Proposition 6.8.1. *A rank-one quadric envelope K has a hyperbundle of double hyperplanes, $\mathfrak{D}(K) = p^*$, $p \in \mathbb{P}_n$, and no simple hyperplane. The polar hyperbundle of any hyperplane H not going through p is p^*.*

Now, for the remaining degenerate quadric envelopes we have:

Theorem 6.8.2. *Let K be a degenerate quadric envelope with rank higher than one and denote by L the kernel of its bundle of singular hyperplanes, $\mathfrak{D}(K) = L^*$. Then:*

(a) *There is a non-degenerate quadric Q of L such that, for any pair of hyperplanes $H, H' \notin L^*$, H and H' are conjugate with respect to K if and only if $H \cap L$ and $H' \cap L$ are (as hyperplanes of L) conjugate with respect to the quadric envelope Q^* or, equivalently, if and only if either of $H \cap L$ and $H' \cap L$ contains the pole of the other with respect to Q. In particular a hyperplane H belongs to K if and only if either $H \supset L$ or $H \cap L$ is a hyperplane (of L) tangent to Q.*

(b) *For each degenerate quadric envelope K, $\mathbf{r}(K) > 1$, there is a unique Q satisfying the conditions of* (a) *above. Mapping $K \mapsto Q$ is a bijection between the set of all quadric envelopes with bundle of singular hyperplanes L^* and the set of all non-degenerate quadrics of L.*

Proof. Take a reference of $\mathbb{P}_n$, $\Delta = (p_0, \dots, p_n, A)$, with $p_0, \dots, p_d$ spanning L and take in L the reference Δ' subordinated by Δ. The hyperplanes $H_i : x_i = 0$, $i = d+1, \dots, n$, are the $n-d$ last vertices of the dual reference $\Delta^\vee$ and, since their intersection is L, they span $L^* = \mathfrak{D}(K)$. It follows then from 6.7.1 that the matrix of K relative to $\Delta^\vee$ has the form

$$\begin{pmatrix} B & 0 \\ 0 & 0 \end{pmatrix}$$

with B a $(d+1) \times (d+1)$ regular matrix. We take Q to be the quadric of L which has matrix relative to Δ' equal to $N = B^{-1}$; the envelope Q^* of Q has thus matrix B, by 6.6.1 (e).

Take any two hyperplanes $H = [u_0, \dots, u_n]$ and $H' = [u'_0, \dots, u'_n]$ not belonging to L^*. The intersections $H \cap L$ and $H' \cap L$ have in L equations $\sum_{i=0}^d u_i x_i = 0$ and $\sum_{i=0}^d u'_i x_i = 0$, and therefore coordinates $u_0, \dots, u_d$ and $u'_0, \dots, u'_d$, respectively. The obvious equality

$$\begin{pmatrix} u_0 & \dots & u_n \end{pmatrix} \begin{pmatrix} B & 0 \\ 0 & 0 \end{pmatrix} \begin{pmatrix} u'_0 \\ \vdots \\ u'_n \end{pmatrix} = \begin{pmatrix} u_0 & \dots & u_d \end{pmatrix} B \begin{pmatrix} u'_0 \\ \vdots \\ u'_d \end{pmatrix}$$

shows that, as claimed, H and H' are conjugate with respect to K if and only if $H \cap L$ and $H' \cap L$ are conjugate with respect to Q^*. We have seen in 6.6.1 that the latter condition is satisfied if and only if either of $H \cap L$, $H' \cap L$ contains the pole of the other relative to Q. Thus $H \in K$ if and only if either $H \in L^* = \mathcal{D}(K)$ or, by the above, $H \cap L$ is conjugate to itself, that is $H \cap L \in Q^*$ or, equivalently, $H \cap K$ is tangent to Q.

Regarding claim (b), it is clear from claim (a) that K determines the conjugation relative to Q^*. By 6.7.13, K determines Q^* and hence, by 6.6.1, also the point-quadric Q. Conversely, two hyperplanes $H, H' \subset \mathbb{P}_n$ are always conjugate with respect to K if one of them contains L, as then it is a singular hyperplane of K. Otherwise, by (a), H and H' are conjugate if and only if $H \cap L$ and $H' \cap L$ are conjugate with respect to Q^*. By 6.7.13, this shows that Q determines K. All together we have seen that $K \mapsto Q$ is an injective map. For its exhaustivity, just note that if a non-degenerate quadric Q has matrix N relative to Δ', then, as seen in the proof of claim (a), it is enough to take K with matrix

$$\begin{pmatrix} N^{-1} & 0 \\ 0 & 0 \end{pmatrix}$$

relative to $\Delta^\vee$ for having K mapped to Q. □

Remark 6.8.3. The polarity relative to Q may be described as follows: any hyperplane of L has the form $H \cap L$ for a hyperplane H of $\mathbb{P}_n$ not containing L, and then, as the reader may easily check, its pole (relative to Q) is the only point shared by all the hyperplanes H' conjugate to H with respect to K.

The quadric envelope K is usually referred to as the *envelope of Q in $\mathbb{P}_n$*. The mention of $\mathbb{P}_n$ is obligatory, as the envelope Q^* of Q (in L) has been already defined in Section 6.5 and it is different from K. In turn, Q is called the quadric *enveloped by K*, or by the elements of K.

Example 6.8.4. If $\mathbf{r}(K) = 2$, Q is a pair of distinct points, $Q = q_1 + q_2$, and conversely. Then the hyperplanes of K are those going through one of the points q_i and K is a pair of distinct hyperbundles (real or imaginary, in case $k = \mathbb{R}$), $K = q_1^* + q_2^*$, as it could be expected from 6.7.8.

Example 6.8.5. The easiest essentially new example of a degenerate quadric envelope is that of an ordinary cone K of $\mathbb{P}_3^\vee$: it has a single singular hyperplane H, on which lies the non-degenerate conic Q. The hyperplanes of K other than H are those whose intersection with H is a line tangent to Q, and the generators of K are the pencils with kernel one of these tangent lines. (See Figure 6.7.)

The relationship between the matrices of Q and its envelope in $\mathbb{P}_n$ shown in the proof of 6.8.2 is worth a separate claim for further reference:

Corollary 6.8.6 (of the proof of 6.8.2). *Assume that Q is a non-degenerate quadric of a linear variety L of $\mathbb{P}_n$, $d = \dim L > 0$, and K its envelope in $\mathbb{P}_n$. If Δ is a*

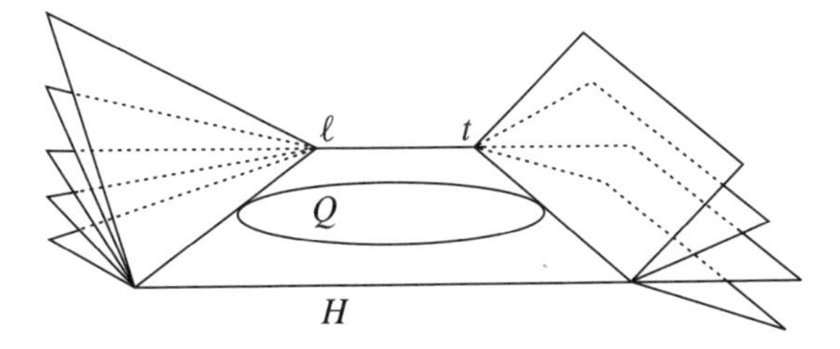

Figure 6.7. An ordinary cone of the dual three-space: there are shown its double hyperplane H, the enveloped conic Q and planes in two generators ℓ^* and t^*.

reference of $\mathbb{P}_n$ whose first $d+1$ vertices span L, Δ' the reference of L subordinated by Δ and N a matrix of Q relative to Δ', then

$$\begin{pmatrix} N^{-1} & 0 \\ 0 & 0 \end{pmatrix}$$

is a matrix of K relative to $\Delta^\vee$.

The next proposition provides a different view of the envelope in $\mathbb{P}_n$ of a non-degenerate quadric of a linear variety, and is useful for effective computations (see Exercises 6.31 and 6.32). The reader may easily give an alternative proof of it using coordinates and 6.8.6 above.

Proposition 6.8.7. *Let L be a linear variety L of $\mathbb{P}_n$, $n > \dim L > 0$, and Q a non-degenerate quadric of L. Assume that Q_0 is a non-degenerate quadric of $\mathbb{P}_n$ and take $\bar{L}$ to be the polar variety of L relative to Q_0. Then $Q = Q_0 \cap L$ if and only if L^* and $\bar{L}^*$ are supplementary and the envelope K of Q in $\mathbb{P}_n$ is the cone over $Q_0^* \cap \bar{L}^*$ with vertex L^*.*

Proof. Assume first that $Q = Q_0 \cap L$. The quadric Q being non-degenerate, Q_0 is not tangent to L; therefore, by 6.5.9, L and $\bar{L}$ are supplementary and hence so are L^* and $\bar{L}^*$. Since we already know that the hyperplanes in L^* are singular hyperplanes of K, by 6.7.2, we need only to check that $Q_0^* \cap \bar{L}^* = K \cap \bar{L}^*$. To this end, using 6.7.13, we will equivalently prove that both induce the same conjugation in $\bar{L}^*$. Assume that $H, H' \in \bar{L}^*$, after which their respective poles q, q' belong to L. H, H' are conjugate with respect to $Q_0^* \cap \bar{L}^*$ if and only if they are conjugate with respect to Q_0^*. By 6.6.1, this is equivalent to having q, q' conjugate with respect to Q_0, which in turn, since $q, q' \in L$, is the same as having q, q' conjugate with respect to $Q_0 \cap L$. Since L and $\bar{L}$ are supplementary, $H \cap L$ and $H' \cap L$ are hyperplanes of L and, clearly, q and q' are their poles relative to $Q = Q_0 \cap L$. Thus, again by 6.6.1, the latter condition is equivalent to having $H \cap L$ and $H' \cap L$ conjugate with respect to Q^* and, by 6.8.2, this is just the condition for H, H' to be conjugate with respect to K or, equivalently, with respect to $K \cap \bar{L}^*$.

Conversely, assume that L^* and $\bar{L}^*$ are supplementary and K is the cone over $Q_0^* \cap \bar{L}^*$ with vertex L^*. The first hypothesis assures that also L and $\bar{L}$ are supplementary, and this in turn that $Q_0 \cap L$ is non-degenerate, by 6.5.9 and 6.4.24. Let T and T' be two hyperplanes of L and q the pole of T (in L) relative to $Q_0 \cap L$. Clearly, $H = T \vee \bar{L}$ and $H' = T' \vee \bar{L}$ are hyperplanes of $\mathbb{P}_n$. The point q being conjugate with respect to Q_0 to all points in $\bar{L}$, it is the pole of H relative to Q_0. The hyperplanes T, T' are conjugate with respect to $(Q_0 \cap L)^*$ if and only if $q \in T'$, which in turn is equivalent to $q \in H'$, and so to being H and H' conjugate with respect to Q_0^*. Since both $H, H' \in \bar{L}^*$ and $Q_0^* \cap \bar{L}^* = K \cap \bar{L}^*$, the latter condition is satisfied if and only if H and H' are conjugate with respect to K and so, by 6.8.2, if and only if $T = H \cap L$ and $T' = H' \cap L$ are conjugate with respect to Q^*. We have thus proved that the conjugations relative to Q^* and $(Q_0 \cap L)^*$ do agree, which in turn forces $Q^* = (Q_0 \cap L)^*$ (by 6.7.12) and hence $Q = Q_0 \cap L$, as wanted. □

6.9 The absolute quadric

The usual setting for developing metric geometry is a real n-dimensional affine space $\mathbb{A}_n$, $n \geq 2$, say with associated vector space F (see Section 3.1), together with a positive-definite symmetric bilinear form K on F called the *scalar product* (or *inner product*). $K(v, w)$ is called the scalar product of the vectors v, w. Often it is written in the more traditional form $K(v, w) = vw$. The triple $(\mathbb{A}_n, F, K)$ is usually called an *Euclidean space* (*Euclidean plane* if $n = 2$). For $n = 3$ it is intended to model our physical ambient space with the added choice of a unit of length. We will see in this section how this setting may be reformulated in order to present metric geometry in terms of projective relations to it.

Instead of the affine space $\mathbb{A}_n$ and its associated vector space F, we will consider the projective closure $\bar{\mathbb{A}}_n$ of $\mathbb{A}_n$ and the improper hyperplane $H_\infty = \mathbb{P}(F)$ as introduced in Section 3.2.

The improper hyperplane H_∞ is a projective space with vector space F; therefore the scalar product K defines a quadric of H_∞ which is named the *absolute quadric* (*conic* if $n = 3$) or just the *absolute*. We will denote it by $\mathbf{K}$. Since the scalar product is assumed to be positive-definite, that is, to satisfy $K(v, v) > 0$ for all non-zero $v \in F$, it is clear that the absolute $\mathbf{K}$ has no real points; in particular it has no double points and therefore the absolute is a non-degenerate quadric of H_∞ with no real point. The (necessarily imaginary) lines joining a proper point and a point of $\mathbf{K}$ are called *isotropic lines*; the isotropic lines through a proper point p are thus the generators of the cone over $\mathbf{K}$ with vertex p, called the *isotropic cone* with vertex p.

By the definition of quadric, the absolute quadric $\mathbf{K} = [K]$ determines the scalar product K up to a non-zero factor $\rho \in \mathbb{R} - \{0\}$, which should be taken positive if one wants ρK to be also positive-definite. Thus taking the triple $(\bar{\mathbb{A}}_n, H_\infty, \mathbf{K})$

instead of $(\mathbb{A}_n, F, K)$ ignores the choice of the unit of length which comes implicit with the latter.

Remark 6.9.1. A positive-dimensional affine linear variety L is as an affine space with associated vector space its director subspace F_L. It is turned into a Euclidean space by taking the restriction $K_{|F_L}$ as the scalar product of L. Thus, if $\dim L \geq 2$, the absolute quadric of L as Euclidean space is the section of the absolute quadric of $\mathbb{A}_n$ by the improper part of L.

An immediate yet very important consequence of the definition of the absolute is the following proposition; it presents orthogonality as a particular case of conjugation relative to a quadric.

Proposition 6.9.2. *Points $p, p' \in H_\infty$ are orthogonal directions if and only if they are conjugate with respect to the absolute quadric.*

Proof. Directions $p = [v]$, $p' = [v']$, $v, v' \in F$ are called orthogonal when so are its representatives v, v', that is, when $K(v, v') = 0$, which is just the condition for p, p' to be conjugate with respect to $\mathbf{K} = [K]$. □

Remark 6.9.3. Pairwise orthogonal directions $p_1, \dots, p_m$ are independent improper points, due to the fact that the absolute has no real point: indeed, the orthogonality conditions assure that each p_i is conjugate to all p_j, $j \neq i$, with respect to K; then, in case $p_i \in \bigvee_{j \neq i} p_j$, p_i would be conjugate to itself and hence $p_i \in \mathbf{K}$.

Positive-dimensional affine linear varieties are called *orthogonal* (or *perpendicular*) if and only if any two vectors chosen in their respective director spaces are orthogonal. When this is the case also the projective closures of the affine linear varieties are called orthogonal. Since (by 3.4.4) the director subspaces of the affine linear varieties define their improper parts, 6.9.2 yields:

Corollary 6.9.4. *Two affine linear varieties of positive dimension are orthogonal if and only if the improper part of either of them is contained in the polar variety, relative to the absolute, of the improper part of the other.*

In particular two affine lines are orthogonal if and only if their improper points are conjugate with respect to the absolute.

The condition of orthogonality depending only on the improper sections, 3.4.10 clearly gives:

Corollary 6.9.5. *If two positive-dimensional affine linear varieties L_1 and L_2 are orthogonal, then any two linear varieties L'_1, L'_2, respectively parallel to L_1 and L_2 and of the same dimensions, are orthogonal too.*

If L is an affine linear variety of dimension $d > 0$, then the polar variety with respect to $\mathbf{K}$ of its improper part has dimension $n - 1 - (d - 1) - 1 = n - d - 1$

and so any linear variety L' orthogonal to L has $\dim L' \leq n - d$. If the equality holds, then the improper part of L' equals the polar variety of the improper part of L and so, in particular,

Corollary 6.9.6. *If L is an affine linear variety of dimension d, $n > d > 0$, then any two affine linear varieties of dimension $n - d$ orthogonal to L are parallel.*

Since $\mathbf{K}$ has no real points, mutually polar linear varieties of H_∞ are disjoint (by 6.5.9). Take any positive-dimensional affine linear variety L, and let S be the polar, relative to $\mathbf{K}$, of its improper part. Then S and the improper part of L are disjoint. This assures that also the projective closure $\overline{L}$ of L and S are disjoint and so, by their dimensions, supplementary linear varieties of $\overline{\mathbb{A}}_n$. The projection from S onto $\overline{L}$ is thus defined: it is called the *orthogonal projection* onto L and maps each $p \in \mathbb{A}_n$ to the intersection point of L and $S \vee p$, the finite part of the latter being, as the reader may easily see, the only linear variety of dimension $n - \dim L$ orthogonal to L and going through p. Since S is improper, the orthogonal projection is a parallel projection.

If $n \geq 3$, the hyperplanes make an exception to the definition of orthogonal varieties recalled above, as two affine hyperplanes H, H' are called *orthogonal* if and only if lines orthogonal to H and H' are themselves orthogonal; the condition obviously does not depend on the choice of the lines due to 6.9.5 and 6.9.6. Since the common improper point of the lines orthogonal to an affine hyperplane H is the pole with respect to $\mathbf{K}$ of the improper part of H, two affine hyperplanes are orthogonal if and only if their improper parts are conjugate with respect the envelope $\mathbf{K}^*$ of $\mathbf{K}$ (see 6.6.1).

In case $n = 2$ the absolute is a pair of imaginary complex-conjugate points of the improper line r_∞. These points are called the *cyclic* or *circular points* of the plane, usually denoted by I, J. For the reason of these names, see 10.1.3.

Remark 6.9.7. The cyclic points being defined as a pair, there are two possible choices of the point to be named I, the other point being then named J. We will refer to each of these choices as a *choice of the cyclic points*. Taking one or the opposite choice is often not relevant because the cyclic points play a symmetric role. However, as we will see in a short while, this is not the case when dealing with angles in an oriented plane.

In the plane case the polarity with respect to the absolute is usually viewed as the involution of the improper line with fixed points I, J (see 6.5.4): it maps each direction to the only direction orthogonal to it, and is called the *involution of orthogonality* (or *perpendicularity*). By either 6.4.11 or 5.6.8, we have the following reformulation of 6.9.2 for $n = 2$:

Corollary 6.9.8. *Lines of a plane are orthogonal if and only if their improper points harmonically divide the cyclic points.*

Fix an affine reference $\Delta = (O, e_1, \dots, e_n)$ of $\mathbb{A}_n$, denote by $\bar{\Delta}$ its associated projective reference and by $\bar{\Delta}_0$ the reference of H_∞ subordinated by $\bar{\Delta}$. The products $g_{i,j} = e_i e_j$ are the entries of the matrix G of the scalar product K, relative to the basis $e_1, \dots, e_n$. Since this basis is adapted to $\bar{\Delta}_0$ (3.3.4), G is also a matrix of $\mathbf{K}$ which therefore has equation relative to $\bar{\Delta}_0$,

$$\sum_{i,j=1}^{n} g_{i,j} x_i x_j = 0.$$

The reader may note that an equality

$$\begin{pmatrix} x_1 & \dots & x_n \end{pmatrix} G \begin{pmatrix} y_1 \\ \vdots \\ y_n \end{pmatrix} = 0$$

may be understood as the condition for the vectors $\sum_{i=0}^{n} x_i e_i$ and $\sum_{i=0}^{n} y_i e_i$ to be orthogonal, and also as the condition for the improper points $[x_1, \dots, x_n]_{\bar{\Delta}_0}$ and $[y_1, \dots, y_n]_{\bar{\Delta}_0}$ to be conjugate with respect to $\mathbf{K}$, in accordance with 6.9.2.

Recall that an affine reference $(O, e_1, \dots, e_n)$ is said to be *orthonormal* if and only if $g_{i,j} = e_i e_j = 0$ for $i \neq j$ and $g_{i,i} = e_i e_i = 1$, $i, j = 1, \dots, n$, or, equivalently, the matrix of the scalar product relative to it is the unit one. In the sequel the projective references associated to orthonormal references will be called orthonormal too. Thus, there holds:

Proposition 6.9.9. *If the reference is orthonormal, then the unit matrix is a matrix of the absolute* $\mathbf{K}$*, or, equivalently,* $\mathbf{K}$ *has equation*

$$\sum_{i=1}^{n} x_i^2 = 0.$$

Note that the converse is not true because the unit matrix is also a matrix of the absolute relative to any affine reference with $e_i e_j = 0$ and just $e_i e_i = e_j e_j$ for $i \neq j$, $i, j = 1, \dots, n$.

Remark 6.9.10. For $n = 2$ the above equation is $x_1^2 + x_2^2 = 0$ and so, once an orthonormal reference Δ of the plane has been fixed, the cyclic points are $[1, \boldsymbol{i}]_{\bar{\Delta}_0} = [0, 1, \boldsymbol{i}]_{\bar{\Delta}}$ and $[1, -\boldsymbol{i}]_{\bar{\Delta}_0} = [0, 1, -\boldsymbol{i}]_{\bar{\Delta}}$.

The angle, taken as a real number defined modulo π that measures the relative position of the directions of two lines, is one of the basic and oldest elements of metric geometry. (Having the angles defined modulo 2π requires that we consider half-lines or oriented lines, which we will not do here.) The theorem below is a cornerstone of the recovering of metric geometry from the frame of projective geometry: it relates the angle to the most important numerical projective invariant, the cross ratio. Before presenting it, let us recall that a Euclidean plane is *oriented* when one of the two possible senses of rotation has been chosen: this is usually done by choosing a certain class of orthonormal references (or orthonormal basis of the associated vector space) which are taken as the positively oriented ones. Changing the orientation turns the angles into their opposites and so, in an unoriented plane, angles are defined up to sign only. Since a Euclidean space $\mathbb{A}_n$, $n > 2$, even if oriented, does not induce orientations on its planes, angles of lines in $\mathbb{A}_n$, $n > 2$, will be always taken up to sign. The angle between two proper lines of $\overline{\mathbb{A}}_n$ will be taken as being the angle between their proper parts.

Theorem 6.9.11 (Laguerre's formula). *In an oriented Euclidean plane $\mathbb{A}_2$ there is a choice of the cyclic points I, J for which the angle between two arbitrary lines ℓ_1, ℓ_2 is*

$$\widehat{\ell_1\ell_2} = \frac{1}{2i}\log(I, J, q_1, q_2),$$

q_1, q_2 being the improper points of ℓ_1, ℓ_2, respectively.

Proof. If the lines are parallel, $q_1 = q_2$ and the claim is obvious. Thus we assume in the sequel that ℓ_1 and ℓ_2 intersect at a point O which, to fix the ideas, will be taken as the origin of an orthonormal reference Δ. To complete it, we take a unitary vector e_1 on ℓ_1 and a second unitary vector e_2, orthogonal to e_1 and such that the basis e_1, e_2 is positively oriented. If $\widehat{\ell_1\ell_2} = \alpha$, then $\cos\alpha\, e_1 + \sin\alpha\, e_2$ is a vector on ℓ_2. Taking the reference subordinated by $\bar{\Delta}$ in the improper line, we have (by 3.3.2 (b)):

$$q_1 = [1, 0], \quad q_2 = [\cos\alpha, \sin\alpha].$$

The reference being orthonormal, by 6.9.10 we may take

$$I = [1, -\boldsymbol{i}], \quad J = [1, \boldsymbol{i}].$$

Then a straightforward computation shows that

$$(I, J, q_1, q_2) = \frac{\cos\alpha + \boldsymbol{i}\,\sin\alpha}{\cos\alpha - \boldsymbol{i}\,\sin\alpha} = e^{2\alpha\boldsymbol{i}}$$

and the claimed equality follows. □

Remark 6.9.12. The logarithm of a complex number being determined up to an integral multiple of $2\pi\boldsymbol{i}$, the expression on the right of Laguerre's formula 6.9.11 is determined up to an integral multiple of π, as is determined, on the other hand, the angle between two lines.

Remark 6.9.13. As one may expect, lines ℓ_1, ℓ_2 are orthogonal if and only if $\widehat{\ell_1\ell_2} = \pi/2$.

Remark 6.9.14. It is clear from Laguerre's formula that $\widehat{\ell_1\ell_2} = -\widehat{\ell_2\ell_1}$ and $\widehat{\ell_1\ell_2} + \widehat{\ell_2\ell_3} = \widehat{\ell_1\ell_3}$, the latter using 2.9.7.

Remark 6.9.15. Swapping over the cyclic points I, J turns the term on the right of Laguerre's formula into its opposite. Therefore taking the opposite choice of the cyclic points would give a correct formula if the orientation of the plane is reversed. In this way each choice of the cyclic points corresponds to one of the two possible orientations of the plane.

Remark 6.9.16. If the plane is not oriented, then Laguerre's formula still holds provided the equality is taken up to sign. In such a case the choice of the cyclic points is, of course, irrelevant.

Corollary 6.9.17. *Assume that ℓ_1, ℓ_2 are two non-parallel lines of a Euclidean space $\mathbb{A}_n$, $n > 2$, with improper points q_1, q_2, respectively. Then, up to sign,*

$$\widehat{\ell_1\ell_2} = \frac{1}{2\boldsymbol{i}} \log(I, J, q_1, q_2),$$

where the pair $I + J$ is the section of the absolute by the line $s = q_1q_2$.

Proof. Since replacing one of the lines with a parallel one does not change its improper point or the angle with the other line, we may assume that ℓ_1 and ℓ_2 have a common point and therefore span a plane L. Then I, J are the cyclic points of L (6.9.1) and therefore the claim follows from Laguerre's formula 6.9.11 in its unoriented version (6.9.16). □

When the angles of the lines of $\mathbb{A}_n$ are taken up to sign, either because $n > 2$ or $n = 2$ and the plane is not oriented, it is useful having a formula symmetric in I, J:

Corollary 6.9.18. *If ℓ_1, ℓ_2 are two non-parallel lines of a Euclidean space $\mathbb{A}_n$, $n \geq 2$, with respective improper points q_1, q_2, then*

$$\widehat{\ell_1\ell_2} = \frac{1}{2} \arccos\left(\frac{(I, J, q_1, q_2) + (J, I, q_1, q_2)}{2}\right),$$

where the pair $I + J$ is the section of the absolute by the line $s = q_1q_2$, or just the pair of cyclic points if $n = 2$.

Proof. Put $\widehat{\ell_1\ell_2} = \alpha$. Up to turning it into its opposite, by 6.9.11 or 6.7.12

$$\cos 2\alpha + \boldsymbol{i} \sin 2\alpha = e^{2\alpha\boldsymbol{i}} = (I, J, q_1, q_2),$$

and so

$$\cos 2\alpha - \boldsymbol{i} \sin 2\alpha = e^{-2\alpha\boldsymbol{i}} = (J, I, q_1, q_2).$$

The claim follows by adding up these equalities. □

The angle between two directions p, p' is taken as the angle of any pair of lines with improper points p, p', the choice of the lines being obviously irrelevant. Next is a useful consequence of Laguerre's formula; being just a particular case of 5.5.16, it requires no proof.

Corollary 6.9.19. *Once $\alpha \in \mathbb{R}$, $0 < \alpha < \pi$, is fixed, mapping each direction of an oriented Euclidean plane $\mathbb{A}_2$ to the direction making an angle α with it, is a projectivity of the improper line of $\mathbb{A}_2$ which has fixed points I, J and invariant $e^{2\alpha i}$.*

Assume again that ℓ_1, ℓ_2 are lines of a Euclidean plane $\mathbb{A}_2$, and assume also that they are not parallel, and so they intersect at a proper point p. It is known, from elementary geometry, that there are two different lines b_1, b_2 through p, called the *bisector lines* (or just the *bisectors*) of ℓ_1, ℓ_2, that satisfy $\widehat{\ell_1 b_i} = \widehat{b_i \ell_2}$, $i = 1, 2$, all angles being taken according to the same orientation of the plane and the bisectors themselves being independent of the choice of the orientation. When working with proper linear varieties instead of the affine ones, we will sometimes refer to the projective closures $\bar{b}_1$, $\bar{b}_2$ as the bisectors of $\bar{\ell}_1$, $\bar{\ell}_2$. Next is a useful characterization of the bisectors that will be proved using Laguerre's formula.

Corollary 6.9.20. *Assume that ℓ_1, ℓ_2 are two different lines of a Euclidean plane meeting at a point p. Two lines b_1, b_2 through p are the bisectors of ℓ_1, ℓ_2 if and only if they are orthogonal and their improper points harmonically divide those of ℓ_1, ℓ_2.*

Proof. Write p_1, p_2, q_1, q_2 the improper points of b_1, b_2, ℓ_1, ℓ_2, respectively. The conditions of the claim are

$$(I, J, p_1, p_2) = -1 \quad \text{and} \quad (q_1, q_2, p_1, p_2) = -1. \tag{6.13}$$

Take σ to be the involution of the complex extension of the improper line that has fixed points p_1, p_2. If the conditions (6.13) are satisfied, then, by 5.6.9, σ has $\{I, J\}$ and $\{q_1, q_2\}$ as corresponding pairs. Therefore, by applying σ and reordering the points,

$$(I, J, q_1, p_i) = (J, I, q_2, p_i) = (I, J, p_i, q_2),$$

$i = 1, 2$, giving the wanted equalities of angles by Laguerre's formula. Furthermore $b_1 \neq b_2$ because they are assumed to be orthogonal.

Conversely, take τ to be the involution of the complex extension of the improper line which has $\{I, J\}$ and $\{q_1, q_2\}$ as corresponding pairs. Using first the equality of angles, applying then τ and reordering,

$$(I, J, q_1, p_i) = (I, J, p_i, q_2) = (J, I, \tau(p_i), q_1) = (I, J, q_1, \tau(p_i)).$$

It follows that $p_i = \tau(p_i)$, $i = 1, 2$. The bisectors b_1 and b_2 being assumed to be different, we have in addition $p_1 \neq p_2$ and so p_1, p_2 is the pair of fixed points of τ: after this, again by 5.6.9, the equalities (6.13) are clear. □

Corollary 6.9.20 may be equivalently stated by saying that a pair of lines $b_1, b_2 \in p^*$ are bisectors of ℓ_1, ℓ_2 if and only if their improper points correspond by both the involution of orthogonality and the involution whose fixed points are the improper points of ℓ_1, ℓ_2 (5.6.9). We will see in the forthcoming chapter 8 that given two different involutions of a real $\mathbb{P}_1$, one at least with imaginary fixed points, there is a unique pair of points corresponding by both involutions (8.2.11). This fact may thus be used to re-prove the existence and uniqueness of the pair of bisectors.

The angles being taken up to sign, affinities of a Euclidean space that preserve angles are named *similarities*. They are characterized by the property of leaving the absolute quadric invariant:

Theorem 6.9.21. *Let $f : \bar{\mathbb{A}}_n \to \bar{\mathbb{A}}_n$ be an affinity of a Euclidean space $\mathbb{A}_n$, $n \geq 2$. The angles being taken up to sign, the following conditions are equivalent:*

(i) *For any two non-parallel lines ℓ_1, ℓ_2 of $\mathbb{A}_n$, $\widehat{\ell_1\ell_2} = \widehat{f(\ell_1)f(\ell_2)}$.*

(ii) *Lines ℓ_1, ℓ_2 of $\mathbb{A}_n$ are orthogonal if and only if $f(\ell_1)$, $f(\ell_2)$ are orthogonal.*

(iii) $f_{|H_\infty}(\mathbf{K}) = \mathbf{K}$.

Proof. (i) $\Rightarrow$ (ii) is clear by 6.9.13. If (ii) is verified, then, by 6.9.2, $f_{|H_\infty}$ preserves the conjugation relative to $\mathbf{K}$ and therefore, by 6.7.13, leaves $\mathbf{K}$ invariant. If (iii) is satisfied, take any two distinct improper points q_1, q_2. If $I + J$ is the section of $\mathbf{K}$ by the line q_1q_2, then $f(I) + f(J)$ is the section of $\mathbf{K}$ by $f(q_1)f(q_2)$. After this, the invariance of the cross ratio (2.9.11) and 6.9.18 above prove (i). □

Remark 6.9.22. In case $n = 2$ and assuming $\mathbb{A}_2$ oriented, the condition (iii) just says that $\{I, J\} = \{f(I), f(J)\}$; so f may either keep invariant or swap over the cyclic points. In the first case, by Laguerre's formula, the angles are kept invariant, while in the other,

$$\widehat{\ell_1\ell_2} = \frac{1}{2i}\log(I, J, q_1, q_2) = \frac{1}{2i}\log(J, I, f(q_1), f(q_2)) = -\widehat{f(\ell_1)f(\ell_2)}.$$

A reflection $(X_1, X_2) \mapsto (X_1, -X_2)$, where X_1, X_2 are affine coordinates relative to an orthonormal reference, provides an easy example of the latter possibility.

Assume that $f : \bar{\mathbb{A}}_n \to \bar{\mathbb{A}}_n$ is an affinity and $\psi : F \to F$ its associated morphism. We know from 3.5.3 that ψ represents $f_{|H_\infty}$ and so, if η is a representative of the absolute $\mathbf{K}$, then $\eta \circ (\psi^{-1} \times \psi^{-1})$ is a representative of $f_{|H_\infty}(\mathbf{K})$. Thus, the affinity f leaves the absolute invariant (that is, f is a similarity) if and only if

$$\eta \circ (\psi^{-1} \times \psi^{-1}) = \lambda\eta \qquad (6.14)$$

or, equivalently,

$$\eta = \lambda\eta \circ (\psi \times \psi),$$

for a representative η of $\mathbf{K}$ and a certain $\lambda \in \mathbb{R} - \{0\}$. If this is the case, then the same condition, with the same λ, is obviously satisfied by all representatives of $\mathbf{K}$. So, by taking $\rho = \lambda^{-1}$, we quote, for future reference:

Lemma 6.9.23. *If an affinity f, with associated morphism ψ, is a similarity, then there is $\rho \in \mathbb{R} - \{0\}$ such that*

$$\eta \circ (\psi \times \psi) = \rho\eta,$$

for any representative η of $\mathbf{K}$. *Conversely, if the above equality is satisfied for certain $\rho \in \mathbb{R} - \{0\}$ and a single representative η of* $\mathbf{K}$, *then f is a similarity.*

The real number ρ of 6.9.23 has an interesting meaning. First note that taking $\eta = K$ and any $v \in F - \{0\}$, the equality of 6.9.23 gives

$$K(\psi(v), \psi(v)) = \rho K(v, v).$$

The scalar product being positive-definite, it follows that $\rho > 0$ and we may take $\delta = \sqrt{\rho}$. On the other hand if p, q are proper points and $v = p - q$, then $f(p) - f(q) = \psi(v)$ and the distance between $f(p)$ and $f(q)$ is

$$d(f(p), f(q)) = K(\psi(v), \psi(v))^{1/2} = \delta K(v, v)^{1/2} = \delta d(p, q).$$

In other words, distances are in general not preserved by a similarity, but are all multiplied by the same positive factor, which is named the *ratio* of the similarity.

Let us go back to Lemma 6.9.23: it shows that the similarities leave invariant the set of representatives of the absolute, but not the representatives themselves: they, according to the equality (6.14), are divided by the square of the ratio of the similarity. The affinities leaving invariant the representatives of $\mathbf{K}$ are the motions. In a more formal way, an affinity f, with associated morphism ψ, is a *motion* if and only if

$$\eta \circ (\psi^{-1} \times \psi^{-1}) = \eta \tag{6.15}$$

for one or, equivalently, for all representatives η of $\mathbf{K}$. Clearly, any motion is a similarity of ratio one and conversely.

Remark 6.9.24. Assume that an affinity f of a Euclidean space $\mathbb{A}_n$ has equations

$$\begin{aligned} Y_1 &= b_1 + a_1^1 X_1 + \cdots + a_n^1 X_n, \\ &\vdots \\ Y_n &= b_n + a_n^1 X_1 + \cdots + a_n^n X_n \end{aligned}$$

relative to an orthonormal reference. Then $N = (a_i^j)_{i,j=1,\dots,n}$ is the matrix of the morphism associated to f and the unit matrix $\mathbf{1_n}$ is the matrix of the scalar product. The equality (6.15) translates thus into

$$(N^{-1})^t \mathbf{1}_n N^{-1} = \mathbf{1}_n,$$

after which there results a usual characterization of motions: f is a motion if and only if $N^t N = \mathbf{1}_n$, that is, N is an orthogonal matrix.

To complete the picture, we will prove the following easy proposition that shows the equivalence between the usual definitions of similarity and motion, and the present ones. As above, $d(p,q)$ indicates the distance between the points p, q.

Proposition 6.9.25. (a) *An affinity f of a Euclidean space is a similarity if and only if for any pair of proper points p, q,*

$$d(f(p), f(q)) = \delta d(p,q)$$

for a fixed $\delta \in \mathbb{R}^+$.

(b) *An affinity f of a Euclidean space is a motion if and only if for any pair of proper points p, q,*

$$d(f(p), f(q)) = d(p,q).$$

Proof. The *only if* parts have been seen above. If an affinity f, with associated morphism ψ, satisfies $d(f(p), f(q)) = \delta d(p,q)$ for all p, q, then it also satisfies $K(\psi(v), \psi(v)) = \delta^2 K(v,v)$ for all $v \in F$. By using the bilinearity of K, in the form

$$K(u,w) = \frac{1}{2}(K(u+w, u+w) - K(u,u) - K(w,w)),$$

one gets $K \circ (\psi \times \psi) = \delta^2 K$ and f is a similarity by 6.9.23. The same argument with $\delta = 1$ completes the proof of claim (b). □

Remark 6.9.26. It is straightforward from the definitions, and also from 6.9.25, that both the set of all similarities and the set of all motions of $\mathbb{A}_n$, are subgroups of the affine group of $\mathbb{A}_n$.

Remark 6.9.27. The similarities and motions of a Euclidean space $\mathbb{A}_n$ remain the same after making a different choice of the unit of length, that is, after taking ρK, $\rho > 0$, instead of K as the scalar product, as their definitions make use of the absolute quadric and its representatives, but not of the scalar product itself.

Affinities between different Euclidean spaces preserving angles or distances are also considered. Affinities $f\colon \bar{\mathbb{A}}_n \to \bar{\mathbb{A}}'_n$ preserving angles are still called *similarities*. If $\mathbb{A}'_n \neq \mathbb{A}_n$, they do not form a group because their composition is not defined. Besides this fact, there is no great difference with the case $\mathbb{A}'_n = \mathbb{A}_n$. In particular the next characterization is similar to 6.9.21 and may be proved by the reader using similar arguments.

Proposition 6.9.28. *Let $f\colon \bar{\mathbb{A}}_n \to \bar{\mathbb{A}}'_n$ be an affinity between Euclidean spaces $\mathbb{A}_n$ and $\mathbb{A}'_n$, $n \geq 2$. The angles being taken up to sign, if $\mathbf{K}$ and $\mathbf{K}'$ are the absolute quadrics of $\mathbb{A}_n$ and $\mathbb{A}'_n$, then the following conditions are equivalent:*

(i) *For any two non-parallel lines* ℓ_1, ℓ_2 *of* $\mathbb{A}_n$, $\widehat{\ell_1\ell_2} = \widehat{f(\ell_1)f(\ell_2)}$.

(ii) *Lines* ℓ_1, ℓ_2 *of* $\mathbb{A}_n$ *are orthogonal if and only if* $f(\ell_1), f(\ell_2)$ *are orthogonal.*

(iii) $f_{|H_\infty}(\mathbf{K}) = \mathbf{K}'$.

However, the condition of leaving invariant one or all representatives of the absolute, already used to define motions in the case $\mathbb{A}'_n = \mathbb{A}_n$, does not apply to the case of different spaces; then, the use of the scalar products K, of $\mathbb{A}_n$, and K', of $\mathbb{A}'_n$ is required: keeping the name *motion* for the case $\mathbb{A}'_n = \mathbb{A}_n$ only, an affinity $f: \overline{\mathbb{A}}_n \to \overline{\mathbb{A}}'_n$ is called an *isometry* if and only if its associated morphism ψ transforms K into K', namely

$$K \circ (\psi^{-1} \times \psi^{-1}) = K'.$$

An isometry between coincident affine spaces is thus a motion, and conversely. Worth noting is the fact that in the case of different spaces, the definition of isometry depends not only on the absolute quadrics, but also on the scalar products themselves: it is clear that then an isometry no longer retains its condition after making different choices of the units of length, but for the cases in which the ratios between new and old unities are the same in both spaces.

Remark 6.9.29. It is clear from the definition that the identity map of the projective closure of any Euclidean space is an isometry, and also that compositions and inverses of isometries are isometries too.

The next proposition asserts that the condition used above to define isometries and the usual one of preserving distances are equivalent. It adds further equivalent conditions that will be useful in the sequel.

Proposition 6.9.30. *If* f *is an affinity between projective closures of Euclidean spaces* $\overline{\mathbb{A}}_n$ *and* $\overline{\mathbb{A}}'_n$, *then the following conditions are equivalent;*

(i) f *is an isometry.*

(ii) *For any* $p, q \in \overline{\mathbb{A}}_n$, $d(p,q) = d(f(p), f(q))$.

(iii) *The image by* f *of any orthonormal reference of* $\overline{\mathbb{A}}_n$ *is orthonormal.*

(iv) *The image by* f *of an orthonormal reference of* $\overline{\mathbb{A}}_n$ *is orthonormal.*

Proof. The equivalence of (i) and (ii) follows from the argument used to prove 6.9.25. (i) $\Rightarrow$ (iii) is clear from the definition of isometry and (iii) $\Rightarrow$ (iv) is obvious. To close, for (iv) $\Rightarrow$ (i), if Δ and $f(\Delta)$ are orthonormal, then, on one hand the unit matrix is a matrix of f relative to these references (by 2.8.1 (b)), while on the other both scalar products have matrix equal to the unit one. After this the equality $K \circ (\psi^{-1} \times \psi^{-1}) = K'$ defining isometries is clearly satisfied. □

Metric geometry may be presented within the projective frame by considering the projective spaces that are projective closures of Euclidean spaces and the objects of projective families belonging to these spaces. Isometries being projectivities between projective closures of Euclidean spaces, their action on these objects is defined, and two of them are called *congruent* if and only if there is an isometry mapping one into the other. Congruence is an equivalence relation because of 6.9.29 above. Invariants by the action of the isometries are called *metric invariants* and classifications by congruence, *metric classifications*. Systems of metric invariants that determine the congruence classes of the objects are called *complete*. Relations and properties kept invariant by the action of the isometries are also called *metric*.

6.10 Exercises

Exercise 6.1. Relative to the quadric of $\mathbb{P}_3$

$$Q : 6x_0^2 - x_0x_1 - 2x_0x_2 - 3x_0x_3 + x_1x_2 + x_1x_3 + x_2x_3,$$

compute:

(1) The tangent planes at the intersection points with the edge $x_2 = x_3 = 0$ of the reference.

(2) The polar plane of the point $[2, 1, 1]$.

(3) The polar line of the line $\ell : x_1 = -2x_2 = 2x_3$, as well as the tangent planes containing ℓ.

(4) The tangent cone with vertex the first vertex of the reference.

Exercise 6.2. Prove that a quadric Q, with equation $\sum_{1,j=1}^{n} a_{i,j}x_ix_j = 0$, goes through the vertices of the reference if and only if $a_{i,i} = 0$, $i = 0, \dots, n$. This being satisfied, compute the tangent hyperplane to Q at each vertex.

Exercise 6.3. Let ℓ and $Q = [\eta]$ be, respectively, a line and a quadric of $\mathbb{P}_n$, and assume $Q \cap \ell = q_1 + q_2$, $q_1, q_2 \in \ell$. Take two different points $p_1 = [v_1]$ and $p_2 = [v_2]$ of ℓ, $p_i \notin Q$, $i = 1, 2$, and $p = [v_1 - v_2]$. Prove that

$$(p_1, p_2, q_1, p)(p_1, p_2, q_2, p) = \frac{\eta(v_1, v_1)}{\eta(v_2, v_2)}.$$

Hint: Use 2.9.7 and equation (6.5).

Exercise 6.4. Assume that $a = (a_0, \dots, a_n)$ and $b = (b_0, \dots, b_n)$ are (row) coordinate vectors of two hyperplanes H and H' of $\mathbb{P}_n$. Prove that $a^t b + b^t a$ is a matrix of the hyperplane-pair $H + H'$.

Exercise 6.5. Write down the equations of all quadrics of $\mathbb{P}_3$ which contain the line ℓ joining the vertices p_0, p_1 of the reference. For any such a quadric Q, study the correspondence $p \mapsto H_{p,Q}$, between ℓ and ℓ^* in terms of the double points of Q lying on ℓ.

Exercise 6.6. Prove that if η is a symmetric bilinear form defining a quadric of $\mathbb{P}_{n,\mathbb{R}}$ with no real points, then η is either positive definite or negative definite (that is, all values $\eta(v,v)$, for v a non-zero vector, are non-zero and have the same sign). *Hint*: Use 6.3.5.

Exercise 6.7 (*Incidence quadric of a correlation*). Let g be a correlation of $\mathbb{P}_n$, not a null system. Assume that, relative to a reference Δ, g has matrix N. Prove that $N + N^t$ is a non-zero symmetric matrix. We associate to g the quadric $I(g)$ with matrix $N + N^t$. $I(g)$ is called the *incidence quadric* of g. Prove that:

(a) $I(g)$ is independent of the choice of the reference Δ and the matrix N of g.

(b) The set of points of $I(g)$ is the incidence set of g. *Hint*: Use that $x^t N^t x = (x^t N^t x)^t = x^t N x$ for any column vector $x = (x_0, \dots, x_n)^t$.

Exercise 6.8. Prove that the polar hyperplane of a point p relative to a pair of distinct hyperplanes $H_1 + H_2$, $p \notin H_1 + H_2$, is the fourth harmonic of $H_1, H_2, p \vee (H_1 \cap H_2)$ (in $(H_1 \cap H_2)^*$).

Exercise 6.9. Prove that if a non-degenerate quadric Q has points, then $\bigcap_{p\in Q} H_p = \emptyset$ *Hint*: Use 6.4.31.

Exercise 6.10. In a projective plane $\mathbb{P}_2$:

(a) Prove that a non-degenerate conic C is tangent to the sides of the reference if and only if its equation may be written in the form

$$a^2x_0^2 + b^2x_1^2 + c^2x_2^2 - 2abx_0x_1 - 2acx_0x_2 - 2bcx_1x_2 = 0,$$

for some $a, b, c \in k - \{0\}$. Compute the contact points with the sides and prove that the lines joining each vertex to the contact point on the opposite side are concurrent. *Warning*: The non-degeneracy condition is essential, degenerate conics satisfying the same tangency conditions have other equations.

(b) Prove that there is a non-degenerate conic tangent to the three sides of a triangle at given contact points if and only if no contact point is a vertex and the lines joining each vertex to the contact point on the opposite side are concurrent (*Ceva*). Write down the dual claim (*Carnot*).

Exercise 6.11. Assume that, in a projective plane, a non-degenerate conic C is tangent to the three sides of a triangle T. Prove that:

(1) No contact point of a side of T is a vertex of T.

(2) The intersection point p, of one of the sides ℓ of T and the line spanned by the contact points of the other two sides of T, is the pole of the line joining the contact point of ℓ and the vertex opposite ℓ.

(3) The above point p and the contact point of ℓ harmonically divide the vertices of T on ℓ.

Exercise 6.12. In $\mathbb{P}_2$, let T be a triangle whose vertices belong to a non-degenerate conic C, and denote by T' the triangle whose sides are the tangents to C at the vertices of T. Prove that there is a homology f of $\mathbb{P}_2$ such that:

(a) $f(T) = T'$.

(b) The axis of f is the polar relative to C of its centre.

Hint: Either compute using Exercise 6.2 or directly apply Exercise 6.10 (b).

Exercise 6.13. Let C be a non-degenerate conic and ℓ, ℓ' two distinct lines of $\mathbb{P}_2$, ℓ' not containing the pole of ℓ. Define a correspondence f between ℓ and ℓ' by taking two points as correspondent if and only if they are conjugate with respect to C. Prove that f is a projectivity, $f : \ell \to \ell'$. Prove that f is a perspectivity if and only if the intersection point of ℓ and ℓ' belongs to C. If this is the case, determine the centre of f.

Exercise 6.14 (*Hesse's theorem*). Use coordinates to prove that if two pairs of opposite vertices of a quadrilateral of $\mathbb{P}_2$ are conjugate with respect to a conic, then so is the third pair. (See Exercise 9.41 for a different proof.) *Hint*: Take two pairs of opposite vertices to compose the reference.

Exercise 6.15. Assume that a point p belongs to a chord of a non-degenerate conic C with ends q_1, q_2, both different from p, and that p' is a point conjugate to p. Assume $p'q_1 \cap C = \{q_1, q_1'\}, q_1 \neq q_1'$, and $p'q_2 \cap C = \{q_2, q_2'\}, q_2 \neq q_2'$. Prove that q_1', q_2' and p are aligned. *Hint*: Use 6.4.27.

Exercise 6.16 (*Seydewitz's theorem*). Assume that the vertices of a triangle T of $\mathbb{P}_2$ belong to a non-degenerate conic C. Prove that for any point p on a side of T, the intersections of the polar of p with the other two sides of T are points conjugate with respect to C. *Hint*: Either compute using Exercise 6.2 or use Exercise 6.13.

Exercise 6.17. Let T and T' be triangles of $\mathbb{P}_2$ sharing no vertex or side. Prove that T and T' are homological triangles if and only if there is a non-degenerate conic C of $\mathbb{P}_2$ relative to which the vertices of $\mathbb{P}_2$ have as polars the sides of T' (*Chasles*). Compare with Exercise 6.12. *Hint*: Take T as the triangle of the reference.

Exercise 6.18. Use 6.4.24 and 6.5.1 to see that if a quadric Q of $\mathbb{P}_3$ has a double line as plane section, then Q is degenerate. Generalize to higher-dimensional quadrics.

Exercise 6.19. If p is a point of a non-degenerate quadric Q of $\mathbb{P}_3$, prove that mapping each line to its polar induces an involution σ of the pencil of lines tangent to Q at p. Prove that the lines fixed by σ are the lines through p contained in Q.

Exercise 6.20. Assume that N is a matrix of an ordinary cone Q of $\mathbb{P}_3$. Prove that any non-zero column of the matrix of cofactors $\bar{N}$ of N is a coordinate vector of the vertex of Q. *Hint*: Prove first that due to being $\det N = 0$, it is $N\bar{N} = 0$.

Exercise 6.21. Prove that the hyperplanes whose coordinate vectors are the non-zero rows of a fixed matrix of a quadric Q, span the bundle $\mathcal{D}(Q)^*$.

Exercise 6.22. Assume that any two different vertices of a tetrahedron T of $\mathbb{P}_3$ are conjugate with respect to a non-degenerate quadric Q. Prove that if ℓ, ℓ' are opposite edges of T, then any chord s of Q meeting ℓ and ℓ' has its ends harmonically dividing $s \cap \ell$, $s \cap \ell'$. *Hint*: Prove first that ℓ and ℓ' are mutually polar lines, and then use 6.4.2

Exercise 6.23. Notations and hypothesis being as in 6.3.3, assume that (λ_1, μ_1) and (λ_2, μ_2) are, up to proportionality, the solutions of equation (6.5).

(a) Prove that the cross ratio of $p = [v]$, $q = [w]$ and the pair of intersections of Q and ℓ, if defined, is

$$\rho = \left(\frac{\lambda_1}{\lambda_2} : \frac{\mu_1}{\mu_2}\right)^{\pm 1},$$

the exponent depending on the ordering on the latter pair.

(b) Express $\rho + 1/\rho$ as a rational function of $\eta(v, v)$, $\eta(v, w)$, $\eta(w, w)$.

(c) Assume $p \notin Q$ fixed and prove that the points q for which the above ρ equals an already fixed value ρ_0, or is undefined, are the points of a quadric which contains $Q \cap H_p$ and is tangent to Q along it. Prove that this quadric is $2H_p$ in case $\rho_0 = -1$, the tangent cone to Q with vertex p in case $\rho_0 = 1$ and Q itself if $\rho_0 = 0, \infty$.

Exercise 6.24. In a projective space $\mathbb{P}_3$:

(a) Prove that a non-degenerate quadric Q is tangent to the edges of the reference if and only if its equation may be written in the form

$$\begin{aligned} a^2x_0^2 + b^2x_1^2 + c^2x_2^2 + d^2x_4^2 - 2abx_0x_1 - 2acx_0x_2 \\ - 2ax_0x_3 - 2bcx_1x_2 - 2bdx_1x_3 - 2cdx_2x_3 = 0, \end{aligned}$$

for some $a, b, c, d \in k - \{0\}$. Compute the contact points with the edges and prove that the lines joining contact points on opposite edges are concurrent. *Hint*: Compute as for Exercise 6.10 and use Exercise 6.18.

(b) Prove that there is a non-degenerate quadric tangent to the six edges of a tetrahedron at given contact points if and only if no contact point is a vertex and the three lines joining pairs of contact points on opposite edges are concurrent.

Exercise 6.25. Prove that the matrix of cofactors $\overline{N}$, of a matrix N of an ordinary cone of $\mathbb{P}_3$ with vertex O, is, in dual coordinates, a matrix of $2O^*$. *Hint*: Use Exercise 6.20 to prove that any $H \in O^*$ is double for the quadric envelope with matrix $\overline{N}$.

Exercise 6.26. Assume that Q is a quadric with $\mathbf{r}(Q) > 1$ and p a point, $p \notin Q$. Prove that $H_{q,Q} = H_{q,TC_p(Q)}$ for all $q \in H_{p,Q}$, and so, in particular, that a point $q \in Q \cap H_{p,Q}$ is simple for Q if and only if it is simple for $TC_p(Q)$ and, in such a case, both quadrics have the same tangent hyperplane at q.

Exercise 6.27. Let Q and Q' be non-degenerate quadrics and H a hyperplane of $\mathbb{P}_n$, $n \geq 2$. Assume that the sections $Q \cap H$ and $Q' \cap H$ are coincident and non-degenerate, H being thus tangent to neither of the quadrics. Put $Q \cap H = Q' \cap H = Q_H$.

(a) Prove that if Q and Q' are tangent along their common section Q_H by H, then they are tangent at all points of Q_H

(b) Use 6.4.32 to prove that the converse is true if Q_H has points.

Exercise 6.28. If Q is a degenerate quadric and $\mathbf{r}(Q) > 1$, prove that the hyperplanes tangent to Q at simple points p, q agree if and only if p and q belong to the same generator of Q.

Exercise 6.29. Let $g = g(x_0, \dots, x_3) = 0$ and $h = h(x_0, \dots, x_3) = 0$ be equations of an ordinary cone K and a plane π of $\mathbb{P}_3$, respectively. Assume that the double point q of K does not belong to π. Prove that $g + h^2 = 0$ is an equation of a non-degenerate quadric Q of $\mathbb{P}_3$ which has K as its tangent cone along $Q \cap \pi$. *Hint*: Assume that p is a double point of Q and use 6.4.30 to prove that either $p \in \pi$ and then $p = q$, or $p \notin \pi$ and $H_{p,K} = \pi$, both against $p \notin \pi$. Example 9.5.22 provides a different approach.

Exercise 6.30. In a projective space $\mathbb{P}_n$, $n \geq 2$, the image of a non-degenerate quadric Q by the polarity relative to a non-degenerate quadric K envelops a third non-degenerate quadric Q' which is called the *polar* of Q relative to (or with *directrix*) K. Compute a matrix of Q' from matrices of Q and K and use it to prove that the points of Q' are the poles relative to K of the hyperplanes tangent to Q, and also that the polar of Q' relative to K is Q.

Exercise 6.31 (*Computing the conic enveloped by a cone of* $\mathbb{P}_3^\vee$). Assume that K is a rank-three quadric envelope of $\mathbb{P}_3$ with singular plane π.

(a) Prove that any non-zero row of the matrix of cofactors of a matrix of K is a coordinate vector of π (see Exercise 6.20).

(b) Fix any $p = [a_0, a_1, a_2, a_3] \in \mathbb{P}_3 - \pi$ and assume that $g = g(u_0, u_1, u_2, u_3) = 0$ is an equation of K. Prove that

$$(a_0u_0 + a_1u_1 + a_2u_2 + a_3u_3)^2 + g = 0$$

is the equation of a non-degenerate quadric envelope Q^* such that $Q \cap \pi$ is the conic enveloped by K. *Hint*: Use 6.8.7 and Exercise 6.29.

(c) Use the above to determine the conic enveloped by the quadric envelope with equation

$$u_0^2 + u_1^2 + u_2^2 + 3u_3^2 + 2u_0u_3 + 2u_1u_3 + 2u_2u_3 = 0.$$

Exercise 6.32 (*Computing the envelope of a conic in* $\mathbb{P}_3$). Assume that a non-degenerate conic C on a plane π of $\mathbb{P}_3$ is $C = Q \cap \pi$, where Q is a non-degenerate quadric of $\mathbb{P}_3$. Prove that the envelope of C is $TC_\pi(Q^*)$. Use this fact to compute an equation of the envelope of the intersection of the quadric

$$x_0^2 + x_1^2 - x_2^2 - x_3^2 = 0$$

and the plane

$$x_1 + x_2 + x_3 = 0.$$

Hint: Use 6.8.7.

Exercise 6.33. Prove that if two non-degenerate quadrics Q, Q' contain a line ℓ and are tangent at three distinct points of ℓ, then they are tangent at all points of ℓ. *Hint*: Prove that taking $p, q \in \ell$ as correspondent if and only if $H_{p,Q} = H_{q,Q'}$ is a projectivity of ℓ.

Exercise 6.34. Use the conjugation relative to the absolute to prove the following claim (*theorem of the three perpendiculars*): If π is a plane of a Euclidian space $\mathbb{A}_3$, ℓ a line perpendicular to π, s a line of π through the foot $q = \ell \cap \pi$ of ℓ, t a line of π perpendicular to s and missing q, and $p = s \cap t$, then for any point $p' \in \ell$, the line $p'p$ is perpendicular to t.

Exercise 6.35. Use Exercises 3.2 and 6.14 to prove that if two pairs of opposite edges of a tetrahedron (with proper vertices) of a Euclidean three-space are orthogonal, then so is the third pair.

Exercise 6.36. Prove that if non-parallel lines of a Euclidean plane have equations relative to an orthonormal reference

$$ax + by + c = 0 \quad \text{and} \quad \bar{a}x + \bar{b}y + \bar{c} = 0,$$

with $a^2 + b^2 = \bar{a}^2 + \bar{b}^2 = 1$, then their bisectors have equations

$$(a + \bar{a})x + (b + \bar{b})y + c + \bar{c} = 0 \quad \text{and} \quad (a - \bar{a})x + (b - \bar{b})y + c - \bar{c} = 0.$$

Exercise 6.37 (*Trilinear coordinates*). Let T be a triangle of a Euclidean plane $\mathbb{A}_2$, with (proper) sides s_i, $i = 0, 1, 2$. Denote by Γ_i and Γ'_i the open half-planes determined by s_i, Γ_i being the one containing the interior of T. Assume to have fixed orthonormal coordinates and, for $i = 0, 1, 2$, let

$$a_i X + b_i Y + c_i = 0$$

be the equation of s_i whose first member takes positive values on Γ_i and has $a_i^2 + b_i^2 = 1$. Let x_0, x_1, x_2 be the homogeneous coordinates associated to the affine coordinates X, Y ($X = x_1/x_0, Y = x_2/x_0$) and, for any $p = [x_0, x_1, x_2]$, take

$$\begin{aligned} y_0 &= c_0 x_0 + a_0 x_1 + b_0 x_2, \\ y_1 &= c_1 x_0 + a_1 x_1 + b_1 x_2, \\ y_2 &= c_2 x_0 + a_2 x_1 + b_2 x_2. \end{aligned} \tag{6.16}$$

(a) Prove that y_0, y_1, y_2 are homogeneous coordinates of p relative to a projective reference of $\bar{\mathbb{A}}_2$ which has T as fundamental triangle and the incentre of T as unit point (see Exercise 2.6). These coordinates are called *trilinear* or *trimetric coordinates* relative to T.

(b) Distances to s_i being taken positive for points of Γ_i, and negative for points of Γ'_i, $i = 0, 1, 2$, prove that the trilinear coordinates of any proper point are proportional to its distances to the sides of T.

(c) Prove that if C_i is the cofactor of c_i in the matrix of the equations (6.16), then the improper line has equation

$$C_0 y_0 + C_1 y_1 + C_2 y_2 = 0.$$

From it, an elementary vector calculus computation shows that an equation of the improper line may also be written

$$\sin \widehat{s_1 s_2}\, y_0 + \sin \widehat{s_2 s_0}\, y_1 + \sin \widehat{s_0 s_1}\, y_2 = 0,$$

the angles being those interior to the triangle.

Exercise 6.38. In computer vision, a singular projectivity κ of a Euclidean space $\mathbb{A}_3$ onto one of its (proper) planes π is called a *camera* if the singular point O of κ does not belong to π, and a *finite camera* if in addition O is a proper point. The point O and the plane π are called the *centre* and the *image plane* of κ. Prove that there is a uniquely determined projectivity $\bar{\kappa}\colon \pi \to \pi$ such that $\kappa = \bar{\kappa} \circ \tau_{O,\pi}$, where $\tau_{O,\pi}$ is the projection from O onto π. A finite camera κ is called a *pinhole camera* if and only if $\bar{\kappa}$ is a similarity, and a *CCD camera* if and only if $\bar{\kappa}$ is an affinity, see [16], Chapter 5.

Chapter 7
Classification and properties of quadrics

7.1 Projective reduced equations of quadrics

As explained in general in Section 2.11, the projective classification of quadrics is induced by the action of the projectivities on them: quadrics Q and Q' belonging to projective spaces $\mathbb{P}_n$ and $\mathbb{P}'_n$ (over the same field $k = \mathbb{R}$ or $\mathbb{C}$) are said to be *projectively equivalent* if and only if there is a projectivity $f : \mathbb{P}_n \to \mathbb{P}'_n$ such that $f(Q) = Q'$. The projective equivalence of quadrics is, indeed, an equivalence relation (by 6.1.4); in the sequel it will be denoted by $\sim_p$. The equivalence class of a quadric by projective equivalence is called the *projective class*, and also the *projective type*, of the quadric. Thus, the projective class of a quadric Q consists of all the images of Q by projectivities. In this section and the next one we will describe all projective classes of quadrics and give a procedure to determine the projective class of any given quadric.

However, we will first fix our attention on an apparently unrelated problem, which is to prove that any quadric Q has, relative to a suitable projective reference, an equation (or matrix) of a particularly simple form. This equation is called the projective reduced equation of Q. The relationship between classifying quadrics and getting their reduced equations will be explained after proving the next lemma. In it, and also in the sequel, equations of quadrics $\sum_{i,j=0}^{n} a_{i,j} x_i x_j = 0$ and $\sum_{i,j=0}^{n} b_{i,j} y_i y_j = 0$, relative to possibly different references, are called equal (or said to be the same) if and only if they have the same coefficients, namely $a_{i,j} = b_{i,j}$, $i, j = 0, \dots, n$, regardless of which letters are used as variables.

Lemma 7.1.1. *Quadrics Q and Q', of projective spaces $\mathbb{P}_n$ and $\mathbb{P}'_n$ over the same field, are projectively equivalent if and only if there exist references Δ, of $\mathbb{P}_n$, and Δ', of $\mathbb{P}'_n$, relative to which Q and Q' have equal equations.*

Proof. If the equations of Q and Q' relative to Δ and Δ' are equal, take the projectivity $f : \mathbb{P}_n \to \mathbb{P}'_n$ such that $f(\Delta) = \Delta'$ (2.8.1). Then $f(Q) = Q'$ by 6.1.14. Conversely, if for a certain projectivity f, $f(Q) = Q'$, again by 6.1.14, any equation of Q relative to a reference Δ equals an equation of Q' relative to $f(\Delta)$. □

Remark 7.1.2. It is clear from 7.1.1 that any fixed $\mathbb{P}_{n,k}$ contains representatives of all projective types of quadrics of n-dimensional projective spaces over k. This follows also from the definition of projective equivalence of quadrics and 2.8.3.

Assume we have proved that each quadric Q has an equation of a certain form, which we call a (projective) reduced equation of Q. If quadrics Q, Q' have equal reduced equations, then, by 7.1.1, they are projectively equivalent. If the converse holds, that is, if any reduced equations of two projectively equivalent quadrics are equal, then there is a unique reduced equation for each quadric (because each quadric is projectively equivalent to itself) and our classification problem is solved. Indeed, on one hand the projective types of quadrics are in one-to-one correspondence with the different reduced equations, and so the different projective classes are described by just listing the reduced equations; on the other, computing the reduced equation of any given quadric determines its projective class.

Of course, asking projectively equivalent quadrics to have equal reduced equations is a strong demand on our reduced equations. It means that the whole of a reduced equation has to have an intrinsic, projectively invariant, character. In other words, the procedure leading from an equation of Q to the reduced one has to drop from the initial equation all but the information intrinsically related to Q. In this section we will prove the existence of projective reduced equations; they will be proved to be the same for projectively equivalent quadrics in Section 7.2.

First of all we introduce a notion that helps to obtain diagonal matrices for quadrics, and also has intrinsic interest. Assume that Q is a quadric of $\mathbb{P}_n$. An n-dimensional simplex S is said to be *self-polar* for (or with respect to) Q if and only if any two different vertices of S are conjugate with respect to Q.

The references whose simplex is self-polar for Q are called *self-polar references* for (or with respect to) Q. They give rise to diagonal matrices:

Lemma 7.1.3. *If a reference Δ is self-polar for a quadric Q, then any matrix of Q relative to Δ is diagonal. Conversely if a matrix of Q relative to Δ is diagonal, then Δ is self-polar for Q.*

Proof. Direct from the definitions and Example 6.4.1. □

The existence of self-polar simplexes is guaranteed by:

Proposition 7.1.4. *For any quadric Q, there is a simplex S which is self-polar for Q.*

Proof. Assume that Q belongs to a projective space $\mathbb{P}_n$. By 6.1.5, we choose any $p_0 \notin Q$ and write $H = H_{p_0}$ its polar hyperplane; since $p_0 \notin Q$, $p_0 \notin H$.

If $n = 1$, then $H = \{p_1\}$, where p_1 is a point different from p_0 and conjugate to it, and so the simplex with vertices p_0, p_1 is self-polar for Q.

If $n > 1$, then, using induction, we assume the claim to be true for any quadric belonging to a space of dimension $n-1$. If $H \subset Q$, we take any independent points $p_1, \dots, p_n \in H$. Otherwise, $Q \cap H$ is a quadric of H and by the induction hypothesis we take $p_1, \dots, p_n$ to be the vertices of a simplex of H self-polar for $Q \cap H$. As noted above, $p_0 \notin H$, from which, in both cases, $p_0, \dots, p_n$ are independent points and we take them as the vertices of S. Also in both cases p_0 is conjugate to all p_i, $i = 1, \dots, n$, because they have been taken in H. In case

$H \subset Q$ any two points p_i, p_j, $i, j = 1, \dots, n$, are conjugate with respect to Q by 6.4.5. If $H \not\subset Q$, by the way they have been chosen, any p_i, p_j, $i \neq j$, $i, j = 1, \dots, n$, are conjugate with respect to $Q \cap H$, and so they are also conjugate with respect to Q by 6.4.5. Thus in both cases S is a self-polar simplex as wanted. □

Remark 7.1.5. We have in fact proved that for any point $p \in \mathbb{P}_n - |Q|$ there is a self-polar simplex with vertex p.

Corollary 7.1.6. *For any quadric Q of P_n there is a reference relative to which Q has an equation of the form*

$$a_0 x_0^2 + \cdots + a_r x_r^2 = 0,$$

where $a_i \neq 0$ for $i = 0, \dots, r$ and $0 \leq r \leq n$.

Proof. By 2.1.3 and 7.1.4 there is a reference Δ which is self-polar for Q and therefore, by 7.1.3, gives rise to an equation of Q with square terms only. Then just renumber the vertices of Δ in order to move the non-zero terms to the first places. □

The above equation of Q can still be improved, but the modifications will be different in the complex and real cases. So, assume first that the base field k is the complex one. Then take new coordinates $y_0, \dots, y_n$, relative to the reference Δ' with the same vertices as Δ and unit point

$$\left[\frac{1}{\sqrt{a_0}}, \dots, \frac{1}{\sqrt{a_r}}, 1, \dots, 1\right].$$

Then

$$\begin{aligned} x_0 &= \frac{1}{\sqrt{a_0}} y_0, \\ &\vdots \\ x_r &= \frac{1}{\sqrt{a_r}} y_r, \\ x_{r+1} &= y_{r+1}, \\ &\vdots \\ x_n &= y_n, \end{aligned}$$

and, by 6.1.13, the new equation of Q is a sum of squares, namely

$$y_0^2 + \cdots + y_r^2 = 0.$$

Assume now that $k = \mathbb{R}$. Up to multiplying it by -1, we assume that the equation already obtained for Q in 7.1.6 has at least as many positive terms as negative. After a suitable renumbering of the coordinates,it has the form

$$b_0 x_0^2 + \cdots + b_j x_j^2 + b_{j+1} x_{j+1}^2 + \cdots + b_r x_r^2 = 0,$$

with $b_0, \ldots, b_j > 0$, $b_{j+1}, \ldots, b_r < 0$, $j + 1 \geq r - j$ and, obviously, $j \leq r \leq n$. Then we take new coordinates $y_0, \ldots, y_n$ relative to the reference Δ' with the same vertices as Δ and unit point

$$\left[\frac{1}{\sqrt{b_0}}, \ldots, \frac{1}{\sqrt{b_j}}, \frac{1}{\sqrt{-b_{j+1}}}, \ldots, \frac{1}{\sqrt{-b_r}}, 1, \ldots, 1\right].$$

The equations relating old and new coordinates are now

$$\begin{aligned}
x_0 &= \frac{1}{\sqrt{b_0}} y_0, \\
&\vdots \\
x_j &= \frac{1}{\sqrt{b_j}} y_j, \\
x_{j+1} &= \frac{1}{\sqrt{-b_{j+1}}} y_{j+1}, \\
&\vdots \\
x_r &= \frac{1}{\sqrt{-b_r}} y_r, \\
x_{r+1} &= y_{r+1}, \\
&\vdots \\
x_n &= y_n,
\end{aligned}$$

and the resulting equation of Q is

$$y_0^2 + \cdots + y_j^2 - y_{j+1}^2 - \cdots - y_r^2 = 0,$$

still with $j + 1 \geq r - j$ and $j \leq r \leq n$.

All together we have proved:

Theorem 7.1.7. *For each quadric Q of a complex projective space $\mathbb{P}_{n,\mathbb{C}}$ there is a reference relative to which Q has equation*

$$x_0^2 + \cdots + x_r^2 = 0, \quad 0 \leq r \leq n.$$

For each quadric Q of a real projective space $\mathbb{P}_{n,\mathbb{R}}$ there is a reference relative to which Q has equation

$$x_0^2 + \cdots + x_j^2 - x_{j+1}^2 - \cdots - x_r^2 = 0,$$

with $j + 1 \geq r - j$ and $j \leq r \leq n$. In both cases such an equation is called a projective reduced equation *of Q, or just a reduced equation if no confusion may result.*

Remark 7.1.8. Actually, we have proved that the vertices of any self-polar simplex may be taken as the vertices of a reference relative to which the equation of the quadric is reduced.

Exercise 7.3 presents an alternative proof of 7.1.6 that allows us an easy computation of both a reduced equation of a quadric and the change of coordinates leading to it.

7.2 Projective classification of quadrics

Throughout this section, we assume that all quadrics belong to projective spaces of a fixed dimension n. We have already established that any quadric has a reduced equation; we will prove now that two quadrics are projectively equivalent if and only if they have equal reduced equations, which is often stated by saying that the reduced equations *projectively classify* quadrics. In the course of the proof we will also get a second and very useful description of the projective types of quadrics by means of numerical invariants.

We will deal with the complex and real cases separately. Thus, to begin with the easier case, we assume first that the base field is the complex onc. By 7.1.7, any quadric has a reduced equation of the form

$$x_0^2 + \cdots + x_r^2 = 0, \quad 0 \leq r \leq n. \tag{7.1}$$

Still denote by $\mathbf{r}(Q)$ the rank of the quadric Q, which, as seen in Section 6.4, is the rank of any matrix of Q. We will prove that for arbitrary complex quadrics Q and Q', the four implications in the diagram

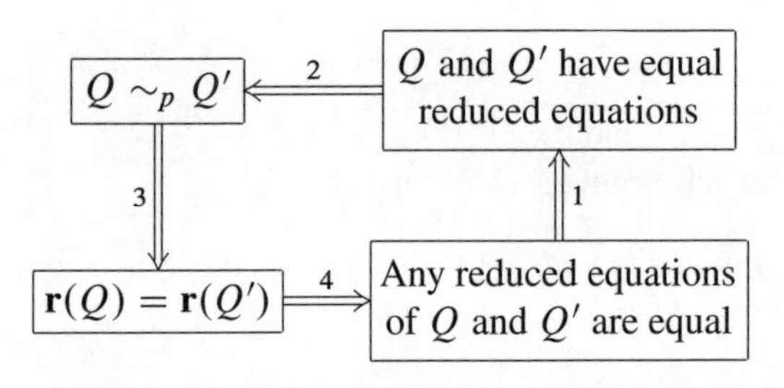

hold true.

Indeed, once it has been proved, in 7.1.7, that any quadric has a reduced equation, the implication 1 is obvious. The implication 2 follows from 7.1.1, while 3 is part of 6.4.13. To prove 4 just note that the reduced equation of a quadric Q, $x_0^2 + \cdots + x_r^2 = 0$, is determined by r, which in turn is $r = \mathbf{r}(Q) - 1$, by the definition of $\mathbf{r}(Q)$. Thus any reduced equations of two quadrics with the same rank are equal.

By taking $Q' = Q$, we have in particular proved that any complex quadric Q has a unique reduced equation, and so we are allowed to call it *the* reduced equation of Q. Note that uniqueness is not claimed for the reference relative to which the quadric has a reduced equation, see 7.1.8 and 7.1.5. All together we have proved a classification theorem for complex quadrics:

Theorem 7.2.1 (Projective classification of complex quadrics). *Any quadric Q of an n-dimensional complex projective space has a unique reduced equation. For any two quadrics Q and Q' of n-dimensional complex projective spaces, the following three conditions are equivalent:*

(i) *Q and Q' are projectively equivalent.*

(ii) *Q and Q' have equal projective reduced equations.*

(iii) *Q and Q' have the same rank.*

We see in particular that $\{\mathbf{r}(Q)\}$ is a complete set of projective invariants for complex quadrics. In other words, complex quadrics are projectively classified by their ranks, which may be understood to measure the degree of degeneration of the quadric. For instance, all non-degenerate quadrics of complex n-dimensional projective spaces compose a single projective class. Checking the projective equivalence of two complex quadrics Q, Q' may thus be done just by comparing their ranks, which in turn may be computed as the ranks of any matrices of Q and Q'. Obviously, the dimension of the variety of double points of Q, $n - \mathbf{r}(Q)$, is a projective invariant carrying the same information as $\mathbf{r}(Q)$, and may be equivalently used instead.

Since any of the reduced equations (7.1) is, clearly, the reduced equation of a quadric, Theorem 7.2.1 assures also that mapping each quadric to its projective reduced equation induces a bijection between the projective types of complex quadrics and the reduced equations listed in (7.1). The next corollary provides a single representative of each projective type of complex quadrics. After the above, it needs no proof.

Corollary 7.2.2. *Fix a complex projective space of dimension n, $\mathbb{P}_{n,\mathbb{C}}$, and a reference Δ in $\mathbb{P}_{n,\mathbb{C}}$. Any quadric of any n-dimensional complex projective space is projectively equivalent to one and only one of the quadrics of $\mathbb{P}_{n,\mathbb{C}}$ whose equations* *relative to* Δ *are:*

$$
\begin{aligned}
x_0^2 + \cdots + x_{n-1}^2 + x_n^2 &= 0, \\
x_0^2 + \cdots + x_{n-1}^2 &= 0, \\
&\vdots \\
x_0^2 + x_1^2 &= 0, \\
x_0^2 &= 0.
\end{aligned}
$$

Remark 7.2.3. Since $x_0^2 + x_1^2 = (x_0 + ix_1)(x_0 - ix_1)$, the reader may note that all complex quadrics of rank two are hyperplane-pairs; the converse has been already seen in 6.4.16. Also, the complex quadrics of rank one are the hyperplanes counted twice, as already seen in 6.4.21.

Now we will turn our attention to the real case. The arguments will be similar to those used in the complex case. However, the more complex form of the reduced equations requires the use of a further projective invariant we will introduce next. Assume that Q is a quadric of a real projective space $\mathbb{P}_{n,\mathbb{R}}$; we say that a linear variety L of $\mathbb{P}_{n,\mathbb{R}}$ is *disjoint* with Q if at only if L has no (real) points belonging to Q, that is, $|Q| \cap L = \emptyset$. The *index* of Q, denoted in the sequel by $\mathbf{j}(Q)$, is defined as being the maximum of the dimensions of the linear varieties of $\mathbb{P}_{n,\mathbb{R}}$ disjoint with Q. Obviously, making a similar definition for complex quadrics would have no interest, as the index of any complex quadric would be 0, due to 6.3.1.

Proceeding as in the complex case, we will prove that the four implications in the following diagram hold true, this time for any pair of real quadrics Q, Q'.

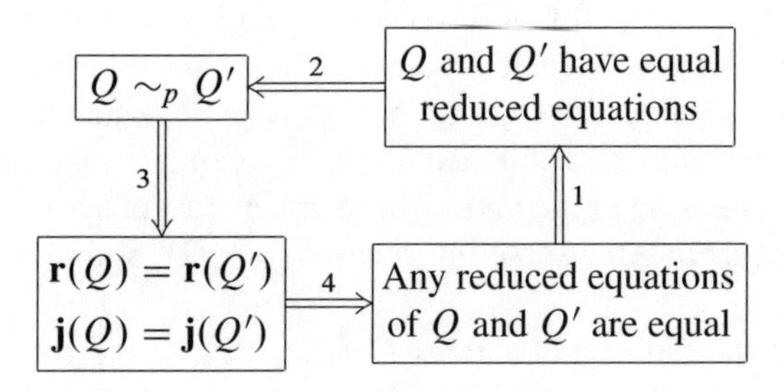

The same arguments used in the complex case prove the validity of 1 and 2. The invariance of the rank has been already proved in 6.4.13. The invariance of the index is direct from its definition: if a linear variety L does not share any point with Q, by 6.1.7 and 6.1.3, neither does $f(L)$ with $f(Q)$ for any projectivity f. Thus $\mathbf{j}(Q) \leq \mathbf{j}(Q')$ and so, by symmetry, $\mathbf{j}(Q) = \mathbf{j}(Q')$. This completes the proof of 3. Now, for 4, note first that a real reduced equation $x_0^2 + \cdots + x_j^2 - x_{j+1}^2 - \cdots - x_r^2 = 0$ is determined by r and j. As in the complex case, $r = \mathbf{r}(Q) - 1$. To relate j to the index $\mathbf{j}(Q)$ we will make use of the next lemma. Its proof requires us to consider also the maximum of the dimensions of the linear varieties contained in the quadric, which, as it will turn out, is a projective invariant enclosing the same information as $\mathbf{j}(Q)$.

Lemma 7.2.4. *If Q is a quadric of a real projective space $\mathbb{P}_{n,\mathbb{R}}$ with equation*

$$x_0^2 + \cdots + x_j^2 - x_{j+1}^2 - \cdots - x_r^2 = 0, \quad j+1 \geq r-j, \tag{7.2}$$

then:

(a) *j is the maximum of the dimensions of the linear varieties disjoint with Q.*

(b) *$n-j-1$ is the maximum of the dimensions of the linear varieties contained in Q.*

Proof. The linear variety L, with equations

$$\begin{aligned} x_{j+1} &= 0, \\ &\vdots \\ x_n &= 0, \end{aligned}$$

obviously has dimension $n-(n-j) = j$. Furthermore, it is disjoint with Q. Indeed, any $p \in L$ is $p = [x_0, \dots, x_j, 0, \dots, 0]$ and $p \in Q$ forces $x_0^2 + \cdots + x_j^2 = 0$ and hence $x_0 = \cdots = x_j = 0$, as the base field is $\mathbb{R}$.

On the other hand, since $j+1 \geq n-j$, we may consider the equations

$$\begin{aligned} x_0 &= x_{j+1}, \\ &\vdots \\ x_{r-j-1} &= x_r, \\ x_{r-j} &= 0, \\ &\vdots \\ x_j &= 0. \end{aligned} \tag{7.3}$$

Clearly, they are independent and so the linear variety T they define has dimension $n-(j+1) = n-j-1$. Furthermore, T is contained in Q, as the equalities (7.3) force the cancellation of all terms in the equation (7.2).

Now if a linear variety L' has dimension strictly greater than j, it has non-empty intersection with T by 1.4.10 and hence $|Q| \cap L' \neq \emptyset$. Similarly, if a linear variety T' has dimension strictly greater than $n-j-1$, again by 1.4.10, it has non-empty intersection with L and hence $T' \not\subset |Q|$. □

Summarizing we have seen:

Proposition 7.2.5. *If Q is a quadric of a real projective space $\mathbb{P}_{n,\mathbb{R}}$ and has projective reduced equation*

$$x_0^2 + \cdots + x_j^2 - x_{j+1}^2 - \cdots - x_r^2 = 0, \quad j+1 \geq r-j,$$

then $r = \mathbf{r}(Q) - 1$ and $j = \mathbf{j}(Q)$.

This shows that the projective invariants $\mathbf{r}(Q)$ and $\mathbf{j}(Q)$ determine the reduced equation and completes the proof of the classification theorem in the real case:

Theorem 7.2.6 (Projective classification of real quadrics). *Any quadric Q of a real projective space $\mathbb{P}_{n,\mathbb{R}}$ has a unique projective reduced equation. For any two quadrics Q and Q' of real n-dimensional projective spaces, the following three conditions are equivalent:*

(i) *Q and Q' are projectively equivalent.*

(ii) *Q and Q' have equal projective reduced equations.*

(iii) *Q and Q' have the same rank and the same index.*

Remark 7.2.7. The pair $\{\mathbf{r}(Q), \mathbf{j}(Q)\}$ is a complete set of projective invariants for the real quadrics.

Remark 7.2.8. Since Lemma 7.2.4 applies to any real quadric (by 7.1.7), the maximum of the dimensions of the linear varieties contained in a real quadric Q is $\mathbf{s}(Q) = n - \mathbf{j}(Q) - 1$. Then the pair $\{\mathbf{r}(Q), \mathbf{s}(Q)\}$ is also a complete set of projective invariants for real quadrics.

Again the reduced equations provide representatives of the projective types:

Corollary 7.2.9. *If $\mathbb{P}_{n,\mathbb{R}}$ is a real projective space of dimension n and Δ a reference of $\mathbb{P}_{n,\mathbb{R}}$, then any quadric of any n-dimensional real projective space is projectively equivalent to one and only one of the quadrics of $\mathbb{P}_{n,\mathbb{R}}$ whose equations relative to Δ are*

$$x_0^2 + \cdots + x_j^2 - x_{j+1}^2 - \cdots - x_r^2 = 0,$$

for $0 \leq r \leq n$ and $(r-1)/2 \leq j \leq r$.

Remark 7.2.10. The ranges of values of the rank (quite obvious) and the index follow from 7.2.9. They are

$$1 \leq \mathbf{r}(Q) \leq n+1 \quad \text{and} \quad \frac{\mathbf{r}(Q)}{2} - 1 \leq \mathbf{j}(Q) \leq \mathbf{r}(Q) - 1.$$

The latter shows that the range of values of $s(Q)$ is

$$\frac{n + \dim \mathcal{D}(Q)}{2} \geq s(Q) \geq \dim \mathcal{D}(Q).$$

Remark 7.2.11. As in the complex case, the rank-one real quadrics are the hyperplanes counted twice and are the representatives of a single projective class, corresponding to the reduced equation $x_0^2 = 0$. There are two projective classes of rank-two real quadrics, namely:

(a) The projective class of all quadrics with reduced equation

$$x_0^2 - x_1^2 = (x_0 + x_1)(x_0 - x_1) = 0,$$

which are pairs of distinct real hyperplanes; any pair of distinct real hyperplanes belongs to this class, as it has rank 2, by 6.4.16, and, obviously, index 0.

(b) The projective class of all quadrics with reduced equation

$$x_0^2 + x_1^2 = (x_0 + ix_1)(x_0 - ix_1) = 0.$$

They are pairs of imaginary hyperplanes, and any pair of imaginary hyperplanes belongs to this class, because it has rank 2 and index 1, by 6.4.18.

Remark 7.2.12. Since any real quadric Q and its complex extension $Q_{\mathbb{C}}$ have equal matrices, they have the same rank. Therefore, projectively equivalent real quadrics have projectively equivalent complex extensions. Obviously, the converse is not true, as real quadrics with the same rank and different indices have projectively equivalent complex extensions.

Next we will examine more closely the projective types of quadrics of spaces $\mathbb{P}_{n,\mathbb{R}}$ for $n \leq 3$. If $n = 1$, then 7.2.11 covers all cases: we have three projective types of point-pairs, namely pairs of real distinct points, pairs of imaginary points, and pairs of coincident points.

For $n = 2$, we find two types of non-degenerate conics. Those with reduced equation $x_0^2 + x_1^2 + x_2^2 = 0$ have index 2 and hence no (real) points: they are called *imaginary conics*. The conics with reduced equation $x_0^2 + x_1^2 - x_2^2 = 0$ have index 1 and hence contain points but no lines: they are called *real non-degenerate conics*. The reader may note that the above use of the adjectives *real* and *imaginary*, although quite usual, is rather misleading, as they refer to the points of the conics rather than to the conics themselves, which in both cases are real conics according to the definitions of Section 6.1. Again as a particular case of 7.2.11, the degenerate conics are either pairs of distinct real lines, or pairs of imaginary lines, or double lines.

In real projective three-spaces we have three projective types of non-degenerate quadrics, corresponding to the reduced equations

$$x_0^2 + x_1^2 + x_2^2 + x_3^2 = 0,$$
$$x_0^2 + x_1^2 + x_2^2 - x_3^2 = 0$$

and

$$x_0^2 + x_1^2 - x_2^2 - x_3^2 = 0.$$

The quadrics of the first type have index 3 and thus no points: as in former cases they are called *imaginary*. Quadrics of the second type have index 2 and therefore contain points but no lines, they are called *non-ruled quadrics*. The quadrics of the third type have index 1 and are called *ruled quadrics*; each contains at least a line. The lines contained in a ruled quadric are called its *generators*; we will see in the forthcoming Section 8.3 that any ruled quadric is covered by its generators (hence their name), which compose two infinite families called the *rulings* of the quadric.

The rank-three quadrics of three-dimensional spaces have each a single double point and, according to 6.7.7, are cones projecting non-degenerate conics from points. They are classified in two projective types corresponding to the reduced equations

$$x_0^2 + x_1^2 + x_2^2 = 0$$

and

$$x_0^2 + x_1^2 - x_2^2 = 0.$$

Each quadric of the first type is the cone projecting the imaginary conic $x_0^2 + x_1^2 + x_2^2 = 0$, of the plane $x_3 = 0$, from the double point $[0, 0, 0, 1]$. It is called an *imaginary cone* of $\mathbb{P}_3$ and clearly has no point other than its double point.

Similarly, each quadric of the second type is the cone projecting the conic $x_0^2 + x_1^2 - x_2^2 = 0$ of $x_3 = 0$, this time real non-degenerate, from the double point $[0, 0, 0, 1]$. These quadrics are called *real cones*, and also *ordinary cones* of $\mathbb{P}_3$. They already appeared in Example 6.7.4.

To close the case $n = 3$, the quadrics of rank ≤ 2 are classified in *pairs of real planes*, *pairs of imaginary planes* and *double planes*, as seen in 7.2.11.

Complete lists of the projective types of real conics and three-space quadrics, with their projective invariants and projective reduced equations, are part of Tables 7.1 and 7.2 below (see p. 292 and p. 293f.).

Remark 7.2.13. The envelope Q^* of a non-degenerate quadric Q is projectively equivalent to Q, as it is defined as the image of Q by the polarity relative to Q (see Section 6.6). After 6.6.1 (e) and the particular form of the reduced equations, it is clear that in both the real and the complex case, if the equation of Q relative to a reference Δ is reduced, then the same equation is the reduced equation of Q^* relative to the dual reference $\Delta^\vee$. This in particular re-proves that $Q \sim_p Q^*$.

The projective equivalence of two real quadrics may be checked by either effectively computing their reduced equations via the method of Exercise 7.3, or by comparing their projective invariants. The computation of the index being not obvious, the following lemma will be useful.

Lemma 7.2.14. *If N is any matrix of a quadric Q of $\mathbb{P}_{n,\mathbb{R}}$ and $c_N = \det(N - X\mathbf{1}_{n+1})$ is the characteristic polynomial of N, then $\mathbf{j}(Q) + 1$ is the maximum of the numbers of positive and negative roots of c_N.*

Proof. The roots of c_N being the eigenvalues of N, their opposites are the eigenvalues of $-N$. Therefore, up to taking $-N$ instead of N, it is not restrictive to assume that the number of positive eigenvalues of N is not less than the number of the negative ones. Denote the former by $s + 1$ and the latter by $r - s$, $r + 1$ being then the number of non-zero roots. Then it is $s + 1 \geq r - s$ and we should prove $\mathbf{j}(Q) = s$.

By its definition, N is the matrix of a representative η of Q relative to a basis $e_0, \dots, e_n$ of the associated vector space E. Let φ be the endomorphism of E which has matrix N relative to the basis $e_0, \dots, e_n$. For any vector $v \in E$, denote by (v) the column matrix whose entries are the components of v relative to $e_0, \dots, e_n$. Define a second bilinear form τ by the rule $\tau(v, w) = (v)^t(w)$. The form τ is obviously symmetric and positive definite, and the basis $e_0, \dots, e_n$ is orthonormal for τ. Furthermore, for any v, w we have

$$\tau(v, \varphi(w)) = (v)^t N(w) = \eta(v, w).$$

Because of this relation, or also because the matrix N of φ, relative to the orthonormal basis $e_0, \dots, e_n$, is symmetric, φ is a symmetric endomorphism with respect to the form τ. A well-known theorem of linear algebra (the *spectral theorem for symmetric morphisms*, see for instance [14], 11.7, or the forthcoming exercise 10.29) assures the existence of eigenvectors $u_0, \dots, u_n$ that are an orthonormal basis for τ. We have thus

$$(u_i)^t(u_i) = 1 \quad \text{and} \quad (u_i)^t(u_j) = 0$$

for $i, j = 0, \dots, n$, $i \neq j$. Up to a suitable reordering we may assume that

$$N(u_i) = a_i(u_i) \text{ where } \begin{cases} a_i > 0 & \text{if } i = 0, \dots, s, \\ a_i < 0 & \text{if } i = s+1, \dots, r, \\ a_i = 0 & \text{if } i = r+1, \dots, n. \end{cases}$$

Then,

$$\eta(u_i, u_i) = \tau(u_i, \varphi(u_i)) = \tau(u_i, a_i u_i) = a_i,$$

while for $i \neq j$,

$$\eta(u_i, u_j) = \tau(u_i, \varphi(u_j)) = \tau(u_i, a_j u_j) = 0.$$

Take

$$w_i = \begin{cases} \frac{1}{\sqrt{a_i}} u_i & \text{if } i = 0, \dots, s, \\ \frac{1}{\sqrt{-a_i}} u_i & \text{if } i = s+1, \dots, r, \\ u_i & \text{if } i = r+1, \dots, n. \end{cases}$$

Then

$$\eta(w_i, w_i) = \begin{cases} 1 & \text{if } i = 0, \dots, s, \\ -1 & \text{if } i = s+1, \dots, r, \\ 0 & \text{if } i = r+1, \dots, n, \end{cases}$$

and

$$\eta(w_i, w_j) = 0 \quad \text{if } i \neq j.$$

Obviously $w_0, \dots, w_n$ is also a basis of E. Let Ω be the reference of $\mathbb{P}_n$ with adapted basis $w_0, \dots, w_n$. By the above equalities, the equation of Q relative to Ω is

$$x_0^2 + \cdots + x_s^2 - x_{s+1}^2 - \cdots - x_r^2 = 0,$$

with $s + 1 \geq r - s$, after which $\mathbf{j}(Q) = s$ by 7.2.5. □

Remark 7.2.15. All roots of the characteristic polynomial c_N of 7.2.14 are real, due to existence of a base of eigenvectors of φ. Therefore, the number of positive roots and the number of negative roots of c_N may be computed using Descartes' rule of signs.

7.3 Determining quadrics by their sets of points

The set of points of a quadric Q does not always determine it. To see some examples take, in a real $\mathbb{P}_n$ and for a fixed r, $1 \leq r \leq n$, the quadrics with equations

$$a_0 x_0^2 + \cdots + a_r x_r^2 = 0,$$

where $a_i > 0$ for $i = 0, \dots, r$. Since a sum of non-negative real numbers is zero if and only if all of them are zero, it is clear that the points of any of these quadrics are those and only those having $x_0 = \cdots = x_r = 0$. The set of these points is thus independent of $a_0, \dots, a_r$; therefore, any two non-proportional equations from the above define different quadrics with the same set of points. Denote by Q any of these quadrics; clearly $\dim \mathcal{D}(Q) = n - r - 1$. On the other hand, the set $|Q|$, described above, is also a linear variety of dimension $n - r - 1$. Therefore $|Q| = \mathcal{D}(Q)$, that is, Q has no points other than its double points. Our goal in this section is to prove that the above are essentially the only examples of quadrics that are not determined by their sets of points. We will see that all complex quadrics, all real quadrics with some simple point and, as already seen in 6.4.20, all real double hyperplanes are determined by their sets of points.

Lemma 7.3.1. *If quadrics Q and Q' have $|Q| = |Q'|$, then $\mathcal{D}(Q) = \mathcal{D}(Q')$.*

Proof. Recall from Section 6.4 that a point $p \in \mathbb{P}_n$ is double for Q if and only if $p \in Q$ and any line ℓ through p is tangent to Q. This in turn is equivalent to having $p \in |Q|$ and either $\ell \subset |Q|$ or $|Q| \cap \ell = \{p\}$, for any line ℓ through p. The latter conditions being formulated in terms of $|Q|$ only, the claim follows. □

Lemma 7.3.2. *If quadrics Q and Q' have $|Q| = |Q'|$ and a line ℓ has non-empty intersection with Q, then $Q \cap \ell = Q' \cap \ell$.*

Proof. Clearly, $\ell \subset Q$ if and only if $\ell \subset Q'$ and in such a case the claimed equality is obvious. Thus we assume both sections to be point-pairs. We have $|Q \cap \ell| = |Q| \cap \ell = |Q'| \cap \ell = |Q' \cap \ell|$. Since by the hypothesis this set is non-empty, neither $Q \cap \ell$ nor $Q' \cap \ell$ is a pair of imaginary points, after which the equality $|Q \cap \ell| = |Q' \cap \ell|$ forces $Q \cap \ell = Q' \cap \ell$ by 6.2.2. □

As already announced, our main claim in this section reads:

Theorem 7.3.3. *Assume that Q and Q' are quadrics of $\mathbb{P}_n$ with equal sets of points. If either the base field k is $k = \mathbb{C}$, or $k = \mathbb{R}$ and $\mathbf{j}(Q) < \mathbf{r}(Q) - 1$, or still $k = \mathbb{R}$ and $\mathbf{r}(Q) = 1$, then $Q = Q'$.*

Proof. The case $\mathbf{r}(Q) = 1$ has been included for the sake of completeness, as the claim has been already proved for rank-one quadrics in 6.4.20. Thus in the sequel we assume $\mathbf{r}(Q) > 1$. Let $\Delta = (p_0, \dots, p_n, A)$ be a reference relative to which Q has reduced equation

$$x_0^2 + \cdots + x_j^2 - x_{j+1}^2 - \cdots - x_r^2 = 0,$$

with $j = r$ in the complex case and $j + 1 \geq r - j$ in the real one. Since we are assuming $\mathbf{r}(Q) > 1$, it is $r > 0$. Furthermore, in the real case, $j < r$ due to the hypothesis $\mathbf{j}(Q) < \mathbf{r}(Q) - 1$. Assume that the equation of Q' relative to Δ is

$$\sum_{i,s=0}^{n} a_{i,s} x_i x_s = 0.$$

It is clear that the vertices $p_{r+1}, \dots, p_n$ are double points of Q. Therefore, by 7.3.1, they are also double points of Q' and so, by 6.7.1,

$$a_{i,s} = 0 \quad \text{if either } i > r \text{ or } s > r. \tag{7.4}$$

Assume first that $k = \mathbb{C}$. Then any line has non-empty intersection with Q and therefore 7.3.2 applies to it. In particular for any line $\ell_{i,s} = p_i \vee p_s$, $i < s \leq r$, we have $Q \cap \ell_{i,s} = Q' \cap \ell_{i,s}$. The equations of these sections being

$$x_i^2 + x_s^2 = 0 \quad \text{and} \quad a_{i,i} x_i^2 + 2a_{i,s} x_i x_s + a_{s,s} x_s^2 = 0$$

we get

$$a_{i,i} = a_{s,s} \quad \text{and} \quad a_{i,s} = 0 \quad \text{for } i < s \leq r.$$

Since we have already seen in (7.4) all the other coefficients to be zero, this shows that Q and Q' have proportional equations and therefore proves the claim in the complex case.

Assume now that the base field is the real one. Still take the lines $\ell_{i,s} = p_i \vee p_s$, $i < s \leq r$. For $i \leq j < s$, the section $Q \cap \ell_{i,s}$ has equation

$$x_i^2 - x_s^2 = 0$$

and therefore is non-empty. Again, 7.3.2 applies and, arguing as in the complex case, we get

$$a_{i,i} = -a_{s,s} \quad \text{for } i \leq j < s \leq r \tag{7.5}$$

and also

$$a_{i,s} = 0 \quad \text{still for } i \leq j < s \leq r. \tag{7.6}$$

Since $0 \leq j < r$, we may collect from (7.5) the equalities $a_{i,i} = -a_{s,s}$, first for $i = 0$ and $s = j+1, \dots, r$ and then for $s = r$ and $i = 0, \dots, j$, thus getting

$$a_{0,0} = \cdots = a_{j,j} = -a_{j+1,j+1} = \cdots = -a_{r,r}. \tag{7.7}$$

In view of equalities (7.4), (7.6) and (7.7), in order to prove that the equations of Q and Q' are proportional, and therefore the claim, it remains only to see that

$$a_{i,s} = 0 \quad \text{if } i < s \text{ and either } i, s \leq j \text{ or } j < i, s.$$

For $i, s \leq j$ take the point with coordinates $x_i = x_s = 1$, $x_r = \sqrt{2}$ and the remaining ones equal to zero. Since it obviously belongs to Q, it belongs to Q' too, and so

$$a_{i,i} + 2a_{i,s} + a_{s,s} + 2a_{r,r} + 2\sqrt{2}\,a_{i,r} + 2\sqrt{2}\,a_{s,r} = 0.$$

Using now (7.6) and (7.7) gives $a_{i,s} = 0$, as wanted. The case $i, s > j$ follows from a similar argument which is left to the reader. □

Remark 7.3.4. In the real case the hypothesis $\mathbf{j}(Q) < \mathbf{r}(Q) - 1$ guarantees that the largest linear varieties included in Q have dimension $n - \mathbf{j}(Q) - 1 > n - \mathbf{r}(Q) = \dim \mathcal{D}(Q)$ and hence that Q has some (real) simple point. Conversely, it is clear from their reduced equations that the real quadrics with $\mathbf{j}(Q) = \mathbf{r}(Q) - 1$ have no simple point.

Remark 7.3.5. The quadrics that do not satisfy the hypothesis of 7.3.3 are the real quadrics with reduced equation of the form

$$x_0^2 + \cdots + x_r^2 = 0, \quad r > 0.$$

We have seen at the beginning of this section that neither of them is determined by its set of points. Note that the set of points of any of these quadrics is a linear variety of codimension higher than one; hence, a heuristic explanation for the fact that these quadrics are not determined by their sets of points is that they have too few points.

Remark 7.3.6. If real quadrics Q, Q' have the same sets of real or imaginary points, that is $|Q_{\mathbb{C}}| = |Q'_{\mathbb{C}}|$, then they have $Q_{\mathbb{C}} = Q'_{\mathbb{C}}$ by 7.3.3 and so, by 6.1.15, $Q = Q'$. Therefore any real quadric Q is determined by the set of points of its complex extension $Q_{\mathbb{C}}$, that is, by the set of the real or imaginary points of Q.

To close this section, we present a result which improves Theorem 7.3.3 in case $\mathbf{r}(Q) > 1$. The reader may see that it does not extend to the case $\mathbf{r}(Q) = 1$. First we will prove a lemma:

Lemma 7.3.7. *Assume that H, H' are two hyperplanes and Q a quadric of $\mathbb{P}_n$. If Q has a simple point and $|Q| \subset H \cup H'$, then $Q = H + H'$.*

Proof. If a point $p \in Q$ belongs to $H \cap H'$, then any line pq, $q \notin H \cup H'$, is tangent to Q because it has no intersection with Q other than p itself, and p is a double point of Q. Thus, if we take p to be a simple point of Q, then, up to swapping over H and H', we may assume $p \in H - H \cap H'$. In particular $H \neq H'$ and clearly $H' \neq H_{p,Q}$, as the former does not contain p. Take any $q \in H' - H \cup H_{p,Q}$; then the line pq is a chord of Q, because $pq \not\subset H_{p,Q}$, and has one end at p: since $|Q| \subset H \cup H'$, the other end needs to be q, and so $q \in Q$. This shows that $H' - H \cup H_{p,Q} \subset |Q|$ and so, by an easy argument using 6.2.5, that $H' \subset Q$. Now, any $q \in H' - H \cup H_{p,Q}$ is a simple point of Q, because pq is a chord through it. Repeating the above argument from it proves that also $H \subset Q$ and hence $Q = H + H'$, by 6.4.17. □

Proposition 7.3.8. *Let Q and Q' be quadrics of a projective space $\mathbb{P}_n$. If Q has a simple point and $|Q| \subset |Q'|$, then $Q = Q'$.*

Proof. The hypothesis assures $\mathbf{r}(Q) \geq 2$ and the case $\mathbf{r}(Q) = 2$ is covered by 6.4.17, so in the sequel we assume $\mathbf{r}(Q) > 2$. Since in case $k = \mathbb{R}$ the hypothesis assures $\mathbf{j}(Q) < r(Q)$ (7.3.4), it will be enough to prove that $|Q| = |Q'|$, by 7.3.3. Assume otherwise, namely that there is $p \in |Q'| - |Q|$. Then, $H_{p,Q}$ is a hyperplane and no line through p is contained in Q. The tangent lines to Q through p have their contact points in $H_{p,Q}$. Thus, a line pq, $q \in |Q| - H_{p,Q}$, intersects Q in two distinct points, each different from p. As a consequence pq contains three different points of Q' and so $pq \subset Q'$, by 6.2.5. If all the lines pq, $q \in |Q| - H_{p,Q}$ were contained in a hyperplane H, then it would be $|Q| \subset H_{p,Q} \cup H$, against 7.3.7 and the assumption $\mathbf{r}(Q) > 2$. It follows thus that there is no hyperplane containing all lines that go through p and are contained in Q'; this assures that p is a double point of Q'.

Now consider the lines pq, $q \in Q$. Each is contained in Q', because it is tangent to Q' at p and contains q, which is different from p and belongs to Q'. Furthermore, we have seen above that no hyperplane contains all these lines. On the other hand, $p \notin Q$, so no one of the lines pq, $q \in Q$, contains more than two points of Q. Therefore each pq, $q \in Q$, may be spanned by points in $|Q'| - |Q|$

and therefore no hyperplane contains $|Q'| - |Q|$. This leads to a contradiction, as the argument already used for p proves that any $p' \in |Q'| - |Q|$ is a double point of Q', after which the set of double points of Q' would be contained in no hyperplane. □

Corollary 7.3.9. *Assume $k = \mathbb{R}$, let Q be a quadric of $\mathbb{P}_n$ and Q' a quadric of $\mathbb{C}\mathbb{P}_n$. If Q has a simple point and $|Q| \subset |Q'|$, then Q' is the complex extension of Q.*

Proof. Fix a reference of $\mathbb{P}_n$ and take it as a real reference of $\mathbb{C}\mathbb{P}_n$. Assume that, relative to it, Q' has equation

$$F = \sum_{i,j=0}^{n} a_{i,j} x_i x_j = 0 \tag{7.8}$$

and take

$$\begin{aligned} F_1 &= \sum_{i,j=0}^{n} (a_{i,j} + \bar{a}_{i,j}) x_i x_j, \\ F_2 &= \boldsymbol{i} \sum_{i,j=0}^{n} (a_{i,j} - \bar{a}_{i,j}) x_i x_j, \end{aligned}$$

the bar meaning complex-conjugate. Both F_1 and F_2 are real polynomials and it is

$$F = \frac{1}{2}(F_1 - \boldsymbol{i} F_2). \tag{7.9}$$

Since the coordinates of any point of Q may be taken real and satisfy the equation (7.8), they satisfy also the equations $F_1 = 0$ and $F_2 = 0$. Thus, for $i = 1, 2$, either the equation $F_i = 0$ is trivial, or it defines a quadric of $\mathbb{P}_n$ which contains all points of Q, and therefore, by 7.3.8, agrees with Q. So, if $G = 0$ is an equation of Q, we have $F_i = \rho_i G$ for some $\rho_i \in \mathbb{R}$. Using the equality (7.8) above, $F = \frac{1}{2}(\rho_1 - \boldsymbol{i}\rho_2)G$; this shows that $G = 0$ may be taken as an equation of Q', and therefore that $Q' = Q_{\mathbb{C}}$, as claimed. □

7.4 Interior and exterior of quadrics

Let Q be a quadric of a real projective space $\mathbb{P}_n$. Its complementary $\mathbb{P}_n - |Q|$ splits into two intrinsically distinct subsets named the interior and the exterior of Q. In this section we will show this splitting and its main properties.

Throughout this section, $\mathbb{P}_n$ will be assumed to be a real projective space of dimension $n > 1$. We will deal with non-degenerate quadrics only; anyway, our considerations may be easily extended to degenerate quadrics of rank greater than 1 by taking them as cones over non-degenerate quadrics in spaces of smaller dimension, as explained in Section 6.7.

We begin by some rather technical considerations. Assume that $Q = [\eta]$ is a non-degenerate quadric of $\mathbb{P}_n$. Once the representative η is fixed, one may assign to each point $p = [v] \notin Q$ the sign, positive or negative, of $\eta(v, v)$. This sign is independent of the choice of v because $\eta(\lambda v, \lambda v) = \lambda^2 \eta(v, v)$ for any $\lambda \in \mathbb{R} - \{0\}$. Obviously the signs of the points depend on the choice of η, but comparing signs will be independent of any choice. Indeed, we will say that points $p, q \notin Q$, $p = [v]$, $q = [w]$, have *the same sign* relative to $Q = [\eta]$ if and only if $\eta(v, v)\eta(w, w) > 0$. If, otherwise, $\eta(v, v)\eta(w, w) < 0$, p and q will be said to have *opposite signs*. Clearly, both conditions are independent of the choice of η because if $\eta' = \lambda\eta$ is another representative of Q, then

$$\eta'(v, v)\eta'(w, w) = \lambda^2 \eta(v, v)\eta(w, w).$$

Having the same sign is an equivalence relation on $\mathbb{P}_n - |Q|$ with at most two equivalence classes, as there are at most two possibilities for the sign of each point $p \in \mathbb{P}_n - |Q|$ relative to a fixed representative of Q.

Remark 7.4.1. Let Δ be a reference of $\mathbb{P}_n$ and assume to have fixed an equation of Q relative to Δ,

$$F(x_0, \dots, x_n) = \sum_{i,s=0}^{n} a_{i,s} x_i x_s = 0.$$

Then there is a representative η of Q and a basis $e_o, \dots, e_n$, adapted to Δ such that $a_{i,s} = \eta(e_i, e_s)$, $i, s = 0, \dots, n$. If $p = [\alpha_0, \dots, \alpha_n]_\Delta$ is any point, $v = \alpha_0 e_0 + \dots + \alpha_n e_n$ is a representative of p and

$$F(\alpha_0, \dots, \alpha_n) = \sum_{i,s=0}^{n} a_{i,s} \alpha_i \alpha_s = \eta(v, v).$$

Thus, for any $p \notin Q$ and any coordinates $\alpha_0, \dots, \alpha_n$ of p, the sign of $F(\alpha_0, \dots, \alpha_n)$ is the sign of p relative to η. As a consequence, the signs of points $p = [\alpha_0, \dots, \alpha_n]$ and $q = [\beta_0, \dots, \beta_n]$ may be compared by just comparing the signs of $F(\alpha_0, \dots, \alpha_n)$ and $F(\beta_0, \dots, \beta_n)$, where $F = 0$ is an arbitrary equation of Q. Using a reduced equation, the reader may easily see that if a quadric Q has index n, then all points have the same sign relative to Q.

In the sequel the signs of the points will be used for comparative purposes only. Thus, they may be understood to be relative to an arbitrarily chosen fixed representative (or equation, by 7.4.1) of Q, all our considerations being independent of such a choice.

Lemma 7.4.2. *Let Q be a non-degenerate quadric. If p, q, $p \neq q$, are conjugate points not in Q, then they have opposite signs if and only if the line $p \vee q$ has two real intersection points with Q.*

Proof. Take $p = [v]$, $q = [w]$, $Q = [\eta]$. The points p and q being conjugate, $\eta(v, w) = 0$. Therefore, according to 6.3.3, the equation of $Q \cap (p \vee q)$ is

$$\eta(v, v)\lambda^2 + \eta(w, w)\mu^2 = 0,$$

with $\eta(v, v) \neq 0$, $\eta(w, w) \neq 0$, from which the claim. □

Lemma 7.4.3. *If Q is a non-degenerate quadric of index $j = \mathbf{j}(Q)$ and Ω a self-polar simplex, then $j + 1$ of the vertices of Ω have one sign, while the remaining $n - j$ have the opposite sign.*

Proof. According to 7.1.8, for a suitable choice of a unit point A, $(p_0, \dots, p_n, A)$ is a reference relative to which Q has reduced equation. Since Q is non-degenerate and has index j, by 7.2.5 and up to a suitable renumbering of the vertices, the reduced equation is

$$x_0^2 + \cdots + x_j^2 - x_{j+1}^2 - \cdots - x_n^2 = 0.$$

As explained in 7.4.1, we may use it to compare signs, after which $p_0, \dots, p_j$ have the same sign and $p_{j+1}, \dots, p_n$ have the opposite sign. □

Next we set the main definitions. Assume we have fixed a non-degenerate quadric Q of $\mathbb{P}_n$, $n > 1$: a point p is called *interior* to Q if and only if its polar hyperplane H_p shares no real point with Q. Since the points of Q lie in their polar hyperplanes, no interior point belongs to Q. By 6.4.11, one may equivalently say that the interior points are those no real line through which is tangent to Q. The set of all interior points is called the *interior* or the *inside* of Q; we will denote it as $\mathrm{Int}(Q)$.

The set $\mathrm{Ext}(Q) = \mathbb{P}_n - \mathrm{Int}(Q) \cup |Q|$ is called the *exterior* or the *outside* of Q. Its points, called *exterior* points, are the points $p \notin Q$ for which $Q \cap H_p$ has some real point or, equivalently, the points $p \notin Q$ through which there is some real line tangent to Q. We have thus a decomposition of $\mathbb{P}_n$ in three pairwise disjoint subsets

$$\mathbb{P}_n = \mathrm{Int}(Q) \cup |Q| \cup \mathrm{Ext}(Q).$$

Remark 7.4.4. Clearly the above splitting has a projective character: if f is any projectivity defined in $\mathbb{P}_n$, by 6.4.9, 6.1.7 and 6.1.3, $f(\mathrm{Int}(Q)) = \mathrm{Int}(f(Q))$ and $f(\mathrm{Ext}(Q)) = \mathrm{Ext}(f(Q))$. In particular, taking f the polarity relative to Q, the hyperplanes interior (resp. exterior) to the envelope Q^* are the polar hyperplanes of the points interior (resp. exterior) to Q.

However, in many cases the above decomposition is trivial:

Proposition 7.4.5. *Assume that Q is a non-degenerate quadric of $\mathbb{P}_n$, $n > 1$:*

(a) *If* $\mathbf{j}(Q) = n$, *then* $\mathrm{Int}(Q) = \mathbb{P}_n$.

(b) *If* $\mathbf{j}(Q) \le n - 2$, *then* $\mathrm{Int}(Q) = \emptyset$.

Proof. If $\mathbf{j}(Q) = n$, then Q has no (real) points and therefore no polar hyperplane shares points with it. If $\mathbf{j}(Q) \le n-2$, then, by the definition of $\mathbf{j}(Q)$, no hyperplane has empty intersection with Q. □

Due to 7.4.5, we will focus our attention on the non-degenerate quadrics Q of $\mathbb{P}_n$, $n > 1$, with $\mathbf{j}(Q) = n - 1$. This includes the most important case of the real non-degenerate conics, and also the case of the non-ruled quadrics of $\mathbb{P}_3$. Note first that if $\mathbf{j}(Q) = n - 1$, then, according to 7.4.3, n vertices of any self-polar simplex Δ have the same sign, while the remaining one has the opposite sign. We will refer to the latter as the *lone vertex* of Δ. Note that it is well determined because the other vertices are $n > 1$ in number.

The key to the properties of the interior points is the next lemma:

Lemma 7.4.6. *If Q is a non-degenerate quadric of index $n - 1$, then any self-polar simplex has its lone vertex interior and all the other vertices exterior.*

Proof. Assume that $p_0, p_1, \dots, p_n$ are the vertices of a self-polar simplex. After a suitable renumbering, by 7.1.8, they may be assumed to be the vertices of a reference relative to which Q has (reduced) equation

$$x_0^2 + \cdots + x_{n-1}^2 - x_n^2 = 0.$$

On one hand p_n clearly is the lone vertex. On the other, since the simplex is self-polar, the face opposite p_i is the polar hyperplane H_{p_i} of p_i. Therefore the equations (relative to the subordinated references) of the sections $Q \cap H_{p_i}$ are

$$x_0^2 + \cdots + \widehat{x_i^2} + \cdots + x_{n-1}^2 - x_n^2 = 0$$

if $i < n$ and

$$x_0^2 + \cdots + x_{n-1}^2 = 0$$

if $i = n$. On one hand, it follows that for $i < n$ it is $\mathbf{j}(Q \cap H_{p_i}) = n-2$ and therefore $|Q \cap H_{p_i}| \neq \emptyset$. On the other, $\mathbf{j}(Q \cap H_{p_n}) = n - 1$ and thus $|Q \cap H_{p_n}| = \emptyset$. This proves that p_n is interior and also that $p_0, \dots, p_{n-1}$ are exterior, as obviously they do not belong to Q. □

Corollary 7.4.7. *If Q is a non-degenerate quadric of index $n - 1$ and p a point interior to Q, then p and any point conjugate to p have opposite signs.*

Proof. Take any $q \in H_p$. $Q \cap H_p$ is a quadric with no points because p is interior. Then $q \notin Q \cap H_p$ and so, by 7.1.5, there is a simplex of H_p, with vertices $q = q_0, \dots, q_{n-1}$, which is self-polar for $Q \cap H_p$. Since $p \notin H_p$, $q = q_0, \dots, q_{n-1}, p$ are the vertices of a simplex of $\mathbb{P}_n$, which clearly is self-polar for Q. Its lone vertex being p, by 7.4.6, and q being another vertex, the claim follows. □

The main property of the interior points follows:

Theorem 7.4.8. *If a point p is interior to a non-degenerate quadric Q of $\mathbb{P}_n$ of index $n-1$, then all lines through p are chords of Q with real ends.*

Conversely, if Q is a quadric of $\mathbb{P}_n$ and p a point any line through which is a chord of Q with real ends, then Q is non-degenerate, has index $n-1$ and p is an interior point of Q.

Proof. Assume that p is interior and ℓ is a line through it. Since $p \notin H_p$, $\ell \cap H_p$ is a point q different from p. Then $\ell = p \vee q$ and, by 7.4.7, p and q have different sign, after which the claim follows from 7.4.2.

Conversely, assume that all lines through p are chords with real ends. Then clearly Q has no double point, as otherwise a line through p and a double point would be tangent. Also $p \notin Q$, as otherwise $p \in H_p$ and any line ℓ, $p \in \ell \subset H_p$, would be tangent. By 7.1.5, there is a self-polar simplex Δ with p as a vertex; let the vertices of Δ be $p = p_0, \dots, p_n$. By the hypothesis, all lines $p \vee p_i$, $i > 0$, are chords with real ends and therefore, by 7.4.2, p and p_i have opposite signs for all $i > 0$. By 7.4.3, $\mathbf{j}(Q) = n-1$. Furthermore p is the lone vertex of Δ and hence an interior point of Q, by 7.4.6. □

Remark 7.4.9. It follows from 7.4.8 that if Q is non-degenerate and has index $n-1$, as above, then the lines that are either tangent to Q or disjoint with Q contain no interior point.

Proposition 7.4.10. *If Q is a non-degenerate quadric of $\mathbb{P}_n$ of index $n-1$, then* $\mathrm{Int}(Q) \neq \emptyset$, $\mathrm{Ext}(Q) \neq \emptyset$, *all points of* $\mathrm{Int}(Q)$ *have the same sign and all points of* $\mathrm{Ext}(Q)$ *have the opposite sign.*

Proof. The first two claims are a direct consequence of the existence of self-polar simplexes and 7.4.6.

If p and p' are interior points, the sign of any $q \in H_p \cap H_{p'}$ is, by 7.4.7, opposite the signs of p and p', which therefore are the same.

If q is exterior, $q \notin Q$ and hence there is a self-polar simplex Δ with vertex q. The lone vertex p of Δ is interior by 7.4.6, and so $q \neq p$. Then, by the definition of lone vertex, the sign of q is opposite the sign of p, and, as already said, p is an interior point. □

Corollary 7.4.11. *If Q is non-degenerate and has index $n-1$, then any chord with real ends contains both interior and exterior points to Q.*

Proof. Let the chord ℓ have ends q_1, q_2, take any $p \in \ell - \{q_1, q_2\}$ and p' the fourth harmonic of q_1, q_2, p. Then p and p' are conjugate, they have opposite sign by 7.4.2 and so, by 7.4.10, one of them is interior and the other is exterior. □

Remark 7.4.12. The proof of 7.4.11 shows that the involution of the chord that leaves fixed its ends, maps interior points to exterior points and conversely.

A direct consequence of 7.4.1 and 7.4.10 is:

Corollary 7.4.13. *Assume, as above, that Q is non-degenerate and has index $n-1$, and let $F(x_0, \dots, x_n) = 0$ be an equation of Q. Then the sign of $F(\alpha_0, \dots, \alpha_n)$ is the same for all coordinates $\alpha_0, \dots, \alpha_n$ of all interior points of Q. The same expression takes the opposite sign for all coordinates of all exterior points of Q.*

As a final note, the reader may easily check that if Q is any non-degenerate quadric with $\mathbf{j}(Q) < n-1$, then $\operatorname{Ext}(Q) = \mathbb{P}_n - |Q|$ splits into two non-empty disjoint subsets according to the sign of the points. This splitting, however, has little interest and in some cases the subsets it gives rise to cannot be projectively distinguished. For instance the projectivity $[x_0, x_1, x_2, x_3] \mapsto [x_0, -x_1, x_2, -x_3]$ leaves invariant the ruled quadric $Q : x_0x_1 - x_2x_3 = 0$ and maps each point $p \notin Q$ to a point with the opposite sign.

7.5 Quadrics of affine spaces

In the remaining sections of this chapter we will deal with affine spaces and their projective closures. We will use the conventions and notations introduced in Chapter 3. Unless otherwise stated, the linear varieties are those of the projective closures: they are called proper if they contain proper points; otherwise, they are called improper. Two proper linear varieties are called parallel when so are their proper parts, or, equivalently (3.4.9), when their improper parts are included in either sense. Affinities are projectivities between projective closures of affine spaces mapping improper points to improper points (see Section 3.5).

The quadrics of an affine space $\mathbb{A}_n$ are, by definition, the quadrics of its projective closure $\overline{\mathbb{A}}_n$. They are called *affine quadrics*, even if such term is somewhat misleading, as it may suggest that the affine quadrics are objects of a different kind than the quadrics of projective spaces, as defined in Chapter 6, and they are not. The image of any affine quadric by any affinity is obviously an affine quadric too, as it is the image of any affine quadric by any projectivity between projective closures of affine spaces.

Affine quadrics are thus just particular samples of quadrics and all we have explained about quadrics till now, such as tangency, conjugation or polarity, does apply to them; in particular the projective classification of quadrics applies to the affine quadrics and it makes sense to consider the projective type, the projective reduced equation and the projective invariants of an affine quadric. Furthermore, by 7.1.2, any projective type of quadrics may be represented by an affine quadric. The only essential difference between quadrics of arbitrary projective spaces and affine quadrics comes from the fact that the latter belong to a particular type of projective spaces – the projective closures of affine spaces – that have a further, non-projective, structure. Such structure allows us to have the affinities as a distinguished class of projectivities, and the affine quadrics may be further studied and classified under the action of affinities, in what is called the *affine geometry of quadrics*. Just to get a preliminary idea of this, pay some attention to the obvious fact that the set of points of an affine quadric Q splits into the disjoint sets of its proper and improper points, namely $|Q| = (|Q| \cap \mathbb{A}_n) \cup (|Q| \cap H_\infty)$. It makes no sense to ask if such a splitting is preserved by projectivities, as it is not defined for quadrics other than the affine ones. Not even the projectivities between projective closures of affine spaces, for which the question makes sense, preserve the properness or improperness of the points of Q (by 2.8.1), but, by its own definition, the affinities do: one says then that splitting the set points of a quadric in those of its proper and improper points belongs to the affine geometry of quadrics, but not to their projective geometry.

We will begin by considering something richer than just the set of the improper points of an affine quadric. To each quadric Q of an n-dimensional affine space, $n \geq 2$, there is associated its section by the improper hyperplane, $Q \cap H_\infty$, which may be either a quadric of H_∞, and then is called the *improper section* of Q, or the whole of the improper hyperplane, in which case it is said that the improper section of Q is *undetermined*. The improper section being not defined for affine point-pairs, in the sequel, unless otherwise stated, we will assume for simplicity that all affine quadrics belong to spaces of dimension greater than one. If wanted, the reader may easily include the affine point-pairs in the forthcoming discussions by taking the improper sections of all point-pairs containing the improper point as undetermined, and the improper sections of all the remaining point-pairs equal to the empty set, with formally assigned rank 0 and, in the real case, index 0.

In the next proposition we will see that the relationship between affine quadrics and their improper sections is preserved by the affinities, which is often stated by saying that the improper section of a quadric Q is an *affine element* of Q.

Proposition 7.5.1. *Let Q be an affine quadric of $\mathbb{A}_n$ and $f : \overline{\mathbb{A}}_n \to \overline{\mathbb{A}}'_n$ an affinity. If H_∞ and H'_∞ denote the improper hyperplanes of $\mathbb{A}_n$ and $\mathbb{A}'_n$, respectively, then $f(Q \cap H_\infty) = f(Q) \cap H'_\infty$.*

Proof. Since f is an affinity, $f(H_\infty) = H'_\infty$, after which the claim is a particular case of 6.1.7. □

Note that 7.5.1 applies to both the cases of determined and undetermined improper sections.

Assume that an affine reference Δ of $\mathbb{A}_n$ has been fixed and denote by $x_0, \dots, x_n$ and $X_1, \dots, X_n$ the homogeneous and affine coordinates, respectively, relative to it. Any quadric Q of $\mathbb{A}_n$ has thus an equation that may be written in the form

$$\sum_{i,j=1}^{n} a_{i,j} x_i x_j + 2 \sum_{i=1}^{n} a_{0,i} x_0 x_i + a_{0,0} x_0^2 = 0, \tag{7.10}$$

where $a_{i,j} = a_{j,i}$ are the coefficients of the corresponding matrix of Q. Since the homogeneous coordinates of a point with affine coordinates $X_1, \dots, X_n$ are $1, X_1, \dots, X_n$, the condition for it to belong to Q may be written

$$\sum_{i,j=1}^{n} a_{i,j} X_i X_j + 2 \sum_{i=1}^{n} a_{0,i} X_i + a_{0,0} = 0. \tag{7.11}$$

This equation is called an *affine equation* of Q relative to Δ, while the former equation (7.10) is called a homogeneous equation of Q, also relative to Δ.

Passing from a homogeneous equation to the corresponding affine one is done by just substituting $1, X_1, \dots, X_n$ for $x_0, x_1, \dots, x_n$ in the homogeneous equation, which is called *dehomogenizing*. The reverse is called *homogenizing* and may be done by substituting x_i/x_0 for X_i, $i = 1, \dots, n$, in the affine equation, and then multiplying the resulting equality by x_0^2.

Obviously there is just a formal difference between the above affine and homogeneous equations, and both carry the same information. Once the affine reference has been fixed, either of these equations determines Q, and is in turn determined by Q up to a non-zero factor.

Remark 7.5.2. The reference Δ being affine, its 0-th face $x_0 = 0$ is the improper hyperplane. If we take in it the reference subordinated by Δ, by 6.1.10, the matrix

$$\begin{pmatrix} a_{1,1} & \dots & a_{n,0} \\ \vdots & & \vdots \\ a_{n,1} & \dots & a_{n,n} \end{pmatrix}$$

is zero if and only if the improper section of Q is undetermined. Otherwise it is a matrix of $Q \cap H_\infty$. Equivalently (6.1.11), the equation

$$\sum_{i,j=1}^{n} a_{i,j} x_i x_j = 0 \tag{7.12}$$

is identically zero if and only if the improper section of Q is undetermined. Otherwise it is an equation of the improper section of Q. The reader may note that the first member of the equation (7.12) is, but for the names of the variables, the part of degree two of the first member of the affine equation (7.11).

The quadrics with undetermined improper section are thus those with homogeneous equations

$$\sum_{i=1}^{n} 2a_{0,i}x_0x_i + a_{0,0}x_0^2 = \Big(\sum_{i=1}^{n} 2a_{0,i}x_i + a_{0,0}x_0\Big)x_0 = 0.$$

In view of their equations, they are the hyperplane-pairs $H + H_\infty$, where H is a proper hyperplane, if $a_{0,i} \neq 0$ for some $i \neq 0$, or $H = H_\infty$ if, otherwise, $a_{0,i} = 0$ for all $i \neq 0$ and therefore $a_{0,0} \neq 0$.

By dehomogenizing the above equations we get the affine equations of the quadrics with undetermined improper section, namely

$$2\sum_{i=1}^{n} a_{0,i}X_i + a_{0,0} = 0,$$

with $a_{0,i} \neq 0$ for some $i \neq 0$. These are equations of the proper parts of the above hyperplanes H which together with H_∞ compose the quadrics. In particular it results in the incompatible equation

$$a_{0,0} = 0$$

in case $H = H_\infty$. This is of course coherent with the fact that the proper points of $H + H_\infty$ are those of H, $2H_\infty$ having in particular no proper points. It is worth noting that, therefore, any hyperplane H' of $\mathbb{A}_n$ may be viewed as the proper part of its projective closure H, and also as the set of proper points of the hyperplane-pair $H + H_\infty$. Corresponding to these two views, an equation of H' may be homogenized either to an equation of H, of degree one, as explained in 3.4.2, or to an equation of $H + H_\infty$, of degree two, as explained above. Obviously, the latter is the product of the former by x_0.

When affine geometry is developed without using the projective closure, quadrics are often defined as being given by equations of the form of (7.11) with the added condition of being $a_{i,j} \neq 0$ for some $i \neq 0$, $j \neq 0$, in order to assure that the defining equation (7.11) has degree two. We will not follow this convention here, as it excludes the affine quadrics with undetermined improper section. These quadrics are not of great interest by themselves, as they are just hyperplane pairs. Nevertheless they appear in families of quadrics and considering them helps to explain some apparently paradoxical situations, such as, for instance, having the line $X_2 = 0$ as a member of the family of conics $X_2 + \lambda X_1^2 = 0$, $\lambda \in k$. Actually, it is the line-pair $x_0x_2 = 0$ – a conic – which belongs to the family $x_0x_2 + \lambda x_1^2 = 0$, $\lambda \in k$.

7.6 Affine reduced equations of quadrics

In this section and the next one we will present a classification that applies to the affine quadrics only, and is finer than the projective one. It is induced by the action of the affinities and called the *affine classification of quadrics*: two affine quadrics Q and Q', of affine spaces $\mathbb{A}_n$ and $\mathbb{A}'_n$, are said to be *affinely equivalent* if and only if there is an affinity $f : \bar{\mathbb{A}}_n \to \bar{\mathbb{A}}'_n$ such that $f(Q) = Q'$. The affine equivalence of Q and Q' will be denoted by $Q \sim_a Q'$. It is a direct consequence of 3.5.4 and 6.1.4 that the affine equivalence is, indeed, an equivalence relation. Its equivalence classes will be called *affine classes* (or *types*) *of quadrics*.

Our line of reasoning will closely follow the one already used for the projective classifications. We start with an affine version of Lemma 7.1.1:

Lemma 7.6.1. *Quadrics Q and Q', of affine spaces $\mathbb{A}_n$ and $\mathbb{A}'_n$ respectively, are affinely equivalent if and only if there exist affine references Δ, of $\mathbb{A}_n$, and Δ', of $\mathbb{A}'_n$, relative to which Q and Q' have equal equations.*

Proof. As in the proof of 7.1.1, if f is the projectivity mapping Δ to Δ', then $f(Q) = Q'$. Now, since Δ and Δ' are affine references, their 0-th faces are the improper hyperplanes H_∞ and H'_∞ of $\mathbb{A}_n$ and $\mathbb{A}'_n$, after which the condition $f(\Delta) = \Delta'$ forces $f(H_\infty) = H'_\infty$ and f is an affinity. For the converse, if f is an affinity and $f(Q) = Q'$, take Δ to be any affine reference of $\mathbb{A}_n$ and $\Delta' = f(\Delta)$: Δ' is an affine reference because f is an affinity, and the equations of Q and Q' relative to Δ and Δ' do agree by 6.1.14. □

Now we will proceed to getting affine reduced equations. As in the projective case, they need to be simplified up to the point of retaining all the information depending of the affine type of the quadric, but no other.

Assume that Q is a quadric of an affine space $\mathbb{A}_n$. If Q contains the improper hyperplane, choose Δ to be any affine reference. Otherwise, let $\Delta' = (p_1, \dots, p_n, A')$ be a reference of H_∞ relative to which the equation of $Q \cap H_\infty$ is reduced. In order to get an affine reference having Δ' as induced reference, we take p_0 to be any proper point, $A \in p_0 \vee A'$, $A \neq p_0, A'$, and $\Delta = (p_0, p_1, \dots, p_n, A)$. Next we check that Δ is a projective reference: $p_0 \notin H_\infty = p_1 \vee \dots \vee p_n$ guarantees the independence of $p_0, \dots, p_n$; since p_0 is proper, $(p_0 \vee A') \cap H_\infty = A'$ and therefore $A \notin p_1 \vee \dots \vee p_n$; if for some $i = 1, \dots, n$ it is $A \in H_i = p_0 \vee \dots \vee \widehat{p_i} \vee \dots \vee p_n$, then $p_0 \vee A \subset H_i$ and therefore

$$A' = (p_0 \vee A) \cap H_\infty \in H_i \cap H_\infty = p_1 \vee \dots \vee \widehat{p_i} \vee \dots \vee p_n$$

against the fact that Δ' is a reference. By its construction, it is clear that Δ is affine and Δ' is the reference of H_∞ induced by it.

Assume that $k = \mathbb{R}$. Then, by 7.5.2, an equation of Q relative to Δ has the form

$$x_1^2 + \dots + x_j^2 - x_{j+1}^2 - \dots - x_r^2 = 2\sum_{i=1}^{n} a_{0,i} x_0 x_i + a_{0,0} x_0^2 \qquad (7.13)$$

with $0 \leq r \leq n$ and $j \geq r - j$; furthermore, in case $r = 0$ it is $a_{0,i} \neq 0$ for some $i = 0, \dots, n$. Note that we include the value $r = 0$, for which the first member is zero, as the set of summands is empty. This covers the case $H_\infty \subset Q$.

Similarly, if $k = \mathbb{C}$, an equation of Q has the form of (7.13) above with $j = r$. For a while we will deal with both the real and complex cases together using the equation (7.13) and assuming that in case $k = \mathbb{C}$ it is $j = r$.

The matrix of the following substitution of variables is obviously regular:

$$\begin{aligned}
x_0 &= y_0, \\
x_1 &= a_{0,1} y_0 + y_1, \\
&\vdots \\
x_j &= a_{0,j} y_0 + y_j, \\
x_{j+1} &= -a_{0,j+1} y_0 + y_{j+1}, \\
&\vdots \\
x_r &= -a_{0,r} y_0 + y_r, \\
x_{r+1} &= y_{r+1}, \\
&\vdots \\
x_n &= y_n,
\end{aligned}$$

and so (2.3.3) $y_0, \dots, y_n$ are new homogeneous coordinates, relative to a certain reference Ω. Furthermore, the first equation, $x_0 = y_0$, assures that the references Δ and Ω to have the same 0-th face, hence Ω is an affine reference too. After substituting, an equation of Q relative to Ω is

$$y_1^2 + \dots + y_j^2 - y_{j+1}^2 - \dots - y_r^2 = 2 \sum_{i=r+1}^{n} a_{0,i} y_0 y_i + b_{0,0} y_0^2, \tag{7.14}$$

where $b_{0,0} = a_{0,0} - \sum_{i=1}^{r} a_{0,i}^2$. To continue reducing the equation we will distinguish three cases:

Case I. $a_{0,r+1} = \dots = a_{0,n} = b_{0,0} = 0$. Note that in such a case $r = 0$ is excluded, as it would give $a_{0,0} = b_{0,0} = 0$, and $a_{0,i} = 0$ for $i = 1, \dots, r$. If $k = \mathbb{C}$, the equation (7.14) takes now the form

$$y_1^2 + \dots + y_r^2 = 0,$$

$1 \leq r \leq n$, while for $k = \mathbb{R}$ it is

$$y_1^2 + \dots + y_j^2 - y_{j+1}^2 - \dots - y_r^2 = 0$$

with $j \geq r - j$ and $1 \leq r \leq n$. We will not modify these equations any further.

Case II. $a_{0,r+1} = \cdots = a_{0,n}$, $b_{0,0} \neq 0$. Now the equation (7.14) is

$$y_1^2 + \cdots + y_j^2 - y_{j+1}^2 - \cdots - y_r^2 = b_{0,0} y_0^2. \tag{7.15}$$

In case $k = \mathbb{R}$ and $b_{0,0} < 0$ we change the sign of the equation and renumber $y_1, \ldots, y_r$ if needed, in order to obtain an equation of the same form as (7.15), but with $b_{0,0} > 0$. Then j may no longer satisfy $j \geq r - j$. After this, in both the real and complex cases, it makes sense to consider the equations

$$y_0 = z_0 \quad \text{and} \quad y_i = \sqrt{b_{0,0}} z_i \quad \text{for } i = 1, \ldots, n,$$

which, as in the former case, define new affine coordinates $z_0, \ldots, z_n$. We divide the resulting equation by $b_{0,0}$ to get

$$z_1^2 + \cdots + z_r^2 = z_0^2,$$

with $0 \leq r \leq n$, if $k = \mathbb{C}$, and

$$z_1^2 + \cdots + z_j^2 - z_{j+1}^2 - \cdots - z_r^2 = z_0^2,$$

with $0 \leq j \leq r$ and $0 \leq r \leq n$, if $k = \mathbb{R}$.

Case III. There is $a_s \neq 0$ with $r + 1 \leq s \leq n$. Note that this may occur only in case $r < n$. Up to renumbering $y_{r+1}, \ldots, y_n$, we may assume without restriction $s = n$. The equations

$$y_0 = z_0, \quad \ldots \quad , y_{n-1} = z_{n-1}, \quad \sum_{i=r+1}^{n} a_{0,i} y_i + \frac{b_{0,0}}{2} y_0 = z_n$$

again define new affine coordinates $z_0, \ldots, z_n$ ($a_{0,n} \neq 0$ is used here) giving

$$z_1^2 + \cdots + z_r^2 = 2 z_0 z_n,$$

with $0 \leq r < n$, in the complex case, and

$$z_1^2 + \cdots + z_j^2 - z_{j+1}^2 - \cdots - z_r^2 = 2 z_0 z_n,$$

with $j \geq r - j$ and $0 \leq r < n$, in the real one.

Summarizing we have proved:

Theorem 7.6.2. *For any quadric Q of an n-dimensional complex affine space there is an affine reference relative to which Q has one of the following equations:*

$$\begin{array}{lll} \text{I} & x_1^2 + \cdots + x_r^2 = 0, & 1 \leq r \leq n, \\ \text{II} & x_1^2 + \cdots + x_r^2 = x_0^2, & 0 \leq r \leq n, \\ \text{III} & x_1^2 + \cdots + x_r^2 = 2x_0 x_n, & 0 \leq r < n. \end{array}$$

Similarly, for any quadric Q of an n-dimensional real affine space there is an affine reference relative to which Q has one of the following equations:

$$\begin{array}{lll} \text{I} & x_1^2+\cdots+x_j^2-x_{j+1}^2-\cdots-x_r^2=0, & j\geq r-j \text{ and } 1\leq r\leq n,\\ \text{II} & x_1^2+\cdots+x_j^2-x_{j+1}^2-\cdots-x_r^2=x_0^2, & 0\leq j\leq r \text{ and } 0\leq r\leq n,\\ \text{III} & x_1^2+\cdots+x_j^2-x_{j+1}^2-\cdots-x_r^2=2x_0x_n, & j\geq r-j \text{ and } 0\leq r< n. \end{array}$$

In both cases such an equation is called an affine reduced equation *of Q.*

7.7 Affine classification of quadrics

We already know from 7.6.1 that quadrics with equal affine reduced equations are affinely equivalent. Now it is time to prove the converse by introducing sets of affine invariants that determine the affine reduced equations. We begin by defining the invariants. Assume that we have affinely equivalent quadrics Q and Q' of affine spaces $\mathbb{A}_n$ and $\mathbb{A}'_n$ with improper hyperplanes H_∞ and H'_∞, respectively. Thus there is an affinity $f\colon \overline{\mathbb{A}}_n \to \overline{\mathbb{A}}'_n$ such that $f(Q) = Q'$. First, since f is in particular a projectivity, Q and Q' are projectively equivalent. Furthermore, by 7.5.1, $f(Q\cap H_\infty) = Q'\cap H'_\infty$ and so either both Q and Q' have undetermined improper section, or the improper sections of Q and Q' are projectively equivalent quadrics. In the sequel we will refer to either of these possibilities by saying that Q and Q' have *projectively equivalent improper sections*. Using this convention, we have proved:

Lemma 7.7.1. *If Q and Q' are affinely equivalent quadrics, then they are projectively equivalent and have projectively equivalent improper sections.*

In particular we see that the affine classification is finer than the projective one. In view of 7.7.1, we take as affine invariants of a quadric its projective invariants together with the projective invariants of its improper section. More precisely:

> If Q is an affine quadric with $Q\cap H_\infty \neq H_\infty$, we define $\mathbf{r}'(Q) = \mathbf{r}(Q\cap H_\infty)$. If Q has undetermined improper section we take $\mathbf{r}'(Q) = 0$. In both cases we call it the *rank of the improper section* of Q.
>
> If $k = \mathbb{R}$ and again Q is an affine quadric with $Q\cap H_\infty \neq H_\infty$, we define $\mathbf{j}'(Q) = \mathbf{j}(Q\cap H_\infty)$. If Q has undetermined improper section we take $\mathbf{j}'(Q) = -1$. We will call $\mathbf{j}'(Q)$ the *index of the improper section* of Q.

It directly follows from 7.7.1 and the above definitions that $\mathbf{r}(Q), \mathbf{r}'(Q)$ in the complex case and $\mathbf{r}(Q), \mathbf{j}(Q), \mathbf{r}'(Q), \mathbf{j}'(Q)$ in the real one, are affine invariants of Q, namely:

Proposition 7.7.2. *If Q and Q' are affinely equivalent quadrics, then*

$$\mathbf{r}(Q) = \mathbf{r}(Q') \quad \textit{and} \quad \mathbf{r}'(Q) = \mathbf{r}'(Q').$$

Furthermore, in case $k = \mathbb{R}$,

$$\mathbf{j}(Q) = \mathbf{j}(Q') \quad \textit{and} \quad \mathbf{j}'(Q) = \mathbf{j}'(Q').$$

Computing the above affine invariants is not difficult. If a matrix of Q relative to an affine reference is

$$N = \begin{pmatrix} a_{0,0} & a_{0,1} & \dots & a_{0,n} \\ a_{1,0} & a_{1,1} & \dots & a_{1,n} \\ \vdots & \vdots & & \vdots \\ a_{n,0} & a_{n,1} & \dots & a_{n,n} \end{pmatrix},$$

then, according to 7.5.2,

$$N' = \begin{pmatrix} a_{1,1} & \dots & a_{1,n} \\ \vdots & & \vdots \\ a_{n,1} & \dots & a_{n,n} \end{pmatrix},$$

is either a matrix of $Q \cap H_\infty$, or zero if the improper section is undetermined. Then, as already known, $\mathbf{r}(Q) = \operatorname{rk} N$ and by its definition $\mathbf{r}'(Q) = \operatorname{rk} N'$. If Q is a real quadric, then 7.2.14 may be used to compute $\mathbf{j}(Q)$ from N and $\mathbf{j}'(Q)$ from N'.

Assume that the base field is the complex one and consider the diagram

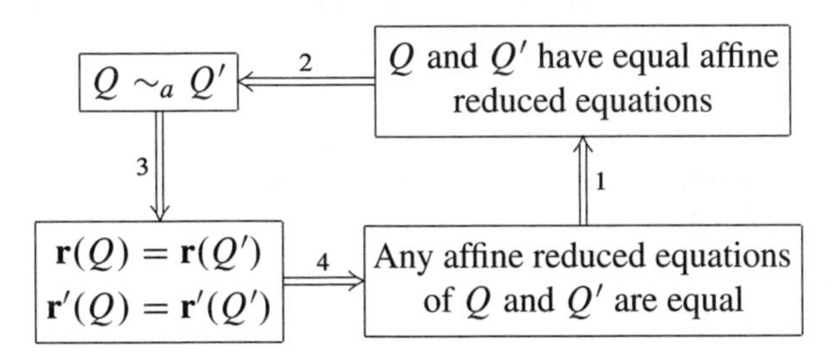

The implication 1 is an obvious consequence of the existence of affine reduced equations (7.6.2), 2 follows from 7.6.1 and 3 has been seen in 7.7.2. Thus, it remains only to prove the implication 4. To this end, we begin by computing the invariants from the reduced equations. Assume that Q has one of the equations

$$\begin{array}{lll} \text{I} & x_1^2 + \dots + x_r^2 = 0, & 1 \le r \le n, \\ \text{II} & x_1^2 + \dots + x_r^2 = x_0^2, & 0 \le r \le n, \\ \text{III} & x_1^2 + \dots + x_r^2 = 2x_0x_n, & 0 \le r < n. \end{array}$$

By writing the corresponding matrices, it easily turns out that:

$$\begin{array}{llll}\mathbf{r}(Q) = r & \text{and} & \mathbf{r}'(Q) = r & \text{if the equation is of type I,}\\ \mathbf{r}(Q) = r+1 & \text{and} & \mathbf{r}'(Q) = r & \text{if the equation is of type II,}\\ \mathbf{r}(Q) = r+2 & \text{and} & \mathbf{r}'(Q) = r & \text{if the equation is of type III.}\end{array}$$

After this we may argue backwards: assume that $\mathbf{r}(Q)$ and $\mathbf{r}'(Q)$ are known: by comparing them we get that the affine reduced equation of Q is of type I if $\mathbf{r}(Q)-\mathbf{r}'(Q) = 0$, of type II if $\mathbf{r}(Q)-\mathbf{r}'(Q) = 1$, and of type III if $\mathbf{r}(Q)-\mathbf{r}'(Q) = 2$. Furthermore, in all cases, the number r of terms in the first member of the equation equals $\mathbf{r}'(Q)$. This shows that the affine invariants of Q determine its affine reduced equation, and hence 4 is proved.

The next theorem summarizes what we have obtained in the complex case:

Theorem 7.7.3 (Affine classification of complex quadrics). *Any quadric Q of an n-dimensional complex affine space has a unique affine reduced equation. For any two quadrics Q and Q' of n-dimensional complex affine spaces, the following three conditions are equivalent:*

(i) *Q and Q' are affinely equivalent,*

(ii) *Q and Q' have equal affine reduced equations,*

(iii) $\mathbf{r}(Q) = \mathbf{r}(Q')$ *and* $\mathbf{r}'(Q) = \mathbf{r}'(Q')$.

We see in particular that $(\mathbf{r}(Q), \mathbf{r}'(Q))$ is a complete system of affine invariants for the quadrics of complex affine spaces. The reader may note that the values of these invariants are not free, but constrained by the inequalities $0 \leq \mathbf{r}(Q)-\mathbf{r}'(Q) \leq 2$. Still $\mathbf{r}(Q)$ measures the degree of degeneration of Q, while $\mathbf{r}'(Q)$ may be interpreted as measuring the degree of degeneration of $Q \cap H_\infty$. Due to 6.4.25, $n - \mathbf{r}'(Q) - 1$ is the dimension of the locus of points at which Q is tangent to the improper hyperplane; this provides a second interpretation of $\mathbf{r}'(Q)$ as a measure of the contact between Q and the improper hyperplane.

Again, representatives of all types are provided by the reduced equations:

Corollary 7.7.4. *Fix any complex affine space $\mathbb{A}_{n,\mathbb{C}}$ and an affine reference Δ of it. Any quadric of any n-dimensional affine space is affinely equivalent to one and only one of the quadrics of $\mathbb{A}_{n,\mathbb{C}}$ that have equations relative to Δ:*

$$\begin{array}{lll}\text{I} & x_1^2 + \cdots + x_r^2 = 0, & 1 \leq r \leq n,\\ \text{II} & x_1^2 + \cdots + x_r^2 = x_0^2, & 0 \leq r \leq n,\\ \text{III} & x_1^2 + \cdots + x_r^2 = 2x_0x_n, & 0 \leq r < n.\end{array}$$

Now we will run for the fourth time the same machinery to complete the affine classification in the real case. Assume thus that $k = \mathbb{R}$. As in the precedent case, from

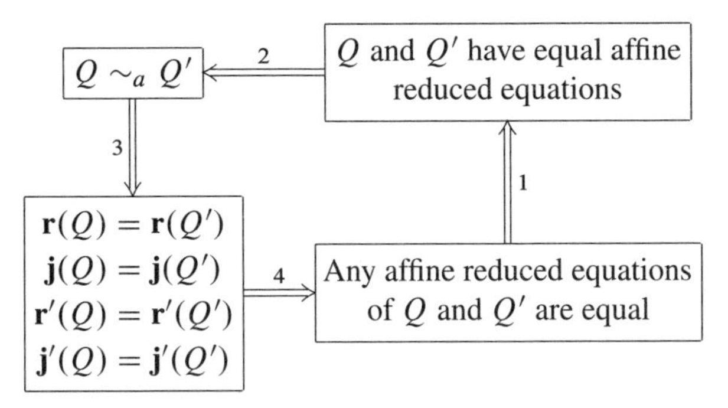

only 4 has not been proved yet. Again we begin by computing the invariants from the reduced equations. Assume that Q has one of the equations

$$\begin{array}{lll} \text{I} & x_1^2 + \cdots + x_j^2 - x_{j+1}^2 - \cdots - x_r^2 = 0, & j \geq r - j \text{ and } 1 \leq r \leq n, \\ \text{II} & x_1^2 + \cdots + x_j^2 - x_{j+1}^2 - \cdots - x_r^2 = x_0^2, & 0 \leq j \leq r \text{ and } 0 \leq r \leq n, \\ \text{III} & x_1^2 + \cdots + x_j^2 - x_{j+1}^2 - \cdots - x_r^2 = 2x_0 x_n, & j \geq r - j \text{ and } 0 \leq r < n. \end{array}$$

Note that an equation of the improper section of Q ($0 = 0$ if the improper section is undetermined) is obtained in all cases by equating to zero the left-hand member of the equation.

Assume first that the equation of Q is of type I. As in the complex case $r = \mathbf{r}'(Q) = \mathbf{r}(Q)$. Since furthermore, in this case, the equations of both Q and its improper section are the projective reduced ones, by 7.2.4 we see that $j - 1 = \mathbf{j}'(Q) = \mathbf{j}(Q)$.

Assume now that the equation of Q is of type II. Then $r = \mathbf{r}'(Q) = \mathbf{r}(Q) - 1$. We will distinguish two subtypes for these equations: subtype IIa if $j > r - j$, and subtype IIb if $j \leq r - j$.

If the equation of Q is of subtype IIa, then the equation of the improper section

$$x_1^2 + \cdots + x_j^2 - x_{j+1}^2 - \cdots - x_r^2 = 0$$

has $j > r - j$. Therefore it is a projective reduced equation of $Q \cap H_\infty$ and, by 7.2.4, $\mathbf{j}'(Q) = j - 1$. The same inequality guarantees that in the complete equation, written

$$x_1^2 + \cdots + x_j^2 - x_{j+1}^2 - \cdots - x_r^2 - x_0^2 = 0,$$

the number of positive terms is not less than the number of negative terms. Thus, such an equation is a projective reduced equation of Q and, again by 7.2.4, $\mathbf{j}(Q) = j - 1$.

If the equation of Q is of subtype IIb and $r > 0$, then, since $j \leq r - j$, the opposites of

$$x_1^2 + \cdots + x_j^2 - x_{j+1}^2 - \cdots - x_r^2 = 0,$$
$$x_1^2 + \cdots + x_j^2 - x_{j+1}^2 - \cdots - x_r^2 - x_0^2 = 0$$

are projective reduced equations of the improper section of Q and Q itself. Therefore this time the counts need to be made from the numbers of negative terms in the above equations and they give $\mathbf{j}'(Q) = r - j - 1$ and $\mathbf{j}(Q) = r - j$. Obviously, these formulas still hold in case $r = 0$, as then $\mathbf{j}(Q) = 0$, $Q \cap H_\infty = H_\infty$ and hence $\mathbf{j}'(Q) = -1$.

Lastly, assume that the equation of Q is of type III. Now $r = \mathbf{r}'(Q) = \mathbf{r}(Q) - 2$. Equating to zero the left-hand member of the equation again gives a projective reduced equation of the improper section, or just $0 = 0$ if the improper section is undetermined. In any case, it results that $\mathbf{j}'(Q) = j - 1$. For equations of type III we will make no use of $\mathbf{j}(Q)$ which, incidentally, is $\mathbf{j}(Q) = j$, as the reader may easily check.

Now we put the machine in reverse to show how the affine reduced equation of Q may be determined from the affine invariants. First, as in the complex case, we compare the invariants $\mathbf{r}(Q)$ and $\mathbf{r}'(Q)$ to check if the reduced equation is of type I, II, or III, and then determine r by the equality $r = \mathbf{r}'(Q)$. This being done, it remains only to determine the value of j in the reduced equation. If the reduced equation has type I or III, then j is determined by the equality $\mathbf{j}'(Q) = j - 1$, which we have seen to hold in both cases. If the reduced equation is of type II, then first we compare $\mathbf{j}(Q)$ and $\mathbf{j}'(Q)$: by the above computations, the subtype is IIa if $\mathbf{j}(Q) = \mathbf{j}'(Q)$, while it is IIb if $\mathbf{j}(Q) = \mathbf{j}'(Q) + 1$. After determining the subtype, we determine j using $\mathbf{j}'(Q) = j - 1$ if the subtype is IIa, or $\mathbf{j}'(Q) = r - j - 1$ if the subtype is IIb. This completes the proof of 4 and hence the following theorem holds:

Theorem 7.7.5 (Affine classification of real quadrics). *Any quadric Q of an n-dimensional real affine space has a unique affine reduced equation. For any two quadrics Q and Q' of real n-dimensional affine spaces, the following three conditions are equivalent:*

(i) *Q and Q' are affinely equivalent,*

(ii) *Q and Q' have equal affine reduced equations,*

(iii) $\mathbf{r}(Q) = \mathbf{r}(Q')$, $\mathbf{r}'(Q) = \mathbf{r}'(Q')$, $\mathbf{j}(Q) = \mathbf{j}(Q')$ *and* $\mathbf{j}'(Q) = \mathbf{j}'(Q')$.

In particular, $(\mathbf{r}(Q), \mathbf{r}'(Q), \mathbf{j}(Q), \mathbf{j}'(Q))$ is a complete system of invariants for the real affine quadrics. As in the complex case, their values are far from being free, see Exercise 7.23. The meaning of $\mathbf{r}'(Q)$ is as in the complex case. By its definition,

$\mathbf{j}'(Q)$ is the highest dimension of a linear variety of H_∞ disjoint with $Q \cap H_\infty$, or, equivalently, the highest dimension of an improper linear variety disjoint with Q. Of course, $\mathbf{j}'(Q)$ also indirectly measures the dimension $\mathbf{s}'(Q) = n - \mathbf{j}'(Q) - 2$ of the largest linear varieties contained in $Q \cap H_\infty$ (see 7.2.8).

Once again the reduced equations give representatives of the affine types:

Corollary 7.7.6. *Fix an n-dimensional real affine space $\mathbb{A}_n$ and an affine reference Δ of it. Any quadric of any real n-dimensional affine space is affinely equivalent to one and only one of the quadrics of $\mathbb{A}_n$ whose equations relative to Δ are:*

I	$x_1^2 + \cdots + x_j^2 - x_{j+1}^2 - \cdots - x_r^2 = 0,$	$j \geq r - j$ *and* $1 \leq r \leq n,$
II	$x_1^2 + \cdots + x_j^2 - x_{j+1}^2 - \cdots - x_r^2 = x_0^2,$	$0 \leq j \leq r$ *and* $0 \leq r \leq n,$
III	$x_1^2 + \cdots + x_j^2 - x_{j+1}^2 - \cdots - x_r^2 = 2x_0x_n,$	$j \geq r - j$ *and* $0 \leq r < n.$

From the affine classification theorems we obtain a nice and very important characterization of the affine equivalence of quadrics, namely:

Corollary 7.7.7. *Quadrics Q and Q', of affine spaces $\mathbb{A}_{n,k}$, $\mathbb{A}'_{n,k}$, $k = \mathbb{R}, \mathbb{C}$, are affinely equivalent if and only if they are projectively equivalent and have projectively equivalent improper sections.*

Proof. The *only if* part has been already seen in 7.7.1. The *if* part is a direct consequence of 7.7.3 and 7.7.5, as the projective equivalence of both the quadrics and their improper sections implies the equality of the affine invariants of the quadrics. □

Corollary 7.7.7 explains how the affine classification refines the projective one: it shows that each projective type of quadrics splits in affine types according to the projective types of the improper sections of the quadrics belonging to it.

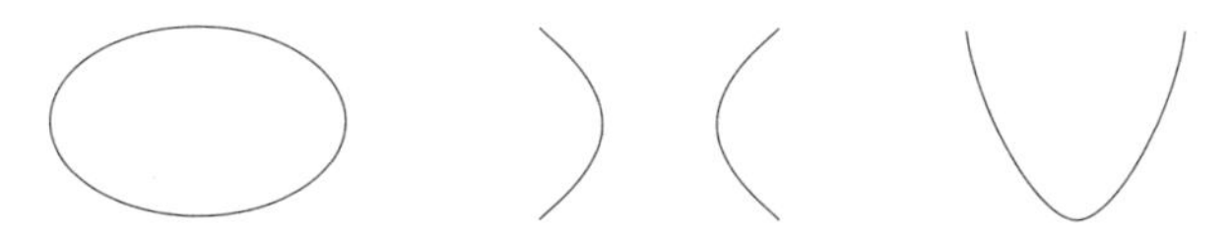

Figure 7.1. Affine non-degenerate conics: from left to right, ellipse, hyperbola and parabola.

After 7.7.6 and the computations of invariants made in the proof of 7.7.5, we have listed in Tables 7.1 and 7.2 (see p. 292, p. 293f) all the affine types of real conics and real quadrics of three-spaces, together with the corresponding values of the invariants $\mathbf{r}, \mathbf{j}, \mathbf{r}', \mathbf{j}'$ used in the classification. We have also included the values of the redundant invariants $\mathbf{s} = n - \mathbf{j} - 1$ and $\mathbf{s}' = n - \mathbf{j}' - 2$. In order to show the splitting of each projective type in affine types, we have grouped together all the affine types of projectively equivalent quadrics; the projective types of the improper sections have been listed too.

Next are some further comments on the contents of Tables 7.1 and 7.2. They include some direct arguments re-proving already justified facts, in order to provide a better insight into the splitting of the projective types in affine types.

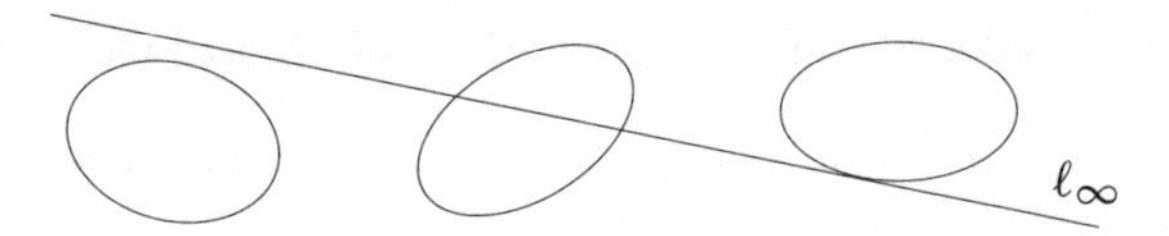

Figure 7.2. A non-affine view of ellipse (left), hyperbola (centre) and parabola (right). ℓ_∞ is the improper line.

We begin with the case of conics. It is shown in Table 7.1 that all non-degenerate imaginary conics have the same affine type. A direct explanation for this fact is that these conics having no (real) points, their improper sections are pairs of imaginary points. Therefore all non-degenerate imaginary conics have projectively equivalent improper sections and so give rise to a unique affine type. By contrast, we see in the table that the real non-degenerate conics give rise to three different affine types: the *ellipses*, which have pairs of imaginary points as improper sections, the *hyperbolas*, which have pairs of real points as improper sections, and the *parabolas*, whose improper sections are pairs of coincident points.

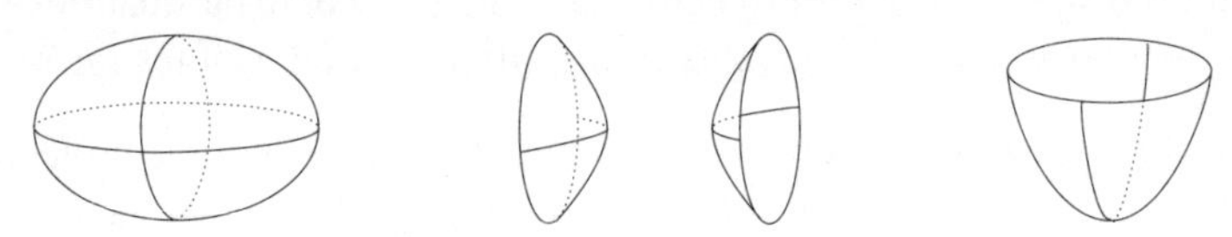

Figure 7.3. Non-ruled quadrics: from left to right, ellipsoid, hyperboloid of two sheets, and elliptic paraboloid.

The reader may easily examine the different relative positions to the improper line of line-pairs and double lines, thus giving a direct confirmation of the remaining part of the list of affine types in Table 7.1.

As for the imaginary conics, the imaginary quadrics give rise to a single affine type. The reason is similar: imaginary quadrics have no (real) points and hence their improper sections must have the only projective type whose conics have no points. Non-ruled quadrics give rise to three affine types: the *ellipsoids*, which have imaginary improper sections, the *hyperboloids of two sheets*, which have real non-degenerate improper sections, and the *elliptic paraboloids*, which have pairs of imaginary lines as improper sections. Since a non-ruled quadric contains no line, no other improper sections can occur. Ruled quadrics split in two affine types, namely the *hyperboloids of one sheet*, also called *ruled hyperboloids*, which have

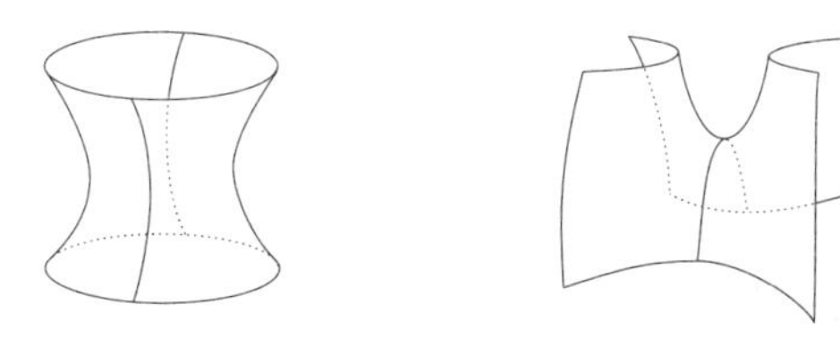

Figure 7.4. Ruled quadrics: hyperboloid of one sheet on the left and hyperbolic paraboloid on the right.

real non-degenerate improper sections, and the *hyperbolic paraboloids*, which have pairs of real lines as improper sections. Of course a ruled quadric cannot have an imaginary improper section because it has index one.

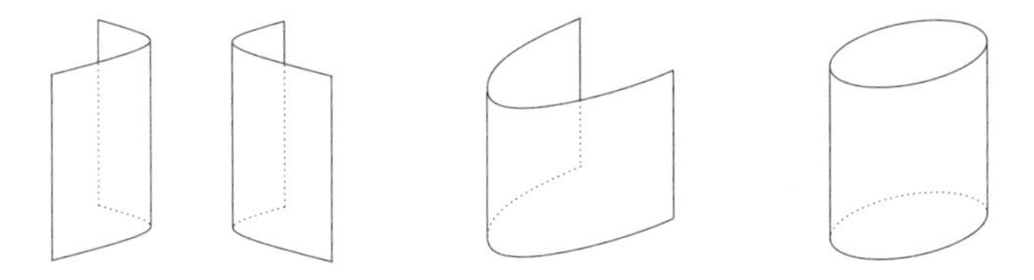

Figure 7.5. Cylinders: from left to right, hyperbolic, parabolic and elliptic.

Cones (real or imaginary) with improper vertex are called *cylinders*, while those with proper vertex keep the name of *cones*; they are called *affine cones* if some confusion may arise. Note that, for a real cone, having improper vertex is equivalent to having all generators parallel. The projective type of the imaginary cones splits into the affine types of the imaginary (affine) cones and the *imaginary cylinders*, according to their vertices being proper or improper. Real cones with proper vertex have all the same affine type, as they all have real non-degenerate improper section (by 6.7.7 and 7.2.4). By contrast, real cones with improper vertex give rise to three affine types of cylinders: *elliptic cylinders*, whose improper section is a pair of imaginary lines, *hyperbolic cylinders*, whose improper section is a pair of real lines, and *parabolic cylinders* whose improper section is a double line. Its affine reduced equation clearly shows that any elliptic cylinder (resp. hyperbolic, parabolic) is a cone with improper vertex over an ellipse (resp. hyperbola, parabola). The examination of the splitting of the projective types of pairs of distinct and coincident planes in affine types is left to the reader.

Table 7.1. Projective and affine types of real conics.

projective characters					*affine non-projective characters*					
projective red. eq.	**r**	**j**	**s**	*projective type*	*affine red. eq.*	**r**′	**j**′	**s**′	*affine type*	*improper section*
$x_0^2 + x_1^2 + x_2^2 = 0$	3	2	−1	imaginary conic	$-y_1^2 - y_2^2 = y_0^2$	2	1	−1	imaginary conic	pair of imaginary points
$x_0^2 + x_1^2 - x_2^2 = 0$	3	1	0	real non-degenerate conic	$y_1^2 + y_2^2 = y_0^2$	2	1	−1	ellipse	pair of imaginary points
					$y_1^2 - y_2^2 = y_0^2$	2	0	0	hyperbola	pair of real points
					$y_1^2 = 2y_0y_2$	1	0	0	parabola	double point
$x_0^2 + x_1^2 = 0$	2	1	0	pair of imaginary lines	$y_1^2 + y_2^2 = 0$	2	1	−1	pair of non-parallel imaginary lines	pair of imaginary points
					$-y_1^2 = y_0^2$	1	0	0	pair of parallel imaginary lines	double point
$x_0^2 - x_1^2 = 0$	2	0	1	pair of real lines	$y_1^2 - y_2^2 = 0$	2	0	0	pair of non-parallel real lines	pair of real points
					$y_1^2 = y_0^2$	1	0	0	pair of parallel real lines	double point
					$0 = 2y_0y_1$	0	−1	1	single proper line	undetermined
$x_0^2 = 0$	1	0	1	double line	$y_1^2 = 0$	1	0	0	double proper line	double point
					$0 = y_0^2$	0	−1	1	the improper line counted twice	undetermined

Table 7.2. Projective and affine types of quadrics of a real three-space.

projective characters					*affine non-projective characters*					
projective red. eq.	$\mathbf{r}$	$\mathbf{j}$	$\mathbf{s}$	*proj. type*	*affine reduced eq.*	$\mathbf{r}'$	$\mathbf{j}'$	$\mathbf{s}'$	*affine type*	*improper section*
$x_0^2+x_1^2+x_2^2+x_3^2=0$	4	3	−1	imaginary	$-y_1^2-y_2^2-y_3^2=y_0^2$	3	2	−1	imaginary	imaginary
$x_0^2+x_1^2+x_2^2-x_3^2=0$	4	2	0	non-ruled	$y_1^2+y_2^2+y_3^2=y_0^2$	3	2	−1	ellipsoid	imaginary
					$y_1^2-y_2^2-y_3^2=y_0^2$	3	1	0	hyperboloid of two sheets	real non-degenerate
					$y_1^2+y_2^2=2y_0y_3$	2	1	0	elliptic paraboloid	pair of im. lines
$x_0^2+x_1^2-x_2^2-x_3^2=0$	4	1	1	ruled	$y_1^2+y_2^2-y_3^2=y_0^2$	3	1	0	hyperboloid of one sheet	real non-degenerate
					$y_1^2-y_2^2=2y_0y_3$	2	0	1	hyperbolic paraboloid	pair of real lines
$x_0^2+x_1^2+x_2^2=0$	3	2	0	imaginary cone	$y_1^2+y_2^2+y_3^2=0$	3	2	−1	imaginary cone	imaginary
					$-y_1^2-y_2^2=y_0^2$	2	1	0	imaginary cylinder	pair of im. lines
$x_0^2+x_1^2-x_2^2=0$	3	1	1	real cone	$y_1^2+y_2^2-y_3^2=0$	3	1	0	real cone	real non-deg.
					$y_1^2+y_2^2=y_0^2$	2	1	0	elliptic cylinder	pair of im. lines
					$y_1^2-y_2^2=y_0^2$	2	0	1	hyperbolic cylinder	pair of r. lines
					$y_1^2=2y_0y_2$	1	0	1	parabolic cylinder	double line

Table 7.2. Continued.

projective characters					*affine non-projective characters*					
projective red. eq.	**r**	**j**	**s**	*proj. type*	*affine reduced eq.*	$\mathbf{r}'$	$\mathbf{j}'$	$\mathbf{s}'$	*affine type*	*improper section*
$x_0^2 + x_1^2 = 0$	2	1	1	pair of imaginary planes	$y_1^2 + y_2^2 = 0$	2	1	0	pair of non-parallel imaginary planes	pair of imaginary lines
					$-y_1^2 = y_0^2$	1	0	1	pair of parallel imaginary planes	double line
$x_0^2 - x_1^2 = 0$	2	0	2	pair of real planes	$y_1^2 - y_2^2 = 0$	2	0	1	pair of non-parallel real planes	pair of real lines
					$y_1^2 = y_0^2$	1	0	1	pair of parallel real planes	double line
					$0 = 2y_0y_1$	0	−1	2	a proper plane plus the improper one	undetermined
$x_0^2 = 0$	1	0	2	double plane	$y_1^2 = 0$	1	0	1	double proper plane	double line
					$0 = y_0^2$	0	−1	2	double impr. plane	undetermined

7.8 Affine elements of quadrics

Affine elements of quadrics are objects associated to quadrics of affine spaces by a relationship invariant by affinities. By 7.5.1, the improper section is an affine element, actually the most important one, as it determines the affine type of a quadric once its projective type has been fixed (7.7.7). We will present in this section other relevant affine elements of quadrics. They are all defined from the quadric, its improper section and the improper hyperplane in terms of conjugation, which makes obvious the affine invariance of their relationship to the quadric. Particular cases of notions introduced in this section already appeared in the former one, while describing the affine types of conics and three-space quadrics.

We will deal with the non-degenerate quadrics first. The *centre* of a non-degenerate quadric Q of $\mathbb{A}_n$ is the pole, relative to Q, of the improper hyperplane; It may, of course, be either a proper or an improper point. A non-degenerate quadric Q with improper centre is called a *paraboloid* (*parabola* if $n = 2$). Paraboloids are characterized in the next proposition, which may be easily proved by the reader using the properties of conjugation and polar hyperplanes seen in Sections 6.4 and 6.5.

Proposition 7.8.1. *Let Q be a non-degenerate quadric and H_∞ the improper hyperplane of $\mathbb{A}_n$. If O is the centre of Q, then the following conditions are equivalent:*

(i) *Q is a paraboloid.*

(ii) *Q is tangent to H_∞.*

(iii) *$Q \cap H_\infty$ is degenerate.*

(iv) *$O \in Q$.*

When they are satisfied, O is the contact point of Q and H_∞, and also the only double point of $Q \cap H_\infty$.

Remark 7.8.2. It follows from 7.8.1, using 7.5.2, that if a non-degenerate quadric Q has matrix $A = (a_{i,j})_{i,j=0,\dots,n}$ relative to an affine reference, then it is a paraboloid if and only if

$$\det(a_{i,j})_{i,j=1,\dots,n} = 0.$$

The non-degenerate quadrics with proper centre are sometimes called *quadrics with centre*. We will not follow this convention, but explicitly write the relevant adjective *proper* and call the non-degenerate quadrics with proper centre *quadrics with proper centre*, to avoid any confusion. A non-degenerate quadric of $\mathbb{A}_n$ is thus either a quadric with proper centre or a paraboloid. It is easy to get a system of equations of the centre (as a zero-dimensional variety). Since the same equations will make sense for degenerate quadrics, in the next lemma we allow Q to be degenerate:

Lemma 7.8.3. *Assume that a quadric Q of $\mathbb{A}_n$ has matrix $A = (a_{i,j})_{i,j=0,\dots,n}$ relative to an affine reference. Then, homogeneous coordinates $x_0, x_1, \dots, x_n$ of a point p satisfy the equations*

$$\begin{aligned} a_{1,0}x_0 + a_{1,1}x_1 + \cdots + a_{1,n}x_n &= 0, \\ &\vdots \\ a_{n,0}x_0 + a_{n,1}x_1 + \cdots + a_{n,n}x_n &= 0 \end{aligned} \tag{7.16}$$

if and only if either p is a double point of Q or, otherwise, the polar hyperplane of p is H_∞.

Proof. The matricial equality

$$A \begin{pmatrix} x_0 \\ x_1 \\ \vdots \\ x_n \end{pmatrix} = \rho \begin{pmatrix} 1 \\ 0 \\ \vdots \\ 0 \end{pmatrix} \tag{7.17}$$

gives rise to $n + 1$ scalar equations from which the first is the only one containing ρ and the other are those of (7.16). The first equation may thus be used to set the value of ρ, in such a way that $x_0, \dots, x_n$ are a solution of (7.16) if and only if they satisfy (7.17) for some $\rho \in k$. Now, the reference being affine, H_∞ has coordinate vector $(1, 0, \dots, 0)^t$; hence, a coordinate vector of a point p satisfies (7.17) for a non-zero ρ if and only if the hyperplane polar of p is H_∞. On the other hand, the non-trivial solutions of (7.17) for $\rho = 0$ clearly are the coordinate vectors of the double points of Q. □

It follows that:

Proposition 7.8.4. *If Q is an affine non-degenerate quadric, then* (7.16) *is a system of equations of the centre of Q.*

Proof. Since Q is non-degenerate, it has no double point and its centre O is the only point with polar hyperplane H_∞. By 7.8.3, O is then the only point whose coordinates satisfy (7.16). Furthermore the matrix A of Q is regular, which assures us that the equations (7.16) are linearly independent, as required. □

The equations (7.16) are called the *equations* (or the *system*) *of the centre*. Of course they depend on the choice of the reference. The equations of the centre may be dehomogenized giving rise to the system

$$\begin{aligned} a_{1,0} + a_{1,1}X_1 + \cdots + a_{1,n}X_n &= 0, \\ &\vdots \\ a_{n,0} + a_{n,1}X_1 + \cdots + a_{n,n}X_n &= 0. \end{aligned} \tag{7.18}$$

It is incompatible if Q is a paraboloid and has the affine coordinates of the centre as its unique solution if Q is a quadric with proper centre.

The lines through the centre of a non-degenerate quadric Q of $\mathbb{A}_n$ are called *diameters* of Q. More generally, any positive-dimensional linear variety containing the centre is called a *diametral variety* of Q.

Remark 7.8.5. If Q is a paraboloid, then its centre O is improper and therefore all proper diameters of Q are parallel. Since, in addition, the improper hyperplane is the tangent hyperplane to Q at O, all improper diameters are tangent to Q, while the proper diameters are not. The proper diameters are thus chords, all sharing the real and improper end O: each proper diameter of a paraboloid has thus a single proper end, which is real.

Assume now that Q is a quadric with proper centre O. By 7.8.1, $Q \cap H_\infty$ is non-degenerate and $O \notin Q$. The next proposition is just an affine specialization of 6.4.11 (b) and therefore needs no proof.

Proposition 7.8.6. *Let Q be a quadric with proper centre O, d a diameter of Q and p the improper point of d. Then one and only one of the following claims is true:*

(1) *$p \in Q$ and d is properly tangent to Q at p.*

(2) *$p \notin Q$, d is a chord of Q with proper ends and O is the midpoint of the ends of d.*

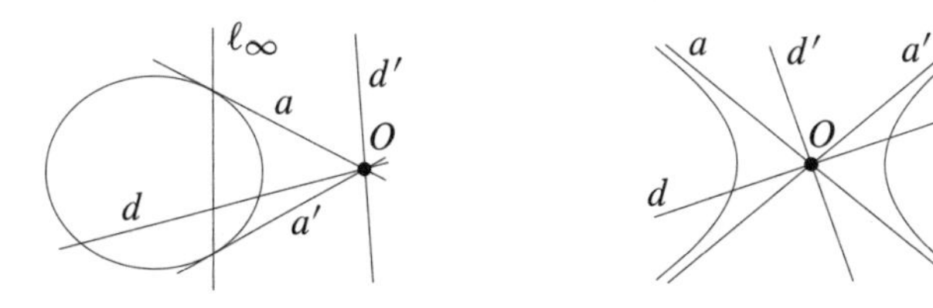

Figure 7.6. Non-affine (left) and affine (right) views of a hyperbola, its centre O, asymptotes a, a' and two diameters d, d'. ℓ_∞ is the improper line.

The diameters tangent to Q are called the *asymptotes* of Q. They describe the tangent cone $TC_O(Q)$, which is called the *asymptotic cone* to Q. By 6.7.5, the asymptotic cone is the cone over $Q \cap H_\infty$ with vertex O and therefore it has rank n. In particular, if $n = 2$, then the asymptotic cone is a pair of distinct lines, and so it is named the *pair of asymptotes*.

Remark 7.8.7. After 7.8.6, the asymptotes are the diameters sharing no proper (real or imaginary, if $k = \mathbb{R}$) point with Q.

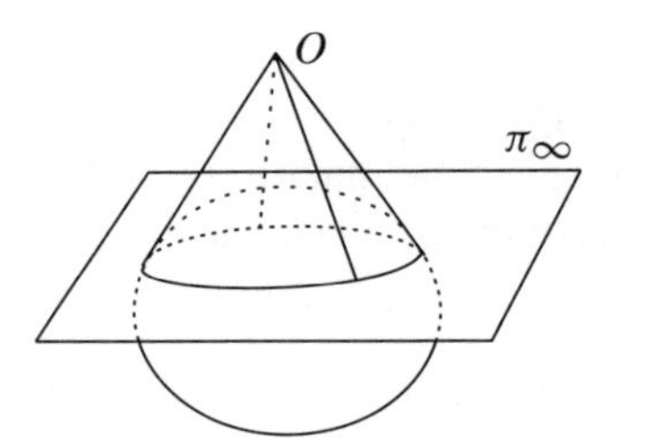

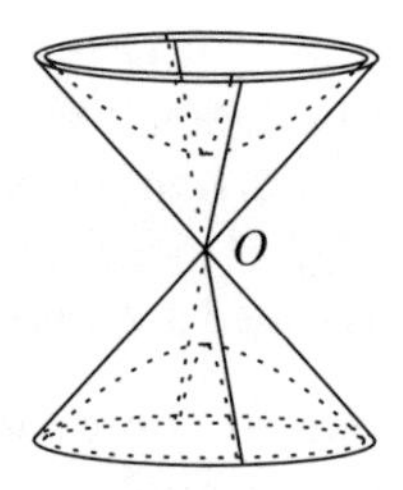

Figure 7.7. Two views of a hyperboloid of two sheets, its centre O and its asymptotic cone, non-affine on the left and affine on the right. π_∞ is the improper plane.

The asymptotic cone being the cone over $Q \cap H_\infty$ with vertex O, it carries the same information as the improper section. When the affine geometry is developed without using improper points, the improper section cannot be considered and the asymptotic cone is used instead. For instance, ellipses may be distinguished from hyperbolas by the fact that the asymptotes of an ellipse are imaginary, while those of a hyperbola are real.

The next corollary justifies the name *centre*. The reader may see also Exercise 7.25.

Corollary 7.8.8. *If a quadric Q has proper centre O, then the set of proper points of Q is invariant by the reflection in O.*

Proof. If q is any proper point of Q, then the diameter Oq is not an asymptote of Q, by 7.8.7. The line qO is thus a chord with proper ends and its other end q' is the symmetric of q with respect to O, by 7.8.6. Since q' is a proper point of Q, the claim follows. □

Two diameters of a quadric Q, with proper centre O, are said to be *conjugate* if and only if their improper points are conjugate with respect to the improper section $Q \cap H_\infty$ (or, equivalently, with respect to Q). The conjugation of diameters is symmetric, because so is the conjugation with respect to $Q \cap H_\infty$. Actually, by its definition and 6.4.4, the conjugation of diameters is the conjugation in the bunch of diameters O^∇ relative to the quadric image of $Q \cap H_\infty$ by the projection map $H_\infty \to O^\nabla$. Of course, the conjugation of diameters and the conjugation relative to $Q \cap H_\infty$ carry the same information; the former provides a way of indirectly handling the latter without using improper points, as does the asymptotic cone with the improper section.

If still d is a diameter of a quadric Q with proper centre O and p the improper point of d, the join of O and the polar hyperplane of p relative to $Q \cap H_\infty$ is a hyperplane H_d of $\mathbb{A}_n$ through O which is called the *diametral hyperplane conjugate to d*. Clearly, a diameter d' is conjugate to d if and only if $d' \subset H_d$. In particular d is an asymptote if and only if $d \subset H_d$.

The next proposition provides a characterization of the conjugation of diameters with no explicit use of improper points:

Proposition 7.8.9. *Let Q be a quadric with proper centre O. Then*

(a) *A diameter of Q is self-conjugate if and only if it is an asymptote.*

(b) *Two different asymptotes d, d' are conjugate if and only if any diameter in the plane they span is also an asymptote.*

(c) *If a diameter d of Q has proper ends q, q', then the tangent hyperplanes H_q, $H_{q'}$, at the ends of d are parallel. Another diameter d' is conjugate to d if and only if it is parallel to H_q (or $H_{q'}$).*

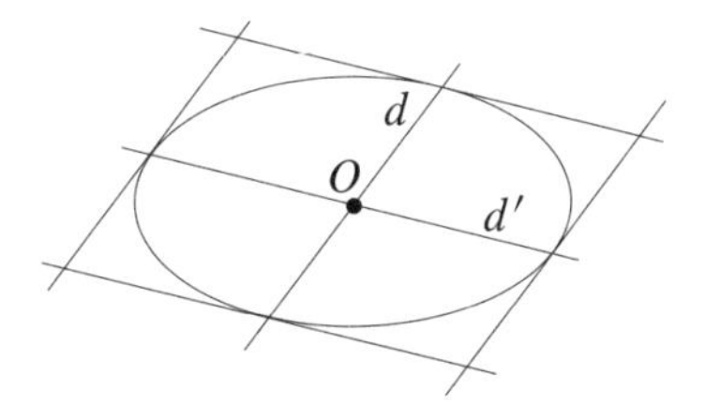

Figure 7.8. Two distinct conjugate diameters d, d'. According to 7.8.9, each has the tangents at its ends parallel to the other, the four tangents making thus a parallelogram.

Proof. Claim (a) follows from 7.8.6, as a diameter d, with improper point p, is self-conjugate if and only if p is self-conjugate with respect to $Q \cap H_\infty$, that is, if and only if $p \in Q \cap H_\infty$.

Regarding claim (b), d, d' being asymptotes, their improper points p, p' belong to $Q \cap H_\infty$ and therefore are conjugate if and only if $p \vee p' \subset Q \cap H_\infty$ (6.4.3), which in turn is the same as saying that any diameter Op'', $p'' \in p \vee p'$ is an asymptote.

For claim (c), note first that the polar variety of d is, by 6.5.6, $H_q \cap H_{q'}$. Then, since $O \in d$, $H_\infty \supset H_q \cap H_{q'}$ (by 6.5.7 (a)) and so $H_q \cap H_{q'} = H_q \cap H_\infty = H_{q'} \cap H_\infty$ proving that H_q and $H_{q'}$ are parallel. Assume that the diameter d' has improper point p': p' and O are conjugate because p' is improper. Then d and d' are conjugate if and only if p' is conjugate to all points in d, which in turn, by the definition of polar variety is equivalent to $p' \in H_q \cap H_{q'} = H_q \cap H_\infty$. □

The reader may note that claim (b) above applies only if $n > 3$. Conjugate diameters will be useful in the sequel mainly due to the next proposition:

Proposition 7.8.10. *A non-degenerate quadric Q of $\mathbb{A}_n$ has diagonal matrix relative to an affine reference Δ if and only if Q has proper centre, the origin of Δ is the centre of Q and any two different axes of Δ are conjugate diameters of Q.*

Proof. By 7.1.3 Q has diagonal matrix if and only if the simplex of (the projective reference associated to) Δ is self-polar. The reference being affine, its 0-th face is the improper hyperplane and so the simplex of the reference is self-polar if and only if the origin of the reference is the centre of C, after which the centre is a proper point, and the improper points of any pair of different axes of Δ are conjugate, hence the claim. □

For $n = 2$, each diameter has a unique conjugate diameter and, as noted in general in 6.5.4, mapping each diameter to its conjugate is an involution of the pencil of diameters: it is named the *involution of conjugate diameters*. Its fixed lines are the asymptotes. The more symmetric situation of the case $n = 2$ is worth the separate claim below, in which the roles of d and d' may obviously be interchanged. Its proof is left to the reader.

Proposition 7.8.11. *If d and d' are distinct diameters of a conic C with proper centre, then the following conditions are equivalent:*

(i) *d and d' are conjugate.*

(ii) *d and d' harmonically divide the asymptotes.*

(iii) *d is the polar of the improper point of d'.*

(iv) *d is a chord with proper ends and d' is parallel to the tangents at the ends of d.*

(v) *The proper points of d are the midpoints of the ends of the proper chords parallel to d' plus the contact points of the tangents parallel to d'.*

(vi) *d, d' and the improper line are the sides of a self-polar triangle.*

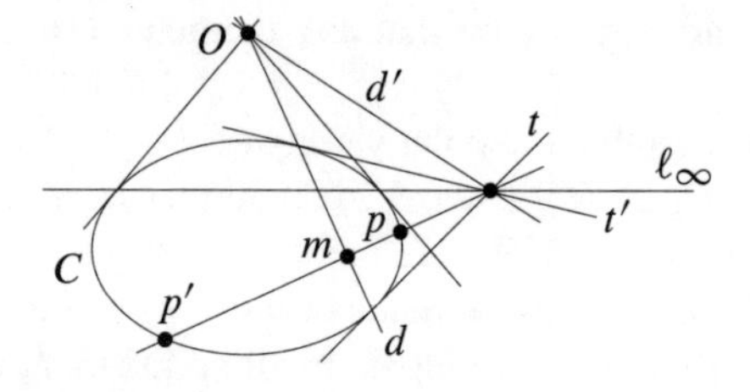

Figure 7.9. A conic C with proper centre O and conjugate diameters d, d'. The tangents t, t' at the ends of d are parallel to d' and m is the midpoint of the ends p, p' of a chord parallel to d'. ℓ_∞ is the improper line.

For the remainder of this section we will assume that Q is a degenerate quadric of an affine space $\mathbb{A}_n$. Since the affine properties of the pairs of distinct or coincident hyperplanes are rather obvious, we will also assume $\mathbf{r}(Q) > 2$. Then the variety of double points $\mathcal{D}(Q)$ of Q has $0 \leq \dim \mathcal{D}(Q) \leq n-3$. As seen in 6.7.7, if T is any supplementary linear variety of $\mathcal{D}(Q)$, Q may be described as the cone over $Q \cap T$ with vertex $\mathcal{D}(Q)$. The section $Q \cap T$ is always a non-degenerate quadric of T and the generators $\mathcal{D}(Q) \vee p$, $p \in Q \cap T$, cover the set of points of Q provided $|Q \cap T| \neq \emptyset$ (6.7.3). Note that $\mathbf{r}(Q) > 2$ assures $\dim T \geq 2$. From the affine viewpoint there are two main possibilities: either the linear variety of double points $\mathcal{D}(Q)$ is improper ($\mathcal{D}(Q) \subset H_\infty$) or it is proper ($\mathcal{D}(Q) \not\subset H_\infty$).

The degenerate quadrics Q of $\mathbb{A}_n$ (with $\mathbf{r}(Q) > 2$) whose variety of double points is improper are called *cylinders*. If Q is a cylinder, then $\mathcal{D}(Q)$ is usually called the *improper vertex* of the cylinder. Any linear variety T supplementary to $\mathcal{D}(Q)$ is proper (otherwise $\bar{\mathbb{A}}_n = \mathcal{D}(Q) \vee T \subset H_\infty$), and so the section $Q \cap T$ is a non-degenerate affine quadric. Any cylinder may thus be described as the cone projecting a non-degenerate quadric of a proper linear variety T, $2 \leq \dim T < n$, from an improper linear variety supplementary to T; its generators $\mathcal{D}(Q) \vee p$, $p \in Q \cap T$, are all parallel.

If Q is a cylinder, then, by the definition, $H_\infty \in \mathcal{D}(Q)^*$ and therefore, by 6.7.10, there exist points whose polar hyperplane is H_∞. In fact 6.7.11 assures that if p is any such point, then the points having H_∞ as polar hyperplane are those of $(\mathcal{D}(Q) \vee p) - \mathcal{D}(Q)$. These points are called the *centres* of the cylinder. By 6.4.11, any centre of a cylinder is the midpoint of the ends of any chord through it, provided these ends are proper. The linear variety $S = \mathcal{D}(Q) \vee p$ is called the *(linear) variety of centres* of Q: its points are the centres and the double points of Q, and its dimension is $1 + \dim \mathcal{D}(Q) = n - \mathbf{r}(Q) + 1$, always positive because Q is degenerate. If the variety of centres of a cylinder Q has dimension one, then it is called the *axis* of Q.

Proposition 7.8.12. *If Q is a cylinder, then any maximal system of independent equations extracted from the equations* (7.16) *of Lemma* 7.8.3 *is a system of equations of the variety of centres of Q.*

Proof. According to the former description of the variety of centres S of Q, Lemma 7.8.3 assures that the points whose coordinates satisfy (7.16) are those lying on S, after which it is enough to recall Remark 2.6.4. □

Since the variety of centers of a cylinder Q is $S = \mathcal{D}(Q) \vee p$, for any centre p of Q, and $\mathcal{D}(Q) \subset H_\infty$, either p is a proper point, in which case $S \cap H_\infty = \mathcal{D}(Q)$ and all centres are proper points, or $p \in H_\infty$, $S \subset H_\infty$ and all centres are improper. In the latter case the cylinder Q is called *parabolic*.

For any choice of the linear variety T, supplementary to $\mathcal{D}(Q)$, $S \cap T$ is, by 1.4.6, a single point p that needs to be a centre because $T \cap \mathcal{D}(Q) = \emptyset$. By 6.4.5, the point p is conjugate, with respect to $\mathcal{D}(Q) \cap T$, of all points of $T \cap H_\infty$ and so it is the centre of the non-degenerate section $Q \cap T$. It follows in particular that all sections $Q \cap T$, for T supplementary to $\mathcal{D}(Q)$, are paraboloids if Q is a parabolic cylinder (hence the name), and quadrics with proper centre if Q is a non-parabolic cylinder.

We leave to the reader to extend part of Proposition 7.8.6 by proving that any centre of a non-parabolic cylinder is the midpoint of the ends of any chord through it.

It remains the case of the degenerated affine quadrics Q, still with $\mathbf{r}(Q) > 2$, whose variety of double points is proper. In the affine context these quadrics keep the name of *cones*, and are often called *affine cones* or *cones with proper vertex* to avoid any confusion with the cylinders, which of course are also cones in the sense of Section 6.7. Not much can be added to the projective description of these quadrics as cones over a non-degenerate section recalled above, but for the fact that its vertex is a proper linear variety and so no two different generators are parallel. Since in this case $H_\infty \notin \mathcal{D}(Q)^*$, by 6.7.10, no point has the improper hyperplane as its polar hyperplane relative to an affine cone. As a consequence, if Q is an affine cone, the points whose coordinates are solutions of the system (7.16) of Lemma 7.8.3 are just the double points of Q. When Q is degenerate, still the names *equations of the centre* and *system of the centre* are used for (7.16).

7.9 Exercises

Exercise 7.1. Use 6.4.27 to prove that if a quadrivertex K has its four vertices on a non-degenerate conic C, then its diagonal triangle is self-polar for C.

Exercise 7.2. Prove that if the vertices of a quadrivertex K belong to a non-degenerate conic C, then the tangents to C at the vertices of K are the sides of a quadrilateral which has the same diagonal triangle as K. *Hint*: Use the polarity relative to C and Exercise 7.1.

Exercise 7.3 (*Effective projective reduction of the equation of a quadric*). Let

$$F = \sum_{i,j=0}^{n} a_{i,j} x_i x_j = 0,$$

be an equation of a quadric Q of a projective space $\mathbb{P}_n$.

(a) Assume that $a_{i,i} = 0$ for $i = 0, \dots, n$. Prove that there are i, j, $0 \leq i < j \leq n$ for which $a_{i,j} \neq 0$. Prove that, for these i, j, the equations

$$x_i = y_i + y_j, \quad x_j = y_i - y_j \quad \text{and} \quad x_s = y_s, \ s = 0, \dots, n, s \neq i, j,$$

define new projective coordinates $y_0, \dots, y_n$ (use 2.3.3) relative to which Q has an equation that effectively contains the monomials in y_i^2 and y_j^2.

(b) Assume that $a_{i,i} \neq 0$ for some $i = 0, \dots, n$. Up to renumbering the coordinates, assume for simplicity $i = 0$. Prove that the left-hand side of the above equation of Q may be rewritten

$$F = a_{0,0}(x_0 + \sum_{0<i\leq n} \frac{a_{0,i}}{a_{0,0}} x_i)^2 + F_0(x_1, \dots, x_n)$$

where F_0 is a homogeneous polynomial in $x_1, \dots, x_n$ of degree two. Prove that the equations

$$y_0 = x_0 + \sum_{0<i\leq n} \frac{a_{0,i}}{a_{0,0}} x_i \quad \text{and} \quad y_i = x_i, i = 1, \dots, n,$$

define new projective coordinates $y_0, \dots, y_n$ relative to which Q has equation

$$a_{0,0} y_0^2 + F_0(y_1, \dots, y_n) = 0.$$

(c) Describe a procedure that starting from an arbitrary equation of a quadric Q, by iterated use of (a) and (b) above, provides the projective reduced equation of Q, as well as equations relating the original coordinates to coordinates relative to which the equation of Q is the projective reduced one. Note that this in particular re-proves the existence of projective reduced equations (7.1.7).

Exercise 7.4. Determine the reduced equations and the projective types of the quadrics of $\mathbb{P}_{3,\mathbb{R}}$ which have equations

$$\begin{aligned} x_0^2 + 2x_0x_1 - 3x_0x_3 - x_2^2 &= 0, \\ x_0^2 + 3x_0x_3 + 2x_0x_1 + x_2^2 + 2x_3^2 &= 0, \\ x_0^2 + 6x_0x_1 + 2x_0x_2 + 9x_1^2 + 5x_1x_2 + x_2^2 + x_2x_3 &= 0. \end{aligned}$$

Exercise 7.5. Determine the projective type of the quadric Q which, relative to a certain reference of $\mathbb{P}_{3,\mathbb{R}}$, has equation

$$x_1x_2 + x_1x_3 + x_2x_3 + x_3^2 - x_0x_3 = 0.$$

Check that it contains lines going through $[1, 0, -1, 0]$ and compute them.

Exercise 7.6. Let Q and π be, respectively, a quadric and a plane of $\mathbb{P}_n$, and assume that both contain a line ℓ. Prove that π is tangent to Q. *Hint*: The case $\pi \subset Q$ being obvious, assume otherwise and prove that $Q \cap \pi$ is degenerate.

Exercise 7.7. Prove that any quadric of a complex projective space $\mathbb{P}_{n,\mathbb{C}}$ contains a linear variety of dimension either $n - \mathbf{r}(Q)/2$, if $\mathbf{r}(Q)$ is even, or $n - (\mathbf{r}(Q)+1)/2$, if $\mathbf{r}(Q)$ is odd. This shows that the maximum allowed by 6.7.9 is reached.

Exercise 7.8. Prove that if a non-degenerate quadric Q of $\mathbb{P}_{n,\mathbb{R}}$ has index j and $j+1 > n-j$, then any section of Q by a non-tangent hyperplane H is a non-degenerate quadric whose index is either j or $j-1$. Prove also that both cases occur. What happens in the excepted case $j+1 = n-j$? *Hint*: Using 7.1.5, take a reference with the pole of H as a vertex and relative to which the equation of Q is reduced.

Exercise 7.9. Prove that the cone Q over a quadric Q' of a linear variety T of $\mathbb{P}_n$, $n > \dim T > 0$, with vertex a supplementary variety of T, has $\mathbf{r}(Q) = \mathbf{r}(Q')$ and, in case $k = \mathbb{R}$, $\mathbf{j}(Q) = \mathbf{j}(Q')$. Deduce that cones over projectively equivalent quadrics with vertices of the same dimension are projectively equivalent.

Exercise 7.10 (*Projective classification of polarities*). Prove that two non-degenerate quadrics are projectively equivalent if and only if their associated polarities are. Deduce that over the field $\mathbb{C}$, the polarities of spaces of a fixed dimension have all the same projective type. Assuming $k = \mathbb{R}$, redefine the index of a non-degenerate quadric in terms of its polarity (use the incidence set of the polarity) and describe the projective types of polarities.

Exercise 7.11. Let Q be a non-degenerate quadric of $\mathbb{P}_{n,\mathbb{R}}$, $n \geq 2$, of index $n-1$. Prove that a linear variety of positive dimension contains no real points of Q if and only if all its points are exterior to Q.

Exercise 7.12. Let Q be a non-degenerate quadric of $\mathbb{P}_{n,\mathbb{R}}$, $n \geq 2$, of index $n-1$ and ℓ a chord of Q with real ends q_1, q_2. Prove that one of the segments of ℓ with ends q_1, q_2 is composed of points exterior to Q, while the other is composed of interior points. *Hint*: Assume that $Q = [\eta]$, take points $p = [v], q = [w]$ in $\ell - \{q_1, q_2\}$ and compare the signs of the roots of $\eta(v,v)X^2 + 2\eta(v,w)X + \eta(w,w)$.

Exercise 7.13. Assume that Q is a non-degenerate quadric of $\mathbb{P}_{n,\mathbb{R}}$, $n \geq 2$, of index $n-1$, and L is a linear variety of $\mathbb{P}_{n,\mathbb{R}}$, of codimension two and composed of points exterior to Q. Prove that:

(1) The polar ℓ of L is a chord of Q with real ends and disjoint with L.

(2) If $q \in \ell$ is interior to Q, then $q \vee L$ is a hyperplane and $Q \cap (q \vee L)$ is a non-degenerate quadric with real points (or, equivalently due to Exercise 7.8, with index $n-2$).

(3) If $q \in \ell$ is exterior to Q, then $q \vee L$ is a hyperplane and $Q \cap (q \vee L)$ is a non-degenerate quadric with no real points (or, equivalently, with index $n-1$).

(4) There are two distinct (real) hyperplanes H_1, H_2 in L^* tangent to Q.

(5) The intersection $Q \cap H$ is a non-degenerate quadric with real points and index $n-2$ if H belongs to one of the segments of L^* with ends H_1, H_2, and is a non-degenerate quadric with no real points if H belongs to the other. *Hint*: Use Exercise 7.12 and project the points of ℓ from L.

Exercise 7.14. Prove that any real non-degenerate conic C of $\mathbb{P}_{2,\mathbb{R}}$ has a chord ℓ_1 with real ends, a chord ℓ_2 with imaginary ends and a tangent ℓ_3. Fix any real affine plane $\mathbb{A}_{2,\mathbb{R}}$ and prove that there are projectivities $f_i\colon \mathbb{P}_{2,\mathbb{R}} \to \bar{\mathbb{A}}_{2,\mathbb{R}}$, $i = 1,2,3$ transforming C into an ellipse, a hyperbola and a parabola, respectively.

Exercise 7.15 (*Effective affine reduction of the equation of a quadric*). Let Q be the quadric of $\mathbb{A}_n$ with equation

$$\sum_{i,j=1}^{n} a_{i,j} X_i X_j + \sum_{i=1}^{n} 2a_i X_i + a = 0 = 0$$

relative to affine coordinates $X_1, \dots, X_n$. Describe a procedure, inspired by the one of 7.3, that re-proves 7.6.2 by effectively providing the affine reduced equation of Q, as well as equations relating $X_1, \dots, X_n$ to affine coordinates relative to which the equation of Q is the affine reduced one. Apply it to the conic

$$C : X_1X_2 - X_1 - X_2 = 0.$$

The quadrics of Exercises 7.16 to 7.22 are assumed to belong to a real affine space $\mathbb{A}_3$ in which an affine reference has been fixed.

Exercise 7.16. Let Q be the quadric with affine equation

$$Y^2 - Z^2 - 2XY - 2XZ + 2X + 2Y + 2 = 0.$$

Determine the affine type of Q, as well as the locus of the centres and double points of the sections of Q by the planes parallel to the plane $x + y + z = 0$.

Exercise 7.17. Determine the affine type of the quadric

$$Q : 2XY + Z^2 + 8Z = 0.$$

Prove that there is a unique plane parallel to the line $X = -2Y = Z$ whose intersection with Q is a pair of parallel lines.

Exercise 7.18. Determine the affine type of the quadric

$$Q : X^2 - Y^2 + 2Z - 1 = 0.$$

Find the planes π for which $C = Q \cap \pi$ is a non-degenerate conic with centre the point $(1, -1, 0)$ and, for each of them, give the affine type of C.

Exercise 7.19. Determine the affine type of

$$Q : X^2 - 2XY - 2XZ - 2Y - 1 = 0.$$

Find the planes π that contain the point (1,-1,-1), are parallel to the X-axis and intersect Q in a parabola.

Exercise 7.20. Determine the affine types of the quadrics

$$Q_a : X^2 - 2XY + aZ^2 + 1 = 0, \quad a \in \mathbb{R}.$$

For which values of a does Q_a contain a pair of lines ℓ_1, ℓ_2 parallel to the line $x + z = y = 0$? Give equations of ℓ_1 and ℓ_2 for each of these values of a.

Exercise 7.21. Determine the affine types of the quadrics with affine equations

$$bX^2 + (2b + a)Y^2 + bZ^2 + 2bXY - 2bYZ - b = 0,$$

for $(a, b) \in \mathbb{R} - \{0, 0\}$.

Exercise 7.22. Prove that all but one of the sections of the quadric

$$Q : X^2 - 4Y^2 + Z^2 + 2XZ - 4YZ + 2X + 3 = 0$$

by the planes $Y + Z = \lambda$, $\lambda \in \mathbb{R}$, are parabolas. *Hint*: Examine the relative position of Q and the improper line of the planes.

Exercise 7.23. Prove that for any quadric Q of $\mathbb{A}_n$, $\mathbf{r}(Q) - 2 \leq \mathbf{r}'(Q) \leq \mathbf{r}(Q)$ and, in case $k = \mathbb{R}$, $\mathbf{j}(Q) - 1 \leq \mathbf{j}'(Q) \leq \mathbf{j}(Q)$.

Exercise 7.24. Assume that π and Q are, respectively, a plane and a quadric of an affine space $\mathbb{A}_{3,\mathbb{R}}$. Denote by π_∞ the improper hyperplane of $A_{3,\mathbb{R}}$, and by $\ell_\infty = \pi \cap \pi_\infty$ the improper line of π. Prove that $(Q \cap \pi) \cap \ell_\infty = (Q \cap \pi_\infty) \cap \ell_\infty$. Use this fact to prove the following claims about plane sections of quadrics:

(1) A non-degenerate plane section of an ellipsoid is either an ellipse or an imaginary conic, and both types occur.

(2) The ordinary cones and the hyperboloids of either type have ellipses, hyperbolas and parabolas as plane sections. The hyperboloids of two sheets have also imaginary plane sections, while the ordinary cones and the hyperboloids of one sheet have not.

(3) All paraboloids have parabolic sections. The non-degenerate and non-parabolic plane sections of a hyperbolic paraboloid are hyperbolas, while those of an elliptic paraboloid are either ellipses or imaginary conics, both cases occurring.

(4) All non-degenerate plane sections of an elliptic (resp. hyperbolic, resp. parabolic) cylinder are ellipses (resp. hyperbolas, resp. parabolas).

Hint: Use 6.4.24 and 6.5.12 to produce non-degenerate plane sections with prescribed improper section.

Exercise 7.25. Let O be a proper point and Q a non-degenerate quadric of $\mathbb{A}_n$. Prove that Q is invariant by the reflection in O if and only if O is the centre of Q.

Exercise 7.26. Prove that the intersection point of the diagonals of a parallelogram is the centre of any non-degenerate conic through its four vertices.

Exercise 7.27. Let T be a triangle (with proper vertices) of an affine plane. Prove that there is a unique non-degenerate conic C going through the vertices of T and having the barycentre of T as centre. Determine the affine type of C in case $k = \mathbb{R}$. *Hint*: Use barycentric coordinates.

Exercise 7.28. Let Q be a non-degenerate quadric of $\mathbb{A}_{3,\mathbb{R}}$ and $p \notin Q$ a point. Assume that p is not the centre of Q and prove that the chords of Q with midpoint p are parallel to the polar plane of p.

Exercise 7.29. Let T be a triangle of an affine plane whose vertices (assumed proper) belong to a non-degenerate conic C. Prove that the barycentre of T is the centre of C if and only if each side of T is parallel to the tangent to C at the opposite vertex.

Exercise 7.30. Let T be a triangle, with proper vertices, of an affine plane $\mathbb{A}_2$. Prove that for a non-degenerate conic C of $\mathbb{A}_2$ the following conditions are equivalent:

(i) C is tangent to the sides of T and its centre is the barycentre of T.

(ii) C is tangent to the sides of T at their midpoints.

Prove that there is a unique non-degenerate conic satisfying them. *Hint*: Use Exercises 6.11 and 6.10.

Exercise 7.31. Prove that if the four sides of a parallelogram are tangent to a non-degenerate conic, then the contact points of the pairs of parallel sides span conjugate diameters.

Exercise 7.32. Prove that in an affine space $\mathbb{A}_{3,\mathbb{R}}$ there is a unique quadric containing two pairs of opposite edges and the barycentre of a given tetrahedron with proper vertices. Determine its affine type. *Hint*: Use barycentric coordinates.

Exercise 7.33. Let Q be a non-degenerate quadric of $\mathbb{A}_3$ and ℓ^*, ℓ an improper line, a pencil of parallel planes. Assume that ℓ is not tangent to Q and prove that the centres of the non-degenerate sections $Q \cap H$, $H \in \ell^*$, lie on a line whose remaining points are the contact points of the planes of ℓ^* tangent to Q. Locate the centres of the non-degenerate sections $Q \cap H$, $H \in \ell^*$ in case where ℓ is tangent to Q.

Exercise 7.34. Let Q be a paraboloid of $\mathbb{A}_{3,\mathbb{R}}$ and ℓ a proper line containing no improper point of Q. Prove that the points of Q at which the tangent plane to Q is parallel to ℓ are the points of a parabola. Prove also that all parabolic sections of Q arise in this way.

Exercise 7.35. Let Q be a quadric of $\mathbb{A}_n$.

(1) Assume that Q has affine equation $f(X_1, \dots, X_n) = 0$ and that its intersection with a line ℓ is a pair of proper points,

$$q = (a_1, \dots, a_n) \quad \text{and} \quad q' = (b_1, \dots, b_n).$$

Prove that for any two different points $p_1, p_2 \in \ell$, $p_1, p_2 \notin Q$, the product of affine ratios $(p_1, p_2, q)(p_1, p_2, q')$ equals $f(a_1, \dots, a_n)/f(b_1, \dots, b_n)$. *Hint*: Use Exercise 6.3.

(2) Use (1) above to prove that if $p_1 = p_{m+1}, p_2, \dots, p_m$ are m distinct proper points of $\mathbb{A}_n$, no one a point of Q, and $Q \cap p_i p_{i+1} = q_i + q'_i$, $i = 1, \dots, m$, are pairs of proper points, then

$$\prod_{i=1}^{m} (p_i, p_{i+1}, q_i)(p_i, p_{i+1}, q'_i) = 1$$

(*Carnot's theorem*).

Exercise 7.36. Assume given four distinct proper points $p_1 = p_5, p_2, p_3, p_4$, of an affine space $\mathbb{A}_3$. Assume that a quadric Q contains no point p_i, $i = 1, \dots, 4$ and has $Q \cap p_{i-1} p_i = \{q_i, q'_i\}$, $q_i \neq q'_i$, $i = 1, \dots, 4$. Prove that any quadric of $\mathbb{A}_3$ containing seven of the points q_i, q'_i, $i = 1, \dots, 4$, contains also the eighth one. *Hint*: Use Carnot's theorem, see Exercise 7.35.

Exercise 7.37. The claim of Exercise 7.36 being projective, prove that it holds if the points p_i, $i = 1, \dots, 4$ and the quadric Q belong to an arbitrary projective space $\mathbb{P}_3$. *Hint*: Use a suitable projectivity $f : \mathbb{P}_3 \to \overline{\mathbb{A}}_3$.

Chapter 8
Further properties of quadrics

This chapter contains the most important results concerning the projective generation and internal structure of conics and three-space quadrics. They have a large number of consequences that include many affine and metric specializations; some of them have been included as exercises.

8.1 Projective generation of conics

In this section and the next one we will deal with non-degenerate complex conics and non-degenerate real conics with points, referred to in the sequel as *non-degenerate conics with points*. Since these conics are determined by their sets of points (by 7.3.3), in the sequel we will often make no distinction between a non-degenerate conic with points and its set of points. We will make frequent use, with no further mention, of the fact that a non-degenerate conic C contains no line (6.5.11), and therefore any line is either a chord of C or a proper tangent to C. The main theorem in this section asserts that the intersection points of corresponding lines in two distinct homographic pencils of $\mathbb{P}_2$ describe a conic. This theorem is named after J. Steiner, who used it to introduce conics without using coordinates. The construction it describes is called the *projective generation* of the conic, and the projectivity f is said to *generate* the conic C.

Theorem 8.1.1 (Steiner's theorem). *If A and B are distinct points of a projective plane $\mathbb{P}_2$, $s = AB$ and $f : A^* \to B^*$ is a projectivity, not a perspectivity, then the intersections $\ell \cap f(\ell)$, $\ell \in A^*$, are the points of a non-degenerate conic C which goes through A and B and has tangents at these points $f^{-1}(s)$ and $f(s)$, respectively.*

Proof. Since f is not a perspectivity, by 5.3.2, the lines $r = f^{-1}(s)$, s and $r' = f(s)$ are distinct. Then $A = r \cap s \neq B = s \cap r'$ and the three lines are not concurrent. Thus we may take $p_0 = A$, $p_1 = B$ and $p_2 = r \cap r'$ as the vertices of a reference of $\mathbb{P}_2$. We choose any line $t \in A^*$, $t \neq r, s$, then, $\bar{t} = f(t) \neq s, r'$. Also $\bar{t} \neq t$, because $t \notin B^*$. This easily gives that $T = t \cap \bar{t}$ is a point and $T \notin r \cup s \cup r'$. We take T as unit point. Then, r, s and t have equations $x_1 = 0$, $x_2 = 0$ and $x_1 - x_2 = 0$, respectively, and so, by 4.4.9, the lines ℓ of A^* have equations

$$\lambda x_1 - \mu x_2 = 0,$$

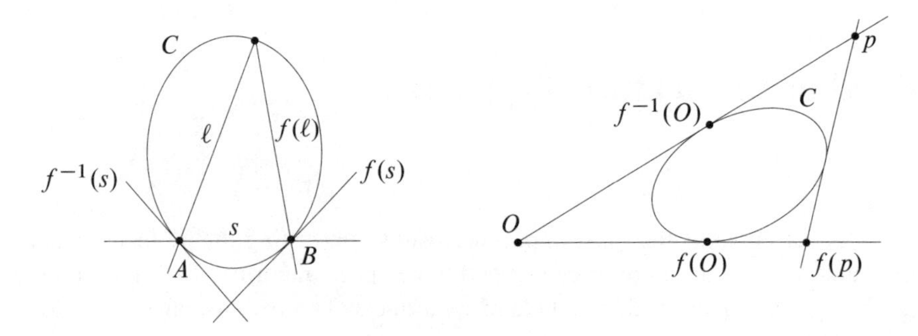

Figure 8.1. Steiner's theorem and its dual.

where λ, μ are homogeneous coordinates of ℓ relative to the reference (r, s, t) of A^*. Similarly, the lines $\bar{\ell} \in B^*$ have equations

$$\alpha x_2 - \beta x_0 = 0,$$

α, β being now coordinates of $\bar{\ell}$ relative to the reference $(s, r', \bar{t})$ of B^*. Since $f(r, s, t) = (s, r', \bar{t})$, by 2.8.1, the image of a line $\lambda x_1 - \mu x_2 = 0$ is $\lambda x_2 - \mu x_0 = 0$ and therefore a point $[x_0, x_1, x_2]$ belongs to both $\ell \in A^*$ and $f(\ell)$ if and only if there exist $\lambda, \mu \in k$, not both zero, such that

$$\lambda x_1 - \mu x_2 = 0,$$
$$\lambda x_2 - \mu x_0 = 0.$$

The above system of equations in λ, μ being compatible if and only if

$$\begin{vmatrix} x_1 & x_2 \\ x_2 & x_0 \end{vmatrix} = 0,$$

we see that the points we are looking for are just those of the conic $C : x_2^2 - x_0 x_1 = 0$. The reader may easily verify now that C satisfies all the remaining properties that have been claimed. □

Remark 8.1.2. All conics given rise to by Steiner's theorem, varying A, B and f, have the same projective type, as they all are non-degenerate and obviously have points.

Remark 8.1.3. As it is clear from the definition of perspectivity (see Section 5.3), if f in 8.1.2 is a perspectivity, then the intersections $\ell \cap f(\ell)$, $\ell \in A^*$ describe the line-pair composed of the axis of f and the line AB, which in this case equals its image.

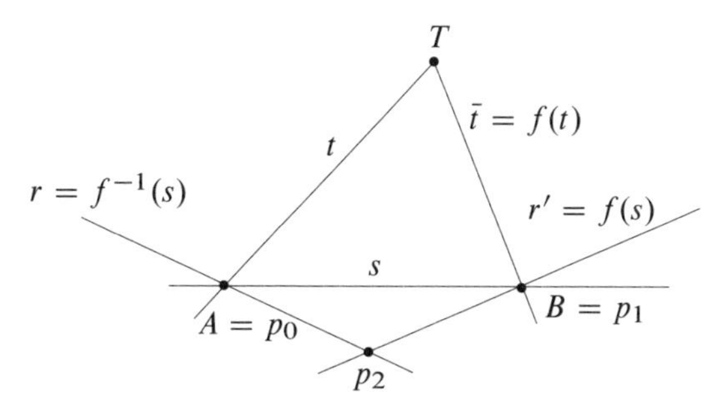

Figure 8.2. The three references used in the proof of Steiner's theorem, namely (p_0, p_1, p_2, T) of $\mathbb{P}_2$, (r, s, t) of A^* and $(s, r', \bar{t})$ of B^*.

The dual of Steiner's theorem provides the projective generation of conic envelopes. By 6.6.1, it reads:

Theorem 8.1.4 (Dual of Steiner's theorem). *If r and s are distinct lines of a projective plane $\mathbb{P}_2$, $O = r \cap s$ and $f : r \to s$ is a projectivity, not a perspectivity, then the joins $p \vee f(p)$, $p \in r$ are the tangents to a non-degenerate conic C which is tangent to r and s at the points $f^{-1}(O)$ and $f(O)$ respectively.*

The projectivity f is said to *generate* the conic envelope C^*.

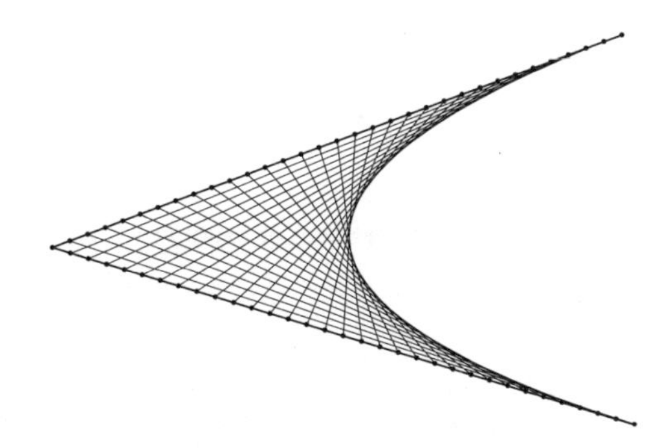

Figure 8.3. Lines joining corresponding points on homographic lines envelop a conic (8.1.4), or tangents to a conic cut on two tangent lines pairs of points corresponding by a homography (8.1.8).

Theorem 8.1.5 (Converse of Steiner's theorem). *Assume that C is a non-degenerate conic and $A, B \in C$, $A \neq B$. If t_A and t_B denote the tangents to C at A and B,* *respectively, then the rules*

$$f(Ap) = Bp \quad \textit{for } p \in C,\ p \notin A,\ p \notin B,$$
$$f(t_A) = AB,$$
$$f(AB) = t_B$$

define a projectivity $f: A^* \to B^*$.

Remark 8.1.6. Once a non-degenerate conic C has been fixed, denoting by pp, for $p \in C$, the tangent line to C at p is a rather usual convention that we will follow in the sequel. According to it, pq, $p, q \in C$, denotes either the chord with ends p, q, if $p \neq q$, or the tangent line at p, if $p = q$. This provides a shorter definition of the projectivity of 8.1.5 by the single rule $f(Ap) = Bp$ for all $p \in C$.

In the proof of 8.1.5 we will make use of the following lemma, which will be useful in many other situations.

Lemma 8.1.7. *For any three distinct points p_0, p_2, T of a non-degenerate conic C of $\mathbb{P}_2$, let p_1 be the pole of p_0p_2. Then (p_0, p_1, p_2, T) is a reference of $\mathbb{P}_2$ and C has equation*

$$x_1^2 - x_0x_2 = 0$$

relative to it.

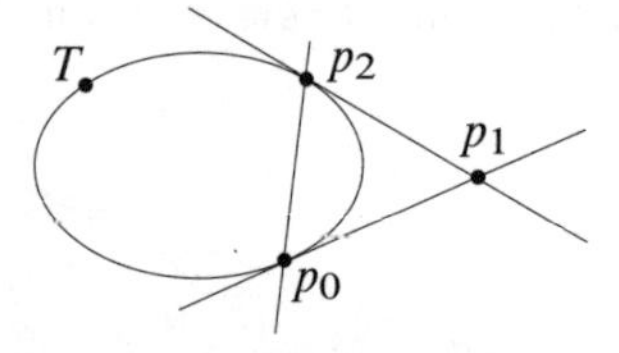

Figure 8.4. The reference of Lemma 8.1.7.

Proof. Since $p_0 \neq p_2$, the line p_0p_2 is not tangent to C, hence it does not contain its pole and p_0, p_1, p_2 are not aligned. The chord p_0p_2 contains no point of C other than its ends p_0, p_2, while, by 6.4.11, p_0p_1 and p_2p_1 are tangent to C at p_0 and p_2, respectively. Therefore no point of C other than p_0, p_2 belongs to p_0p_2, p_0p_1 or p_1p_2. This proves that for any choice of $T \in C, T \neq p_0, p_2$, $\Delta = (p_0, p_1, p_2, T)$ is a reference of $\mathbb{P}_2$. If $a_{i,j}$, $i, j = 0, 1, 2$, are the coefficients of an equation of C relative to Δ, then $a_{0,0} = a_{2,2} = a_{0,1} = a_{1,2} = 0$ because p_0 and p_2 are conjugate to themselves and both conjugate to p_1 (6.4.1). The equation has thus the form $a_{1,1}x_1^2 + 2a_{0,2}x_0x_2 = 0$. The further condition $T = [1, 1, 1] \in C$ gives $a_{1,1} + 2a_{0,2} = 0$, from which the last part of the claim follows. □

Proof of 8.1.5. Take a reference as in 8.1.7 with $p_0 = A$, $p_2 = B$. Then t_A and AB have equations $x_2 = 0$ and $x_1 = 0$ respectively and C is $C : x_1^2 - x_0x_2 = 0$. Any line ℓ through A has an equation of the form $\lambda x_1 - \mu x_2 = 0$, λ, μ being homogeneous coordinates of ℓ relative to the reference (AB, Ap_1, AT) of A^* (4.4.9). Assume that $\ell \neq t_A, AB$: then, in the above equation, $\lambda \neq 0$ and $\mu \neq 0$. An easy computation shows that there is a single point p other than A in $C \cap \ell$, namely $p = [\mu^2, \lambda\mu, \mu^2] \neq B = [0, 0, 1]$. Thus $\ell = Ap$ for a single point $p \in C$, $p \neq A, B$, and so ℓ has a well-defined image $f(\ell) = pB$ which, clearly, has equation $\lambda x_0 - \mu x_1 = 0$.

The tangent t_A is $x_2 = 0$. According to the claim its image is $AB : x_1 = 0$, while the image of AB is in turn $t_B : x_0 = 0$. We see thus that in all cases the image by f of $\ell : \lambda x_1 - \mu x_2 = 0$ is $f(\ell) : \lambda x_0 - \mu x_1 = 0$. The coordinates of ℓ being also coordinates of $f(\ell)$ in B^*, by 2.8.5, f is a projectivity as claimed. □

The dual of 8.1.5 reads:

Theorem 8.1.8 (Dual of the converse of Steiner's theorem). *Assume that r and s, $r \neq s$, are lines of $\mathbb{P}_2$ tangent to a non-degenerate conic C at points p and q respectively, and take $O = r \cap s$. Then the rules*

$$\begin{aligned} f(t \cap r) &= t \cap s \quad \text{for } t \text{ tangent to } C,\ t \neq r, s, \\ f(p) &= O, \\ f(O) &= q \end{aligned}$$

define a projectivity $f : r \to s$.

Next are six corollaries about the existence and uniqueness of a non-degenerate conic of which points and tangents, in a total number of five, have been already given in certain conditions. In forthcoming Sections 9.4 and 9.5, we will come back to the problem of determining conics by prescribing points and tangents, using a deeper and more fruitful viewpoint.

Corollary 8.1.9. *There is one and only one non-degenerate conic going through five given points of $\mathbb{P}_2$ no three of which are aligned.*

Proof. Assume that the given points are q_1, q_2, q_3, q_4, q_5. By the hypothesis $q_1q_3, q_1q_4, q_1q_5 \in q_1^*$ are three distinct lines, and so are $q_2q_3, q_2q_4, q_2q_5 \in q_2^*$. Thus, we may take $f : q_1^* \to q_2^*$ the projectivity that maps q_1q_i to q_2q_i, $i = 3, 4, 5$. It is not a perspectivity because q_3, q_4, q_5 are not aligned. Therefore, Steiner's Theorem 8.1.1 applies and gives rise to a non-degenerate conic C through the five points. For the uniqueness, assume that a non-degenerate conic C' goes through the same five points. According to 8.1.5, C' induces a projectivity $g : q_1^* \to q_2^*$ for which $g(q_1q_i) = q_2q_i = f(q_1q_i)$, $i = 3, 4, 5$. Therefore $f = g$ and both $|C|$ and $|C'|$ are the locus of intersection points of lines corresponding by $f = g$. It follows that $|C| = |C'|$ and, by 7.3.3, $C = C'$. □

In old books, the next corollary is often presented as a limit case of 8.1.9, when one of the points becomes infinitely near to another along a given line, and then vaguely justified by a continuity argument.

Corollary 8.1.10. *Given four points q_1, q_2, q_3, $q_4 \in \mathbb{P}_2$, no three aligned, and a line r through q_1 containing no one of the remaining points, there is one and only one non-degenerate conic of $\mathbb{P}_2$ that goes through q_1, q_2, q_3, q_4 and is tangent to r at q_1.*

Proof. Proceed as in the proof of 8.1.9 but for mapping $r \mapsto q_1q_2$ instead of $q_1q_5 \mapsto q_2q_5$. □

A further limit case may be proved similarly by the reader, this time by mapping q_1q_2 to r':

Corollary 8.1.11. *Given three non-aligned points $q_1, q_2, q_3 \in \mathbb{P}_2$ and two lines r and r' through q_1 and q_2 respectively, each not going through any of the other two points, there is one and only one non-degenerate conic of $\mathbb{P}_2$ that goes through q_1, q_2, q_3 and is tangent to r and r' at q_1 and q_2, respectively.*

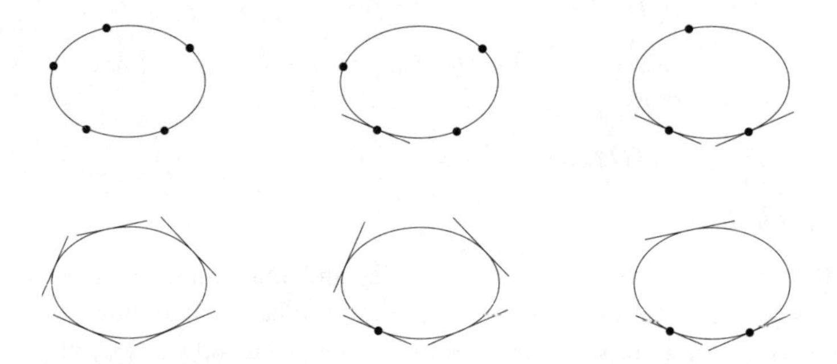

Figure 8.5. Determining a non-degenerate conic by points and tangents, according to Corollaries 8.1.9 through 8.1.14.

The dual claims determine the conic by five tangents, or four tangents and one contact point, or three tangents and two contact points. Being just the duals of 8.1.9, 8.1.10 and 8.1.11, they need no proof. Anyway the reader may prove them using 8.1.5 and 8.1.8.

Corollary 8.1.12. *There is one and only one non-degenerate conic tangent to five given lines of $\mathbb{P}_2$, no three of which are concurrent.*

Corollary 8.1.13. *Given four lines $r_1, r_2, r_3, r_4 \in \mathbb{P}_2$, no three concurrent, and a point q on r_1 belonging to no one of the remaining lines, there is one and only one non-degenerate conic of $\mathbb{P}_2$ that has tangents r_1, r_2, r_3, r_4 and contact point q with r_1.*

Corollary 8.1.14. *Given three non-concurrent lines $r_1, r_2, r_3 \in \mathbb{P}_2$ and two points q and q' on r_1 and r_2 respectively, each lying not on any of the other two lines, there is one and only one non-degenerate conic of $\mathbb{P}_2$ that is tangent to r_1, r_2, r_3 and has contact points q and q' with r_1 and r_2, respectively.*

Remark 8.1.15. Since all degenerate conics are line-pairs, it is clear that no degenerate conic goes through the five points of Corollary 8.1.9, which therefore still holds without the non-degeneracy hypothesis on the conic. Similar remarks apply to 8.1.10 and 8.1.11, but not to 8.1.12, 8.1.13 or 8.1.14, because double lines are tangent to any line.

Remark 8.1.16. Assuming the lines r_i to be already given on the drawing plane, there is a construction allowing us to draw as many tangents as wanted to the conic C of 8.1.12, which is called *constructing the conic by tangents*. Indeed, according to the proof of 8.1.12, the intersections of r_1 and r_2 with r_3, r_4 and r_5 are three pairs of points corresponding by the projectivity f. For each choice of $p \in r_1$, its image $f(p) \in r_2$ may be constructed following 5.3.6, after which the line $pf(p)$ is tangent to C. The same construction, with minor variations, applies to the conics of 8.1.13 and 8.1.14. If the data are reasonably placed, just a few tangents provide a surprisingly appealing picture of the conic. The reader is advised to try this construction at least once.

Remark 8.1.17. The construction dual to that of 8.1.16 provides as many points as wanted of either of the conics of 8.1.9, 8.1.10 and 8.1.11, which is called *constructing the conic by points*. Constructing conics by tangents gives, however, nicer pictures.

8.2 Projective structure on a conic

In this section we will see that any non-degenerate conic with points inherits from its ambient plane a structure of one-dimensional projective space which is, up to equivalence, independent of any choice. The existence of such a projective structure has interesting consequences, not only relative to conics, but also to the geometry of abstract one-dimensional projective spaces, which, by means of a projectivity onto a conic, may be immersed in a plane in a very useful manner. The reader may see for instance the proof of Proposition 8.2.11 below.

Still we make no distinction between the set of points of a non-degenerate conic with points and the conic itself; so, in particular, a projective structure on the set of points of a non-degenerate conic with points will be just called a projective structure on the conic.

Assume that C is a non-degenerate conic and A a point of C. We have the map

$$\begin{aligned} \delta_A \colon C &\longrightarrow A^*, \\ p &\longmapsto Ap. \end{aligned}$$

Note that $\delta_A(A) = AA = t_A$ according to the convention of 8.1.6. In the opposite sense, we have

$$\sigma_A \colon A^* \longrightarrow C,$$
$$\ell \longmapsto p,$$

where p is the only point for which $C \cap \ell = \{A, p\}$. As it is clear, δ_A and σ_A are inverses of each other, and so they are in particular bijective.

The map δ_A is usually called a *projection map* (or just a *projection*), and σ_A a *section map* (or just a *section*); they should not be confused with the projection and section maps introduced in Section 1.9. The reader may note that the hypothesis $A \in C$ is essential in the definitions of δ_A and σ_A.

Now, we will fix a point $A_1 \in C$ and use the corresponding section map σ_{A_1} to translate the projective structure of A_1^* into a projective structure on C. The important fact about such a projective structure is that it turns all projection and section maps δ_A, σ_A, for $A \in C$, into projectivities. This property determines the projective structure on C up to equivalence, and in particular makes irrelevant the choice of the point A_1. To be precise, we have:

Theorem 8.2.1. *If C is a non-degenerate conic with points, there is a one-dimensional projective structure on C such that, for any point $A \in C$, the projection map*

$$\delta_A \colon C \longrightarrow A^*,$$
$$p \longmapsto Ap,$$

and the section map

$$\sigma_A \colon A^* \longrightarrow C,$$
$$\ell \longmapsto p,$$

$C \cap \ell = \{A, p\}$, are reciprocal projectivities. This property determines the projective structure on C up to equivalence.

Proof. Fix $A_1 \in C$ and denote by $\pi_1 \colon F_1 - \{0\} \to A_1^*$ the map defining the projective structure on the pencil A_1^*. As seen in Section 4.2, $F_1 = \langle v \rangle^{\perp}$, v a representative of A_1, and π is the restriction to $F_1 - \{0\}$ of the structure map $\pi^{\vee}$ of the dual plane $\mathbb{P}_2^{\vee}$, although we will make no use of these facts now. Since the section map

$$\sigma_1 = \sigma_{A_1} \colon A_1^* \longrightarrow C$$

is bijective, it is obvious that the composite map $\bar{\pi}_1 = \sigma_1 \circ \pi_1$, satisfies the conditions of the definition of projective spaces in Section 1.2. Therefore it defines a structure of one-dimensional projective space on (the set of points of) the conic

C, the representatives of any point $p \in C$ being just the representatives of the line $A_1 p \in A_1^*$.

Assume that A is any point of C. The composite map

$$A_1^* \xrightarrow{\sigma_1} C \xrightarrow{\delta_A} A^*$$

is a projectivity. Indeed, this is obvious if $A = A_1$ and has been seen in 8.1.5 if $A \neq A_1$. Denote by $\pi : F - \{0\} \to A^*$ the map giving the projective structure on A^* and by φ any representative of $\delta_A \circ \sigma_1$. We have the diagram

$$\begin{array}{ccccc} & & F_1 - \{0\} & \xrightarrow{\varphi} & F - \{0\} \\ & \swarrow^{\pi_1} & \downarrow{\scriptstyle \bar{\pi}_1} & & \downarrow{\scriptstyle \pi} \\ A_1^* & \xrightarrow{\sigma_1} & C & \xrightarrow{\delta_A} & A^* \end{array}$$

in which, just because $\delta_A \circ \sigma_1 = [\varphi]$, it holds that

$$\delta_A \circ \sigma_1 \circ \pi_1 = \pi \circ \varphi.$$

Therefore, since $\bar{\pi}_1 = \sigma_1 \circ \pi_1$,

$$\delta_A \circ \bar{\pi}_1 = \pi \circ \varphi.$$

This proves that if C is taken with the structure defined by $\bar{\pi}_1$, then δ_A is a projectivity, the one represented by φ. Of course, the section map σ_A being the inverse of δ_A, it is a projectivity too.

To close, the property of the claim determines the projective structure due to 1.6.3 (b). □

In the sequel any non-degenerate conic C with points will be endowed with the projective structure of Theorem 8.2.1, and therefore considered as a one-dimensional projective space. Rather than using the associated vector space and representatives, this projective structure is usually handled through projectivities δ_A and σ_A, using projective invariance. The next proposition shows some examples:

Proposition 8.2.2. *Assume that C is a non-degenerate conic with points and take any $A \in C$. Then:*

(a) *The cross ratio of four points $q_1, q_2, q_3, q_4 \in C$, three at least distinct, is the cross ratio of the lines $Aq_1, Aq_2, Aq_3, Aq_4 \in A^*$.*

(b) *Three points $p_0, p_1, T \in C$ compose a projective reference of C if and only if the lines Ap_0, Ap_1, AT compose a projective reference of A^*. Of course this occurs if and only if p_0, p_1, T are three distinct points.*

(c) *A point $p \in C$ has homogeneous coordinates α_0, α_1 relative to a reference p_0, p_1, T of C if and only if the line Ap has homogeneous coordinates α_0, α_1 relative to the reference Ap_0, Ap_1, AT of A^*.*

(d) *$f: C \to \mathbb{P}_1$ is a projectivity if and only if $f \circ \sigma_A$ is.*

(e) *$g: \mathbb{P}_1 \to C$ is a projectivity if and only if $\delta_A \circ g$ is.*

Proof. The claims directly follow from being δ_A and σ_A projectivities (8.2.1), using 2.9.11 for (a), 1.7.5 for (b), 2.8.1 (b) for (c), and 1.6.2 (b) for (d) and (e). □

Remark 8.2.3. The existence of a projective structure on it (or just the existence of the bijective maps δ_A, σ_A) shows that any non-degenerate conic with points contains infinitely many points.

After fixing references in C and $\mathbb{P}_2$, any $p \in \mathbb{C}$ has coordinates in the conic and coordinates in the plane: our first task is to relate them. This will provide a second way (the first one is 8.2.1) of relating the internal geometry of C and the geometry of $\mathbb{P}_2$.

Proposition 8.2.4 (Coordinates on a conic). *Let C be a non-degenerate conic of $\mathbb{P}_2$ with points. Let p_0, p_2, T be three distinct points on C and take p_1 to be the pole of $p_0 p_2$. Then $\Omega = (p_0, p_2, T)$ is a reference of C and $\Delta = (p_0, p_1, p_2, T)$ is a reference of $\mathbb{P}_2$, C has equation $x_1^2 = x_0 x_2$ relative to Δ and any point $p = [a_0, \alpha_1]_\Omega \in C$ has coordinates x_0, x_1, x_2 relative to Δ given by*

$$\begin{aligned} x_0 &= \alpha_0^2, \\ x_1 &= \alpha_0\alpha_1, \\ x_2 &= \alpha_1^2. \end{aligned} \tag{8.1}$$

Proof. That Ω is a reference of C is clear from 8.2.2 (b). We have already seen in Lemma 8.1.7 that Δ is a reference of $\mathbb{P}_2$ relative to which an equation of C is $x_1^2 = x_0 x_2$. Take any $p = [\alpha_0, \alpha_1]_\Omega \in C$. By 8.2.2 (c), the coordinates of p relative to $\Omega = (p_0, p_2, T)$ are coordinates of pp_0 relative to the reference $(p_0 p_0, p_2 p_0, T p_0)$ of p_0^*. Since $p_0 p_0$, $p_2 p_0$ and $T p_0$ have equations $x_2 = 0$, $x_1 = 0$ and $x_2 - x_1 = 0$, respectively, an equation of pp_0 is $\alpha_0 x_2 - \alpha_1 x_1 = 0$ (by 4.4.9). Once we have equations of C and pp_0, it is enough to compute their intersection points to see that $p = [\alpha_0^2, \alpha_0\alpha_1, \alpha_1^2]$, as claimed. □

Remark 8.2.5. Using the absolute coordinate $\alpha = \alpha_0/\alpha_1$ on C turns the equations (8.1) into

$$\begin{aligned} x_0 &= \alpha^2, \\ x_1 &= \alpha, \\ x_2 &= 1, \end{aligned} \tag{8.2}$$

which for $\alpha = \infty$ are intended to give $x_0 = 1$ and $x_1 = x_2 = 0$, the result of taking $\alpha_1 = 0$ in the equations (8.1).

The equations (8.1) may be taken as *parametric equations* of the conic: the free variation of the pair of homogeneous parameters α_0, α_1 in $k^2 - \{(0,0)\}$ gives the coordinates of all points of C. Furthermore each point determines its corresponding parameters up to a common factor, because they are its coordinates in C. The reader may compare with 2.5.3. A similar comment may be made regarding equations (8.2): in this case the absolute parameter α varies in $k \cup \{\infty\}$ and is determined by its corresponding point.

In the sequel, we will say that the reference Δ of 8.2.4 is the reference *associated* to the reference Ω of C. One may of course ask for the coordinates of the points of C relative to another reference Δ' of $\mathbb{P}_2$. They are given by the equations

$$\begin{aligned} y_0 &= a_0^0\alpha_0^2 + a_1^0\alpha_0\alpha_1 + a_2^0\alpha_1^2, \\ y_1 &= a_0^1\alpha_0^2 + a_1^1\alpha_0\alpha_1 + a_2^1\alpha_1^2, \\ y_2 &= a_0^2\alpha_0^2 + a_1^2\alpha_0\alpha_1 + a_2^2\alpha_1^2, \end{aligned} \tag{8.3}$$

where (a_i^j) is the matrix changing coordinates relative to Δ into coordinates relative to Δ'.

Having their sets of points given by polynomial parameterizations such as (8.1) and (8.3), causes the non-degenerate conics with points to be *rational curves*, see the forthcoming Section 9.2.

If C is a non-degenerate conic with points, next we will obtain equations of its chords and tangents pq, $p, q \in C$:

Lemma 8.2.6. *If C is a non-degenerate conic with points and Ω a reference of C, then the line pq, $p, q \in C$, $p = [\alpha_0, \alpha_1]_\Omega$, $q = [\beta_0, \beta_1]_\Omega$, has equation*

$$\alpha_1\beta_1 x_0 - (\alpha_0\beta_1 + \alpha_1\beta_0)x_1 + \alpha_0\beta_0 x_2 = 0$$

relative to the reference of $\mathbb{P}_2$ associated to Ω.

Proof. Assume first $p \neq q$. According to 8.2.4, the equation of the chord pq is

$$\begin{vmatrix} x_0 & x_1 & x_2 \\ \alpha_0^2 & \alpha_0\alpha_1 & \alpha_1^2 \\ \beta_0^2 & \beta_0\beta_1 & \beta_1^2 \end{vmatrix} = 0.$$

After development it may be written

$$(\alpha_0\beta_1 - \alpha_1\beta_0)\big(\alpha_1\beta_1 x_0 - (\alpha_0\beta_1 + \alpha_1\beta_0)x_1 + \alpha_0\beta_0 x_2\big) = 0,$$

which, after dropping the non-zero factor $\alpha_0\beta_1 - \alpha_1\beta_0$, is the claimed equation. The reader may easily check the case $p = q$, by direct computation of an equation of the tangent pp to $C : x_1^2 - x_0x_2 = 0$ at $p = [\alpha_0^2, \alpha_0\alpha_1, \alpha_1^2]$. □

Remark 8.2.7. As usual, a lighter version of the above equation is obtained by using absolute coordinates $\alpha = \alpha_0/\alpha_1$ and $\beta = \beta_0/\beta_1$: if both α and β are finite, then an equation of pq is

$$x_0 - (\alpha + \beta)x_1 + \alpha\beta x_2 = 0.$$

Suitable conventions for the cases in which one or both of the absolute coordinates are infinity may be set by the reader.

Theorem 8.2.8 (Involutions of a conic). *If $\tau: C \to C$ is an involution of a non-degenerate conic with points C of $\mathbb{P}_2$, then there is a point $O \in \mathbb{P}_2 - C$ such that all lines $p\tau(p)$ go through O. Conversely, given C as above and any $O \in \mathbb{P}_2 - C$, the pairs $p, \bar{p} \in C$ for which $O \in p\bar{p}$ are the pairs of an involution τ of C.*

Proof. Take a reference of C and its associated one in $\mathbb{P}_2$. Fix any $\bar{a}, \bar{b}, \bar{c} \in k$ such that $\bar{b}^2 - \bar{a}\bar{c} \neq 0$. The equality

$$\bar{a}\alpha_0\beta_0 + \bar{b}(\alpha_0\beta_1 + \alpha_1\beta_0) + \bar{c}\alpha_1\beta_1 = 0 \tag{8.4}$$

is, on one hand, the necessary and sufficient condition for the points $p = [\alpha_0, \alpha_1]$ and $q = [\beta_0, \beta_1]$ to correspond by the involution that has equation $\bar{a}x_0x_0^* + \bar{b}(x_0x_1^* + x_1x_0^*) + \bar{c}x_1x_1^* = 0$ (see Section 5.6). On the other, according to 8.2.6, the same equality (8.4) is the necessary and sufficient condition for the line pq to go through the point $O = [\bar{c}, -\bar{b}, \bar{a}]$, which does not belong to C because $\bar{b}^2 - \bar{a}\bar{c} \neq 0$. From the identity between these two conditions, the claim obviously follows. □

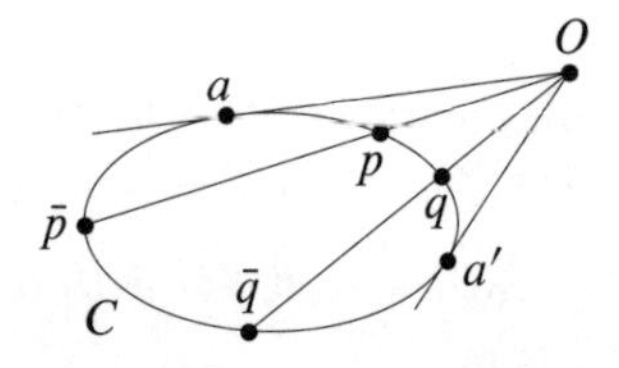

Figure 8.6. Pairs of corresponding points $\{p, \bar{p}\}$ and $\{q, \bar{q}\}$ and fixed points a, a' by the involution of C with centre O.

Remark 8.2.9. It is clear from the claim of 7.5.6 that τ and O determine each other. This follows also from the proof, by just comparing the coordinates of O and the coefficients of the equation (8.4) of τ. The point O is called the *centre* of τ.

A direct consequence of 8.2.8 is that the fixed points of an involution τ of a conic C are the contact points of the tangents to C through the centre of τ, that is:

Corollary 8.2.10. *C being as above, the fixed points of the involution of C with centre O are those of $C \cap H_O$.*

After 8.2.8, some facts that we have already proved for involutions of one-dimensional projective spaces in Section 5.6 are obvious for involutions of conics. For instance it is clear that two disjoint pairs $\{p, \bar{p}\}$, $\{q, \bar{q}\}$ of points of C belong to one and only one involution of C, namely the one with centre $O = p\bar{p} \cap q\bar{q}$. Note that the condition $\{p, \bar{p}\} \cap \{q, \bar{q}\} = \emptyset$ assures that O is a point and also that it does not belong to C. Furthermore, once C and the points $p, \bar{p}, q, \bar{q}$ are drawn on a plane, it suffices a straight edge to construct first the centre O and then, from it, the images of arbitrary points on C. Similarly, by 8.2.10, an involution of a conic has either two distinct fixed points or, only in case $k = \mathbb{R}$, none. In fact, if $k = \mathbb{R}$, by the definition of interior point, an involution has no fixed points if and only if its centre is an interior point. The construction of 6.4.29 may be used to decide if an involution of a conic has real fixed points and to construct them in the affirmative case.

An arbitrary one-dimensional projective space $\mathbb{P}_1$ may be immerged into a projective plane $\mathbb{P}_2$ by means of a projectivity from $\mathbb{P}_1$ onto a non-degenerate conic with points of $\mathbb{P}_2$ (see also Exercise 9.6 for an intrinsic immersion). Then results of one-dimensional projective geometry that may be easily proved on a non-degenerate conic with points using plane projective geometry, are extended to any $\mathbb{P}_1$ by means of such an immersion. As an example, we will prove in this way the next result, which has not been included in Section 5.6 and has interesting consequences relative to the metric properties of conics (see Proposition 9.2.1). The reader may provide an alternative direct proof in an arbitrary $\mathbb{P}_1$ using coordinates.

Proposition 8.2.11. *If τ_1 and τ_2 are two different involutions of the same $\mathbb{P}_1$ and either $k = \mathbb{C}$ or $k = \mathbb{R}$ and one of the involutions has no (real) fixed points, then there is a unique pair $\{p, \bar{p}\}$ belonging to both involutions, that is, such that $\bar{p} = \tau_1(p) = \tau_2(p)$. Furthermore, in the real case $p \neq \bar{p}$.*

Proof. Let C be a non-degenerate conic with points. We will first reduce ourselves to the case of $\mathbb{P}_1 = C$. To this end, fix any projectivity $f : \mathbb{P}_1 \to C$ (2.8.1 provides many choices). It is clear that $\tilde{\tau}_i = f \circ \tau_i \circ f^{-1}$, $i = 1, 2$, are two different involutions of C. If $q, \bar{q} \in C$ are such that $\bar{q} = \tilde{\tau}_1(q) = \tilde{\tau}_2(q)$, then $p = f^{-1}(q)$ and $\bar{p} = f^{-1}(\bar{q})$ satisfy $\bar{p} = \tau_1(p) = \tau_2(p)$, and conversely. Furthermore, regarding the last claim, obviously $p \neq \bar{p}$ if and only if $q \neq \bar{q}$.

Now, the problem of finding the pairs of points corresponding by two different involutions has an easy solution on C (see Figure 8.7): if O_1, O_2 are the centres of different involutions $\tilde{\tau}_1, \tilde{\tau}_2$ of C, points $q, \bar{q} \in C$ correspond by both $\tilde{\tau}_1$ and $\tilde{\tau}_2$ if and only if $q\bar{q}$ contains both O_1 and O_2. Since $\tilde{\tau}_1 \neq \tilde{\tau}_2$, $O_1 \neq O_2$ and O_1O_2 is a line. Therefore the unique solution is to take $q + \bar{q} = C \cap O_1O_2$. This solves the problem in the case $k = \mathbb{C}$. If $k = \mathbb{R}$, the centre of the involution with no fixed points is an interior point, after which $q, \bar{q}$ are real and different by 7.4.8. □

Remark 8.2.12. The uniqueness of the pair $p, \bar{p}$ in 8.2.11 follows also from 5.6.10.

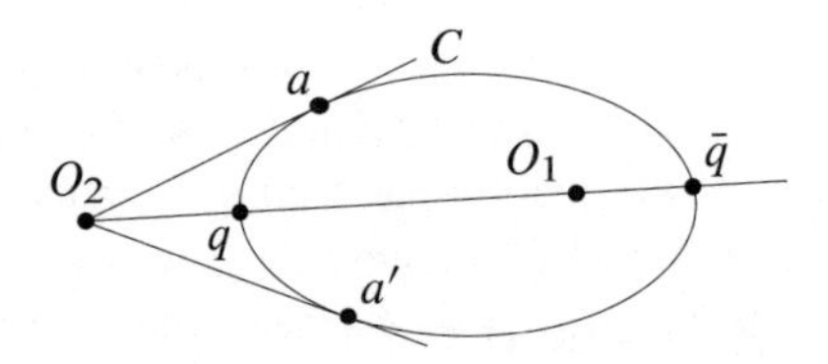

Figure 8.7. The centres O_1 and O_2 of two involutions τ_1 and τ_2 of a conic C. τ_1 has no fixed points, while those of τ_2 are a, a'. The pair common to both involutions is $\{q, \bar{q}\}$.

Like the involutions, also the homographies of a conic may be graphically described, the description being in this case a bit more complicated.

Theorem 8.2.13 (Cross-axis theorem for conics). *Let C be a non-degenerate conic with points of $\mathbb{P}_2$ and $f : C \to C$ a non-identical homography of C. Then there is a uniquely determined line s of $\mathbb{P}_2$ such that for any two points $p, q \in C$, the lines $pf(q)$, $qf(p)$ and s are concurrent.*

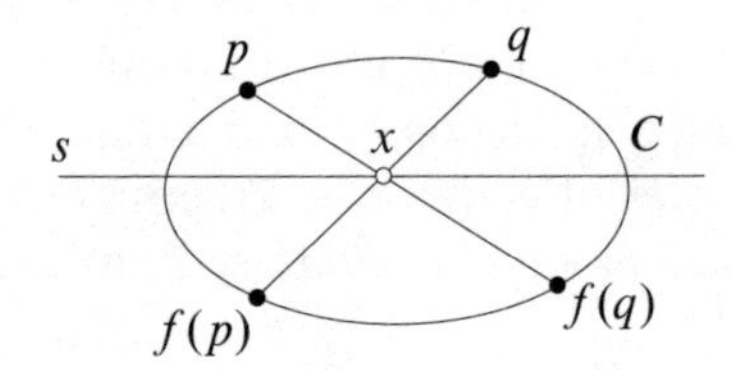

Figure 8.8. The cross-axis of a projectivity f of a conic C: while p and q vary on C, the point x lies on the cross-axis s.

Proof. Fix a reference of C and take the associated one in $\mathbb{P}_2$. Choose points $p = [\alpha_0, \alpha_1], q = [\beta_0, \beta_1] \in C$. Since the claim is obviously satisfied for any choice of s if $p = q$, we assume in the sequel $p \neq q$. If

$$\begin{pmatrix} a & b \\ c & d \end{pmatrix}, \quad ad - bd \neq 0,$$

is a matrix of f, then

$$f(p) = [a\alpha_0 + b\alpha_1, c\alpha_0 + d\alpha_1],$$
$$f(q) = [a\beta_0 + b\beta_1, c\beta_0 + d\beta_1].$$

Lemma 8.2.6 provides an equation of $pf(q)$, which after an easy computation may be written

$$\alpha_0\beta_0(ax_0+cx_1)+\alpha_1\beta_1 b(x_1+dx_2)+\alpha_0\beta_1(bx_0+dx_1)+\alpha_1\beta_0(ax_1+cx_2) = 0.$$

Similarly, an equation of $qf(p)$ is

$$\alpha_0\beta_0(ax_0+cx_1)+\alpha_1\beta_1 b(x_1+dx_2)+\alpha_1\beta_0(bx_0+dx_1)+\alpha_0\beta_1(ax_1+cx_2)=0.$$

Subtracting these equations cancels the parts symmetric in p, q and gives

$$(\alpha_0\beta_1-\alpha_1\beta_0)\big(bx_0-(a-d)x_1+cx_2\big)=0$$

which, after dropping the non-zero factor $(\alpha_0\beta_1-\alpha_1\beta_0)$ yields

$$bx_0-(a-d)x_1+cx_2=0. \tag{8.5}$$

This equation is obviously independent of the choice of p and q, and is not identically zero because $f \neq \mathrm{Id}_C$. We define s as the line with equation (8.5). As it is clear from the way this equation has been obtained, either $pf(q)=qf(p)=s$ or $pf(q) \neq qf(p)$ and s belongs to the pencil generated by $pf(q)$ and $qf(p)$, as required. For the uniqueness, suppose that a second line s' satisfies the same property and put $z=s\cap s'$. Take any non-fixed point $p \in C$ and any $p' \in C$, $p' \neq f(p)$. If $pp' = f(p)f^{-1}(p')$, then either $p=f(p)$ or $p=f^{-1}(p')$, that is $f(p)=p'$, both already excluded. Thus $pp' \cap f(p)f^{-1}(p')$ is a point which, by the property satisfied by s and s', belongs to both lines, and therefore is the point z. We have thus $z \in pp'$. The arbitrariness in the choice of p' forces $z=p$, which makes no sense because p may in turn be taken any non-fixed point of C. □

The line s of 8.2.13 is called the *cross-axis* of f.

Remark 8.2.14. If $p \in C$ is not fixed by f and $q \in C$ is an arbitrary point other than p, then $p \neq q$ and $p \neq f(p)$, which assures $pf(q) \neq qf(p)$; therefore, $pf(q) \cap qf(p)$ is a point which by 8.2.13 belongs to s.

Corollary 8.2.15. *Hypothesis and notations being as in* 8.2.13, *the fixed points of f are those of $C \cap s$.*

Proof. Since f has at most two fixed points and $C \cap s$ contains no more than two points, we may take $p \in C$ a non-fixed point with $p \notin s$, $f(p) \notin s$. If $q \in C \cap s$, then, since $f(p) \notin s$, $q \neq f(p)$ and $qf(p) \neq s$; it follows that q is the only point in $qf(p) \cap s$. By 8.2.13, $q \in pf(q)$ and since $q \neq p$ because $p \notin s$, $q = f(q)$ as claimed. Conversely, if $q \in C$ and $f(q)=q$, then $q \neq p$, and so, by 8.2.14, $q = pf(q) \cap qf(p) \in s$. □

In case $k=\mathbb{R}$ we see that the projectivity is parabolic (resp. hyperbolic, resp. elliptic) if and only if its cross-axis s is a tangent to C (resp. a chord of C with real ends, resp. a chord of C with imaginary ends). The reader may note that 8.2.15 re-proves 5.5.1 for projectivities of a conic. The argument may be extended to an arbitrary $\mathbb{P}_1$ as in the proof of 8.2.11.

Corollary 8.2.16. *For any non-fixed point $p \in C$ and any two different points $q_1, q_2 \in C$, $pf(q_1) \cap q_1 f(p)$ and $pf(q_2) \cap q_2 f(p)$ are points spanning the cross-axis s of f.*

Proof. Take $p, q_1, q_2 \in C$ as in the claim. The point p being not fixed, $f(p)$ is not fixed either and therefore, by 8.2.15, $f(p) \notin s$. Since $q_1 \neq q_2$, the lines $q_1 f(p)$ and $q_2 f(p)$ are different; since $f(p) \notin s$ they meet s at different points, which therefore span s. By 8.2.14, these points are $pf(q_1) \cap q_1 f(p)$ and $pf(q_2) \cap q_2 f(p)$. □

The cross-axis and the image of a non-fixed point determine the projectivity:

Corollary 8.2.17. *If f, g are two projectivities of C which have the same cross-axis and both map a certain point p to the same image $\bar{p} = f(p) = g(p) \neq p$, then $f = g$.*

Proof. Take any $q \in C$ different from p. Since p is not fixed, $p, \bar{p} \notin s$, after which $q\bar{p} \cap s$ is a point different from p. Since by 8.2.13 and 8.2.14 both $pf(q)$ and $pg(q)$ go through it, $pf(q) = pg(q)$ and therefore $f(q) = g(q)$. □

From the equation of s obtained in the proof of 8.2.13, it is clear that the cross-axis s alone does not determine f. Next is a reciprocal of 8.2.13 that shows the existence of a projectivity of which the cross-axis s and a pair of different corresponding points, not on s, have been given:

Proposition 8.2.18. *Assume that C is a non-degenerate conic with points, s a line and $p, \bar{p} \in C$ a pair of different points, no one on s. Mapping $q \mapsto \bar{q}$, $q, \bar{q} \in C$, if and only if $p\bar{q} \cap q\bar{p} \cap s \neq \emptyset$ is a projectivity f of C that maps p to $\bar{p}$ and has cross-axis s.*

Proof. The condition $p\bar{q} \cap q\bar{p} \cap s \neq \emptyset$ just says that $q\bar{p}$ is the image of $p\bar{q}$ by the perspectivity $h_s \colon \bar{p}^* \to p^*$ of axis s. Thus the correspondence f of the claim is the composition of projectivities

$$C \xrightarrow{\delta_{\bar{p}}} \bar{p}^* \xrightarrow{h_s} p^* \xrightarrow{\sigma_p} C,$$

$\delta_{\bar{p}}$ and σ_p the projection and section maps. That f has cross-axis s and maps p to $\bar{p}$ is clear. □

Assume that the conic C is already drawn on a plane and that the images by f of three different points of C are given. Since at least one of the points, let us name it p, is not fixed, 8.2.16 provides an easy construction of the cross-axis of f. Once the cross-axis has been drawn, the fixed points are evident, by 8.2.15, and the construction of the image of an arbitrary point $q \neq p$, from the cross-axis and the pair of corresponding points p, $f(p)$, $p \neq f(p)$, follows from 8.2.13 (see Figure 8.9).

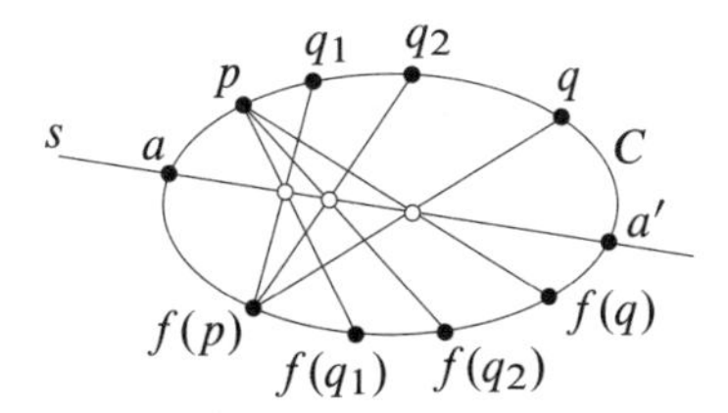

Figure 8.9. The cross-axis s of a projectivity f of a conic C determined from the images of the points p, q_1, q_2, according to 8.2.16. The cross-axis allows us in turn to determine the fixed points a, a' (8.2.15) and the image $f(q)$ of an arbitrary further point q (8.2.13).

As the reader may have noticed, the above constructively re-proves the existence and uniqueness of a projectivity of a conic C mapping three given distinct points to three given distinct points, already seen in 2.8.1. An extension to projectivities of an arbitrary $\mathbb{P}_1$ follows as in former cases.

Remark 8.2.19. Using the coordinates of the centre and the equation of the cross-axis that appeared in the proofs of 8.2.8 and 8.2.13, it is direct to check that the cross-axis of any involution of a conic is the polar of its centre. See also Exercise 8.25. The involutions of conics being easily described from their centres (8.2.8), their axes are seldom considered.

Part of the following definitions will be used next. If $m \geq 3$, an *m-gon* $\mathcal{M}$ is the figure that consists of m cyclically ordered points $p_0, p_1, \dots, p_m = p_0 \in \mathbb{P}_2$ – its *vertices* – and the lines $a_i = p_i p_{i+1}, i = 0, \dots, m-1$ – its *sides* – in such a way that no three vertices are collinear and no three sides are concurrent. Reversing the ordering of the vertices does not change the m-gon. The 3-gons are just triangles. A 4-gon is a quadrivertex with two distinguished pairs of opposite sides, which are the sides of the 4-gon. For $m = 5, 6, \dots$, m-gons are called *pentagons*, *hexagons*, etc. Obviously an m-gon is determined by its cyclically ordered vertices, and also by its cyclically ordered sides, and the notion of m-gon is self-dual. If m is even, by definition, $\{p_0, p_{m/2}\}, \{p_1, p_{(m/2)+1}\}, \dots, \{p_{(m/2)-1}, p_{m-1}\}$ are the pairs of *opposite vertices* and $\{a_0, a_{m/2}\}, \{a_1, a_{(m/2)+1}\}, \dots, \{a_{(m/2)-1}, a_{m-1}\}$ the pairs of *opposite sides* of $\mathcal{M}$. An m-gon $\mathcal{M}$ (in particular a triangle) is said to be *inscribed* in a non-degenerate conic C if and only if all its vertices belong to C, in which case it is equivalently said that C is *circumscribed* to $\mathcal{M}$. If all sides of $\mathcal{M}$ are tangent to a non-degenerate conic C, then it is said that $\mathcal{M}$ is *circumscribed* to C and also that C is *inscribed* in $\mathcal{M}$.

One of the oldest and nicest theorems of projective geometry, due to B. Pascal, characterizes the hexagons inscribed in conics. It is essentially equivalent to the cross-axis theorem 8.2.13, so its proof will be easy.

Theorem 8.2.20 (Pascal). *A hexagon is inscribed in a non-degenerate conic C if and only if the intersection points of its pairs of opposite sides are collinear.*

Proof. Assume that the vertices of the hexagon are, in their cyclic order, $p_1, \dots, p_6$, and belong to the non-degenerate conic C. Take the homography f of C that maps $p_1 \mapsto p_4$, $p_5 \mapsto p_2$, $p_3 \mapsto p_6$. By 8.2.13, the intersections of opposite sides $p_1p_2 \cap p_4p_5$, $p_2p_3 \cap p_5p_6$ and $p_3p_4 \cap p_6p_1$ belong to the cross-axis of f, hence the direct claim (see Figure 8.10).

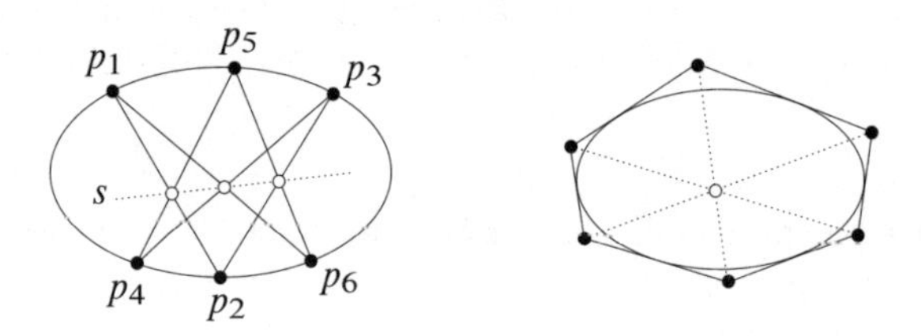

Figure 8.10. Pascal and Brianchon theorems.

For the converse, let C be the non-degenerate conic that, by 8.1.9, goes through the points $p_1, \dots, p_5$. The reader may easily see that the line s which, by the hypothesis, joins the intersections of opposite sides cannot contain any vertex. Thus we may take the projectivity of C that maps p_1 to p_4 and has cross-axis s (8.2.18). Since the point $p_1p_2 \cap p_4p_5$ belongs to s, by 8.2.18, $f(p_5) = p_2$ and similarly $f(p_3)$ is the intersection point of C and p_1p_6 other than p_1, let us call it p'. Now, by 8.2.13, p_5p' meets s at the same point as p_2p_3, which in turn, by the hypothesis, meets s at the same point as p_5p_6. It follows that $p_5p_6 = p_5p'$ and therefore p' belongs to both p_5p_6 and p_1p_6. This gives $p' = p_6$ and therefore $p_6 \in C$. □

The reader may compare Pascal's theorem above with Pappus' theorem 1.8.5: heuristically, the latter may be seen as a limit case of the former when the conic degenerates into a line-pair.

Six points of the plane $\mathbb{P}_2$ need not lie on a non-degenerate conic. Indeed, take any five points $p_1, \dots, p_5 \in \mathbb{P}_2$, no three aligned; we have seen (8.1.9) that there is a unique non-degenerate conic C containing them. Thus, it is enough to choose a sixth point p_6 lying not on C for having no non-degenerate conic through $p_1, \dots, p_6$. Pascal's theorem characterizes in terms of an easy linear construction the sets of six points of $\mathbb{P}_2$ that belong all to the same non-degenerate conic. The dual of Pascal's theorem, which of course needs no proof, is also a nice one:

Theorem 8.2.21 (Brianchon). *A hexagon is circumscribed to a non-degenerated conic C if and only if the three lines joining its pairs of opposite vertices are concurrent.*

If C is a non-degenerate conic with points of a projective plane $\mathbb{P}_2$, we have on one hand the projectivities of C, as one-dimensional projective space: they compose the projective group $\mathrm{PG}(C)$ of C. On the other, we have the projectivities of $\mathbb{P}_2$ that leave C invariant: they obviously form a subgroup of $\mathrm{PG}(\mathbb{P}_2)$, and so in particular a group we will denote by $\mathrm{PG}(\mathbb{P}_2)_C$. The interesting fact is that any projectivity of C is the restriction of a uniquely determined projectivity of $\mathbb{P}_2$, as claimed next:

Proposition 8.2.22. *Mapping each projectivity of $\mathbb{P}_2$ leaving C invariant to its restriction to C is a group isomorphism*

$$\begin{aligned} \theta\colon \mathrm{PG}(\mathbb{P}_2)_C &\longrightarrow \mathrm{PG}(C), \\ f &\longmapsto f_{|C}. \end{aligned}$$

Proof. First we will check that $f_{|C}$ is a projectivity. Fix a point $q \in C$. Call g the projectivity induced by f,

$$\begin{aligned} g = (f^{\vee})^{-1}_{|q^*}\colon q^* &\longrightarrow f(q)^*, \\ \ell &\longmapsto f(\ell). \end{aligned}$$

By composing it with the projection and section maps δ_q and $\sigma_{f(q)}$ we get a projectivity

$$\begin{array}{ccccccc} C & \xrightarrow{\delta_q} & q^* & \xrightarrow{g} & f(q)^* & \xrightarrow{\sigma_{f(q)}} & C, \\ p & \longmapsto & qp & \longmapsto & f(q)f(p) & \longmapsto & f(p), \end{array}$$

that clearly agrees with $f_{|C}$. We have thus seen that θ is well defined. It obviously is a group homomorphism. For its injectivity, pick any four distinct points of C: they compose a reference Δ of $\mathbb{P}_2$ because no three points of C are aligned. If two projectivities f, f' of $\mathbb{P}_2$ have $f_C = f'_C$, then $f(\Delta) = f'(\Delta)$ and therefore $f = f'$ by 2.8.1. To prove the exhaustivity, assume that $h\colon C \to C$ is a projectivity: we will give an explicit determination of a projectivity f of $\mathbb{P}_2$ that leaves C invariant and has $f_{|C} = h$. Take a projective reference Ω of C and its associated one Δ in $\mathbb{P}_2$, as described in 8.2.4; assume that h has matrix

$$\begin{pmatrix} a & b \\ c & d \end{pmatrix},$$

with $ad - bc \neq 0$, and consider the matrix

$$\begin{pmatrix} a^2 & 2ab & b^2 \\ ac & ad+bc & bd \\ c^2 & 2cd & d^2 \end{pmatrix}. \tag{8.6}$$

It has determinant $(ad - bc)^3$ and therefore is regular. The projectivity f of $\mathbb{P}_2$ that it defines maps an arbitrary point

$$p = [\alpha_0, \alpha_1]_\Omega = [\alpha_0^2, \alpha_0\alpha_1, \alpha_1^2]_\Delta \in C$$

to the point

$$\begin{aligned}[][(a\alpha_0 + b\alpha_1)^2, (a\alpha_0 + b\alpha_1)(c\alpha_0 + d\alpha_1), (c\alpha_0 + d\alpha_1)^2]_\Delta \\ = [a\alpha_0 + b\alpha_1, c\alpha_0 + d\alpha_1]_\Omega = h(p),\end{aligned}$$

which shows that f leaves C invariant and $f_{|C} = h$, thus ending the proof. For a less explicit argument, see Exercise 8.16. For more about the relationship between h and f, see Exercise 9.7. □

8.3 Lines on quadrics

In this section we will study the lines contained in the non-degenerate quadrics of projective three-spaces, these lines being called the *generators* of the quadric. We will consider only the non-degenerate quadrics containing lines, namely the non-degenerate quadrics of $\mathbb{P}_{3,\mathbb{C}}$ and the non-degenerate quadrics of $\mathbb{P}_{3,\mathbb{R}}$ of index one. For simplicity, these quadrics will be called *ruled* quadrics in the sequel. All ruled quadrics being non-degenerate, they contain no planes (by 6.5.11), and therefore all their plane sections are conics; this fact will be used in the sequel without further mention.

First of all we prove the existence of a particularly simple equation for any ruled quadric:

Lemma 8.3.1. *For any ruled quadric Q there is a projective reference relative to which Q has equation*

$$z_0z_1 - z_2z_3 = 0.$$

Proof. Assume first $k = \mathbb{C}$. Any non-degenerate quadric has reduced equation

$$x_0^2 + x_1^2 + x_2^2 + x_3^2 = 0.$$

Taking new coordinates z_r, $r = 0, \dots, 3$, defined by the rules

$$\begin{aligned} z_0 &= x_0 + \boldsymbol{i}x_1, \\ z_1 &= x_0 - \boldsymbol{i}x_1, \\ z_2 &= x_2 + \boldsymbol{i}x_3, \\ z_3 &= -x_2 + \boldsymbol{i}x_3, \end{aligned}$$

the resulting equation is as wanted. If $k = \mathbb{R}$, then the reduced equation is

$$x_0^2 + x_1^2 - x_2^2 - x_3^2 = 0,$$

in which case we use the change of coordinates

$$\begin{aligned} z_0 &= x_0 + x_2, \\ z_1 &= x_0 - x_2, \\ z_2 &= x_1 + x_3, \\ z_3 &= -x_1 + x_3 \end{aligned}$$

to get the same result. □

The next theorem describes the generators of a ruled quadric.

Theorem 8.3.2. *If Q is a ruled quadric of $\mathbb{P}_3$, then there are two sets S_1, S_2, of generators of Q, called the* **systems of generators** *or* **rulings** *of Q, such that:*

(1) *Any generator of Q belongs to one and only one ruling of Q.*

(2) *Any two different generators of the same ruling are skew.*

(3) *Any two generators of different rulings meet.*

(4) *There is one generator of each ruling going through each point of Q.*

(6) *If two sets of generators S_1', S_2' satisfy the above properties* (1) *to* (3)*, then $S_1' = S_i$, $S_2' = S_j$, where $\{i, j\} = \{1, 2\}$.*

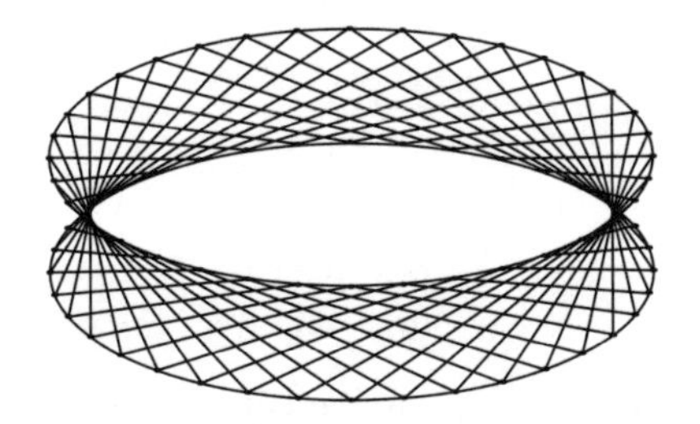

Figure 8.11. Generators of a ruled quadric.

Proof. First we will define S_1 and S_2. As allowed by 8.3.1, take the coordinates such that Q has equation

$$z_0 z_1 - z_2 z_3 = 0.$$

For any $\alpha_0, \alpha_1 \in k$, not both zero, the equations

$$\begin{aligned} \alpha_0 z_0 &= \alpha_1 z_2, \\ \alpha_1 z_1 &= \alpha_0 z_3 \end{aligned} \tag{8.7}$$

obviously define a line s_α that depends only on the ratio $\alpha = \alpha_0/\alpha_1 \in k \cup \{\infty\}$. Assume that $\alpha \neq 0, \infty$. Then the coordinates of any point $p \in s_\alpha$ satisfy the above equations and hence also their product

$$\alpha_0\alpha_1 z_0 z_1 = \alpha_0\alpha_1 z_2 z_3.$$

Since $\alpha_0\alpha_1 \neq 0$, it turns out that

$$z_0 z_1 - z_2 z_3 = 0$$

and therefore $s_\alpha \subset Q$ for $\alpha \neq 0, \infty$. Checking that the same is true for $\alpha = 0, \infty$ is direct: the equations of s_0 are $z_1 = z_2 = 0$, which obviously imply $z_0 z_1 - z_2 z_3 = 0$, and similarly for s_∞. Thus we take $S_1 = \{s_\alpha\}_{\alpha \in k \cup \{\infty\}}$ and we have seen that all the elements of S_1 are generators of Q.

Similarly, all lines ℓ_β, with equations

$$\begin{aligned} \beta_0 z_0 &= \beta_1 z_3, \\ \beta_1 z_1 &= \beta_0 z_2, \end{aligned} \tag{8.8}$$

$\beta = \beta_0/\beta_1 \in k \cup \{\infty\}$, are generators of Q and we take $S_2 = \{\ell_\beta\}_{\beta \in k \cup \{\infty\}}$.

The coordinates of the points of $s_\alpha \cap s_{\alpha'}$ are the solutions of the system of equations

$$\begin{aligned} \alpha_0 z_0 - \alpha_1 z_2 &= 0, \\ \alpha_1 z_1 - \alpha_0 z_3 &= 0, \\ \alpha_0' z_0 - \alpha_1' z_2 &= 0, \\ \alpha_1' z_1 - \alpha_0' z_3 &= 0, \end{aligned}$$

which has determinant

$$-\begin{vmatrix} \alpha_0 & \alpha_1 \\ \alpha_0' & \alpha_1' \end{vmatrix}^2.$$

Therefore $s_\alpha \cap s_{\alpha'} = \emptyset$ if $\alpha \neq \alpha'$. A similar computation shows that $\ell_\beta \cap \ell_{\beta'} = \emptyset$ if $\beta \neq \beta'$. This proves (2), and also that $s_\alpha \neq s_{\alpha'}$ if $\alpha \neq \alpha'$ and $\ell_\beta \neq \ell_{\beta'}$ if $\beta \neq \beta'$. In particular both S_1 and S_2 are infinite.

The system of equations

$$\begin{aligned} \alpha_0 z_0 - \alpha_1 z_2 &= 0, \\ \alpha_1 z_1 - \alpha_0 z_3 &= 0, \\ \beta_0 z_0 - \beta_1 z_3 &= 0, \\ \beta_1 z_1 - \beta_0 z_2 &= 0 \end{aligned}$$

may be easily checked to have its determinant equal to zero for all values of α, β. It follows that any two generators s_α and ℓ_β meet, which proves (3).

The equalities 8.7 give rise to a system of equations in α_0, α_1,

$$z_0\alpha_0 - z_2\alpha_1 = 0,$$
$$z_3\alpha_0 - z_1\alpha_1 = 0,$$

whose solutions are the parameters α_0, α_1 of the lines s_α going through the point $p = [z_0, z_1, z_2, z_3] \in \mathbb{P}_3$. This system has a non-trivial solution if and only if $z_0z_1 - z_2z_3 = 0$, which is just the equation of Q. We see thus that for any point $p \in Q$ there is a line $s_\alpha \in S_1$ through p. The same argument applies to the equations (8.8) and completes the proof of (4).

Now, for claim (1), $S_1 \cap S_2 = \emptyset$, because a generator $s \in S_1 \cap S_2$ would simultaneously meet and be skew to the infinitely many generators of, say, S_1 other than s itself, by (2) and (3). If r is any generator of Q, fix any point $p \in r$ and take s_α and ℓ_β to be the generators of S_1 and S_2 going through p. Clearly $s_\alpha \neq \ell_\beta$ because we have seen that $S_1 \cap S_2 = \emptyset$. We have the generators r, s_α and ℓ_β going through the same point p, all contained in the tangent plane H_p, by 6.4.23, and therefore in $Q \cap H_p$. Since by 6.4.17 no conic contains three different lines, either $r = s_\alpha \in S_1$ or $r = \ell_\beta \in S_2$.

To close, we prove the uniqueness (6). For any pair of generators r, r' define $r \sim r'$ if and only if $r = r'$ or $r \cap r' = \emptyset$, which obviously does not depend on S_1 and S_2. Nevertheless it is clear from (1), (2) and (3) that $r \sim r'$ if and only if r and r' belong to the same S_i. From this it easily follows that $\sim$ is an equivalence relation whose equivalence classes are S_1 and S_2. Since the same argument applies to S_1', S_2', (6) follows. □

Remark 8.3.3. Some further information directly follows from 8.3.2:

(a) The intersection of any two generators of different rulings is a point, because the generators meet by (3) and are different by (1).

(b) For any $p \in Q$, there is one and only one generator of each ruling going through p; the existence has been seen in (4) and the uniqueness follows from (2).

The two rulings of a ruled quadric are said to be *opposite* each other. There is no intrinsic difference between them: taking the coordinates as in 8.3.1, the projectivity of $\mathbb{P}_3$,

$$[z_0, z_1, z_2, z_3] \longmapsto [z_0, z_1, z_3, z_2],$$

leaves the quadric invariant and swaps over its rulings. For identification purposes, it is usual to arbitrarily choose one of the rulings and call it the *first ruling*, while the other is then called the *second ruling*.

The points on a generator parameterize the ruling that does not contain it:

Corollary 8.3.4. *If s is a generator of a ruled quadric Q, for each $q \in s$ write ℓ_q the generator through q other than s itself. The rules $q \mapsto \ell_q$ and $\ell \mapsto \ell \cap s$ define a pair of reciprocal bijections between s and the ruling of Q that does not contain s.*

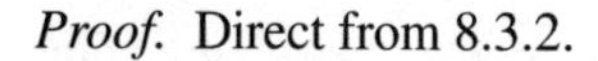

Proof. Direct from 8.3.2. □

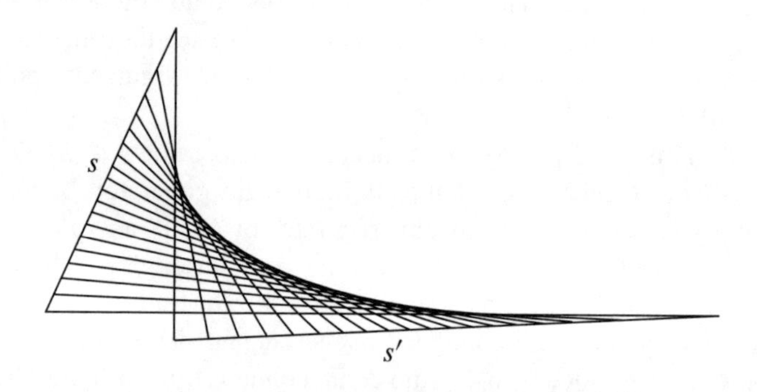

Figure 8.12. Generators of a ruling of a ruled quadric meeting two generators s, s' of the opposite ruling.

Two of the above maps put the set of points of a ruled quadric in one-to-one correspondence with a product of projective lines:

Corollary 8.3.5. *Assume that s and ℓ are generators of Q of different rulings. As above denote by ℓ_q the generator through $q \in s$ other than s and, similarly, by $s_{q'}$ the generator through $q' \in \ell$ other than ℓ. Then the correspondence*

$$\begin{aligned} \Phi \colon s \times \ell &\longrightarrow Q, \\ (q, q') &\longmapsto \ell_q \cap s_{q'}, \end{aligned}$$

is a bijection.

Proof. Assume that s belongs to the first ruling S_1. The map of the claim is the composition of

$$\begin{aligned} s \times \ell &\longrightarrow S_2 \times S_1 \longrightarrow Q, \\ (q, q') &\longmapsto (\ell_q, s_{q'}) \longmapsto \ell_q \cap s_{q'}. \end{aligned}$$

The map on the left is bijective by 8.3.4, while the one on the right is well defined by 8.3.3 (a) and bijective by 8.3.2 (4). □

A map Φ as above may be explicitly written using coordinates, which provides a parametric representation of the ruled quadric. Take the coordinates of 8.3.1.

Choose s to be the line $z_1 = z_2 = 0$, its points have the form $q = [\beta_1, 0, 0, \beta_0]$ where β_0, β_1 are projective coordinates on s. Similarly, we take ℓ the line $z_0 = z_2 = 0$ and its points in the form $q' = [0, \alpha_0, 0, \alpha_1]$. Then ℓ_q has equations

$$\begin{aligned}\beta_0 z_0 &= \beta_1 z_3,\\ \beta_1 z_1 &= \beta_0 z_2,\end{aligned}$$

while those of $s_{q'}$ are

$$\begin{aligned}\alpha_0 z_0 &= \alpha_1 z_2,\\ \alpha_1 z_1 &= \alpha_0 z_3.\end{aligned}$$

The reader may recognize the equations (8.8) and (8.7) used in the proof of 8.3.2 to define the rulings S_2 and S_1. An easy computation shows that the point

$$[\alpha_1\beta_1, \alpha_0\beta_0, \alpha_0\beta_1, \alpha_1\beta_0]$$

belongs to both ℓ_q and $s_{q'}$ and therefore

$$\Phi(q, q') = [\alpha_1\beta_1, \alpha_0\beta_0, \alpha_0\beta_1, \alpha_1\beta_0].$$

In particular, the equalities

$$\begin{aligned}z_0 &= \alpha_1\beta_1,\\ z_1 &= \alpha_0\beta_0,\\ z_2 &= \alpha_0\beta_1,\\ z_3 &= \alpha_1\beta_0\end{aligned} \tag{8.9}$$

provide a parameterization of the ruled quadric Q: (α_0, α_1), (β_0, β_1) are two pairs of homogeneous parameters, each pair determined up to a non-zero factor clearly irrelevant to the above equations. The map Φ being bijective, the free variation of these homogeneous parameters makes the point $p = [z_0, z_1, z_2, z_3]$ to describe the whole of $|Q|$, while in turn each $p \in Q$ determines its pairs of parameters up to non-zero factors.

It is often preferred to use the absolute coordinates $\alpha = \alpha_0/\alpha_1, \beta = \beta_0/\beta_1 \in k \cup \{\infty\}$ as ordinary – non-homogeneous – parameters, which give rise to the equivalent equations

$$\begin{aligned}z_0 &= 1,\\ z_1 &= \alpha\beta,\\ z_2 &= \alpha,\\ z_3 &= \beta.\end{aligned}$$

As in former cases, the suitable conventions for $\alpha = \infty$ or $\beta = \infty$ are obtained from the homogeneous equations, (8.9) in this case. As an heuristic rule that the reader may easily check, when one or both parameters take value ∞, correct values of the z_i are obtained by first dividing the equations by the parameter(s) – which does not affect $p = [z_0, z_1, z_2, z_3]$ – and then giving to the parameter(s) the value ∞ using the rule $1/\infty = 0$.

Surfaces that may be given by homogeneous polynomial parameterizations like (8.9) are called *rational surfaces*. The planes and the ruled quadrics of a three-space provide the easiest examples of rational surfaces. A nice book on rational surfaces is [5].

The tangent planes play an important role in the geometry of a ruled quadric. We begin by describing their intersections with the quadric:

Corollary 8.3.6. *The section of a ruled quadric by a tangent plane is the pair of generators through the contact point.*

Proof. If $p \in Q$, by 6.4.23, the generators through p are contained in the plane section $Q \cap H_p$ which therefore, by 6.4.17, equals the pair of generators. □

Remark 8.3.7. Assume $k = \mathbb{R}$. If a non-degenerate quadric Q is not ruled and has points, then its section by any tangent plane is conic with a double point at the contact point (by 6.4.24) and containing no (real) line. In view of the listing of projective types, it necessarily is a pair of imaginary lines with the contact point as its only real point. Thus for any such Q and any $p \in Q$, $|Q| \cap H_p = |Q \cap H_p| = \{p\}$, as it is intuitively clear, for instance, for a sphere. To complete the picture, the reader may easily see using 6.7.2 that if Q is an ordinary real cone and p one of its simple points, then the tangent plane at p is tangent to the cone at all points of the generator through p. Again by 6.4.24, in this case the section by the tangent plane is two times the generator through the contact point.

Remark 8.3.8. If s is a generator of a ruled quadric, it follows from 6.5.10 that s equals its polar line and therefore the tangent planes at the points of s are the planes of the pencil s^*, of the planes through s.

Remark 8.3.9. The bijections of 8.3.4 may now be described as follows: While the point q describes s, the tangent hyperplanes H_q describe s^* and then, by 8.3.6, the plane sections $Q \cap H_q$ have as fixed part s and as variable part the image ℓ_q of q. Conversely the point q is the pole of $\ell_q \vee s$ and so the map $\ell \mapsto \ell \cap s$ is the composition of projecting the generators from s, $\ell \mapsto \ell \vee s$, and the restriction to s^* of the inverse of the polarity.

Remark 8.3.10. By 8.3.6 or the above Remark 8.3.9, the inverse of the bijection Φ of 8.3.5, $(q, q') \mapsto p = \ell_q \cap s_{q'}$, consists of taking the poles q, q' of the planes projecting the point $p \in Q$ from the generators s, ℓ, provided the projection from s of $p \in s$ is taken to be the tangent plane H_p, and similarly for ℓ and $p \in \ell$.

Proposition 8.3.11. *If Q is ruled, its envelope Q^* is ruled too. The generators of Q^* are the pencils whose kernel is a generator of Q and its rulings are the families of pencils*

$$\{s^* | s \in S_i\}, \quad i = 1, 2,$$

S_1, S_2 the rulings of Q.

Proof. By its definition, Q^* is the image $\mathcal{P}_Q(Q)$ of Q by its own polarity $\mathcal{P}_Q$. It is thus ruled, and has as generators and rulings the images of the generators and rulings of Q by the polarity. Just add that if s is a generator, then, as seen in 8.3.8, $\mathcal{P}_Q(s) = s^*$. □

Remark 8.3.12. Being a generator of a ruled quadric is thus a self-dual condition: the points of s belong to Q if and only if the planes through s belong to Q^*. Also the conditions of two generators to belong to the same or to different rulings are self-dual.

Next we will see that composing two of the bijections of 8.3.4, relative to two different generators and taken in opposite senses, gives an already well-known projectivity:

Corollary 8.3.13. *Assume that s, s' are different generators belonging to the same ruling of Q. Mapping $q \in s$ to $q' \in s'$ if and only if they belong to the same generator (necessarily of the other ruling) of Q is the perspectivity with centre any generator $\bar{s}$ other than s, s' and of the same ruling. Any ruling of any ruled quadric is the set of all lines joining corresponding points of two perspective skew lines of $\mathbb{P}_3$.*

Proof. Take the ruling containing s, s' and $\bar{s}$ as the first one. If points q, q' correspond by the perspectivity, then, for some $H \in \bar{s}^*$, $q, q' \in H$ or, equivalently, $q, q' \in Q \cap H$. By 8.3.8 and 8.3.6, $Q \cap H = \bar{s} + \ell$ where ℓ belongs to the second ruling. Different generators of the same ruling being disjoint, it follows that $q, q' \in \ell$. Conversely, if q and q' belong to the same generator ℓ of the second ruling, then they belong to the plane $\bar{s} \vee \ell$ and therefore they correspond by the perspectivity. The second claim obviously follows from the first one. □

Remark 8.3.14. As the reader may easily check, the lines joining corresponding points on two perspective skew lines $r_1, r_2 \subset \mathbb{P}_3$ meet r_1, r_2 and the centre of the perspectivity, and conversely. Thus, after 8.3.13, any ruling of any ruled quadric is the set of all lines meeting three fixed pairwise skew lines of $\mathbb{P}_3$.

Remark 8.3.15. Any ruling S of a ruled quadric may be endowed with a one-dimensional projective structure that turns the bijections of 8.3.4, for all choices of the generator $s \notin S$, into projectivities. Moreover, such a condition determines the projective structure on S up to equivalence. Indeed, it is enough to proceed as for the projective structure on a conic in the proof of 8.2.1: one of the bijections of 8.3.4 is used to define the projective structure, after which the other bijections become projectivities by 8.3.13, and the structure is uniquely determined by 1.6.3 (b). The details are left to the reader. Note that, by 8.3.9, if this structure is taken on S, then also projecting the generators of S from any fixed generator of the other ruling becomes a projectivity.

Next is the dual of 8.3.13.

Corollary 8.3.16. *Let s, s' be different generators belonging to the same ruling of Q. Take planes $H \supset s$ and $H' \supset s'$ as corresponding if and only if they contain the same generator (necessarily of the other ruling) of Q. This correspondence is the perspectivity with axis any generator $\bar{s}$ other than s, s' and of the same ruling. Any ruling of any ruled quadric is the set of all intersection lines of corresponding planes of two perspective disjoint pencils of planes of $\mathbb{P}_3$.*

The two propositions that follow are dual of each other. They generalize Steiner's theorems to ruled quadrics and are the converses of the second halves of 8.3.13 and 8.3.16. We will prove the second one.

Proposition 8.3.17 (Projective generation of ruled quadrics I). *The lines joining corresponding points of two homographic skew lines of $\mathbb{P}_3$ describe a ruling of a ruled quadric.*

Proposition 8.3.18 (Projective generation of ruled quadrics II). *The intersection lines of corresponding planes of two homographic disjoint pencils of planes of $\mathbb{P}_3$ describe a ruling of a ruled quadric.*

Proof. The pencils being disjoint, all the intersections of pairs of corresponding planes are lines. Let

$$f : r^* \longrightarrow s^*$$

be the homographic pencils of planes. The lines r and s being skew, we take a reference of $\mathbb{P}_3$ with the vertices $p_0, p_1 \in r$ and $p_2, p_3 \in s$. The planes of r^* have thus equations $\lambda x_2 + \mu x_3 = 0$ where λ, μ are homogeneous coordinates on r^* (4.4.9), and similarly for the planes $\lambda' x_0 + \mu' x_1 = 0$ of s^*. If f has matrix

$$\begin{pmatrix} a & b \\ c & d \end{pmatrix},$$

the image of $\lambda x_2 + \mu x_3 = 0$ is

$$(a\lambda + b\mu)x_0 + (c\lambda + d\mu)x_1 = 0$$

and therefore $p = [x_0, x_1, x_2, x_3]$ belongs to the intersection of two corresponding planes if and only if there exist λ, μ, not both zero, such that

$$\lambda x_2 + \mu x_3 = 0,$$
$$\lambda(a x_0 + c x_1) + \mu(b x_0 + d x_1) = 0.$$

Since this is equivalent to having

$$x_2(b x_0 + d x_1) - x_3(a x_0 + c x_1) = 0, \tag{8.10}$$

we see that the points of the intersection lines of corresponding planes are those of the quadric Q with equation (8.10) above, and thus matrix

$$\begin{pmatrix} 0 & 0 & b & -a \\ 0 & 0 & d & -c \\ b & d & 0 & 0 \\ -a & -c & 0 & 0 \end{pmatrix}.$$

The condition of regularity of the matrix of f, $ad - bc \neq 0$, assures that Q is non-degenerate and hence, by the way we got it, ruled. We may thus apply Theorem 8.3.2. The line r has equations $x_2 = x_3 = 0$ and so clearly is a generator of Q. All intersections of corresponding planes being coplanar with r, they all belong to the ruling S that does not contain r. Conversely, since there is one intersection of corresponding planes through each point of r (by an easy direct argument or just because $r \subset Q$), the intersections of corresponding planes cover the whole of S and the proof is complete. □

Corollary 8.3.19. *There is a unique quadric Q containing three given pairwise skew lines r_1, r_2, r_3 of $\mathbb{P}_3$. The quadric Q is ruled, the lines meeting r_1, r_2 and r_3 compose one of its rulings and the points of these lines are the points of Q.*

Proof. For the existence of a ruled quadric containing r_1, r_2, r_3, let f be the perspectivity from r_1 onto r_2 with centre r_3. By 8.3.17, assume that Q is a ruled quadric one of whose rulings consist in all joins of points corresponding by f. Clearly, all points of r_1 and r_2 belong to Q. If $p \in r_3$, then the line $(p \vee r_1) \cap (p \vee r_2)$ meets r_1 and r_2 in points that correspond by f, from which $p \in Q$ and hence $r_3 \subset Q$.

For the uniqueness and the remaining claims, assume that Q is a quadric containing r_1, r_2, r_3. Since no plane-pair contains three pairwise skew lines, $\mathbf{r}(Q) \geq 3$. If $\mathbf{r}(Q) = 3$, one of the lines r_i, say r_1, does not contain the double point O of Q. Then a plane π through r_1 and missing O would have $r_1 \subset Q \cap \pi$ and hence $Q \cap \pi$ degenerate, against 6.7.7. Thus, Q is non-degenerate and therefore ruled.

Since r_1, r_2, r_3 are pairwise skew, they belong to the same ruling of Q; we take it as the first ruling. Every line meeting r_1, r_2, r_3 shares with Q three different points and therefore (by 6.2.5) is a generator of Q, obviously of the second ruling. Conversely, by 8.3.2, any generator of the second ruling of Q meets r_1, r_2, r_3, because they belong to the first ruling. This proves that the second ruling of Q equals the set of lines meeting r_1, r_2, r_3: in particular (again using 8.3.2) the points of Q are the points of the lines meeting r_1, r_2, r_3, as claimed, and this in turn uniquely determines Q by 7.3.3. □

As the reader may have noticed, regarding projective generation and projective structure, the rulings of the non-degenerate quadrics play the same role as the sets of points of (or tangents to) the non-degenerate conics. It is worth comparing 8.3.17, 8.3.18, 8.3.16 and 8.3.15 to 8.1.4, 8.1.1, 8.1.5 and 8.2.1, respectively.

Projecting the points of a non-degenerate quadric Q from a fixed $p \in Q$ defines a map $\delta\colon |Q| - \{p\} \to p^{\nabla}$ which is the easiest way of relating the points of the quadric to the elements of a two-dimensional projective space. The projection δ is usually composed with a section map $\sigma: p^{\nabla} \to H$, H a plane missing p, in order to have points of the same space as images of the points of Q. The composition $\psi = \sigma \circ \delta$ is called the *stereographic projection of Q onto H from* (or *with centre*) p. There is of course no essential difference between ψ and δ, because σ is a projectivity. The next proposition describes stereographic projections. Its easy proof is left to the reader, who may also compare it to 8.3.5.

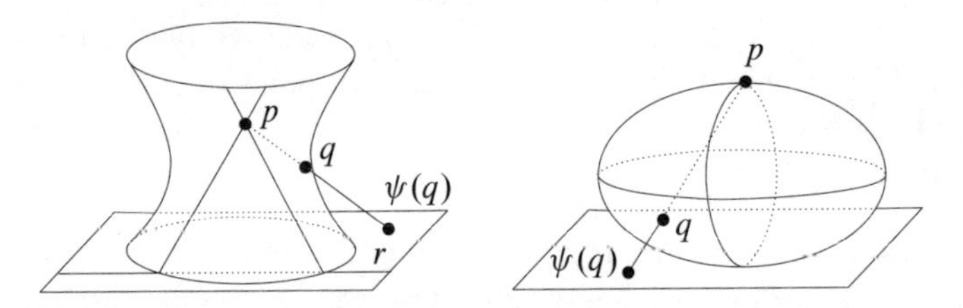

Figure 8.13. Stereographic projections: the ruled and non-ruled cases.

Proposition 8.3.20 (Plane representation of quadrics). *Assume that Q is a non-degenerate quadric of $\mathbb{P}_3$, p a point of Q, $H \subset \mathbb{P}_3$ a plane missing p and r the line intersection of H and the tangent plane to Q at p. Let ψ be the map* (*stereographic projection*)

$$\begin{aligned} \psi\colon |Q| - \{p\} &\longrightarrow H, \\ q &\longmapsto pq \cap H. \end{aligned}$$

Then the following holds:

(1) *If Q is not ruled then ψ is injective and has image $H - r$.*

(2) *If Q is ruled and s, ℓ are the generators of Q through p, then ψ contracts s and ℓ to their respective traces on H, $\psi(s-\{p\}) = s \cap H$, $\psi(\ell-\{p\}) = \ell \cap H$, and restricts to a bijection*

$$|Q| - s \cup \ell \longrightarrow H - r.$$

8.4 Lines of $\mathbb{P}_3$

We know that both the set of points and the set of hyperplanes of a projective space $\mathbb{P}_n$ have a structure of n-dimensional projective space. The set of all linear varieties of $\mathbb{P}_n$ of a fixed intermediate dimension d, $0 < d < n - 1$, may also receive a geometric structure which, however, is not so simple as in the extremal cases. This section is devoted to the case of the lines of $\mathbb{P}_3$; we will see that they may be represented as the points of a certain quadric of a five-dimensional projective space. For $n > 3$ the linear varieties of intermediate dimension d are represented as the points of a certain projective algebraic variety, called a Grassmannian variety: presenting Grassmannian varieties is beyond the scope of this book; the interested reader is referred to [18], book II, chapter VII, or [15], Chapter 6.

Throughout this section we assume to have fixed a three-dimensional projective space $\mathbb{P}_3$ and a reference frame Δ in it. Let ℓ be a line of $\mathbb{P}_3$, spanned by distinct points $a = [a_0, a_1, a_2, a_3]$ and $b = [b_0, b_1, b_2, b_3]$, and take

$$p_{i,j} = \begin{vmatrix} a_i & a_j \\ b_i & b_j \end{vmatrix}, \quad i, j = 0, \dots, 3.$$

Obviously, there is at least one $p_{i,j} \neq 0$, $i \neq j$, because we are assuming $a \neq b$. Furthermore, a different choice of the coordinates of a and b results in the multiplication of all the $p_{i,j}$ by the same non-zero factor. The first interesting point is:

Lemma 8.4.1. *The above $p_{i,j}$, $i, j = 0, \dots, 3$, taken up to a non-zero common factor, do not depend on the choice of the points a, b spanning ℓ.*

Proof. Assume that ℓ is also spanned by points $a' = [a'_0, a'_1, a'_2, a'_3]$ and $b' = [b'_0, b'_1, b'_2, b'_3]$. Then there are $\alpha, \beta, \gamma, \delta \in k$, $\alpha\delta - \beta\gamma \neq 0$, such that

$$(a'_0, a'_1, a'_2, a'_3) = \alpha(a_0, a_1, a_2, a_3) + \beta(b_0, b_1, b_2, b_3),$$
$$(b'_0, b'_1, b'_2, b'_3) = \gamma(a_0, a_1, a_2, a_3) + \delta(b_0, b_1, b_2, b_3).$$

Then

$$\begin{pmatrix} a'_i & a'_j \\ b'_i & b'_j \end{pmatrix} = \begin{pmatrix} \alpha & \beta \\ \gamma & \delta \end{pmatrix} \begin{pmatrix} a_i & a_j \\ b_i & b_j \end{pmatrix}$$

and therefore

$$p'_{i,j} = \begin{vmatrix} a'_i & a'_j \\ b'_i & b'_j \end{vmatrix} = \begin{vmatrix} \alpha & \beta \\ \gamma & \delta \end{vmatrix} p_{i,j} \quad \text{for } i, j = 0, \dots, 3.$$

Of course, $p_{i,j} = -p_{j,i}$ and $p_{i,i} = 0$, $i, j = 0, \dots, 3$. So, to avoid redundancies, we will only consider the $p_{i,j}$ with $0 \leq i < j \leq 3$. Taken in the lexicographical order, namely

$$p_{0,1}, \quad p_{0,2}, \quad p_{0,3}, \quad p_{1,2}, \quad p_{1,3}, \quad p_{2,3},$$

and up to a common non-zero factor, they are called the *Plücker coordinates*, and also the *line coordinates*, of ℓ (relative to the reference Δ). The redundant $p_{i,j}$, $i \geq j$, are sometimes used to give a more symmetric form to some expressions.

Fix any five-dimensional projective space $\mathbb{P}_5$ and a reference of it. If $\mathcal{G}$ denotes the set of all lines of $\mathbb{P}_3$, then 8.4.1 assures that mapping each line ℓ of $\mathbb{P}_3$ to the point of $\mathbb{P}_5$ whose coordinates are the Plücker coordinates of ℓ is a well-defined map

$$\begin{aligned} \mathfrak{p}\colon \mathcal{G} &\longrightarrow \mathbb{P}_5, \\ \ell &\longmapsto [p_{i,j}]_{0 \leq i < j \leq 3}, \end{aligned}$$

which is called the *Plücker map*.

Theorem 8.4.2. *The Plücker map is a bijection between the set of lines of $\mathbb{P}_3$ and the set of points of the quadric*

$$\mathbb{G}: p_{0,1}p_{2,3} - p_{0,2}p_{1,3} + p_{0,3}p_{1,2} = 0. \tag{8.11}$$

Proof. Assume that $p_{i,j} = a_i b_j - a_j b_i, 0 \leq i < j \leq 3$, are the Plücker coordinates of the line spanned by $a = [a_0, a_1, a_2, a_3]$ and $b = [b_0, b_1, b_2, b_3]$. After expanding the determinant in terms of its first two rows, the obvious equality

$$\begin{vmatrix} a_0 & a_1 & a_2 & a_3 \\ b_0 & b_1 & b_2 & b_3 \\ a_0 & a_1 & a_2 & a_3 \\ b_0 & b_1 & b_2 & b_3 \end{vmatrix} = 0$$

gives

$$p_{0,1}p_{2,3} - p_{0,2}p_{1,3} + p_{0,3}p_{1,2} = 0$$

and therefore proves that all the images of lines by the Plücker map are points of $\mathbb{G}$.

Conversely, assume that a point of $p \in \mathbb{P}_5$ has coordinates $p_{i,j}, 0 \leq i < j \leq 3$, satisfying equation (8.11) and consider the matrix

$$\begin{pmatrix} 0 & p_{0,1} & p_{0,2} & p_{0,3} \\ -p_{0,1} & 0 & p_{1,2} & p_{1,3} \\ -p_{0,2} & -p_{1,2} & 0 & p_{2,3} \\ -p_{0,3} & -p_{1,3} & -p_{2,3} & 0 \end{pmatrix}.$$

The other cases being dealt with similarly, assume $p_{0,1} \neq 0$ and take a and b to be the points of $\mathbb{P}_3$ with coordinate vectors the rows of the above matrix involving $p_{0,1}$, namely

$$a = [0, p_{0,1}, p_{0,2}, p_{0,3})],$$
$$b = [-p_{0,1}, 0, p_{1,2}, p_{1,3}].$$

Clearly, $a \neq b$. A direct computation gives as Plücker coordinates of $\ell = a \vee b$,

$$p_{0,1}^2, \quad p_{0,1}p_{0,2}, \quad p_{0,1}p_{0,3}, \quad p_{0,1}p_{1,2}, \quad p_{0,2}p_{1,3} - p_{0,3}p_{1,2} = p_{0,1}p_{2,3},$$

and so $\mathfrak{p}(\ell) = p$.

For the injectivity, assume as above that ℓ is the line of $\mathbb{P}_3$ spanned by distinct points $a = [a_0, a_1, a_2, a_3]$ and $b = [b_0, b_1, b_2, b_3]$. Then a point $[x_0, x_1, x_2, x_3]$ belongs to ℓ if and only if

$$\operatorname{rk}\begin{pmatrix} x_0 & x_1 & x_2 & x_3 \\ a_0 & a_1 & a_2 & a_3 \\ b_0 & b_1 & b_2 & b_3 \end{pmatrix} < 3,$$

which may be equivalently stated by annihilating all order-three minors of the matrix, that is

$$\begin{aligned} p_{1,2}x_0 - p_{0,2}x_1 + p_{0,1}x_2 &= 0, \\ p_{1,3}x_0 - p_{0,3}x_1 + p_{0,1}x_3 &= 0, \\ p_{2,3}x_0 - p_{0,3}x_2 + p_{0,2}x_3 &= 0, \\ p_{2,3}x_1 - p_{1,3}x_2 + p_{1,2}x_3 &= 0. \end{aligned} \tag{8.12}$$

The points of ℓ are thus the solutions of the above system of equations, which in turn is determined by the Plücker coordinates of ℓ. Hence ℓ itself is determined by its Plücker coordinates and so the Plücker map is injective. □

Remark 8.4.3. Although not needed in the argument, the reader may easily check that the above system of equations (8.12) has rank two, as it could not be otherwise (see 2.6.4). Two independent equations extracted from (8.12) compose thus a system of equations of ℓ.

Remark 8.4.4. The quadric $\mathbb{G}$ of 8.4.2 is called the *Plücker quadric*, and also the *Klein quadric*. As it is clear from its equation (8.11), it is non-degenerate and, in case $k = \mathbb{R}$, has index 2.

It is often said that the Plücker map *represents* the lines of $\mathbb{P}_3$ as the points of the Plücker quadric; therefore, images by the Plücker map are sometimes called *representations*. The remainder of this section is devoted to relating properties of lines of $\mathbb{P}_3$ to projective properties of their representations as points of $\mathbb{G}$.

Proposition 8.4.5. *If ℓ and r are two distinct lines of $\mathbb{P}_3$, then the following conditions are equivalent:*

(i) *ℓ and r are coplanar.*

(ii) *$\mathfrak{p}(\ell)$ and $\mathfrak{p}(r)$ are conjugate with respect to $\mathbb{G}$.*

(iii) *The line $\mathfrak{p}(\ell) \vee \mathfrak{p}(r)$ of $\mathbb{P}_5$ is contained in $\mathbb{G}$.*

Proof. Assume as above that ℓ is spanned by

$$a = [a_0, a_1, a_2, a_3] \quad \text{or} \quad b = [b_0, b_1, b_2, b_3],$$

and so has Plücker coordinates

$$p_{i,j} = \begin{vmatrix} a_i & a_j \\ b_i & b_j \end{vmatrix}, \quad 0 \leq i < j \leq 3.$$

Assume also that r is spanned by $c = [c_0, c_1, c_2, c_3]$ and $d = [d_0, d_1, d_2, d_3]$, and so has Plücker coordinates

$$q_{i,j} = \begin{vmatrix} c_i & c_j \\ d_i & d_j \end{vmatrix}, \quad 0 \leq i < j \leq 3.$$

The lines ℓ and r are coplanar if and only if a, b, c, d are linearly dependent, that is, if and only if

$$\begin{vmatrix} a_0 & a_1 & a_2 & a_3 \\ b_0 & b_1 & b_2 & b_3 \\ c_0 & c_1 & c_2 & c_3 \\ d_0 & d_1 & d_2 & d_3 \end{vmatrix} = 0.$$

After expanding the determinant in terms of its first two rows, the above equality reads

$$p_{0,1}q_{2,3} - p_{0,2}q_{1,3} + p_{0,3}q_{1,2} + p_{1,2}q_{0,3} - p_{1,3}q_{0,2} + p_{2,3}q_{0,1} = 0$$

which, as the reader may easily check, is just the condition of conjugation of $\mathfrak{p}(\ell) = [p_{i,j}]_{0 \leq i < j \leq 3}$ and $\mathfrak{p}(r) = [q_{i,j}]_{0 \leq i < j \leq 3}$ with respect to $\mathbb{G}$.

Both $\mathfrak{p}(\ell)$ and $\mathfrak{p}(r)$ being self-conjugate, the equivalence (ii) $\Longleftrightarrow$ (iii) is clear after 6.4.6. □

Let ℓ, r be distinct coplanar lines of $\mathbb{P}_3$ and put $\pi = \ell \vee r$, $O = \ell \cap r$. The lines on π through O compose the pencil of lines of π spanned by ℓ and r: in a three-dimensional context, it is often referred to as the *flat pencil* spanned by ℓ and r. The next proposition assures that the lines contained in $\mathbb{G}$ are just the representations of the flat pencils of $\mathbb{P}_3$:

Proposition 8.4.6. *If ℓ and r are distinct coplanar lines of $\mathbb{P}_3$, then the line spanned by $\mathfrak{p}(\ell)$ and $\mathfrak{p}(r)$ is the image by $\mathfrak{p}$ of the flat pencil spanned by ℓ and r. Any line of $\mathbb{P}_5$ contained in $\mathbb{G}$ is the image by $\mathfrak{p}$ of the flat pencil spanned by two coplanar distinct lines of $\mathbb{P}_3$.*

Proof. Take $O = \ell \cap r$. Since ℓ and r are coplanar, by 8.4.5, $L = \mathfrak{p}(\ell) \vee \mathfrak{p}(r) \subset \mathbb{G}$. Thus any point $T \in L, T \neq \mathfrak{p}(\ell), \mathfrak{p}(r)$, is $T = \mathfrak{p}(t)$, t a line of $\mathbb{P}_3, t \neq \ell, r$. Again by 8.4.5, t is coplanar with, and hence meets, ℓ and r. Let s be any line of $\mathbb{P}_3$ that meets both ℓ and r. By 8.4.5, $\mathfrak{p}(s)$ is conjugate to both $\mathfrak{p}(\ell)$ and $\mathfrak{p}(r)$ and therefore also to $\mathfrak{p}(t)$. Once again by 8.4.5, t is coplanar with s and therefore meets it. If s is chosen such that $O \notin s \subset \ell \vee r$, then the points $t \cap \ell, t \cap r, t \cap s$ belong to $\ell \vee r$ and at least two of them are distinct: it follows that $t \subset \ell \vee r$. Now choose s such that $O \in s \not\subset \ell \vee r$. Then, by the above, $t \cap s \subset (\ell \vee r) \cap s = O$, hence $O \in t$ and, all together, t belongs to the flat pencil spanned by ℓ and r.

For the converse, assume that for a line t of $\mathbb{P}_3, t \neq \ell, r$, it is $O \in t \subset \ell \vee r$. Choose three aligned points b_1, b_2, b_3 on ℓ, r, and t, respectively, all different from O: then the coordinate vectors of b_1, b_2, b_3 are linearly dependent. If Plücker coordinate vectors of ℓ, r and t are computed from the points O, b_1, O, b_2 and O, b_3, respectively, they easily turn out to be linearly dependent too. This proves that $\mathfrak{p}(\ell), \mathfrak{p}(r), \mathfrak{p}(t)$ are aligned points.

For the second claim, just note that any line contained in $\mathbb{G}$ is spanned by two points $\mathfrak{p}(\ell)$, $\mathfrak{p}(r)$, with ℓ, r distinct and coplanar, by 8.4.2 and 8.4.5. Then the first claim applies. □

Proposition 8.4.7. (1) *For any plane π of $\mathbb{P}_3$, the image by the Plücker map of the set of all lines on π is a plane of $\mathbb{P}_5$ contained in $\mathbb{G}$.*

(2) *For any point O of $\mathbb{P}_3$, the image by the Plücker map of the bunch of all lines through O is a plane of $\mathbb{P}_5$ contained in $\mathbb{G}$.*

(3) *There are no planes of $\mathbb{P}_5$ contained in $\mathbb{G}$ other than those described in* (1) *and* (2) *above.*

Proof. Pick three non-concurrent lines $\ell_i, i = 1, 2, 3$, on π. Their representations $\mathfrak{p}(\ell_i), i = 1, 2, 3$, are three distinct points of $\mathbb{G}$. Furthermore, they are non-aligned, as otherwise the line they span would be contained in $\mathbb{G}$ by 6.2.5 against 8.4.6 and the assumption of the lines being not concurrent. The points $\mathfrak{p}(\ell_i), i = 1, 2, 3$, are conjugate to each other because the lines are coplanar (8.4.5) and so, by 6.4.6, span a plane Π of $\mathbb{P}_5$ contained in $\mathbb{G}$. Now, a point of $q = \mathfrak{p}(\ell) \in \mathbb{G}$ belongs to Π if and only if it is collinear with $\mathfrak{p}(\ell_1)$ and a point $\mathfrak{p}(\ell') \in \mathfrak{p}(\ell_2) \vee \mathfrak{p}(\ell_3)$. Again by 8.4.6, this occurs if and only if ℓ belongs to the flat pencil spanned by ℓ_1 and a line ℓ' of the flat pencil spanned by ℓ_2 and ℓ_3. The latter condition being clearly equivalent to $\ell \subset \pi$, we have seen that $\Pi = \mathfrak{p}(\{\ell \mid \ell \subset \pi\})$, as wanted.

Proving claim (2) by a quite similar argument, using three non-coplanar lines through O, is left to the reader.

For claim (3), assume that Π is a plane contained in $\mathbb{G}$ and take three points $\mathfrak{p}(\ell_i)$, $i = 1, 2, 3$, spanning it. Since each two of these points span a line contained in $\mathbb{G}$ that does not contain the third point, by 8.4.5 and 8.4.6, any two of the lines ℓ_i, $i = 1, 2, 3$, are coplanar and no one belongs to the flat pencil spanned by the other two. Then either ℓ_1, ℓ_2 and ℓ_3 belong to a plane π and are non-concurrent, or they all go through a point O and are non-coplanar. Since we already know from claims (1) and (2) that the representations of either the lines on π, or the lines through O, according to the case, describe a plane Π' of $\mathbb{P}_5$ that obviously contains the points $\mathfrak{p}(\ell_i)$, $i = 1, 2, 3$, $\Pi = \Pi'$ and Π is as claimed. □

Remark 8.4.8. The planes contained in $\mathbb{G}$ are thus of two different types: they are either the set of points representing the lines on a plane $\pi \subset \mathbb{P}_3$, or the set of points representing the lines through a point $O \in \mathbb{P}_3$. The reader may easily check that two different planes of the same type meet at a single point, while planes of different types either share a line or are disjoint (according to $O \in \pi$ or not).

Remark 8.4.9. Due to 6.5.11, there are no linear varieties of dimension higher than two contained in $\mathbb{G}$. Therefore 8.4.2, 8.4.6 and 8.4.7 describe all the non-empty linear varieties contained in $\mathbb{G}$.

Proposition 8.4.10. *The images by the Plücker map of the lines meeting three pairwise skew lines ℓ_1, ℓ_2, ℓ_3 of $\mathbb{P}_3$ are the points of a non-degenerate plane section of $\mathbb{G}$. Any non-degenerate plane section of $\mathbb{G}$ with points arises in this way.*

Proof. Let S be the set of all lines meeting ℓ_1, ℓ_2 and ℓ_3. The reader may note first that, either by an easy argument using 5.3.8, or directly from 8.3.19, S contains infinitely many lines, any two of which, if different, are skew.

By 8.4.5, a line r meets ℓ_1, ℓ_2 and ℓ_3 if and only if $\mathfrak{p}(r)$ is conjugate to $\mathfrak{p}(\ell_1)$, $\mathfrak{p}(\ell_2)$) and $\mathfrak{p}(\ell_3)$. This occurs if and only if $\mathfrak{p}(r)$ belongs to the intersection of polar hyperplanes relative to $\mathbb{G}$,

$$V = H_{\mathfrak{p}(\ell_1)} \cap H_{\mathfrak{p}(\ell_2)} \cap H_{\mathfrak{p}(\ell_3)},$$

or, which is the same, to $\mathbb{G} \cap V$. Thus $\mathfrak{p}(S) = \mathbb{G} \cap V$.

Next we will see that V is a plane and $\mathbb{G} \cap V$ a non-degenerate conic. Since the points $\mathfrak{p}(\ell_1), \mathfrak{p}(\ell_2)), \mathfrak{p}(\ell_3)$ are different, a line containing them would be contained in $\mathbb{G}$, against 8.4.6. It follows that $\mathfrak{p}(\ell_1), \mathfrak{p}(\ell_2)), \mathfrak{p}(\ell_3)$ are independent, and therefore so are their polar hyperplanes (because $\mathbb{G}$ is non-degenerate); hence V is a plane. Since, as noted above, S contains infinitely many lines and any two different lines in S are skew, $\mathbb{G} \cap V$ contains infinitely many points and no line, as any two different points in $\mathbb{G} \cap V$ span a line not contained in $\mathcal{G}$ by 8.4.6. It follows that $\mathbb{G} \cap V$ is a non-degenerate conic.

For the converse, assume that V is a plane and $\mathbb{G} \cap V$ is a non-degenerate conic with points. Pick three different, necessarily non-aligned, points $\mathfrak{p}(s_1), \mathfrak{p}(s_2), \mathfrak{p}(s_3)$

of $\mathbb{G} \cap V$. The conic $\mathbb{G} \cap V$ being non-degenerate, each two of these points span a line not contained in $\mathbb{G}$ and therefore (again by 8.4.6) s_1, s_2, s_3 are pairwise skew. Using 5.3.8, take three different lines ℓ_1, ℓ_2, ℓ_3, each meeting s_1, s_2 and s_3. The lines ℓ_1, ℓ_2, ℓ_3 are pairwise skew too, as otherwise the lines s_i, $i = 1, 2, 3$, would be coplanary. By the part already proved, the representations of the lines that meet ℓ_1, ℓ_2 and ℓ_3, are the points of a non-degenerate plane section $\mathbb{G} \cap V'$ of $\mathbb{G}$. Since each s_i, $i = 1, 2, 3$, meets ℓ_1, ℓ_2 and ℓ_3, $\mathfrak{p}(s_1), \mathfrak{p}(s_2), \mathfrak{p}(s_3) \in V'$. Then $V' = V$ and $\mathbb{G} \cap V = \mathbb{G} \cap V'$ is as claimed. □

As already noted, if $k = \mathbb{R}$, then the index of $\mathbb{G}$ is 2 and therefore $\mathbb{G}$ has imaginary plane sections.

Corollary 8.4.11. *The image by the Plücker map of a ruling of a ruled quadric is the set of points of a non-degenerate plane section of $\mathbb{G}$, and any non-degenerate plane section of $\mathbb{G}$ with points arises in this way.*

Proof. Follows from 8.4.10 and 8.3.14. □

For any hyperplane H of $\mathbb{P}_5$, the set F_H of all lines of $\mathbb{P}_3$ whose image by the Plücker map belongs to H is called a *linear line-complex*. It is thus $F_H = \mathfrak{p}^{-1}(H)$. Equivalently, one may define the linear line-complexes as the sets of all lines of $\mathbb{P}_3$ whose Plücker coordinates satisfy a non-trivial homogeneous linear relation

$$\sum_{0 \leq i < j \leq 3} A_{i,j} p_{i,j} = 0,$$

the coefficients $A_{i,j}$ being those of an equation of H. Allowing the above relation to be homogeneous of arbitrary degree in the $p_{i,j}$, instead of linear, gives rise to more general families called *line-complexes*; we will not study them here.

The easier examples of linear line-complexes are obtained by fixing a line ℓ of $\mathbb{P}_3$ and taking all the lines of $\mathbb{P}_3$ that meet ℓ. For, according to 8.4.5, two lines meet if and only if their images by the Plücker map are conjugate with respect to $\mathbb{G}$, and so the representations of the lines that meet ℓ are the points of the section of $\mathbb{G}$ by its tangent hyperplane at $\mathfrak{p}(\ell)$. These line-complexes are called *special*. The remaining linear line-complexes will be called *non-special*. To describe them, we need to set a couple of facts relative to null-systems of $\mathbb{P}_3$ not already seen in Section 5.8. If wanted, the reader may easily extend them to arbitrary dimension following the lines of Section 6.5.

Assume that ℓ is a line and g a null-system of $\mathbb{P}_3$. The images $g(p)$, for $p \in \ell$, compose a pencil of planes $\bar{\ell}^*$; its kernel $\bar{\ell}$ is a line which is called the *polar line*, or just the *polar*, of ℓ relative to g. Obviously it is $\bar{\ell} = g(p_1) \cap g(p_2)$ for any two points p_1, p_2 spanning ℓ. If q_1, q_2 span $\bar{\ell}$, then $q_1, q_2 \in g(p_i)$, $i = 1, 2$, and, by the reciprocity, $p_1, p_2 \in g(q_i)$, $i = 1, 2$. It follows that the polar of $\bar{\ell}$ is ℓ. Lines equating their own polars are called *self-polar*. Clearly from the above, the line ℓ, spanned by p_1, p_2, is self-polar with respect to g if and only if

$$p_1 \in g(p_1), \quad p_1 \in g(p_2), \quad p_2 \in g(p_1), \quad p_2 \in g(p_1).$$

The first and fourth conditions are always satisfied because g is a null-system, while the second and third ones are equivalent by the reciprocity. It follows thus:

Lemma 8.4.12. *The line of $\mathbb{P}_3$ spanned by two distinct points p_1, p_2 is self-polar with respect to a correlation g of $\mathbb{P}_3$ if and only if $p_2 \in g(p_1)$.*

The theorem below provides a description of the non-special line-complexes in terms of null-systems.

Theorem 8.4.13. *There is a one-to-one correspondence between non-special line-complexes and null-systems of $\mathbb{P}_3$, such that the lines composing a non-special line complex are those which are self-polar with respect to the corresponding null-system.*

Proof. Take $q = [q_{i,j}]_{0 \leq i < j \leq 3} \in \mathbb{P}_5 - |\mathbb{G}|$ and denote by F_q the linear line-complex determined by the polar hyperplane H_q of q relative to $\mathbb{G}$, that is

$$F_q = \mathfrak{p}^{-1}(|\mathbb{G} \cap H_q|).$$

Clearly F_q determines $\mathfrak{p}(F_q) = |\mathbb{G} \cap H_p|$. Since $\mathbb{G}$ is non-degenerate and contains planes, H_q contains a simple (real if $k = \mathbb{R}$) point q' of $\mathbb{G}$. Furthermore, H_q is not tangent to $\mathbb{G}$ at q' because $q \notin \mathbb{G}$ and therefore $q \neq q'$. Thus Corollary 6.4.33 applies, showing that $|\mathbb{G} \cap H_q|$ determines H_q. Since obviously H_q determines in turn its pole q, we have seen that mapping q to F_q defines a bijection θ between $\mathbb{P}_5 - |\mathbb{G}|$ and the set of all non-special line-complexes of $\mathbb{P}_3$.

On the other hand, consider the matrix

$$N = \begin{pmatrix} 0 & q_{0,1} & q_{0,2} & q_{0,3} \\ -q_{0,1} & 0 & q_{1,2} & q_{1,3} \\ -q_{0,2} & -q_{1,2} & 0 & q_{2,3} \\ -q_{0,3} & -q_{1,3} & -q_{2,3} & 0 \end{pmatrix}.$$

By direct computation it is easy to check that

$$\det N = (q_{0,1}q_{2,3} - q_{0,2}q_{1,3} + q_{0,3}q_{1,2})^2,$$

and so, the point q being assumed not to belong to $\mathbb{G}$, the matrix N is regular. Since it is also skew-symmetric, N defines a null-system g_q of $\mathbb{P}_3$ which is obviously determined by q. It is then clear that $q \mapsto g_q$ defines a bijection θ' between $\mathbb{P}_5 - |\mathbb{G}|$ and the set of all null-systems of $\mathbb{P}_3$: next we will prove that the bijection $\theta' \circ \theta^{-1}$ fulfills the condition of the claim, thus ending the proof.

Take any pair of distinct points of $\mathbb{P}_3$, $a = [a_0, a_1, a_2, a_3]$ and $b = [b_0, b_1, b_2, b_3]$, and

$$p_{i,j} = \begin{vmatrix} a_i & a_j \\ b_i & b_j \end{vmatrix}, \quad 0 \leq i < j \leq 3,$$

as Plücker coordinates of the line ℓ they span. According to Lemma 8.4.12, a necessary and sufficient condition for ℓ to be self-polar with respect to g_q is

$$\begin{pmatrix} a_0 & a_1 & a_2 & a_3 \end{pmatrix} \begin{pmatrix} 0 & q_{0,1} & q_{0,2} & q_{0,3} \\ -q_{0,1} & 0 & q_{1,2} & q_{1,3} \\ -q_{0,2} & -q_{1,2} & 0 & q_{2,3} \\ -q_{0,3} & -q_{1,3} & -q_{2,3} & 0 \end{pmatrix} \begin{pmatrix} b_0 \\ b_1 \\ b_2 \\ b_3 \end{pmatrix} = 0.$$

A bit of computation allows us to turn this equality into the equivalent one

$$p_{0,1}q_{2,3} - p_{0,2}q_{1,3} + p_{0,3}q_{1,2} + p_{1,2}q_{0,3} - p_{1,3}q_{0,2} + p_{2,3}q_{0,1} = 0,$$

which, as already seen in the proof of 8.4.5, is the condition for being $\mathfrak{p}(\ell)$ and q conjugate with respect to $\mathbb{G}$, and so also for being $\ell \in F_q$. □

8.5 Exercises

Exercise 8.1. Describe procedures for constructing by points:

(1) A parabola, from three proper points and a diameter.

(2) A hyperbola, from three proper points and two lines parallel to the asymptotes.

(3) A hyperbola, from three proper points and one asymptote.

(4) A hyperbola, from a proper point and the asymptotes.

Exercise 8.2. Describe procedures for constructing by tangents:

(1) A parabola, from four proper tangents.

(2) A hyperbola, from three tangents and one asymptote.

(3) A hyperbola, from a proper tangent and the asymptotes.

Exercise 8.3. Given a quadrivertex T, with vertices A, B, C, D, in a projective plane $\mathbb{P}_2$, prove that the intersection points of the lines $\ell \in A^*$ and $s \in B^*$ whose intersection points with CD harmonically divide the points C, D, lie on a non-degenerate conic of $\mathbb{P}_2$ circumscribed to T. *Hint*: Use 8.1.1.

Exercise 8.4. Let C be a non-degenerate conic of $\mathbb{P}_2$ and $q_1, q_2 \in \mathbb{P}_2$ two points non-conjugate with respect to C and spanning a chord of C. Prove that the points $p \neq q_1, q_2$ for which the line pq_2 contains the pole of pq_1 relative to C belong to a non-degenerate conic C'. Prove also that the poles of q_1q_2 relative to C and C' are coincident.

Exercise 8.5. In a real affine plane let ℓ, s be two proper and non-parallel lines, $O = \ell \cap s$ and P a proper point, $P \notin \ell \cup s$. Consider all the parallelograms that have sides ℓ and s, one vertex at O and the diagonal missing O going through p. Prove that they have their vertices opposite O on a hyperbola whose asymptotes are the lines through P parallel to ℓ and s.

Exercise 8.6. Assume given two proper points q_1, q_2 and two proper lines ℓ, s of an affine plane $\mathbb{A}_{2,\mathbb{R}}$ such that ℓ and s are not parallel and $q_1 \notin s$. Prove that the quadrivertices with vertices q_1, q_2, p', p, for which $p' \in s$, pq_2 is parallel to $p'q_1$ and pp' is parallel to ℓ, have their vertices p on a conic C which is degenerate if and only if ℓ is parallel to q_1q_2. Determine the affine type of C.

Exercise 8.7. Assume given in a real affine plane a proper triangle with vertices A, B, C. Consider pairs of lines ℓ, s, parallel to AB and AC, respectively, whose intersections with BC harmonically divide B, C. Prove that the intersections $\ell \cap s$ lie on a hyperbola with centre the midpoint of B, C and asymptotes parallel to AB and AC.

Exercise 8.8. Let C be a hyperbola of an affine plane $\mathbb{A}_2$, with asymptotes a, a', and p, p' two proper and different points of C. Consider the triangles with vertices q, q', q'' for which q and q' belong to pp' and are conjugate with respect to C, and the sides qq'' and $q'q''$ are parallel to a and a' respectively. Prove that their vertices q'' lie on a hyperbola which has the midpoint of p, p' as centre and asymptotes parallel to a and a'. Compare with Exercise 8.7.

Exercise 8.9. Let C be a hyperbola of a real affine plane, and p a proper point other than the centre O of C. Prove that the midpoints of the ends of the chords of C through p lie on a conic C' that contains O and P, and has the same improper points as C. Determine for which positions of p the conic C' is non-degenerate and therefore is a hyperbola. *Hint*: For each $\ell \in p^*$, consider the polar relative to C of its improper point.

Exercise 8.10. Let C be a non-degenerate conic of $\mathbb{P}_2$ and ℓ and s two different lines tangent to C at points p, q. Prove that the lines joining conjugate points (with respect to C) of ℓ and s envelop a non-degenerate conic which is tangent to ℓ and s at p and q respectively. *Hint*: Use 8.1.4 and Exercise 6.13.

Exercise 8.11. Prove that the tangents to a hyperbola determine with its asymptotes triangles which have all the same area. *Hint*: Choose an affine reference with origin the centre of the hyperbola and unitary vectors on the asymptotes, and write the equation of the projectivity of 8.1.8 between the asymptotes using the abscissa and the ordinate as absolute coordinates.

Exercise 8.12. Assume given a hyperbola C of an affine plane $\mathbb{A}_2$, with asymptotes a, a'. Consider the triangles with vertices q, q', q'' for which $q \in a$, $q' \in a'$, qq'

is tangent to C, qq'' parallel to a' and $q'q''$ parallel to a. Prove that their vertices q'' lie on a hyperbola which also has asymptotes a, a'. *Hint*: Use the projectivity $a \to a'$ defined by C (8.1.8).

Exercise 8.13. Let p and ℓ be a point and a line of a Euclidean plane, $p \notin \ell$. Prove that if a pair of distinct, non-parallel lines $r + s$ varies so that $p \in r$, $r \cap s \in \ell$ and the angle $\widehat{rs}$ is constant, then the line s remains tangent to a parabola which is in turn tangent to ℓ.

Exercise 8.14. Notations being as in Section 8.2, use 8.1.5 and the diagram

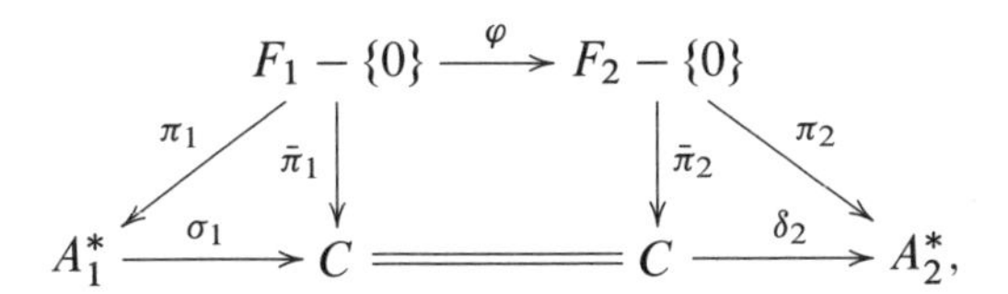

$$\begin{array}{ccccccc} & & F_1 - \{0\} & \xrightarrow{\varphi} & F_2 - \{0\} & & \\ & \swarrow^{\pi_1} & \downarrow{\bar{\pi}_1} & & \downarrow{\bar{\pi}_2} & \searrow^{\pi_2} & \\ A_1^* & \xrightarrow{\sigma_1} & C & = & C & \xrightarrow{\delta_2} & A_2^*, \end{array}$$

where φ is a representative of the projectivity $\delta_2 \circ \sigma_1$, to directly prove that the projective structures on C defined using different points $A_1, A_2 \in C$ are equivalent.

Exercise 8.15. Assume we are given two non-degenerate conics with points C and C', of planes $\mathbb{P}_2$ and $\mathbb{P}'_2$, and three distinct points on each conic, say $p_1, p_2, p_3 \in C$ and $p'_1, p'_2, p'_3 \in C'$. Prove that there is a unique projectivity $f : \mathbb{P}_2 \to \mathbb{P}'_2$ such that $f(C) = C'$ and $f(p_i) = p'_i$ for $i = 1, 2, 3$. *Hint*: Consider the projectivity mapping the reference of $\mathbb{P}_2$ associated to (p_1, p_2, p_3) to the reference of $\mathbb{P}'_2$ associated to (p'_1, p'_2, p'_3) and use 8.1.11.

Exercise 8.16. Prove that if C is a non-degenerate conic with points of $\mathbb{P}_2$, any projectivity $f : \mathbb{P}_2 \to \mathbb{P}'_2$ restricts to a projectivity $\bar{f} : C \to f(C)$. Conversely, if C' is a non-degenerate conic with points of a projective plane $\mathbb{P}'_2$, prove that any projectivity $g : C \to C'$ is the restriction of a uniquely determined projectivity $f : \mathbb{P}_2 \to \mathbb{P}'_2$ for which $f(C) = C'$. *Hint*: Use Exercise 8.15.

Exercise 8.17. Let C be a non-degenerate conic with points of $\mathbb{P}_2$, Ω a reference of C and Δ the reference of $\mathbb{P}_2$ associated to Ω. Prove that the images $\tilde{\Omega}$ and $\tilde{\Delta}$ of Ω and Δ by the polarity relative to C are, respectively, a reference of the envelope C^* and its associated reference of $\mathbb{P}_2^\vee$. Prove that $\tilde{\Delta}$ is not the reference dual $\Delta^\vee$ of Δ, although it has the same vertices. Write a matrix changing coordinates relative to $\tilde{\Delta}$ into coordinates relative to $\Delta^\vee$. Prove that the tangent at $p = [\alpha_0, \alpha_1]_\Omega \in C$ has coordinates α_0, α_1 relative to $\tilde{\Omega}$, compute its coordinates relative to $\tilde{\Delta}$ and compare with 8.2.6 for $p = q$.

Exercise 8.18 (*Chasles' definition of conics*). Assume we are given points $p_1, p_2, p_3, p_4 \in \mathbb{P}_2$, no three aligned, and $\lambda \in k - \{0, 1\}$. Prove that p_1, p_2, p_3, p_4 and the points $p \neq p_i$, $i = 1, 2, 3, 4$, for which $(pp_1, pp_2, pp_3, pp_4) = \lambda$ compose the set of points of a non-degenerate conic. Prove that the set of points of any non-degenerate conic with points may be obtained in this way for a suitable choice of p_1, p_2, p_3, p_4 and λ. *Hint*: Take the line ℓ through p_4 such that $(p_4p_1, p_4p_2, p_4p_2, \ell) = \lambda$ and consider the conic through p_1, p_2, p_3, p_4 tangent to ℓ at p_4.

Exercise 8.19. Assume given on a line ℓ of the drawing plane two different pairs of points corresponding by an involution σ of ℓ. Describe a graphic procedure allowing us to determine:

(1) The image $\sigma(p)$ of an arbitrary point $p \in \ell$.

(2) If σ has real or imaginary fixed points.

(3) The fixed points of σ in case of being real.

Hint: Draw an auxiliary circle C, fix a point $A \in C$ and, as in the proof of 8.2.11, reduce the problem to a similar one on C by projecting the points of ℓ from A and then taking section with C.

Exercise 8.20. Same as in Exercise 8.19, assuming this time that σ is a projectivity of ℓ and three different ordered pairs of points corresponding by σ have been given.

Exercise 8.21. Describe a graphic construction allowing to decide if a non-degenerate conic of the drawing plane, given by five of its points, is a ellipse, a parabola or a hyperbola. *Hint*: Consider the projectivity of the improper line induced by a projectivity between pencils of lines generating the conic, and use the construction of Exercise 8.20.

Exercise 8.22. Let p_1, p_2 and q_1, q_2 be two disjoint pairs of different points of a non-degenerate conic C of $\mathbb{P}_{2,\mathbb{R}}$. Prove that (p_1, p_2, q_1, q_2) is negative (the pairs separate each other on C) if and only if the chords p_1p_2 and q_1q_2 meet at an interior point. *Hint*: Use Exercise 5.9.

Exercise 8.23. Prove that two different involutions of a non-degenerate conic with points C commute if and only if their centres are conjugate with respect to C.

Back to Exercise 5.10, assume that $\mathbb{P}_1 = C$, C a non-degenerate conic with points, and prove that the three equivalent conditions (i), (ii), (iii) of 5.10 are also equivalent to

(iv) the centres of τ_1, τ_2 and τ_3 are the vertices of a self-polar triangle for C.

Prove that if $k = \mathbb{R}$ and the above conditions are satisfied, then one of the involutions has imaginary fixed points, while the other two have real fixed points. Extend the result to an arbitrary $\mathbb{P}_1$.

Exercise 8.24. Describe a procedure based on Pascal's theorem 8.2.20 allowing us to construct by points the conic through five given points, no three aligned. Describe also its dual and compare both with the procedures described at the end of Section 8.1.

Exercise 8.25. Use 8.2.13 and 6.4.27 to prove that if $\{p, \bar{p}\}$ and $\{q, \bar{q}\}$ are two different pairs of an involution τ of a conic C, then the lines $p\bar{p}$ and $q\bar{q}$ meet at the pole of the cross-axis of τ, thus re-proving 8.2.8.

Exercise 8.26. Let C be a non-degenerate conic of $\mathbb{P}_2$, p_1, p_2 two different points of C and p the pole of the line p_1p_2. Fix a line ℓ through p, $\ell \neq pp_1, pp_2$, put $\ell \cap C = \{q_1, q_2\}$ and prove that for any $q \in C$, $q \neq q_1, q_2$, the points $qq_1 \cap p_1p_2$ and $qq_2 \cap p_1p_2$ are conjugate with respect to C. *Hint*: The cases $q = q_1, q_2$ being clear, use the involution of C with centre p and projection from q to prove that $(p_1, p_2, q_1, q_2) = -1$.

Exercise 8.27. Let C be a non-degenerate conic of $\mathbb{P}_2$ and q_1, q_2, $q_3 \in \mathbb{P}_2$ three independent points, no one on C. Prove that there are either infinitely many or at most two triangles inscribed in C whose sides contain, one each, the points q_1, q_2, q_3. Prove that the triangles are infinitely many if and only if q_1, q_2, q_3 are the vertices of a self-polar triangle, and also that, if this is the case, any point of C other than those aligned with two of the points q_i, $i = 1, 2, 3$, is a vertex of such a triangle.

Assume $k = \mathbb{R}$ and that C and q_1, q_2, q_3 are already drawn on a drawing plane. Describe a graphic procedure allowing to recognize the case of infinitely many solutions and to construct the solutions in case of them being finitely many. *Hint*: Consider the fixed points of the composition of the involutions of C with centres q_1, q_2, q_3, and use Exercise 8.23.

Exercise 8.28. Let T and T' be triangles of $\mathbb{P}_2$, no vertex of either triangle lying on a side of the other. Prove that if T and T' are inscribed in the same non-degenerate conic, then they are both circumscribed to a second one, and conversely (*Brianchon*). *Hint*: Use the conic to define a projectivity between sides of T and T' in such a way that the remaining sides join corresponding points; then use 8.1.4.

Exercise 8.29 (*Poncelet's porism*). Prove that if C and C' are non-degenerate conics of $\mathbb{P}_2$ and there exists a triangle T inscribed in C and circumscribed to C', then any point p of C, $p \notin C'$, is a vertex of a triangle T_p which is also inscribed in C and circumscribed to C'. (The name *porism* refers to problems which, depending on the data, have either no solution or infinitely many.) *Hint*: Take T_p with vertices p, p_1, p_2, $p_1, p_2 \in C$ and pp_1, pp_2 tangent to C'; prove that the conic which, according to Exercise 8.28, is inscribed in T and T_p, is C'.

Exercise 8.30. Let T and T' be two triangles of $\mathbb{P}_2$, no vertex of either triangle lying on a side of the other. Prove that if T and T' are self-polar with respect to the same non-degenerate conic, then they are both inscribed in a second non-degenerate conic and, by duality (or by Exercise 8.28), circumscribed to a third one. *Hint*: If the triangles have vertices A, B, C and A', B', C', use section by $C'B'$ and the polarity to show that AB', AC', AB, AC and $A'B'$, $A'C'$, $A'B$, $A'C$ have equal cross ratios, and then apply 8.1.1.

Exercise 8.31 (*Projective generation of cones*). Let ℓ_1, ℓ_2 be coplanary distinct lines of $\mathbb{P}_3$ and assume we are given a homography $f : \ell_1^* \to \ell_2^*$ such that $f(\ell_1 \vee \ell_2) \neq \ell_1 \vee \ell_2$. Prove that the intersections of the pairs of corresponding planes by f are the generators of an ordinary cone with generators ℓ_1, ℓ_2, and therefore vertex $\ell_1 \cap \ell_2$. Prove also that any ordinary cone of $\mathbb{P}_3$ arises in this way.

Exercise 8.32. Prove that if a non-degenerate quadric of an affine space $\mathbb{A}_{3,\mathbb{R}}$ contains two distinct parallel lines, then it is a hyperboloid of one sheet, and conversely.

Exercise 8.33. Prove that all proper generators in a ruling of a hyperbolic paraboloid are parallel to a plane.

Exercise 8.34. Let ℓ, s be skew and non-perpendicular lines of a Euclidean space $\mathbb{A}_3$. Prove that the lines that join points of ℓ and s and are perpendicular to s are generators of a hyperbolic paraboloid that contains ℓ and s. Examine the case in which ℓ and s, still skew, are perpendicular.

Exercise 8.35. Let Q be a ruled quadric of $\mathbb{P}_3$, s and ℓ generators in different rulings of Q, $p = s \cap \ell$ and H a plane of $\mathbb{P}_3$ missing p. Consider the correspondence $s \times \ell \leftrightarrow H$ defined by composing the bijection Φ of 8.3.5 and the stereographic projection of Q onto H with centre p. Determine the elements on either side which have no correspondent on the other, and also those which have more than one correspondent.

Exercise 8.36. Conventions and notations being as in Section 8.4, prove that the images by the Plücker map of the edges of the simplex of the reference of $\mathbb{P}_3$ are the vertices of the reference of $\mathbb{P}_5$.

Exercise 8.37. Assume that ℓ is a line in $\mathbb{P}_3$, take $p_{i,j}$, $i, j = 0, \dots, 3$, as defined at the beginning of Section 8.4 and $M = (p_{i,j})_{i,j=0,\dots,3}$ the matrix with entries the $p_{i,j}$. Prove that the i-th row of M is zero if and only if ℓ is contained in the i-th face of the reference of $\mathbb{P}_3$, and also that, otherwise, the i-th row is a coordinate vector of the intersection of ℓ and the i-th face of the reference. Use this to re-prove the injectivity of the Plücker map.

Exercise 8.38. Let π, π' be planes of $\mathbb{P}_5$ whose intersections with the Plücker quadric $\mathbb{G}$ are non-degenerate conics with points $\Sigma = \mathbb{G} \cap \pi$ and $\Sigma' = \mathbb{G} \cap \pi'$. Prove that Σ and Σ' represent the two rulings of a ruled quadric of $\mathbb{P}_3$ if and only if $\Sigma \cap \Sigma' = \emptyset$ and $p \vee p' \subset \mathbb{G}$ for any $p \in \Sigma$ and $p' \in \Sigma'$.

Exercise 8.39. Let T be a tetrahedron and Q a non-degenerate quadric of $\mathbb{P}_3$ such that no vertex of T is the pole relative to Q of its opposite face. Prove that the polar hyperplanes relative to Q of the vertices of T are the faces of a tetrahedron T'. Prove also that each vertex p of T and the vertex of T' opposite $H_{p,Q}$ span a line, and that the four lines arising in this way, if pairwise skew, belong to the same ruling of a non-degenerate ruled quadric (*Chasles*). *Hint*: Take T as the tetrahedron of the reference, compute the Plücker coordinates of the lines in terms of the cofactors of the entries of a matrix of Q and use 8.4.11.

Exercise 8.40 (*The dual way to Plücker coordinates*). Assume, that ℓ is the line of $\mathbb{P}_3$ spanned by $a = [a_0, a_1, a_2, a_3]$ and $b = [b_0, b_1, b_2, b_3]$, and therefore has Plücker coordinates

$$p_{i,j} = \begin{vmatrix} a_i & a_j \\ b_i & b_j \end{vmatrix}, \quad 0 \leq i < j \leq 3.$$

Assume that ℓ is the intersection of the planes which have equations

$$\sum_{i=0}^{3} \alpha_i x_i = 0 \quad \text{and} \quad \sum_{i=0}^{3} \beta_i x_i = 0$$

and take

$$\pi_{i,j} = \begin{vmatrix} \alpha_i & \alpha_j \\ \beta_i & \beta_j \end{vmatrix}, \quad 0 \leq i < j \leq 3.$$

Prove that

$$[\pi_{2,3}, \pi_{1,3}, \pi_{1,2}, \pi_{0,3}, \pi_{0,2}, \pi_{0,1}] = [p_{0,1}, -p_{0,2}, p_{0,3}, p_{1,2}, -p_{1,3}, p_{2,3}].$$

Hint: Eliminate a_0 from the conditions expressing $a \in \ell$,

$$\sum_{i=0}^{3} \alpha_i a_i = 0, \quad \sum_{i=0}^{3} \beta_i a_i = 0,$$

to get $\pi_{0,1}a_1 + \pi_{0,2}a_2 + \pi_{0,3}a_3 = 0$ and, similarly, $\pi_{0,1}b_1 + \pi_{0,2}b_2 + \pi_{0,3}b_3 = 0$. Proceed in the same way for indices other than 0.

Exercise 8.41. Examine the intersection of the Plücker quadric $\mathbb{G}$ and the plane spanned by the representations of three different lines, in the cases in which just one or two pairs of these lines are coplanar.

Exercise 8.42. Let T a tetrahedron and Q a quadric, both of a projective space $\mathbb{P}_3$. Assume that the vertices of T are simple points of Q and prove that if the intersections of the faces of T with the planes tangent to Q at the opposite vertices are four pairwise skew lines, then they belong to a ruling of a non-degenerate quadric of $\mathbb{P}_3$. *Hint*: Take T as the tetrahedron of the reference, compute Plücker coordinates of the lines using Exercises 6.2 and 8.40, and use 8.4.11.

Exercise 8.43 (*Regarding the intrinsic character of the Plücker map and the Plücker quadric*). Conventions and notations still being as in Section 8.4, take another reference $\bar{\Delta}$ of $\mathbb{P}_3$ and another projective space $\overline{\mathbb{P}}_5$ to define Plücker coordinates $\bar{p}_{i,j}$ and a second Plücker map

$$\bar{\mathfrak{p}}\colon \mathcal{G} \longrightarrow \overline{\mathbb{P}}_5.$$

Assume that $(\alpha_t^s)_{s,t=0,1,2,3}$ is the matrix that changes coordinates relative to Δ into coordinates relative to $\bar{\Delta}$ and prove:

(1) The Plücker coordinates $p_{i,j}$ and $\bar{p}_{i,j}$ of any line of $\mathbb{P}_3$ are related by the equalities

$$\bar{p}_{i,j} = \sum_{0\leq s<t\leq 3} \begin{vmatrix} \alpha_i^s & \alpha_i^t \\ \alpha_j^s & \alpha_j^t \end{vmatrix} p_{s,t}, \quad 0 \leq i < j \leq 3.$$

(2) The 6×6 matrix whose entry in row i, j and column s, t is

$$\begin{vmatrix} \alpha_i^s & \alpha_i^t \\ \alpha_j^s & \alpha_j^t \end{vmatrix},$$

for $0 \leq i < j \leq 3$ and $0 \leq s < t \leq 3$, defines either a projectivity or a singular projectivity $\varphi\colon \mathbb{P}_5 \to \overline{\mathbb{P}}_5$ that makes commutative the diagram

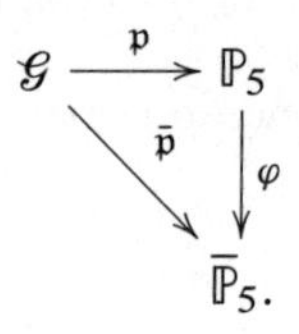

(3) The image of φ contains the Plücker quadric of $\overline{\mathbb{P}}_5$, after which, by 8.4.2 and 6.4.31, φ is a projectivity.

This proves that, taken up to a projectivity, both the Plücker map and the Plücker quadric are independent of the choices of the reference in $\mathbb{P}_3$ and the projective space $\mathbb{P}_5$.

Exercise 8.44. If π is a plane of $\mathbb{P}_3$, prove that the Plücker map restricts to a map $\pi^\vee \to \mathbb{P}_5$ which is a projectivity from $\pi^\vee$ onto its image (see 8.4.8). *Hint*: Take a reference of $\mathbb{P}_3$ with π as a face, and relate the line-coordinates and the Plücker coordinates of each line of π, using Exercise 8.40.

Exercise 8.45. If O is a point of $\mathbb{P}_3$, prove that the Plücker map restricts to a map $O^\nabla \to \mathbb{P}_5$, which is a projectivity from O^∇ onto its image (see 8.4.8). *Hint*: Take a reference of $\mathbb{P}_3$ with O as vertex and relate the coordinates in O^∇ and the Plücker coordinates of any line $\ell \in O^\nabla$.

Exercise 8.46. Prove that the image by the Plücker map of the set of generators of an ordinary cone of $\mathbb{P}_3$ is the set of points of a non-degenerate conic of a plane Π of $\mathbb{P}_5$ contained in the Plücker quadric. Same for the image of the set of lines tangent to a non-degenerate conic with points of a plane of $\mathbb{P}_3$. Prove also that the set of points of any non-degenerate conic with points of a plane contained in the Plücker quadric arises in one and only one of these ways, the cone or the conic being uniquely determined in each case. *Hint*: Use 6.7.14, 6.6.1 and Exercises 8.44 and 8.45, or just the computations made to solve the latter.

Chapter 9
Projective spaces of quadrics

As already noted, one of the most important features of projective geometry is the existence of intrinsic projective structures on sets of geometric objects, other than points, belonging to a fixed projective space. In this chapter we will study the projective structure on the set of the quadrics of a projective space $\mathbb{P}_n$. Considering the quadrics of $\mathbb{P}_n$ as the points of a new projective space offers a radically different view on the geometry of quadrics and provides a far deeper understanding of them. Incidentally, as far as $n \geq 2$, the space of the quadrics of $\mathbb{P}_n$ has dimension higher than three, which is an important reason for studying projective geometry in spaces of arbitrary dimension.

The first two sections of this chapter contain some auxiliary definitions and results which are needed later on. The study of the spaces of quadrics will begin in Section 9.3.

9.1 Effective divisors on projective lines

This section presents homogeneous versions of basic facts regarding algebraic equations in one variable which will be needed in the sequel. Assume to have fixed a one-dimensional projective space $\mathbb{P}_1$ and a projective reference Δ of it. Consider a non-zero homogeneous polynomial of degree d, $d \geq 0$,

$$P = P(X_0, X_1) = a_d X_0^d + a_{d-1} X_0^{d-1} X_1 + \cdots + a_1 X_0 X_1^{d-1} + a_0 X_1^d,$$

with $a_i \in k$ for $i = 0, \dots, d$, and the points $q = [\alpha_0, \alpha_1] \in \mathbb{P}_1$ for which $P(\alpha_0, \alpha_1) = 0$. As in the case $d = 2$, the latter condition is independent of the arbitrary factor involved in the coordinates of q because, due to the homogeneity of P, $P(\lambda\alpha_0, \lambda\alpha_1) = \lambda^d P(\alpha_0, \alpha_1)$. When $P(\alpha_0, \alpha_1) = 0$, it is said that the point q *satisfies*, or is a *solution* of, the equation $P = 0$. The next lemma partially generalizes 5.4.2:

Lemma 9.1.1. *If $P(X_0, X_1)$ is a non-zero homogeneous polynomial of degree $d \geq 0$ and $q = [\alpha_0, \alpha_1] \in \mathbb{P}_1$, then:*

(a) *$P(\alpha_0, \alpha_1) = 0$ if and only if $P = (\alpha_1 X_0 - \alpha_0 X_1)P'$, with P' homogeneous of degree $d - 1$.*

(b) *The number of points q that satisfy $P = 0$ is at most d.*

(c) *Assume $d > 0$. If either $k = \mathbb{C}$, or $k = \mathbb{R}$ and d is odd, then there is at least one point q satisfying $P(X_0, X_1) = 0$.*

Proof. The *if* part of the first claim is clear. Regarding the converse, assume first $\alpha_1 = 0$: then $\alpha_0 \neq 0$, $P(\alpha_0, 0) = a_d \alpha_0^d = 0$ forces $a_d = 0$ and thus

$$P(X_0, X_1) = X_1 \sum_{j=0}^{d-1} a_j X_0^j X_1^{d-j-1} = (0\,X_0 - \alpha_0 X_1)\Big(-\frac{1}{\alpha_0}\sum_{j=0}^{d-1} a_j X_0^j X_1^{d-j-1}\Big).$$

Otherwise, $q = [\alpha_0/\alpha_1, 1]$ and so we have

$$P(\alpha_0/\alpha_1, 1) = \sum_{j=0}^{d} a_j (\alpha_0/\alpha_1)^j = 0.$$

This shows that the polynomial $P(X, 1) \in k[X]$ has the root α_0/α_1 and therefore is a multiple of $X - \alpha_0/\alpha_1$. Then

$$\sum_{j=1}^{d} a_j X^j = P(X, 1) = (X - \alpha_0/\alpha_1) \sum_{j=1}^{d-1} b_j X^j,$$

for some polynomial $\sum_{j=1}^{d-1} b_j X^j \in K[X]$. After substituting X_0/X_1 for X in the above equality and multiplying by X_1^d we obtain

$$P(X_0, X_1) = \sum_{j=0}^{d} a_j X_0^j X_1^{d-j} = (\alpha_1 X_0 - \alpha_0 X_1)\Big(\frac{1}{\alpha_1}\sum_{j=0}^{d-1} b_j X_0^j X_1^{d-j-1}\Big)$$

as claimed.

Claim (b) is obvious for $d = 0$. For $d > 0$ it follows by induction using the already proved factorization of P and the fact that $q = [\alpha_0, \alpha_1]$ is the only point which satisfies the equation

$$\alpha_1 X_0 - \alpha_0 X_1 = \begin{vmatrix} X_0 & X_1 \\ \alpha_0 & \alpha_1 \end{vmatrix} = 0.$$

If $a_d = 0$, claim (c) is clearly satisfied by taking $q = [1, 0]$. Otherwise the polynomial $P(X, 1)$, in the single variable X, has degree $d > 0$. Therefore, if either $k = \mathbb{C}$, or $k = \mathbb{R}$ and d is odd, $P(X, 1)$ has a root $\alpha \in k$. Then it is enough to take $q = [\alpha, 1]$. □

Iterated use of 9.1.1 (a) shows that $q = [\alpha_0, \alpha_1]$ is a solution of $P = 0$ if and only if P may be written $P = (\alpha_1 X_0 - \alpha_0 X_1)^\nu P_1$, where $\nu > 0$, P_1 is homogeneous of degree $d - \nu$ and has no factor $\alpha_1 X_0 - \alpha_0 X_1$ (or, equivalently, $P_1 = 0$ has no solution q). The positive integer ν is the multiplicity of $\alpha_1 X_0 - \alpha_0 X_1$ as a prime factor of P: it is also called the *multiplicity* of q, or of the pair (α_0, α_1), as a solution of $P = 0$. Clearly, solutions of $P = 0$ other than q are solutions of $P_1 = 0$ and conversely. So, proceeding inductively with the other solutions of $P = 0$ and using 9.1.1 (c) for $k = \mathbb{C}$, we have:

Theorem 9.1.2. *Let $P(X_0, X_1)$ be a non-zero homogeneous polynomial of degree $d \geq 0$. The solutions of $P = 0$ are $q_j = [\alpha_{j,0}, \alpha_{j,1}] \in \mathbb{P}_1$, $j = 1, \dots, m$, $q_i \neq q_j$ for $i \neq j$, with respective multiplicities ν_j, $j = 1, \dots, m$, if and only if*

$$P = Q \prod_{j=1}^{m} (\alpha_{j,1} X_0 - \alpha_{j,0} X_1)^{\nu_j}, \tag{9.1}$$

where Q is homogeneous of degree $d - \sum_{j=0}^{m} \nu_j$ and $Q = 0$ has no solutions. If $k = \mathbb{C}$, then Q has degree zero.

Remark 9.1.3. If $k = \mathbb{R}$, then either $\deg Q = 0$, or $\deg Q \geq 2$, as, obviously, if $\deg Q = 1$, then $Q = 0$ has a solution.

It follows in particular from 9.1.2 that $\sum_{j=0}^{m} \nu_j \leq d$, with equality if $k = \mathbb{C}$. The sum $\sum_{j=0}^{m} \nu_j$ is the number of solutions of $P = 0$ if each solution is counted as many times as indicated by its multiplicity, which is called *counting solutions according to multiplicities.* Using these terms, we have:

Corollary 9.1.4. *If P is a non-zero homogeneous polynomial in two variables, then the number of solutions of the equation $P = 0$ in $\mathbb{P}_1$, counted according to multiplicities, cannot exceed the degree of P, and always equals it if $k = \mathbb{C}$.*

The next proposition provides a way of measuring the multiplicity of a solution. In it, to make the induction easier, we take the points which are not solutions of $P = 0$ as *solutions of multiplicity zero* and P itself as its only 0-th derivative.

Proposition 9.1.5. *The point $q = [\alpha_0, \alpha_1]$ is a solution of multiplicity $\nu \geq 0$ of an equation $P = 0$, $P \in k[X_0, X_1]$ homogeneous and non-zero, if and only if*

$$\frac{\partial^{\nu-1} P}{\partial^i X_0 \partial^{\nu-1-i} X_1}(\alpha_0, \alpha_1) = 0 \quad \textit{for all } i = 0, \dots, \nu - 1, \tag{9.2}$$

provided $\nu > 0$, and

$$\frac{\partial^{\nu} P}{\partial^i X_0 \partial^{\nu-i} X_1}(\alpha_0, \alpha_1) \neq 0 \quad \textit{for some } i = 0, \dots, \nu. \tag{9.3}$$

Proof. Assume that q is a solution of multiplicity ν. The case $\nu = 0$ being clear, we assume $\nu > 0$ and use induction on ν. According to the definition of multiplicity,

$$P = (\alpha_1 X_0 - a_0 X_1)^{\nu} Q \quad \text{with } Q(\alpha_0, \alpha_1) \neq 0.$$

Then,

$$\frac{\partial P}{\partial X_0} = (\alpha_1 X_0 - \alpha_0 X_1)^{\nu-1} \Big(\nu \alpha_1 Q + (\alpha_1 X_0 - \alpha_0 X_1) \frac{\partial P}{\partial X_0} \Big)$$

and

$$\frac{\partial P}{\partial X_1} = (\alpha_1 X_0 - \alpha_0 X_1)^{\nu-1}\Big(- \nu\alpha_0 Q + (\alpha_1 X_0 - \alpha_0 X_1)\frac{\partial P}{\partial X_1}\Big),$$

which show that q is a solution of multiplicity $\geq \nu - 1$ of both $\partial P/\partial X_0 = 0$ and $\partial P/\partial X_1 = 0$, the multiplicity being equal to $\nu - 1$ for at least one of them (depending on which α_i is non-zero). The direct claim follows then from the induction hypothesis. The converse is clear, as an integer ν satisfying the conditions (9.2) and (9.3) is necessarily unique, and therefore, by the direct claim, equal to the multiplicity. □

A *group of points* of $\mathbb{P}_1$, or *effective divisor* of $\mathbb{P}_1$, is a finite set of pairwise different points $p_1, \dots, p_r \in \mathbb{P}_1$, each with an assigned positive integer μ_j which is called its *multiplicity*. Such a group is usually written as a formal sum of repeated points, $G = \mu_1 q_1 + \cdots + \mu_m q_m$. The *degree* of G is defined as $\deg G = \mu_1 + \cdots + \mu_r$. The definition includes the group with no points (and hence no multiplicities), which has degree 0 and is denoted by 0 (the value usually assigned to an empty sum). *Divisors* are defined as the effective divisors, but for allowing their multiplicities μ_i to be negative; we will make no use of them here. The image of the group of points $\mu_1 q_1 + \cdots + \mu_m q_m$ of $\mathbb{P}_1$ by a projectivity $f\colon \mathbb{P}_1 \to \overline{\mathbb{P}}_1$ is taken to be $\mu_1 f(q_1) + \cdots + \mu_m f(q_m)$, the properties of Section 1.10 being then obviously satisfied.

The notations being as in 9.1.2, the solutions of $P = 0$, together with their multiplicities, compose the group $D = \nu_1 q_1 + \cdots + \nu_m q_m$, which is called the *group of solutions* or *divisor of solutions* of $P = 0$. Conversely, it is clear from 9.1.2 that an arbitrary group of points $D = \nu_1 q_1 + \cdots + \nu_m q_m$, with $q_j = [\alpha_{j,0}, \alpha_{j,1}]$, is the group of solutions of an equation $P = 0$, with P homogeneous and $\deg P = \deg D$, if and only if

$$P = \alpha \prod_{j=1}^{m} (\alpha_{j,1} X_0 - \alpha_{j,0} X_1)^{\nu_j},$$

for $\alpha \in k - \{0\}$. Furthermore the condition on the degrees may be dropped if $k = \mathbb{C}$. When D is the group of solutions of $P = 0$ and $\deg P = \deg D$, $P = 0$ is called an *equation* of D. Note that an equation of D determines D and is in turn is determined by it up to a constant factor.

All the above is relative to the previous choice of the reference Δ. Regarding the use other coordinates we have:

Proposition 9.1.6. *Assume we have a second reference Ω of $\mathbb{P}_1$, coordinates y_0, y_1 relative to Ω being related to the former coordinates by the equalities*

$$x_0 = a y_0 + b y_1,$$
$$x_1 = c y_0 + d y_1,$$

with $ad - bc \neq 0$. Assume also that, using coordinates relative to Δ, the equation $P(X_0, X_1) = 0$, with P homogeneous, has a group of solutions D. Then, using coordinates relative to Ω, the equation

$$\bar{P}(Y_0, Y_1) = P(aY_0 + bY_1, cY_0 + dY_1) = 0$$

has a group of solutions D. In particular $P = 0$ is an equation relative to Δ of a group of points D if and only if $\bar{P} = 0$ is an equation relative to Ω of the same group D.

Proof. Assume that $D = \nu_1 q_1 + \cdots + \nu_m q_m$, with $q_i \neq q_j$ for $i \neq j$ and $q_j = [\alpha_{j,0}, \alpha_{j,1}]$, $j = 1, \ldots, m$. Then, by 9.1.2,

$$P = Q \prod_{j=1}^{m} (\alpha_{j,1} X_0 - \alpha_{j,0} X_1)^{\nu_j},$$

the factor Q having no solutions. On the other hand, by reversing the equations (9.1.6) using the matrix of cofactors, it results that

$$\begin{aligned} \beta_{j,0} &= d\alpha_{j,0} - b\alpha_{j,1}, \\ \beta_{j,1} &= -c\alpha_{j,0} + a\alpha_{j,1} \end{aligned}$$

are coordinates of q_j relative to Ω. Then, performing the substitution of coordinates in one of the linear factors of P gives

$$\alpha_{j,1}(aY_0 + bY_1) - \alpha_{j,0}(cY_0 + dY_1) = \beta_{j,1} Y_0 - \beta_{j,0} Y_1$$

and so

$$\bar{P}(Y_0, Y_1) = P(aY_0 + bY_1, cY_0 + dY_1) = \bar{Q}(Y_0, Y_1) \prod_{j=1}^{m} (\beta_{j,1} Y_0 - \beta_{j,0} Y_1)^{\nu_j},$$

where, Q having no solution, $\bar{Q}(Y_0, Y_1) = Q(aY_0 + bY_1, cY_0 + dY_1)$ has no solution either. Indeed, $\bar{Q}(\beta_0, \beta_1) = 0$ would give $Q(a\beta_0 + b\beta_1, c\beta_0 + d\beta_1) = 0$ and $(a\beta_0 + b\beta_1, c\beta_0 + d\beta_1) \neq (0, 0)$ because $ad - bc \neq 0$. This proves the first claim. From it, the second one follows by just noticing that $\deg P = \deg \bar{P}$. □

9.2 Rational curves

Roughly speaking, rational curves are subsets of projective spaces whose points have their coordinates given as homogeneous polynomial functions of two homogeneous parameters (see the precise definition below). The sets of points of the non-degenerate conics with points were presented in this way in Section 8.2 (equations (8.1) and (8.3). Since some other rational curves will appear in the sequel, we devote this section mainly to set the definition of rational curve. We will make a projective study of the rational curves of a special kind, named *normal rational curves*, in forthcoming Sections 9.11 and 9.12. The general study of rational curves belongs to algebraic geometry.

In the sequel, when polynomials $P_0, \dots, P_n$ are said to be coprime, or share no factor, we in particular assume that at least one of them is not zero.

Fix references in spaces $\mathbb{P}_1$ and $\mathbb{P}_n$, $n > 1$ and let $P_0, \dots, P_n \in k[T_0, T_1]$ be homogeneous polynomials of the same degree d sharing no factor. Let us see first that

$$\begin{aligned} \Phi\colon \mathbb{P}_1 &\longrightarrow \mathbb{P}_n, \\ p = [\alpha_0, \alpha_1] &\longmapsto [P_0(\alpha_0, \alpha_1), \dots, P_n(\alpha_0, \alpha_1)], \end{aligned} \tag{9.4}$$

is a well-defined map. Indeed, if

$$P_0(\alpha_0, \alpha_1) = \dots = P_n(\alpha_0, \alpha_1) = 0,$$

then, by 9.1.1, each P_i is a multiple of $\alpha_1 T_0 - \alpha_0 T_1$, against the hypothesis. On the other hand, $P_i(\lambda\alpha_0, \lambda\alpha_1) = \lambda^d P_i(\alpha_0, \alpha_1)$ for any $\lambda \in k$ and $i = 0, \dots, n$, which makes irrelevant the undetermined factor up to which the coordinates α_0, α_1 are determined by p. A map such as Φ above is called a *rational map*, from $\mathbb{P}_1$ to $\mathbb{P}_n$, *of degree* d. The coordinates being fixed, first we will see that the rational map Φ determines the polynomials P_i, $i = 0, \dots, n$, up to a non-zero common scalar factor. This will in particular prove that the degree of a rational map is well defined, as it depends only on the map itself, and not on the set of polynomials defining it.

Lemma 9.2.1. *If two ordered sets of coprime homogeneous polynomials in two variables $P_0, \dots, P_n$ and $Q_0, \dots, Q_n$ define the same rational map, namely*

$$[P_0(\alpha_0, \alpha_1), \dots, P_n(\alpha_0, \alpha_1)] = [Q_0(\alpha_0, \alpha_1), \dots, Q_n(\alpha_0, \alpha_1)]$$

for any $(\alpha_0, \alpha_1) \in k^2 - \{(0,0)\}$, then there is $\rho \in k - \{0\}$ such that

$$\rho P_i = Q_i, \quad i = 0, \dots, n.$$

Proof. By 2.2.5, the equalities of the hypothesis force

$$P_i(\alpha_0, \alpha_1) Q_j(\alpha_0, \alpha_1) = P_j(\alpha_0, \alpha_1) Q_i(\alpha_0, \alpha_1)$$

for $0 \le i < j \le n$ and any $(\alpha_0, \alpha_1) \in k^2 - \{(0,0)\}$, and hence

$$P_i Q_j = P_j Q_i, \quad 0 \le i < j \le n, \tag{9.5}$$

as polynomials. If $P_i = 0$, then, clearly, $Q_i = 0$, as at least one P_j is not zero. Assume $P_i \neq 0$ and let F be a prime divisor of multiplicity ν of P_i. The polynomials P_s, $s = 0, \dots, n$, being coprime, there is $j \neq i$ for which P_j has no factor F. Then the equality $P_i Q_j = P_j Q_i$ easily gives $F^\nu | Q_i$. The prime factor F of P_i

being arbitrary, it follows that P_i divides Q_i, say $Q_i = D_i P_i$. Furthermore, if both $P_i \neq 0$ and $P_j \neq 0$, then the equalities (9.5) give $D_i = D_j$: rename $D = D_i$ for $P_i \neq 0$, so that we will have $Q_i = DP_i$ for $i = 0, \dots, n$. Since the polynomials $Q_i, i = 0, \dots, n$, are coprime too, it follows that $D \in k - \{0\}$, and hence the claim by taking $\rho = D$. □

We need also to prove that the above definition of rational map of degree d is independent of the choices of coordinates in $\mathbb{P}_1$ and $\mathbb{P}_n$. This is done next:

Lemma 9.2.2. *Notations being as above, assume that Ω and Δ are arbitrary references of $\mathbb{P}_1$ and $\mathbb{P}_n$, respectively. Then there are $\bar{P}_0, \dots, \bar{P}_n \in k[T_0, T_1]$, coprime and homogeneous of degree d, such that*

$$\Phi([\beta_0, \beta_1]_\Omega) = [\bar{P}_0(\beta_0, \beta_1), \dots, \bar{P}_n(\beta_0, \beta_1)]_\Delta$$

for any $p = [\beta_0, \beta_1]_\Omega \in \mathbb{P}_1$.

Proof. Let M be the matrix that changes the present coordinates of $\mathbb{P}_n$ into coordinates relative to Δ. Then the polynomials $\widehat{P}_0, \dots, \widehat{P}_n$, defined by the equality

$$(\widehat{P}_0, \dots, \widehat{P}_n)^t = M(P_0, \dots, P_n)^t$$

define the map Φ using in $\mathbb{P}_n$ the coordinates relative to Δ. Clearly they are homogeneous of degree d. Furthermore, they are coprime, because so are $P_0, \dots, P_n$ and

$$(P_0, \dots, P_n)^t = M^{-1}(\widehat{P}_0, \dots, \widehat{P}_n)^t.$$

Then a change of coordinates in $\mathbb{P}_1$ results in a linear, homogeneous and invertible substitution of the variables X_0, X_1 in $\widehat{P}_0, \dots, \widehat{P}_n$: obviously the resulting polynomials $\bar{P}_0, \dots, \bar{P}_n$ still are homogeneous of degree d and coprime. □

Clearly, a degree-zero rational map is constant. The image Γ of a rational map $\Phi \colon \mathbb{P}_1 \to \mathbb{P}_n$, of degree $d > 0$, is called a *rational curve* of $\mathbb{P}_n$, a *plane rational curve* if $n = 2$. Φ is called a *parameterization map* of Γ.

Assume that coordinates in $\mathbb{P}_1$ and $\mathbb{P}_n$ have been fixed and Φ, as above, is

$$\begin{aligned} \Phi \colon \mathbb{P}_1 &\longrightarrow \mathbb{P}_n, \\ [\alpha_0, \alpha_1] &\longmapsto [P_0(\alpha_0, \alpha_1), \dots, P_n(\alpha_0, \alpha_1)], \end{aligned} \tag{9.6}$$

where $P_0, \dots, P_n$ are coprime homogeneous polynomials of the same positive degree. The rational curve Γ is then the set of points with coordinates

$$x_0 = P_0(\alpha_0, \alpha_1), \ \dots \ , x_n = P_n(\alpha_0, \alpha_1),$$

for some $(\alpha_0, \alpha_1) \in k^2 - \{(0, 0)\}$. The above are called *parametric equations* of Γ. The scalars α_0, α_1 are called *homogeneous parameters*, or just *parameters*, of the point $[x_0, \dots, x_n]$.

Of course, once the coordinates are fixed, the expression (9.6) of the parameterization map and the parametric equations carry equivalent information, the relevant data being in both cases the polynomials $P_0, \dots, P_n$ taken up to a common non-zero scalar factor.

The lines are the easiest examples of rational curves. As already said, the parametric equations (8.1) of Section 8.2 show that any non-degenerate conic with points is a rational curve. The corresponding parameterization map has

$$P_0 = T_0^2, \quad P_1 = T_0T_1, \quad P_2 = T_1^2,$$

and therefore has degree two.

Proposition 9.2.3. *A rational curve has infinitely many points.*

Proof. Let the rational curve have a parameterization map Φ given, as above, by

$$[\alpha_0, \alpha_1] \longmapsto [P_0(\alpha_0, \alpha_1), \dots, P_n(\alpha_0, \alpha_1)].$$

Since $\mathbb{P}_1$ is an infinite set (2.4.3), if $\mathrm{Im}(\Phi)$ is finite, then at least one of its points q is the image of infinitely many points. Assume $q = [c_0, \dots, c_n]$: then there are infinitely many points $[\alpha_0, \alpha_1]$ satisfying

$$c_j P_i(\alpha_0, \alpha_1) - c_i P_j(\alpha_0, \alpha_1) = 0, \quad 0 \leq i < j \leq n,$$

and so, by 9.1.1,

$$c_j P_i - c_i P_j = 0, \quad 0 \leq i < j \leq n,$$

as polynomials in T_0, T_1. Picking $c_j \neq 0$, these qualities give

$$P_i = \frac{c_i}{c_j} P_j, \quad 0 \leq j \leq n,$$

against the fact that $P_0, \dots, P_n$ are coprime and of positive degree. □

A rational curve may be given by different rational maps which may even have different degrees. For instance, the rational curve which has parametric equations

$$x_i = P_i(\alpha_0, \alpha_1), \quad i = 0, \dots, n$$

also has parametric equations

$$x_i = P_i(\alpha_0^3, \alpha_1^3), \quad i = 0, \dots, n.$$

The *degree* of a rational curve is defined as the minimum of the degrees of its parameterization maps. Both the parameterization maps whose degree equals the degree of the rational curve and their corresponding parametric equations, are called *proper*. The reader may easily see that all rational maps of degree one define lines, after which it is clear that, as rational curves, the non-degenerate conics with points have degree two and so the parameterization (8.1) of Section 8.2 is proper.

A rational curve Γ of $\mathbb{P}_n$ is said *to span* $\mathbb{P}_n$ (in old books it is said that Γ *belongs to* $\mathbb{P}_n$) if and only if it is contained in no hyperplane of $\mathbb{P}_n$. Clearly, if Γ does not span $\mathbb{P}_n$, then it may be seen as a rational curve of a projective space of dimension smaller than n. We have:

Proposition 9.2.4. *If a rational curve is given by the parameterization map*

$$\begin{aligned}\Phi\colon \mathbb{P}_1 &\longrightarrow \mathbb{P}_n,\\ p = [\alpha_0, \alpha_1] &\longmapsto [P_0(\alpha_0, \alpha_1), \dots, P_n(\alpha_0, \alpha_1)],\end{aligned}$$

then it spans $\mathbb{P}_n$ if and only if the polynomials $P_0, \dots, P_n$ are linearly independent. No rational curve of degree less than n spans $\mathbb{P}_n$.

Proof. The polynomials are linearly dependent if and only if there are scalars $a_0, \dots, a_n$, not all zero, for which

$$a_0 P_0(\alpha_0, \alpha_1) + \dots + a_n P_n(\alpha_0, \alpha_1) = 0$$

for any $(\alpha_0, \alpha_1) \in k^2$, and this is in turn equivalent to having Γ contained in the hyperplane $a_0 x_0 + \dots + a_n x_n = 0$. If Γ has degree $d < n$, then the polynomials P_i, $i = 0, \dots, n$, may be assumed to have degree d, after which they cannot be linearly independent because the dimension of the space of homogeneous polynomials of degree d in $k[T_0, T_1]$ is $d + 1$. □

Proposition 9.2.5. *Any plane rational curve of degree two spanning P_2 is a non-degenerate conic with points.*

Proof. Assume the parameterization map of degree two to be given by polynomials P_0, P_1, P_2. By 9.2.4 they are linearly independent. If they are written

$$P_i = a_0^i T_0^2 + a_1^i T_0 T_1 + a_2^i T_1^2,$$

$i = 0, 1, 2$, then the matrix $A = (a_j^i)$ is regular and parametric equations of the curve may be written

$$\begin{pmatrix} x_0 \\ x_1 \\ x_2 \end{pmatrix} = A \begin{pmatrix} \alpha_0^2 \\ \alpha_0 \alpha_1 \\ \alpha_1^2 \end{pmatrix}.$$

Since these equations are turned into the equations (8.1) by a change of coordinates of $\mathbb{P}_2$ with matrix A^{-1} (see the proof of 9.2.2), the claim follows. □

9.3 Linear systems of quadrics

Let $\mathbb{P}_n$ be a projective space with associated vector space E. Denote by $S^2(E)$ the vector space of the symmetric bilinear forms on E. The quadrics of $\mathbb{P}_n$ were defined in Section 6.1 as being the points of the projective space $\mathbb{P}(S^2(E))$. In the sequel we will write $\mathbb{P}(S^2(E)) = \mathbb{Q}(\mathbb{P}_n)$ and call $\mathbb{Q}(\mathbb{P}_n)$ the *space of quadrics* (*of conics* if $n = 2$, *of pairs of points* if $n = 1$) of $\mathbb{P}_n$. The projective space $\mathbb{Q}(\mathbb{P}_n)$ is usually handled using the coordinates of quadrics introduced in the next proposition:

Proposition 9.3.1. *Assume we have fixed a projective reference Δ of $\mathbb{P}_n$. There is a projective reference Δ' of $\mathbb{Q}(\mathbb{P}_n)$ such that, for any quadric Q of $\mathbb{P}_n$, Q has equation*

$$\sum_{i,j=0}^{n} a_{i,j} x_i x_j = 0$$

relative to Δ if and only if it has coordinate vector $(a_{i,j})_{0 \leq i \leq j \leq n}$ relative to Δ'.

Proof. Let $e_0, \dots, e_n$ be a basis adapted to Δ. It is known that the bilinear forms $\eta_{i,j}$, $0 \leq i \leq j \leq n$, determined by the conditions

$$\eta_{i,j}(e_s, e_t) = \begin{cases} 1 & \text{if } \{i, j\} = \{s, t\}, \\ 0 & \text{otherwise} \end{cases}$$

compose a basis of $S^2(E)$. Furthermore, for any $\eta \in S^2(E)$,

$$\eta = \sum_{0 \leq i \leq j \leq n} \eta(e_i, e_j) \eta_{i,j}. \tag{9.7}$$

Let Δ' be the reference of $\mathbb{Q}(\mathbb{P}_n)$ with adapted basis $\{\eta_{i,j}\}_{0 \leq i \leq j \leq n}$. If $Q = [\eta]$ is a quadric of $\mathbb{P}_n$, we have seen in Section 6.1 that Q has equation

$$\sum_{i,j=0}^{n} a_{i,j} x_i x_j = 0$$

if and only there is $\rho \in k - \{0\}$ such that $a_{i,j} = \rho \eta(e_i, e_j)$, $i, j = 0, \dots, n$. The claim follows thus from the definition of projective coordinates and the equality (9.7). □

Remark 9.3.2. The reference Δ' of 9.3.1 is the one with adapted basis $\eta_{i,j}$, $0 \leq i \leq j \leq n$ (2.1.3). It is seldom explicitly mentioned in practice: after fixing a reference of $\mathbb{P}_n$, one just takes the coefficients of any equation (or matrix) of a quadric Q as projective coordinates of Q and conversely. Usually, equations or matrices of quadrics are used instead of their coordinate vectors, as they carry the same information and linear operations with coordinate vectors are equally performed using equations or matrices. For instance, assume that quadrics Q_i have equations $f_i = 0$ and matrices M_i, $i = 0, \dots, m$. Then, by 2.2.7 and 9.3.1, $Q_1, \dots, Q_m$ are linearly independent, as elements of $\mathbb{Q}(\mathbb{P}_n)$, if and only if the f_i, $i = 1, \dots, m$, are linearly independent polynomials, and also if and only if the M_i, $i = 0, \dots, m$, are linearly independent matrices.

From 9.3.1, by just counting the coordinates, we get:

Corollary 9.3.3. *The dimension of* $\mathbb{Q}(\mathbb{P}_n)$ *is* $(n^2 + 3n)/2$.

In particular the space of the pairs of points of a projective line has dimension two, the space of the conics of a projective plane has dimension five and the space of the quadrics of a three-dimensional projective space has dimension nine.

Transforming quadrics by a projectivity is a projectivity between spaces of quadrics:

Proposition 9.3.4. *Assume that* $f : \mathbb{P}_n \to \mathbb{P}'_n$ *is a projectivity. Mapping each quadric of* $\mathbb{P}_n$ *to its image* $f(Q)$ *is a projectivity*

$$\mathbb{Q}(f) : \mathbb{Q}(\mathbb{P}_n) \longrightarrow \mathbb{Q}(\mathbb{P}'_n).$$

Proof. Let E and E' be the vector spaces associated to $\mathbb{P}_n$ and $\mathbb{P}'_n$, respectively. Assume that $f = [\varphi]$ and $Q = [\eta]$. It is direct to check that the map between spaces of symmetric bilinear forms

$$\begin{aligned} S^2(E) &\longrightarrow S^2(E'), \\ \eta &\longmapsto \eta \circ (\varphi^{-1} \times \varphi^{-1}), \end{aligned}$$

is linear and has as inverse the map $\eta' \mapsto \eta' \circ (\varphi \times \varphi)$, $\eta' \in S^2(E')$. It is thus an isomorphism of vector spaces which, by the definition of $f(Q)$ in Section 6.1, induces $\mathbb{Q}(f)$, hence the claim. □

Remark 9.3.5. Equations of $\mathbb{Q}(f)$ have been implicitly written in 6.1.14 (b). The fact that they are linear, homogeneous and invertible provides a second, not essentially different, proof of 9.3.4.

Remark 9.3.6. With the present notations, Proposition 6.1.4 reads $\mathbb{Q}(\mathrm{Id}_{\mathbb{P}_n}) = \mathrm{Id}_{\mathbb{Q}(\mathbb{P}_n)}$, $\mathbb{Q}(g \circ f) = \mathbb{Q}(g) \circ \mathbb{Q}(f)$ and $\mathbb{Q}(f^{-1}) = \mathbb{Q}(f)^{-1}$, for any projective spaces $\mathbb{P}_n$, $\mathbb{P}'_n$, $\mathbb{P}''_n$, and projectivities $f : \mathbb{P}_n \to \mathbb{P}'_n$, $g : \mathbb{P}'_n \to \mathbb{P}''_n$.

When no confusion may arise, it is usual to write just f for $\mathbb{Q}(f)$, which we will do in the sequel.

The m-dimensional linear varieties of $\mathbb{Q}(\mathbb{P}_n)$ are called *m-dimensional linear systems* (or *families*) *of quadrics of* $\mathbb{P}_n$. The one-dimensional linear systems of quadrics are called *pencils of quadrics*, while the two-dimensional ones are called *nets of quadrics*. The pencils and the nets of quadrics are thus, respectively, the lines and the planes of the spaces of quadrics. Of course one says *pencil of conics* and *net of conics* if $n = 2$.

Assume that $f : \mathbb{P}_n \to \mathbb{P}'_n$ is a projectivity. The images by f of the quadrics of a d-dimensional linear system of quadrics $\mathcal{L} \subset \mathbb{Q}(\mathbb{P}_n)$ compose the set $\mathbb{Q}(f)(\mathcal{L})$, which is a d-dimensional linear system of quadrics of $\mathbb{P}'_n$ by 9.3.4. We have thus an action of projectivities on d-dimensional linear systems of quadrics. Since the properties required in Section 1.10 are a direct consequence of 6.1.4 (or 9.3.6), this action turns the class of all d-dimensional linear systems of quadrics into a projective class.

The linear systems of quadrics may be thought of in two different ways, corresponding to the parametric and implicit representations of linear varieties.

Assume first that $\mathcal{L} \subset \mathbb{Q}(\mathbb{P}_n)$ is a d-dimensional linear system of quadrics and that

$$f_0 = 0, \dots, f_d = 0$$

are equations of $d + 1$ independent quadrics of $\mathcal{L}$. Using equations instead of coordinate vectors, as explained above, this means that $f_0, \dots, f_d$ are linearly independent polynomials and, by 2.5.1, $\mathcal{L}$ is the family of all quadrics Q having equation

$$\lambda_0 f_0 + \cdots + \lambda_d f_d = 0,$$

for $(\lambda_0, \dots, \lambda_d) \in k^{d+1} - \{(0, \dots, 0)\}$. Furthermore, by 2.5.3, the parameters $\lambda_0, \dots, \lambda_d$ are determined by Q up to a non-zero scalar factor. In the sequel we will refer to $\mathcal{L}$ above as the linear system $\mathcal{L} : \lambda_0 f_0 + \cdots + \lambda_d f_d = 0$ or just $\lambda_0 f_0 + \cdots + \lambda_d f_d = 0$. On the other hand, once $d + 1$ linearly independent symmetric polynomials of degree two, $f_0, \dots, f_d \in k[x_0, \dots, x_n]$, are given, the family of all quadrics with equation $\lambda_0 f_0 + \cdots + \lambda_d f_d = 0$, $(\lambda_0, \dots, \lambda_m) \in k^{d+1} - \{(0, \dots, 0)\}$, is, by 2.5.2, a linear system of quadrics of dimension d.

Example 9.3.7. The pencil spanned by two different quadrics $f = 0$ and $g = 0$ of $\mathbb{P}_n$ is $\mathcal{P} : \lambda f + \mu g = 0$.

By 2.6.1 and 2.6.2 a set $\mathcal{L} \subset \mathbb{Q}(\mathbb{P}_n)$ is an $(n-m)$-dimensional linear system of quadrics if and only if there is system of independent homogeneous linear equations

$$\sum_{0 \leq i \leq j \leq n} b^r_{i,j} a_{i,j} = 0, \tag{9.8}$$

$r = 1, \dots, m$, in the coordinates $a_{i,j}$ of an undetermined quadric, such that a quadric $Q : \sum_{i,j=0,\dots,n} a_{i,j} x_i x_j = 0$ belongs to $\mathcal{L}$ if and only if its coordinates $a_{i,j}$, $0 \leq i \leq j \leq n$, are a non-trivial solution of it. The system (9.8) may then be seen as a condition to be imposed on (the coordinates of) a quadric, which is satisfied by the quadrics of $\mathcal{L}$, and only by them. The system (9.8) may be just the algebraic translation of an equivalent geometrical condition imposed on quadrics: conditions imposed on the quadrics of a fixed projective space P_n that translate into a rank-m homogeneous linear system of equations in the coordinates of the quadric are called *linear conditions* of *order* m. Equivalently, a condition on quadrics of $\mathbb{P}_n$ is linear of order m if and only if the quadrics satisfying it compose a linear system of quadrics of $\mathbb{P}_n$ of codimension m. Order-one linear conditions are called *simple*. The quadrics satisfying a simple linear condition describe a hyperplane of $\mathbb{Q}(\mathbb{P}_n)$. Let us see some examples:

Example 9.3.8 (Going through a given point). If $p \in \mathbb{P}_n$ is a point, then *going through* p is a simple linear condition on the quadrics of $\mathbb{P}_n$, or, equivalently, the quadrics going through p are those of a hyperplane of $\mathbb{Q}(\mathbb{P}_n)$. Indeed, if, after fixing a coordinate frame in $\mathbb{P}_n$, $p = [\alpha_0, \dots, \alpha_n]$, then the quadric $Q : \sum_{i,j=0}^{n} a_{i,j} x_i x_j = 0$ goes through p if and only if

$$\sum_{i,j=0}^{n} a_{i,j}\alpha_i\alpha_j = 0,$$

a homogeneous linear equation in the coordinates $a_{i,j}$, $0 \le i \le j \le n$, of Q, which is not trivial because $\alpha_i \neq 0$ for at least one i. The condition of going through a point $p \in \mathbb{P}_n$ is often referred to as the *condition imposed by* p (on the quadrics of $\mathbb{P}_n$).

Example 9.3.9 (Containing a given linear variety). Fix a linear variety $L \subset \mathbb{P}_n$ of dimension d: containing it is a linear condition of order $(d+1)(d+2)/2$ on the quadrics of $\mathbb{P}_n$. To check this, take the notations as above, the reference being chosen with its first $d+1$ vertices spanning L. Then by 6.1.10, the quadric Q contains L if and only if

$$a_{i,j} = 0, \quad 0 \le i \le j \le d,$$

which clearly is a system of $(d+1)(d+2)/2$ independent homogeneous linear equations in the $a_{i,j}$.

Obviously 9.3.9 reduces to 9.3.8 if $d = 0$.

Example 9.3.10 (Having a pair of points as a conjugate pair). Assume $p, q \in \mathbb{P}_n$: *having* p, q *as conjugate pair* is a simple linear condition on the quadrics of $\mathbb{P}_n$. Indeed, proceeding as in Example 9.3.8, with the same notations, if $q = [\beta_0, \dots, \beta_n]$ we get

$$\sum_{i,j=0}^{n} a_{i,j}\beta_i\alpha_j = 0,$$

again a non-trivial homogeneous linear equation in the $a_{i,j}$. With some abuse of language, the quadrics satisfying this condition are said *to harmonically divide* p *and* q, due to 6.4.2.

Also 9.3.10 reduces to 9.3.8, in this case when $p = q$.

Example 9.3.11 (Having a pair of imaginary and complex-conjugate points as a conjugate pair). Assume $k = R$ and let $p, \bar{p}$ be a pair of imaginary and complex-conjugate points: *having* $p, \bar{p}$ *conjugate with respect to* Q (that is, with respect to $Q_{\mathbb{C}}$) is a simple linear condition on the quadrics Q of $\mathbb{P}_n$. For, proceeding as in the former cases, if still $p = [\alpha_0, \dots, \alpha_n]$, the condition translates into

$$\sum_{i,j=0}^{n} a_{i,j} |\alpha_j|^2 = 0.$$

By definition, the quadrics of $\mathbb{P}_n$ satisfying a linear condition describe a linear system, often called the linear system *given rise to* by the condition. Conversely, as seen above, any linear system of quadrics $\mathcal{L}$ may be seen as the linear system given rise to by a linear condition, namely a system of equations of $\mathcal{L}$. In fact, speaking about linear conditions imposed on quadrics is rather tautological, because there is no formal difference between a linear system of quadrics and the class of all logically equivalent conditions giving rise to it. Nevertheless, it makes a very appealing and useful language, and we will use it in the sequel. The reader may easily avoid its use, if wanted, by translating all claims in terms of the linear families given rise to by the conditions.

Of course, not all subsets of $\mathbb{Q}(\mathbb{P}_n)$ are linear varieties, and not all conditions on quadrics are linear: the conditions which translate into a homogeneous polynomial system of equations in the coordinates $a_{i,j}$ of an undetermined quadric,

$$P_1 = 0, \dots, P_m = 0,$$

where each P_s, $s = 1, \dots, m$, is a homogeneous polynomial in the $a_{i,j}$, $0 \leq i \leq j \leq n$, are called *algebraic conditions*. Those which translate into a system with a single equation are called *simple*. Linear conditions are, in particular, algebraic. A couple of examples of algebraic, non-linear, conditions follow:

Example 9.3.12 (Being tangent to a line). On the quadrics of $\mathbb{P}_n$, $n \geq 2$, the condition of *being tangent to a given line* is a simple, non-linear, algebraic condition given by an equation of degree two. In other words, the quadrics of $\mathbb{P}_n$ tangent to a line ℓ describe the set of points of a quadric $\mathcal{Q}_\ell$ of $\mathbb{Q}(\mathbb{P}_n)$ and $|\mathcal{Q}_\ell|$ is not a linear variety. For, take a coordinate frame with its two first vertices on the given line ℓ. Then, by 6.3.4, $Q : \sum_{i,j=0}^{n} a_{i,j} x_i x_j = 0$ is tangent to ℓ if and only if $a_{0,0}a_{1,1} - a_{0,1}^2 = 0$, which is the equation of a quadric (of quadrics!) $\mathcal{Q}_\ell$ of $\mathbb{Q}(\mathbb{P}_n)$. Since $\mathcal{Q}_\ell$ shares with the pencil of quadrics $\lambda x_0^2 + \mu x_1^2 = 0$ the quadrics $x_0^2 = 0$

and $x_1^2 = 0$ and no other, by 1.5.5, $|\mathcal{Q}_\ell|$ is not a linear variety. Incidentally, note that $\mathbf{r}(\mathcal{Q}_\ell) = 3$.

Example 9.3.13 (Being degenerate). The condition of *being degenerate* is a simple, non-linear, algebraic condition on the quadrics of $\mathbb{P}_n$, given by an equation of degree $n+1$. Indeed, the quadric $Q : \sum_{i,j=0}^n a_{i,j}x_ix_j = 0$ is degenerate if and only if $\det(a_{i,j}) = 0$, which is clearly homogeneous of degree $n+1$ in the $a_{i,j}$. As the reader may easily check, the pencil spanned by the degenerate quadrics $x_0^2 + \cdots + x_{n-1}^2 = 0$ and $x_1^2 + \cdots + x_n^2 = 0$ contains non-degenerate quadrics, which, as in the former example, proves that the condition is not linear.

In the sequel we will deal with linear conditions mainly, but as shown by the two preceding examples, non-linear conditions cannot be avoided. Collecting examples of linear conditions is useful, next are some further ones.

Example 9.3.14 (Being tangent to a line at a given point). If p is a point of a line ℓ of $\mathbb{P}_n$, $n \geq 2$, *being tangent to ℓ at p* is a linear condition of order two on the quadrics of $\mathbb{P}_n$. To check this, recall first, from the definition of contact point, that a quadric is tangent to ℓ at p if and only if it is tangent to ℓ and goes through p. As for 9.3.12, take a reference of $\mathbb{P}_n$ with its two first vertices on ℓ, this time the point p being taken as the first one. Then $Q : \sum_{i,j=0}^n a_{i,j}x_ix_j = 0$ is tangent to ℓ and goes through p if and only if $a_{0,0}a_{1,1} - a_{0,1}^2 = 0$ (by 9.3.12) and $a_{0,0} = 0$. These equalities are equivalent to $a_{0,1}^2 = 0$ and $a_{0,0} = 0$, and so to $a_{0,0} = 0$ and $a_{0,1} = 0$, hence the claim.

There is no contradiction in the fact that the non-linear condition of being tangent to ℓ is turned into a linear one by adding the further condition of going through $p \in \ell$. What we have seen in 9.3.14 is that the intersection of the quadric $\mathcal{Q}_\ell$: $a_{0,0}a_{1,1} - a_{0,1}^2 = 0$ and the hyperplane $\mathcal{H} : a_{0,0} = 0$, both of $\mathbb{Q}(\mathbb{P}_n)$, is two times a hyperplane of $\mathcal{H}$, and its set of points is, indeed, a linear variety.

Example 9.3.15 (Having a given point as a double point). If $p \in \mathbb{P}_n$, *having p as a double point* is a linear condition of order $n+1$ on the quadrics of $\mathbb{P}_n$. Indeed, if p is taken as the first vertex of the coordinate frame, then it is a double point of the quadric $Q : \sum_{i,j=0}^n a_{i,j}x_ix_j = 0$ if and only if $a_{0,i} = 0$, $i = 0, \dots, n$.

Example 9.3.16 (Having a given simplex as self-polar). Fix an n-dimensional simplex Δ of $\mathbb{P}_n$. *Having Δ as a self-polar simplex* is a linear condition of order $n(n+1)/2$ on the quadrics of $\mathbb{P}_n$. For, if Δ itself is taken as simplex of reference, then Δ is self-polar for a quadric $Q : \sum_{i,j=0}^n a_{i,j}x_ix_j = 0$ if and only if $a_{i,j} = 0$ for $i < j$, $i,j = 0, \dots, n$ (7.1.3).

Our first use of the projective structure of $\mathcal{Q}(\mathbb{P}_n)$ will be a useful characterization of the quadrics for which a given simplex is self-polar. The reader may note that no mention of $\mathbb{Q}(\mathbb{P}_n)$ is made in the claim. The particular case in which Δ is the simplex of the reference has been, indeed, already seen in 7.1.3.

Proposition 9.3.17. *Assume that homogeneous coordinates have been fixed in a projective space $\mathbb{P}_n$. Let Ω be an n-dimensional simplex of $\mathbb{P}_n$ whose $(n-1)$-dimensional faces have equations $F_i = 0$, $i = 0, \dots, n$. Then Ω is self-polar for a quadric $Q : f = 0$ if and only if there exist $\lambda_0, \dots, \lambda_n \in k$ for which*

$$f = \lambda_0 F_0^2 + \dots + \lambda_n F_n^2.$$

Proof. Note first that the $(n-1)$-dimensional faces of Ω counted twice, $Q_i : F_i^2 = 0$, $i = 0, \dots, n$, are linearly independent elements of $\mathbb{Q}(\mathbb{P}_n)$: otherwise there would be a non-trivial relation $\sum_{i=0}^n \mu_i F_i^2 = 0$ and one of the faces would go through the vertex shared by the remaining ones. On the other hand, we know from 9.3.16 that the quadrics relative to which Ω is self-polar describe a linear system $\mathcal{L}$ of dimension n. Since, clearly, $Q_0, \dots, Q_n \in \mathcal{L}$, they span $\mathcal{L}$ and the claim follows. □

Let $L \subset \mathbb{P}_n$ be a positive-dimensional linear variety and Q' a quadric of L. We will say that a quadric Q of $\mathbb{P}_n$ *contains*, or *goes through*, Q', denoted by $Q \supset Q'$, if and only if either $Q \supset L$ or $Q \cap L = Q'$. The reader may easily check that in case of L being a line and Q' a pair of distinct, maybe imaginary, points, $Q' = p_1 + p_2$, then Q contains Q' if and only if p_1 and p_2 are points of Q. Similarly, in case $Q' = 2p$, Q contains Q' if and only if Q is tangent to L at p.

Proposition 9.3.18 (Going through a quadric of a linear variety). *If L is a linear variety of $\mathbb{P}_n$ of dimension $d > 0$ and Q' a quadric of L, then containing Q' is a linear condition of order $(d^2 + 3d)/2$ on the quadrics of $\mathbb{P}_n$.*

Proof. Choose a coordinate frame $\Delta = (p_0, \dots, p_n, A)$ with $L = p_0 \vee \dots \vee p_d$, take on L the coordinates subordinated by those of $\mathbb{P}_n$ and assume that $f = f(x_0, \dots, x_d) = 0$ is an equation of Q'. A quadric $Q : \sum_{i,j=0}^n a_{i,j} x_i x_j = 0$ contains Q' if and only if $\sum_{i,j=0}^d a_{i,j} x_i x_j$ is either identically zero or proportional to f (6.1.11), that is, if and only if the equation of Q has the form

$$\sum_{i=d+1}^{n} a_{i,i} x_i^2 + 2 \sum_{\substack{d<j\le n \\ i<j}} a_{i,j} x_i x_j + \lambda f = 0$$

for arbitrary values of the coefficients $a_{i,j}$ and λ, not all zero. Now, since $f \in k[x_0, \dots, x_d]$, f and the monomials appearing in the above expression are linearly independent polynomials. Then the quadrics containing Q' describe a linear system which, by a straightforward counting, has dimension $(n^2 + 3n - d^2 - 3d)/2$, hence the claim. □

Corollary 9.3.19 (Containing an imaginary point). *Assume $k = \mathbb{R}$. Containing an imaginary point $p \in \mathbb{C}\mathbb{P}_n - \mathbb{P}_n$ is a linear condition of order two on the quadrics of $\mathbb{P}_n$.*

Proof. Let $\bar{p}$ be the point complex-conjugate to p. Since any quadric of $\mathbb{P}_n$ containing p contains also $\bar{p}$ (6.1.16), a quadric of $\mathbb{P}_n$ contains p if and only if it contains the point-pair $p + \bar{p}$ of the real line spanned by $p, \bar{p}$. Then 9.3.18 applies. □

9.4 Independence of linear conditions on quadrics

Assume that $\mathcal{C}_1, \dots, \mathcal{C}_m$ are linear conditions on the quadrics of $\mathbb{P}_n$ and denote by $\mathcal{L}_i$ the linear system of the quadrics satisfying $\mathcal{C}_i$. Imposing on quadrics to satisfy all the conditions $\mathcal{C}_1, \dots, \mathcal{C}_m$ is also a linear condition because the quadrics satisfying it are those in $\mathcal{L}_1 \cap \dots \cap \mathcal{L}_m$; we will denote this condition by $\mathcal{C}_1 \wedge \dots \wedge \mathcal{C}_m$. By 1.4.13, we have

Lemma 9.4.1. *It holds that*

$$\operatorname{ord}(\mathcal{C}_1 \wedge \dots \wedge \mathcal{C}_m) \leq \operatorname{ord} \mathcal{C}_1 + \dots + \operatorname{ord} \mathcal{C}_m. \tag{9.9}$$

The conditions $\mathcal{C}_1, \dots, \mathcal{C}_m$ are called *independent* if and only if the inequality (9.9) is an equality. If all conditions are simple, then (9.9) reads

$$n - \dim(\mathcal{L}_1 \cap \dots \cap \mathcal{L}_m) = \operatorname{ord}(\mathcal{C}_1 \wedge \dots \wedge \mathcal{C}_m) \leq m$$

and so the independence of the conditions $\mathcal{C}_1, \dots, \mathcal{C}_m$ is equivalent to the linear independence of their corresponding linear systems $\mathcal{L}_1, \dots, \mathcal{L}_m$ as hyperplanes of $\mathbb{Q}(\mathbb{P}_n)$, by 4.2.6.

The following are useful criteria for the independence of linear conditions on quadrics. They are just translations of 1.4.8 and 4.2.7:

Lemma 9.4.2. *If $\mathcal{C}$ and $\mathcal{C}'$ are linear conditions on the quadrics of $\mathbb{P}_n$ and $\mathcal{C}'$ is simple, then they are independent if and only if there is a quadric Q that satisfies $\mathcal{C}$ and does not satisfy $\mathcal{C}'$.*

Lemma 9.4.3. *Simple linear conditions on the quadrics of $\mathbb{P}_n$, $\mathcal{C}_1, \dots, \mathcal{C}_m$, $m \geq 2$, are independent if and only if for each $i = 2, \dots, m$ there is a quadric Q_i of $\mathbb{P}_n$ that satisfies $\mathcal{C}_1, \dots, \mathcal{C}_{i-1}$ and does not satisfy $\mathcal{C}_i$.*

Exercises 7.36 and 9.37 (4) provide examples of dependent conditions. The next theorem will be useful in many cases.

Theorem 9.4.4. *Assume that $n \geq 2$, H is a hyperplane of $\mathbb{P}_n$, Q' is a quadric of H and $p_0, \dots, p_m$, $-1 \leq m \leq n$, are independent points of $\mathbb{P}_n$, no one in H. Then the quadrics of $\mathbb{P}_n$ going through both Q' and all points $p_0, \dots, p_m$ describe a linear system of dimension $n - m$.*

Proof. The case of no point has been proved in 9.3.18, thus we assume $m \geq 0$ and use induction on m. By the induction hypothesis, the quadrics through Q' and

$p_0, \dots, p_{m-1}$ describe a linear system of dimension $n-m+1$. By the independence of the points, there is a hyperplane H' going through all points $p_i, i = 0, \dots, m-1$, and missing p_m (4.2.10). Then, since $p_m \notin H$, the pair of hyperplanes $H + H'$ is a quadric going through Q' and $p_0, \dots, p_{m-1}$, and not going through p_m. Using 9.4.2 completes the proof. □

Remark 9.4.5. In case $m = n$, 9.4.4 above assures the existence and uniqueness of a quadric through Q' and $p_0, \dots, p_n$.

The next corollary and 8.1.9 are very similar: it is worth comparing their proofs, as they are quite different.

Corollary 9.4.6. *There is one and only one conic of $\mathbb{P}_2$ going through five different given points, provided no four of them are aligned.*

Proof. First we prove that among five different points $q_i \in \mathbb{P}_2$, $i = 1, \dots, 5$, no four aligned, there are three independent points none of which belongs to the line spanned by the remaining two points. Indeed, the claim is obvious if no three of the five points are aligned. Otherwise assume that q_1, q_2 and q_3 are aligned. Then the remaining points q_4, q_5 may be aligned with at most one of the former, say with none or with q_1. Then, since q_4 cannot be aligned with q_1, q_2, q_3, the points q_1, q_2 and q_4 are independent. Furthermore, the line q_3q_5 does not contain q_4, because q_3 is assumed not to be aligned with q_4 and q_5, and neither contains q_1 or q_2, as in such a case q_5 would be aligned with q_1, q_2, q_3. Now, to prove the claim, just apply 9.4.4 (or 9.4.5), taking the three independent points as p_1, p_2, p_3 and the remaining ones to compose the pair of points Q' of the line H they span. □

Remark 9.4.7. The claim of 9.4.6 is false if four of the five points are aligned, as then, obviously, infinitely many pairs of lines are going through them.

Unfortunately, we cannot be so precise about determining quadrics by points in higher dimensions. Since going through a point is a simple linear condition, it is clear that there is always a quadric of $\mathbb{P}_n$ going through $m \leq (n^2+3n)/2$ given different points, as by 9.4.1 and 9.3.3, the quadrics satisfying such a number of simple linear conditions describe a linear system of dimension at least $(n^2+3n)/2-m$. Having a unique quadric through the m given points needs thus $m = (n^2+3n)/2$ and, this being fulfilled, is equivalent to the independence of the conditions imposed by the points. The next corollary of 9.4.4 assures the existence of a set of $(n^2+3n)/2$ different points imposing independent conditions.

Corollary 9.4.8. *There is always one quadric going through $(n^2+3n)/2$ given points of $\mathbb{P}_n$. Furthermore, there is a choice of the points for which such a quadric is unique.*

Proof. The first part has been seen above. For the second one we will use induction on n, the case $n = 1$ being already proved in 6.2.2. If $n > 1$, choose a set A of $n + 1$ independent points of $\mathbb{P}_n$. Then take a hyperplane H missing all points in A: let it be $H : x_0 + \cdots + x_n = 0$, the coordinates being relative to any reference whose vertices are the points in A. Using the induction hypothesis, take a set B of $((n-1)^2 + 3(n-1))/2$ different points in H such that there is a unique quadric Q' of H containing them. The points of $A \cup B$ are $(n^2 + 3n)/2$ in number. Assume that Q is a quadric of $\mathbb{P}_n$ going through them. Then either $Q \supset H$ or $Q \cap H$ is a quadric of H going through all points of B and hence $Q \cap H = Q'$. In any case $Q \supset Q'$ and so, by 9.4.5, Q is the only quadric containing Q' and all points of A. □

Actually, in the above proof $Q \supset H$ cannot occur, as, by 6.4.22, this would force Q to be a hyperplane-pair and therefore all points of A to lie on a hyperplane.

In fact 9.4.8 falls rather short, as after the case $n = 2$ (9.4.6) one may expect to have a unique quadric through any given set of $(n^2+3n)/2$ points except for the sets whose points are in some sort of special positions. The next corollary is a precise statement in this sense, but, unlike 9.4.6, it does not give any explicit description of the special positions in which the points do not determine the quadric. Roughly speaking, it says that there is a unique quadric of $\mathbb{P}_n$ through any given $(n^2+3n)/2$ points, except for the case in which the coordinates of the points satisfy a certain non-trivial algebraic relation, which is not explicitly given. The reader may see also the forthcoming Example 9.11.29.

Corollary 9.4.9. *There exists a non-zero polynomial $D(X_i^s)$ in the variables X_i^s, $i = 0, \dots, n$, $s = 1, \dots, (n^2 + 3n)/2$, homogeneous in the variables $X_0^s, \dots, X_n^s$ for each fixed s, such that, after fixing any coordinate frame in $\mathbb{P}_n$, $D(\alpha_i^s) \neq 0$ guarantees that there is a unique quadric of $\mathbb{P}_n$ going through the points $p_s = [\alpha_0^s, \dots, \alpha_n^s]$, $s = 1, \dots, (n^2 + 3n)/2$.*

Proof. Put $\delta = (n^2 + 3n)/2$. Fix any total order on the set of pairs (i, j), $0 \leq i \leq j \leq n$. For each fixed s take the $(\delta + 1)$-dimensional row vector v^s whose (i, j)-component is

$$v_{i,j}^s = \begin{cases} (X_{i,i}^s)^2 & \text{for } 0 \leq i = j \leq n, \\ 2X_i^s X_j^s & \text{for } 0 \leq i < j \leq n. \end{cases}$$

and let $M(X_i^s)$ be the matrix whose rows are the vectors v^s. The matrix $M(X_i^s)$ is thus a $\delta \times (\delta + 1)$ matrix whose entries are polynomials in the X_i^s. Let $D_r(X_i^s)$, $r = 1, \dots, \delta+1$, be the maximal minors of $M(X_i^s)$, $D_r(X_i^s)$ being the determinant of the matrix obtained from $M(X_i^s)$ by dropping its r-th column. Each $D_r(X_i^s)$ is a polynomial in the X_i^s, homogeneous in $X_0^s, \dots, X_n^s$ for each fixed s, due to the fact that the entries of v_s are homogeneous in the variables X_i^s, $i = 0, \dots, n$, and these variables do not appear in the other rows.

A quadric

$$Q : \sum_{i,j=0}^{n} a_{i,j} x_i x_j = 0$$

goes through given points $p_s = [\alpha_0^s, \dots, \alpha_n^s]$, $s = 1, \dots, \delta$, if and only if

$$\sum_{i=0}^{n} a_{i,i} (\alpha_i^s)^2 + \sum_{0 \le i < j \le n} 2a_{i,j} \alpha_i^s \alpha_j^s = \sum_{i,j=0}^{n} a_{i,j} \alpha_i^s \alpha_j^s = 0, \quad s = 1, \dots, \delta.$$

The above is a system of homogeneous linear equations in the $a_{i,j}$. Its matrix, if the variables are suitably ordered, is $M(\alpha_i^s)$, the result of substituting α_i^s for X_i^s in the entries of $M(X_i^s)$. The existence of a unique quadric through the points p_i, $i = 1, \dots, \delta$, is thus equivalent to rk $M(\alpha_i^s) = \delta$, that is, to being $D_r(\alpha_i^s) \neq 0$ for at least one r. Since by 9.4.8 there exist points determining the quadric, their coordinates are values of the α_i^s for which $D_r(\alpha_i^s) \neq 0$ for a certain r. Then the corresponding polynomial $D_r(X_i^s)$ is not zero and taking it as $D(X_i^s)$ fulfills the claim. □

Requiring the polynomial $D(X_i^s)$ to be homogeneous in each set of variables with fixed s is usual: this guarantees that the condition $D(\alpha_i^s) \neq 0$ does not depend on the arbitrary factor up to which are determined the homogeneous coordinates of each point p_s.

9.5 Pencils of quadrics

In the sequel, the pencils of quadrics will be often called just pencils if no confusion may result. As defined in Section 9.3, the pencils of quadrics of $\mathbb{P}_n$ are the lines of $\mathbb{Q}(\mathbb{P}_n)$. Therefore any two different quadrics Q, Q' span a pencil (as a line of $\mathbb{Q}(\mathbb{P}_n)$) and any pencil is spanned by any two different quadrics of it.

Once a reference of $\mathbb{P}_n$ has been fixed, by 9.3.2, the pencil $\mathcal{P}$ spanned by different quadrics $Q : f = 0$ and $Q' : g = 0$ is the linear family of quadrics

$$\mathcal{P} = \{Q_{\lambda,\mu} : \lambda f + \mu g = 0 \mid (\lambda, \mu) \in k^2 - (0,0)\}, \tag{9.10}$$

where $f = 0$ and $g = 0$ are equations of Q and Q'. Any pencil $\mathcal{P}$ may thus be represented in this form, $f = 0$ and $g = 0$ being equations of any two different quadrics of $\mathcal{P}$. As already agreed for linear systems of quadrics, the pencil $\mathcal{P}$ of (9.10) will be denoted by $\mathcal{P} : \lambda f + \mu g = 0$, or just $\lambda f + \mu g = 0$. One may equivalently use matrices M of Q and N of Q', the quadric $Q_{\lambda,\mu}$ being then the one with matrix $\lambda M + \mu N$.

Once the equations (or matrices) of Q and Q' are fixed, λ, μ, are referred to as *parameters* of $Q_{\lambda,\mu}$. They are homogeneous coordinates of $Q_{\lambda,\mu}$ relative to the reference of $\mathcal{P}$ with vertices Q, Q' and unit point $f + g = 0$ (2.5.3).

We will make frequent use of the next lemma. It needs no proof, as it is just the translation of the easy fact that any linear variety sharing two different points with a line does contain it (1.5.5).

Lemma 9.5.1. *If a linear condition is satisfied by two different quadrics of a pencil $\mathcal{P}$, then it is satisfied by all the quadrics of $\mathcal{P}$.*

In particular by 9.3.8 and 9.3.19, a (real or imaginary, in case $k = \mathbb{R}$) point belongs to two different quadrics of a pencil $\mathcal{P}$ if and only if it belongs to all quadrics of $\mathcal{P}$. If this is the case, the point is called a *base point* of $\mathcal{P}$. The set of all base points is called the *base locus* of the pencil. In the case $k = \mathbb{R}$ a base point may be *real* or *imaginary*.

Remark 9.5.2. By 6.1.16, the complex-conjugate to an imaginary base point of a pencil of real quadrics is also an imaginary base point of the pencil.

The fact that a line is either contained in a hyperplane or meets it at a single point, obviously translates into:

Lemma 9.5.3. *If a simple linear condition $\mathcal{C}$ is not satisfied by all quadrics of a pencil $\mathcal{P}$, then there is exactly one quadric in $\mathcal{P}$ satisfying $\mathcal{C}$.*

The sections of the quadrics of a pencil by a linear variety do not always describe a pencil, namely:

Proposition 9.5.4. *Assume that L is a positive-dimensional linear variety and $\mathcal{P}$ a pencil of quadrics, both of $\mathbb{P}_n$. Then either*

(a) *all quadrics of $\mathcal{P}$ contain L and therefore $Q \cap L = L$ for all $Q \in \mathcal{P}$, or*

(b) *a single quadric $Q_0 \in \mathcal{P}$ contains L and then all sections $Q \cap L$ equal the same quadric of L for $Q \in \mathcal{P}$, $Q \neq Q_0$, or*

(c) *no quadric of $\mathcal{P}$ contains L, the sections $Q \cap L$, $Q \in \mathcal{P}$ describe a pencil $\mathcal{P}'$, of quadrics of L, and $Q \mapsto Q \cap L$ defines a projectivity from $\mathcal{P}$ onto $\mathcal{P}'$.*

Proof. Fix coordinates $x_0, \dots, x_n$ relative to a projective reference whose vertices $p_0, \dots, p_d$ span L, and take $x_0, \dots, x_d$ as coordinates on L (2.2.6). For any equation $h = 0$ of a quadric Q of $\mathbb{P}_n$ we write $\bar{h} = h(x_0, \dots, x_d, 0, \dots, 0)$. If $f = 0$ and $g = 0$ are equations of two different quadrics of $\mathcal{P}$, then the quadrics of $\mathcal{P}$ are $Q_{\lambda,\mu} : \lambda f + \mu g = 0$, for $\lambda, \mu \in k$ and not both zero. By 6.1.11, the polynomial

$$\lambda \bar{f} + \mu \bar{g} = \overline{\lambda f + \mu g}$$

is zero if and only if $Q_{\lambda,\mu} \supset L$; otherwise $Q_{\lambda,\mu} \cap L$ is a quadric of L and has equation

$$\lambda \bar{f} + \mu \bar{g} = 0.$$

After this, as the reader may easily check, case (a) occurs if and only if $\bar{f} = \bar{g} = 0$, case (b) occurs if and only if $\bar{f}$ and $\bar{g}$ are linearly dependent but not both zero, and case (c) occurs if and only if $\bar{f}$ and $\bar{g}$ are linearly independent. In the last case mapping each quadric in $\mathcal{P}$ to its section by L is a projectivity because both the quadric and its section have the same parameters in the pencils $\mathcal{P}$ and $\mathcal{P}'$, respectively. □

The degenerate quadrics of a pencil $\mathcal{P}$ may be determined in the following way: fix matrices $M = (a_{i,j})$ and $N = (b_{i,j})$ of two quadrics $Q : \sum_{i,j=0}^{n} a_{i,j} x_i x_j = 0$ and $Q' : \sum_{i,j=0}^{n} b_{i,j} x_i x_j = 0$ spanning $\mathcal{P}$; then the quadric of $\mathcal{P}$ with matrix $\lambda M + \mu N$ is degenerate if and only if

$$\det(\lambda M + \mu N) = \begin{vmatrix} \lambda a_{0,0} + \mu b_{0,0} & \dots & \lambda a_{n,0} + \mu b_{n,0} \\ \vdots & & \vdots \\ \lambda a_{0,n} + \mu b_{0,n} & \dots & \lambda a_{n,n} + \mu b_{n,n} \end{vmatrix} = 0. \tag{9.11}$$

The first member of this equation is a homogeneous polynomial in λ, μ of degree $n+1$, as all entries of the determinant are homogeneous of degree one in λ, μ. If it is zero, then all the quadrics of the pencil are degenerate. Otherwise, 9.1.1 applies, after which we have:

Proposition 9.5.5. *Either all or at most $n+1$ of the members of a pencil of quadrics of $\mathbb{P}_n$ are degenerate. If $k = \mathbb{C}$, every pencil of quadrics contains a degenerate quadric. The same is true for $k = \mathbb{R}$ if the dimension n is even.*

So, in particular, if a pencil of quadrics of $\mathbb{P}_n$ contains a non-degenerate quadric, then it contains at most $n + 1$ degenerate quadrics. The pencils containing non-degenerate quadrics will be called *regular pencils* in the sequel.

The equation (9.11) has a deeper meaning: if $\mathcal{P}$ is regular and λ, μ are taken as coordinates in $\mathcal{P}$, then the divisor of solutions of (9.11) is an effective divisor of $\mathcal{P}$ named the *characteristic divisor* of $\mathcal{P}$. Its points are the degenerate quadrics of $\mathcal{P}$. We will see next that the characteristic divisor depends only on $\mathcal{P}$, and not on any of the choices we made to write its equation (9.11).

Theorem 9.5.6. *Let $\mathcal{P}$ be a regular pencil of quadrics of $\mathbb{P}_n$. Choose a reference Δ of $\mathbb{P}_n$, two different quadrics $Q, Q' \in \mathcal{P}$ and matrices M, N of Q and Q' relative to Δ. Take λ, μ as homogeneous coordinates in $\mathcal{P}$ of the quadric of matrix $\lambda M + \mu N$. Then the divisor of solutions of*

$$\det(\lambda M + \mu N) = 0$$

is independent of the choices of Δ, Q, Q', M and N.

Proof. We have seen above that $P(\lambda, \mu) = \det(\lambda M + \mu N)$ is a non-zero homogeneous polynomial in λ, μ and so $\det(\lambda M + \mu N) = 0$ has a divisor of solutions.

Matrices of any other pair of different conics of $\mathcal{P}$ are $\bar{M} = aM + cN$ and $\bar{N} = bM + dN$ for suitable $a, b, c, d \in k$ with $ad - bc \neq 0$. Then

$$\begin{aligned} 0 &= \det(\bar{\lambda}\bar{M} + \bar{\mu}\bar{N}) = \det((a\bar{\lambda} + b\bar{\mu})M + (c\bar{\lambda} + d\bar{\mu})N) \\ &= P(a\bar{\lambda} + b\bar{\mu}, c\bar{\lambda} + d\bar{\mu})) \end{aligned}$$

has, by 9.1.6, the same divisor of solutions as $P(\lambda, \mu) = 0$. This proves that the characteristic divisor does not depend on the choice of Q, Q', or their matrices M, N. Relative to other coordinates of $\mathbb{P}_n$, Q and Q' have matrices of the form $T^t MT$ and $T^t NT$, respectively, where T is a regular $(n+1) \times (n+1)$ matrix (6.1.12). Then, clearly,

$$0 = \det(\lambda T^t MT + \mu T^t NT) = (\det T)^2 \det(\lambda M + \mu N)$$

defines the same divisor as $\det(\lambda M + \mu N) = 0$. □

Remark 9.5.7. A polynomial $\det(\lambda M + \mu N)$ (resp. equation $\det(\lambda M + \mu N) = 0$) as above is called a *characteristic polynomial* (resp. *characteristic equation*) of $\mathcal{P}$. When using $\theta = \mu/\lambda$ as an absolute coordinate in $\mathcal{P}$, it is usual to take the non-homogeneous polynomial $\det(M + \theta N))$ instead, and still call it characteristic polynomial. The proof of 9.5.6 shows that a characteristic polynomial of a regular pencil $\mathcal{P}$ depends on the choice of the coordinates λ, μ in $\mathcal{P}$, but not, up to a non-zero constant factor, on the choice of coordinates in $\mathbb{P}_n$.

Remark 9.5.8. Since any characteristic equation has degree $n + 1$, by 9.1.4, the degree of the characteristic divisor is at most $n + 1$, and the equality is always reached in the case $k = \mathbb{C}$. Equivalently, the number of degenerate quadrics in a regular pencil of quadrics of $\mathbb{P}_n$, each degenerate quadric counted according to its multiplicity in the characteristic divisor, does not exceed $n + 1$, and equals $n + 1$ if $k = \mathbb{C}$; this improves the bound of 9.5.5.

Corollary 9.5.9 (Projective invariance of the characteristic divisor). *Let $f : \mathbb{P}_n \to \mathbb{P}'_n$ be a projectivity and $\mathcal{P}$ a regular pencil of quadrics of $\mathbb{P}_n$. The images by f of the quadrics of $\mathcal{P}$ describe a regular pencil of quadrics of $\mathbb{P}'_n$ whose characteristic divisor is the image by f of the characteristic divisor of $\mathcal{P}$.*

Proof. We already know from 9.3.4 that the images of the quadrics of $\mathcal{P}$ describe a pencil $f(\mathcal{P})$, and obviously $F(\mathcal{P})$ is regular if $\mathcal{P}$ is. Fix any reference Δ in $\mathbb{P}_n$. As above, let Q, Q' be quadrics spanning $\mathcal{P}$, assume that they have matrices M, N, respectively, and denote by $Q_{\lambda,\mu}$ the quadric with matrix $\lambda M + \mu N$. Then, using λ, μ as homogeneous coordinates of $Q_{\lambda,\mu}$ in $\mathcal{P}$, the divisor of solutions of $\det(\lambda M + \mu N) = 0$ is the characteristic divisor of $\mathcal{P}$.

Take $f(\Delta)$ as reference in $\mathbb{P}'_n$. Then the matrix of f is the unit one (2.8.1) and so, for any quadric K of $\mathbb{P}_n$, K and $f(K)$ have the same matrices. The quadrics

$f(Q)$ and $f(Q')$, which span $f(\mathcal{P})$, have thus matrices M and N, and furthermore, for any $(\lambda,\mu) \in k^2 - \{(0,0)\}$, $f(Q_{\lambda,\mu})$ has matrix $\lambda M + \mu N$. Thus, we proceed as in $\mathcal{P}$: we take λ, μ as coordinates of $f(\mathcal{P})$ in $\mathcal{P}$, and using these coordinates the divisor of solutions of $\det(\lambda M + \mu N) = 0$ is the characteristic divisor of $f(\mathcal{P})$.

Now a quadric $Q_{a,b} \in \mathcal{P}$ belongs with multiplicity ν to the characteristic divisor of $\mathcal{P}$ if and only if $b\lambda - a\mu$ is a prime factor of multiplicity ν of $\det(\lambda M + \mu N)$. Since $f(Q_{a,b})$ also has coordinates a, b, the latter condition is satisfied if and only if $f(Q_{a,b})$ belongs with multiplicity ν to the characteristic divisor of $f(\mathcal{P})$, hence the claim. □

The dimension of the variety of double points of a degenerate quadric Q provides a lower bound of the multiplicity of Q in the characteristic divisor of any regular pencil of quadrics containing Q.

Proposition 9.5.10. *Assume that Q is a degenerate quadric and belongs to a regular pencil $\mathcal{P}$ of quadrics of $\mathbb{P}_n$. If ν is the multiplicity of Q in the characteristic divisor of $\mathcal{P}$, then $\nu \geq \dim \mathcal{D}(Q) + 1$, where $\mathcal{D}(Q)$ is the linear variety of double points of Q.*

Proof. Put $d = \dim \mathcal{D}(Q)$. Take a reference of $\mathbb{P}_n$ whose vertices $p_0, \dots, p_d$ are double points of Q. If M is a matrix of Q and N a matrix of any $Q' \in \mathcal{P}$, by 9.5.6, the divisor of solutions of $\det(\lambda M + \mu N) = 0$ is the characteristic divisor of $\mathcal{P}$. The choice of the reference causes all entries on the first $d + 1$ rows of M to be zero, after which the polynomial $\det(\lambda M + \mu N)$ clearly has a factor μ^{d+1}. Since Q has parameters $\lambda = 1$, $\mu = 0$ the claim follows. □

In the remainder of this section we examine some noteworthy examples of pencils of quadrics. The next lemma will be useful to handle some pencils with imaginary base points.

Lemma 9.5.11. *Assume $k = \mathbb{R}$ and let $Q : f = 0$, $Q' : g = 0$ and $Q'' : h = 0$ be quadrics of $\mathbb{P}_n$, $Q \neq Q'$. If the complex extension of Q'' belongs to the pencil of quadrics of $\mathbb{C}\mathbb{P}_n$ spanned by the complex extensions of Q and Q', then Q'' belongs to the pencil spanned by Q and Q', and conversely.*

Proof. The converse being clear, we prove the direct claim only. By the hypothesis and 6.1.15, there exist complex numbers λ, μ such that $h = \lambda f + \mu g$. The equations f, g, h being real, taking complex-conjugates gives $h = \bar{\lambda} f + \bar{\mu} g$, after which $2h = (\lambda + \bar{\lambda}) f + (\mu + \bar{\mu}) g$, with $\lambda + \bar{\lambda}, \mu + \bar{\mu} \in \mathbb{R}$, which proves the claim. □

Assume that $\mathcal{P}$ is a pencil of quadrics of a real projective space $\mathbb{P}_n$. If Q and Q' are any quadrics spanning $\mathcal{P}$, in the sequel we will denote by $\mathcal{P}_{\mathbb{C}}$ the pencil of quadrics of $\mathbb{C}\mathbb{P}_n$ spanned by the complex extensions $Q_{\mathbb{C}}$ and $Q'_{\mathbb{C}}$ of Q and Q'. By 9.5.11, $\mathcal{P}_{\mathbb{C}}$ is the only pencil of quadrics of $\mathbb{C}\mathbb{P}_n$ that contain the complex extensions of all quadrics of $\mathcal{P}$. In particular, it is independent of the choice of Q, Q' in $\mathcal{P}$. The pencil $\mathcal{P}_{\mathbb{C}}$ will be called the *complex extension* of $\mathcal{P}$.

Example 9.5.12 (Conics through four points, three aligned). If four different points of $\mathbb{P}_2$ are aligned, then any conic through them contains the line they span (by 6.2.5) and conversely. Therefore (by 6.4.22) the conics through four aligned points describe the net

$$v(\lambda_0 x_0 + \lambda_1 x_1 + \lambda_2 x_2) = 0,$$

where $v = 0$ is an equation of the line spanned by the four points. If just three of the points are aligned, then the same argument shows that any conic through the four points is composed of the line spanned by the three aligned points and a second line that this time needs to go through the fourth point. The converse being clear, the conics through four non-aligned points, three of which are aligned, describe the pencil

$$v(\lambda u_1 + \mu u_2) = 0,$$

where $v = 0$ is the line spanned by the three aligned points and $u_1 = 0, u_2 = 0$ are any two different lines through the fourth point. Note that the conics of the pencil are composed of a fixed and a variable line. Clearly, the pencil is not regular and has infinitely-many base points.

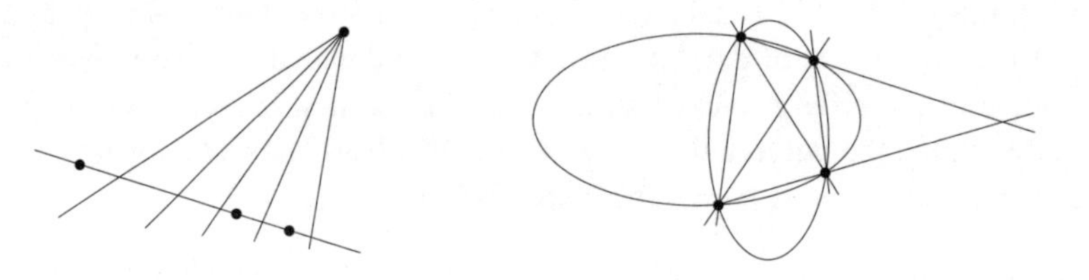

Figure 9.1. Conics through four points, just three aligned on the left and no three aligned on the right.

Example 9.5.13 (Conics through four points, no three aligned). Fix four points p_1, p_2, p_3, p_4 of $\mathbb{P}_2$, no three aligned. The conics through them describe a pencil, as taking two of the points as composing a pair of points of the line they span, 9.4.4 applies. The degenerate conics of the pencil are the pairs of lines going through the four points, and these are just the pairs of opposite sides of the quadrivertex with vertices p_1, p_2, p_3, p_4; there are three of them, so the bound of 9.5.5 is reached and, by 9.5.8, each pair of opposite sides has multiplicity one in the characteristic divisor. Since any two of the pairs of opposite sides span the pencil, it is clear that the base points are just p_1, p_2, p_3, p_4.

A quadrivertex having all its vertices on a conic C is said to be *inscribed* in C; then it is equivalently said that C is *circumscribed* to the quadrivertex. Dually for non-degenerate conics, a quadrilateral is *circumscribed* to a non-degenerate conic

C if and only if all its sides are tangent to C; in such a case it is equivalently said that C is *inscribed* in the quadrilateral.

In 9.5.13 we have seen:

Lemma 9.5.14. *The conics circumscribed to a quadrivertex describe a pencil whose degenerate members are the pairs of opposite sides of the quadrivertex; each has multiplicity one in the characteristic divisor.*

Remark 9.5.15. In particular, the three pairs of opposite sides of a quadrivertex belong to the same pencil of conics. This is a remarkable fact that underlies some elementary theorems involving quadrivertices, see for instance Exercises 9.15, 9.17 and 9.41.

Lemma 9.5.14 provides a quite different proof of a claim which was already proposed as Exercise 7.1:

Proposition 9.5.16. *The diagonal triangle of a quadrivertex is self-polar with respect to any conic circumscribed to the quadrivertex.*

Proof. By 9.5.13 the conics circumscribed to a quadrivertex describe a pencil. Since the condition of having two given points as a conjugate pair is linear (9.3.10), by 9.5.1 it is enough to see that any two vertices q, q' of the diagonal triangle are conjugate with respect to two different circumscribed conics. Take Q and Q' to be the pairs of sides of K that meet at q and q', respectively. Then q, q' are conjugate with respect to Q because q is a double point of Q, and with respect to Q' because q' is a double point of Q'. □

Corollary 9.5.17. *If K is a quadrivertex of $\mathbb{P}_2$, then there is a reference of $\mathbb{P}_2$ whose vertices are those of the diagonal triangle of K and relative to which the conics circumscribed to K have equations*

$$(\lambda + \mu)x_0^2 - \lambda x_1^2 - \mu x_2^2 = 0,$$

$(\lambda, \mu) \in k^2 - \{(0,0)\}$.

Proof. Take any reference with vertices those of the diagonal triangle. By 9.5.16 and 7.1.3, all circumscribed conics have diagonal matrices. In particular the pairs of opposite sides having $[0, 1, 0]$ and $[0, 0, 1]$ as double points span the pencil and have equations

$$x_0^2 - ax_2^2 = 0 \quad \text{and} \quad x_0^2 - bx_1^2 = 0$$

with $a, b \in k - \{0\}$, and $a, b > 0$ if $k = \mathbb{R}$, as in such a case the pairs of opposite sides are pairs of real lines. The conics of the pencil have thus equations

$$(\lambda + \mu)x_0^2 - \lambda b x_1^2 - \mu a x_2^2 = 0, \quad (\lambda, \mu) \in k^2 - \{(0,0)\}.$$

Taking $[1, 1/\sqrt{b}, 1/\sqrt{a}]$ as a new unit point turns these equations into the claimed ones. □

Example 9.5.18 (Conics through three points with a prescribed tangent at one of them). Let p_1, p_2, p_3 be three different points and ℓ a line of $\mathbb{P}_2$, $p_3 \in \ell$. In order to describe the conics through p_1, p_2, p_3 tangent to ℓ at p_3, we assume p_1, p_2, p_3 non-aligned and $p_1, p_2 \notin \ell$, as the other cases are similar to those of Example 9.5.12 and may be dealt with by the reader. By taking $2p_3$ as a quadric of ℓ, 9.4.4 applies and shows that the conics through p_1, p_2, p_3 tangent to ℓ at p_3 describe a pencil. Its degenerate conics are pairs of lines through p_1, p_2 and tangent to ℓ at p_3. By the latter condition, they either contain ℓ or, otherwise, have p_3 as double point. This gives $p_1p_3 + p_2p_3$ and $\ell + p_1p_2$ as the only degenerate conics of the pencil, and in particular proves that the base points of the pencil are just p_1, p_2, p_3.

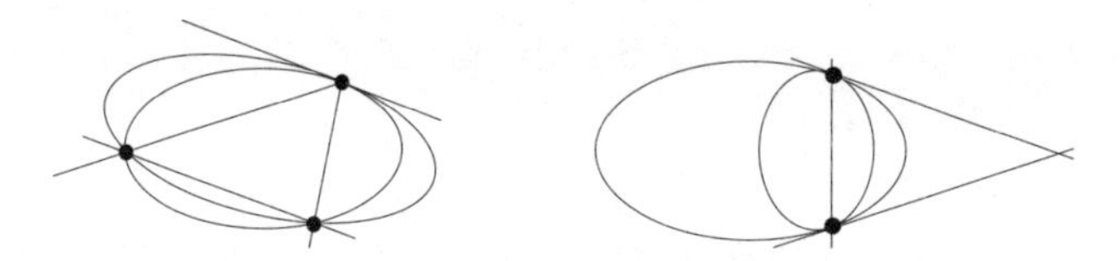

Figure 9.2. Conics through three points with a fixed tangent at one of them, on the left, and bitangent conics, on the right.

Example 9.5.19 (Bitangent conics). Assume to have fixed in $\mathbb{P}_2$ a point p and a pair of distinct points $q_1 + q_2$ of a line $\ell \subset \mathbb{P}_2$ not going through p. In case $k = \mathbb{R}$, the points q_1, q_2 may be either real or imaginary and complex-conjugate. We are interested in the family $\mathcal{P}$ of all conics C of $\mathbb{P}_2$ such that:

(1) $C \supset q_1 + q_2$, and

(2) p and any $q \in \ell$ are conjugate with respect to C.

After fixing any two different points $p_1, p_2 \in \ell$, condition (2) above is clearly equivalent to being p conjugate to p_1 and p_2.

We take a reference of $\mathbb{P}_2$ with first vertex $p_0 = p$, two distinct points $p_1, p_2 \in \ell$, conjugate with respect to $q_1 + q_2$, as second and third vertices, and any suitable unit point.

The choice of p_1, p_2 assures that an equation of $q_1 + q_2$, relative to the subordinated reference of ℓ, has the form $x_1^2 + bx_2^2 = 0$. Then a conic C : $\sum_{i,j=0}^{2} a_{i,j}x_ix_j = 0$ satisfies $C \supset q_1 + q_2$ if and only if

$$\sum_{i,j=1}^{2} a_{i,j}x_ix_j = \mu(x_1^2 + bx_2^2), \quad \mu \in k,$$

(by 6.1.10). On the other hand, $p = p_0$ is conjugate to p_1 and p_2 if and only if $a_{0,1} = a_{0,2} = 0$ (6.4.1). All together, after renaming $a_{0,0} = -\lambda$, the conics of $\mathcal{P}$

are those with equation

$$\lambda x_0^2 + \mu(x_1^2 + bx_2^2) = 0, \quad (\lambda, \mu) \in k^2 - \{0, 0\} \tag{9.12}$$

and hence $\mathcal{P}$ is a pencil: it is called the *pencil of bitangent conics* or just the *bitangent pencil* determined by p and $q_1 + q_2$. The reason for the name will become clear in a short while.

Clearly, the base points of $\mathcal{P}$ are q_1, q_2. Its degenerate conics are the pair of lines $pq_1 + pq_2$ ($\lambda = 0$) and the double line 2ℓ ($\mu = 0$); by 9.5.10 and 9.5.8 (or by just a direct computation), their multiplicities in the characteristic divisor are one and two, respectively. As the reader may easily check, a suitable choice of a new unit point allows us to take $b = \pm 1$ in the equation (9.12) above, the minus sign being only for the case $k = \mathbb{R}$ and q_1, q_2 real.

Obviously, any bitangent pencil $\mathcal{P}$ may be generated by its double line and any other conic in $\mathcal{P}$. Conversely, as the reader may easily check, in a projective plane, a double line 2ℓ and a conic C not tangent to ℓ span a bitangent pencil: it is the bitangent pencil determined by p and $C \cap L$, where p is either the double point of C if C is degenerate, or the pole of ℓ relative to C otherwise.

Assume $k = \mathbb{C}$ or $k = \mathbb{R}$ and q_1, q_2 real, after which $t_1 = pq_1$ and $t_2 = pq_2$ are lines of $\mathbb{P}_2$. Then, by 6.4.2, a conic C belongs to $\mathcal{P}$ if and only if C is tangent to t_1 and t_2 at q_1 and q_2. In case $k = \mathbb{R}$ and q_1, q_2 imaginary, using the above in $\mathbb{C}\mathbb{P}_2$ and 9.5.11, $C \in \mathcal{P}$ if and only if $C_{\mathbb{C}}$ is tangent to t_1 and t_2 at q_1 and q_2.

Two different conics of $\mathbb{P}_2$ are said to be *bitangent at* q_1, q_2 if and only if they span a bitangent pencil with base points q_1, q_2. We have seen above that two different conics are bitangent at q_1, q_2 if and only if they (or their complex extensions, if q_1, q_2 are imaginary) are tangent at q_1 and q_2.

Example 9.5.20 (Quadrics through a skew quadrilateral). A *skew quadrilateral* of $\mathbb{P}_3$ consists of four independent points $p_0, p_1, p_2, p_3 \in \mathbb{P}_3$ (its *vertices*) together with four lines cyclically joining the vertices (its *sides*). Up to renumbering the vertices, the sides will be taken to be p_0p_1, p_1p_2, p_2p_3, p_3p_0. Take a reference with vertices the vertices p_0, p_1, p_2, p_3 of a skew quadrilateral K. It is direct to check that a quadric

$$Q : \sum_{i,j=0}^{3} a_{i,j} x_i x_j = 0$$

contains the four sides of K if and only if

$$a_{0,0} = a_{1,1} = a_{2,2} = a_{3,3} = a_{0,1} = a_{1,2} = a_{2,3} = a_{0,3} = 0.$$

It follows that the quadrics containing the four sides of K are those with equations

$$\lambda x_0 x_2 + \mu x_1 x_3 = 0,$$

$(\lambda, \mu) \in k^2 - \{0, 0\}$, and hence describe a pencil. The reader may easily check that its base points are those of the sides of K and its degenerate quadrics the plane-pairs $x_0x_2 = 0$ and $x_1x_3 = 0$, each having multiplicity two in the characteristic divisor.

Example 9.5.21 (Quadrics through two conics). Assume to have fixed two conics C_1 and C_2 of two different planes π_1 and π_2, respectively, of $\mathbb{P}_3$. Put $\ell = \pi_1 \cap \pi_2$. Assume furthermore that C_2 is non-degenerate and $C_1 \cap \ell = C_2 \cap \ell$, which is a pair of points of ℓ because C_2 is non-degenerate. Choose three distinct points, p_1, p_2, p_3, of C not on ℓ. Since p_1, p_2, p_3 are non-aligned points of π_2 and no one lies on ℓ, 9.4.4 applies and C_2 is the only conic of π_2 that goes through p_1, p_2, p_3 and contains $C_1 \cap \ell$. If a quadric Q contains C_1 and p_1, p_2, p_3, then either it contains π_2 or it intersects π_2 in a conic that contains both p_1, p_2, p_3 and $C_1 \cap \ell$, and therefore is C_2. So in both cases Q contains C_1 and C_2. The converse being obvious, the quadrics of $\mathbb{P}_3$ containing C_1 and C_2 are exactly those containing C_1 and p_1, p_2, p_3. Since the latter describe a pencil (again by 9.4.4), we see that the quadrics that contain C_1 and C_2 are the elements of a pencil $\mathcal{P}$ of quadrics of $\mathbb{P}_3$.

Example 9.5.22 (Quadrics tangent along a common hyperplane section). Assume given in a projective space $\mathbb{P}_n$ a point p, a hyperplane H missing p and a non-degenerate quadric $\bar{Q}$ of H. Let K be the cone over $\bar{Q}$ with vertex p and $\mathcal{P}$ the pencil spanned by $2H$ and K. Obviously the base points of $\mathcal{P}$ are the points of $\bar{Q}$. By 9.5.10 and 9.5.8, $2H$ and K, have, respectively, multiplicities $n-1$ and 1 in the characteristic divisor of $\mathcal{P}$, and so they are the only degenerate quadrics of $\mathcal{P}$.

We will show next that $\mathcal{P}$ is the set of all quadrics Q of $\mathbb{P}_n$ that contain $\bar{Q}$ and for which $H_{p,Q} \supset H$. These conditions being satisfied by both $2H$ and K, they are satisfied by all quadrics of $\mathcal{P}$ by 9.5.1. For the converse, assume that Q satisfies the above conditions. Assume first that $Q \supset H$: then any $q \in H$ is conjugate to all points of H and, by the hypothesis, also to p; this shows that all points of H are double points of Q and therefore that $Q = 2H$, which obviously belongs to $\mathcal{P}$.

Assume thus $Q \not\supset H$ and so $Q \cap H = \bar{Q}$. Since $\bar{Q}$ is non-degenerate, Q is not tangent to H and in particular has no double point in H. If $p \in Q$, then p is a double point of Q, because it is conjugate to itself and to all points in H: it follows that $Q = K$ (by 6.7.2) and so $Q \in \mathcal{P}$.

We are now reduced to the case in which $p \notin Q$. Then $H_{p,Q} = H$ and so, in particular, H contains all double points of Q (6.7.10). Since we have already noted that H contains no double point of Q, Q is non-degenerate. By 6.7.5, K is the tangent cone to Q with vertex p. The equation of $TC_p(Q)$ obtained in Section 6.3 (Equation (6.7)) directly shows that if $TC_p(Q)$ is defined, then Q belongs to the pencil spanned by $2H_{p,Q}$ and $TC_p(Q)$, that is, in our case, that $Q \in \mathcal{P}$.

Actually, the arguments above have proved a bit more, namely that $\mathcal{P}$ is composed of $2H$, K and all non-degenerate quadrics Q with $Q \cap H = \bar{Q}$ and $H_{p,Q} = H$. Equivalently, $\mathcal{P}$ is composed of $2H$, K and all non-degenerate quadrics with $\bar{Q}$ as hyperplane section and K as tangent cone along $\bar{Q}$. Obviously

$\mathcal{P}$ may be spanned by any of its non-degenerate members Q and either $2H$ or K, the tangent cone to Q along $Q \cap H$. The reader may have noticed that Example 9.5.19 is a particular case of the present one.

9.6 Pencils of conics

In this section we will describe all the different types of regular pencils of conics. For the non-regular ones, see Exercise 9.23.

Let $\mathcal{P}$ be a regular pencil of conics and let C be a non-degenerate conic of $\mathcal{P}$. By 9.5.5, there is a degenerate conic $D = \ell_1 + \ell_2 \in \mathcal{P}$. The conics C and D are different and therefore span $\mathcal{P}$. If either $k = \mathbb{C}$ or $k = \mathbb{R}$ and the lines ℓ_1, ℓ_2 are real, we write $C \cap \ell_1 = p_1 + p_1'$ and $C \cap \ell_2 = p_2 + p_2'$. Otherwise, the lines ℓ_1, ℓ_2 are imaginary and we put $C_{\mathbb{C}} \cap \ell_1 = p_1 + p_1'$ and $C_{\mathbb{C}} \cap \ell_2 = p_2 + p_2'$. In any case, the base points of $\mathcal{P}$ are p_1, p_1', p_2, p_2'; no three of them may be different and aligned, because C is non-degenerate; in case $k = \mathbb{R}$, some of them may be imaginary, the whole set $\{p_1, p_1', p_2, p_2'\}$ being invariant by complex conjugation (9.5.2). In the sequel, when needed, we will swap over the points in any of the pairs $\{p_1, p_1'\}$ and $\{p_2, p_2'\}$ and also the pairs themselves, in order to fix the notations.

We will distinguish between the different types of pencils by the coincidences between their base points. The reader may easily give examples of each of the types described below by taking suitable conics as C and D. Some details of the case $k = \mathbb{R}$ will also be left to the reader.

9.6.1. General pencils. *The points p_1, p_1', p_2, p_2' are all different.* If either $k = \mathbb{C}$, or $k = \mathbb{R}$ and the four base points are real, we have seen in Example 9.5.13 that the conics through p_1, p_1', p_2, p_2' describe a pencil which contains $\mathcal{P}$ and therefore, by 1.3.6 (b), equals it. The pencil $\mathcal{P}$ is thus a pencil of conics through four points, no three aligned; we have already described these pencils in Example 9.5.13. Proposition 9.5.16 as well as Corollary 9.5.16 apply to them.

If $k = \mathbb{R}$ and some of the base points are imaginary, still $\mathcal{P}$ is the pencil of conics through p_1, p_1', p_2, p_2'. Indeed, if a conic C' of $\mathbb{P}_2$ goes through p_1, p_1', p_2, p_2', then so does its complex extension $C'_{\mathbb{C}}$ which, therefore, as seen above, belongs to the pencil spanned by $C_{\mathbb{C}}$ and $D_{\mathbb{C}}$. Then, by 9.5.22, C' belongs to the pencil spanned by C and D, that is, to $\mathcal{P}$, as wanted.

Still in the real case, there may be one or two pairs of complex conjugate imaginary points among p_1, p_1', p_2, p_2'. In the first case it is easy to check that the only degenerate conic of $\mathcal{P}$ is $r_1 + r_2$, r_1 the line spanned by the real base points and r_2 the line spanned by the imaginary ones. Obviously, both r_1 and r_2 are real.

Assume now that the four base points are imaginary: then they are the vertices of a quadrivertex K (of $\mathbb{C}\mathbb{P}_2$) all of whose pairs of opposite sides are real conics. For, joining pairs of conjugate vertices gives a pair of real lines, while the other two choices give pairs of complex-conjugate imaginary lines. Thus, in this case, the degenerate members of $\mathcal{P}$ are three pairs of distinct lines, from which a single one is of real lines. Their double points, which are the vertices of the diagonal triangle of K, are all real and are the vertices a self-polar triangle, as the arguments used in the proof of 9.5.16 apply to the present case without changes. Proceeding as in the proof of 9.5.17, it easily turns out that the conics of $\mathcal{P}$ have equations

$$(\lambda + \mu)x_0^2 + \lambda x_1^2 + \mu x_2^2 = 0, \quad (\lambda, \mu) \in k^2 - \{(0,0)\}$$

relative to a suitable reference whose vertices are those of the diagonal triangle of K.

9.6.2. Simple contact pencils. *There is a single pair of coincident points among* p_1, p_1', p_2, p_2'. Rename p the coincident points. They may or may not belong to the same line ℓ_i. So, after reordering, we have either $p_1 = p_1' = p$ (ℓ_1 is tangent to C or $C_{\mathbb{C}}$ at p and ℓ_2 is a chord that does not contain p) or $p_1 = p_2 = p$ ($p = \ell_1 \cap \ell_2$ and both ℓ_1 and ℓ_2 are chords).

Assume that $k = \mathbb{C}$ or $k = \mathbb{R}$ and all three base points are real. Name t the tangent line to C at p. No matter if p is the double point of D or $t = \ell_1$, D is tangent to t at p and therefore (by 9.3.14 and 9.5.1) all conics of $\mathcal{P}$ go through p and the two remaining base points, and are tangent to t at p. Clearly p is the only base point on t, as it is the only point of C on t, so we have seen in Example 9.5.18 that the conics through the three base points and tangent to t at p describe a pencil which, again by 1.3.6 (b), equals $\mathcal{P}$. In this case the pencil $\mathcal{P}$ is thus as those described in Example 9.5.18.

If $k = \mathbb{R}$ and there is some imaginary base point, then two base points compose a pair of imaginary and complex-conjugate points and the remaining base point needs to be real. We will see first that the real base point is p, and therefore the tangent t to C at p is real too. Indeed, p is clearly real if it is the double point of D. Otherwise all base points are simple points of D, D has a real simple point and therefore is a pair of real lines. Since one of these lines is tangent to C at p, p is a real point (by 6.7.16). Now, the same argument used in Example 9.6.1 shows that also in this case the conics through the base points and tangent to t at p describe a pencil which equals $\mathcal{P}$. The pencil $\mathcal{P}$ is then the set of all conics going through the real point p with tangent t and going also through a pair of complex-conjugate imaginary points lying not on t. The degenerate conics are as in the case of real base points, but for being imaginary the lines composing the line-pair with double point p.

9.6.3. Bitangent pencils. *There are two different pairs of coincident points among* p_1, p_1', p_2, p_2'. Rename the points so that $\{p_1, p_1', p_2, p_2'\} = \{q_1, q_2\}$. Then, up to renumbering q_1, q_2 or swapping over p_2, p_2', it is either $p_1 = p_2 = q_1$ and $p_1' = p_2' = q_2$, or $p_1 = p_1' = q_1$ and $p_2 = p_2' = q_2$. In the first case $\ell_1 = \ell_2 = q_1q_2$ and $D = 2q_1q_2$, after which $\mathcal{P}$ is a bitangent pencil, as described in 9.5.19.

In the second case assume $k = \mathbb{C}$ or $k = \mathbb{R}$ and q_1, q_2 real. Then ℓ_1, ℓ_2 are the tangents to C at q_1, q_2. It directly follows that $\mathcal{P}$ is the bitangent pencil determined by $q_1 + q_2$ and the double point of D. The same holds if $k = \mathbb{R}$ and q_1, q_2 are imaginary, arguing with the complex extensions of C and D as in the preceding cases.

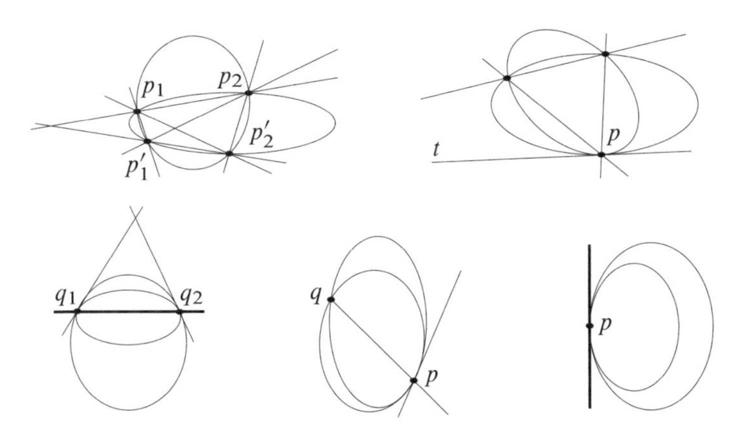

Figure 9.3. Pencils of conics: from left to right and from top to bottom, general, simple contact, bitangent, double contact and triple contact. Some of the base points pictured may be imaginary.

9.6.4. Double contact pencils. *Three points among p_1, p_1', p_2, p_2' are coincident, but not the fourth one.* After reordering we assume $p_1 = p_1' = p_2 \neq p_2'$. We write $p_1 = p$ and $p_2' = q$, after which $p \neq q$, $C \cap \ell_1 = 2p$ and $C \cap \ell_2 = p + q$. The double point of D is p. In particular, in case $k = \mathbb{R}$, the point p is real and therefore so is q. The line ℓ_1 is the tangent to C at p and ℓ_2 is the chord pq. By 9.5.1, all conics of $\mathcal{P}$ go through p and q and are tangent to ℓ_1 at p. Nevertheless, these conditions do not determine a pencil, but at most a net. Furthermore, there are no base points other than p and q, and the conics of $\mathcal{P}$ do not have a common tangent at q (because C and D have not): it follows that in the present case the pencil $\mathcal{P}$ cannot be defined by conditions of going through given points with given tangents at some of them.

Name t the tangent to C at q. A conic $C' \in \mathcal{P}$, $C' \neq C$, cannot intersect C at a point other than p or q, as then such a point would be a base point. Neither can C' be tangent to t at q as then t would be tangent to all conics of $\mathcal{P}$ and it is not tangent to D. After this the reader may easily check that D is the only degenerate conic of $\mathcal{P}$.

9.6.5. Triple contact pencils. *All four points p_1, p_1', p_2, p_2' are coincident.* Write $p_1 = p_1' = p_2 = p_2' = p$, which is of course real if $k = \mathbb{R}$. Then both ℓ_1 and ℓ_2

are tangent to C at p and so they agree. Write $\ell_1 = \ell_2 = t$, the tangent to C at p, after which $D = 2t$. All conics of $\mathcal{P}$ go through p and are tangent to t at p, but, clearly, these conditions are not enough to determine the pencil. Since p is the only base point of $\mathcal{P}$, as in the former case, $\mathcal{P}$ cannot be defined by conditions imposed by points and tangents. Furthermore the only degenerate conic of $\mathcal{P}$ is D, as any pair of lines through p other than $2t$ has at least one further intersection with C.

Since we have covered all possible coincidences between the points of the pairs $p_1 + p_1'$ and $p_2 + p_2'$, any regular pencil belongs to one of the five types listed above. Calling *general* the pencils with four base points, no three aligned, is classical: the reason for such a name is that the pencil spanned by two different conics belongs to this type unless the conics have some special relative position (they should be at least tangent at a point) and therefore the other pencils are scarcer (see Exercise 9.9 for a more precise claim). In classical books ([25], for instance) the pencils other than the general ones are often described as limits of general pencils when some of the base points become "infinitely near" to other base points. This viewpoint leads to nice and very suggestive descriptions, but making it rigorous is far beyond the scope of this book. Heuristically, one may think of a simple contact pencil as the result of moving one base point q of a general pencil $\mathcal{P}$ to another base point p along a conic of the pencil. At the limit, the conics of the resulting pencil $\mathcal{P}'$ have as common tangent at p the limit of the varying chord pq and $\mathcal{P}'$ is a simple contact pencil. The "limit position" $\bar{q}$ of q is somewhat understood as a first-order infinitely near point to p, shared by all the conics of $\mathcal{P}'$. If now another base point q' moves to p along a conic of $\mathcal{P}'$, one gets a double contact pencil $\mathcal{P}''$. The limit position $\bar{q}'$ of q' is understood as a second-order infinitely near point to p, which is shared, together with p and $\bar{q}$, by all conics of $\mathcal{P}''$. The term *second-order* refers to the fact that $\bar{q}'$ does not come immediately after p, but after the first-order infinitely near point $\bar{q}$. Sharing $\bar{q}'$ means that the conics of the pencil have a higher contact, not just being tangent, at p. Similarly, moving to p the remaining base point q'' of $\mathcal{P}''$, gives rise to a triple contact pencil: its conics share a further infinitely near point, of the third order, which is the limit position of q'', thus having a higher contact at p. The *intersection multiplicity*, introduced in algebraic geometry, is an integer, associated to each pair of curves at a common point, which precisely measures what we have intuitively called contact. For precise definitions of intersection multiplicity and infinitely near points, the reader is referred to [3].

Corollary 9.6.6 (of the preceding description). *If two different conics of a real projective plane share no real point, then the pencil they span is either a general or a bitangent pencil, in both cases with no real base point.*

Proof. After the hypothesis, all base points of the spanned pencil are imaginary, and so they compose either a single or two different pairs of imaginary complex-conjugate points. According to the preceding description, in the first case the pencil is a bitangent pencil, while in the second case it is a general pencil. □

We close this section by proving, by elementary means, that the projective types of regular pencils of complex conics are the five types of pencils described above. The projective classification of regular pencils of quadrics of a complex projective space of arbitrary dimension demands a stronger algebraic background. It is the subject of the forthcoming Section 11.6 and of course covers the plane case we will deal with next, see Example 11.6.17. First we need:

Lemma 9.6.7. *If C and C' are non-degenerate conics of projective planes $\mathbb{P}_2$ and $\mathbb{P}'_2$, q_0, q_1, q_2 are three distinct points of C and q'_0, q'_1, q'_2 three distinct points of C', then there is a projectivity $f : \mathbb{P}_2 \to \mathbb{P}'_2$ such that $f(C) = C'$ and $f(q_i) = q'_i$ for $i = 0, 1, 2$.*

Proof. According to 8.1.7 q_0, the pole of q_0q_1 relative to C, q_1 and q_2 compose a reference Δ of $\mathbb{P}_2$ relative to which C has equation

$$x_1^2 - x_0x_2 = 0.$$

Similarly, q'_0, the pole of $q'_0q'_1$ relative to C', q'_1 and q'_2 compose a reference Δ' of $\mathbb{P}'_2$ relative to which the equation of C' is the same as the equation of C above. The projectivity mapping Δ to Δ' (2.8.1) obviously maps q_i to q'_i, $i = 0, 1, 2$, and also C to C' because their equations are equal. □

Theorem 9.6.8 (Projective classification of regular pencils of complex conics). *Assume that $\mathcal{P}$ and $\mathcal{P}'$ are regular pencils of conics of complex projective planes $\mathbb{P}_2$ and $\mathbb{P}'_2$, respectively. Then the following conditions are equivalent:*

(i) *There is a projectivity $f : \mathbb{P}_2 \to \mathbb{P}'_2$ for which $f(\mathcal{P}) = \mathcal{P}'$.*

(ii) *$\mathcal{P}$ and $\mathcal{P}'$ belong to the same of the five types of pencils of conics described in* 9.6.1 *to* 9.6.5 *above.*

(iii) *$\mathcal{P}$ and $\mathcal{P}'$ contain the same number of double lines and the same number of pairs of distinct lines.*

Proof. That (i) ⇒ (iii) is clear, while (iii) ⇒ (ii) follows from the above descriptions, as they show that pencils of different type have either different numbers of double lines or different numbers of pairs of distinct lines. Next we will prove (ii) ⇒ (i) type by type, by showing that there is a projectivity mapping two different conics of $\mathcal{P}$ to conics of $\mathcal{P}'$.

If both $\mathcal{P}$ and $\mathcal{P}'$ are general pencils, we take f to be the projectivity mapping the base points p_0, p_1, p_2, p_3 of $\mathcal{P}$ to the base points of p'_0, p'_1, p'_2, p'_3 of $\mathcal{P}'$, in the given order (2.8.1). Then f maps $p_0p_1 + p_2p_3$ to $p'_0p'_1 + p'_2p'_3$ and $p_0p_2 + p_1p_3$ to $p'_0p'_2 + p'_1p'_3$.

If $\mathcal{P}$ and $\mathcal{P}'$ are simple contact pencils, assume that the conics of $\mathcal{P}$ go through the points q_0, q_1, q_2 all with the same tangent at q_0 and, similarly, that the conics of

$\mathcal{P}'$ go through points q_0', q_1', q_2', with common tangent at q_0'. If $C \in \mathcal{P}$ and $C' \in \mathcal{P}'$ are non-degenerate, by 9.6.7 there is a projectivity f such that $f(C) = C'$ and, furthermore. $f(q_i) = q_i'$ for $i = 0, 1, 2$, which assures that $p_0p_1 + p_0p_2 \in \mathcal{P}$ is mapped to $p_0'p_1' + p_0'p_2' \in \mathcal{P}'$.

The case of two bitangent pencils is dealt with similarly: again by 9.6.7 there is a projectivity mapping a non-degenerate $C \in \mathcal{P}$ to a non-degenerate $C' \in \mathcal{P}'$ and the base points q_0, q_1 of $\mathcal{P}$ to the base points q_0', q_1' of $\mathcal{P}'$. Then the double line $2q_0q_1 \in \mathcal{P}$ is mapped to $2q_0'q_1' \in \mathcal{P}'$

Assume that $\mathcal{P}$ and $\mathcal{P}'$ are double contact pencils, with, respectively, base points q_0, q_1 and q_0', q_1', and common tangents t at q_0 and t' at q_0'. Take f mapping a non-degenerate $C \in \mathcal{P}$ to $C' \in \mathcal{P}$, q_0 to q_0' and q_1 to q_1': then $f(t) = t'$ and so the image of $q_0q_1 + t \in \mathcal{P}$ is $q_0'q_1' + t' \in \mathcal{P}'$.

To close, in the case of $\mathcal{P}$ and $\mathcal{P}'$ being triple contact pencils with base points q_0 and q_0', and common tangents t and t', any projectivity mapping a non-degenerate $C \in \mathcal{P}$ to $C' \in \mathcal{P}$ and q_0 to q_0' maps t to t' and so $2t \in \mathcal{P}$ to $2t' \in \mathcal{P}'$. □

Remark 9.6.9. It follows from 9.6.8 that the projective type of a pencil of complex conics, not all degenerate, may be determined by just examining its degenerate conics.

9.7 Desargues' theorem on pencils of quadrics

We devote this section to presenting and proving the theorem of Desargues on pencils of quadrics, which is among the oldest theorems of projective geometry: it appeared in Desargues' foundational *Brouillon project* of 1639, see the Introduction and [9]). Desargues' theorem shows an important relationship between pencils of quadrics and involutions of projective lines.

First we pay some attention to the pencils of quadrics of $\mathbb{P}_1$. If such a pencil has a base point p, then its pairs of points are composed of p and a variable point. If, otherwise, the pencil has no base point, any two distinct pairs in the pencil are disjoint (by 9.5.1). We have already dealt with pencils of pairs of points with no base point when presenting the parametric form of the involutions of $\mathbb{P}_1$ in 5.6.14 and 5.6.15. We have in part seen:

Proposition 9.7.1. *The pairs of points (either real or imaginary and complex-conjugate, if $k = \mathbb{R}$) corresponding by an involution τ of $\mathbb{P}_1$ describe a pencil of quadrics of $\mathbb{P}_1$ with no base point. Conversely, the pairs of points of any pencil $\mathcal{P}$ of quadrics of $\mathbb{P}_1$ with no base point are the pairs of points (real or imaginary and complex-conjugate, if $k = \mathbb{R}$) corresponding by an involution of $\mathbb{P}_1$.*

Proof. The first claim has been proved in 5.6.14 and 5.6.15. For the converse, if $k = \mathbb{C}$, just take $p + p'$ and $q + q'$ to be two different pairs in $\mathcal{P}$. If $k = \mathbb{R}$, in order to assure that all four points are real, choose first any real point p and take

$p + p'$ the only point-pair of $\mathcal{P}$ containing p: it obviously is a pair of real points. Then choose any real point $q, q \neq p, q \neq p'$ and let $q + q'$ be the only point-pair of $\mathcal{P}$ containing q: it is also a pair of real points and, clearly, $q + q' \neq p + p'$.

In both the real and complex cases, since $\mathcal{P}$ has no base point, $\{p, p'\} \cap \{q, q'\} = \emptyset$. Therefore there is an involution τ of $\mathbb{P}_1$ for which $\tau(p) = p'$ and $\tau(q) = q'$ (5.6.10); by 5.6.14 the pairs of points corresponding by τ describe the pencil spanned by $p + p'$ and $q + q'$, which is $\mathcal{P}$. □

Remark 9.7.2. The pencil of quadrics and the involution of 9.7.1 determine each other: given the involution τ, its corresponding pencil is spanned by any two different pairs $p + \tau(p), q + \tau(q)$. Conversely, once the pencil $\mathcal{P}$ is given, for each $p \in \mathbb{P}_1$ there is a unique pair of points $Q \in \mathcal{P}$ containing p: if $Q = p + p'$, then the image of p by the involution corresponding to $\mathcal{P}$ is p'.

Remark 9.7.3. It is worth listing the many different ways an involution of $\mathbb{P}_1$ may be thought of. We have seen it to be:

(1) a map $\tau\colon \mathbb{P}_1 \to \mathbb{P}_1$ satisfying $\tau^2 = \mathrm{Id}_{\mathbb{P}_1}, \tau \neq \mathrm{Id}_{\mathbb{P}_1}$ (the definition),

(2) taking the fourth harmonic with respect to two fixed points (5.6.8),

(3) the conjugation relative to a non-degenerate quadric of $\mathbb{P}_1$ (6.5.4) and

(4) mapping to each other the points composing each of the pairs of a pencil of quadrics of $\mathbb{P}_1$ with no base point (9.7.1).

Theorem 9.7.4 (Desargues' on pencils of quadrics). *Assume that $\mathcal{P}$ is a pencil of quadrics of $\mathbb{P}_n$ and $\ell \subset \mathbb{P}_n$ a line containing no (real or imaginary, in case $k = \mathbb{R}$) base point of $\mathcal{P}$. Then the quadrics of $\mathcal{P}$ intersect on ℓ the pairs of points (either real or imaginary and complex-conjugate, in case $k = \mathbb{R}$) corresponding by an involution of ℓ.*

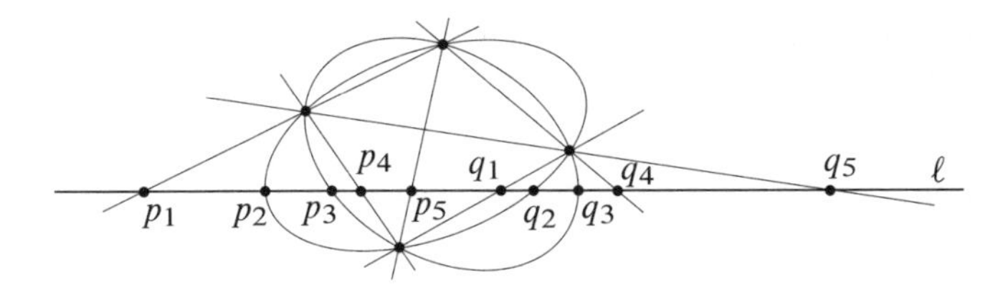

Figure 9.4. Desargues' theorem for a pencil of conics: all pairs $\{p_i, q_i\}$ belong to the same involution of ℓ.

Proof. First note that no quadric of the pencil contains ℓ. For, if $Q \supset \ell$ and $Q \in \mathcal{P}$, take $Q' \in \mathcal{P}$, $Q' \neq Q$ and p a (maybe imaginary) point of $Q' \cap \ell$. Then the point p belongs to both Q and Q' and therefore is a base point which belongs to ℓ, against the hypothesis.

Now, by 9.5.4, the sections $Q \cap \ell$, $Q \in \mathcal{P}$, describe a pencil $\mathcal{P}'$ of pairs of points of ℓ. Since a base point of $\mathcal{P}'$ would obviously be a base point of $\mathcal{P}$ on ℓ, $\mathcal{P}'$ has no base point. After this, the claim follows from 9.7.1. □

Corollary 9.7.5. *If $\mathcal{P}$ is a pencil of quadrics of $\mathbb{P}_n$ and $\ell \subset \mathbb{P}_n$ a line containing no (real or imaginary if $k = \mathbb{R}$) base point of $\mathcal{P}$, then there are exactly two quadrics of $\mathcal{P}$ tangent to ℓ or, only in case $k = \mathbb{R}$, none.*

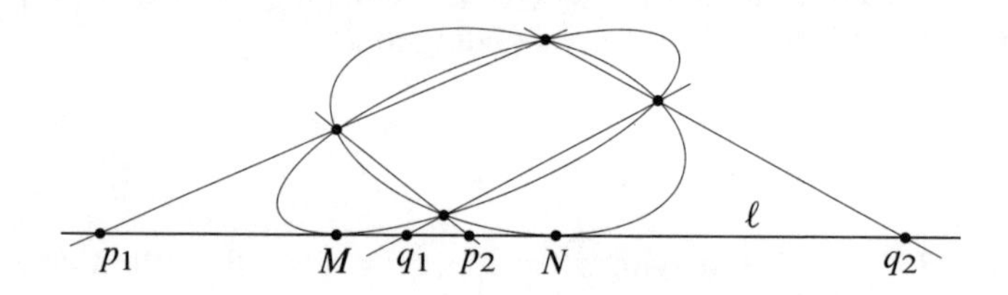

Figure 9.5. Desargues' theorem for a pencil of conics: the contact points M, N are the fixed points of the involution of ℓ determined by the pairs $\{p_1, q_1\}$ and $\{p_2, q_2\}$.

Proof. Since the line ℓ contains no base points, for any $p \in \ell$ there is one and only one quadric of $\mathcal{P}$ going through p. According to 9.7.4, such a quadric is tangent at p if and only if p is a fixed point of the involution whose pairs are cut on ℓ by the quadrics of $\mathcal{P}$. Any involution being elliptic or hyperbolic (5.6.6), the claim follows. □

Remark 9.7.6. Corollary 9.7.5 confirms that the condition of being tangent to a fixed line is not linear. It also shows that the pencils of quadrics with no base point on ℓ are chords of the quadric $\mathcal{Q}_\ell$ of 9.3.12. Actually, there is no other chord, because a pencil of quadrics with a base point on ℓ either is contained in $\mathcal{Q}_\ell$ or shares with $\mathcal{Q}_\ell$ a single quadric, by 9.3.14.

9.8 Spaces of quadric envelopes, ranges

Throughout this section we assume $n > 1$. Together with the space of quadrics of $\mathbb{P}_n$, $\mathbb{Q}(\mathbb{P}_n)$, we have the space of quadrics of $\mathbb{P}_n^\vee$, $\mathbb{Q}(\mathbb{P}_n^\vee)$, called the *space of quadric envelopes* of $\mathbb{P}_n$. The m-dimensional linear varieties of $\mathbb{Q}(\mathbb{P}_n^\vee)$ are called *m-dimensional linear systems* (or *linear families*) *of quadric envelopes*. Denote by $\mathbb{D}(\mathbb{P}_n)$ and $\mathbb{D}(\mathbb{P}_n^\vee)$ the sets of degenerate quadrics and degenerate quadric envelopes of $\mathbb{P}_n$, respectively. The main relationship between the spaces of quadrics and quadric envelopes is the bijection

$$\begin{aligned} \Upsilon\colon \mathbb{Q}(\mathbb{P}_n) - \mathbb{D}(\mathbb{P}_n) &\longrightarrow \mathbb{Q}(\mathbb{P}_n^\vee) - \mathbb{D}(\mathbb{P}_n^\vee), \\ Q &\longmapsto Q^*, \end{aligned}$$

that maps each non-degenerate quadric to its envelope (6.6.1). The first point is that Υ cannot be extended to a projectivity:

Lemma 9.8.1. *There is no projectivity from $\mathbb{Q}(\mathbb{P}_n)$ to $\mathbb{Q}(\mathbb{P}_n^\vee)$ mapping each non-degenerate quadric of $\mathbb{P}_n$ to its envelope.*

Proof. Fix any reference of $\mathbb{P}_n$. The non-degenerate quadrics

$$\begin{aligned} x_0^2 + x_1^2 + x_2^2 + \sum_{i=3}^n x_i^2 &= 0, \\ 3x_0^2 + 2x_1^2 + x_2^2 + \sum_{i=3}^n x_i^2 &= 0, \\ 4x_0^2 + 3x_1^2 + 2x_2^2 + 2\sum_{i=3}^n x_i^2 &= 0 \end{aligned}$$

obviously belong to the same pencil. A straightforward computation shows that their envelopes are linearly independent, after which the claim follows from 1.7.5. ☐

The bijection Υ, often seen as a correspondence between $\mathbb{Q}(\mathbb{P}_n)$ and $\mathbb{Q}(\mathbb{P}_n^\vee)$, is easily described using coordinates. Indeed, fix any reference of $\mathbb{P}_n$ and assume that a quadric Q has matrix $A = (a_{i,j})$ relative to it. As seen in Section 6.6, $(\det A)A^{-1}$ may be taken as a matrix of Q^* and therefore the entries $b_{s,t}$ of a matrix $B = (b_{s,t})$ of Q^* may be taken to be

$$b_{i,j} = A_{i,j}, \quad 0 \le i \le j \le n, \tag{9.13}$$

where $A_{i,j}$ is the cofactor of $a_{i,j}$ in A. Coordinates of Q^* appear thus as homogeneous polynomial functions of degree n of the coordinates of Q. This is an example of a rational map between projective spaces. We have already dealt with rational maps $\mathbb{P}_1 \to \mathbb{P}_n$ in Section 8.2. A *rational map of degree d* from a projective space $\overline{\mathbb{P}}_m$ to a second one $\mathbb{P}_n$ is defined similarly: it maps the point $[y_0, \dots, y_n] \in \overline{\mathbb{P}}_m$ to the point $[x_0, \dots, x_n] \in \mathbb{P}_n$ given by equalities (the *equations* of the rational map)

$$x_j = P_j(y_0, \dots, y_n), \quad j = 0, \dots, n,$$

where the P_j are homogeneous polynomials, all of the degree d, sharing no factor. Still the arbitrary factor involved in the coordinates $y_0, \dots, y_m$ is irrelevant and the definition is independent of the choice of coordinates in both spaces. Nevertheless, unlike the case $m = 1$ and in spite of its name, a rational map such as the above is in general not defined on the whole of $\overline{\mathbb{P}}_m$ but only on the subset

$$\overline{\mathbb{P}}_m - \{[y_0, \dots, y_m] \mid P_j(y_0, \dots, y_m) = 0,\ j = 0, \dots, n\}.$$

The cofactors of a square matrix whose entries are free variables, are polynomials with no common factor, as each entry fails to appear in one of the cofactors. Therefore the equations (9.13) define a rational map of degree n,

$$\mathbb{Q}(\mathbb{P}_n) \longrightarrow \mathbb{Q}(\mathbb{P}_n^\vee),$$

$$Q \sum_{i,j=0}^{n} a_{i,j} x_1 x_j = 0 \longmapsto Q' : \sum_{i,j=0}^{n} A_{i,j} u_1 u_j = 0,$$

defined on the set of quadrics of rank at least n, which maps each non-degenerate quadric Q to its envelope Q^* and therefore extends the bijection Υ.

Arguing by duality using 6.6.4, coordinates $a_{i,j}$ of any non-degenerate quadric Q are computed from those of its envelope Q^* by equations similar to (9.13), namely:

$$a_{i,j} = B_{i,j}, \quad 0 \le i \le j \le n, \tag{9.14}$$

where $B_{i,j}$ is the cofactor of $b_{i,j}$ in the matrix B of Q^*. Hence also Υ^{-1} extends to a rational map of degree n, this time the one given by the equations (9.14), from $\mathbb{Q}(\mathbb{P}_n^\vee)$ to $\mathbb{Q}(\mathbb{P}_n)$.

The above explains some apparent violations of the principle of duality that appear when linear conditions are imposed on quadric envelopes. For an easy example take $n = 2$ and consider the conditions on conics *going through a point* p and *being tangent to a line* ℓ: the second condition appears to be the dual of the first one, although we have seen that *going through a point* is linear, while *being tangent to a line* is not (9.3.8 and 9.3.12). Example 9.3.8 may, indeed, be read in $\mathbb{P}_n^\vee$ and then it says that *containing* ℓ is a linear condition on conic envelopes. To have a closer look, take a reference of $\mathbb{P}_2$ such that ℓ has equation $x_2 = 0$, and hence coordinate vector $(0, 0, 1)$. Then the conic envelope $\sum_{i,j=0}^{2} b_{i,j} u_i u_j = 0$ contains ℓ if and only if $b_{2,2} = 0$, a linear equation in the coordinates of the conic envelope, which confirms that the condition is linear on conic envelopes. Nevertheless, this equation is turned into a non-linear one if point-conic coordinates are introduced through the equations (9.13): the last of these equations being $b_{2,2} = a_{0,0}a_{1,1} - a_{0,1}^2$, from $b_{2,2} = 0$ it results in $a_{0,0}a_{1,1} - a_{0,1}^2 = 0$, just the equation of the quadric $\mathcal{Q}_\ell$ already seen in Example 9.3.12.

The point is that neither Υ nor Υ^{-1} preserve linearity, and so, while a non-degenerate point-quadric (resp. quadric envelope) varies in a linear system, its envelope (resp. enveloped quadric) need not stay in a linear system of the same dimension. As a consequence, regarding its linearity, the conditions need to be clearly specified if they are imposed on point-quadrics or on quadric envelopes. Then duality leads to no contradiction, as it just says that the dual of a linear condition on point-quadrics is a linear condition on quadric envelopes, and conversely.

After the above, one cannot expect the point-quadrics enveloped by the non-degenerate members of a regular pencil of quadric envelopes, to belong in turn to a pencil of point-quadrics, even if in some cases they do (see Proposition 9.8.4 below). The family of all non-degenerate point-quadrics whose envelopes belong to a fixed regular pencil of quadric envelopes is called a *range of quadrics*, or just a *range*. In other words, a range of quadrics of $\mathbb{P}_n$ is any set of the form

$$\mathcal{R} = \{Q \in \mathbb{Q}(\mathbb{P}_n) \mid Q \text{ is non-degenerate and } Q^* \in \mathcal{P}\}$$

where $\mathcal{P} \subset \mathbb{Q}(\mathbb{P}_n^*)$ is a regular pencil of conic envelopes. In the sequel we will refer to $\mathcal{P}$ as the *pencil associated to* the range $\mathcal{R}$. By 9.5.5, any range of quadrics $\mathcal{R}$ is an infinite set and, clearly, its associated pencil is spanned by the envelopes of any two different members of $\mathcal{R}$. In particular the associated pencil is determined by the range.

Example 9.8.2 (Non-degenerate conics inscribed in a quadrilateral). By the definition, a non-degenerate conic C is inscribed in the quadrilateral with sides ℓ_1, ℓ_2, ℓ_3, ℓ_4 if and only if its envelope C^* contains ℓ_1, ℓ_2, ℓ_3, ℓ_4. The non-degenerate conics inscribed in a quadrilateral describe thus a range $\mathcal{R}$ whose associated pencil $\mathcal{P}$ is a general pencil as those described in Example 9.5.13. Applying 9.5.16 to $\mathcal{P}$ (and using 4.6.7), there is a reference relative to which the conic envelopes of $\mathcal{P}$ have equations

$$(\lambda + \mu)u_0^2 - \lambda u_1^2 - \mu u_2^2 = 0, \quad (\lambda, \mu) \in k^2 - \{(0,0)\},$$

and hence the quadrics of $\mathcal{R}$ are

$$\lambda\mu x_0^2 - \mu(\lambda + \mu)x_1^2 - \lambda(\lambda + \mu)x_2^2 = 0, \tag{9.15}$$

with $(\lambda, \mu) \in k^2$, $\lambda \neq 0$, $\mu \neq 0$ and $\lambda + \mu \neq 0$, in order to avoid the values of the projective parameters giving rise to the degenerate envelopes of $\mathcal{P}$.

The above equation (9.15) may of course be equivalently written

$$\frac{x_0^2}{\lambda + \mu} - \frac{x_1^2}{\lambda} - \frac{x_2^2}{\mu} = 0, \tag{9.16}$$

but in neither form does it depend linearly on the parameters.

The fact that a range does not contain quadrics corresponding to the degenerate envelopes of its associated pencil causes it to be a somewhat incomplete family, a bit tricky to deal with. Assume for instance we are looking for the conics of the range $\mathcal{R}$ of Example 9.8.2 that are tangent to a given line ℓ: for a conic $C \in \mathcal{R}$, being tangent to ℓ is equivalent to having $\ell \in C^*$, which in turn is a linear condition on conic envelopes (9.3.8). Clearly, all envelopes of the associated pencil $\mathcal{P}$ contain ℓ if $\ell = \ell_i$ for some $i = 1, \dots, 4$, and therefore, in such a case, all conics of $\mathcal{R}$ are tangent to ℓ (as it is obvious from the definition of the range). Otherwise there is a single $E \in \mathcal{P}$ containing ℓ (9.5.3). If E is non-degenerate, then the conic it envelops is the only conic of $\mathcal{R}$ tangent to ℓ, but if E is degenerate, then no conic of $\mathcal{R}$ is tangent to ℓ. As the reader may easily see, the latter situation occurs if and only if ℓ goes through one of the vertices of the quadrilateral. For instance there is no quadric of R tangent to $\ell : x_0 - x_1 = 0$.

When dealing with a range it is usually preferred to argue, as above, with its associated pencil rather than with the range itself, as this is often easier. The reader may prove in this way part (2) of the next proposition, using 9.7.5. Part (1) has been seen above.

Proposition 9.8.3. *Assume that K is a quadrilateral of $\mathbb{P}_2$:*

(1) *If ℓ is a line missing all vertices of K, then there is exactly one conic inscribed in K and tangent to ℓ.*

(2) *If p is a point of $\mathbb{P}_2$ belonging to no side of K or its diagonal triangle, then there are either two non-degenerate conics inscribed in K and going through p, or none, the last possibility being for $k = \mathbb{R}$ only.*

Even if the bijections Υ and Υ^{-1} do not preserve collinearity, they map some collinear points to collinear points, namely:

Proposition 9.8.4. *The non-degenerate quadrics of a bitangent pencil compose a range whose associated pencil is a bitangent one. Any range with bitangent associated pencil is the set of non-degenerate quadrics of a bitangent pencil.*

Proof. If, according to 9.5.19, the non-degenerate conics of the bitangent pencil are written

$$\lambda x_0^2 + \mu(x_1^2 \pm x_2^2) = 0, \quad \lambda, \mu \neq 0,$$

their envelopes are

$$\pm\mu^2 u_0^2 \pm \lambda\mu u_1^2 + \lambda\mu u_2^2 = 0, \quad \lambda, \mu \neq 0,$$

or, after discarding the factor μ,

$$\pm\mu u_0^2 \pm \lambda u_1^2 + \lambda u_2^2 = 0, \quad \lambda, \mu \neq 0,$$

which are the non-degenerate envelopes of a bitangent pencil, hence the first claim. The computation dual of the above proves the second claim. If preferred, the reader may also argue directly from the fact that the conditions determining a bitangent pencil are self-dual. $\square$

The bitangent pencils are thus (but for their degenerate members) also ranges. By this reason, in old books, they are called *pencil-ranges*. This is misleading though, as the non-degenerate conics of a triple contact pencil also compose a range and a claim similar to 9.8.4 does hold for triple contact pencils (see Exercise 9.35).

9.9 Apolarity

This section is devoted to explaining a relation between point-quadrics and quadric envelopes, named *apolarity*, which allows us to describe all simple linear conditions on quadrics by identifying the space of hyperplanes of $\mathbb{Q}(\mathbb{P}_n)$ to $\mathbb{Q}(\mathbb{P}_n^\vee)$. Assume we have fixed a reference Δ of $\mathbb{P}_n$. Let Q be a point-quadric with matrix $A = (a_{i,j})$ relative to Δ and T a quadric envelope with matrix $B = (b_{s,t})$ relative to $\Delta^\vee$: they are said to be *apolar*, or apolar to each other, if and only if

$$\sum_{i,j=0}^{n} a_{i,j} b_{i,j} = 0. \tag{9.17}$$

A point-quadric Q is said to be *outpolar* to a non-degenerate point-quadric Q' if and only if Q is apolar to the envelope of Q'. The condition defining apolarity is clearly independent of the arbitrary scalar factors up to which the matrices are determined. Checking that it is also independent of the choice of the coordinates will be a bit harder:

Proposition 9.9.1. *The condition* (9.17) *defining apolarity does not depend on the choice of the reference.*

Proof. We will perform some matrix computations. First of all, assume we have fixed a total order on the set of ordered pairs of indices $\Gamma = \{(i,j) \mid i,j = 0,\dots,n\}$ in order to use it as a totally ordered set of indices. Take for instance the inverse lexicographical order: $(i,j) \leq (i',j')$ if and only if either $j < j'$ or $j = j'$ and $i \leq i'$. Let $A = (a_{i,j})$ be any $(n+1) \times (n+1)$ matrix, $a_{i,j}$ being its entry on row i and column j. We associate to A the column vector $\widehat{A}$, indexed by Γ, with entries $a_{i,j}$, $(i,j) \in \Gamma$.

Assume on the other hand that $C = (c_j^i)$ is an $(n+1) \times (n+1)$ regular matrix, c_j^i being its entry on row i and column j. An equality

$$D = C^t A C, \tag{9.18}$$

holds if and only if the entry on row r and column s of D is

$$d_{r,s} = \sum_{(i,j)\in\Gamma} a_{i,j} c_i^r c_j^s, \tag{9.19}$$

for any $(r,s) \in \Gamma$. Now we define the $\Gamma \times \Gamma$-indexed matrix $\widetilde{C}$ as being the one whose entry on row (r,s) and column (i,j) is $c_i^r c_j^s$. Then, in view of (9.19) the equality (9.18) may be equivalently written

$$\widehat{C^t A C} = \widehat{D} = \widetilde{C}\widehat{A}. \tag{9.20}$$

We need the following lemma:

Lemma 9.9.2. *For any $(n+1) \times (n+1)$ regular matrix C,*

$$\widetilde{C^t} = \widetilde{C}^t \quad \textit{and} \quad \widetilde{C^{-1}} = \widetilde{C}^{-1}.$$

Proof. The first equality is direct from the definition of $\widetilde{C}$. For the second one note that, using (9.20), the obvious equality

$$(C^{-1})^t (C^t A C) C^{-1} = A$$

translates into

$$\widetilde{C^{-1}}\widetilde{C}\widehat{A} = \widehat{A},$$

which in turn, A being arbitrary, assures that $\widetilde{C^{-1}}\widetilde{C}$ is the unit matrix, as claimed. □

End of the proof of 9.9.1. If the point-quadric Q has matrix A relative to Δ and the quadric envelope T has matrix B relative to $\Delta^\vee$, the condition of apolarity may be written

$$0 = \widehat{B}^t \widehat{A}.$$

After a change of reference with matrix C, Q has matrix $C^t A C$ and the new matrix of T is $C^{-1} B (C^{-1})^t$ (4.4.12). Thus the condition of apolarity using the new coordinates reads

$$0 = (\widehat{C^{-1} B (C^{-1})^t})^t (\widehat{C^t A C}) = ((\widetilde{C^{-1}})^t \widehat{B})^t (\widetilde{C}\widehat{A}) = \widehat{B}^t \widetilde{C}^{-1} \widetilde{C} \widehat{A} = \widehat{B}^t \widehat{A},$$

from which the claim. □

Theorem 9.9.3. *Mapping each quadric envelope $T \in \mathbb{Q}(\mathbb{P}_n^\vee)$ to the set of point-quadrics apolar to it, is a projectivity $\Xi : \mathbb{Q}(\mathbb{P}_n^\vee) \to \mathbb{Q}(\mathbb{P}_n)^\vee$.*

Proof. Fix any reference and assume that T has matrix $B = (b_{i,j})$ and hence coordinates $b_{i,j}$, $0 \le i \le j \le n$. The condition for a quadric Q, of matrix $A = (a_{i,j})$, to be apolar to T may be written

$$0 = \sum_{i,j=0}^{n} a_{i,j} b_{i,j} = \sum_{i=0}^{n} a_{i,i} b_{i,i} + 2 \sum_{0 \le i < j \le n} a_{i,j} b_{i,j},$$

clearly a non-trivial linear equation in the coordinates $a_{i,j}$, $0 \le i \le j \le n$, of Q. The set of quadrics apolar to T is thus the hyperplane of $\mathbb{Q}(\mathbb{P}_n)$ with coordinates

$$\beta_{i,j} = \begin{cases} b_{i,j} & \text{if } 0 \le i = j \le n, \\ 2b_{i,j} & \text{if } 0 \le i < j \le n, \end{cases}$$

and Ξ is, as claimed, a projectivity. □

The projectivity Ξ is independent of any choice and may therefore be used to identify $\mathbb{Q}(\mathbb{P}_n)^\vee$ and $\mathbb{Q}(\mathbb{P}_n^\vee)$, see Exercise 9.45. It follows from Theorem 9.9.3 that *being apolar to a given quadric envelope* is a simple linear condition on quadrics, and also that any simple linear condition on quadrics may be equivalently presented in this form. In old books *being apolar to a given quadric envelope* is said to be the *most general* simple linear condition on quadrics, while the conditions *going through a given point* or *having two given points as a conjugate pair* are qualified as *particular*, this meaning that any simple linear condition on quadrics may be equivalently presented as a condition of apolarity to a given conic, but not as having as conjugate a given pair of points or as going through a given point. This is clear after 9.9.3 and the next proposition, which reformulates the latter two conditions in terms of apolarity.

Proposition 9.9.4. *A quadric Q of $\mathbb{P}_n$ goes through a point p if and only if it is apolar to $2p^*$, while it has two points p, q as a conjugate pair if and only if it is apolar to $p^* + q^*$.*

Proof. The first claim has been included for the sake of clarity only, as it clearly follows from the second one. To prove the latter, fix a reference and assume that $(a_{i,j})_{i,j=0,\dots,n}$ is the matrix of Q, while $(\alpha_i)_{i=0,\dots,n}$ and $(\beta_j)_{j=0,\dots,n}$ are coordinate vectors of p and q. Then the condition of conjugation reads

$$\sum_{i,j=0}^{n} a_{i,j}\alpha_i\beta_j = 0. \tag{9.21}$$

Since on the other hand an equation of $p^* + q^*$ is

$$0 = \Big(\sum_{i=0}^{n} \alpha_i u_i\Big)\Big(\sum_{j=0}^{n} \beta_j u_j\Big) = \sum_{i,j=0}^{n} \alpha_i\beta_j u_i u_j,$$

the condition for being Q apolar to $p^* + q^*$ is just (9.21), hence the claim. □

9.10 Pencils and polarity

In this section we will examine the family of polar hyperplanes of a fixed point with respect to the quadrics of a pencil, as well as the locus of poles of a fixed hyperplane relative to the non-degenerate quadrics of a pencil. First we will see that the polar hyperplanes of a fixed point relative to the quadrics of a pencil may either describe a pencil of hyperplanes, or remain constant, or be all undefined:

Proposition 9.10.1. *Let $\mathcal{P}$ be a pencil of quadrics and p a point of $\mathbb{P}_n$. The polar hyperplanes of p relative to the quadrics of $\mathcal{P}$ either*

(a) *describe a pencil of hyperplanes $\mathcal{W}$ and the map*

$$\begin{aligned} \mathcal{P} &\longrightarrow \mathcal{W}, \\ Q &\longmapsto H_{p,Q}, \end{aligned}$$

is a projectivity, or

(b) *are all coincident but for one quadric $\bar{Q} \in \mathcal{P}$, which has p as a double point and therefore $H_{p,\bar{Q}}$ undetermined, or*

(c) *are all undetermined, and therefore p is a double point of all quadrics of $\mathcal{P}$.*

Proof. After fixing coordinates, assume that M and N are matrices of two different quadrics of $\mathcal{P}$. Then the quadrics of $\mathcal{P}$ are the quadrics $Q_{\lambda,\mu}$ with matrix $\lambda M + \mu N$, for $(\lambda, \mu) \in k^2 - \{(0,0)\}$. Furthermore (λ, μ) may be taken as projective coordinates of $Q_{\lambda,\mu}$ in $\mathcal{P}$ (see the beginning of Section 9.5). Assume that $p = [a_0, \dots, a_n]$; then the column vector

$$v_{\lambda,\mu} = \lambda M \begin{pmatrix} a_0 \\ \vdots \\ a_n \end{pmatrix} + \mu N \begin{pmatrix} a_0 \\ \vdots \\ a_n \end{pmatrix} = (\lambda M + \mu N) \begin{pmatrix} a_0 \\ \vdots \\ a_n \end{pmatrix}$$

is either zero, in which case p is a double point of $Q_{\lambda,\mu}$ and $H_{p,Q_{\lambda,\mu}}$ is undetermined, or a coordinate vector of $H_{p,Q_{\lambda,\mu}}$.

If the vectors

$$v_{1,0} = M \begin{pmatrix} a_0 \\ \vdots \\ a_n \end{pmatrix}, \quad v_{0,1} = N \begin{pmatrix} a_0 \\ \vdots \\ a_n \end{pmatrix}$$

are linearly independent, then no $v_{\lambda,\mu}$ is zero and the hyperplanes $H_{p,Q_{\lambda,\mu}}$ describe a pencil of hyperplanes. Furthermore, mapping $Q_{\lambda,\mu}$ to $H_{p,Q_{\lambda,\mu}}$ is a projectivity because both the quadric and its image have coordinates λ, μ in the pencils of quadrics and hyperplanes they respectively belong to.

If the vectors $v_{1,0}, v_{0,1}$ are linearly dependent but not both zero, we assume $v_{1,0} \neq 0$, the case $v_{0,1} \neq 0$ being dealt with similarly. Then $v_{0,1} = \alpha v_{1,0}$ for a uniquely determined $\alpha \in k$ and therefore all polar hyperplanes $H_{p,Q_{\lambda,\mu}}$ have proportional coordinate vectors

$$v_{\lambda,\mu} = \lambda v_{1,0} + \mu v_{0,1} = (\lambda + \alpha\mu) v_{1,0},$$

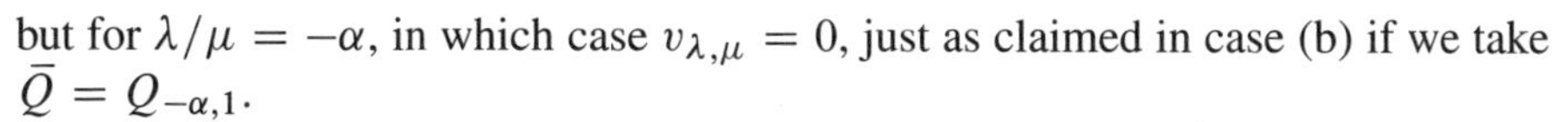

but for $\lambda/\mu = -\alpha$, in which case $v_{\lambda,\mu} = 0$, just as claimed in case (b) if we take $\bar{Q} = Q_{-\alpha,1}$.

To close, if $v_{1,0} = v_{0,1} = 0$, obviously $v_{\lambda,\mu} = 0$, and so p is a double point of $Q_{\lambda,\mu}$, for all λ, μ. □

In case (a) of 9.10.1, it is usually said that the point p has *variable* polar hyperplane with respect to the quadrics of the pencil, while in case (b) it is said that p has *constant* polar hyperplane.

A pencil of quadrics $\mathcal{P}$ is said to have *self-polar simplex* Δ (or that Δ is self-polar for $\mathcal{P}$) if and only if Δ is a self-polar simplex for all quadrics of $\mathcal{P}$.

Corollary 9.10.2. *If a simplex Δ is self-polar for a pencil of quadrics $\mathcal{P}$, then each vertex of Δ is a double point of a degenerate quadric of $\mathcal{P}$.*

Proof. Direct from 9.10.1, as the vertices of a self-polar simplex do not have variable polar hyperplanes with respect to the quadrics of the pencil. □

Corollary 9.10.2 poses a strong obstruction to the existence of a self-polar simplex for a pencil $\mathcal{P}$ of quadrics of $\mathbb{P}_n$: it makes necessary the existence of $n+1$ independent points among the double points of the degenerate quadrics of $\mathcal{P}$, which is not always the case. Next are three examples:

Example 9.10.3. Simple contact, double contact and triple contact pencils of conics have no self-polar triangle.

Example 9.10.4. If $\mathcal{P}$ is a general pencil which has either four or (in case $k = \mathbb{R}$ only) no base points, we have seen in 9.6.1 that the degenerate conics of $\mathcal{P}$ are three pairs of lines whose double points are the vertices of a self-polar triangle of $\mathcal{P}$; by 9.10.2, there is no other. If $k = \mathbb{R}$, general pencils with two (real) base points have a single degenerate conic and hence no (real) self-polar triangle.

Example 9.10.5. According to its description in 9.5.19, a bitangent pencil of conics $\mathcal{P}$ has a double line, 2ℓ, and a pair of different lines, $r + s$, as its only degenerate members. Furthermore the point $p = r \cap s$ does not belong to ℓ. Thus any self-polar triangle needs to have one vertex at p and the remaining ones on ℓ. As the reader may easily check, such a triangle actually is self-polar for $\mathcal{P}$ if and only if its vertices on ℓ are conjugate with respect to the pair of base points $q_1 + q_2$ of $\mathcal{P}$. This gives infinitely many self-polar triangles, a self-polar triangle being determined by the arbitrary choice of one of its vertices in $\ell - \{q_1, q_2\}$.

Corollary 9.10.6. *The poles of a fixed hyperplane H with respect to the quadrics of a range $\mathcal{R}$ of $\mathbb{P}_n$ either are all coincident or describe the set of all but at most $n+1$ points of a line of $\mathbb{P}_n$. The first situation occurs if and only if H is a singular hyperplane of one of the (necessarily degenerate) envelopes of the pencil $\mathcal{P}$ associated to $\mathcal{R}$.*

Proof. Follows from 9.10.1 applied to $\mathcal{P}$. Note that, by its definition, the pencil associated to a range is regular, and therefore 9.10.1 (c) does not occur. □

Example 9.10.7. The line joining the base points of a bitangent pencil of conics $\mathcal{T}$ has the same pole with respect to all non-degenerate conics of $\mathcal{T}$, which, as seen in 9.8.4, describe a range.

Assume now that $\mathcal{P}$ is a regular pencil of quadrics of $\mathbb{P}_n$. After fixing coordinates, assume that $\mathcal{P}$ is spanned by quadrics Q and Q', with matrices A and B, and let H be the hyperplane with coordinate vector $(a) = (a_0, \dots, a_n)$. For $(\lambda, \mu) \in k^2 - \{(0,0)\}$, let $Q_{\lambda,\mu}$ be the quadric with matrix $M(\lambda, \mu) = \lambda A + \mu B$ and write $\overline{M(\lambda, \mu)}$ the matrix of cofactors of $M(\lambda, \mu)$. If $Q_{\lambda,\mu}$ is non-degenerate, the vector

$$\begin{pmatrix} P_0(\lambda, \mu) \\ \vdots \\ P_n(\lambda, \mu) \end{pmatrix} = \overline{M(\lambda, \mu)} \begin{pmatrix} a_0 \\ \vdots \\ a_n \end{pmatrix}$$

is a coordinate vector of the pole of H relative to $Q_{\lambda,\mu}$, in particular $P_i(\lambda, \mu) \neq 0$ for at least one i.

If for variable X, Y we take

$$\begin{pmatrix} P_0 \\ \vdots \\ P_n \end{pmatrix} = \overline{M(X, Y)} \begin{pmatrix} a_0 \\ \vdots \\ a_n \end{pmatrix},$$

the $P_i, i = 0, \dots, n$, are homogeneous polynomials in X, Y of degree n. Take $D = \gcd(P_0, \dots, P_n)$ and $P_i' = P_i/D$. Then, for $Q_{\lambda,\mu}$ non-degenerate, $D(\lambda, \mu) \neq 0$ (otherwise $P_i(\lambda, \mu) = 0$ for all i) and also $(P_0'(\lambda, \mu), \dots, P_n'(\lambda, \mu))$ is a coordinate vector of the pole of H relative to $Q_{\lambda,\mu}$. It follows that

$$\begin{aligned} \psi \colon \mathcal{P} &\longrightarrow \mathbb{P}_n, \\ Q_{\lambda,\mu} &\longmapsto [P_0'(\lambda, \mu), \dots, P_n'(\lambda, \mu)], \end{aligned}$$

is a rational map, defined in $\mathcal{P}$, that maps each non-degenerate quadric of $\mathcal{P}$ to the pole of H relative to it. If the common degree of the P_i' is zero, then ψ is a constant map; otherwise its image is a rational curve of $\mathbb{P}_n$. Therefore, we have proved:

Proposition 9.10.8. *Assume that $\mathcal{P}$ is a regular pencil of quadrics of $\mathbb{P}_n$ and $H \subset \mathbb{P}_n$ a hyperplane. The poles of H relative to the non-degenerate quadrics of $\mathcal{P}$ either are all coincident or describe a set consisting of all but finitely many points of a rational curve of $\mathbb{P}_n$.*

Example 9.10.9. If $\mathcal{P}$ is the pencil of the conics circumscribed to a quadrivertex K, ℓ a side of the diagonal triangle of K and q its opposite vertex, then, by 9.5.16, the pole of ℓ relative to any non-degenerate quadric of $\mathcal{P}$ is q, and therefore the above rational map ψ is constant.

We will prove a more precise claim for pencils of conics:

Theorem 9.10.10. *Assume that $\mathcal{P}$ is a regular pencil of conics and ℓ a line of a projective plane $\mathbb{P}_2$. If $k = \mathbb{C}$, assume that no double point of a conic of $\mathcal{P}$ belongs to ℓ. If $k = \mathbb{R}$, assume that no double point of a conic of $\mathcal{P}_{\mathbb{C}}$, the complex extension of $\mathcal{P}$, is a (real or imaginary) point of ℓ. Then the poles of ℓ relative to the non-degenerate conics of $\mathcal{P}$ belong to a non-degenerate conic whose other points are the double points of the degenerate conics of $\mathcal{P}$.*

Proof. Pick two different points p_1, p_2 in ℓ. Since neither of them is a double point of a conic in $\mathcal{P}$, according to 9.10.1, the polars of $p_i, i = 1, 2$, relative to the conics of $\mathcal{P}$ describe a pencil of lines q_i^* and the map

$$\begin{aligned} g_i : \mathcal{P} &\longrightarrow q_i^*, \\ C &\longmapsto H_{p_i,C}, \end{aligned}$$

is a projectivity. Note that q_i is conjugate to p_i with respect to all conics of $\mathcal{P}$.

Our first task is to prove that $q_1 \neq q_2$. Assume otherwise and put $q = q_1 = q_2$: then q is conjugate to both p_1 and p_2, and therefore to all points of ℓ, with respect to all conics of $\mathcal{P}$. In particular, by 9.10.1, q is a double point of a conic of $\mathcal{P}$ and so, by the hypothesis, $q \notin \ell$. Assume first that $k = \mathbb{C}$. Then, by 9.3.12, the conics tangent to ℓ are the elements of a quadric of $\mathbb{Q}(\mathbb{P}_2)$ which has non-empty intersection with any line of $\mathbb{Q}(\mathbb{P}_2)$ (by 6.3.1), and in particular with $\mathcal{P}$. Thus there is one conic C_0 in $\mathcal{P}$ which is tangent to ℓ. Take p to be a contact point of ℓ and C_0: p is then conjugate with respect to C_0 of all points of ℓ, and also of $q \notin \ell$. This means that p is a double point of C_0, against the hypothesis.

If $k = \mathbb{R}$ we will apply a similar argument after placing ourselves on the complex field. The pencil $\mathcal{P}_{\mathbb{C}}$ being spanned by the complex extensions of two different conics of $\mathcal{P}$, still q is conjugate to p_1 and p_2, and hence to all points of $\ell_{\mathbb{C}}$, with respect to all conics of $\mathcal{P}_{\mathbb{C}}$. Then, as above, there is a conic C_0 in $\mathcal{P}_{\mathbb{C}}$ tangent to $\ell_{\mathbb{C}}$ and any contact point of $\ell_{\mathbb{C}}$ and C_0 is a (maybe imaginary) double point of C_0 on $\ell_{\mathbb{C}}$, against the hypothesis.

Now, once we know that $q_1 \neq q_2$, the projectivity

$$f = g_2 \circ g_1^{-1} : q_1^* \longrightarrow q_2^*$$

is not a perspective, as otherwise the line q_1q_2 would be the polar of both p_1 and p_2 with respect to a certain conic of $\mathcal{P}$ which, by 6.7.11, would have a double point on ℓ. By Steiner's theorem (8.1.1) the intersections of corresponding lines, $r \cap f(r)$, $r \in q_1^*$, are the points of a non-degenerate conic K. We end the proof by showing that K is the conic we want. By the definition of f, the points of K are $p_C = g_1(C) \cap g_2(C) = H_{p_1,C} \cap H_{p_2,C}$, for $C \in \mathcal{P}$. Thus, for C non-degenerate, p_C is the pole of $\ell = p_1p_2$ with respect to C. If C is degenerate, p_C being the intersection of two distinct polar lines relative to C, it is the only double point of C, by 6.7.10. After this the proof is complete. □

Remark 9.10.11. Theorem 9.10.10 does not apply to pencils of conics containing a double line $2r$, as obviously any line contains a double point of $2r$. Nevertheless these pencils are either bitangent pencils or triple contact pencils, their non-degenerate conics describe ranges (9.8.4 and Exercise 9.35) and so the loci of poles they give rise to have been described in 9.10.6.

Example 9.10.12. In a real projective plane take a general pencil $\mathcal{P}$ with two real base points (9.6.1). The two non-real degenerate conics of $\mathcal{P}_{\mathbb{C}}$ have imaginary and complex-conjugate double points p_1, p_2 which therefore span a real line ℓ. Since these points together with the double point p of the only degenerate conic of $\mathcal{P}$ are the vertices of a self-polar triangle of $\mathcal{P}_{\mathbb{C}}$ (by 9.5.16), p is the pole of ℓ with respect to any non-degenerate conic of $\mathcal{P}$. Thus in this case the locus of poles is not a conic, yet no conic of $\mathcal{P}$ has a double point on ℓ.

In the particular case in which the hyperplane H, or the line ℓ, is the improper hyperplane of an affine space, 9.10.6, 9.10.8, and 9.10.10 describe locus of centres, see for instance Exercises 9.28 and 9.29.

9.11 Rational normal curves of $\mathbb{P}_n$

In this section we will study a particular type of rational curves that generalize the non-degenerate conics with points, as plane rational curves, to projective spaces of dimension $n > 2$.

Assume we have fixed a coordinate frame in a projective space $\mathbb{P}_n$, $n \geq 2$. A subset $\Gamma \subset \mathbb{P}_n$ is called a *rational normal curve* of $\mathbb{P}_n$ if and only if there are $n+1$ linearly independent homogeneous polynomials of degree n, $P_0, \dots, P_n \in k[T_0, T_1]$, such that Γ is the set of all points of $\mathbb{P}_n$ which have coordinates $x_0, \dots, x_n$ satisfying

$$\begin{aligned} x_0 &= P_0(t_0, t_1), \\ &\vdots \\ x_n &= P_n(t_0, t_1), \end{aligned} \tag{9.22}$$

for some $(t_0, t_1) \in k^2 - (0, 0)$. Rational normal curves of $\mathbb{P}_3$ are also called *twisted cubics*.

First of all, note that the linear independence of $P_0, \dots, P_n$ guarantees that they compose a basis of the subspace of the homogeneous polynomials of degree n in $k[T_0, T_1]$. In particular they are coprime, as otherwise all homogeneous polynomials of degree n in $k[T_0, T_1]$ would share a factor.

Take $\mathbb{P}_1$ to be a one-dimensional projective space with a fixed coordinate frame. Then, by the above,

$$\begin{aligned} \varphi \colon \mathbb{P}_1 &\longrightarrow \mathbb{P}_n, \\ [t_0, t_1] &\longmapsto [P_0(t_0, t_1), \dots, P_n(t_0, t_1)], \end{aligned} \tag{9.23}$$

is a rational map with image Γ and therefore, according to the definitions of Section 9.2, Γ is, indeed, a rational curve of $\mathbb{P}_n$ with parameterization map (9.23) and parametric equations (9.22). As for any rational curve, t_0, t_1 are called *homogeneous parameters* of the point $x_0, \dots, x_n$. Sometimes the parametric equations are written with an arbitrary non-zero factor ρ, in the form

$$x_i = \rho P_i(t_0, t_1), \quad i = 0, \dots, n,$$

in order to assure that they provide all the coordinates of the points of Γ (see Exercise 9.46).

Since $P_0, \dots, P_n$ are assumed to be linearly independent, by 9.2.4, Γ spans $\mathbb{P}_n$. On the other hand, Γ has no parameterization map of degree less than n, as, again by 9.2.4, no rational curve defined by a parameterization map of degree less than n spans $\mathbb{P}_n$. It follows that Γ has degree n.

Once we know that the degree of Γ in n, according to the definitions of Section 9.2, (9.22) and φ, defined by (9.23), are proper parametric equations and a parameterization map of Γ. Conversely, any proper parametric equations (resp. proper parameterization map) of Γ has the form of (9.23) (resp. (9.22)) for some homogeneous polynomials $P_0, \dots, P_n \in k[T_0, T_1]$ of degree n, and these polynomials necessarily are linearly independent because Γ spans $\mathbb{P}_n$, (9.2.4). Only proper parametric equations and proper parameterization maps of rational normal curves will be considered in the sequel.

The term *normal* in the name of the rational normal curves is classical: it refers to the fact that they cannot be obtained as projections of rational curves of the same degree spanning spaces of higher dimension (actually, these curves do not exist, by 9.2.4). This use of the term *normal* is obsolete, as in modern algebraic geometry it has a somewhat related but definitively different meaning.

Assume as above that

$$x_0 = P_0(t_0, t_1), \ \dots, \ x_n = P_n(t_0, t_1)$$

are proper parametric equations of a rational normal curve Γ of $\mathbb{P}_n$. If the polynomials P_i are

$$P_i = \sum_{j=0}^{n} a^i_j T_0^j T_1^{n-j}, \quad i = 0, \dots, n,$$

then the same parametric equations may be rewritten in matricial form

$$\begin{pmatrix} x_0 \\ x_1 \\ \vdots \\ x_n \end{pmatrix} = \begin{pmatrix} a^0_0 & a^0_1 & \dots & a^0_n \\ a^1_0 & a^1_1 & \dots & a^1_n \\ \vdots & \vdots & & \vdots \\ a^n_0 & a^n_1 & \dots & a^n_n \end{pmatrix} \begin{pmatrix} t_0^n \\ t_0^{n-1} t_1 \\ \vdots \\ t_1^n \end{pmatrix} \tag{9.24}$$

and the linear independence of $P_0, \dots, P_n$ is equivalent to the regularity of the matrix $A = (a^i_j)$. If a new coordinate frame of $\mathbb{P}_n$ is used and B is a matrix changing old into new coordinates, then Γ is the set of points which have new coordinates $y_0, \dots, y_n$ satisfying

$$\begin{pmatrix} y_0 \\ y_1 \\ \vdots \\ y_n \end{pmatrix} = BA \begin{pmatrix} t_0^n \\ t_0^{n-1} t_1 \\ \vdots \\ t_1^n \end{pmatrix},$$

for some $(t_0, t_1) \in k^2 - (0, 0)$, the matrix BA being also regular. The above are thus proper parametric equations of Γ relative to the new coordinates of $\mathbb{P}_n$. This shows in particular that the definition of rational normal curve is independent of the choice of coordinates in $\mathbb{P}_n$.

In particular we may take the new coordinates in such a way that $B^{-1} = A$ (2.3.3), after which we have:

Lemma 9.11.1. *Any proper parametric equations of a rational normal curve Γ of $\mathbb{P}_n$ may be turned into*

$$\begin{aligned} x_0 &= t_0^n, \\ x_1 &= t_0^{n-1} t_1, \\ &\vdots \\ x_{n-1} &= t_0 t_1^{n-1}, \\ x_n &= t_1^n \end{aligned} \tag{9.25}$$

by a suitable change of coordinates in $\mathbb{P}_n$.

The equations (9.25) are called *canonical equations* of Γ. They are often written using the absolute parameter $t = t_0/t_1 \in k \cup \{\infty\}$ in the form

$$\begin{aligned} x_0 &= t^n, \\ x_1 &= t^{n-1}, \\ &\vdots \\ x_{n-1} &= t, \\ x_n &= 1. \end{aligned}$$

Remark 9.11.2. If $n = 2$, then the canonical equations are just the equations (8.1) of Section 8.2. Thus the rational normal curves of $\mathbb{P}_2$ are just the non-degenerate conics with points of $\mathbb{P}_2$.

Theorem 9.11.3. *Any two rational normal curves of projective spaces of the same dimension are projectively equivalent.*

Proof. Assume that Γ and Γ' are rational normal curves of projective spaces $\mathbb{P}_n$ and $\mathbb{P}'_n$, respectively. If Δ and Δ' are references relative to which Γ and Γ' have canonical equations, then, by 2.8.1, the projectivity $f : \mathbb{P}_n \to \mathbb{P}'_n$ mapping Δ to Δ', satisfies $f(\Gamma) = \Gamma'$. □

Rational normal curves have their proper parameterization maps injective. In other words:

Proposition 9.11.4. *The homogeneous parameters of a point p in a proper parameterization of a rational normal curve of $\mathbb{P}_n$ are determined by p up to a common non-zero factor.*

Proof. Up to a change of coordinates in $\mathbb{P}_n$, we are allowed to use the canonical equations (9.25). After them, if $x_0 = 0$, then $t_0 = 0$. Otherwise $t_0 \neq 0$ and $t_1/t_0 = x_1/x_0$. □

Remark 9.11.5. It follows from Proposition 9.11.4 that any rational normal curve is an infinite set, which was already proved for any rational curve in 9.2.3.

Proper parametric equations of a given rational normal curve Γ, relative to fixed coordinates of $\mathbb{P}_n$, are not unique: for instance, the reader may easily check that if

$$x_i = P_i(t_0, t_1), \quad i = 0, \dots, n,$$

are proper parametric equations of Γ, then so are

$$x_i = P_i(az_0 + bz_1, cz_0 + dz_1), \quad i = 0, \dots, n,$$

for any fixed $a, b, c, d \in k$ for which $ad - bc \neq 0$. It is important to see that there are no other examples, namely:

Proposition 9.11.6. *Assume we have two systems of proper parametric equations of the same rational normal curve Γ of $\mathbb{P}_n$, relative to the same reference of $\mathbb{P}_n$:*

$$x_i = P_i(t_0, t_1), \quad P_i \in k[T_0, T_1], \quad i = 0, \dots, n$$

and

$$x_i = Q_i(z_0, z_1), \quad Q_i \in k[Z_0, Z_1], \quad i = 0, \dots, n.$$

Then there exist $a, b, c, d \in k$, with $ad - bc \neq 0$, and $\rho \in k - \{0\}$ such that

$$Q_i(Z_0, Z_1) = \rho P_i(aZ_0 + bZ_1, cZ_0 + dZ_1), \quad i = 0, \dots, n.$$

Proof. By performing a change of coordinates in $\mathbb{P}_n$, it is not restrictive to assume that the first equations are canonical, namely that $P_i = T_0^{n-i} T_1^i$, $i = 0, \dots, n$.

Then, any coordinates of any point of Γ being $x_i = \delta t_0^{n-i} t_1^i$, $i = 0, \dots, n$, $\delta \in k - \{0\}$, they satisfy the equalities

$$x_i^2 = x_{i-1}x_{i+1}, \quad i = 1, \dots, n-1.$$

Thus we have

$$Q_i(z_0, z_1)^2 = Q_{i-1}(z_0, z_1)Q_{i+1}(z_0, z_1), \quad i = 1, \dots, n-1,$$

for all $(z_0, z_1) \in k - \{(0, 0)\}$, and so, as polynomials in two variables,

$$Q_i^2 = Q_{i-1}Q_{i+1}, \quad i = 1, \dots, n-1. \tag{9.26}$$

Assume that F is a prime factor of Q_0. The equalities (9.26) inductively show that F is also a prime factor of $Q_1, \dots, Q_{n-1}$. Therefore F does not divide Q_n, because $Q_0, \dots, Q_n$ are coprime. Assume that the multiplicity of F as a prime factor of Q_{n-1} is ν: arguing inductively backwards, from the last of the equalities (9.26), we see that F has multiplicity $(n-i)\nu$ as a factor of Q_i and so, eventually, multiplicity $n\nu$ as a factor of Q_0. Since Q_0 has degree n it follows that $\nu = 1$ and $\deg F$=1. Take now a prime factor $\bar{F}$ of Q_n; it is coprime with F because F does not divide Q_n. Arguing as above, it results that $\bar{F}$ also has degree one and appears with multiplicity i as a prime factor of Q_i, $i = 0, \dots, n$. The degree n of the Q_i allowing no further factors, we get

$$Q_i = a_i F^{n-i} \bar{F}^i, \quad i = 0, \dots, n,$$

for some $a_i \in k - \{0\}$. Up to multiplying $\bar{F}$ by a suitable non-zero constant factor, it is not restrictive to assume that $a_0 = a_1$, after which, using again the equalities (9.26), $a_0 = a_1 = a_2 = \dots = a_n$ and so

$$Q_i = a_0 F^{n-i} \bar{F}^i, \quad i = 0, \dots, n.$$

The polynomials F and $\bar{F}$ are homogeneous, as otherwise either $Q_0 = F^n$ or $Q_n = \bar{F}^n$ would be non-homogeneous, against the hypothesis. Since F and $\bar{F}$ have degree one and are coprime, they are

$$F = aZ_0 + bZ_1,$$
$$\bar{F} = cZ_0 + dZ_1,$$

with $ad - bc \neq 0$. After this, renaming $a_0 = \rho$,

$$Q_i = \rho(aZ_0 + bZ_1)^{n-i}(cZ_0 + dZ_1)^i = \rho P_i(aZ_0 + bZ_1, cZ_0 + dZ_1),$$

for $i = 0, \dots, n$, as claimed. $\square$

Remark 9.11.7. If Γ is seen as the image of a proper parameterization map

$$\varphi\colon \mathbb{P}_1 \longrightarrow \mathbb{P}_n,$$

then 9.11.6 assures that any proper parameterization map of Γ,

$$\varphi'\colon \mathbb{P}'_1 \longrightarrow \mathbb{P}_n,$$

is $\varphi' = \varphi \circ \tau$, where $\tau\colon \mathbb{P}'_1 \to \mathbb{P}_1$ is a projectivity, namely the one with matrix

$$\begin{pmatrix} a & b \\ c & d \end{pmatrix}$$

if the parameters are those of 9.11.6.

Fix a rational normal curve Γ of $\mathbb{P}_n$ and take the coordinates of $\mathbb{P}_n$ such that Γ has canonical equations $x_i = t_0^i t_1^{n-i}$, $i = 0, \dots, n$. If H is the hyperplane with equation $u_0 x_0 + \dots + u_n x_n = 0$, then the point of Γ with parameters t_0, t_1 belongs to H if and only if

$$u_0 t_0^n + u_1 t_0^{n-1} t_1 + \dots + u_n t_1^n = 0. \tag{9.27}$$

Then, by 9.1.1 and 9.11.4:

Proposition 9.11.8. *The intersection of a rational normal curve and a hyperplane of $\mathbb{P}_n$ contains at most n points.*

Two interesting facts directly follow:

Corollary 9.11.9. *If Γ is a rational normal curve of $\mathbb{P}_n$, then*

(a) *any $n+1$ different points of Γ are independent, and*

(b) *any ordered set of $n+2$ different points of Γ is a projective reference of $\mathbb{P}_n$.*

Remark 9.11.10. Also any r different points $p_0, \dots, p_{r-1} \in \Gamma$, $0 < r \leq n+1$, are independent. Indeed, by 9.11.5, the set $\{p_0, \dots, p_{r-1}\}$ can be enlarged to a set of $n+1$ different points of Γ and then 1.5.4 applies.

Assume we have fixed coordinates in $\mathbb{P}_n$ and let Γ be a rational normal curve of $\mathbb{P}_n$ with proper parametric equations

$$x_i = P_i(t_0, t_1), \quad i = 0, \dots, n.$$

We proceed as above: if a hyperplane H has coordinates $u_0, \dots, u_n$, then, the homogeneous polynomial $u_0 P_0 + \dots + u_n P_n$ is non-zero, due to the independence of the P_i, and has degree n; the equation

$$u_0 P_0(T_0, T_1) + \dots + u_n P_n(T_0, T_1) = 0, \tag{9.28}$$

has as solutions the pairs of parameters of the points of $\Gamma \cap H$. Assume that $p \in \Gamma \cap H$ has parameters t_0, t_1: they are determined by p up to a scalar factor (9.11.4) and so we may attach to p the multiplicity ν of (t_0, t_1) as a solution of the equation (9.28) (see Section 9.1). In the sequel we will call ν the *multiplicity* of p in the intersection of Γ and H; we will equivalently say that p counts with multiplicity ν in $\Gamma \cap H$. Our first job is to see that ν does not depend on the choices made:

Lemma 9.11.11. *The multiplicity of p in the intersection of Γ and H, as defined above, is independent of the choice of the coordinates of $\mathbb{P}_n$ and also of the choice of the proper parametric equations of Γ.*

Proof. We will see first that using different coordinates in $\mathbb{P}_n$ and suitable proper parametric equations, the resulting multiplicity is the same.

The notations being as above, we write the equation (9.28) in matricial form

$$\begin{pmatrix} u_0 & \dots & u_n \end{pmatrix} \begin{pmatrix} P_0(T_0, T_1) \\ \vdots \\ P_n(T_0, T_1) \end{pmatrix} = 0. \tag{9.29}$$

Assume that we take in $\mathbb{P}_n$ new coordinates $y_0, \dots, y_n$ and let B be the matrix changing old into new coordinates. Then the hyperplane H has new coordinates

$$\begin{pmatrix} v_0 \\ \vdots \\ v_n \end{pmatrix} = (B^{-1})^t \begin{pmatrix} u_0 \\ \vdots \\ u_n \end{pmatrix},$$

while one may take

$$\begin{pmatrix} y_0 \\ \vdots \\ y_n \end{pmatrix} = B \begin{pmatrix} P_0(t_0, t_1) \\ \vdots \\ P_n(t_0, t_1) \end{pmatrix}$$

as proper parametric equations of Γ relative to the new coordinates. Using these equations and the new coordinates, the definition of multiplicity leads us to consider the equation

$$\begin{pmatrix} v_0 & \dots & v_n \end{pmatrix} B \begin{pmatrix} P_0(T_0, T_1) \\ \vdots \\ P_n(T_0, T_1) \end{pmatrix} = \begin{pmatrix} u_0 & \dots & u_n \end{pmatrix} B^{-1} B \begin{pmatrix} P_0(T_0, T_1) \\ \vdots \\ P_n(T_0, T_1) \end{pmatrix} = 0,$$

which is just (9.29).

Now we will check that the multiplicity is the same no matter which proper parametric equations, relative to the same coordinates of $\mathbb{P}_n$, are used.

Assume we have proper parametric equations of Γ,

$$x_i = P_i(t_0, t_1), \quad i = 0, \dots, n.$$

According to 9.11.6, any other proper parametric equations of Γ, relative to the same reference of $\mathbb{P}_n$ are

$$x_i = P_i(az_0 + bz_1, cz_0 + dz_1), \quad i = 0, \dots, n,$$

for some $a, b, c, d \in k$ with $\delta = ad - bc \neq 0$. Therefore, if a point $p \in \Gamma$ has parameters β_0, β_1 relative to the latter equations, then it has parameters

$$\alpha_0 = a\beta_0 + b\beta_1, \quad \alpha_1 = c\beta_0 + d\beta_1$$

relative to the former equations. For any given coordinates $u_0, \dots, u_n$ of a hyperplane, we have to see that the multiplicities of $[\alpha_0, \alpha_1]$ and $[\beta_0, \beta_1]$, as solutions of

$$F(T_0, T_1) = u_0 P_0(T_0, T_1) + \cdots + u_n P_n(T_0, T_1) = 0$$

and

$$\begin{aligned} G(Z_0, Z_1) \\ &= F(aZ_0 + bZ_1, cZ_0 + dZ_1) \\ &= u_0 P_0(aZ_0 + bZ_1, cZ_0 + dZ_1) + \cdots + u_n P_n(aZ_0 + bZ_1, cZ_0 + dZ_1) \\ &= 0, \end{aligned}$$

respectively, do agree, and this follows from a direct computation. For, assume that the multiplicity of $[\alpha_0, \alpha_1]$ is ν, that is,

$$F(T_0, T_1) = (\alpha_1 T_0 - \alpha_0 T_1)^\nu Q(T_0, T_1),$$

with $Q(\alpha_0, \alpha_1) \neq 0$. Then

$$\begin{aligned} G(Z_0, Z_1) &= \big(\alpha_1(aZ_0 + bZ_1) - \alpha_0(cZ_0 + dZ_1)\big)^\nu Q(aZ_0 + bZ_1, cZ_0 + dZ_1) \\ &= \big((-c\alpha_0 + a\alpha_1)Z_0 - (d\alpha_0 - b\alpha_1)Z_1\big)^\nu Q'(Z_0, Z_1) \\ &= (\beta_1 Z_0 - \beta_0 Z_1)^\nu \delta^\nu Q'(Z_0, Z_1), \end{aligned}$$

where $Q'(Z_0, Z_1) = Q(aZ_0 + bZ_1, cZ_0 + dZ_1)$ and so $Q'(\beta_0, \beta_1) \neq 0$. □

Theorem 9.11.12. *Assume that Γ is a rational normal curve of $\mathbb{P}_n$. The following holds:*

(a) *For any hyperplane H of $\mathbb{P}_n$, the sum of the multiplicities of the points $p \in \Gamma \cap H$ in the intersection of Γ and H is at most n. It always equals n in case $k = \mathbb{C}$.*

(b) *Given points* $p_1, \dots, p_m \in \Gamma$ *and positive integers* $\nu_1, \dots, \nu_m$ *such that* $\sum_{i=1}^m \nu_i = n$*, there is a unique hyperplane* H *of* $\mathbb{P}_n$ *for which* $\Gamma \cap H = \{p_1, \dots, p_m\}$ *and the multiplicity of* p_i *in the intersection of* Γ *and* H *is* ν_i*, for* $i = 1, \dots, m$.

Proof. In view of the definition of multiplicity, part (a) is a direct consequence of 9.1.4. Regarding part (b), take the coordinates such that Γ has canonical equations and assume that p_i has parameters $\alpha_{i,0}, \alpha_{i,1}$. Then a hyperplane $u_0x_0+\dots+u_nx_n = 0$ satisfies the condition of the claim if and only if the equation

$$u_0T_0^n + u_1T_0^{n-1}T_1 + \dots + u_nT_1^n = 0$$

has solutions $[\alpha_{1,0}, \alpha_{1,1}], \dots, [\alpha_{m,0}, \alpha_{m,1}]$, with respective multiplicities $\nu_1, \dots, \nu_m$. By 9.1.2, this is equivalent to having

$$u_0T_0^n + u_1T_0^{n-1}T_1 + \dots + u_nT_1^n = \alpha \prod_{i=1}^m (\alpha_{i,1}T_0 - \alpha_{i,0}T_1)^{\nu_i},$$

for $\alpha \in k - \{0\}$, which in turn is satisfied by the coordinates $u_0, \dots, u_n$ of a single hyperplane of $\mathbb{P}_n$. □

Corollary 9.11.13 (of the proof of 9.11.12). *If* Γ *is given by canonical equations, the points* p_i *have parameters* $\alpha_{i,0}$, $\alpha_{i,1}$ *and*

$$\prod_{i=1}^m (\alpha_{i,1}T_0 - \alpha_{i,0}T_1)^{\nu_i} = u_0T_0^n + u_1T_0^{n-1}T_1 + \dots + u_{n-1}T_0T_1^{n-1} + u_nT_1^n,$$

then $u_0, \dots, u_n$ *may be taken as coordinates of the hyperplane* H *of* 9.11.12 (b).

If $k = \mathbb{C}$ and each point $p \in \Gamma \cap H$ is counted as many times as its multiplicity in the intersection of Γ and H, then 9.11.12 (a) asserts that the number of intersections of Γ and an arbitrary hyperplane of $\mathbb{P}_n$ is the degree n of Γ. This is the way in which the degree of a (non-necessarily rational) algebraic curve of $\mathbb{P}_n$ is usually defined in algebraic geometry, see [22], Chapter 5, for instance. The next lemma adds further information in the real case; it will be useful later on.

Lemma 9.11.14. *Assume* $k = \mathbb{R}$ *and let* Γ *and* H *be a rational normal curve and a hyperplane of* $\mathbb{P}_n$*, respectively. If the sum of the multiplicities of the points of* $\Gamma \cap H$ *in the intersection of* Γ *and* H *is not less than* $n-1$*, then it equals* n.

Proof. As for part (a) of 9.11.12, it follows from the definition of multiplicity and 9.1.4, this time using also 9.1.3. □

It follows from 9.11.12 (b) that for each $p \in \Gamma$ there is a uniquely determined hyperplane H_p whose intersection with Γ is p, counted with multiplicity n: H_p is called the hyperplane *osculating* Γ at p. If Γ is given by canonical equations and the parameters of p are α_0, α_1, then, by 9.11.13, the coordinates $u_0, \dots, u_n$ of H_p may be taken such that

$$u_0 T_0^n + u_1 T_0^{n-1} T_1 + \cdots + u_n T_1^n = (\alpha_1 T_0 - \alpha_0 T_1)^n,$$

that is,

$$\begin{aligned} u_0 &= \binom{n}{0} \alpha_1^n, \\ u_1 &= -\binom{n}{1} \alpha_0 \alpha_1^{n-1}, \\ &\vdots \\ u_{n-1} &= (-1)^{n-1} \binom{n}{n-1} \alpha_0^{n-1} \alpha_1, \\ u_n &= (-1)^n \binom{n}{n} \alpha_0^n. \end{aligned} \tag{9.30}$$

We see thus that the osculating hyperplanes of Γ are the points of a rational normal curve Γ^* of the space of hyperplanes $\mathbb{P}_n^\vee$, namely the one with proper parametric equations (9.30) above.

Rational curves of $\mathbb{P}_n^\vee$ are called *rational developables* of $\mathbb{P}_n$. Thus Γ^* above is a rational normal developable: it is called the *envelope* of Γ.

In case $n = 2$, Γ is a non-degenerate conic with points (9.11.2). The reader may easily check that then the osculating hyperplane at $p \in \Gamma$ is just the tangent line. Therefore, the envelope of Γ as a rational normal curve is its envelope as a conic and the double use of the term *envelope* causes no confusion.

Theorem 9.11.15 (Clifford). *Given a rational normal curve Γ of $\mathbb{P}_n$, there is a unique correlation $\mathcal{P}_\Gamma$ of $\mathbb{P}_n$ mapping each point $p \in \Gamma$ to the hyperplane osculating Γ at p. $\mathcal{P}_\Gamma$ is a polarity if n is even, and a null-system if n is odd.*

Proof. If Γ is given by canonical equations, then, according to equations (9.30), the correlation with matrix

$$\begin{pmatrix} & & & & \binom{n}{0} \\ & & & -\binom{n}{1} & \\ & & \cdot^{\cdot^{\cdot}} & & \\ & (-1)^{n-1}\binom{n}{n-1} & & & \\ (-1)^n \binom{n}{n} & & & & \end{pmatrix}$$

maps p to H_p for every $p \in \Gamma$ and clearly is a polarity or a null-system depending on the parity of n, as claimed. Since any $n + 2$ different points chosen on Γ may be taken as the points of a reference (9.11.9) and have their images determined, the uniqueness follows from 2.8.1. $\square$

For $n = 2$ we recover the polarity with respect to a non-degenerate conic. In spite of being not a polarity for n odd, $\mathscr{P}_\Gamma$ is sometimes called the *polarity relative to* Γ.

Arguing by duality, if Ω is a rational normal developable of $\mathbb{P}_n$, for each hyperplane H of Ω there is a unique point $p_H \in \mathbb{P}_n$ for which $\Omega \cap (p_H^*)$ consists of H counted with multiplicity n. Such a p_H is called the *focal point* of Ω on H. The focal points of Ω describe a rational normal curve of $\mathbb{P}_n$ which is called the *cuspidal edge* of Ω. We leave to the reader to prove, as an exercise, that the cuspidal edge of the envelope of a rational normal curve Γ is Γ itself.

The lines joining two different points $p, q \in \Gamma$ are called *chords* of Γ, the points p, q being called the *ends* of the chord.

Remark 9.11.16. No point of a chord other than its ends may belong to Γ, as otherwise there would be three linearly dependent points on Γ, against 9.11.10.

Clearly the hyperplanes through two different points $p, q \in \Gamma$ describe an $(n-2)$-dimensional bundle whose kernel is the chord with ends p, q. Considering a single point $p \in \Gamma$ and the hyperplanes H for which p has multiplicity at least two in $\Gamma \cap H$, will give rise to the tangent line to Γ at p. To this end we need:

Proposition 9.11.17. *If p is a point of a rational normal curve Γ, then the set of all hyperplanes H, for which p has multiplicity at least two in the intersection of Γ and H, is an $(n-2)$-dimensional bundle of hyperplanes ℓ_p^*.*

Proof. Fix a canonical parameterization of Γ and assume that p has parameters α_0, α_1. Proceeding as in the proof of 9.11.12, the hyperplanes we are interested in are those with coordinates $(u_0, \dots, u_n)$ for which the equation

$$u_0 T_0^n + u_1 T_0^{n-1} T_1 + \dots + u_n T_1^n = 0$$

has the solution $[\alpha_0, \alpha_1]$ with multiplicity at least two. According to 9.1.5, this is equivalent to the vanishing of the two first derivatives at α_0, α_1, namely to

$$\begin{aligned} u_0 n \alpha_0^{n-1} + u_1 (n-1) \alpha_0^{n-2} \alpha_1 + \dots + u_{n-1} \alpha_1^{n-1} &= 0, \\ u_1 \alpha_0^{n-1} + \dots + u_{n-1} (n-1) \alpha_0 \alpha_1^{n-2} + u_n n \alpha_1^{n-1} &= 0. \end{aligned}$$

Since the above are two homogeneous linear equations in the hyperplane coordinates $u_0, \dots, u_n$, and they are independent because either

$$\begin{vmatrix} n\alpha_0^{n-1} & (n-1)\alpha_0^{n-2}\alpha_1 \\ 0 & \alpha_0^{n-1} \end{vmatrix} = n\alpha_0^{2n-2} \neq 0$$

or

$$\begin{vmatrix} \alpha_1^{n-1} & 0 \\ (n-1)\alpha_0\alpha_1^{n-2} & n\alpha_1^{n-1} \end{vmatrix} = n\alpha_1^{2n-2} \neq 0,$$

the proof is complete. □

The kernel ℓ_p of the bundle ℓ_p^* of 9.11.17 is a line through p: it is called the *tangent line* to Γ at p, while p is in turn called the *contact point* of ℓ_p. Heuristically the tangents to Γ may be understood as a sort of chords with coincident ends, and so in many classical books the tangent lines are taken as chords; we do not follow this convention here. While proving 9.11.17, we have seen:

Corollary 9.11.18 (of the proof of 9.11.17). *If the rational normal curve Γ is given by canonical equations, then the points*

$$[n\alpha_0^{n-1}, (n-1)\alpha_0^{n-2}\alpha_1, \dots, \alpha_1^{n-1}, 0]$$

and

$$[0, \alpha_0^{n-1}, \dots, (n-1)\alpha_0\alpha_1^{n-2}, n\alpha_1^{n-1}]$$

span the tangent line to Γ at the point with parameters α_0, α_1.

Like the chords, the tangent lines have no further intersection with the curve:

Proposition 9.11.19. *A tangent line ℓ_p to a rational normal curve Γ shares with it no point other than its contact point p.*

Proof. Assume that $q_1 \in \ell_p \cap \Gamma$ and $q_1 \neq p$. Γ being an infinite set (9.11.5), enlarge $\{p, q_1\}$ to a set of n different points $p, q_1, q_2, \dots, q_{n-1} \in \Gamma$. Then there is a hyperplane H through all these points. Since $H \supset p \vee q_1 = \ell_p$, by the definition of ℓ_p, the point p has multiplicity at least two in the intersection of Γ and H. This, together with the fact that H contains $n-1$ different points of Γ other than p, contradicts 9.11.12 (a). □

Remark 9.11.20. Since p has multiplicity $n \geq 2$ in the intersection of Γ and its osculating hyperplane H_p at p, it is clear that H_p contains the tangent line ℓ_p. Of course, they agree if $n = 2$.

Fix $n-1$ different points $p_1, \dots, p_{n-1}$ of a rational normal curve of $\mathbb{P}_n$. Since they are independent (by 9.11.10), they span an $(n-2)$-dimensional linear variety L: its corresponding pencil of hyperplanes L^* is the set of the hyperplanes through $p_1, \dots, p_{n-1}$. Take $H \in L^*$. In both the real and complex cases, the sum of the multiplicities of the points of $\Gamma \cap H$ in the intersection of Γ and H is n. Indeed, in the complex case this is clear from 9.11.12 (a), while in the real one it follows from 9.11.14, because we already have $p_1, \dots, p_{n-1} \in H$. Thus, we have two possibilities: either

(1) $\Gamma \cap H = \{p_1, \dots, p_{n-1}, p\}$ with $p \neq p_i, i = 1, \dots, n-1$, or

(2) $\Gamma \cap H = \{p_1, \dots, p_{n-1}\}$ and all points p_i have multiplicity one in the intersection of Γ and H but for a single one, p_j, which has multiplicity two.

If in case (2) we write $p = p_j$, then, in both cases, $\Gamma \cap H = \{p_1, \dots, p_{n-1}, p\}$, with the points repeated according to their multiplicities in the intersection, and we will refer to p as the *further intersection point* of Γ and H.

Conversely, according to 9.10.10 (b), for any $p \in \Gamma$ there is a unique hyperplane H such that the points of $\Gamma \cap H$, repeated according to their multiplicities in the intersection, are $p_1, \dots, p_{n-1}, p$. We will write $H = pp_1 \dots p_{n-1}$ and name it the *projection of p from $p_1, \dots, p_{n-1}$*, even in the cases in which $p = p_i$ for some $i = 1, \dots, n-1$. Then reader may then easily prove:

Proposition 9.11.21. *If $p_1, \dots, p_{n-1}$ are different points of a rational normal curve Γ of $\mathbb{P}_n$, then the hyperplanes through $p_1, \dots, p_{n-1}$ describe $(p_1 \vee \dots \vee p_{n-1})^*$, which is a pencil of hyperplanes. Mapping each hyperplane through $p_1, \dots, p_{n-1}$ to its further intersection point with Γ, and each $p \in \Gamma$ to its projection from $p_1, \dots, p_{n-1}$, are reciprocal bijections between $(p_1 \vee \dots \vee p_{n-1})^*$ and Γ.*

We will refer to the map

$$\begin{aligned} \Gamma &\longrightarrow (p_1 \vee \dots \vee p_{n-1})^*, \\ p &\longmapsto pp_1 \dots p_{n-1}, \end{aligned}$$

as the *projection* (of the points of Γ) *from* $p_1, \dots, p_{n-1}$. The following lemma will prove very useful.

Lemma 9.11.22. *If $p_1, \dots, p_{n-1}$ are different points of a rational normal curve Γ of $\mathbb{P}_n$, then the composition of any proper parameterization map of Γ and the projection from $p_1, \dots, p_{n-1}$, is a projectivity from $\mathbb{P}_1$ to $(p_1 \vee \dots \vee p_{n-1})^*$.*

Proof. Let $\varphi\colon \mathbb{P}_1 \to \Gamma$ be a proper parameterization map of Γ and name θ its composition with the projection from $p_1, \dots, p_n$. By 9.11.1, choose the coordinates such that φ is given by canonical equations, namely by the rule $[\alpha_0, \alpha_1] \mapsto [\alpha_0^{n-j} \alpha_1^j]_{j=0,\dots,n}$. Assume that p_i has parameters $\alpha_{i,0}, \alpha_{i,1}$, $i = 1, \dots, n-1$. If $q = [\alpha_0, \alpha_1] \in \mathbb{P}_1$, we have seen in 9.11.13 that the coefficients of the polynomial

$$(\alpha_1 T_0 - \alpha_0 T_1) \prod_{j=1}^{n-1} (\alpha_{j,1} T_0 - \alpha_{j,0} T_1)$$

are coordinates of $\theta(q) = \varphi(q) p_1 \dots p_{n-1}$. In particular the coefficients of

$$T_1 \prod_{j=1}^{n-1} (\alpha_{j,1} T_0 - \alpha_{j,0} T_1) \quad \text{and} \quad T_0 \prod_{j=1}^{n-1} (\alpha_{j,1} T_0 - \alpha_{j,0} T_1)$$

are coordinates of $\theta([1, 0])$ and $\theta([0, 1])$. Since these polynomials are linearly independent, $\theta([1,0])$ and $\theta([0,1])$ span the pencil $(p_1 \vee \cdots \vee p_{n-1})^*$. The equality

$$\begin{aligned}(\alpha_1 T_0 - \alpha_0 T_1) &\prod_{j=1}^{n-1}(\alpha_{j,1}T_0 - \alpha_{j,0}T_1)\\ &= \alpha_1 T_0 \prod_{j=1}^{n-1}(\alpha_{j,1}T_0 - \alpha_{j,0}T_1) - \alpha_0 T_1 \prod_{j=1}^{n-1}(\alpha_{j,1}T_0 - \alpha_{j,0}T_1)\end{aligned}$$

shows then that $\theta([\alpha_0, \alpha_1])$ has coordinates α_0, α_1 relative to the reference $(\theta([1,0]), \theta([0,1]), \theta([1,1,]))$ of $(p_1 \vee \cdots \vee p_{n-1})^*$, hence the claim. □

The reader may compare the next theorem to the converse of Steiner's theorem (8.1.5):

Theorem 9.11.23. *If Γ is a rational normal curve of $\mathbb{P}_n$ and $\{p_1, \dots, p_{n-1}\}$, $\{q_1, \dots, q_{n-1}\}$ are two sets of $n-1$ different points of Γ, then taking two hyperplanes as corresponding if and only if they are the projections the same point of Γ defines a pair of reciprocal projectivities*

$$\begin{aligned}(p_1 \vee \cdots \vee p_{n-1})^* &\longleftrightarrow (q_1 \vee \cdots \vee q_{n-1})^*,\\ pp_1 \dots p_{n-1} &\longleftrightarrow pq_1 \dots q_{n-1}.\end{aligned}$$

Proof. Call ψ_1 and ψ_2 the projection maps

$$\begin{aligned}\psi_1 \colon \Gamma &\longrightarrow (p_1 \vee \cdots \vee p_{n-1})^*,\\ p &\longmapsto pp_1 \dots p_{n-1},\end{aligned}$$

and

$$\begin{aligned}\psi_2 \colon \Gamma &\longrightarrow (q_1 \vee \cdots \vee q_{n-1})^*,\\ p &\longmapsto pq_1 \dots q_{n-1},\end{aligned}$$

and fix any proper parameterization map $\varphi \colon \mathbb{P}_1 \to \Gamma$. We have seen in 9.11.22 that both $\psi_1 \circ \varphi$ and $\psi_2 \circ \varphi$ are projectivities: hence so are $\psi_2^{-1} \circ \psi_1 = (\varphi \circ \psi_2)^{-1} \circ (\varphi \circ \psi_1)$ and $\psi_1^{-1} \circ \psi_2 = (\varphi \circ \psi_1)^{-1} \circ (\varphi \circ \psi_2)$, which are the maps of the claim. □

The rational normal curves have a well-determined one-dimensional projective structure. The reader may check that it is the one of Theorem 8.2.1 in the case $n = 2$.

Theorem 9.11.24. *If Γ is a rational normal curve of $\mathbb{P}_n$, there is a one-dimensional projective structure on Γ such that*

(a) *for any choice of $n-1$ different points $p_1, \dots, p_{n-1} \in \Gamma$, the projection map*

$$\begin{aligned} \Gamma &\longrightarrow (p_1 \vee \cdots \vee p_{n-1})^*, \\ p &\longmapsto pp_1 \dots p_{n-1}, \end{aligned}$$

is a projectivity, and

(b) *any proper parameterization map of Γ is a projectivity.*

Furthermore such a projective structure is uniquely determined up to equivalence by either of the above properties (a), (b).

Proof. Fix $n-1$ different points $q_1, \dots, q_{n-1} \in \Gamma$. Arguing as in the proof of 8.2.1, use the map sending each $H \in (q_1 \vee \cdots \vee q_{n-1})$ to its further intersection with Γ to define a projective structure on Γ, and then 9.11.23 to prove that such a projective structure satisfies property (a).

Regarding (b), assume that φ is a proper parameterization map. Take ψ to be any of the projection maps of part (a). Both ψ and $\psi \circ \varphi$ are projectivities, the latter by 9.11.22. Then $\varphi = \psi^{-1} \circ (\psi \circ \varphi)$ is also a projectivity.

The uniqueness of the projective structure follows from 1.6.3 (b). □

Remark 9.11.25. By 9.11.24 (b) and 2.8.1, after fixing proper parametric equations of Γ, the parameters of the points of Γ are their homogeneous coordinates relative to the reference of Γ whose vertices and unit point have pairs of parameters $(1,0)$, $(0,1)$ and $(1,1)$, respectively. As for non-degenerate conics with points, the proper parametric equations of Γ may be understood as relating coordinates in Γ and coordinates in $\mathbb{P}_n$ of the same point $p \in \Gamma$.

To close this section, we will pay some attention to the quadrics that contain a fixed rational normal curve. Assume that Γ is a rational normal curve of $\mathbb{P}_n$, given by canonical equations

$$x_i = t_0^{n-i} t_1^i, \quad i = 0, \dots, n.$$

It is clear from these equations that the coordinates of all points of Γ satisfy

$$x_i x_{j+1} = x_{i+1} x_j, \quad 0 \le i < j < n,$$

and therefore all points of Γ belong to the quadrics

$$Q_{i,j} : x_i x_{j+1} - x_{i+1} x_j = 0, \quad 0 \le i < j < n.$$

We leave to the reader to check that the above quadrics, in number of $\frac{1}{2}n(n-1)$, are linearly independent. Next we will prove that their intersection is Γ. Assume

$$p = [a_0, \dots, a_n] \in \bigcap_{0 \le i < j < n} Q_{i,j}.$$

Among the quadrics $Q_{i,j}$ we have

$$Q_{i,i+1} : x_i x_{i+2} - x_{i+1}^2 = 0, \quad 0 \le i \le n-2,$$

and from their equations it is easy to see by decreasing induction that if p has $a_n = 0$, then $p = [1, 0, \dots, 0]$, which clearly belongs to Γ. We may assume thus that p has $a_n = 1$ and call $a_{n-1} = \alpha$. Assume inductively, that $x_{n-r} = \alpha^r$, $r = 0, \dots, s$, $0 < s < n$. Then the equation $x_{n-s-1}x_n - x_{n-s}x_{n-1} = 0$, of $Q_{n-s-1,n-1}$, forces $a_{n-s-1} = \alpha^s \alpha = \alpha^{s+1}$ and so, at the end, $p = [\alpha^n, \dots, \alpha, 1]$ and therefore $p \in \Gamma$. Since for $n > 3$ some of the quadrics $Q_{i,j}$ have played no role in the above argument, actually we have proved a bit more, namely:

Lemma 9.11.26. *The curve Γ is the intersection of the quadrics*

$$Q_{0,2}, Q_{1,3}, \dots, Q_{n-2,n}$$

and

$$Q_{0,n-1}, Q_{1,n-1}, \dots, Q_{n-2,n-1}.$$

The reader may note that only $Q_{n-3,n-1}$ has been listed twice, so the above quadrics number $2n - 3$.

On the other hand we have:

Lemma 9.11.27. *If a quadric Q contains $2n + 1$ different points of a rational normal curve Γ of $\mathbb{P}_n$, then it contains Γ.*

Proof. Assume that Γ is given by canonical equations and that Q has equation $\sum_{i,j=0,\dots,n} a_{i,j}x_i x_j = 0$. Then the parameters of the points of Γ in Q are the solutions of the equation

$$\sum_{i,j=0,\dots,n} a_{i,j} T_0^{2n-i-j} T_1^{i+j} = 0 \tag{9.31}$$

which therefore, by the hypothesis, has more solutions than degree. By 9.1.4 the equation is trivial and so all points of Γ belong to Q, as claimed. □

Pick any $2n + 1$ different points $q_0, \dots, q_{2n} \in \Gamma$. By 9.11.27, the quadrics through $q_0, \dots, q_{2n}$ are the quadrics containing Γ, and so they describe a linear system of dimension at least $\frac{1}{2}(n^2 + 3n) - (2n + 1) = \frac{1}{2}n(n-1) - 1$. Since, on the other hand, we already have the $\frac{1}{2}n(n-1)$ quadrics $Q_{i,j} : x_i x_{j+1} - x_{i+1}x_j = 0$, $0 \le i < j < n$, which are independent and contain Γ, the dimension of the linear system of the quadrics containing Γ is just $\frac{1}{2}n(n-1) - 1$ and we have proved:

Theorem 9.11.28. *The set of the quadrics through a rational normal curve Γ of $\mathbb{P}_n$ is a linear system $\mathcal{L}_\Gamma$ of dimension $\binom{n}{2} - 1$. If Γ is given by canonical equations, then $\mathcal{L}_\Gamma$ is spanned by the quadrics*

$$Q_{i,j} : x_i x_{j+1} - x_{i+1}x_j = 0, \quad 0 \le i < j < n.$$

Furthermore, the intersection of these quadrics is Γ itself.

We may now show by an example that points imposing dependent conditions to quadrics need not have special linear relations between them, which gives further insight into 9.4.8 and 9.4.9. The reader may compare with the plane case 9.4.6.

Example 9.11.29. Take $n = 3$. According to 9.11.28 the quadrics through a twisted cubic describe a net, and so any nine different points on a twisted cubic impose dependent conditions to the quadrics of $\mathbb{P}_3$. Nevertheless, by 9.11.9 (a), any four of these points are linearly independent.

9.12 Twisted cubics

The twisted cubics have been already defined as being the rational normal curves of projective spaces of dimension three, so all the results of the preceding Section 9.11 apply to them. In the present section we will deal with properties of twisted cubics that are either specific to them or not worth generalizing to higher-dimensional spaces. We will assume $k = \mathbb{C}$ along the whole section. Most of the results (essentially those depending on Lemma 9.12.3) would need to be modified allowing chords with imaginary ends in order to continue to hold for $k = \mathbb{R}$, which can be done by the reader without major difficulty.

Let Γ be a twisted cubic of a projective space $\mathbb{P}_3$. As for conics, if $p, q \in \Gamma$, pq will denote the chord with ends p, q if $p \neq q$, and the tangent at p if, otherwise, $p = q$. In both cases we will refer to pq as the line *spanned* by the pair p, q. We will always assume that the coordinates in $\mathbb{P}_3$ have been taken such that Γ is given by canonical equations, namely by

$$\begin{aligned} x_0 &= t_0^3, \\ x_1 &= t_0^2 t_1, \\ x_2 &= t_0 t_1^2, \\ x_3 &= t_1^3. \end{aligned} \tag{9.32}$$

We have seen in 9.11.28 that the quadrics through Γ describe a net $\mathcal{L}_\Gamma$, which is spanned by

$$Q_{0,1} : x_0x_2 - x_1^2 = 0, \quad Q_{0,2} : x_0x_3 - x_1x_2 = 0, \quad Q_{1,2} : x_1x_3 - x_2^2 = 0,$$

and also that

$$Q_{0,1} \cap Q_{0,2} \cap Q_{1,2} = \Gamma.$$

Given points $q_1, q_2, q_3 \in \Gamma$, with respective pairs of parameters (α_0, α_1), (β_0, β_1) and (δ_0, δ_1), the equation of the plane $q_1q_2q_3$ whose intersection with Γ consists of q_1, q_2, q_3, repeated according to multiplicities, is (9.11.13)

$$\begin{aligned} &\alpha_1\beta_1\delta_1 x_0 - (\alpha_0\beta_1\delta_1 + \alpha_1\beta_0\delta_1 + \alpha_1\beta_1\delta_0)x_1 \\ &\quad + (\alpha_1\beta_0\delta_0 + \alpha_0\beta_1\delta_0 + \alpha_0\beta_0\delta_1)x_2 - \alpha_0\beta_0\delta_0 x_3 = 0. \end{aligned} \tag{9.33}$$

Since this equation may be written

$$\begin{aligned}&\delta_0(-\alpha_1\beta_1x_1 + (\alpha_1\beta_0 + \alpha_0\beta_1)x_2 - \alpha_0\beta_0x_3)\\&\qquad + \delta_1(\alpha_1\beta_1x_0 - (\alpha_1\beta_0 + \alpha_0\beta_1)x_1 + \alpha_0\beta_0x_2) = 0,\end{aligned}$$

while q_3 varies freely in Γ, these planes describe the pencil through q_1q_2. The reader may note that the parameters of q_3 appear as coordinates of $q_1q_2q_3$ in the pencil, and compare with 9.11.22. Note also that in the present case we are allowing $q_1 = q_2$.

In particular we may take as equations of q_1q_2,

$$\begin{aligned}-\alpha_1\beta_1x_1 + (\alpha_1\beta_0 + \alpha_0\beta_1)x_2 - \alpha_0\beta_0x_3 &= 0,\\ \alpha_1\beta_1x_0 - (\alpha_1\beta_0 + \alpha_0\beta_1)x_1 + \alpha_0\beta_0x_2 &= 0.\end{aligned} \tag{9.34}$$

For $q_1 \neq q_2$, the above are equations of the chord with ends q_1, q_2, while for $q_1 = q_2$ they are equations of the tangent line at q_1. In such a case they may be rewritten

$$\begin{aligned}-\alpha_1^2x_1 + 2\alpha_0\alpha_1x_2 - \alpha_0^2x_3 &= 0,\\ \alpha_1^2x_0 - 2\alpha_0\alpha_1x_1 + \alpha_0^2x_2 &= 0.\end{aligned}$$

The next two lemmas are different versions of the same fact. Both will be of use below.

Lemma 9.12.1. *If Γ is a twisted cubic, $p_1, q_1, p_2, q_2 \in \Gamma$ and $\{p_1, q_1\} \cap \{p_2, q_2\} = \emptyset$, then the lines $\ell_1 = p_1q_1$ and $\ell_2 = p_2q_2$ are not coplanar.*

Proof. Assume that a plane π contains both lines. If for some $i = 1, 2$, the line ℓ_i is tangent to Γ, then, by the definition of tangent line in Section 9.11, the point $p_i = q_i$ has multiplicity at least two in the intersection of Γ and π. Otherwise $p_i \neq q_i$. Thus, in any case, the sum of the multiplicities of the points in $\Gamma \cap \pi$ is at least four, against 9.11.12. □

Lemma 9.12.2. *Two different lines, each a tangent to or a chord of the same twisted cubic Γ, cannot meet at a point $p \notin \Gamma$.*

Proof. Let the lines be $\ell_1 = p_1q_1$ and $\ell_2 = p_2q_2$, $p_1, q_1, p_2, q_2 \in \Gamma$. If $p \in \ell_1 \cap \ell_2$, then ℓ_1 and ℓ_2 are coplanar and therefore, by 9.12.1, $\{p_1, q_1\} \cap \{p_2, q_2\} \neq \emptyset$. Then ℓ_1 and ℓ_2 share the point p and also a point $q \in \{p_1, q_1\} \cap \{p_2, q_2\}$. Since $p \neq q$ because $p \notin \Gamma$, it follows that $\ell_1 = \ell_2$. □

As allowed by 9.11.1, 9.11.24 and 9.11.25, we consider Γ as a one-dimensional projective space, choose a reference of $\mathbb{P}_n$ relative to which Γ has canonical equations and take the parameters of the points of Γ as their homogeneous coordinates. Just as a tool, it will be useful to consider a class of correspondences on Γ slightly larger than the class of all involutions. Fix $a, b, c \in \mathbb{C}$, one at least non-zero, and consider the correspondence σ on Γ through which the points with parameters α_0, α_1 and β_0, β_1 correspond if and only if

$$a\alpha_1\beta_1 + b(\alpha_1\beta_0 + \alpha_0\beta_1) + c\alpha_0\beta_0 = 0,$$

the condition obviously depending on the points only, and not on the choice of their parameters. We will say that a, b, c, taken up to a non-zero factor, are the coefficients of σ. Obviously σ is symmetric. If $ac - b^2 \neq 0$, then σ is an involution of Γ (see Section 5.6). If, otherwise, $ac - b^2 = 0$, σ is called a *degenerate involution* of Γ. Degenerate involutions are not involutions; their definition may obviously be extended to arbitrary one-dimensional projective spaces.

Assume that σ is a degenerate involution with coefficients a, b, c. Due to being $b^2 = ac$, it is either $a \neq 0$ or $c \neq 0$. Take $O \in \Gamma$ to be the point with parameters either $a, -b$, if $a \neq 0$, or $b, -c$, if $c \neq 0$; in case $a \neq 0$ and $c \neq 0$, both possibilities give the same point because $ac - b^2 = 0$. The point O is called the *centre* of σ. If $a \neq 0$, then

$$a\alpha_1\beta_1 + b(\alpha_1\beta_0 + \alpha_0\beta_1) + c\alpha_0\beta_0 = \frac{1}{a}(b\alpha_0 + a\alpha_1)(b\beta_0 + a\beta_1),$$

while for $c \neq 0$,

$$a\alpha_1\beta_1 + b(\alpha_1\beta_0 + \alpha_0\beta_1) + c\alpha_0\beta_0 = \frac{1}{c}(c\alpha_0 + b\alpha_1)(c\beta_0 + b\beta_1).$$

These equalities show that two points correspond by a degenerate involution if and only if one of them is its centre, which provides a complete description of the degenerate involutions.

The reader may easily check that σ determines its coefficients up to proportionality, the case of an involution being already seen in Section 5.2. This follows also from the next lemma, which will have more important consequences:

Lemma 9.12.3. *Assume that σ is either an involution or a degenerate involution of Γ which has coefficients a, b, c. Then the points of the lines spanned by the pairs of points corresponding by σ are the points of the quadric*

$$Q_{a,b,c} : a(x_1x_3 - x_2^2) + b(x_0x_3 - x_1x_2) + c(x_0x_2 - x_1^2) = 0,$$

of the net of quadrics through Γ.

Proof. Using the equations (9.34), the point $[x_0, x_1, x_2, x_3]$ belongs to a line q_1q_2, $q_1, q_2 \in \Gamma$ and corresponding by σ, if and only if there exist pairs of parameters α_0, α_1 and β_0, β_1 for which

$$\begin{aligned} -\alpha_1\beta_1x_1 + (\alpha_1\beta_0 + \alpha_0\beta_1)x_2 - \alpha_0\beta_0x_3 = 0,\\ \alpha_1\beta_1x_0 - (\alpha_1\beta_0 + \alpha_0\beta_1)x_1 + \alpha_0\beta_0x_2 = 0 \end{aligned} \tag{9.35}$$

and

$$a\alpha_1\beta_1 + b(\alpha_1\beta_0 + \alpha_0\beta_1) + c\alpha_0\beta_0 = 0. \tag{9.36}$$

If this is the case, the system of equations in Z_1, Z_2, Z_3,

$$\begin{aligned} -x_1Z_1 + x_2Z_2 - x_3Z_3 &= 0,\\ x_0Z_1 - x_1Z_2 + x_2Z_3 &= 0,\\ aZ_1 + bZ_2 + cZ_3 &= 0 \end{aligned} \tag{9.37}$$

has the non-trivial solution

$$Z_1 = \alpha_1\beta_1, \quad Z_2 = (\alpha_1\beta_0 + \alpha_0\beta_1), \quad Z_3 = \alpha_0\beta_0$$

and therefore

$$\begin{vmatrix} -x_1 & x_2 & -x_3 \\ x_0 & -x_1 & x_2 \\ a & b & c \end{vmatrix} = 0,$$

or, equivalently,

$$a(x_1x_3 - x_2^2) + b(x_0x_3 - x_1x_2) + c(x_0x_2 - x_1^2) = 0.$$

Conversely, if the above equality is satisfied, then the system (9.37) has a non-trivial solution z_1, z_2, z_3. Since $k = \mathbb{C}$, by 9.1.2 we may take $[\alpha_0, \alpha_1]$ and $[\beta_0, \beta_1]$ to be the solutions (repeated according to multiplicities) of

$$z_1T_0^2 + z_2T_0T_1 + z_3T_1^2 = 0.$$

Then, again by 9.1.2, the triples $\alpha_1\beta_1$, $\alpha_1\beta_0 + \alpha_0\beta_1$, $\alpha_0\beta_0$ and z_1, z_2, z_3 are proportional, the former is also a solution of (9.37) and so α_0, α_1 and β_0, β_1 satisfy both (9.35) and (9.36), as wanted. □

Theorem 9.12.4. *Assume that Γ is a twisted cubic of a complex projective space $\mathbb{P}_3$. Then:*

(a) *If σ is an involution of Γ, then the lines $q\sigma(q)$, for $q \in \Gamma$, describe a ruling of a non-degenerate quadric of $\mathbb{P}_3$ that contains Γ.*

(b) *If $p \in \Gamma$, then the lines pq, $q \in \Gamma$, are the generators of an ordinary cone of $\mathbb{P}_3$ that has vertex p and contains Γ.*

(c) *Any quadric of $\mathbb{P}_3$ containing Γ may be obtained from a unique involution of Γ, as described in* (a) *above, if it is non-degenerate, or from a unique $p \in \Gamma$, as described in* (b), *if it is degenerate.*

Proof. According to 9.12.3, the points of the lines $q\sigma(q)$ are those of a quadric Q through Γ. Different pairs $\{q, \sigma(q)\}$, $\{q', \sigma(q')\}$ give non-coplanar lines $q\sigma(q)$, $q'\sigma(q')$, by 9.12.1. Since Γ shares at most three points with an arbitrary plane, obviously Q cannot be a pair of distinct or coincident planes. Ordinary cones containing no pair of skew lines, Q cannot be an ordinary cone either, and so it is non-degenerate. Any two different lines $q\sigma(q)$ and $q'\sigma(q')$ being pairwise skew, all lines $q\sigma(q)$, $q \in \Gamma$, belong to the same ruling S of Q. Conversely, if $\ell \in S$, pick $p \in \ell$; since $p \in Q$, by 9.12.3, p belongs to some generator $q\sigma(q) \in S$: it follows that $\ell = q\sigma(q)$ because they belong to the same ruling and share a point.

Regarding part (b), the lines pq, $q \in \Gamma$ are those spanned by the pairs of points corresponding by the degenerate involution with centre p. By 9.12.3 their points are those of a quadric C through Γ. As argued above, C cannot be a pair of planes. On the other hand, p is a double point of C, as otherwise the polar plane of p should contain all lines pq, $q \in \Gamma$ and in particular the curve Γ itself. Thus C is an ordinary cone with vertex p. Obviously, all lines pq, $q \in \Gamma$ are generators of C. If ℓ is any generator of C, pick $q' \in \ell$, $q' \neq p$. Then $q' \in C$ and by 9.12.3 $q' \in pq$ for some $q \in \Gamma$, after which $\ell = pq$.

To close, 9.12.3 assures that any quadric Q through Γ comes from a unique correspondence σ, as the coefficients of σ are the parameters of the quadric in the net. If σ is an involution, then Q is as described in claim (a), while if σ is a degenerate involution, then Q is the cone of claim (b) with p the centre of σ, hence claim (c). □

The cone of 9.12.4 (b) is called the *cone projecting* Γ *from* p. After 9.12.4 (c), these cones are the only degenerate quadrics containing Γ. If Q is a non-degenerate quadric and contains Γ, then, by 9.12.4 (c), the generators of one of its rulings are the tangents and chords of Γ spanned by the pairs of points corresponding by an involution σ on Γ. The generators of the opposite ruling have a quite different behaviour:

Proposition 9.12.5. *If Q is a non-degenerate quadric of $\mathbb{P}_3$ containing a twisted cubic Γ, then all generators of one of the rulings of Q have a single intersection point with Γ and no one is tangent to Γ.*

Proof. By 9.12.4, there is an involution σ of Γ whose pairs span the generators of one of the rulings of Q: name this ruling S_1 and let S_2 be the opposite ruling. First we will see that no $s \in S_2$ is a tangent to or a chord of Γ. Indeed, assume otherwise: $s \cap \Gamma$ being at most two points (by 9.11.16), we may choose two different points q_1, q_2 of Γ, corresponding by σ and lying not on s; then $q_1q_2 \in S_1$ and the plane $\pi = q_1q_2 \vee s$ contains either q_1, q_2 and the ends of s, if s is a chord, or q_1, q_2 and the contact point of s, which has multiplicity at least two in $\Gamma \cap \pi$, if s is tangent. Since $q_1, q_2 \notin s$ this shows the sum of the multiplicities of the points in $\Gamma \cap \pi$ to be at least four, against 9.11.12 (a).

Thus, the proof will be complete after showing $\Gamma \cap s \neq \emptyset$ for any $s \in S_2$, which we will do next. Fix any chord $q_1q_2 \in S_1$. Note that, by 9.11.21, there are in the pencil $(q_1q_2)^*$ just two planes having less than three different intersection points with Γ, namely $q_1q_1q_2$ and $q_2q_1q_2$. We examine first the generators $s_i \in S_2$ through the points q_i, $i = 1, 2$. Of course they meet Γ, as wanted, but we need something more about them. Take $\pi_i = q_1q_2 \vee s_i$, $i = 1, 2$, which obviously are two different planes. Since neither s_i, as seen above, nor q_1q_2, by 9.11.16, have any intersection with Γ other than q_1, q_2,

$$\pi_i \cap \Gamma = \pi_i \cap Q \cap \Gamma = (s_i \cup q_1q_2) \cap \Gamma = \{q_1, q_2\}.$$

It follows that any plane $\pi \in (q_1q_2)^*$ other than π_1, π_2 has three different intersections with Γ. Assume now that $s \in S_2$, $s \neq s_1, s_2$: then $s \vee q_1q_2 \neq \pi_1, \pi_2$ and therefore

$$(s \cup q_1q_2) \cap \Gamma = (s \vee q_1q_2) \cap Q \cap \Gamma = (s \vee q_1q_2) \cap \Gamma = \{q_1, q_2, q\},$$

with $q \neq q_1, q_2$. Using again 9.11.16, $q \in s$ and the proof is complete. □

Remark 9.12.6. The lines that meet Γ at a single point and are not tangent to Γ are called *unisecants* of Γ. Combining 9.12.4 and 9.11.4, we see that if a non-degenerate quadric Q contains a twisted cubic Γ, then there is an involution σ of Γ whose pairs span the generators in one of the rulings of Q, while the opposite ruling consists of unisecants of Γ. The reader may easily check that any of the unisecants s contained in Q determines the above involution σ in the following way: if $\Gamma \cap s = p$, then two points $q_1, q_2 \in \Gamma$ correspond by σ if and only if $pq_1q_2 \in s^*$.

The above leads to the following characterization of the involutions of a twisted cubic as the correspondences whose pairs of corresponding points are cut on the cubic (besides the obvious fixed point) by the planes through a unisecant. The reader may compare it with 8.2.8 and 9.7.4:

Corollary 9.12.7. *Let Γ be a twisted cubic, s a unisecant of Γ and $p = \Gamma \cap s$. Taking points $q_1, q_2 \in \Gamma$ as corresponding if and only if the plane pq_1q_2 contains s, defines an involution of Γ. Any involution of Γ may be obtained in this way by a suitable choice of the unisecant s.*

Proof. Choose any two different points $p_1, p_2 \in s$, both different from p. Since the quadrics through Γ describe a net, one of them, Q, goes through p_1 and p_2 and therefore, since it contains also p, $Q \supset s$ (by 6.2.5). By 9.12.4 no degenerate quadric through Γ contains a unisecant. Therefore Q is non-degenerate, s belongs to the ruling of Q composed of unisecants and the arguments of 9.12.6 apply. For the converse, once the involution σ is given, take Q to be the non-degenerate quadric through Γ it gives rise to by 9.12.4, and s any of the unisecants contained in Q. □

Each of the unisecants s of 9.12.7 is called a *directrix* of σ.

The net of quadrics through a twisted cubic Γ being, as above,

$$\mathcal{L}_\Gamma = \{Q_{a,b,c} : a(x_1x_3 - x_2^2) + b(x_0x_3 - x_1x_2) + c(x_0x_2 - x_1^2) = 0\}$$

we have seen in 9.12.4 that $Q_{a,b,c}$ is degenerate if and only if $b^2 - ac = 0$ (see also Exercise 9.48). The equation $b^2 - ac = 0$ defines a non-degenerate conic of $\mathcal{L}_\Gamma$ ($\mathcal{L}_\Gamma$ seen as a two-dimensional projective space) whose points are the degenerate quadrics through Γ. The pencils of quadrics through Γ being the lines of $\mathcal{L}_\Gamma$, we obtain a useful result that will be improved in a short while (see 9.12.11 below):

Lemma 9.12.8. *Any pencil of quadrics of $\mathbb{P}_3$ through a twisted cubic has at most two degenerate members.*

We already know that any twisted cubic is the intersection of three quadrics. Next we will see that no twisted cubic is the intersection of two quadrics.

Corollary 9.12.9. *If two different quadrics Q, $\bar{Q}$ contain a twisted cubic Γ, then $Q \cap \bar{Q} = \Gamma \cup \ell$ where ℓ is either a tangent to or a chord of Γ.*

Proof. By 9.5.1, any two different quadrics of the pencil spanned by Q and $\bar{Q}$ have the same intersection as Q and $\bar{Q}$. So 9.12.8 allows us to assume without restriction that both Q and $\bar{Q}$ are non-degenerate. Then, by 9.12.4, Q and $\bar{Q}$ correspond to two different involutions $\sigma, \bar{\sigma}$ of Γ. If q_1, q_2 is the common pair of these involutions (8.2.11), then $\ell = q_1q_2$ is either a tangent to or a chord of Γ, and also a common generator of Q and $\bar{Q}$, after which $Q \cap \bar{Q} \supset \Gamma \cup \ell$. Conversely, assume that $p \in Q \cap \bar{Q} - \Gamma$. Then p belongs to a generator s of Q of spanned by a pair of points p_1, p_2 corresponding by σ, and also to a generator $\bar{s}$ of $\bar{Q}$ spanned by a pair of points $\bar{p}_1, \bar{p}_2$ corresponding by $\bar{\sigma}$: by 9.12.2, $s = \bar{s}$, hence $\{p_1, p_2\} = \{\bar{p}_1, \bar{p}_2\}$ is the common pair of σ_1 and σ_2. It turns out that $s = \bar{s} = \ell$ and hence $p \in \ell$, thus ending the proof. □

Corollary 9.12.10. *If Γ is a twisted cubic of $\mathbb{P}_3$, then through each point $p \in \mathbb{P}_3 - \Gamma$ there is a unique line ℓ which is either a chord of Γ or a tangent to Γ.*

Proof. The uniqueness has been already seen in 9.12.2. For the existence note that, the quadrics through Γ making a net, one may select a quadric Q such that $Q \supset \Gamma$ and $p \in Q$. Then, by 8.12.2, one of the generators of Q through p may be taken as ℓ. □

We end this section, and the chapter, with a description of the pencils of quadrics through a twisted cubic. These are new samples of pencils of quadrics that did not appear in Section 9.5.

Corollary 9.12.11. *If Γ is a twisted cubic of $\mathbb{P}_3$ and ℓ a tangent to or a chord of Γ, then:*

(a) *The quadrics containing* $\Gamma \cup \ell$ *describe a pencil* $\mathcal{P}_{\Gamma,\ell}$.

(b) *The set of base points of* $\mathcal{P}_{\Gamma,\ell}$ *is* $\Gamma \cup \ell$.

(c) *The degenerate quadrics of* $\mathcal{P}_{\Gamma,\ell}$ *are the cones projecting* Γ *from the ends of* ℓ, *if* ℓ *is a chord, or the cone projecting* Γ *from the contact point of* ℓ, *if* ℓ *is tangent to* Γ.

(d) *Any pencil* $\mathcal{P}$ *of quadrics through* Γ *is* $\mathcal{P} = \mathcal{P}_{\Gamma,\ell}$ *for a uniquely determined line* ℓ *which is a tangent to or a chord of* Γ.

Proof. Notations being as above, let

$$Q_{a,b,c} : a(x_1x_3 - x_2^2) + b(x_0x_3 - x_1x_2) + c(x_0x_2 - x_1^2) = 0$$

be an arbitrary quadric through Γ. Assume that $q_1, q_2 \in \Gamma$ are points spanning ℓ, with parameters α_0, α_1 and β_0, β_1, respectively. By 9.12.4 and 9.12.6, $Q_{a,b,c}$ contains ℓ if and only if q_1 and q_2 correspond by an involution or degenerate involution σ of Γ. Since, by 9.12.3, σ has coefficients a, b, c, the last condition is equivalent to having

$$a\alpha_1\beta_1 + b(\alpha_1\beta_0 + \alpha_0\beta_1) + c\alpha_0\beta_0 = 0,$$

a non-trivial linear condition on the coordinates a, b, c of $Q_{a,b,c}$ in the net of quadrics through Γ. This proves claim (a). Claim (b) is direct from 9.12.9. The degenerate quadrics through Γ are the cones projecting Γ from each $p \in \Gamma$, by 9.12.4; one of them contains ℓ if and only if $p \in \ell$, hence claim (c). Lastly, for claim (d), take two different $Q, \bar{Q} \in \mathcal{P}$: by 9.12.9, $Q \cap \bar{Q} = \Gamma \cup \ell$ where ℓ is a tangent to or a chord of Γ. Then all the quadrics of $\mathcal{P}$ contain $\Gamma \cup \ell$, after which $\mathcal{P} \subset \mathcal{P}_{\Gamma,\ell}$. So $\mathcal{P} = \mathcal{P}_{\Gamma,\ell}$ and ℓ is uniquely determined by $\mathcal{P}$ due to claim (b). □

9.13 Exercises

Exercise 9.1. Use 6.2.5 to show that four aligned points impose dependent conditions on quadrics.

Exercise 9.2. Re-prove 9.3.19 by translating the conditions $p \in Q$ and $\bar{p} \in Q$ into two linear equations in the coefficients of an equation of Q, and proving then that these equations are linearly independent if p is imaginary.

Exercise 9.3. Prove that for a fixed $p \in \mathbb{P}_n$, mapping Q to $H_{p,Q}$ is a singular projectivity $f : \mathbb{Q}(\mathbb{P}_n) \to \mathbb{P}_n^\vee$ whose singular points are the quadrics of which p is a double point. Use this fact and Exercise 5.34 to re-prove 9.10.1.

Exercise 9.4. If $L \subset \mathbb{P}_n$ is a positive-dimensional linear variety, prove that mapping $Q \mapsto Q \cap L$ (for $Q \not\supset L$) is a singular projectivity $f : \mathbb{Q}(\mathbb{P}_n) \to \mathbb{Q}(L)$ whose singular points are the quadrics containing L. Use this fact and Exercise 5.34 to re-prove 9.5.4.

Exercise 9.5. Let $\mathbb{P}_2$ be a real projective plane and C a non-degenerate conic of $\mathbb{C}\mathbb{P}_2$. Prove that if C contains five real points, then it is the complex extension of a real non-degenerate conic with points. Prove by an example that the same claim is false if the number of real points is assumed to be four. *Hint*: Prove that the five real points determine a non-degenerate real conic and consider its projective closure.

Exercise 9.6 (*Constructing a projective plane in which an abstract projective line is immerged as a conic*). Fix a projective line $\mathbb{P}_1$ and consider $\mathbb{Q}(\mathbb{P}_1)$, the two-dimensional projective space of the point-pairs of $\mathbb{P}_1$. Prove that mapping each point $p \in \mathbb{P}_1$ to $2p$ is a map

$$\theta : \mathbb{P}_1 \longrightarrow \mathbb{Q}(\mathbb{P}_1)$$

whose image is the set of points of a non-degenerate conic (with points) C of $\mathbb{Q}(\mathbb{P}_1)$. Prove also that θ induces a projectivity from $\mathbb{P}_1$ to C.

Exercise 9.7. Let C be a non-degenerate conic with points of a projective plane $\mathbb{P}_2$, endowed with its one-dimensional projective structure, and consider the two-dimensional projective space $\mathbb{Q}(C)$ of the point-pairs of C.

(1) Prove that mapping each point $p \in \mathbb{P}_2$ to the point-pair $C \cap H_{p,C}$, section of C by the polar of p, is a projectivity

$$\Gamma : \mathbb{P}_2 \longrightarrow \mathbb{Q}(C)$$

that maps each $p \in C$ to $2p$. (Note that Γ does not depend on any choice. If $\mathbb{Q}(C)$ is identified with $\mathbb{P}_2$ through Γ, then it appears as the ambient plane of C recovered from the internal geometry of C.)

(2) Prove also that if h is any projectivity of C, then $f = \Gamma^{-1} \circ \mathbb{Q}(h) \circ \Gamma$ is a projectivity of $\mathbb{P}_2$ that leaves C invariant and has $f_{|C} = h$ (the only one according to 8.2.22).

(3) After taking coordinates on C and $\mathbb{P}_2$ as in 8.2.4, compute a matrix of f from a matrix of h and compare it with the matrix (8.6) appearing in the proof of 8.2.22.

Exercise 9.8. Prove that two distinct pairs of (distinct or coincident) lines of $\mathbb{P}_2$ sharing no double point, span a regular pencil of conics. *Hint*: Prove that they span a pencil as described in one of the Examples 9.5.13, 9.5.18 or 9.5.19.

Exercise 9.9. (*Requires using the discriminant of a polynomial*) Prove that there exists a non-zero polynomial $P(X_{i,j}, Y_{s,t})$, separately homogeneous in each set of variables $\{X_{i,j}\}_{0 \leq i \leq j \leq 2}$ and $\{Y_{s,t}\}_{0 \leq s \leq t \leq 2}$, such that the pencil spanned by two different conics with matrices $(a_{i,j})$ and $(b_{s,t})$ is general if $P(a_{i,j}, b_{s,t}) \neq 0$. This is a precise form of claiming that most pairs of different conics span a general pencil, which is in turn the reason for the name of these pencils. *Hint*: Use that regular pencils with three degenerate members are general and impose a characteristic equation to be non-trivial and have three distinct solutions.

Exercise 9.10. Prove that the quadrics of Exercise 6.23 (c) are the members of the pencil of quadrics tangent to Q along $Q \cap H_p$. Write their matrices in terms of ρ, coordinates of p and a matrix of Q.

Exercise 9.11. Let $\mathcal{P}$ a pencil of quadrics of $\mathbb{A}_3$ containing a quadric with proper centre. Prove that:

(1) $\mathcal{P}$ contains at most three paraboloids.

(2) If $\mathcal{P}$ contains a cylinder or a pair of planes, then there are at most two paraboloids in $\mathcal{P}$.

(3) If $\mathcal{P}$ contains a parabolic cylinder or a pair of parallel or coincident planes, then there is at most one paraboloid in $\mathcal{P}$.

Give examples in which the bounds of (1), (2) and (3) above are attained. Give also an example of a pencil of quadrics of $\mathbb{A}_{3,\mathbb{C}}$ containing a quadric with proper centre and no paraboloid. *Hint*: Consider the improper sections of the quadrics of $\mathcal{P}$ and use 9.5.4, 9.5.10 and 7.8.1.

Exercise 9.12. Hypothesis and notations being as in Example 9.5.21, assume that C_1 is non-degenerate and $p_1 \neq p_2$. Take any $p \in \pi_1 \cap \pi_2$, $p \neq p_1, p_2$, and $\pi = H_{p,Q}$, for any $Q \in \mathcal{P} - \{\pi_1 + \pi_2\}$, the latter being independent of Q due to 9.10.1. Prove that

(1) The sections $Q \cap \pi$, $Q \in \mathcal{P}$ compose a general pencil of conics of π.

(2) A point p is double for a $Q \in \mathcal{P} - \{\pi_1 + \pi_2\}$ if and only if it is double for $Q \cap \pi$.

(3) The degenerate quadrics of $\mathcal{P}$ other than $\pi_1 + \pi_2$ are either two or none if $k = \mathbb{R}$, and just two if $k = \mathbb{C}$; they have rank three.

Reformulate the above for the case $p_1 = p_2$.

Exercise 9.13. Hypothesis and notations being as in Example 9.5.21, prove that if C_1 is a double line, say $C_1 = 2s$, then the quadrics through C_1 and C_2 are $\pi_1 + \pi_2$ and the cones projecting C_2 from the points of s other than $s \cap \pi_2$. $\mathcal{P}$ is thus a non-regular pencil in which different quadrics share no double point.

Exercise 9.14 (*Ordinary quadratic transformations*). Fix a quadrivertex K of $\mathbb{P}_2$. Assume that its diagonal triangle has vertices p_0, p_1, p_2, their opposite sides being ℓ_0, ℓ_1, ℓ_2, respectively. Let $\mathcal{P}$ be the pencil of the conics circumscribed to K. For any point $p \in \mathbb{P}_2$ set

$$\varphi(p) = \bigcap_{C \in \mathcal{P}} H_{p,C},$$

the intersection of all polars of p relative to the conics of $\mathcal{P}$.

(1) Prove that $\varphi(p)$ is a point, but for $p = p_i, i = 0, 1, 2$.

(2) Let A be a vertex of K and take as reference $\Delta = (p_0, p_1, p_2, A)$. Prove that for any $p = [x_0, x_1, x_2] \neq p_0, p_1, p_2$ it is $\varphi(p) = [\bar{x}_0, \bar{x}_1, \bar{x}_2]$, where

$$\begin{aligned} \bar{x}_0 &= x_1 x_2, \\ \bar{x}_1 &= x_0 x_2, \\ \bar{x}_2 &= x_0 x_1. \end{aligned} \tag{9.38}$$

(3) Prove that

$$\begin{aligned} \varphi\colon \mathbb{P}_2 - \{p_0, p_1, p_2\} &\longrightarrow \mathbb{P}_2 - \{p_0, p_1, p_2\}, \\ p &\longmapsto \varphi(p), \end{aligned}$$

maps the points of ℓ_i to p_i and restricts to a bijection

$$\mathbb{P}_2 - \ell_0 \cup \ell_1 \cup \ell_2 \longrightarrow \mathbb{P}_2 - \ell_0 \cup \ell_1 \cup \ell_2$$

whose square is the identical map.

(4) Prove that for any line ℓ through p_0, $\ell \neq \ell_1, \ell_2$, the images of the points of ℓ (other than a vertex) span a line. It is called the transform of ℓ and denoted by $\varphi(\ell)$. Take $\varphi(\ell_1) = \ell_2$ and $\varphi(\ell_2) = \ell_1$ and prove that then

$$\begin{aligned} \tau\colon p_0^* &\longrightarrow \ell_0, \\ \ell &\longmapsto \varphi(\ell) \cap \ell_0, \end{aligned}$$

is a projectivity.

(5) Assume we have an analytic arc of curve at p_0,

$$\gamma = \{p(t) = [1, x(t), y(t)] \mid t \in k, |t| < \varepsilon\},$$

where $\varepsilon \in \mathbb{R}^+$ and $x(t), y(t)$ are analytic functions defined for $|t| < \varepsilon$, $x(0) = y(0) = 0$. If $\ell : (dy/dt)_{t=o} x_1 - (dx/dt)_{t=o} x_2 = 0$ is the tangent line to γ at $p_0 = p(0)$, prove that

$$\lim_{t \to 0, t \neq 0} \varphi(p(t)) = \tau(\ell).$$

Claims similar to (4) and (5) are also true for p_1, p_2. The map φ, seen as a such or as a correspondence of $\mathbb{P}_2$, is called an *ordinary quadratic transformation* of $\mathbb{P}_2$. It is often defined by the equations (9.38). The points p_0, p_1, p_2, at which the transformation is not defined are called its *fundamental points*, and the triangle they determine, its *fundamental triangle*. Heuristically, each fundamental point is intended to be blown up by φ in such a way that the infinitesimal directions at it materialize through τ into the points of the opposite side of the fundamental triangle.

Exercise 9.15 (*Desargues' theorem on quadrivertices*). Let K be a quadrivertex of $\mathbb{P}_2$ and $\ell \subset \mathbb{P}_2$ a line containing no vertex of K. Prove that the three pairs of opposite sides of K cut on ℓ three pairs of points belonging to the same involution.

Exercise 9.16. Use Desargues' theorem 9.7.4 to prove that a parabola cannot be circumscribed to a parallelogram.

Exercise 9.17. Give two proofs of the theorem of the complete quadrilateral (2.10.3), one using 9.5.16 and the other using Desargues' theorem 9.7.4.

Exercise 9.18. Assume that K is a complete quadrilateral of a Euclidean plane $\mathbb{A}_2$ and $p \in \mathbb{A}_2$ a point lying on no side of K. Use the dual of Desargues' theorem on quadrivertices (Exercise 9.15) to prove that if two of the pairs of lines projecting the pairs of opposite vertices of K from p are of orthogonal lines, then so is the third.

Exercise 9.19. Fix a projective line $\mathbb{P}_1$ and a non-identical projectivity $f : \mathbb{P}_1 \to \mathbb{P}_1$.

(1) Prove that the point-pairs $p + f(p)$ are the points of a conic C_f of $\mathbb{Q}(\mathbb{P}_1)$.

(2) Prove that C_f is degenerate if and only if f is an involution, and that in such a case C_f is two times the pencil of point-pairs corresponding to f by 9.7.1.

(3) Let q be a fixed point of f. Prove that the only pencil of point-pairs which has q as base point is, as a line ℓ_q of $\mathbb{Q}(\mathbb{P}_1)$, tangent to C_f at $2q$.

(4) Prove that for f hyperbolic with given fixed points q_1, q_2, all conics C_f belong to a bitangent pencil whose other conics are $\ell_{q_1} + \ell_{q_2}$ and the conic C of Exercise 9.6.

(5) Fix a point $p \in \mathbb{P}_2$ and prove that for f parabolic with fixed point q, all conics C_f belong to a triple contact pencil whose other conics are $2\ell_q$ and the conic C of Exercise 9.6.

Hint: If M, α and β are, respectively, a matrix of f and coordinate vectors of p and p', write the condition of being $p + p' \in C_f$ in the form $\det(\alpha, M\beta)\det(\beta, M\alpha) = 0$ and introduce the coordinates of $p + p'$ using 6.2.3. The computations are easier using the references of 5.5.6, 5.5.14 and 5.5.17.

Exercise 9.20. Let C be a non-degenerate conic with points of a projective plane $\mathbb{P}_2$ and $f: C \to C$ a projectivity which is not identical or involutive.

(1) Prove that the poles relative to C of the lines $pf(p)$, $p \in C$, are the points of a non-degenerate conic of $\mathbb{P}_2$ which is tangent to C at the fixed points of f.

(2) Prove that the lines $pf(p)$, $p \in C$, envelop a non-degenerate conic of $\mathbb{P}_2$ which is tangent to C at the fixed points of f.

Hint: Either compute using 8.2.6, or use Exercises 9.19 and 9.7.

Exercise 9.21. If H_1 and H_2 are distinct hyperplanes of $\mathbb{P}_n$ and $L = H_1 \cap H_2$, prove that the quadrics of the pencil $\mathcal{P}$, spanned by $2H_1$ and $2H_2$ are the pairs of hyperplanes $H + \sigma(H)$, where $H \in L^*$ and σ is the involution of L^* which leaves fixed H_1 and H_2. In particular the members of $\mathcal{P}$ other than $2H_1$ and $2H_2$ are the pairs of (necessarily distinct) hyperplanes that harmonically divide H_1, H_2.

Exercise 9.22. Prove that if C and C' are tangent cones to a non-degenerate quadric Q of $\mathbb{P}_3$ with vertices p and p', $p, p' \notin Q$, $p \neq p'$, then the intersection of C and C' is a pair of conics whose planes harmonically divide $H_{q,C}$ and $H_{q,C'}$ (in $(H_{q,C} \cap H_{q,C'})^*$). *Hint*: In the net spanned by Q, C and C', consider the pencils spanned by C and C' and by $2H_{q,C}$ and $2H_{q,C'}$, and use 1.4.9 (c) and Exercise 9.21.

Exercise 9.23. Assume that $\mathcal{P}$ is a pencil of conics, all degenerate. Prove that either there are a line ℓ and a point p such that $\mathcal{P} = \{\ell + \ell' \mid \ell' \in p^*\}$ (the conics of $\mathcal{P}$ are composed of a fixed line and a line varying in a pencil), or there is a point p and an involution τ of p^* such that $\mathcal{P} = \{\ell + \tau(\ell) \mid \ell \in p^*\}$ (the conics of $\mathcal{P}$ are the pairs of an involution of a pencil of lines). *Hint*: Use Exercise 9.8.

Exercise 9.24. Re-prove part (b) of Exercise 6.10 using the pencil of conics tangent to two sides of the triangle at the given contact points, and Desargues' theorem 9.7.4.

Exercise 9.25. Prove that a hyperbola and its asymptotes cut on any line two pairs of points with the same midpoint. In particular, the contact point of any tangent to a hyperbola is the midpoint of its intersections with the asymptotes. *Hint*: Consider the pencil of conics spanned by the hyperbola and its asymptotes and use Desargues' theorem 9.7.4.

Exercise 9.26. Given on the drawing plane the vertices of a quadrivertex T and a line ℓ missing them, describe a procedure allowing us to construct by points the two conics circumscribed to T and tangent to ℓ. *Hint*: Determine first the contact points as the fixed points of the involution whose pairs are cut on ℓ by the conics circumscribed to T.

Exercise 9.27. Let $\mathcal{P}$ be a general pencil of conics of $\mathbb{P}_2$ and C a non-degenerate conic with points going through exactly two base points p, p' of $\mathcal{P}$. Prove that the intersections of the conics of $\mathcal{P}$ with C other than p, p' span lines through a fixed point $q \neq C$ (and so are the pairs of an involution of C). *Hint*: Take a point p'' distinct from p, p' and collinear with them. In the net of conics spanned by $\mathcal{P}$ and C consider the pencil $\mathcal{P}'$ of the conics through p'' and show that the conics of $\mathcal{P}'$ have the same intersections with C as those of $\mathcal{P}$.

Exercise 9.28 (*Nine-points conic*). Let G be a quadrivertex of a projective plane $\mathbb{P}_2$ and ℓ a line going through no diagonal point of G. Prove that there is a non-degenerate conic K that contains the three diagonal points of G and whose other points are the poles of ℓ relative to the non-degenerate conics circumscribed to G. Prove that for each side s_i of G, $i = 1, \dots, 6$, K goes through the fourth harmonic of $\ell \cap s_i$ with respect to the two vertices on s_i. In particular, if G is a quadrivertex of an affine plane with no improper diagonal point, then the centres and double points of the conics circumscribed to G are the points of a non-degenerate conic that goes through the diagonal points of G and the midpoints of the six pairs of distinct vertices of G (the *nine-points conic* of G)

Exercise 9.29. Let G be a quadrivertex and ℓ a line of a projective plane $\mathbb{P}_2$. Assume that ℓ is not a side of G and goes through exactly one diagonal point p of G, say the one at which opposite sides r and s meet. Use the arguments of the proof of 9.10.10 to see that the locus of the points that are conjugate to all points of ℓ with respect to a conic circumscribed to G is composed of two lines, namely the fourth harmonic of r, s, ℓ, and the line t spanned by the diagonal points of G other than p. Prove that the poles of ℓ relative to the non-degenerate conics circumscribed to G are the points of t other than the diagonal points spanning it.

Exercise 9.30. Describe the locus of the centres of the non-degenerate conics circumscribed to a quadrivertex of $\bar{\mathbb{A}}_2$ which has no improper side and just one pair of opposite sides parallel. *Hint*: Use Exercise 9.29.

Exercise 9.31. Assume given two non-degenerate conics C_1, C_2 of planes π_1, π_2 of $\mathbb{A}_3$, respectively. Assume also that π_1, π_2 are proper, distinct and non-parallel planes, and that C_1, C_2 share two distinct proper points. Name $\mathcal{P}$ the pencil of the quadrics of $\mathbb{A}_3$ going through C_1 and C_2. Let p be the improper point of $\pi_1 \cap \pi_2$ and $\pi = H_{p,Q}$, $Q \in \mathcal{P}$, $Q \neq \pi_1 + \pi_2$, which is independent of Q due to 9.5.4. Take $\bar{\mathcal{P}} = \{Q \cap \pi \mid Q \in \mathcal{P}\}$, which has been seen to be a general pencil of conics in Exercise 9.12.

(1) Prove that the centre of any non-degenerate $Q \in \mathcal{P}$ is the centre of $Q \cap \pi$.

(2) Prove that the centres of the non-degenerate quadrics of $\mathcal{P}$ either are all coincident, or lie on a line, or lie on a non-degenerate conic of π. Characterize each case in terms of the data π_1, π_2, C_1, C_2.

Exercise 9.32. Let Q be a non-degenerate quadric of an affine space $\mathbb{A}_n$, $n \geq 2$. Assume that it has proper centre O and take $p \notin Q$, $p \neq O$. Prove that the centres of all the non-degenerate quadrics tangent to Q along its section by $H_{p,Q}$, belong to the line Op. *Hint*: Take an affine reference with origin O, the first $n-1$ axes parallel to $H_{p,Q}$ and with $p = (0, \dots, 0, 1)$, and use 9.5.22.

Exercise 9.33. In $\mathbb{P}_2$, notations being as in Section 9.8, prove that:

(1) The rational map extending Υ maps any pair of distinct lines $\ell + \ell'$ to the conic envelope $2p^*$, $p = \ell \cap \ell'$, and is undefined on any double line.

(2) The rational map extending Υ^{-1} maps any pair of distinct pencils of lines $p^* + q^*$ to the point-conic $2pq$ and is undefined on any pencil of lines counted twice.

Note that neither of the extensions of Υ and Υ^{-1} is injective.

Exercise 9.34. Let Q be a quadric of $\mathbb{P}_n$, $n > 1$, of rank n and with matrix $A = (a_{i,j})$. Use the properties of the determinants to prove that $(a_{i,j})(A_{i,j}) = 0$, $(A_{i,j}) = \tilde{A}$ being the matrix of cofactors of A. Deduce that the non-zero columns of $\tilde{A}$ are all coordinate vectors of the double point p of Q, and therefore that the image $\tilde{Q}$ of Q by the rational map extending Υ (see Section 9.8) has rank one. Write thus

$$\sum_{i,j=0}^{n} A_{i,j} u_i u_j = \rho\Big(\sum_{s=0}^{n} \alpha_s u_s\Big)^2, \quad \rho \neq 0,$$

and prove that also $(\alpha_0, \dots, \alpha_n)$ is a coordinate vector of p. Conclude that $\tilde{Q}$ is $2p^*$, the bundle of hyperplanes through p counted twice.

Exercise 9.35. Prove that the pencil of conics $\lambda x_0^2 + \mu(x_1^2 - x_0 x_2) = 0$ is a triple contact pencil, and also that any triple contact pencil may be written in this form using a suitable reference. Prove that the non-degenerate conics of a triple contact pencil compose a range whose associated pencil is also a triple contact pencil. Prove also that any range of conics with triple contact associated pencil is the set of non-degenerate conics of a triple contact pencil.

Exercise 9.36. Determine a plane π of $\mathbb{Q}(\mathbb{P}_2)$, a non-degenerate conic κ of π (a conic of conics, thus) and three elements D_1, D_2, D_3 of κ such that the elements of κ other than D_1, D_2, D_3 are the members of the range of Example 9.8.2.

Exercise 9.37. Assume given a triangle T of a projective plane $\mathbb{P}_2$. Lemma 7.1.3 proves that the conics of $\mathbb{P}_2$ with self-polar triangle T describe a net $\mathcal{N}$.

(1) Determine the degenerate conics of $\mathcal{N}$, proving in particular that they compose three non-concurrent (as lines of $\mathcal{N}$) pencils.

(2) If $p \in \mathbb{P}_2$ lies on no side of T, prove that the conics of $\mathcal{N}$ going through p describe a general pencil whose base points are the vertices of a quadrivertex of $\mathbb{P}_2$ with diagonal triangle T and one vertex at p. Use this to re-prove the existence and uniqueness of such a quadrivertex, see Exercises 2.18 and 2.19.

(3) If p is now assumed to belong to a side of T and not to be a vertex, prove that the conics of $\mathcal{N}$ through p describe a bitangent pencil one of whose members is two times the side of T containing p.

(4) Prove that two quadrivertices of $\mathbb{P}_2$ with the same diagonal triangle are both inscribed in a non-degenerate conic. State the dual claim.

(5) If $\mathbb{P}_2$ is assumed to be the projective closure of a real affine plane and the vertices of T to be proper points, prove that the parabolas in $\mathcal{N}$ describe a range. Determine the degenerate envelopes of the associated pencil.

Exercise 9.38. Let T be a quadrilateral of an affine plane with proper sides, no two parallel. Prove that the centres of the non-degenerate conics inscribed in T belong to a line ℓ, which contains also the midpoints of the three pairs of opposite vertices of T and no other point. This in particular proves that these midpoints are collinear. *Hint* use 9.10.1 in the dual plane.

Exercise 9.39 (*Involves some tensor calculus*). Prove that a point quadric $Q = [\psi]$ and a quadric envelope $T = [\delta]$ are apolar if and only if the result of contracting the indices $1, 1$ and $2, 2$ of the two times contravariant and two times covariant tensor $\psi \otimes \delta$, is zero. This provides an alternative proof of 9.9.1.

Exercise 9.40. Prove that, in a projective plane $\mathbb{P}_2$, a double line 2ℓ is apolar to a conic envelope K if and only if ℓ belongs to K. Use this to prove that given six lines of $\mathbb{P}_2$, $\ell_i : h_i = 0, i = 1, \dots, 6$, the polynomials h_i^2, $i = 1, \dots, 6$, are linearly dependent if and only if the lines ℓ_i, $i = 1, \dots, 6$, either are all tangent to a non-degenerate conic, or each goes through one of the points of a certain pair of distinct or coincident points of $\mathbb{P}_2$.

Exercise 9.41. Re-prove Hesse's theorem (see Exercise 6.14) using that the envelopes of the three pairs of opposite vertices of a quadrivertex belong to the same pencil of conic envelopes (9.5.15), and that being a given conic apolar to a conic envelope K is a linear condition on K.

Exercise 9.42. Combine 9.3.17 and Exercise 9.40 to prove that two triangles of $\mathbb{P}_2$, each with no vertex on a side of the other, are both self-polar with respect to a non-degenerate conic C if and only if they are both circumscribed to a second non-degenerate conic C'. Compare with Exercise 8.30. Write the dual claim and re-prove the claim of Exercise 8.28.

Exercise 9.43. Let Q_1 and Q_2 be different quadrics of a projective space $\mathbb{P}_n$ and assume that Q_1 is non-degenerate. Write a characteristic polynomial of the pencil they span in the form

$$\mathcal{C}(\lambda) = \det(A + \lambda B) = c_0 + c_1\lambda + \cdots + c_{n+1}\lambda^{n+1},$$

where A and B are matrices of Q_1 and Q_2, respectively. Prove that $c_1 = 0$ if and only if Q_2 is outpolar to Q_1. *Hint*: Use that c_1 is the value of the first derivative of $\mathcal{C}(\lambda)$ for $\lambda = 0$.

Exercise 9.44. Let K be a quadric of $\mathbb{P}_{n,\mathbb{R}}$ without points. Prove that a quadric Q of $\mathbb{P}_{n,\mathbb{R}}$ contains the vertices of a simplex S, self-polar for K, if and only if Q is outpolar to K. Prove also that if this is the case, then one vertex of S may be arbitrarily chosen in Q. *Hint*: For the *if* part, prove that Q has points, pick any $p \in Q$ as a vertex of a reference relative to which the equation of K is reduced and use induction on n.

Exercise 9.45. Assume that $f : \mathbb{P}_n \to \overline{\mathbb{P}}_n$ is a projectivity. If Ξ and $\overline{\Xi}$ are the morphisms of 9.9.3 corresponding to $\mathbb{P}_n$ and $\overline{\mathbb{P}}_n$, respectively, prove that $\mathbb{Q}(f^\vee) \circ \Xi = \overline{\Xi} \circ \mathbb{Q}(f)^\vee$.

Exercise 9.46. Prove that for $k = \mathbb{R}$ and n even, the canonical equations of a rational normal curve of $\mathbb{P}_n$ (9.25) do not give all the coordinates of the points of Γ.

Exercise 9.47. Assume that a twisted cubic Γ of $\mathbb{P}_{3,\mathbb{C}}$ has canonical equations relative to a reference (p_0, p_1, p_2, p_3, A), and prove that

(a) $p_0, p_1, A \in \Gamma$,

(b) the tangent line to Γ at p_0 is $p_0 \vee p_1$,

(c) the osculating plane to Γ at p_0 is $p_0 \vee p_1 \vee p_2$,

(d) the tangent line to Γ at p_3 is $p_2 \vee p_3$,

(e) the osculating plane to Γ at p_3 is $p_1 \vee p_2 \vee p_3$.

Conversely, assume that Γ is a twisted cubic given by proper parametric equations relative to a reference (p_0, p_1, p_2, p_3, A). Prove that if the above conditions (a) to (d) are satisfied and, furthermore, the pairs of parameters of p_0, p_1, and A are $(1, 0)$, $(0, 1)$ and $(1, 1)$, respectively, then the parametric equations of Γ are canonical.

Exercise 9.48. The hypothesis and notations being as in 9.12.3, check by direct computation from the equation of $Q_{a,b,c}$ that, as implicitly said in Theorem 9.12.4, σ is an involution if and only if $Q_{a,b,c}$ is non-degenerate.

Exercise 9.49. Let p be a point of a twisted cubic Γ of $\mathbb{P}_{3,\mathbb{C}}$ and π a plane missing p. Prove that mapping $q \mapsto pq \cap \pi$ is a projectivity between Γ and the section of the cone projecting Γ from p by π.

Exercise 9.50. Prove that Theorem 9.11.23 still holds true for a twisted cubic if the points of one or both of the pairs $\{p_1, p_2\}, \{q_1, q_2\}$ are coincident.

Exercise 9.51 (*Projective generation of twisted cubics*). Prove that any twisted cubic of $\mathbb{P}_{3,\mathbb{C}}$ may be obtained as the locus of the intersection points of triples of corresponding planes $\pi, f(\pi), g(\pi)$ of three pencils of planes related by two projectivities

$$\ell^* \xleftarrow{\;f\;} r^* \xrightarrow{\;g\;} s^*.$$

Hint: Take r, s, ℓ spanned by three pairwise disjoint pairs of different points of Γ not corresponding by the same involution.

Exercise 9.52. Let Γ be a twisted cubic of $\mathbb{P}_{3,\mathbb{C}}$, $\mathcal{L}_\Gamma$ the net of quadrics through Γ, seen as a projective plane, and K the conic of $\mathcal{L}_\Gamma$ whose elements are the cones through Γ (see the argument preceding 9.12.8).

(a) Determine the pencils of quadrics through Γ that are, as lines of $\mathcal{L}_\Gamma$, tangent to K.

(b) If Q is a non-degenerate quadric through Γ, determine the cones through Γ which, in $\mathcal{L}_\Gamma$, are the contact points of the tangent lines to K through Q, and also the pencil which is the polar line of Q relative to K.

Chapter 10
Metric geometry of quadrics

In Section 6.9 we have dealt with the basic points of linear metric geometry in terms of the absolute quadric. In this chapter we will present the metric properties and metric classification of quadric hypersurfaces. This will complete our study of quadrics, already done at the projective and affine levels in preceding chapters. As needed for metric geometry, throughout this chapter we take $k = \mathbb{R}$. Unless otherwise said, we will work in a Euclidean n-dimensional space $\mathbb{A}_n$, $n \geq 2$, with improper hyperplane H_∞, scalar product K and absolute quadric $\mathbf{K} = [K]$ (see the beginning of Section 6.9). If Q is a quadric of $\mathbb{A}_n$ and ℓ a chord of Q with proper and real ends, then the *length* of ℓ is taken to be the distance between its ends. This is reminiscent of elementary geometry, in which chords are segments rather than lines. If p is a simple point of Q, the *normal* to Q at p is the line through p orthogonal to the hyperplane tangent to Q at p. Two quadrics of n-dimensional Euclidean spaces $\mathbb{A}_n$, $\mathbb{A}'_n$ are called *congruent* if and only if there is an isometry $f : \mathbb{A}_n \to \mathbb{A}'_n$ mapping one quadric into the other. Congruence of quadrics is an equivalence relation and the classification of quadrics of Euclidean spaces modulo congruence is the *metric classification of quadrics* (see Section 6.9). We will proceed separately with the plane and three-space cases, as they are richer in details not always worth generalizing to higher dimension, and furthermore they allow a lighter treatment.

10.1 Circles and spheres

The metric properties of quadrics depending on their relationship with the absolute, we begin by studying the quadrics of a Euclidean space $\mathbb{A}_n$ that contain the absolute. We know from 9.3.18 that the quadrics of $\mathbb{A}_n$ containing the absolute describe a linear system of dimension $n + 1$, but anyway this will be re-proved by the computations below. Fix an orthonormal reference and use homogeneous coordinates relative to it. As seen in Section 6.9, an equation of the absolute K is

$$x_1^2 + \cdots + x_n^2 = 0$$

and so, computing as in the proof of 9.3.18, a quadric Q contains K if and only if it has an equation of the form

$$\lambda(x_1^2 + \cdots + x_n^2) + 2\sum_{i=1}^{n} \bar{a}_i x_0 x_i + \bar{a} x_0^2 = 0 \tag{10.1}$$

for $\lambda, \bar{a}_1, \ldots, \bar{a}_n, \bar{a} \in \mathbb{R}$, not all zero.

If $\lambda = 0$ we obtain all the hyperplane-pairs $H + H_\infty$, for H an arbitrary hyperplane of $\bar{\mathbb{A}}_n$. In particular we get $2H_\infty : x_0^2 = 0$.

If, otherwise, $\lambda \neq 0$, then the quadric defined by (10.1) does not contain H_∞. By dividing (10.1) by λ and taking $a = \bar{a}/\lambda$ and $a_i = -\bar{a}_i/\lambda$, $i = 1, \dots, n$, we get

$$x_1^2 + \dots + x_n^2 - 2\sum_{i=1}^{n} a_i x_0 x_i + a x_0^2 = 0, \tag{10.2}$$

and so a quadric has the absolute as improper section if and only if it has an equation such as (10.2).

After switching to affine coordinates $X_i = x_i/x_0$, $i = 1, \dots, n$, and completing squares, equation (10.2) reads

$$(X_1 - a_1)^2 + \dots + (X_n - a_n)^2 = D, \tag{10.3}$$

with $D = \sum_{i=0}^{n} a_i^2 - a$. The quadric Q it defines is degenerate if and only if $D = 0$, and in such a case it is the cone over the absolute with vertex the proper point $(a_1, \dots, a_n)$. Thus we get all the cones over the absolute with vertex a proper point, which have been already named *isotropic cones*.

If $D \neq 0$, then the quadric Q is non-degenerate and it is clear that it has proper centre $(a_1, \dots, a_n)$. If $D > 0$, the equation (10.3) is just the condition for a proper point $(X_1, \dots, X_n)$ to be at distance $\sqrt{D}$ from $(a_1, \dots, a_n)$: then, according to the elementary definition, the quadric Q is called the sphere (circle if $n = 2$) with centre $(a_1, \dots, a_n)$ and radius $\sqrt{D}$; it clearly contains points. If $D < 0$, then Q is a non-degenerate quadric with no real point and improper section $\mathbf{K}$. Also this quadric is determined by its centre $(a_1, \dots, a_n)$ and D: it is called the *imaginary sphere* (*imaginary circle* if $n = 2$) with centre $(a_1, \dots, a_n)$ and radius $\sqrt{D} = \boldsymbol{i}\sqrt{-D}$.

All together we have seen:

Theorem 10.1.1. *The quadrics of a Euclidean space $\mathbb{A}_n$ that contain the absolute quadric are:*

- *the spheres,*
- *the imaginary spheres,*
- *the isotropic cones, and*
- *the pairs of hyperplanes $H + H_\infty$, where H is any hyperplane of $\bar{\mathbb{A}}_n$.*

An orthonormal reference being arbitrarily fixed in $\mathbb{A}_n$, a quadric of $\mathbb{A}_n$ is a sphere (resp. an imaginary sphere, an isotropic cone) if and only if it has an equation of the form

$$(X_1 - a_1)^2 + \dots + (X_n - a_n)^2 = D,$$

with $a_1, \dots, a_n, D \in \mathbb{R}$ and $D > 0$ (resp. $D < 0$, $D = 0$).

A direct consequence is:

Corollary 10.1.2. *A quadric of a Euclidean space is a sphere if and only if it is non-degenerate, has real points and contains the absolute quadric.*

Remark 10.1.3. In particular the circles are the non-degenerate conics with points that contain the cyclic points.

The next proposition shows that the spheres, the imaginary spheres and the isotropic cones have a high degree of symmetry.

Proposition 10.1.4. *Let Q be either a sphere, an imaginary sphere or an isotropic cone of a Euclidean space $\mathbb{A}_n$. If O is the centre or the double point of Q, according to the case, then any motion of $\mathbb{A}_n$ leaving O invariant, leaves Q invariant too.*

Proof. Let f be any motion of $\mathbb{A}_n$ leaving O invariant and take an orthonormal reference with origin O. According to the computations above and 6.9.24 the matrices of Q and f may be taken as, in blocks,

$$\begin{pmatrix} -D & 0 \\ 0 & \mathbf{1}_n \end{pmatrix} \quad \text{and} \quad \begin{pmatrix} 1 & 0 \\ 0 & N \end{pmatrix},$$

with $N^t N = \mathbf{1}_n$. Then a direct computation proves that $f(Q) = Q$. □

The affine types of the quadrics containing the absolute are clear from their descriptions and equations above. In particular the spheres are non-degenerate quadrics with proper centre containing points but no lines or improper points ($\mathbf{r} = n + 1$, $\mathbf{j} = n - 1$, $\mathbf{r}' = n$, $\mathbf{j}' = n - 1$). Thus, the spheres of a three-space are ellipsoids and the circles are ellipses.

Proposition 10.1.5. *If Q is either a sphere or an imaginary sphere and has centre O, then for any point $p \neq O$ the diameter Op and the polar hyperplane H_p are orthogonal. Conversely, any non-degenerate quadric with proper centre O, relative to which the polar hyperplane of any point $p \neq O$ is orthogonal to Op, is either a sphere or an imaginary sphere.*

Proof. The polar hyperplane H_p is proper because $p \neq O$. Let p_∞ be the improper point of the diameter Op. Conjugation being relative to Q, since H_∞ is the polar hyperplane of O, all points of $H_p \cap H_\infty$ are conjugate to O and p, and therefore also to any point of the diameter Op. They are in particular conjugate to p_∞ and so (by 6.4.5), $H_p \cap H_\infty$ is the polar hyperplane of p_∞ relative to $Q \cap H_\infty$. Since $Q \cap H_\infty = \mathbf{K}$, the direct claim is clear after 6.9.4. For the converse, assume that p is any improper point. Then, still $H_p \cap H_\infty$ is the polar hyperplane of p relative to $Q \cap H_\infty$. On the other hand, since by the hypothesis H_p is orthogonal to Op, $H_p \cap H_\infty$ is also the polar hyperplane of p relative to the absolute $\mathbf{K}$: their polarities being coincident, $Q \cap H_\infty = \mathbf{K}$ by 6.5.2, and then 10.1.1 applies. □

As a direct consequence we obtain a well-known property of spheres, namely:

Corollary 10.1.6. *If S is a sphere, then, for any $p \in S$, the diameter through p and the tangent hyperplane at p are orthogonal.*

If Q is a sphere and its equation relative to an orthonormal reference has the form of (10.3), then the term D equals the square of the distance from the centre to any point of Q. Therefore D does not depend on the choice of the reference. The next proposition shows a different intrinsic characterization of D which applies also to the imaginary spheres:

Proposition 10.1.7. *If, relative to an orthonormal reference, the equation of a sphere or imaginary sphere Q, with centre O, is written, as in* 10.1.1,

$$(X_1 - a_1)^2 + \cdots + (X_n - a_n)^2 = D,$$

then, for any proper point $p \neq O$, the product of distances from O to p and its polar hyperplane equals $|D|$. The radius of Q, $\sqrt{D}$, is independent of the choice of the orthonormal reference.

Proof. A direct computation from the equation of Q shows that if $p = (\alpha_1, \dots, \alpha_n)$, then H_p has equation

$$\sum_{i=1}^{n}(\alpha_i - a_i)X_i - \sum_{i=1}^{n}(\alpha_i - a_i)a_i - D = 0.$$

After this, the centre of Q being $O = (a_1, \dots, a_n)$, the first claim follows by a direct computation. For the second one just note that the sign of D is already determined by the projective type of Q. □

Corollary 10.1.8. *An isometry between Euclidean spaces transforms spheres (resp. imaginary spheres) into spheres (resp. imaginary spheres) of the same radius.*

Proof. Let $f\colon \mathbb{A}_n \to \mathbb{A}'_n$ be an isometry. If $\mathbf{K}$ is the absolute quadric of $\mathbb{A}_n$, then, by the definition of isometry in Section 6.9, $f(\mathbf{K})$ is the absolute quadric of $\mathbb{A}'_n$. If Q is a sphere (resp. an imaginary sphere) of $\mathbb{A}_n$, then $Q \supset \mathbf{K}$ and therefore $f(Q) \supset f(\mathbf{K})$. On the other hand, since f is a projectivity, $f(Q)$ is non-degenerate and has real points (resp. no real points). Thus, by 10.1.1, $f(Q)$ is a sphere (resp. an imaginary sphere) of $\mathbb{A}'_n$. Now, for the equality of radii, let O be the centre of Q. Pick any $p \in \mathbb{A}_n$, $p \neq O$, and let H_p be the polar hyperplane of p relative to Q. Since f is an affinity, $f(O)$ is the centre of $f(Q)$ and $f(H_p)$ is the polar hyperplane of $f(p)$ relative to $f(Q)$. After this it is enough to use 10.1.7 and the preservation of distances by f. □

10.2 Metric properties of conics

In this section we will be concerned with non-degenerate conics of Euclidean planes. The case of parabolas being quite different, we begin by considering a conic C of a Euclidean plane $\mathbb{A}_2$ with proper centre O. Thus, $C \cap \ell_\infty$ is a pair of distinct points and C may be either an imaginary conic, an ellipse, or a hyperbola. We have two involutions on the improper line ℓ_∞ of $\mathbb{A}_2$, namely the conjugation with respect $C \cap \ell_\infty$ and the conjugation with respect to the pair of cyclic points $I + J$. By projecting the improper points from the centre O, these involutions give rise to two involutions on the pencil of diameters O^*, which are, in this order, the conjugation of diameters (see Section 7.8) and the orthogonality of diameters. We are looking for a pair of simultaneously conjugate and orthogonal diameters. If C is a circle or an imaginary circle, then $C \cap \ell_\infty = I + J$, the two involutions coincide and therefore any two orthogonal diameters are conjugate, and conversely. Otherwise, by 6.5.2, the two involutions on ℓ_∞ are distinct. By 8.2.11, they share a unique pair of corresponding points, which, I, J being imaginary, is a pair of distinct real points. Projecting them from the centre of C gives rise to a unique pair of orthogonal and conjugate diameters and therefore we have proved:

Proposition 10.2.1. *Let C be a conic with proper centre of a Euclidean plane. If C is a circle or an imaginary circle, then any two orthogonal diameters of C are conjugate, and conversely. Otherwise there is a unique pair of diameters of C which are simultaneously orthogonal and conjugate.*

A pair of orthogonal and conjugate diameters of a conic with proper centre C is called a pair of *principal axes* (or just *axes*) of C. What is claimed above is that a conic with proper centre has a unique pair of principal axes, but for the circles and the imaginary circles, for which any pair of orthogonal diameters is a pair of principal axes.

A principal axis is orthogonal to its conjugate and hence different from it. By 7.8.9, it is not an asymptote and therefore has two distinct ends, which may be real or imaginary: the ends of the principal axes are called the *vertices* of the conic. As the reader may easily check any point of a circle or an imaginary circle is a vertex, while any other conic with proper centre has four distinct vertices. We have:

Proposition 10.2.2. *A real point p of a conic C with proper centre O is a vertex of C if and only if the tangent to C at p is orthogonal to the diameter Op.*

Proof. The tangent to C at p is parallel to the diameter conjugate to Op by 7.8.11, hence it is orthogonal to Op if and only if Op is a principal axis. □

Assume that the conic C, with proper centre O, has principal axes e, e'. Denote by p, p' the improper points of e, e', respectively. Define an affine reference Δ by taking the centre O as origin and a couple of unitary vectors on the principal axes

e, e' as basis: e, e' being orthogonal, Δ is an orthonormal reference. On the other hand, by 7.8.10, the matrix of C is diagonal. It is of course regular because C is non-degenerate, so an equation of C relative to Δ may be written in the form

$$\frac{X^2}{\lambda} + \frac{Y^2}{\mu} = 1,$$

where X, Y are the affine coordinates relative to Δ and $\lambda, \mu \in \mathbb{R} - \{0\}$. The signs of λ, μ determine the affine type of C. If λ and μ have opposite signs then C is a hyperbola and, up to swapping over the coordinates, we assume $\lambda > 0 > \mu$.

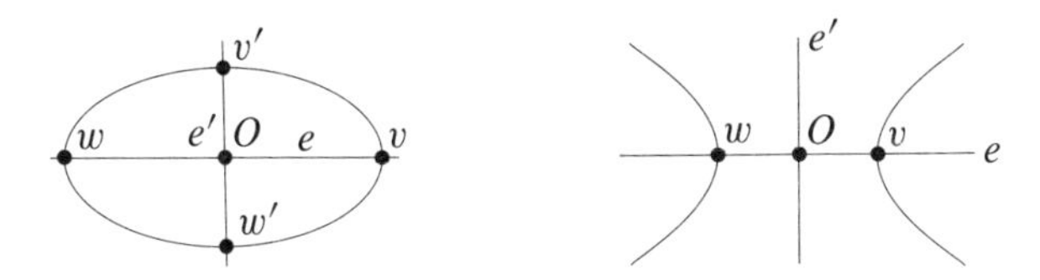

Figure 10.1. Centre O, principal axes e, e' and vertices v, w, v', w' of an ellipse (left) and (minus v', w') a hyperbola (right).

Otherwise, C is imaginary if both λ and μ are negative, or is an ellipse if λ and μ are positive. In these cases we assume, again up to swapping over the coordinates, that $|\lambda| \geq |\mu|$. By taking $\mathbf{a} = \sqrt{|\lambda|}$ and $\mathbf{b} = \sqrt{|\mu|}$ we have seen:

Theorem 10.2.3. *If C is a conic with proper centre of a Euclidean plane $\mathbb{A}_2$, there is an orthonormal reference of $\mathbb{A}_2$ relative to which*

- *if C is an imaginary conic, it has equation*

$$-\frac{X^2}{\mathbf{a}^2} - \frac{Y^2}{\mathbf{b}^2} = 1, \tag{10.4}$$

  with $\mathbf{a} \geq \mathbf{b} > 0$;

- *if C is an ellipse, it has equation*

$$\frac{X^2}{\mathbf{a}^2} + \frac{Y^2}{\mathbf{b}^2} = 1, \tag{10.5}$$

  with still $\mathbf{a} \geq \mathbf{b} > 0$;

- *if C is a hyperbola, it has equation*

$$\frac{X^2}{\mathbf{a}^2} - \frac{Y^2}{\mathbf{b}^2} = 1, \tag{10.6}$$

  with $\mathbf{a}, \mathbf{b} > 0$.

Remark 10.2.4. Note that, conversely, if a conic C has an equation of the form of (10.4), (10.5) or (10.6) above, relative to an orthonormal reference, then it is, respectively, an imaginary conic, an ellipse or a hyperbola.

An equation of a conic C with proper centre is called a *metric reduced equation* of C if and only if it is relative to an orthonormal reference and has the form of (10.4), (10.5) or 10.6); Theorem 10.2.3 assures thus that any conic with proper centre of a Euclidean plane has a metric reduced equation. The orthonormality of the reference is essential to the usefulness of the reduced equations in the metric context.

Proposition 10.2.5. *Each conic C with proper centre, of a Euclidean plane, has a unique metric reduced equation. If Δ is an orthonormal reference relative to which C takes its metric reduced equation, then the origin of Δ is the centre of C and its axes are principal axes of C.*

Proof. Let us prove the second claim first. If the equation is reduced, then the matrix of C is diagonal and therefore, by 7.8.10, the origin of Δ is the centre of C and the axes of Δ are conjugate diameters. Since the axes of Δ are by hypothesis orthogonal, they are principal axes of C.

The existence of reduced equations has been already proved in 10.2.4. Its uniqueness is direct from 10.1.7 if C is a circle or an imaginary circle; we assume thus that C is not, after which it has a unique pair of principal axes, by 10.2.1. Assume that C has reduced equations

$$\alpha X^2 + \beta Y^2 = 1 \quad \text{and} \quad \bar{\alpha}\bar{X}^2 + \bar{\beta}\bar{Y}^2 = 1,$$

relative to orthonormal references Δ, $\bar{\Delta}$ respectively; we have already seen that both references have as origin the centre of C and as coordinate axes the principal axes of C. Since both references are orthonormal, the vectors of their basis are pairwise equal or opposite and so the corresponding coordinates are related either by the equalities

$$\bar{X} = \pm X, \quad \bar{Y} = \pm Y$$

or by the equalities

$$\bar{X} = \pm Y, \quad \bar{Y} = \pm X.$$

In the first case, after substitution, $\bar{\alpha}X^2 + \bar{\beta}Y^2 = 1$ is also an equation of C relative to Δ; this forces $\bar{\alpha} = \alpha$, $\bar{\beta} = \beta$ and so the two reduced equations agree. In the second case, proceeding similarly, $\bar{\beta}X^2 + \bar{\alpha}Y^2 = 1$ is an equation of C relative to Δ and so $\bar{\alpha} = \beta$, $\bar{\beta} = \alpha$. For each of the three affine types, these equalities are incompatible with the inequalities set for the coefficients of the reduced equations. Thus, the second case does not occur and, as claimed, the reduced equations agree. □

A number of direct consequences of 10.2.3 will we presented next, case by case. There is not much to say about imaginary conics but for the clear fact that one of them is an imaginary circle if and only if its reduced equation has $\mathbf{a} = \mathbf{b}$. Similarly, an ellipse is a circle if and only if its reduced equation has $\mathbf{a} = \mathbf{b}$. As seen above, ellipses other than circles have a single pair of principal axes, which are the coordinate axes of any reference giving a metric reduced equation (10.2.5). Computing from the reduced equation, the vertices of an ellipse are the points $(\mathbf{a}, 0)$ and $(-\mathbf{a}, 0)$, at distance $\mathbf{a}$ from the centre, and $(0, \mathbf{b})$ and $(0, -\mathbf{b})$, at distance $\mathbf{b}$ from the centre. The principal axis spanned by the two vertices laying furthest from the centre (the X-axis if the coordinates are named as in 10.2.3) is called the *major axis* of the ellipse; the other axis is called the *minor axis*. Hyperbolas always have a single pair of principal axes, which again need to be taken as the coordinate axes to get a metric reduced equation (10.2.5). Two of the four vertices of a hyperbola, namely $(\mathbf{a}, 0)$ and $(-\mathbf{a}, 0)$, are real, while the other, $(0, \boldsymbol{i}\,\mathbf{b})$ and $(0, -\boldsymbol{i}\,\mathbf{b})$, are imaginary. The principal axis spanned by the real vertices is called the *transverse axis* of the hyperbola, the other being called the *conjugate* (or *non-transverse*) *axis*. Hyperbolas with orthogonal asymptotes (or, equivalently, with improper points harmonically dividing the cyclic ones) are called *equilateral hyperbolas*: the reader may easily check that they are characterized by having $\mathbf{a} = \mathbf{b}$ in their reduced equations.

For an ellipse (resp. hyperbola) $\mathbf{2a}$ is called the *length of the major axis* (resp. *transverse axis*) and $\mathbf{2b}$ the *length of the minor axis* (resp. *conjugate axis*). Similarly $\mathbf{a}$ and $\mathbf{b}$ are called the *lengths of the semi-axes* (major or transverse and minor or conjugate, according to the case). This is reminiscent of elementary geometry, in which the axes and *semi-axes* are defined as the segments of the principal axes joining the two vertices and the centre to the vertices, respectively, but for the conjugate axes and semi-axes of the hyperbolas, which are conventionally defined to suit.

Getting metric reduced equations for parabolas is easier. Assume that C is now a parabola and O its (improper) centre. The improper line ℓ_∞ is thus tangent to C at O. Take $O' \in \ell_\infty$ to be the conjugate to O with respect to the pair of cyclic points (the direction orthogonal to O, thus) and e the polar of O' relative to C: the line e is called the *axis* of C. The centre O being conjugate to all the improper points, $O \in e$, while $e \neq \ell_\infty$ because $O' \neq O$. The axis e is thus a proper diameter and therefore (7.8.5) it is a chord with a single proper end V, which is real. The point V is called the *vertex* of C. The axis being the polar of O', the tangent t to C at V goes through O' and so, by the definition of O', e and t are orthogonal. We leave to the reader to check that V is the only point of the parabola at which the tangent is orthogonal to the diameter. We take an affine reference Δ with origin V and basis a couple of unitary vectors on e and t: it is an orthonormal reference because, as seen above, t and e are orthogonal. Furthermore the projective reference associated to Δ has two vertices, V and O, on C, while the remaining one, O' is just the pole

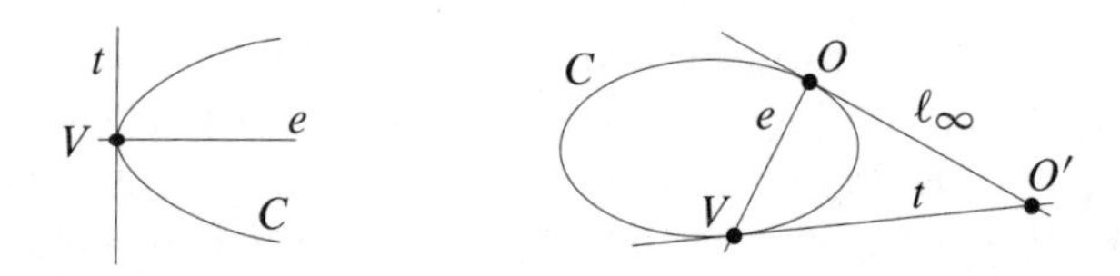

Figure 10.2. Euclidean (left) and non-affine (right) views of a parabola C, its axis e, vertex V and tangent at the vertex t. ℓ_∞ is the improper line, O the improper centre and O' the direction orthogonal to O.

of $e = VO$. Thus the vertices V and O are self-conjugate and both are conjugate to O', after which the equation of C in homogeneous coordinates has the form

$$2a_{0,2}x_0x_1 + a_{1,1}x_2^2 = 0$$

with $a_{0,2}, a_{1,1} \neq 0$ because C is non-degenerate. Using affine coordinates, C has thus equation

$$Y^2 = -2\frac{a_{0,2}}{a_{1,1}}X.$$

Up to replacing the first vector of the basis of the reference with its opposite, one may assume $\mathbf{p} = -a_{0,2}/a_{1,1} > 0$, after which we obtain:

Theorem 10.2.6. *If C is a parabola of a Euclidean plane $\mathbb{A}_2$, there is an orthonormal reference of $\mathbb{A}_2$ relative to which C has equation*

$$Y^2 = 2\mathbf{p}X, \tag{10.7}$$

with $\mathbf{p} > 0$.

Conversely, it is clear that any equation $Y^2 = 2\mathbf{p}X$ defines a parabola. An equation of a parabola C of the form $Y^2 = 2\mathbf{p}X$, with $\mathbf{p} > 0$, and relative to an orthonormal reference, is called a *metric reduced equation* of C. Also for parabolas the metric reduced equation is unique:

Proposition 10.2.7. *Each parabola C of a Euclidean plane has a unique metric reduced equation. If Δ is an orthonormal reference relative to which the equation of C is reduced, then the origin of Δ is the vertex of C, and the first and second axis of Δ are the axes of C and the tangent at the vertex of C, respectively.*

Proof. The second claim follows by a direct computation of the axis, the vertex and the tangent at the vertex from the reduced equation of C. The existence of a metric reduced equation has been seen in 10.2.6 above. Regarding its uniqueness, if C has metric reduced equations

$$Y^2 = 2\mathbf{p}X \quad \text{and} \quad \bar{Y}^2 = 2\mathbf{q}\bar{X},$$

$\mathbf{p}, \mathbf{q} > 0$, relative to orthonormal references Δ and $\bar{\Delta}$, then, by the already proved second claim, Δ and $\bar{\Delta}$ have the same origin and the same first and second axis. The coordinates relative to them are thus related by equalities

$$\bar{X} = \pm X, \quad \bar{Y} = \pm Y.$$

After substitution, we see that $Y^2 = \pm 2\mathbf{q}X$ is also an equation of C relative to Δ, from which $\mathbf{q} = \pm\mathbf{p}$ and so, both being positive, $\mathbf{q} = \mathbf{p}$, as claimed. □

The positive real number $\mathbf{p}$ appearing in the reduced equation $Y^2 = 2\mathbf{p}X$ of a parabola is called the *parameter* of the parabola.

The orthonormal references relative to which a non-degenerate conic C takes its metric reduced equation, are called *principal references* of C. The metric reduced equations of the non-degenerate conics provide an easy description of their metric classification.

Theorem 10.2.8 (Metric classification of non-degenerate conics). *Two non-degenerate conics of Euclidean planes are congruent if and only if they have equal metric reduced equations.*

Proof. Let C and C' be non-degenerate conics of Euclidean planes $\mathbb{A}_2$ and $\mathbb{A}'_2$, respectively. If $f : \mathbb{A}_2 \to \mathbb{A}'_2$ is an isometry and $f(C) = C'$, take Δ to be a principal reference of C. Then, $f(\Delta)$ is an orthonormal reference of $\mathbb{A}'_2$ relative to which (by 2.8.1 (b)) the equation of C' agrees with the equation of C; therefore the latter is the metric reduced equation of C' and both reduced equations agree. Conversely, if C and C' have equal reduced equations, relative to orthonormal references Δ and Δ', then the affinity mapping Δ to Δ' is an isometry by 6.9.30 and maps C to C' because both have the same equation. □

In particular, the lengths of the semi-axes $\mathbf{a}$, $\mathbf{b}$ and the parameter $\mathbf{p}$ are metric invariants. Note that their range is the set of all positive real numbers. This makes a substantial difference with the invariants used in the projective and affine classifications, all of which have a finite range. Numerical invariants whose range, as the above, is a subset of $\mathbb{R}$ or $\mathbb{C}$ with non-empty interior, are called *continuous*. As we will see, the metric classification of quadrics involves many continuous invariants.

10.3 Focal properties of conics

In this section we will introduce the foci of conics and present the main properties of conics related to them, often called *focal properties*. The foci are very important metric elements of the non-degenerate conics, by far the most important ones regarding the applications: let us just mention their role in Kepler's first planetary motion law or in parabolic reflectors. The foci directly appear in some elementary presentations of conics, for instance an ellipse is often taken as the locus of the points whose sum of distances to two given points is constant, these points being the foci of the ellipse (see 10.3.7 and 10.3.10 below). Nevertheless, the relationship between a non-degenerate conic and its foci is not elementary: as we will see next, it quite naturally involves pencils of conic envelopes.

Since the absolute $\mathbf{K} = I + J$ is a non-degenerate quadric of the improper line ℓ_∞, its envelope in $\bar{\mathbb{A}}_2$ is defined; we will denote it as $\mathbf{K}^\circ$. According to 6.8.4, $\mathbf{K}^\circ$ is a rank-two quadric of $\bar{\mathbb{A}}_2^\vee$ with singular line ℓ_∞ containing no other (real) line, because $\mathbf{K}$ has no (real) point. In other words, $\mathbf{K}^\circ$ is a pair of imaginary lines of $\bar{\mathbb{A}}_2^\vee$ whose only real point (line of $\bar{\mathbb{A}}_2$) is ℓ_∞.

Next is a nice characterization of the perpendicularity of lines using $\mathbf{K}^\circ$; it will be useful in the sequel.

Proposition 10.3.1. *Two proper lines of $\bar{\mathbb{A}}_2$ are orthogonal if and only if they are conjugate with respect to $\mathbf{K}^\circ$.*

Proof. By 6.8.2 (a), lines r and s are conjugate with respect to $\mathbf{K}^\circ$ if and only if $r \cap \ell_\infty$ and $s \cap \ell_\infty$ are conjugate with respect to the envelope $\mathbf{K}^*$ of $\mathbf{K}$ (in ℓ_∞). By 6.6.2, this occurs if and only if the improper points of r and s are conjugate with respect to $\mathbf{K}$, which, by 6.9.4, is equivalent to the orthogonality of r and s. □

This causing no confusion, in the sequel we will refer to $\mathbf{K}^\circ$ just as the envelope of $\mathbf{K}$.

If Δ is the projective reference associated to an orthonormal reference of $\mathbb{A}_2$, we have already seen that the equation of the absolute $\mathbf{K}$ relative to the reference of $\ell_\infty : x_0 = 0$ subordinated by Δ is

$$x_1^2 + x_2^2 = 0$$

and therefore, if u_0, u_1, u_2 are line coordinates relative to Δ, by 6.8.6, the equation of $\mathbf{K}^\circ$ is

$$u_1^2 + u_2^2 = 0. \tag{10.8}$$

This equation confirms, indeed, that $\mathbf{K}^\circ$ is a pair of imaginary lines of $\bar{\mathbb{A}}_2^\vee$ whose double element is the improper line $\ell_\infty = [1.0, 0]$ of $\bar{\mathbb{A}}_2$. An easy alternative proof of 10.3.1 using the equation (10.8) may be provided by the reader.

Next we will have a closer look at $\mathbf{K}^\circ$ by examining the lines composing it. They are the lines of $\mathbb{C}(\bar{\mathbb{A}}_2^\vee)$ with equations

$$u_1 + \boldsymbol{i} u_2 = 0, \quad u_1 - \boldsymbol{i} u_2 = 0,$$

relative to $\Delta^\vee$ taken as a reference of $\mathbb{C}(\bar{\mathbb{A}}_2^\vee)$ (5.1.7). By 5.1.20, these lines may be identified, through the projectivity g of 5.1.19, to the lines of $(\mathbb{C}\bar{\mathbb{A}}_2)^\vee$ which have the same equations, this time the u_i being taken as line coordinates relative to Δ as a reference of $\mathbb{C}\bar{\mathbb{A}}_2$. The lines of $(\mathbb{C}\bar{\mathbb{A}}_2)^\vee$ being pencils of lines of $\mathbb{C}\bar{\mathbb{A}}_2$, the ones we have in hand are just the pencils of lines through the points $[0, 1, \boldsymbol{i}]$ and $[0, 1, -\boldsymbol{i}]$

of $\mathbb{C}\bar{\mathbb{A}}_2$, which are the cyclic points. Summarizing, if the projectivity g of 5.1.19 is used to identify $\mathbb{C}(\bar{\mathbb{A}}_2^\vee)$ to $(\mathbb{C}\bar{\mathbb{A}}_2)^\vee$, then the lines composing the envelope $\mathbf{K}^\circ$ of the absolute $\mathbf{K}$ are identified with the pencils I^* and J^*, of the lines of $\mathbb{C}\bar{\mathbb{A}}_2$ through the cyclic points I and J, respectively.

Assume now that C is a non-degenerate conic of the Euclidean plane $\mathbb{A}_2$. We have two conic envelopes of $\bar{\mathbb{A}}_2$: the envelope C^* of C and the envelope $\mathbf{K}^\circ$ of the absolute $\mathbf{K}$. They span a pencil of conic envelopes $\mathcal{F}_C$ whose degenerate members are pairs of (real or imaginary, maybe coincident) lines of $\bar{\mathbb{A}}_2^\vee$. When these lines are real, they are pencils of lines of $\bar{\mathbb{A}}_2$ and their kernels, taken in pairs, are the foci of C according to the following definition:

A pair of points F, F' of $\bar{\mathbb{A}}_n$ is a *pair of foci* of C if and only if the degenerate conic envelope $F^* + F'^*$ belongs to the pencil of conic envelopes $\mathcal{F}_C$, spanned by the envelopes of C and the absolute.

A pair of imaginary points G, G' for which $G^* + G'^* \in \mathcal{F}_C$ is sometimes referred to as a *pair of imaginary foci*. We will not make any further mention of them in the sequel, but the reader may note after the computations below that ellipses other than circles and hyperbolas have a single pair of imaginary foci.

According to the above definition, the foci naturally come in pairs, rather than as individual points, which is relevant (see the definition of focal conic in next Section 10.4). Anyway, each of the points F, F' is called a focus of C. The *directrices* of C are the polars, relative to C, of the foci of C. The line joining a focus F of C to a point $p \in C$, $p \neq F$, is called a *focal radius* of p.

The following is a characterization of the foci of a non-degenerate conic, as individual points, which makes no use of the dual plane:

Proposition 10.3.2. *A point p of a Euclidean plane $\mathbb{A}_2$ is a focus of a non-degenerate conic C of $\mathbb{P}_2$ if and only if the pair of tangents to C through p and the pair of isotropic lines through p do agree.*

Proof. The pair of tangents to C through p is $C^* \cap p^*$, while the pair of isotropic lines through p is $\mathbf{K}^\circ \cap p^*$. If $C^* \cap p^* = \mathbf{K}^\circ \cap p^*$, by 9.5.4, one quadric of the pencil spanned by C^* and $\mathbf{K}^\circ$ contains p^* and therefore, according to the definition, p is a focus. The converse follows by just reversing the argument. □

Next we examine the foci of the conics by direct computation from the metric reduced equations. Parabolas will be given a separate treatment and, for simplicity, we will drop the less interesting case of the imaginary conics; the reader may easily cover it by slightly modifying the computations and arguments below.

Assume that C is either an ellipse or a hyperbola. In the computations that follow we will use the double signs $\pm$ and $\mp$, in order to cover both cases simultaneously: unless otherwise stated, the upper sign applies to ellipses and the lower sign to hyperbolas.

Assume that C has metric reduced equation

$$\frac{X^2}{\mathbf{a}^2} \pm \frac{Y^2}{\mathbf{b}^2} = 1,$$

with $\mathbf{a}, \mathbf{b} > 0$ and, in case of C being an ellipse, $\mathbf{a} \geq \mathbf{b}$. The reference is thus orthonormal, has the centre O of C as origin and its axes are a pair of principal axes of C (10.2.5). An equation of C^* is

$$-u_0^2 + \mathbf{a}^2 u_1^2 \pm \mathbf{b}^2 u_2^2 = 0,$$

while, the reference being orthonormal, $\mathbf{K}^\circ$ has equation $u_1^2 + u_2^2 = 0$. Therefore the pencil $\mathcal{F}_C$ is

$$-\lambda u_0^2 + (\lambda \mathbf{a}^2 + \mu) u_1^2 + (\pm \lambda \mathbf{b}^2 + \mu) u_2^2 = 0. \tag{10.9}$$

If C is a circle, then the above equation is

$$-\lambda u_0^2 + (\lambda \mathbf{a}^2 + \mu)(u_1^2 + u_2^2) = 0$$

and $\mathcal{F}_C$ is a bitangent pencil whose only degenerate member other than $\mathbf{K}^\circ$ is $2O^* : u_0^2 = 0$, two times the pencil of diameters of C. Hence:

Proposition 10.3.3. *The centre of a circle is its only focus.*

Remark 10.3.4. If C is a circle, then $C \cap \ell_\infty = \mathbf{K}$ and so 6.8.7 assures that $\mathbf{K}^\circ$ is the cone with vertex ℓ_∞ over $C^* \cap O^*$. Since O^* is the polar of ℓ_∞ relative to C^*, this provides an alternative proof of the fact that $\mathcal{F}_C$ is a bitangent pencil with degenerate members $\mathbf{K}^\circ$ and $2O^*$.

If C is not a circle, an easy computation from equation (10.9) shows that the degenerate members of $\mathcal{F}_C$ other than $\mathbf{K}^\circ$ are

$$u_0^2 + \mathbf{c}^2 u_2^2 = 0 \quad \text{and} \quad u_0^2 - \mathbf{c}^2 u_1^2 = 0,$$

where $\mathbf{c} = \sqrt{\mathbf{a}^2 \mp \mathbf{b}^2}$. Only the last of them is composed of real lines of $\bar{\mathbb{A}}_2$, and these lines are

$$u_0 + \mathbf{c} u_1 = 0 \quad \text{and} \quad u_0 - \mathbf{c} u_1 = 0.$$

Therefore C has foci $F = [1, \mathbf{c}, 0]$ and $F' = [1, -\mathbf{c}, 0]$ or, using affine coordinates, $F = (\mathbf{c}, 0)$ and $F' = (-\mathbf{c}, 0)$. Clearly, they lie on the first coordinate axis, their midpoint is the origin and they are at distance $\mathbf{c}$ from it. Since the origin is the centre and the first coordinate axis is either the major or the transverse axis of C, we have proved:

Proposition 10.3.5. *If C is an ellipse, not a circle, then it has two different foci F, F'; they are proper points, belong to the major axis and their midpoint is the centre of C. If C has reduced equation*

$$\frac{X^2}{\mathbf{a}^2} + \frac{Y^2}{\mathbf{b}^2} = 1,$$

then F and F' are at distance $\mathbf{c} = \sqrt{\mathbf{a}^2 - \mathbf{b}^2}$ from the centre of C.

If C is a hyperbola, then it has two different foci F, F'; they are proper points, belong to the transverse axis and their midpoint is the centre of C. If C has reduced equation

$$\frac{X^2}{\mathbf{a}^2} - \frac{Y^2}{\mathbf{b}^2} = 1,$$

then F and F' and are at distance $\mathbf{c} = \sqrt{\mathbf{a}^2 + \mathbf{b}^2}$ from the centre of C.

The positive real number $\mathbf{c}$ above is called the *half-focal separation* of the conic. The half-focal separation of a circle is taken to be $\mathbf{c} = 0$: This extends the equality $\mathbf{c} = \sqrt{\mathbf{a}^2 - \mathbf{b}^2}$ to circles and is consistent with the fact that the circles have $F = F' = O$. For both ellipses and hyperbolas, the *eccentricity* is defined as being $\mathbf{e} = \mathbf{c}/\mathbf{a}$. Clearly, circles have $\mathbf{e} = 0$, ellipses have $0 \leq \mathbf{e} < 1$ and hyperbolas have $1 < \mathbf{e}$.

The next lemma is the key to many focal properties of the conics with proper centre. In it we assume to have taken orthonormal coordinates relative to which the equation of C is reduced and the notations are as above.

Lemma 10.3.6. *If C is an ellipse, then the distances from a point $p \in C$, with abscissa x, to the foci of C are*

$$\mathbf{r} = \mathbf{a} - \mathbf{e}x \quad \textit{and} \quad \mathbf{r}' = \mathbf{a} + \mathbf{e}x.$$

If C is a hyperbola, then the same distances are

$$\mathbf{r} = \pm(\mathbf{e}x - \mathbf{a}) \quad \textit{and} \quad \mathbf{r}' = \pm(\mathbf{e}x + \mathbf{a}),$$

where the upper sign applies to the points p with $x > 0$ and the lower sign to those with $x < 0$.

Proof. The square of the distance from $p = (x, y)$ to $F = (\mathbf{c}, 0)$ is

$$\mathbf{r}^2 = (x - \mathbf{c})^2 + y^2.$$

After replacing y^2 by the value obtained from the reduced equation, easy computations give

$$\mathbf{r}^2 = (\mathbf{e}x - \mathbf{a})^2;$$

proceeding similarly with F',

$$\mathbf{r}'^2 = (\mathbf{e}x + \mathbf{a})^2.$$

If C is an ellipse, $0 \leq \mathbf{e} < \mathbf{1}$ and it is clear from its reduced equation that $-\mathbf{a} \leq x \leq \mathbf{a}$, after which $-\mathbf{a} \leq \mathbf{e}x \leq \mathbf{a}$ and the claimed equalities follow. For a hyperbola it is $\mathbf{e} > \mathbf{1}$ and either $x > \mathbf{a}$ or $x < -\mathbf{a}$, which force $\mathbf{e}x - \mathbf{a}$ and $\mathbf{e}x + \mathbf{a}$ to have both the sign of x, hence the remaining two equalities. □

Directly from 10.3.6 we have:

Proposition 10.3.7. *The sum of distances from any point of an ellipse to the foci equals the length of the major axis.*

The absolute value of the difference of distances from any point of a hyperbola to the foci equals the length of the transverse axis.

Taking in account that the polars of the foci $(\mathbf{c}, 0)$, $(-\mathbf{c}, 0)$ are, respectively, $X = \mathbf{a}/\mathbf{e}$ and $X = -\mathbf{a}/\mathbf{e}$, also:

Proposition 10.3.8. *If C is an ellipse or a hyperbola, then the ratio of the distances from any point $p \in C$ to a focus and its corresponding directrix is independent of p and equals the eccentricity of C.*

The converses of 10.3.7 and 10.3.8 do hold, see Exercises 10.17 and 10.18.

Next we will present a different, synthetic, view on part of the above. It re-proves part of 10.3.5 and provides a nice interpretation of the metric elements of ellipses and hyperbolas (see Figure 10.3), but carries no new information. To avoid confusion, the line-pairs of $\bar{\mathbb{A}}_2^\vee$ will be called *pairs of pencils*.

Assume that C is a conic with proper centre, not a circle or an imaginary circle. Then C^* shares no (real) lines with $\mathbf{K}^\circ$ and so, by 9.6.6, $\mathcal{F}_C$ is either a general or a bitangent pencil of $\bar{\mathbb{A}}_2^\vee$ with no base lines. In the second case, by 6.8.7, $C \cap \ell_\infty = \mathbf{K}$ and so C would be a circle or an imaginary circle, which has been excluded. $\mathcal{F}_C$ is thus general and, according to 9.6.1, it contains three different pairs of distinct pencils, from which a single one is composed of real pencils. If this one is $F^* + F'^*$ then, by the definition of foci, F, F' is the pair of foci of C. From the remaining two pairs of pencils of $\mathcal{F}_C$, one obviously is $\mathbf{K}^\circ$. The singular lines of $F^* + F'^*$ and $\mathbf{K}^\circ$ are, respectively, the join of F and F', which we write $e = FF'$, and the improper line ℓ_∞. We denote by e' the singular line of the third pair of pencils of $\mathcal{P}_C$. Again by 9.6.1, e, ℓ_∞ and e' are the vertices of a triangle T of $\bar{\mathbb{A}}_2^\vee$ which is self-polar with respect to all conic envelopes of $\mathcal{F}_C$.

In particular, since the centre O of C is the pole of ℓ_∞ with respect to C, in the dual plane O^* is the polar of ℓ_∞ with respect to C^*: by the above, $O^* = e^* \vee e'^*$. Equivalently, $O = e \cap e'$ and so e and e' are different diameters of C.

We have seen in 10.3.5 that e is a principal axis of C; we will re-prove this fact next, and see also that e' is the other principal axis. Using again that T is self-polar with respect to all conic envelopes of $\mathcal{F}_C$, e and e' are conjugate with respect to $\mathbf{K}^\circ$ and hence orthogonal, by 10.3.1. On the other hand, e and e' are conjugate with respect to C^*. Either of them contains thus the pole of the other with respect to C, by 6.6.1 (c), and this assures that e and e' are conjugate diameters, by 7.8.11.

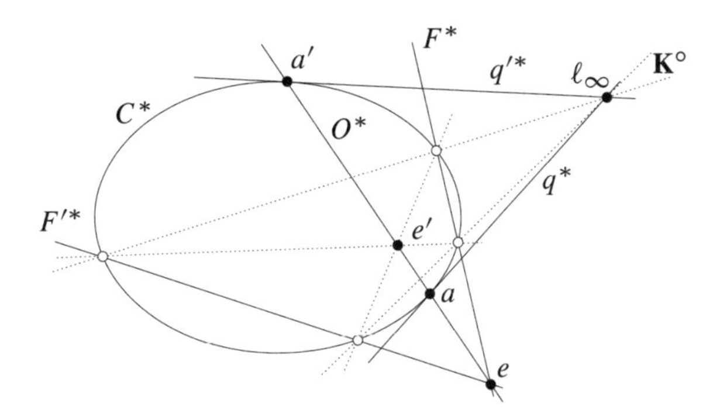

Figure 10.3. A dual view of the main affine and metric elements of a hyperbola C: lines are pictured as points and pencils of lines as lines. Pairs of imaginary lines of $\bar{\mathbb{A}}_2^\vee$ are dotted. ℓ_∞ is the improper line and $\mathbf{K}^\circ$ the envelope of the absolute. q, q' are the improper points of C, a, a' the asymptotes, O the centre, F, F' the pair of foci and e, e' the principal axis, e being the transverse one. Details are in the text. The picture for an ellipse other than a circle would be similar, but for having a, a' imaginary.

Assume now that C is a parabola and take its reduced equation to be $Y^2 = 2\mathbf{p}X$, $\mathbf{p} > 0$. Its envelope has equation $\mathbf{p}u_2^2 - 2u_0u_1 = 0$, after which the pencil $\mathcal{F}_C$ is

$$\mathcal{F}_C : (\lambda\mathbf{p} + \mu)u_2^2 + \mu u_1^2 - 2\lambda u_0 u_1 = 0.$$

It direct to check from the above equation that $\mathcal{F}_C$ is a simple contact pencil (as one could expect, because the singular line ℓ_∞ of $\mathbf{K}^\circ$ belongs to C^*). Its degenerate members are $\mathbf{K}^\circ$ and

$$2u_0u_1 + \mathbf{p}u_1^2 = 0.$$

The latter splits in $u_1 = 0$, the pencil of lines through the improper centre $O = [0, 1, 0]$ of C, and

$$2u_0 + \mathbf{p}u_1 = 0,$$

the pencil of lines through the point $F = (\mathbf{p}/2, 0)$. The polar of F being $X = -\mathbf{p}/2$, we have seen:

Proposition 10.3.9. *Any parabola C of a Euclidean plane has a single pair of foci. One of the foci is the improper centre of C, while the other is a proper point that belongs to the axis of C, is interior to C and, if C has reduced equation $Y^2 = 2\mathbf{p}X$, lies at distance $\mathbf{p}/2$ from the vertex of C. The directrix corresponding to the proper* *focus does not meet C, is orthogonal to the axis and lies at distance $\mathbf{p}/2$ from the vertex.*

The proper focus of a parabola and its corresponding directrix are often referred to as *the focus* and *the directrix* of the parabola. Using these terms, the next proposition states a well-known metric property of parabolas, often used to define them. Its proof follows from an elementary computation which is left to the reader.

Proposition 10.3.10. *All points of a parabola lie at equal distance from the focus and the directrix.*

The converse is also true, see Exercise 10.18. The *eccentricity* of a parabola is taken, by definition, equal to one, which is the only non-negative value not reached by the eccentricity of the conics with proper centre. After this 10.3.10 above extends 10.3.8 to parabolas.

As already done in the case of conics with proper centre, next we present a different view on the metric elements of a parabola (see Figure 10.4); it re-proves most of 10.3.9 but adds no new information.

Since the singular line ℓ_∞ of $\mathbf{K}^\circ$ belongs to C^*, we know from 9.5.1 that $\mathcal{F}_C$ is a single contact pencil and also that its degenerate members are $\mathbf{K}^\circ$ and a degenerate envelope composed of the pencil of lines O^*, which is tangent to C^* at ℓ_∞, and another pencil of lines F^* that does not contain ℓ_∞: F is thus a proper point and O, F is the pair of foci of C. If $e = OF$, then e is the singular line of $O^* + F^*$. Clearly, $e \neq \ell_\infty$, as F is proper. In particular $e \notin C^*$, as $C^* \cap O^* = \ell_\infty$, and so, if O'^* is the polar pencil of e relative to C^*, $C^* \cap O'^*$ consists of two different lines, one of which is ℓ_∞, say $C^* \cap O'^* = \{\ell_\infty, t\}$. Since e is the singular line of $O^* + F^*$, e and t are conjugate with respect to two, and hence all, conic envelopes in $\mathcal{F}_C$. In particular they are conjugate with respect to $\mathbf{K}^\circ$ and therefore are orthogonal, by 10.3.1. The improper points O of e and O' of t are thus orthogonal directions and so e, which is the polar of O' relative to C by the definition of O', is the axis of C. Since by its definition $e \in F^*$, we have seen that the proper focus belongs to the axis of the parabola. If V^* is the pencil spanned by e and t, V^* contains e and is tangent to C^* at t, which means that V is the proper end of the axis, hence the vertex of C, and t is the tangent to C at its vertex.

The next proposition states one of the most relevant focal properties of conics.

Proposition 10.3.11. *If C is a non-degenerate conic, not a circle, with foci F, F', and p is a proper point of C, then the tangent and the normal to C at p are the bisectors of the pair of focal radii pF, pF'.*

Proof. The proof will be done in the dual plane $\bar{\mathbb{A}}_2^\vee$. Call t and t' the tangent and normal to C at p, respectively: since they are orthogonal lines, they are conjugate with respect to $\mathbf{K}^\circ$ (by 10.3.1). Denote by $i + j$ the intersection of the conic envelope $\mathbf{K}^\circ$ and the pencil p^* (as a line of $\bar{\mathbb{A}}_2^\vee$): since $\ell_\infty \notin p^*, i, j$ are imaginary elements

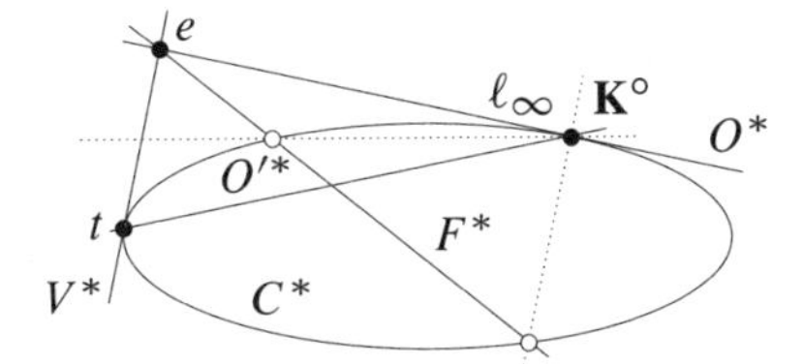

Figure 10.4. Another dual view, this time of the main affine and metric elements of a parabola C. Again, lines are pictured as points and pencils of lines as lines, ℓ_∞ is the improper line and $\mathbf{K}^\circ$ the envelope of the absolute. Now O is the improper centre of C and O' the direction orthogonal to O; F is the proper focus of C, e its axis, V its vertex and t the tangent to C at V.

of p^*, that is, they belong to $(p^*)_{\mathbb{C}} \subset \mathbb{C}(\bar{\mathbb{A}}_2^\vee)$. By 6.4.11, $(i, j, t, t') = -1$. The conic envelopes C^*, $F^* + F'^*$ and $\mathbf{K}^\circ$ all belong to the pencil $\mathcal{F}_C$, and their respective intersections with p^* are $2t$, $pF + pF'$ and $t + t'$. Desargues' theorem on pencils of quadrics 9.7.4 applies and shows that these three pairs of lines are corresponding pairs of an involution τ of p^*. The tangent t is thus fixed by τ, and then so is the normal t' because i, j correspond by τ and $(i, j, t, t') = -1$. After this, since also pF and pF' correspond by τ, $(pF, pF', t, t') = -1$. The orthogonality of t and t' being clear, the claim follows from 6.9.20. □

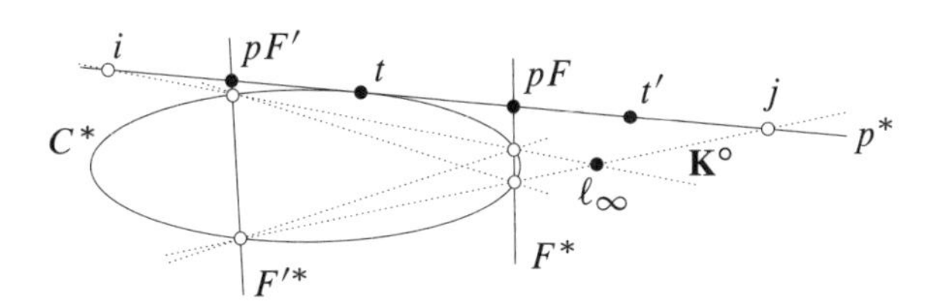

Figure 10.5. The proof of 10.3.11 for a conic with proper centre, not a circle: a dual view.

Non-degenerate conics of a Euclidean plane $\mathbb{A}_2$ with the same pair of foci are called *confocal*. According to the next proposition, the non-degenerate conics confocal with a given one describe a range; these ranges are called *confocal ranges*.

Proposition 10.3.12. *Let C be a non-degenerate conic of a Euclidean plane. Another non-degenerate conic E is confocal with C if and only if it belongs to the range with associated pencil $\mathcal{F}_C$.*

Proof. Since the pencils $\mathcal{F}_C$ and $\mathcal{F}_E$ share the conic envelope $\mathbf{K}^\circ$, $E^* \in \mathcal{F}_C$ if and only if $\mathcal{F}_C = \mathcal{F}_E$, which in turn clearly implies that C and E are confocal.

Conversely, if both conics have the same foci F, F', then $\mathcal{F}_C = \mathcal{F}_E$ because they share the two clearly different conic envelopes $F^* + F'^*$ and $\mathbf{K}^\circ$. □

Since the conics with proper centre have two proper foci, while the parabolas have a single one, either all conics in a confocal range have proper centre, or all are parabolas. It is clear that a confocal range is determined by any of its conics, and also by the common pair of foci. It is easy to make this determination explicit, because a conic in the range, or the pair or foci, provide a conic envelope which, together with $\mathbf{K}^\circ$, spans the associated pencil. Assume for instance to have been given a pair of proper points F, F' as a foci. By taking orthonormal coordinates with origin the midpoint of F, F' and the first axis containing them, the foci will be $F = (\mathbf{c}, 0)$ and $F' = (-\mathbf{c}, 0)$, where $\mathbf{c} \geq 0$ is the one half of the distance between them. Then $F^* + F'^*$ is $u_0^2 - \mathbf{c}^2 u_1^2 = 0$ which, together with the envelope of the absolute $u_1^2 + u_2^2 = 0$, spans the pencil

$$\lambda u_0^2 + (\mu - \lambda \mathbf{c}^2) u_1^2 + \mu u_2^2 = 0.$$

Since assuming $\lambda \neq 0$ just drops the degenerate envelope $F^* + F'^*$, we take $\delta = -\mu/\lambda$ and write the non-degenerate conic envelopes in the above pencil as

$$u_0^2 - (\delta + \mathbf{c}^2) u_1^2 - \delta u_2^2 = 0, \quad \delta \neq 0, -\mathbf{c}^2.$$

After this, the equations of the conics with foci $F = (\mathbf{c}, 0)$ and $F' = (-\mathbf{c}, 0)$ are

$$x_0^2 - \frac{x_1^2}{\delta + \mathbf{c}^2} - \frac{x_2^2}{\delta} = 0, \quad \delta \neq 0, -\mathbf{c}^2,$$

or, using affine coordinates,

$$\frac{X^2}{\delta + \mathbf{c}^2} + \frac{Y^2}{\delta} = 1, \quad \delta \neq 0, -\mathbf{c}^2. \tag{10.10}$$

In case $F = F'$, it is $\mathbf{c} = 0$ and the reader may recognize the family of all circles and imaginary circles with centre the origin.

A similar computation assuming F proper and F' improper gives

$$Y^2 + 2\delta X - \delta^2 = 0, \quad \delta \neq 0 \tag{10.11}$$

as the equations of all parabolas with foci F, F', the equations being relative to an orthonormal reference with origin at the proper focus and first axis containing the improper one.

There are no other possibilities for a confocal range, as we have already seen that any non-degenerate conic has at least one proper focus. Anyway the reader may try a similar computation starting from improper points F, F': it will result in a pencil of degenerate conic envelopes giving rise to no range of conics.

Confocal ranges are very nice families of conics. Their main properties are summarized in the next proposition. Its proof is left to the reader, who may use the equations (10.10) and (10.11) above.

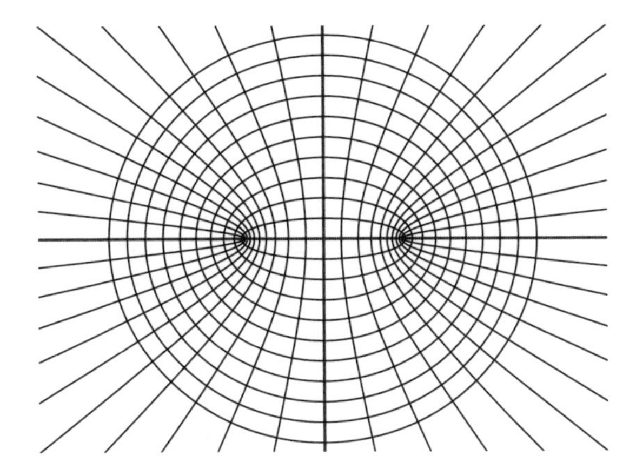

Figure 10.6. Confocal ellipses and hyperbolas.

Proposition 10.3.13. *If $\mathcal{R}$ is a confocal range, then one of the following three possibilities occurs:*

(a) *$\mathcal{R}$ is the family of all circles and imaginary circles with a given centre.*

(b) *No circle belongs to $\mathcal{R}$. All conics in $\mathcal{R}$ have the same proper centre and the same principal axes. If p is a proper point lying on no one of the common principal axes, there are two different conics of $\mathcal{R}$ through p: one of them is an ellipse, the other is a hyperbola and their tangents at p are orthogonal.*

(c) *All conics of $\mathcal{R}$ are parabolas and have the same axis. If p is a proper point and does not lie on the common axis, then there are two different parabolas of $\mathcal{R}$ through p; these parabolas have orthogonal tangents at p.*

10.4 Metric properties of three-space quadrics

In this section we present metric properties that are specific to quadrics of three-dimensional Euclidean spaces. Other properties of these quadrics, in particular their metric classification, will appear as particular cases of the general treatment of Sections 10.5, 10.6 and 10.7. We will not consider the metric properties of plane-pairs. We have seen in Section 10.1 that the quadrics with improper section equal to the absolute are the spheres, the imaginary spheres and the isotropic cones. If Q is not one of these quadrics, we will consider the pencil $\mathcal{P}_Q$, of conics of the improper plane π_∞, spanned by the absolute $\mathbf{K}$ and the improper section Q^∞ of Q. Since $\mathbf{K}$ has no real points, according to 9.6.6, $\mathcal{P}_Q$ is either a general or a bitangent pencil, and in no case has real base points.

The quadrics Q invariant by all rotations with a prescribed axis are called *quadrics of revolution*. The spheres, imaginary spheres and isotropic cones are obvious examples of quadrics of revolution due to 10.1.4. The remaining ones will be characterized in terms of the pencil $\mathcal{P}_Q$ as a consequence of the next theorem, which provides some further information:

Theorem 10.4.1. *Assume that Q is a quadric of a three-dimensional Euclidean space, other than a sphere, an imaginary sphere or an isotropic cone, and has $\mathbf{r}(Q) \geq 3$. If Q is invariant by a rotation of $\mathbb{A}_3$ of angle $\alpha \neq 0, \pm\pi/2, \pi$, then Q is not a parabolic cylinder and the pencil $\mathcal{P}_Q$ is a bitangent one. Conversely, if Q is not a parabolic cylinder and $\mathcal{P}_Q$ is a bitangent pencil, then there is a uniquely determined proper line e such that Q is invariant by any motion leaving all points of e invariant.*

Proof. Assume that Q is invariant by a rotation $\mathcal{R}$, with axis e and angle $\alpha \neq 0, \pm\pi/2, \pi$. Let $\bar{e}$ be the kernel of the pencil of planes orthogonal to e. We have seen in Example 5.7.12, that $\bar{e}$ is invariant by $\mathcal{R}$ and that the points q_1, q_2 of $\mathbf{K} \cap \bar{e}$ (the cyclic points of all planes orthogonal to e) are imaginary isolated fixed points of $\mathcal{R}$; they are in particular the only fixed points of $\mathcal{R}$ on $\bar{e}$. Since $\alpha \neq \pm\pi/2$, either by a direct argument, or by computing the invariant of $\mathcal{R}_{|\bar{e}}$ from the eigenvalues given in 5.7.12, it is clear that $\mathcal{R}_{|\bar{e}}$ is not an involution.

Let us check first that $\mathbf{K} \cap \bar{e} = q_1 + q_2$ is the only (real) point-pair of $\bar{e}$ invariant by $\mathcal{R}$. Indeed, take $p_1 + p_2 \neq q_1 + q_2$. Since q_1, q_2 are imaginary and two real point-pairs sharing an imaginary point share also the other, neither p_1 nor p_2 are fixed points. So, if $p_1 + p_2$ is invariant, then its points are distinct and mapped to each other by $\mathcal{R}_{\mathbb{C}}$. By 5.6.4 or 5.6.5, this would force $\mathcal{R}_{|\bar{e}}$ to be an involution, and we have noted above that it is not.

Both Q and $\bar{e}$ being invariant by $\mathcal{R}$, so is $Q^\infty \cap \bar{e} = Q \cap \bar{e}$. Thus, by the above, either $Q^\infty \cap \bar{e} = \bar{e}$, or $Q^\infty \cap \bar{e} = \mathbf{K} \cap \bar{e}$; in either case $Q^\infty \supset \mathbf{K} \cap \bar{e}$. Take $\bar{p} = e \cap \pi_\infty$. Since $\bar{p}$ and Q are invariant by $\mathcal{R}$, either $H_{\bar{p},Q} = \bar{\mathbb{A}}_3$ or $H_{\bar{p},Q}$ is a plane invariant by $\mathcal{R}$. We already know $\bar{e}^*$ to be the only real fundamental bundle of $\mathcal{R}$ (5.7.12), so in any case $H_{\bar{p},Q} \supset \bar{e}$. It follows that $\bar{p}$ is conjugate to all points of $\bar{e}$ with respect to Q^∞, and also with respect to $\mathbf{K}$, by its definition. All together, the pencil $\mathcal{P}_Q$, spanned by Q^∞ and $\mathbf{K}$, is the bitangent pencil determined by $q_1 + q_2$ and $\bar{p}$ (see 9.5.19).

If Q is a parabolic cylinder, then Q^∞ is a double line which, being a member of $\mathcal{P}_Q$, is $Q^\infty = 2\bar{e}$. In such a case, the double point of Q would be an obviously invariant real point on $\bar{e}$, and we know that there is none.

The above completes the proof of the first claim. Before proving the converse we will show that, in the above situation, the axis e of $\mathcal{R}$ is determined by Q, as this will be useful below. First, it is clear that Q determines $2\bar{e}$ as the only rank-one conic in $\mathcal{P}_Q$, and therefore determines also the line $\bar{e}$. The argument already used to prove that $H_{\bar{p},Q} \supset \bar{e}$ applies to any $p \in e$ and proves that $H_{p,Q} \supset \bar{e}$ for any $p \in e$;

hence, $H_{q,Q} \supset e$ for any $q \in \bar{e}$. On the other hand either Q is non-degenerate, or it has a single double point that, being invariant by $\mathcal{R}$, cannot belong to $\bar{e}$, as already said. Thus, in any case, the $H_{q,Q}$, $q \in \bar{e}$, are planes and describe a pencil (by 6.5.5 or 5.10.6). Then $e = \bigcap_{q\in\bar{e}} H_{q,Q}$, which determines e.

Now, for the converse, assume that Q is not a parabolic cylinder, $\mathbf{r}(Q) \geq 3$ and $\mathcal{P}_Q$ is bitangent. Take $\bar{e}$ to be the only line of π_∞ for which $2\bar{e} \in \mathcal{P}_Q$. Since $Q^\infty \neq 2\bar{e}$, as otherwise Q would be a parabolic cylinder, and $\mathcal{P}_Q$ has no real base point, no real point of Q belongs to $\bar{e}$. In particular $\bar{e} \not\subset Q$. We may thus choose a proper plane π through $\bar{e}$ which either is non-tangent to Q if Q is non-degenerate, or misses the double point of Q if Q is degenerate. In both cases $\bar{Q} = Q \cap \pi$ is a non-degenerate conic of π. Let $\mathcal{T}$ be the pencil of quadrics spanned by Q and 2π: it is a pencil of quadrics tangent along $\bar{Q}$, as described in Example 9.5.22; in particular its degenerate quadrics are 2π and a single rank-three quadric. Since $Q \cap \pi_\infty = Q^\infty$ and $2\pi \cap \pi_\infty = 2\bar{e}$, the improper sections of the quadrics of Q are the conics of $\mathcal{P}_Q$ and there is a unique quadric $S \in \mathcal{T}$ with improper section $\mathbf{K}$ (9.6.5). S is thus either a sphere, or an imaginary sphere, or an isotropic cone.

Since no real point of Q belongs to $\bar{e}$, in case of Q being degenerate, its double point does not belong to $\bar{e}$. Thus, by 6.5.5 or 5.10.6, the $H_{q,Q}$, $q \in \bar{e}$, describe a pencil of planes: we take its kernel to be the line e. Any point of e is thus conjugate to all points of $\bar{e}$ with respect to Q. This in particular assures that e is a proper line, as otherwise a real point in $e \cap \bar{e}$ would belong to Q and $\bar{e}$. Furthermore, since any point $p \in e$ is also conjugate to all points of $\bar{e}$ with respect to 2π, any point $p \in e$ is conjugate to all points of $\bar{e}$ with respect to any quadric of $\mathcal{T}$. This in particular implies that the improper point of e is conjugate to all points of $\bar{e}$ with respect to S, and so, also with respect to $\mathbf{K}$: the line e is thus orthogonal to π. Take any $Q' \in \mathcal{T}$, $Q' \neq 2\pi$. Still Q' has no real point on $\bar{e}$, because $Q' \cap \pi_\infty$ is not $2\bar{e}$ and belongs to $\mathcal{P}_Q$, which has no real base point. Arguing as above, Q' has no double point on $\bar{e}$, the $H_{q,Q'}$, $q \in \bar{e}$, describe a pencil of planes and $e = \bigcap_{q\in\bar{e}} H_{q,Q'}$. In particular the centre or double point of any $Q' \in \mathcal{T}$, $Q' \neq 2\pi$ belongs to e, because it belongs to the polar plane of any improper point.

Let $\mathcal{R}$ be any motion of $\mathbb{A}_3$ leaving invariant all points of e. On one hand, the plane π is orthogonal to e and hence invariant by $\mathcal{R}$; therefore so is the quadric 2π. On the other, the centre or double point of S has been seen to belong to e: hence, by 10.1.4, S is also invariant by $\mathcal{R}$. The whole pencil $\mathcal{T}$, which is spanned by 2π and S, is thus invariant by $\mathcal{R}$ and $\mathcal{R}$ induces the projectivity

$$\mathbb{Q}(\mathcal{R})_{|\mathcal{T}} : \mathcal{T} \to \mathcal{T},$$

which has 2π and S as fixed points. Denote by D the only pair of distinct lines in $\mathcal{P}_Q$, and by Q_1 the quadric of $\mathcal{T}$ which has improper section D. Since D is the only degenerate conic of $\mathcal{P}_Q$ besides $2\bar{e}$, Q_1 is the only quadric of $\mathcal{T}$ besides 2π which is tangent to π_∞ (and so a paraboloid or a non-parabolic cylinder). As a such, it is necessarily invariant by $\mathcal{R}$. In view of their improper sections, 2π, S

and Q_1 are three different members of $\mathcal{T}$; since we have seen each of them to be invariant by $\mathcal{R}$, it is $\mathbb{Q}(\mathcal{R})_{|\mathcal{T}} = \mathrm{Id}_{\mathcal{T}}$, all quadrics of $\mathcal{T}$ are invariant by $\mathcal{R}$, and so is Q, as claimed.

To close, for the uniqueness of the line e, just note that a line satisfying the property of the claim may be taken as the axis of a rotation of angle $\alpha \neq 0, \pm\pi/2, \pi$ leaving Q invariant; we have already seen that such an axis is uniquely determined by Q. □

Now, the already announced characterization of the quadrics of revolution directly follows:

Corollary 10.4.2 (von Staud). *Let Q be a quadric of a three-dimensional Euclidean space, other than a sphere, an imaginary sphere or an isotropic cone, and with $\mathbf{r}(Q) \geq 3$. Then Q is invariant by all rotations with axis a certain line e if and only if Q is not a parabolic cylinder and $\mathcal{P}_Q$ is a bitangent pencil. If this is the case, then the axis e is uniquely determined by Q.*

The line e of 10.4.2 is called the *axis of revolution* of Q. By 10.1.4, the spheres, imaginary spheres and isotropic cones have infinitely many axes of revolution.

The following is a more computable characterization of the quadrics of revolution:

Corollary 10.4.3. *Let Q be a quadric of a three-dimensional Euclidean space with rank at least three and affine equation*

$$\sum_{i,j=1}^{3} a_{i,j} X_i X_j + 2\sum_{i=1}^{n} a_i X_i + a = 0.$$

Take $A = (a_{i,j})$ as a matrix of the improper section of Q, call G the matrix of the scalar product and consider the polynomial in λ,

$$D(z) = \det(A - \lambda G).$$

Then Q is a quadric of revolution other than a sphere, an imaginary sphere or an isotropic cone if and only if $D(\lambda)$ has a non-zero double (and non-triple) root.

Proof. If Q is not a sphere, an imaginary sphere or an isotropic cone, then the pencil $\mathcal{P}_Q$ is defined. Taking λ as the absolute parameter of the conic of $\mathcal{P}_Q$ with matrix $A - \lambda G$, $\mathcal{P}_Q$ has characteristic polynomial $D(\lambda)$. If in addition Q is a quadric of revolution, then, by 10.4.2, it is not a parabolic cylinder and $\mathcal{P}$ is a bitangent pencil. Then the parameter of the only double line of $\mathcal{P}_Q$ is not zero, because Q is not a parabolic cylinder, and is a double root of $D(\lambda)$.

Conversely, assume that $D(\lambda)$ has a non-zero double root. First note that Q cannot be a sphere, an imaginary sphere or an isotropic cone, as for any of these quadrics A and G are proportional and as a consequence $D(\lambda)$ has a triple root. Since $D(\lambda)$ has degree three, either $D(0) \neq 0$ or 0 is a simple root of $D(\lambda)$. Then Q^∞, which has parameter zero in $\mathcal{P}_Q$, is either non-degenerate or has a single double point, by 9.5.10; therefore Q is not a parabolic cylinder. Furthermore, since $\mathcal{P}_Q$ is either general or bitangent and the characteristic equations of general pencils have all their roots simple, $\mathcal{P}_Q$ is a bitangent pencil. Corollary 10.4.2 shows then that Q is a quadric of revolution. □

Still assume that Q is a quadric of a Euclidean three-dimensional space other than a plane-pair. It is interesting to obtain triples of independent directions (that is, improper points) which are pairwise orthogonal and conjugate with respect to Q. These triples are called *triples of principal directions* of Q. In other words, a triple of principal directions is the set of vertices of a triangle of π_∞ self-polar with respect to both $\mathbf{K}$ and $Q \cap \pi_\infty$.

Proposition 10.4.4. *Any quadric Q of a Euclidean three-space, with rank at least three, has a triple of principal directions.*

Proof. If Q is either a sphere, or an imaginary sphere, or an isotropic cone, then $Q \cap \pi_\infty = \mathbf{K}$ and any triple of pairwise orthogonal directions is a triple of principal directions. Otherwise the pencil $\mathcal{P}_Q$ is defined, and by 9.6.6 it is either a general pencil with imaginary base points or a bitangent pencil. Therefore, as seen in 9.6.1 and 9.10.5, it has a self-polar triangle. □

Remark 10.4.5. As seen in the above proof, any three pairwise orthogonal directions compose a triple of principal directions of any sphere, imaginary sphere or isotropic cone. When $\mathcal{P}_C$ is bitangent, there are many choices of a self-polar triangle: according to 9.10.5, the triples of principal directions of a quadric of revolution are all the triples of pairwise orthogonal directions one of which is the direction of the axis. Similarly, the triples of principal directions of a parabolic cylinder Q are the triples of pairwise orthogonal directions two of which belong to the improper section of Q. In the remaining cases, $\mathcal{P}_Q$ is general and by 9.10.4, the triple of principal directions is unique.

Three unitary vectors representing a triple of principal directions of a quadric Q compose an orthonormal basis. By adding to it a suitable origin, it results in an orthonormal reference relative to which Q has a particularly simple equation that will be taken as its metric reduced equation. The choice of the origin depending on the affine type of Q, we will proceed case by case:

10.4.6. Quadrics with proper centre. The centre O of Q is taken as origin, after which the vertices of the associated projective reference are the vertices of a self-polar tetrahedron and the equation of any of these quadrics may be written in the form

$$\lambda_1 X^2 + \lambda_2 Y^2 + \lambda_3 Z^2 = 1, \tag{10.12}$$

with $\lambda_i \neq 0$, $i = 1, 2, 3$. Clearly the quadric is a sphere or an imaginary sphere if and only if $\lambda_1 = \lambda_2 = \lambda_3$. By 10.4.3, it is a quadric of revolution if and only if at least two of the coefficients λ_i agree.

Any triple of diameters projecting from the centre a triple of principal directions – such as the axes of the above reference – is called a *triple of principal axes*, and the diameters themselves *principal axes* (or just *axes*) of Q. As for conics, the proper (real or imaginary) ends of the principal axes are called *vertices*: Arguing as in the proof of 10.2.2, the reader may easily see that the vertices are the points of Q at which the tangent plane is orthogonal to the diameter.

After renaming the coefficients in the equation (10.12) to make evident their signs – and hence the affine type of the quadric – and reordering the monomials, the metric reduced equations are as follows.

Imaginary quadrics:

$$\frac{X^2}{\mathbf{a}^2} + \frac{Y^2}{\mathbf{b}^2} + \frac{Z^2}{\mathbf{c}^2} = -1,$$

with $\mathbf{a} \geq \mathbf{b} \geq \mathbf{c} > 0$. The imaginary spheres are characterized by having $\mathbf{a} = \mathbf{b} = \mathbf{c}$, and the imaginary quadrics of revolution by having either $\mathbf{a} = \mathbf{b} \geq \mathbf{c}$ or $\mathbf{a} \geq \mathbf{b} = \mathbf{c}$.

Ellipsoids:

$$\frac{X^2}{\mathbf{a}^2} + \frac{Y^2}{\mathbf{b}^2} + \frac{Z^2}{\mathbf{c}^2} = 1,$$

with $\mathbf{a} \geq \mathbf{b} \geq \mathbf{c} > 0$. The spheres are the ellipsoids with $\mathbf{a} = \mathbf{b} = \mathbf{c}$, while the ellipsoids of revolution are the ellipsoids with either $\mathbf{a} = \mathbf{b} \geq \mathbf{c}$ or $\mathbf{a} \geq \mathbf{b} = \mathbf{c}$.

Hyperboloids of two sheets:

$$\frac{X^2}{\mathbf{a}^2} - \frac{Y^2}{\mathbf{b}^2} - \frac{Z^2}{\mathbf{c}^2} = 1,$$

with $\mathbf{a} > 0$ and $\mathbf{b} \geq \mathbf{c} > 0$. They are quadrics of revolution if and only if $\mathbf{b} = \mathbf{c}$.

Hyperboloids of one sheet:

$$\frac{X^2}{\mathbf{a}^2} + \frac{Y^2}{\mathbf{b}^2} - \frac{Z^2}{\mathbf{c}^2} = 1,$$

with $\mathbf{a} \geq \mathbf{b} > 0$ and $\mathbf{c} > 0$. They are quadrics of revolution if and only if $\mathbf{a} = \mathbf{b}$.

As for the conics with proper center, the positive real numbers $\mathbf{a}, \mathbf{b}, \mathbf{c}$ appearing in the above reduced equations are called the *lengths of the semi-axes* of the quadric.

10.4.7. Cones. If Q is a cone, then we take its double point as origin, after which Q has an equation of the form

$$\lambda_1 X^2 + \lambda_2 Y^2 + \lambda_3 Z^2 = 0,$$

with $\lambda_i \neq 0, i = 1, 2, 3$. Clearly Q is isotropic if and only if $\lambda_1 = \lambda_2 = \lambda_3$, while, as above, it is easy to check that Q is a quadric of revolution if and only if at least two of the coefficients λ_i agree. Multiplying the equation by -1, if needed, and proceeding as for the quadrics with proper centre, we obtain the following metric reduced equations.

Imaginary cones:

$$\frac{X^2}{\mathbf{a}^2} + \frac{Y^2}{\mathbf{b}^2} + \frac{Z^2}{\mathbf{c}^2} = 0,$$

with $\mathbf{a} \geq \mathbf{b} \geq \mathbf{c} > 0$. An imaginary cone is isotropic if and only if $\mathbf{a} = \mathbf{b} = \mathbf{c}$, while it is a quadric of revolution if and only if $\mathbf{a} = \mathbf{b} \geq \mathbf{c}$ or $\mathbf{a} \geq \mathbf{b} = \mathbf{c}$.

Real cones:

$$\frac{X^2}{\mathbf{a}^2} + \frac{Y^2}{\mathbf{b}^2} - \frac{Z^2}{\mathbf{c}^2} = 0,$$

with $\mathbf{a} \geq \mathbf{b} > 0$ and $\mathbf{c} > 0$. They are cones of revolution if and only if $\mathbf{a} = \mathbf{b}$.

10.4.8. Paraboloids. The improper section of any paraboloid Q is a pair of distinct lines whose double point O is the centre of the paraboloid. If p_1, p_2, p_3 is a triple of principal directions, then, by 9.10.4 or 9.10.5, one of them is O. By renumbering the principal directions, we assume $p_3 = O$. Then the line $\ell = p_1 \vee p_2$ – the direction of the planes orthogonal to O – does not contain O and therefore is not tangent to Q. If e is the polar line of ℓ with respect to Q, then e is supplementary to ℓ (6.5.12) and so, in particular, since $\ell \subset \pi_\infty$, e is a proper line. Furthermore $O \in e$, because O, the centre of Q, is conjugate to all improper points: then O is the improper point of e. The line e going through O and being proper, it is not tangent to Q and therefore is a chord of Q, one of whose ends is O: its other end V is a real and proper point: it is called the *vertex* of Q, while e is called the *axis* of Q. By 6.5.13, $V \vee \ell$ is the plane tangent to Q at V, and clearly it is orthogonal to e. The reader may easily check that the vertex V is the only proper point of Q at which the tangent plane is orthogonal to the diameter.

We take an orthonormal reference with origin V and unitary vectors e_1, e_2, e_3 representing p_1, p_2, p_3. The points p_1, p_2, p_3 are pairwise conjugate and, by its definition, V is conjugate to itself, p_1 and p_2. An equation of Q may thus be written in the form

$$\mu_1 X^2 + \mu_2 Y^2 = 2\mu_3 Z,$$

with $\mu_i \neq 0, i = 1, 2, 3$, because Q is non-degenerate. After dividing by μ_3 the equation has the form

$$\lambda_1 X^2 + \lambda_2 Y^2 = 2Z.$$

Up to swapping over the first and second axis, it is non-restrictive to assume that $|\lambda_1| \leq |\lambda_2|$. Furthermore, λ_1 may be turned into its opposite by multiplying the equation by -1 and then reversing the orientation of the third axis (that is, taking $-Z$ as new third coordinate) to recover a right-hand side equal to $2Z$. Thus we may in addition assume $\lambda_1 > 0$. Again it is direct to check that Q is a paraboloid of revolution if and only if $\lambda_1 = \lambda_2$. As in former cases, the coefficients are renamed $\lambda_1 = 1/\mathbf{a}^2$ and $\lambda_2 = \pm 1/\mathbf{b}^2$, to give the following metric reduced equations.

Elliptic paraboloids:

$$\frac{X^2}{\mathbf{a}^2} + \frac{Y^2}{\mathbf{b}^2} = 2Z,$$

with $\mathbf{a} \geq \mathbf{b} > 0$. An elliptic paraboloid is a quadric of revolution if and only if $\mathbf{a} = \mathbf{b}$.

Hyperbolic paraboloids:

$$\frac{X^2}{\mathbf{a}^2} - \frac{Y^2}{\mathbf{b}^2} = 2Z,$$

with $\mathbf{a} \geq \mathbf{b} > 0$. In no case is a hyperbolic paraboloid a quadric of revolution.

10.4.9. Non-parabolic cylinders. If Q is a non-parabolic cylinder, then its improper section is a pair of distinct lines whose double point is the vertex O of the cylinder. As for paraboloids, the vertex O belongs to any triple of principal directions and, up to reordering, we take it as its third member. We take an orthonormal reference with origin a centre p_0 of Q and, as in former cases, unitary vectors representing the members p_1, p_2, p_3 of a triple of principal directions. Then p_0, p_1, p_2, p_3 are the vertices of a self-polar tetrahedron and p_3 is in addition self-conjugate, which gives an equation of the form

$$\lambda_1 X^2 + \lambda_2 Y^2 = 1.$$

As in former cases, the above cylinder is of revolution if and only if $\lambda_1 = \lambda_2$. The metric reduced equations are as follows.

Imaginary cylinders:

$$\frac{X^2}{\mathbf{a}^2} + \frac{Y^2}{\mathbf{b}^2} = -1,$$

with $\mathbf{a} \geq \mathbf{b} > 0$. The imaginary cylinders of revolution are those with $\mathbf{a} = \mathbf{b}$.

Elliptic cylinders:

$$\frac{X^2}{\mathbf{a}^2} + \frac{Y^2}{\mathbf{b}^2} = 1,$$

with $\mathbf{a} \geq \mathbf{b} > 0$. An elliptic cylinder is of revolution if and only if $\mathbf{a} = \mathbf{b}$.

Hyperbolic cylinders:

$$\frac{X^2}{\mathbf{a}^2} - \frac{Y^2}{\mathbf{b}^2} = 1,$$

with $\mathbf{a}, \mathbf{b} > 0$. No hyperbolic cylinder is a quadric of revolution.

10.4.10. Parabolic cylinders. The improper section of parabolic cylinder Q is a double line 2ℓ; ℓ is then a chord of $\mathbf{K}$ with imaginary ends and so $\mathcal{P}_Q$ is a bitangent pencil with imaginary base points (see 9.5.19). Using again 9.10.5, we choose a triple of principal directions containing the double point O of Q: we take p_1, p_2, p_3 such that $p_3 = O$, $p_2 \in \ell$ is conjugate to O with respect to the pair of base points of $\mathcal{P}_Q$, and p_1 is the pole of ℓ relative to any non-degenerate conic of $\mathcal{P}_Q$. Since $p_1 \notin \ell$, its polar plane π, relative to Q, is $\pi \neq \pi_\infty$; therefore $Q \cap \pi$ a pair of different lines, one of which is ℓ, say $Q \cap \pi = \ell + \ell'$. Of course ℓ' is a proper line, as otherwise $\pi = \pi_\infty$. Thus, we choose the origin p_0 of the reference to be a proper point of ℓ' and take a triple of unitary vectors representing p_1, p_2, p_3 as basis, this making, as in the preceding cases, an orthonormal reference. After our choices, p_1, p_2, p_3 are pairwise conjugate, p_0 and p_3 are self-conjugate and p_0 is also conjugate to p_2, p_3, conjugation being always relative to Q. After this, Q has an equation of the form

$$\lambda X^2 = 2\mu Y,$$

with $\lambda, \mu \neq 0$ because Q has rank three. After dividing by λ and reversing the orientation of the second axis if needed, we get as metric reduced equation

$$X^2 = 2\mathbf{p}Y,$$

with $\mathbf{p} > \mathbf{0}$. By 10.4.2, no parabolic cylinder is a quadric of revolution.

The orthonormal references relative to which the quadric has a metric reduced equation are called *principal references*. As in the case of conics, they are not determined by the quadric. By contrast, the metric reduced equation is determined by the quadric, but for the cones, which determine their metric reduced equations up to proportionality only. The metric reduced equation may be used to classify the three-space quadrics modulo congruence. We will not give a proof of these facts here, as using arguments similar to those used for conics would give rise to too many cases. Instead, the reader is referred to the forthcoming Sections 10.5,10.6 and 10.7, where the same facts are proved for quadrics of Euclidean spaces of arbitrary dimension.

Generalizing the notion of foci to quadrics of Euclidean spaces of dimension greater than two is not straightforward. We will fix our attention on the three-dimensional case. Let $\mathbf{K}^\circ$ be the envelope in $\bar{\mathbb{A}}_3$ of the absolute conic $\mathbf{K}$: $\mathbf{K}^\circ$ is a quadric envelope of rank three, whose singular plane is the improper one. Still the envelope of a non-degenerate quadric Q, other than a sphere or an imaginary sphere, and $\mathbf{K}^\circ$ span a pencil $\mathcal{F}_Q$, but now a degenerate member of $\mathcal{F}_Q$ other than $\mathbf{K}^\circ$ is expected to have rank three: if this is the case, it envelops a non-degenerate conic which is said to be a *focal conic* of Q. Only in exceptional cases (see Example 10.4.11 below) $\mathcal{F}_Q$ does contain a rank-two envelope: then such a rank-two envelope envelops a pair of points, which are called *foci* of the quadric. The focal conics of quadrics are far less interesting objects than the foci of conics. For quadrics, the attention is rather fixed on the range of the non-degenerate quadrics whose envelope belongs to $\mathcal{F}_Q$, which is called a *confocal range* of quadrics. Next, as a representative example, we will examine the case of the ellipsoids by direct computation. The reader may deal similarly with other affine types of non-degenerate quadrics.

Example 10.4.11 (Focal conics and foci of ellipsoids). Let Q be an ellipsoid and assume that it has metric reduced equation

$$\frac{X^2}{\mathbf{a}^2} + \frac{Y^2}{\mathbf{b}^2} + \frac{Z^2}{\mathbf{c}^2} = 1,$$

with $\mathbf{a} \geq \mathbf{b} \geq \mathbf{c}$. Its envelope has equation

$$\mathbf{a}^2 u_1^2 + \mathbf{b}^2 u_2^2 + \mathbf{c}^2 u_3^2 - u_0^2 = 0,$$

while, the reference being orthonormal, $\mathbf{K}^\circ$ has equation

$$u_1^2 + u_2^2 + u_3^2 = 0.$$

The quadric envelopes of $\mathcal{F}_Q$, $\mathbf{K}^\circ$ excluded, have thus equations

$$(\mathbf{a}^2 - \lambda)u_1^2 + (\mathbf{b}^2 - \lambda)u_2^2 + (\mathbf{c}^2 - \lambda)u_3^2 - u_0^2 = 0, \tag{10.13}$$

for $\lambda \in \mathbb{R}$; they are degenerate if and only if $\lambda = \mathbf{a}^2, \mathbf{b}^2, \mathbf{c}^2$.

Assume first that $\mathbf{a} > \mathbf{b} > \mathbf{c}$. In case $\lambda = \mathbf{a}^2$, the degenerate quadric envelope is

$$(\mathbf{b}^2 - \mathbf{a}^2)u_2^2 + (\mathbf{c}^2 - \mathbf{a}^2)u_3^2 - u_0^2 = 0$$

which envelops the imaginary conic

$$\frac{Y^2}{\mathbf{b}^2 - \mathbf{a}^2} + \frac{Z^2}{\mathbf{c}^2 - \mathbf{a}^2} = 1, \quad X = 0.$$

Similarly $\lambda = \mathbf{b}^2$ gives as second focal conic the hyperbola

$$\frac{X^2}{\mathbf{a}^2 - \mathbf{b}^2} + \frac{Z^2}{\mathbf{c}^2 - \mathbf{b}^2} = 1, \quad Y = 0.$$

and $\lambda = \mathbf{c}^2$ the ellipse

$$\frac{X^2}{\mathbf{a}^2 - \mathbf{c}^2} + \frac{Y^2}{\mathbf{b}^2 - \mathbf{c}^2} = 1, \quad Z = 0.$$

Assume now that Q is an ellipsoid of revolution with $\mathbf{a} > \mathbf{b} = \mathbf{c}$. Then $\lambda = \mathbf{a}^2$ still gives a focal conic, this time the imaginary circle

$$Y^2 + Z^2 = \mathbf{b}^2 - \mathbf{a}^2, \quad X = 0.$$

By contrast, the degenerate envelope with $\lambda = \mathbf{b}^2$ has equation

$$(\mathbf{a}^2 - \mathbf{b}^2)u_1^2 - u_0^2 = 0.$$

It has thus rank two and envelops the pair of real points

$$X^2 = \mathbf{a}^2 - \mathbf{b}^2$$

of the axis $Y = Z = 0$. These points are the foci of the ellipsoid. The case $\mathbf{a} = \mathbf{b} > \mathbf{c}$ is similar, but for giving a real focal circle and a pair of imaginary foci.

If Q is a sphere ($\mathbf{a} = \mathbf{b} = \mathbf{c}$), it has centre $O = (0,0,0)$ and the same computation gives $2O^*$ as the only degenerate envelope: the centre O is thus the only focus of Q.

To close the example, it follows from the equation (10.13) above that the quadrics of the confocal range determined by Q have equations

$$\frac{X^2}{\mathbf{a}^2 - \lambda} + \frac{Y^2}{\mathbf{b}^2 - \lambda} + \frac{Z^2}{\mathbf{c}^2 - \lambda} = 1,$$

for $\lambda \in \mathbb{R}$, $\lambda \neq \mathbf{a}^2, \mathbf{b}^2, \mathbf{c}^2$.

10.5 Metric reduced equations of quadrics

This section contains the first step towards the metric classification of quadrics of n-dimensional Euclidean spaces, which consists of proving the existence of metric reduced equations. Essentially we will proceed as for three-dimensional quadrics, the next theorem providing enough pairwise orthogonal and conjugate directions.

Theorem 10.5.1. *Assume that $\mathcal{P}$ is a pencil of quadrics of a real projective space $\mathbb{P}_n$ containing a quadric with no real point; then $\mathcal{P}$ is a regular pencil. If $Q_1, \dots, Q_m$ are the degenerate quadrics of $\mathcal{P}$ and L_i the linear variety of double points of Q_i, $i = 1, \dots, m$, then the following holds:*

(a) *The degree of the characteristic divisor of $\mathcal{P}$ is $n + 1$.*

(b) *The multiplicity of Q_i in the characteristic divisor of $\mathcal{P}$ is* $\dim L_i + 1$.

(c) *For each i=1,...,m,*

$$\emptyset = L_i \cap \Big(\bigvee_{j \neq i} L_j\Big).$$

Proof. A quadric $Q' \in \mathcal{P}$ with no real points has in particular no double points. Hence it is non-degenerate and $\mathcal{P}$ is regular.

We will make a special proof of claim (a) for $n = 1$. Let $Q' \in \mathcal{P}$ be a pair of imaginary points. $\mathcal{P}$ being regular, it may be spanned by Q' and a second pair of distinct points Q. The conjugations with respect to Q and Q' are then two different involutions of $\mathbb{P}_1$ and Proposition 8.2.11 applies to them, showing the existence of two different real points q_1, q_2 that are conjugate with respect to both Q and Q', and therefore also with respect to any point-pair in $\mathcal{P}$. Let Q'' be the point-pair of $\mathcal{P}$ containing q_1: q_1 is conjugate to itself and to q_2 with respect to Q''; it is thus a double point of Q'' and so $Q'' = 2q_1$ is degenerate. The same argument shows that $2q_2 \in \mathcal{P}$, after which the degree of characteristic divisor is two, by 9.5.8.

Now, for the case $n > 1$, fix coordinates in $\mathbb{P}_n$ and take Q and Q' spanning $\mathcal{P}$, with matrices M and N, respectively. If all prime factors of $\det(\lambda M + \mu N)$ in $\mathbb{R}[\lambda, \mu]$ are linear, then we are done. Assume otherwise: then, by 9.1.2, $\det(\lambda M + \mu N)$ is a product of linear factors in $\mathbb{C}[\lambda, \mu]$, and $\det(\lambda M + \mu N) = 0$ has some non-real solution $(1, \alpha)$, $\alpha \in \mathbb{C} - \mathbb{R}$. We will see that this hypothesis leads us to the existence of a real base point of $\mathcal{P}$, which contradicts the existence of a quadric in $\mathcal{P}$ with no real points. If $\bar{\alpha}$ is the complex-conjugate of α, then $\bar{\alpha} \neq \alpha$, we have $\det(M + \alpha N) = \det(M + \bar{\alpha} N) = 0$ and so $A = M + \alpha N$ and $\bar{A} = M + \bar{\alpha} N$ are matrices of two different degenerate quadrics $T, \bar{T}$ of $\mathbb{C}\mathbb{P}_n$. Since

$$M = -\frac{\bar{\alpha}}{\alpha - \bar{\alpha}} A + \frac{\alpha}{\alpha - \bar{\alpha}} \bar{A} \quad \text{and} \quad N = \frac{1}{\alpha - \bar{\alpha}} A - \frac{1}{\alpha - \bar{\alpha}} \bar{A},$$

the pencil of quadrics $\bar{\mathcal{P}}$ of $\mathbb{C}\mathbb{P}_n$ spanned by T and $\bar{T}$ obviously contains the complex extensions of all the quadrics of $\mathcal{P}$. Any real base point of $\bar{\mathcal{P}}$ is thus a base point of $\mathcal{P}$ and therefore it will be enough to see that T and $\bar{T}$ share a real point.

Let p be a double point of T. Since the matrices $M + \alpha N$ and $M + \bar{\alpha} N$ are complex-conjugate of each other, the complex-conjugate $\bar{p}$ of p is a double point of $\bar{T}$. If $p = \bar{p}$, then p is a real point that belongs to both T and $\bar{T}$ and we are done. Otherwise $\ell = p\bar{p}$ is a real line; Choose a real plane π of $\mathbb{C}\mathbb{P}_n$ containing ℓ. Since p is a double point of T, $T \cap \pi$ is a line-pair and hence contains a line s; The complex-conjugate $\bar{s}$ of s is then contained in π, and also in $\bar{T}$, using again that T and $\bar{T}$ have complex-conjugate matrices. If $s = \bar{s}$, then s is a real line contained in both T and $\bar{T}$ and it is enough to pick any of its real points. Otherwise s and $\bar{s}$ are different complex-conjugate lines in a plane and therefore they intersect at a real point which, clearly, belongs to both T and $\bar{T}$.

For claim (b), take a degenerate quadric Q_i of $\mathcal{P}$ and let $Q' \in \mathcal{P}$ be a quadric with no real points. As already said, Q' is non-degenerate. In particular Q_i and Q' are different and therefore span $\mathcal{P}$. Take $d = \dim L_i$, and let L' be the polar variety of L_i relative to Q'. Since Q' has no real points, by 6.5.9, $L_i \cap L' = \emptyset$ and so L_i and L' are supplementary linear varieties. Choose $d + 1$ independent points $p_0, \dots, p_d \in L_i$ and $n - d$ independent points $p_{d+1}, \dots, p_n \in L'$. Since L_i and L' are supplementary, the points $p_0, \dots, p_n$ are independent and we take them as the vertices of a reference Δ of $\mathbb{P}_n$. By 6.7.1 and due to being p_s and p_j conjugate with respect to Q' for any s, j, $0 \leq s \leq d < j \leq n$, Q and Q' have matrices relative to Δ, in blocks,

$$M = \begin{pmatrix} 0 & 0 \\ 0 & M'' \end{pmatrix} \quad \text{and} \quad N = \begin{pmatrix} N' & 0 \\ 0 & N'' \end{pmatrix},$$

where M'' and N'' are $(n-d) \times (n-d)$ matrices, M'' is regular, and so is N' because N is. A characteristic equation of $\mathcal{P}$ is thus

$$0 = \begin{vmatrix} \mu N' & 0 \\ 0 & \lambda M'' + \mu N'' \end{vmatrix} = \mu^{d+1} \det N' \det(\lambda M'' + \mu N''),$$

where $\det(\lambda M'' + \mu N'')$ has no factor μ because for $\mu = 0$ it gives $\lambda^{n-d} \det M'' \neq 0$. This completes the proof of claim (b).

Regarding claim (c), assume that $p \in L_i \cap (\bigvee_{j \neq i} L_j)$. For $j \neq i$, $j = 1, \dots, m$, take points $q_{j,0}, \dots, q_{j,d_j}$ spanning L_j. The point p is conjugate to each of the $q_{j,s}$ with respect to Q_i, because p is double for Q_i, and also with respect to Q_j because $q_{j,s}$ is double for Q_j. By 9.5.1, p is then conjugate to all the $q_{j,s}$, $j \neq i$, $j = 1, \dots, m$, $s = 0, \dots, d_j$ with respect to all quadrics of $\mathcal{P}$. Since the whole of the above points $q_{j,s}$ span $\bigvee_{j \neq i} L_j$, p is conjugate to all points of $\bigvee_{j \neq i} L_j$, and so in particular to itself, with respect to all quadrics of $\mathcal{P}$. This proves that p is a base point, against the existence in $\mathcal{P}$ of a quadric with no real points. □

Corollary 10.5.2. *Hypothesis and notations being as in* 10.5.1, $\dim L_1 + \dots + \dim L_m + m - 1 = n$ *and* $L_1 \vee \dots \vee L_m = \mathbb{P}_n$.

Proof. By 10.5.1 (a), $n+1$ equals the degree of the characteristic divisor which in turn, by 10.5.1 (b), equals $\sum_{i=1}^{m} (\dim L_i + 1)$, hence the first equality. Using 10.5.1 (c) inductively, it is easy to check that $\dim L_1 \vee \dots \vee L_m = \dim L_1 + \dots + \dim L_m - m + 1$, after which the second equality follows from the first one. □

Corollary 10.5.3. *If a pencil $\mathcal{P}$, of quadrics of a real projective space $\mathbb{P}_n$, contains a quadric with no real point, then there is a self-polar simplex for $\mathcal{P}$.*

Proof. Take the notations as in 10.5.1 and still call Q' the quadric with no real points which is assumed to belong to $\mathcal{P}$. For each $i = 1, \dots, m$, $Q' \not\supset L_i$ and so $Q' \cap L_i$ is a quadric of L_i. Let Ω_i be a simplex of L_i self-polar for $Q' \cap L_i$. Collecting the vertices of all the Ω_i, $i = 1 \dots, m$, we obtain $n+1$ points spanning $\mathbb{P}_n$, due to 10.5.2: these points are thus independent and may be taken as the vertices of an n-dimensional simplex Ω of $\mathbb{P}_n$. We will see next that Ω is self-polar for $\mathcal{P}$. Indeed, if different vertices p, q of Ω belong to the same Ω_i, then they are double points of Q_i, and hence conjugate with respect to it, while they are also conjugate with respect to Q' by the definition of Ω_i: they are thus conjugate with respect to any quadric of $\mathcal{P}$. If, otherwise, p and q are vertices of Ω_i and Ω_j, respectively, and $i \neq j$, then p is a double point of Q_i and q a double point of Q_j, after which they are conjugate with respect to both Q_i and Q_j, and hence are conjugate with respect to all quadrics of $\mathcal{P}$. □

Remark 10.5.4. If a simplex Ω is self-polar for $\mathcal{P}$, then, by 9.10.2, all its vertices are double points of quadrics of $\mathcal{P}$. Furthermore, for each $i = 1, \dots, m$, the vertices of Ω which are double points of Q_i are in number of $\dim L_i + 1$, as obviously their number cannot be higher, and in case of being lower for one i, the total number of vertices cannot reach $\sum_{i=1}^{m}(\dim L_i + 1)$, which by 10.5.2 equals $n + 1$.

Back to our metric setting, as before, assume that we have fixed a Euclidean space $\mathbb{A}_n$, $n \geq 2$, with improper hyperplane H_∞ and absolute quadric $\mathbf{K}$. For any quadric Q of $\mathbb{A}_n$, $Q \not\supset H_\infty$, we write $Q^\infty = Q \cap H_\infty$ the improper section of Q. We have:

Corollary 10.5.5. *If Q is a quadric of $\mathbb{A}_n$, then there are $p_1, \dots, p_n \in H_\infty$ pairwise orthogonal and conjugate with respect to Q.*

Proof. If $Q \supset H_\infty$ or $Q^\infty = \mathbf{K}$, then obviously any set of n pairwise orthogonal directions does the job. Otherwise Q^∞ and $\mathbf{K}$ span a pencil $\mathcal{P}$. Since $\mathbf{K}$ has no real points, 10.5.3 applies to $\mathcal{P}$ providing a simplex Δ of H_∞, self-polar for $\mathcal{P}$. Then it is enough to take $p_1, \dots, p_n$ the vertices of Δ. □

Remark 10.5.6. A set of n improper points pairwise orthogonal and conjugate with respect to Q is called a set of *principal directions* of Q. Note that the orthogonality conditions assure that the points of any set of principal directions are independent (6.9.3). The cases in which the improper section Q^∞ is undefined or equal to $\mathbf{K}$ excepted, a set of principal directions of Q is thus the set of vertices of a self-polar simplex for the pencil spanned by Q^∞ and $\mathbf{K}$, and conversely.

Next we will associate to each quadric of a Euclidean space a metric reduced equation. We will consider first the quadrics Q for which there exist a proper point O which is conjugate with respect to Q to all the improper points. By 10.5.5, there is a set $p_1, \dots, p_n$ of principal directions of Q. We take the point O, as origin, and unitary vectors representing $p_1, \dots, p_n$ to compose an orthonormal reference. Then the simplex with vertices $O, p_1, \dots, p_n$ is self-polar for Q and so, after renumbering the basis if needed, Q has an affine equation of the form

$$a_{0,0} + a_{1,1}X_1^2 + \dots + a_{s,s}X_s^2 = 0$$

with $1 \leq s \leq n$ and $a_{i,i} \neq 0$ for $i = 1, \dots, s$. If $a_{0,0} = 0$ we just rename $a_{i,i} = \lambda_i$, $i = 1, \dots, s$, and take

$$\lambda_1 X_1^2 + \dots + \lambda_s X_s^2 = 0$$

as metric reduced equation of Q, with $1 \leq s \leq n$ and $\lambda_i \neq 0$ for $i = 1, \dots, s$. If $a_{0,0} \neq 0$, then we put $\lambda_i = -a_{i,i}/a_{0,0}$ and take the resulting equation

$$\lambda_1 X_1^2 + \dots + \lambda_s X_s^2 = 1$$

this time with $0 \leq s \leq n$ and $\lambda_i \neq 0$ for $i = 1, \dots, s$, as metric reduced equation of Q.

Assume now that there is no proper point conjugate with respect to Q to all the improper points. We will consider first the case in which Q is non-degenerate: then it is a paraboloid and its centre, denoted in the sequel by $\bar{O}$, is (by 6.4.24) the only double point of the improper section Q^∞. Choose a set $p_1, \dots, p_n$ of principal directions of Q: by 10.5.6 and 10.5.4, $\bar{O}$ belongs to it: up to reordering, we assume $\bar{O} = p_n$. Take $T = p_1 \vee \dots \vee p_{n-1}$, which has dimension $n-2$, and ℓ the polar variety of T relative to Q: ℓ is thus a line. Since T does not contain the double point p_n of Q^∞, no improper point has hyperplane polar relative to Q^∞ equal to T, and therefore p_n is the only improper point conjugate with respect to Q^∞ (or Q, it is the same) to all points of T. This proves that $\ell \cap H_\infty = p_n$, and so ℓ goes through p_n, is a proper line, and is not tangent to Q at p_n. In other words, ℓ is a proper chord of Q one of whose ends is p_n: we take its other end O as origin and unitary vectors representing the principal directions $p_1, \dots, p_n$ to compose an orthonormal reference. The origin O is self-conjugate because it belongs to Q, and conjugate to $p_1, \dots, p_{n-1}$ because it belongs to ℓ. On the other hand, any two different $p_i, p_j, i, j = 1, \dots, n$ are conjugate because they are part of a set of principal directions, and furthermore p_n is self-conjugate because it belongs to Q. This causes an equation of Q to have the form

$$a_{1,1}X_1^2 + \dots + a_{n-1,n-1}X_n^2 + 2a_{0,n}X_n = 0$$

with $a_{i,i} \neq 0, i = 1, \dots, n-1$ and $a_{0,n} \neq 0$ because Q is non-degenerate. Taking $\lambda_i = -a_{i,i}/a_{0,n}, i = 1, \dots, n-1$, gives as metric reduced equation

$$\lambda_1 X_1^2 + \dots + \lambda_{n-1}X_{n-1}^2 = 2X_n,$$

with $\lambda_i \neq 0, i = 1, \dots, n-1$.

Assume Q degenerate and still that there is no proper point conjugate to all the improper ones. Then the variety of double points $D = \mathcal{D}(Q)$ is improper and different from H_∞. Therefore, if we take $d = \mathbf{r}(Q) - 1$, then it is $0 \leq n - d - 1 \leq n - 2$, or, equivalently, $1 \leq d \leq n-1$. In case $d = 1$, Q is a pair of planes with improper intersection, say $Q = H + H'$, $T = H \cap H' \subset H_\infty$. If both $H, H' \neq H_\infty$ then the fourth harmonic of H_∞ with respect to H, H' (in T^*) is a proper plane and any of its proper points is conjugate to all improper points against the hypothesis. Q is thus $Q = H + H_\infty$, H a proper hyperplane, and after an obvious choice of the orthonormal reference we may take as metric reduced equation of Q just $0 = 2X_n$. Next we deal with the more interesting case $d \geq 2$. In H_∞ we take $\bar{D}$ to be the polar variety of D with respect to the absolute, or, equivalently, the set of all directions orthogonal to all the directions which are double points of Q. Since the absolute has no real points, $D \cap \bar{D} = \emptyset$ (6.5.9) and so, if for a fixed proper point p we take $L = p \vee \bar{D}$, then $L \cap D = \emptyset$

and $\dim L = 1 + \dim \bar{D} = n - \dim D - 1$: this assures that L and D are supplementary varieties, and also that $\dim L = d \geq 2$. By 6.7.7, $Q' = Q \cap L$ is a non-degenerate quadric of L. Furthermore no proper point $q \in L$ is conjugate to all points of $L \cap H_\infty = \bar{G}$ with respect to Q', as otherwise, any point being obviously conjugate with respect to Q to all points of G, q would be conjugate with respect to Q to all points of $H_\infty = G \vee \bar{G}$, against the hypothesis. What we have proved for Q non-degenerate applies now to Q' showing the existence of an orthonormal reference $(O, e_1, \dots, e_d)$ of (the proper part of) L relative to which Q' has equation

$$\lambda_1 X_1^2 + \cdots + \lambda_{d-1} X_{d-1}^2 = 2X_d,$$

with $\lambda_i \neq 0, i = 1, \dots, d-1$. Since any two directions in D and $\bar{D}$ are orthogonal, we enlarge Δ' to an orthonormal reference Δ of $\mathbb{A}_n$ by adding to $e_1, \dots, e_d$, unitary vectors $e_{d+1}, \dots, e_n$ representing pairwise orthogonal directions in D. In this way, the last $n-d$ vertices of the projective reference associated to Δ are double points of Q and therefore, by 6.7.1, the above equation of Q' may be taken as an equation of Q relative to Δ. In order to follow the usual conventions, we reorder the coordinates by swapping over the d-th and n-th ones, and take $s = d - 1$. After this the metric reduced equation of any Q for which no proper point is conjugate to all the improper ones will be

$$\lambda_1 X_1^2 + \cdots + \lambda_s X_s^2 = 2X_n,$$

with $\lambda_i \neq 0, i = 1, \dots, s-1, 0 \leq s \leq n-1$. Note that we are allowing the value $s = n - 1$ to include the case of Q non-degenerate, and also $s = 0$ to include the case $d = 1$, already dealt with separately. All together we have proved:

Theorem 10.5.7 (Metric reduced equations of quadrics). *For any quadric Q of an n-dimensional Euclidean space, $n \geq 2$, there is an orthonormal reference relative to which Q has one of the following equations.*

Type I:

$$\lambda_1 X_1^2 + \cdots + \lambda_s X_s^2 = 0,$$

with $1 \leq s \leq n$ and $\lambda_i \neq 0$ for $i = 1, \dots, s$.

Type II:

$$\lambda_1 X_1^2 + \cdots + \lambda_s X_s^2 = 1,$$

with $0 \leq s \leq n$ and $\lambda_i \neq 0$ for $i = 1, \dots, s$.

Type III:

$$\lambda_1 X_1^2 + \cdots + \lambda_s X_s^2 = 2X_n,$$

with $0 \leq s \leq n-1$ and $\lambda_i \neq 0$ for $i = 1, \dots, s$.

Such an equation is called a ***metric reduced equation*** *of Q.*

By its definition, a metric reduced equation of a quadric is always relative to an orthonormal reference, which is called a *principal reference*, and its axes *principal axes*, or just *axes*, of the quadric. For $n = 2, 3$ the reader may recognize, with minor variations, the metric reduced equations already found in Sections 10.2 and 10.4 for conics and three-space quadrics.

The above reduced equations are not fully determined by the quadric: besides the obvious possibility of renumbering the monomials of their left-hand side, equations of type I may be multiplied by an arbitrary non-zero factor, and equations of type III have their coefficients λ_i turned into their opposites by just replacing the n-th vector of the base with its opposite. It is usual to take the metric reduced equations up to these indeterminacies: to be precise, in the sequel we will call *congruent* two metric reduced equations whose left-hand sides are $\sum_{i=1}^{s} \lambda_i X_i^2$ and $\sum_{i=1}^{r} \lambda_i' X_i^2$, if and only if $s = r$ and for a certain permutation σ of $\{1, \dots, n\}$, they either

(a) are both of type I and have $\lambda_i' = \rho\lambda_{\sigma(i)}$, $i = 1, \dots, s$, for some $\rho \in \mathbb{R} - \{0\}$, or

(b) are both of type II and have $\lambda_i' = \lambda_{\sigma(i)}$, $i = 1, \dots, s$, or

(c) are both of type III and have $\lambda_i' = \varepsilon\lambda_{\sigma(i)}$, $i = 1, \dots, s$, for $\varepsilon = \pm 1$.

The next lemma allows the metric reduced equations to play a role similar to that of the projective and affine reduced equations.

Lemma 10.5.8. *Let Q and Q' be quadrics of Euclidean spaces of the same dimension. If Q and Q' have congruent reduced equations, then they are congruent.*

Proof. Assume that Q and Q' are quadrics of Euclidean spaces $\mathbb{A}_n$ and $\mathbb{A}_n'$, respectively, and have congruent metric reduced equations relative to orthonormal references Δ and Δ'. First of all, after reordering the basis of Δ', it is not restrictive to assume that the permutation σ involved in the congruence of the equations is the identity. The projectivity $f : \mathbb{A}_n' \to \mathbb{A}_n$ mapping Δ' to Δ is an isometry by 6.9.30 and $f(Q')$ is thus congruent to Q'. By 2.8.1, the equation of $f(Q')$ relative to Δ equals the equation of Q' relative to Δ'. We may thus replace Q', $\mathbb{A}_n'$ and Δ' with $f(Q')$, $\mathbb{A}_n$ and Δ and make the proof assuming that $\mathbb{A}_n' = \mathbb{A}_n$ and $\Delta' = \Delta$.

Assume thus that Q and Q' are quadrics of the same Euclidean space $\mathbb{A}_n$ that have congruent metric reduced equations relative to the same reference, and that the permutation of indices involved in the congruence of the equations is the identity. If the reduced equations are of type I or II, then clearly $Q = Q'$. If the reduced equations are of type III, then either $\varepsilon = 1$ and again $Q = Q'$, or $\varepsilon = -1$ and then the reflection in the coordinate hyperplane $X_n = 0$,

$$(X_1, \dots, X_{n-1}, X_n) \mapsto (X_1, \dots, X_{n-1}, -X_n),$$

maps Q to Q'. □

10.6 Metric invariants of quadrics

If Q is a quadric of a Euclidean space $\mathbb{A}_n$ and q a proper point of $\mathbb{A}_n$, in this section we associate to the pair (Q, q) an ordered set of real numbers that are metric invariants of the pair. It will turn out that part of these invariants are independent of the choice of the point q and compose a complete system of metric invariants for quadrics of Euclidean spaces.

For any quadric Q we still denote by $\mathcal{D}(Q)$ the linear variety of double points of Q and, if Q belongs to an affine space, by Q^∞ the improper section of Q. Assume we have fixed a quadric Q of a Euclidean space $\mathbb{A}_n$, $n \geq 2$, and a proper point q. Let S_q be the imaginary sphere with centre q and radius $\boldsymbol{i} = \sqrt{-1}$. We choose representatives η of Q and ψ of S_q, and define two finite unordered sequences of real numbers

$$\{\tilde{\mu}_1, \dots, \tilde{\mu}_{n+1}\} \quad \text{and} \quad \{\mu_1, \dots, \mu_n\}$$

in the following way: If $Q = S_q$, then $\eta = \delta\psi$ for some $\delta \in \mathbb{R}$ and we take

$$\tilde{\mu}_1 = \dots = \tilde{\mu}_{n+1} = \mu_1 = \dots = \mu_n = \delta.$$

Otherwise Q and S_q span a pencil $\mathcal{P}$ and we take the reference of $\mathcal{P}$ with vertices S_q and Q and unit point the quadric $[\eta - \psi]$: relative to it, the quadric $Q_\mu = [\eta - \mu\psi]$ has absolute coordinate μ.

(1) We define $\tilde{\mu}_1, \dots, \tilde{\mu}_{n+1}$ to be the values of μ for which the quadric Q_μ is degenerate, each repeated as many times as indicated by $\dim \mathcal{D}(Q_\mu) + 1$. Since S_q is non-degenerate and has no real points, 10.5.1 applies showing that $\tilde{\mu}_1, \dots, \tilde{\mu}_{n+1}$ are the absolute coordinates of the elements of the characteristic divisor of $\mathcal{P}$ repeated according to multiplicities, and that their number is, indeed, $n + 1$.

(2) If there is a quadric Q_δ containing H_∞, then all the other quadrics in $\mathcal{P}$ have the same improper section (the absolute $\mathbf{K} = S_q^\infty$, in fact), by 9.5.4: then we take $\mu_1 = \dots = \mu_n = \delta$. Otherwise $\mu_1, \dots, \mu_n$ are taken to be the values of μ for which Q_μ^∞ is degenerate, each repeated as many times as indicated by $\dim \mathcal{D}(Q_\mu^\infty) + 1$. In this case, again by 9.5.4, the improper sections of the quadrics of $\mathcal{P}$ describe a pencil $\mathcal{P}^\infty$ in which μ may be taken as the absolute coordinate of Q_μ^∞ relative to the reference $(S_q^\infty, Q^\infty, Q_1^\infty)$. As above, S_q^∞ is non-degenerate and has no real points, after which, using 10.5.1, $\mu_1, \dots, \mu_n$ are the absolute coordinates of the elements of the characteristic divisor of $\mathcal{P}^\infty$, repeated according to multiplicities, and their number is n.

The reader may note that no use of coordinates of $\mathbb{A}_n$ has been made in the above definitions. A different choice of the representatives of Q and S_q (or, equivalently, of the unit point of the reference of $\mathcal{P}$) causes the multiplication of all the $\tilde{\mu}_i$ and μ_j by the same non-zero constant factor. Because of this, $\tilde{\mu}_1, \dots, \tilde{\mu}_{n+1}, \mu_1, \dots, \mu_n$

will be taken as defined up to a common non-zero factor $\rho \in \mathbb{R} - \{0\}$ and usually written

$$\rho\tilde{\mu}_1, \dots, \rho\tilde{\mu}_{n+1}, \rho\mu_1, \dots, \rho\mu_n$$

to recall this fact.

Proposition 10.6.1. *The unordered sequences*

$$\rho\tilde{\mu}_1, \dots, \rho\tilde{\mu}_{n+1} \quad \textit{and} \quad \rho\mu_1, \dots, \rho\mu_n,$$

taken up to the indeterminacy of the common factor ρ, *are metric invariants of the pair* (Q, q).

Proof. Assume that f is an isometry and write $Q' = f(Q), q' = f(q)$. By 10.1.8, $f(S_q)$ is the imaginary sphere $S_{q'}$ with centre q' and radius $\boldsymbol{i}$. If $Q = S_q$, then $Q' = S_{q'}$ and the claim is obvious. Otherwise $Q' \neq S_{q'}$ and if $\mathcal{P}'$ is the pencil they span, by 9.3.4, the isometry f induces the projectivity

$$\begin{aligned} \mathbb{Q}(f)_{|\mathcal{P}} \colon \mathcal{P} &\longrightarrow \mathcal{P}', \\ T &\longmapsto f(T). \end{aligned}$$

Assume that the $\tilde{\mu}_i$ and μ_j associated to (Q, q) have been taken as absolute coordinates relative to a reference Ω of $\mathcal{P}$ with vertices S_q, Q, and similarly that the $\tilde{\mu}'_i$ and μ'_j associated to (Q', q') have been taken as absolute coordinates relative to a reference Ω' of $\mathcal{P}'$ with vertices $S_{q'}$, Q'. Since $\mathbb{Q}(f)_{|\mathcal{P}}$ maps vertices to vertices, in their order, it has diagonal matrix. Therefore, for a certain $\rho \in \mathbb{R} - \{0\}$ and all $\mu \in \mathbb{R} \cup \{\infty\}$, $\mathbb{Q}(f)_{|\mathcal{P}}$ maps the quadric Q_μ, with absolute coordinate μ in $\mathcal{P}$, to the quadric $Q'_{\rho\mu}$, with absolute coordinate $\rho\mu$ in $\mathcal{P}'$.

Now, Q_μ is degenerate if and only if $f(Q_\mu) = Q'_{\rho\mu}$ is, and furthermore $\dim \mathcal{D}(Q_\mu) = \dim \mathcal{D}(f(Q_\mu))$: this proves that, up to renumbering, $\tilde{\mu}'_i = \rho\tilde{\mu}_i$, $i = 1, \dots, n+1$.

If there is Q_μ containing the improper hyperplane, then so does $f(Q_\mu) = Q'_{\rho\mu}$, and conversely. Otherwise Q_μ has degenerate improper section if and only if the improper section of $f(Q_\mu) = Q'_{\rho\mu}$ is degenerate, and furthermore $\dim \mathcal{D}(Q_\mu^\infty) = \dim \mathcal{D}(f(Q_\mu)^\infty)$. This proves that, in both cases and up to renumbering $\mu'_j = \rho\mu_j$, $j = 1, \dots, n$. □

Assume to have fixed a system of affine, non-necessarily orthonormal, coordinates in $\mathbb{A}_n$, denote by $\widetilde{M} = (a_{i,j})_{i,j=0,\dots,n}$ a matrix of Q and by M the submatrix $M = (a_{i,j})_{i,j=1,\dots,n}$, which is a matrix of Q^∞ relative to the reference subordinated on H_∞. Take similarly $\widetilde{N} = (b_{i,j})_{i,j=0,\dots,n}$ a matrix of S_q and $N = (b_{i,j})_{i,j=1,\dots,n}$, which is a matrix of $S_q^\infty = \mathbf{K}$. Both $\widetilde{N}$ and N being regular, we define the polynomials in z:

$$\widetilde{D}(z) = (\det \widetilde{N})^{-1} \det(\widetilde{M} - z\widetilde{N})$$

and

$$D(z) = (\det N^{-1}) \det(M - zN).$$

We have:

Proposition 10.6.2. *For a certain* $\rho \in \mathbb{R} - \{0\}$,

$$\tilde{D} = \prod_{i=1}^{n+1} (\rho\tilde{\mu}_i - z)$$

and

$$D = \prod_{i=1}^{n} (\rho\mu_i - z).$$

Proof. We will prove that if the $\tilde{\mu}_i$ and μ_j are defined using the representatives of Q and S_q that have matrices $\widetilde{M}$ and $\widetilde{N}$, then the claim is satisfied for $\rho = 1$.

If $Q = S_q$, then, for some $\delta \in \mathbb{R} - \{0\}$, $\widetilde{M} = \delta\widetilde{N}$, and by definition $\tilde{\mu}_i = \mu_j = \delta$ for $i = 1, \dots, n+1$ and $j = 1, \dots, n$. On the other hand

$$\widetilde{D}(z) = (\det \widetilde{N})^{-1} \det(\widetilde{M} - z\widetilde{N}) = (\delta - z)^{n+1}$$

and

$$D(z) = (\det N)^{-1} \det(M - zN) = (\delta - z)^n,$$

hence the claim.

Otherwise, the quadrics with matrices $\widetilde{M} - \tilde{\mu}_i\widetilde{N}$, $i = 1, \dots, n+1$, are the elements of the characteristic divisor of $\mathcal{P}$, each repeated according to its multiplicity, and so $[1, -\tilde{\mu}_i]$, $i = 1, \dots, n+1$, are the solutions, repeated according to multiplicities, of the characteristic equation $\det(\lambda\widetilde{M} + \mu\widetilde{N}) = 0$. Therefore, by 9.1.2,

$$\det(\lambda\widetilde{M} + \mu\widetilde{N}) = a \prod_{i=1}^{n+1} (\tilde{\mu}_i\lambda + \mu)$$

with $a \in \mathbb{R}$. By taking $\lambda = 0$ and $\mu = 1$, it follows that $a = \det \widetilde{N}$, and so

$$\widetilde{D}(z) = (\det \widetilde{N})^{-1} \det(\widetilde{M} - z\widetilde{N}) = \prod_{i=1}^{n+1} (\tilde{\mu}_i - z),$$

as claimed.

If Q_δ^∞ is undetermined, then $M - \delta N = 0$ and so

$$D(z) = (\det N)^{-1} \det(M - zN) = (\delta - z)^n,$$

according to the claim, because in such a case $\mu_1 = \dots = \mu_n = \delta$.

If no quadric of $\mathcal{P}$ contains the improper hyperplane, then, arguing as above, $[1, -\mu_j]$, $j = 1, \ldots, n$, are the solutions, repeated according to multiplicities, of the characteristic equation $\det(\lambda M + \mu N) = 0$ of the pencil of improper sections. Consequently,

$$\det(\lambda M + \mu N) = b \prod_{j=1}^{n} (\mu_j \lambda + \mu)$$

with $b \in \mathbb{R}$. By taking $\lambda = 0$ and $\mu = 1$, it follows that $b = \det N$. Hence,

$$D(z) = (\det N)^{-1} \det(M - zN) = \prod_{j=1}^{n} (\mu_j - z),$$

which completes the proof. □

Remark 10.6.3. It follows in particular from 10.6.2 that $\tilde{D}$ has degree $n+1$ and leading term $(-1)^{n+1} z^{n+1}$, while D has degree n and leading term $(-1)^n z^n$.

After 10.6.2, the roots of the polynomials $\tilde{D}$ and D, repeated according to multiplicities, may be taken as the invariants $\tilde{\mu}_i$, $i = 1, \ldots, n+1$, and μ_j, $j = 1, \ldots, n$, respectively. Conversely, if the polynomials are written

$$\tilde{D}(z) = (-1)^{n+1} \tilde{d}_0 z^{n+1} + (-1)^n \tilde{d}_1 z^n + \cdots - \tilde{d}_n z + \tilde{d}_{n+1}$$

and

$$D(z) = (-1)^n d_0 z^n + (-1)^{n-1} d_1 z^{n-1} + \cdots - d_{n-1} z + d_n,$$

then $\tilde{d}_0 = d_0 = 1$ and the other coefficients appear as elementary symmetric functions of the $\tilde{\mu}_i$ and μ_j; still writing the indeterminacy factor ρ, it is:

$$\rho^s \tilde{d}_s = \sum_{i_1 < \cdots < i_s} \rho^s \tilde{\mu}_{i_1} \ldots \tilde{\mu}_{i_s}, \quad s = 1, \ldots, n+1, \tag{10.14}$$

and

$$\rho^s d_s = \sum_{j_1 < \cdots < j_s} \rho^s \mu_{j_1} \ldots \mu_{j_s}, \quad s = 1, \ldots, n. \tag{10.15}$$

So, the ordered sequences

$$\rho \tilde{d}_1, \ldots, \rho^n \tilde{d}_n \quad \text{and} \quad \rho d_1, \ldots, \rho^{n-1} d_n$$

are metric invariants of (Q, q), defined up to the indeterminacy of ρ and carrying the same information as $\{\rho \tilde{\mu}_i\}_{i=1,\ldots,n+1}$ and $\{\rho \mu_i\}_{i=1,\ldots,n}$. The reader may note that the r-th power of the indeterminacy parameter ρ appears as a factor of $\tilde{d}_r$ and d_r. Because of this $\tilde{d}_r$ and d_r are said to be invariants of degree r, while, by the same reason, the $\tilde{\mu}_i$ and μ_j are called invariants of degree one. Invariants of positive degree may be combined to obtain degree-zero invariants, which are invariants involving no indeterminacy, often called *absolute invariants*. To this end let us recall that a polynomial $\Phi \in \mathbb{R}[Z_1, \dots, Z_m]$ is called isobaric of weight r if and only if it has the form

$$\Phi = \sum_{s_1+\dots+s_\ell=r} a_{s_1,\dots,s_\ell} Z_{s_1} \dots Z_{s_\ell},$$

that is, the sum of subscripts of any of its non-zero monomials equals r. If Φ is isobaric of weight r and $\nu_1, \dots, \nu_m$ are invariants of degrees equal to their indices, then, clearly $\Phi(\nu_1, \dots, \nu_m)$ is an invariant of degree r. So, if both $\Phi_1, \Phi_2 \in \mathbb{R}[Z_1, \dots, Z_m]$ are isobaric of the same weight, then

$$\frac{\Phi_1(\nu_1, \dots, \nu_m)}{\Phi_2(\nu_1, \dots, \nu_m)}$$

is an invariant of degree zero. In Section 10.7 the coefficients of the metric reduced equations will appear as degree-zero invariants of the quadric in this way.

So far, we have dealt with metric invariants of the pair (Q, q). The next theorem proves that some of them are independent of the choice of q and therefore compose a system of metric invariants of Q. This system will be proved to be complete in Section 10.7.

Theorem 10.6.4. *Let Q be a quadric and q a point, both of a Euclidean space $\mathbb{A}_n$, and consider the metric invariants of the pair (Q, p)*

$$\rho d_1, \dots, \rho^n d_n, \rho \tilde{d}_1, \dots, \rho^{n+1} \tilde{d}_{n+1}$$

defined above. We have:

(a) *If s is defined by the conditions $d_s \neq 0$ and $d_i = 0$ for $i > s$, then s is the rank of Q^∞ and $\tilde{d}_{n+1} = \dots = \tilde{d}_{s+3} = 0$.*

(b) *If $s < n$, then*

$$\rho d_1, \dots, \rho^n d_n, \rho^{s+2} \tilde{d}_{s+2}$$

are, but for the indeterminacy of $\rho \in \mathbb{R} - \{0\}$, independent of the choice of q and therefore compose a set of metric invariants of Q. If Q has a metric reduced equation of type III, then $\tilde{d}_{s+2} \neq 0$.

(c) *If either $s = n$, or $s < n$ and $\tilde{d}_{s+2} = 0$, then*

$$\rho d_1, \dots, \rho^n d_n, \rho^{s+1} \tilde{d}_{s+1}$$

are, but for the indeterminacy of $\rho \in \mathbb{R} - \{0\}$, independent of the choice of q and therefore compose a set of metric invariants of Q. If Q has a metric reduced equation of type II, then either $s = n$ or $\tilde{d}_{s+2} = 0$ and in both cases $\tilde{d}_{s+1} \neq 0$. If Q has a metric reduced equation of type I, then still $s = n$ or $\tilde{d}_{s+2} = 0$, but now, in both cases, $\tilde{d}_{s+1} = 0$.

Proof. First of all, note that the invariants $\rho d_1, \dots, \rho^{n-1} d_n$ depend only on Q^∞ and the improper section of S_q, which is the absolute: they are thus independent of the choice of q.

Regarding the first half of claim (a), by the definition of s, $n-s$ is the multiplicity of 0 as root of $D(z)$. By 10.6.2, $n-s$ is then the number of times that 0 appears among the $\rho\mu_j$, $j = 1, \dots, n$, and so, by the definition of the latter, $n - s = \dim \mathcal{D}(Q^\infty) + 1$, from which $s = \mathbf{r}(Q^\infty)$.

Now we will compute the invariants from a metric reduced equation of Q. The reference being orthonormal, if $q = (\alpha_1, \dots, \alpha_n)$, an equation of S_q is

$$\sum_{j=1}^{n} (X_j - \alpha_j)^2 + 1 = 0$$

and therefore S_q has matrix

$$\begin{pmatrix} \alpha & -\alpha_1 & \dots & -\alpha_n \\ -\alpha_1 & 1 & \dots & 0 \\ \vdots & \vdots & \ddots & \vdots \\ -\alpha_n & 0 & \dots & 1 \end{pmatrix},$$

where we have taken $\alpha = \sum_{j=1}^{n} \alpha_j^2 + 1$.

Assume first that Q has a metric reduced equation of type III. Since, as already proved, $s = \mathbf{r}(Q^\infty)$, a reduced equation of Q has the form

$$\lambda_1 X_1^2 + \dots + \lambda_s X_s^2 = 2X_n,$$

with $0 \le s \le n-1$ and $\lambda_i \neq 0$ for $i = 1, \dots, s$. Hence Q has matrix

$$\begin{pmatrix} 0 & 0 & \dots & 0 & 0 & \dots & -1 \\ 0 & \lambda_1 & & & & & 0 \\ \vdots & & \ddots & & & & \vdots \\ 0 & & & \lambda_s & & & 0 \\ 0 & & & & 0 & & 0 \\ \vdots & & & & & \ddots & \vdots \\ -1 & 0 & \dots & 0 & 0 & \dots & 0 \end{pmatrix}$$

and so,

$$\widetilde{D}(z) = \begin{vmatrix} -\alpha z & \alpha_1 z & \dots & \alpha_s z & \alpha_{s+1} z & \dots & \alpha_n z - 1 \\ \alpha_1 z & \lambda_1 - z & & & & & 0 \\ \vdots & & \ddots & & & & \vdots \\ \alpha_s z & & & \lambda_s - z & & & 0 \\ \alpha_{s+1} z & & & & -z & & 0 \\ \vdots & & & & & \ddots & \vdots \\ \alpha_n z - 1 & 0 & \dots & 0 & 0 & \dots & -z \end{vmatrix}.$$

By adding to the first column the products of the $(s+2)$-th, ..., n-th ones by, respectively, $\alpha_{s+1}, \dots, \alpha_{n-1}$, we get

$$\widetilde{D}(z) = \begin{vmatrix} -\alpha' z & \alpha_1 z & \dots & \alpha_s z & \alpha_{s+1} z & \dots & \alpha_{n-1} z & \alpha_n z - 1 \\ \alpha_1 z & \lambda_1 - z & & & & & & 0 \\ \vdots & & \ddots & & & & & \vdots \\ \alpha_s z & & & \lambda_s - z & & & & 0 \\ 0 & & & & -z & & & 0 \\ \vdots & & & & & \ddots & & \vdots \\ 0 & & & & & & -z & 0 \\ \alpha_n z - 1 & 0 & \dots & 0 & 0 & \dots & 0 & -z \end{vmatrix}$$

$$= (-1)^{n-s-1} z^{n-s-1} \begin{vmatrix} -\alpha' z & \alpha_1 z & \dots & \alpha_s z & \alpha_n z - 1 \\ \alpha_1 z & \lambda_1 - z & & & 0 \\ \vdots & & \ddots & & \vdots \\ \alpha_s z & & & \lambda_s - z & 0 \\ \alpha_n z - 1 & 0 & \dots & 0 & -z \end{vmatrix},$$

where $\alpha' = \alpha - \sum_{j=s+1}^{n-1} \alpha_j^2$. Thus, $\rho^{n+1}\tilde{d}_{n+1} = \dots = \rho^{s+3}\tilde{d}_{s+3} = 0$. After dividing by z^{n-s-1} and taking $z = 0$, it turns out that

$$\rho^{s+2}\tilde{d}_{s+2} = \begin{vmatrix} 0 & 0 & \dots & 0 & -1 \\ 0 & \lambda_1 & & & 0 \\ \vdots & & \ddots & & \vdots \\ 0 & & & \lambda_s & 0 \\ -1 & 0 & \dots & 0 & 0 \end{vmatrix} = -\lambda_1 \dots \lambda_s.$$

In particular $\rho^{s+2}\tilde{d}_{s+2}$ is in this case, as claimed, non-zero and independent of the choice of q, because the coordinates α_i of q do not appear in the above expression.

Assume now that a metric reduced equation of Q is of type I or II, and so write it in the form

$$\lambda_1 X_1^2 + \dots + \lambda_s X_s^2 = u, \quad u = 0, 1,$$

where $0 \leq s \leq n$ if $u = 1$ and $1 \leq s \leq n$ if $u = 0$. Computing as above,

$$\widetilde{D}(z) = \begin{vmatrix} -u - \alpha z & \alpha_1 z & \dots & \alpha_s z & \alpha_{s+1} z & \dots & \alpha_n z \\ \alpha_1 z & \lambda_1 - z & & & & & 0 \\ \vdots & & \ddots & & & & \vdots \\ \alpha_s z & & & \lambda_s - z & & & 0 \\ \alpha_{s+1} z & & & & -z & & 0 \\ \vdots & & & & & \ddots & \vdots \\ \alpha_n z & 0 & \dots & 0 & 0 & \dots & -z \end{vmatrix}$$

$$= \begin{vmatrix} u - \alpha'' z & \alpha_1 z & \dots & \alpha_s z & \alpha_{s+1} z & \dots & \alpha_n z \\ \alpha_1 z & \lambda_1 - z & & & & & 0 \\ \vdots & & \ddots & & & & \vdots \\ \alpha_s z & & & \lambda_s - z & & & 0 \\ 0 & & & & -z & & 0 \\ \vdots & & & & & \ddots & \vdots \\ 0 & 0 & \dots & 0 & 0 & \dots & -z \end{vmatrix}$$

$$= (-1)^{n-s} z^{n-s} \begin{vmatrix} -u - \alpha'' z & \alpha_1 z & \dots & \alpha_s z \\ \alpha_1 z & \lambda_1 - z & & 0 \\ \vdots & & \ddots & \vdots \\ \alpha_s z & 0 & \dots & \lambda_s - z \end{vmatrix}.$$

It follows that $\rho^{n+1}\tilde{d}_{n+1} = \dots = \rho^{s+2}\tilde{d}_{s+2} = 0$. In particular, still $\rho^{s+2}\tilde{d}_{s+2}$ is independent of the choice of q. Furthermore,

$$\rho^{s+1}\tilde{d}_{s+1} = \begin{vmatrix} -u & 0 & \dots & 0 \\ 0 & \lambda_1 & & 0 \\ \vdots & & \ddots & \vdots \\ 0 & 0 & \dots & \lambda_s \end{vmatrix} = -\lambda_1 \dots \lambda_s u$$

and so, in this case, $\rho^{s+1}\tilde{d}_{s+1}$ is independent of the choice of q, $\rho^{s+1}\tilde{d}_{s+1} \neq 0$ if Q has a metric reduced equation of type II, and $\rho^{s+1}\tilde{d}_{s+1} = 0$ if Q has a metric reduced equation of type I. This completes the proof. □

Remark 10.6.5. As seen while proving Theorem 10.6.4, exactly s of the invariants $\rho\mu_1, \dots, \rho\mu_n$ are non-zero; up to renumbering them, assume they are $\mu_1, \dots, \mu_s$. Then $\rho\mu_1, \dots, \rho\mu_s$ may obviously replace $\rho d_1, \dots, \rho^s d_s$ in the sets of invariants of claims (b) and (c), giving rise to sets of metric invariants carrying the same information.

It is worth retaining part of the output of the above computations:

Corollary 10.6.6 (of the proof of 10.6.4). *Computing from the metric reduced equations of* 10.5.7 *it follows that*

(a) *if the reduced equation is of type* III, *then*

$$\rho^{s+2}\tilde{d}_{s+2} = -\lambda_1 \dots \lambda_s,$$

(b) *if the reduced equation is of type* II, *then*

$$\rho^{s+1}\tilde{d}_{s+1} = -\lambda_1 \dots \lambda_s, \textit{and}$$

(c) *in all three cases we have*

$$\rho\mu_1 = \lambda_1, \dots, \rho\mu_s = \lambda_s, \quad \rho\mu_{s+1} = \dots = \rho\mu_n = 0.$$

Proof. The first two equalities have been obtained while proving 10.6.4. The remaining ones are clear, because computing as above,

$$D(z) = \begin{vmatrix} \lambda_1 - z & \dots & 0 & 0 & \dots & 0 \\ \vdots & \ddots & & & & \vdots \\ 0 & & \lambda_s - z & & & 0 \\ 0 & & & -z & & 0 \\ \vdots & & & & \ddots & \vdots \\ 0 & \dots & 0 & 0 & \dots & -z \end{vmatrix}.$$ □

As allowed by 10.6.4, we will take as *metric invariants* of a quadric Q of an n-dimensional Euclidean space, up to the indeterminacy of $\rho \in -\{0\}$:

$\rho^n d_n, \dots, \rho d_1$ or, equivalently, $\rho\mu_1, \dots, \rho\mu_n$

and furthermore, for $s = \max\{i \mid d_i \neq 0\}$,

$\rho^{s+2}\tilde{d}_{s+2}$ in case $s < n$, and

$\rho^{s+1}\tilde{d}_{s+1}$ only in case $s = n$ or $\tilde{d}_{s+2}$=0.

These invariants are computed from the already introduced polynomials

$$\begin{aligned} \tilde{D}(z) &= (\det \tilde{N})^{-1} \det[\tilde{M} - z\tilde{N}] \\ &= (-1)^{n+1} z^{n+1} + (-1)^n \tilde{d}_1 z^n + \dots - \tilde{d}_n z + \tilde{d}_{n+1} \end{aligned}$$

and

$$\begin{aligned} D(z) &= (\det N)^{-1} \det[M - zN] \\ &= (-1)^n z^n + (-1)^{n-1} d_1 z^{n-1} + \dots - d_{n-1} z + d_n, \end{aligned}$$

where $\widetilde{M} = (a_{i,j})_{i,j=0,\dots,n}$ and $\widetilde{N} = (b_{i,j})_{i,j=0,\dots,n}$ are matrices of Q and an imaginary sphere S_q of radius $\boldsymbol{i}$, relative to any affine, non-necessarily orthonormal, reference, and M, N are the submatrices $M = (a_{i,j})_{i,j=1,n}$ and $N = (b_{i,j})_{i,j=1,n}$ corresponding to their improper sections. The choice of the centre q of S_q being irrelevant, it is usually taken to be the origin of coordinates, after which the matrix $\widetilde{N}$ may be taken to be

$$\widetilde{N} = \begin{pmatrix} 1 & 0 & \dots & 0 \\ 0 & g_{1,1} & \dots & g_{1,n} \\ \vdots & \vdots & & \vdots \\ 0 & g_{n,1} & \dots & g_{n,n} \end{pmatrix},$$

where $(g_{i,j}) = N$ is the matrix of the scalar product. If in particular the reference is orthonormal, then both $\widetilde{N}$ and N are unit matrices and

$$\widetilde{D}(z) = \det(\widetilde{M} - zI_{n+1}), \quad D(z) = \det(M - zI_n).$$

Invariants which are rational functions of the coefficients of the quadric are usually called *rational invariants*. As the reader may have noticed, the invariants $\rho^n d_n, \dots, \rho d_1, \rho^{s+2}\tilde{d}_{s+2}$ and, when suitable, $\rho^{s+1}\tilde{d}_{s+1}$, are rational, while, as far as $s > 1$, $\rho\mu_1, \dots, \rho\mu_n$ are not. For this reason, whenever possible, the use of $\rho^n d_n, \dots, \rho d_1$ instead of $\rho\mu_1, \dots, \rho\mu_n$ is preferred.

10.7 Metric classification of quadrics

Now we have all we need in order to describe the congruence classes of quadrics both in terms of invariants and of reduced equations. As in former cases, consider the diagram:

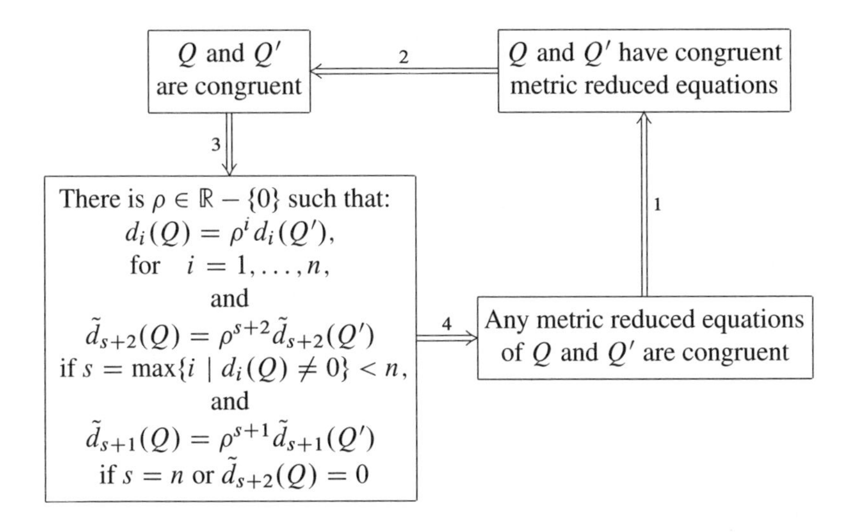

From it, assertion 1 is clear once the existence of metric reduced equations (10.5.7) has been proved. Assertions 2 and 3 have been proved in 10.5.8 and 10.6.4, respectively. It remains to prove assertion 4, that is, that the metric invariants determine the metric reduced equation up to congruence, which will be done next, mainly using 10.6.4 and 10.6.6.

First of all, $s = \max\{i \mid d_i \neq 0\}$ is obviously determined by the $\rho^i d_i$, $i = 1, \dots, n$; by 10.6.4 (a), it equals the number of monomials of the left-hand side of the metric reduced equation. The type of the metric reduced equation is determined through the next lemma:

Lemma 10.7.1. (a) *Q has a metric reduced equation of type III if and only if $s < n$ and $\rho^{s+2}\tilde{d}_{s+2}(Q) \neq 0$.*

(b) *Q has a metric reduced equation of type II if and only if either $s = n$ or $s < n$ and $\rho^{s+2}\tilde{d}_{s+2}(Q) = 0$, and, in addition, in either case, $\rho^{s+1}\tilde{d}_{s+1}(Q) \neq 0$.*

(c) *Q has a metric reduced equation of type I if and only if either $s = n$ or $s < n$ and $\rho^{s+2}\tilde{d}_{s+2}(Q) = 0$, and, in addition, in either case, $\rho^{s+1}\tilde{d}_{s+1}(Q) = 0$.*

Proof. The *only if* parts are part of 10.6.4. The converses follow from them and the fact that each of the assertions on the right is incompatible with any of the remaining two. □

Once the type of the metric reduced equation and the length s of its left-hand member have been determined, it remains to determine the coefficients $\lambda_1, \dots, \lambda_s$. From 10.6.6 (c) we have:

$$\rho\mu_1 = \lambda_1, \quad \dots, \quad \rho\mu_s = \lambda_s.$$

If the metric reduced equation is of type I, then its coefficients $\lambda_1, \dots, \lambda_s$ have to be determined up to proportionality and hence the above equalities suffice.

If the metric equation is of type II, from 10.6.6 (b) we have

$$\rho^{s+1}\tilde{d}_{s+1} = -\lambda_1 \dots \lambda_s,$$

while from the equalities (10.15) and 10.6.6 (c),

$$\rho^s d_s = \rho^s \mu_1 \dots \mu_s = \lambda_1 \dots \lambda_s.$$

It results that

$$\rho = -\frac{d_s}{\tilde{d}_{s+1}},$$

and so

$$\lambda_i = -\frac{\mu_i d_s}{\tilde{d}_{s+1}}, \quad i = 1, \dots, s,$$

which are degree-zero invariants.

Lastly, if the equation is of type III one has $\rho^{s+2}\tilde{d}_{s+2} = -\lambda_1 \dots \lambda_s$ from 10.6.6 (a), which, using again that $\rho^s d_s = \lambda_1 \dots \lambda_s$, gives

$$\rho^2 = -\frac{d_s}{\tilde{d}_{s+2}},$$

and so

$$\lambda_i = \mu_i \sqrt{-\frac{d_s}{\tilde{d}_{s+2}}}, \quad i = 1, \dots, s,$$

where the sign of the square root is irrelevant because the metric reduced equation is taken up to congruence.

This completes the proof of the metric classification theorem:

Theorem 10.7.2 (Metric classification of quadrics). *Any quadric Q of a Euclidean space $\mathbb{A}_n$ determines its metric reduced equation up to congruence. For any two quadrics Q and Q', of Euclidean n-dimensional spaces, the following three conditions are equivalent:*

(i) *Q and Q' are congruent.*

(ii) *Q and Q' have congruent metric reduced equations.*

(iii) *For a $\rho \in \mathbb{R} - \{0\}$, it holds that*

(a) *$d_i(Q) = \rho^i d_i(Q')$, $\quad i = 1, \dots, n$, and*

(b) *$\tilde{d}_{s+2}(Q) = \rho^{s+2}\tilde{d}_{s+2}(Q')$ if $s = \max\{i \mid d_i(Q) \neq 0\} < n$, and*

(c) *$\tilde{d}_{s+1}(Q) = \rho^{s+1}\tilde{d}_{s+1}(Q')$ if either $s = n$ or $s < n$ and $\tilde{d}_{s+2}(Q) = \tilde{d}_{s+2}(Q') = 0$.*

The way the metric reduced equation may be computed from an arbitrary equation is summarized next.

Corollary 10.7.3 (of the proof of 10.7.2). *Let a quadric Q of a Euclidean space $\mathbb{A}_n$ have matrix $\tilde{M} = (a_{i,j})_{i,j=0,\dots,n}$ relative to an affine reference. Write $M = (a_{i,j})_{i,j=1,\dots,n}$ for the corresponding matrix of Q^∞, let N be the matrix of the scalar product and, in blocks,*

$$\tilde{N} = \begin{pmatrix} 1 & 0 \\ 0 & N \end{pmatrix}.$$

Take

$$\begin{aligned} \tilde{D}(z) &= (\det \tilde{N})^{-1} \det(\tilde{M} - z\tilde{N}) \\ &= (-1)^{n+1} z^{n+1} + (-1)^n \tilde{d}_1 z^n + \dots - \tilde{d}_n z + \tilde{d}_{n+1} \end{aligned}$$

and

$$\begin{aligned} D(z) &= (\det N^{-1}) \det(M - zN) \\ &= (-1)^n z^n + (-1)^{n-1} d_1 z^{n-1} + \cdots - d_{n-1} z + d_n. \end{aligned}$$

If the non-zero roots of $D(z)$, repeated according to multiplicities, are $\mu_1, \dots, \mu_s$, then $s = \min\{i \geq 0 \mid d_j = 0 \text{ for } j > i\}$ and the metric reduced equation of Q may be taken to be either

I

$$\mu_1 X_1^2 + \cdots + \mu_s X_s^2 = 0$$

if either $s = n$, $\tilde{d}_{s+1} = 0$ or $s < n$, $\tilde{d}_{s+2} = \tilde{d}_{s+1} = 0$; or

II

$$-\frac{\mu_1 d_s}{\tilde{d}_{s+1}} X_1^2 - \cdots - \frac{\mu_s d_s}{\tilde{d}_{s+1}} X_s^2 = 1$$

if either $s = n$, $\tilde{d}_{s+1} \neq 0$ or $s < n$, $\tilde{d}_{s+2} = 0$, $\tilde{d}_{s+1} \neq 0$; or

III

$$\mu_1 \sqrt{-\frac{d_s}{\tilde{d}_{s+2}}} X_1^2 + \cdots + \mu_s \sqrt{-\frac{d_s}{\tilde{d}_{s+2}}} X_s^2 = 2X_n$$

if $s < n$, $\tilde{d}_{s+2} \neq 0$.

10.8 Exercises

Exercise 10.1. Prove that if a triangle of a Euclidean plane is self-polar with respect to a circle, then its orthocentre is the centre of the circle. Show by an example that the converse is false.

Exercise 10.2. Prove that the non-degenerate sections of a sphere by linear varieties of dimension at least two are spheres or imaginary spheres, while the degenerate sections are isotropic cones.

Exercise 10.3. Let A, A' be the ends of a diameter of a circle C. Use Exercise 8.26 to prove that for any $p \in C$, $p \neq A, A'$, the lines pA and pA' are orthogonal.

Exercise 10.4. Let $\mathcal{P}$ be a pencil of conics spanned by two circles C, C' of a Euclidean plane. Prove that all the non-degenerate quadrics with points of $\mathcal{P}$ are circles. Prove also that $\mathcal{P}$ contains a single pair of lines $\ell + \ell_\infty$, containing the improper line ℓ_∞. The line ℓ is called the *radical axis* of C, C'. Conversely, prove that any pencil spanned by a circle and a line-pair $\ell + \ell_\infty$, as above, may be spanned by two circles. Consider all possible positions of the radical axis and prove that the pencil spanned by two circles either:

(1) is general, in which case it has exactly two proper base points, real or imaginary and complex-conjugate, and these base points span the radical axis; or

(2) is a simple contact pencil, in which case the contact point is its only proper base point and the common tangent is the radical axis; or

(3) is a bitangent pencil, in which case its base points are the cyclic points, the radical axis is the improper line and all the non-degenerate conics of the pencil have the same centre.

Prove also that all these cases occur.

Exercise 10.5. Assume that C and C' are two different circles of a Euclidean plane and prove that the following conditions are equivalent:

(i) C and C' have orthogonal tangents at a common (proper) point.

(ii) C and C' span a general pencil $\mathcal{P}$ with two real base points, and, in the complex extension $\mathcal{P}_{\mathbb{C}}$ of $\mathcal{P}$, C and C' harmonically divide the two degenerate conics other than the line-pair containing the improper line (see Exercise 10.4).

When these conditions are satisfied, the circles C and C' are called *orthogonal*. The condition (ii) being independent of the choice of the common point, two orthogonal circles share two proper points and have orthogonal tangents at both of them. *Hint*: Use the projectivity of 9.10.1 (a), p being the common point.

Exercise 10.6. Prove that two orthogonal circles intersect any diameter of either of them in two pairs of points that, if disjoints, harmonically divide each other. *Hint*: If the circles are C and C' and d is a diameter of C, consider the circles through one of the proper points shared by C and C' and orthogonal to C, and apply 9.7.4.

Exercise 10.7. Use 9.4.4 to prove that given an n-dimensional simplex T, with proper vertices, of a Euclidean space $\mathbb{A}_n$, $n \geq 2$, there is one and only one sphere containing all vertices of T (the sphere *circumscribed* to T). *Hint*: Prove first that a quadric of $\mathbb{P}_n$ with index $n-1$ and at least two real points is non-degenerate.

Exercise 10.8. Let T be an n-dimensional simplex of a Euclidean space $\mathbb{A}_n$, with (proper) vertices $p_0, \dots, p_n$. If S is the sphere circumscribed to T (see Exercise 10.7), for each edge $p_i p_j$, $i \neq j$, of T, prove that the hyperplane orthogonal to $p_i p_j$ through the midpoint of p_i, p_j is the polar hyperplane relative to S of the improper point of $p_i p_j$. Deduce that all these hyperplanes are concurrent at the centre of S.

Exercise 10.9. Assume that C is a non-degenerate conic of a Euclidean plane and e is either the axis of C, if C is a parabola, or a principal axis of C, if C has proper centre. Use 7.8.11 to prove that e contains the midpoints of the ends of the chords of C orthogonal to e and the contact points of the tangents to C orthogonal to e. Re-prove the same fact using a metric reduced equation of C.

Exercise 10.10. Let C be a non-degenerate conic with points of a Euclidean plane, and p a proper point of C. Consider all the right triangles that are inscribed in C and have right angle at p. Prove that their hypotenuses are all going through a point O that does not belong to C and lies on the normal to C at p. The point O is called the *Frégier point* of p (relative to C). Prove that the Frégier point is improper if and only if C is an equilateral hyperbola, and that it is the centre of C if and only if C is a circle. *Hint*: Prove that the other two vertices of the triangles correspond by an involution of C and use 8.2.8. For the last claim, locate the fixed points of the involution.

Exercise 10.11 (*Locus of the Frégier points* I). Assume that C is an equilateral hyperbola of a Euclidean plane $\mathbb{A}_2$. Prove that while a point p describes the set of proper points of C, its Frégier point relative to C describes the set of all the improper points of $\mathbb{A}_2$ other than the improper points of C.

Exercise 10.12 (*Locus of the Frégier points* II). Let C be a non-degenerate conic with points of a Euclidean plane, not a circle or an equilateral hyperbola.

(1) Let τ_I and τ_J be the involutions of $C_{\mathbb{C}}$ with centres the cyclic points I, J. Prove that for any proper point $p \in C$ the line $\tau_I(p)\tau_J(p)$ is the polar of the Frégier point of p relative to C.

(2) Use Exercises 9.20 and 9.5 to prove that while p describes the set of proper points of C, the Frégier point of p relative to C describes the set of proper points of a non-degenerate conic C' which is tangent to C at its improper points.

Exercise 10.13. Let C be a conic in a Euclidean plane that does not contain the improper line. Prove that the following conditions are equivalent:

(i) The invariant d_1 of C is zero.

(ii) The cyclic points are conjugate with respect to C.

(iii) C is either a pair of orthogonal (real) lines or an equilateral hyperbola.

Hint: Use Exercise 5.7.

Exercise 10.14. Let T be a non-right triangle of a Euclidean plane.

(1) Prove that the conics that are circumscribed to T and have as conjugate the cyclic points compose a pencil $\mathcal{P}$ whose members are either equilateral hyperbolas or pairs of real lines (see Exercise 10.13; $\mathcal{P}$ is usually referred to as the pencil of equilateral hyperbolas circumscribed to T).

(2) Determine the degenerate conics of $\mathcal{P}$, prove that it is a general pencil and determine its fourth base point, re-proving on the way that the altitudes of T are concurrent.

(3) Prove that the centres of the non-degenerate conics of $\mathcal{P}$ lie on a circle that contains the feet of the altitudes of T, the midpoints of its sides and the midpoints of the orthocentre of T and each of its vertices (the *nine-points circle* or *Feuerbach's circle* of T). *Hint*: Use Exercise 9.28, determine the centres of the parabolas that belong to the complex extension $\mathcal{P}_{\mathbb{C}}$ of $\mathcal{P}$, and use 7.3.9 to relate the loci of centres of $\mathcal{P}$ and $\mathcal{P}_{\mathbb{C}}$.

Exercise 10.15. Reexamine the pencil of conics of Exercise 10.14, as well as its degenerate members and its locus of centres, in case of T being a right triangle.

Exercise 10.16. Let T be a triangle in a Euclidean plane.

(1) Prove that the conics relative to which T is self-polar and the cyclic points conjugate, compose a pencil $\mathcal{P}$ whose non-degenerate members are equilateral hyperbolas (see Exercise 10.13).

(2) Determine the degenerate conics of $\mathcal{P}$, prove that it is a general pencil and determine its base points. Prove on the way that any two bisectors of two different pairs of sides of T are concurrent with a bisector of the third pair of sides.

(3) Prove that the centres of the non-degenerate conics of $\mathcal{P}$ lie on the circle circumscribed to T (see Exercise 10.7). *Hint*: As for Exercise 10.14.

Exercise 10.17. Assume given two different points F, F' of a Euclidean plane $\mathbb{A}_2$, and $\delta \in \mathbb{R}$, $\delta > 0$.

(1) Assume $\delta > d(F, F')$ and prove that the points of $\mathbb{A}_2$ whose sum of distances from F and F' is equal to δ, are the points of the only ellipse with foci F, F' and major axis of length δ.

(2) Assume $\delta < d(F, F')$ and prove that the points of $\mathbb{A}_2$ whose difference of distances from F and F' is equal to $\pm\delta$, are the points of the only hyperbola with foci F, F' and major axis of length δ.

Exercise 10.18. Let F be a proper point and d a proper line of a Euclidean plane $\mathbb{A}_2$, $F \notin d$. Prove that the points of $\mathbb{A}_2$ whose distances from F and d are in constant ratio $\varepsilon > 0$, are the points of a non-degenerate conic C with focus F, directrix d and eccentricity ε.

Exercise 10.19. Prove that two tangents to a parabola C are orthogonal if and only if they meet on the directrix of C.

Exercise 10.20. Let C be a non-degenerate conic with foci F, F', $P \neq F, F'$ a point, and t, t' the tangents to C through P, assumed to be real. Prove that $\widehat{PFt} = \widehat{t'PF'}$. Compare with 10.3.11. *Hint*: Use Desargues' theorem 9.7.4 and Exercise 5.8.

Exercise 10.21. Prove that if F is a focus of a conic C, then it is also a focus of the image of C by any homology with centre F. *Hint*: Use 10.3.2.

Exercise 10.22. Prove that the polar of a circle relative to a second circle with centre O (see Exercise 6.30) is a non-degenerate conic with one focus at O. Prove also that, conversely, any non-degenerate conic with points, with a focus at O, is the polar of a circle with directrix a circle with centre O. *Hint*: Use 10.3.2.

Exercise 10.23. Let Q be a quadric of a three-dimensional Euclidean space other than a sphere, an imaginary sphere or an isotropic cone. Assume that Q has rank at least two and is not a parabolic cylinder.

(1) Prove that the intersection of Q and a proper plane π is a circle if and only if π contains real points of Q, is not tangent to Q, and the improper line of π is part of one of the degenerate conics of the pencil spanned by the absolute $\mathbf{K}$ and the improper section of Q.

(2) For each possible affine type of Q and starting from its metric reduced equation, compute the equations of all planes whose intersection with Q is a circle.

Hint: Use Exercise 7.13 to discard imaginary and degenerate sections.

Exercise 10.24. Extend the validity of 10.4.4 to pairs of distinct or coincident planes.

Exercise 10.25. Prove that the altitudes of a tetrahedron of $\mathbb{A}_3$, if pairwise disjoint, are generators of a ruled hyperboloid. *Hint*: Consider a sphere with centre one of the vertices of the tetrahedron and use 10.1.5 and Exercise 8.39.

Exercise 10.26. Prove that all quadrics tangent to a sphere along one of its non-degenerate plane sections are quadrics of revolution.

Exercise 10.27. Assume that a pencil $\mathcal{P}$, of quadrics of a Euclidean space $\mathbb{A}_3$, contains a sphere S. Prove that the non-degenerate quadrics of $\mathcal{P}$ other than S either:

(1) are all spheres or imaginary spheres, or

(2) are all quadrics of revolution and have their axes of revolution parallel, or

(3) are not quadrics of revolution and those which have proper centre have their principal axes parallel.

Give an example of a pencil in the situation of (2) above and containing a parabolic cylinder.

Exercise 10.28. Let Q be a quadric of revolution of a Euclidean space $\mathbb{A}_3$, other than a sphere, an imaginary sphere or an isotropic cone and with $\mathbf{r}(Q) \geq 3$. Let $C = Q \cap \pi$ be a non-degenerate section with points of Q by a plane π orthogonal to the axis of revolution (such a C is called a *parallel* of Q). Prove that C is a circle and that there is a sphere tangent to Q along C.

Exercise 10.29 (*Spectral theorem for symmetric morphisms*). Let F be a finite-dimensional real vector space equipped with a positive definite symmetric bilinear form K (which is usually called an *Euclidean vector space*). An endomorphism φ of F is called *symmetric* if and only if $K(\varphi(v), w) = K(v, \varphi(w))$ for any $v, w \in F$. Prove that if φ is symmetric, then there is a K-orthonormal basis of F composed of eigenvectors of φ. *Hint*: Let aside the obvious case $\varphi = \lambda \mathrm{Id}_F$, $\lambda \in \mathbb{R}$; prove that the rule $\eta(v, w) = K(\varphi(v), w)$ defines a non-zero symmetric bilinear form and apply 10.5.3 to the pencil of quadrics of $\mathbb{P}(F)$ spanned by $[K]$ and $[\eta]$.

Exercise 10.30. A calculus computation shows the area of the plane region limited by the ellipse with reduced equation $X^2/\mathbf{a}^2 + Y^2/\mathbf{b}^2 = 1$ to be $\pi\mathbf{ab}$. Similarly, the volume of the region limited by the ellipsoid with reduced equation $X^2/\mathbf{a}^2 + Y^2/\mathbf{b}^2 + Z^2/\mathbf{c}^2 = 1$ is $4\pi\mathbf{abc}/3$. Write their squares as degree-zero rational metric invariants, in terms of $d_i, \tilde{d}_j$.

Exercise 10.31. Prove that the points of a Euclidean plane from which the tangents to an ellipse or hyperbola C are orthogonal describe a circle with centre the centre of C and radius $\sqrt{\mathbf{a}^2 \pm \mathbf{b}^2}$, $\mathbf{a}$ and $\mathbf{b}$ being the lengths of the semi-axes of C. *Hint*: Take C given by its metric reduced equation, compute the equation of the tangent cone from an arbitrary point and use Exercise 10.13.

Exercise 10.32. Prove that a quadric Q of a Euclidean space $\mathbb{A}_n$, that does not contain the improper hyperplane, has $d_1(Q)$ =0 if and only if its improper section is outpolar to the absolute quadric. *Hint*: Either compute from the reduced equation of Q, or use Exercise 9.43.

Exercise 10.33. Let Q be either a ruled hyperboloid or an ordinary cone of a Euclidean three-dimensional space. Prove that Q contains three pairwise perpendicular generators if and only if $d_1(Q) = 0$. Prove also that if this is the case, then any generator of Q is part of a triple of pairwise perpendicular generators. *Hint*: Compute, using for the *only if* part, an orthonormal reference whose axis are generators, and for the *if* part, an orthonormal reference with one generator as axis; as an alternative, Exercises 9.44 and 10.32 may be used.

Exercise 10.34. Let C be an ellipse of a Euclidean plane, O the centre of C and ℓ, ℓ' a pair of conjugate diameters of C. Choose ends p, p' of ℓ and ℓ' and name δ, δ' the distances $\delta = d(O, p)$, $\delta' = d(O, p')$ (the *lengths of the semi-diameters*). Prove that

$$\delta^2 + \delta'^2 = -\frac{d_1 \tilde{d}_3}{d_2^2}$$

and

$$\delta^2 \delta'^2 \sin^2(\widehat{pOp'}) = \frac{\tilde{d}_3^2}{d_2^3}.$$

Note that this proves in particular that both the sum of the squares of the lengths of two conjugate semi-diameters and the area of the triangle they determine are independent of the choice of the conjugate semi-diameters (*Apollonius*).*Hint*: Use $(O, p - O, p' - O)$ as affine reference.

Exercise 10.35. Let Q be an ordinary cone of a Euclidean three-dimensional space. Prove that there are three pairwise perpendicular planes tangent to Q if and only if $d_2(Q) = 0$. Prove also that if this is the case, then any tangent plane to Q is part of a triple of pairwise perpendicular tangent planes. *Hint*: Relate the characteristic polynomials of the pencils spanned by Q_∞ and $\mathbf{K}$, and by Q_∞^* and $\mathbf{K}^*$, and then argue in $\pi_\infty^\vee$ using Exercises 9.43 and 9.44.

Exercise 10.36. Let Q be an ellipsoid of a Euclidean space $\mathbb{A}_3$. Prove that the points from which there are three pairwise orthogonal planes tangent to Q are the points of a sphere with the same centre as Q (the *Monge sphere* of Q). Compute its radius from the lengths of the semi-axes of Q. *Hint*: Use the reduced equation of Q, compute the equation of the tangent cone to Q from an arbitrary point and use Exercises 6.26 and 10.35.

Exercise 10.37. Compute the metric reduced equations of the quadrics of a Euclidean three-dimensional space which relative to an orthonormal reference have equations:

$$2x^2 + 2xy + \sqrt{2}xz + 2y^2 + \sqrt{2}yz + 3z^2 - 2x - 4y - \sqrt{2}z + 1 = 0,$$
$$2x^2 - 2\sqrt{2}xy + 2\sqrt{2}xz + y^2 - 2yz + z^2 + 2\sqrt{2}x + 6y - 6z + 5 = 0,$$
$$2x^2 - y^2 + 2\sqrt{3}yz - 3z^2 + 2y - 2\sqrt{3}z - 3 = 0,$$
$$6x^2 + 2\sqrt{2}xy - 2\sqrt{2}xz + y^2 - 10yz + z^2 + 2\sqrt{2}x + 10y - 2z - 3 = 0,$$
$$2x^2 - 2\sqrt{2}xy + 2\sqrt{2}xz + 5y^2 + 6yz + 5z^2 + 2\sqrt{2}x - 2y - 14z + 9 = 0,$$
$$4x^2 - y^2 - 6\sqrt{3}yz + 5z^2 + 6\sqrt{3}y - 10z + 9 = 0,$$
$$-2x^2 + 3y^2 + 2\sqrt{3}yz + z^2 - 4y - 4\sqrt{3}z + 1 = 0.$$

Chapter 11
Three projective classifications

This chapter is devoted to the projective equivalence and projective classification of collineations, regular pencils of quadrics and correlations. All these classifications are strongly based on the algebraic theory of equivalence of polynomial matrices, whose more relevant parts are explained in Sections 11.1 and 11.2. For further information about polynomial matrices, the reader may see [11], [18] and [23], from which we have taken many of the arguments.

11.1 Polynomial matrices

As in preceding chapters, the base field k is $\mathbb{R}$ or $\mathbb{C}$. *Polynomial matrices* are matrices whose entries are polynomials over k in a single variable which, following usual conventions, will be denoted by λ. A polynomial matrix has thus the form

$$M(\lambda) = \big(a_i^j(\lambda)\big)_{\substack{i=1,\dots,n\\ j=1,\dots,m}}, \quad a_i^j(\lambda) \in k[\lambda].$$

Its *degree* is taken to be

$$\deg M(\lambda) = \max_{i,j}\{\deg a_i^j(\lambda)\}.$$

The usual matrices with entries in k are thus the polynomial matrices M with $\deg M \leq 0$: they will be called *constant matrices*, in order to distinguish them from the positive-degree polynomial matrices. In the sequel we will often omit mentioning the variable λ by writing $M(\lambda) = M$ and $a_i^j(\lambda) = a_i^j$.

Since polynomial matrices may be always considered to be matrices with entries in the field of rational fractions $k(\lambda)$, all usual definitions and properties of matrix calculus do apply to polynomial matrices, but those involving division by a non-zero polynomial, which of course is in general not possible in $k[\lambda]$. In particular, addition and multiplication of polynomial matrices, as well as the product of a polynomial matrix by a polynomial are defined by the same rules as for matrices with entries in a field, and so are the transpose of a polynomial matrix and the determinant of a square polynomial matrix, the latter being a polynomial. The *rank* of a polynomial matrix is defined as the maximum of the orders of its non-zero minors. In the sequel we will call *regular* the square matrices with non-zero determinant (or, equivalently, with maximal rank), while we call a square matrix M *invertible* if and only if it has an inverse, that is, a (necessarily unique) matrix M^{-1} for which $M^{-1}M = MM^{-1} = \mathbf{1}$. Both conditions are equivalent if the matrices have their entries in a field, but not for polynomial matrices. In fact we have:

Lemma 11.1.1. *A square polynomial matrix M is invertible if and only if its determinant is a non-zero element of k.*

Proof. If a square polynomial matrix M has $\det M \in k - \{0\}$, then division by $\det M$ is possible in $k[\lambda]$ and the usual rule $M^{-1} = (\det M)^{-1}\widetilde{M}$, $\widetilde{M}$ the matrix of cofactors of M, provides a polynomial matrix which is the inverse of M. Conversely, if M has an inverse polynomial matrix M^{-1}, then both $\det M$ and $\det(M^{-1})$ are polynomials and

$$\det M \det(M^{-1}) = \det(MM^{-1}) = 1,$$

from which $\deg(\det M) = 0$ and so $\det M \in k - \{0\}$. □

Remark 11.1.2. Any $m \times n$ polynomial matrix $M = (a_i^j)$, of degree d, may be written in the form

$$M = M_0\lambda^d + M_1\lambda^{d-1} + \cdots + M_d,$$

where M_r is the $m \times n$ constant matrix with entries the coefficients of degree $d - r$ of the a_i^j. In particular $M_0 \neq 0$.

Remark 11.1.3. If

$$\begin{aligned} M &= M_0\lambda^d + M_1\lambda^{d-1} + \cdots + M_d, \\ M' &= M_0'\lambda^{d'} + M_1'\lambda^{d'-1} + \cdots + M_{d'}' \end{aligned}$$

are non-zero n-dimensional square polynomial matrices of degrees d and d', written as in 11.1.2, and either M_0 or M_0' is invertible, then

$$MM' = M_0M_0'\lambda^{d+d'} + \text{ terms of lower degree in } \lambda$$

and $M_0M_0' \neq 0$ because one of the factors is invertible and the other is non-zero. It follows that

$$\deg MM' = d + d' = \deg M + \deg M',$$

which may fail to be true if both M_0 and M_0' are non-invertible.

We will make use of a sort of Euclidean division for polynomial matrices which is described next.

Lemma 11.1.4 (Division of polynomial matrices). *Assume that*

$$M = M_0\lambda^d + M_1\lambda^{d-1} + \cdots + M_d$$

and

$$N = N_0\lambda^s + N_1\lambda^{s-1} + \cdots + N_s$$

are n-dimensional square polynomial matrices of degrees d and s, written as in 11.1.2. Assume also that N_0 is invertible. Then:

1 ***Dividing M by N on the right.*** *There exist uniquely determined n-dimensional square polynomial matrices Q and R (**quotient** and **remainder**) such that*

$$M = QN + R \quad \textit{and} \quad \deg R < \deg N.$$

2 ***Dividing M by N on the left.*** *There exist uniquely determined n-dimensional square polynomial matrices $\bar{Q}$ and $\bar{R}$ (**quotient** and **remainder**) such that*

$$M = N\bar{Q} + \bar{R} \quad \textit{and} \quad \deg \bar{R} < \deg N.$$

Proof. We will prove the first claim; similar arguments prove the second one. For the existence, note first that in case $d < s$ it suffices to take $Q = 0$, $R = M$. Thus we assume $d \geq s$. Write

$$M = M_0 N_0^{-1} N \lambda^{d-s} + (M - M_0 N_0^{-1} N \lambda^{d-s}),$$

where, clearly, the term of degree d in $M' = M - M_0 N_0^{-1} N \lambda^{d-s}$ is canceled and so $\deg M' < d$. If $d = s$ we take $Q = M_0 N_0^{-1} \lambda^0$ and $R = M'$. Otherwise, using induction on $d - s$, it suffices to divide M' by N on the right and substitute the resulting expression for M' in the last displayed equality.

As regards the uniqueness, having a quotient Q' and a remainder R' gives $(Q - Q')N = R - R'$. If $Q - Q' \neq 0$, then, by 11.1.3, $\deg N \leq \deg(Q - Q')N = \deg(R - R') \leq \deg R$, against what has been proved for R. Thus, $Q = Q'$, and so $R = R'$, as claimed. □

Unless otherwise stated, all matrices appearing from now on in this section will be assumed to be square and of fixed dimension $n + 1$, $n \geq 1$; their rows and columns will be numbered from 0 to n. Performing an *elementary operation* on such a matrix M consists of either

– multiplying one row (resp. column) by a non-zero $\alpha \in k$ (*elementary operation of type* 1), or

– adding to a row (resp. column) the result of multiplying another row (resp. column) by an arbitrary polynomial $a(\lambda) \in k[\lambda]$ (*elementary operation of type* 2).

We say that two polynomial matrices are *equivalent* when one may be obtained from the other by a finite, maybe empty, sequence of elementary operations. The reader may easily check that the inverse of any elementary operation is itself an elementary operation, after which it is clear that the equivalence of polynomial matrices is, indeed, an equivalence relation.

Lemma 11.1.5. *The matrix obtained from a polynomial matrix M by interchanging two different rows (or two different columns) is equivalent to M.*

Proof. Assume the rows to be the i-th one, denoted by a, and the j-th one, denoted by b. It is enough to consider the sequence of elementary operations

$$\begin{pmatrix} \cdots \\ a \\ \cdots \\ b \\ \cdots \end{pmatrix} \mapsto \begin{pmatrix} \cdots \\ a \\ \cdots \\ b+a \\ \cdots \end{pmatrix} \mapsto \begin{pmatrix} \cdots \\ -b \\ \cdots \\ b+a \\ \cdots \end{pmatrix} \mapsto \begin{pmatrix} \cdots \\ -b \\ \cdots \\ a \\ \cdots \end{pmatrix} \mapsto \begin{pmatrix} \cdots \\ b \\ \cdots \\ a \\ \cdots \end{pmatrix}$$

which are, in the given order: adding the i-th row to the j-th one; adding the product of the j-th row by -1 to the i-th row; adding the i-th row to the j-th one; multiplying the i-th row by -1. A similar argument proves the claim for columns. □

Any polynomial matrix may be reduced to a very simple form by performing a sequence of elementary operations, namely:

Theorem 11.1.6. *Any square polynomial matrix M is equivalent to a diagonal matrix of the form*

$$\begin{pmatrix} I_0(\lambda) & & & & & & 0 \\ & I_1(\lambda) & & & & & \\ & & \ddots & & & & \\ & & & I_\ell(\lambda) & & & \\ & & & & 0 & & \\ & & & & & \ddots & \\ 0 & & & & & & 0 \end{pmatrix},$$

where $n \geq \ell \geq -1$, the $I_j(\lambda)$ are monic polynomials and each $I_{j-1}(\lambda)$ divides $I_j(\lambda)$ for $j = 1, \dots, \ell$.

Proof. For an arbitrary non-zero polynomial matrix N, denote by $\delta(N)$ the minimal degree of the non-zero entries of N; clearly $\delta(N) \geq 0$. Assume $M \neq 0$, as otherwise the claim is obvious. Choose a non-zero entry a of M with $\deg a = \delta(M)$. By performing an elementary operation of type 1, it is not restrictive to assume that a is monic. Then use 11.1.5 to turn M into an equivalent matrix $M_1 = (b_i^j)$ with $b_0^0 = a$ and still $\delta(M_1) = \deg a$. Perform the divisions of polynomials $b_0^i = aq_i + r_i, i = 1, \dots, n$, and turn M_1 into an equivalent matrix M_2 by adding to the i-th row the result of multiplying the first one by $-q_i$, for $i = 1, \dots, n$. The elements on the first column of M_2 are thus $a, r_1, \dots, r_n$. If $\delta(M_2) < \delta(M)$, then take M_2 in the place of M and repeat the procedure from scratch. Since the values of δ cannot decrease indefinitely, after finitely many steps we will be in the case $\delta(M_2) = \delta(M)$. Then since $\deg r_i < \deg a$, necessarily $r_i = 0$ for $i = 1, \dots, n$, after which the matrix M_2 has first column $(a, 0, \dots, 0)^t$ and $\deg a = \delta(M_2)$.

This achieved, as done above with the first column, divide by a the other entries on the first row and perform an elementary operation of type 2, this time using the first column, in order to get these entries replaced with their remainders. If the resulting matrix M_3 has $\delta(M_3) < \delta(M_2)$, then perform the whole procedure from the beginning. Again, δ cannot decrease indefinitely and so, eventually, the case $\delta(M_3) = \delta(M_2)$ will occur. Then, as for the first column, all remainders on the first row are zero, $\delta(M_3) = \deg a$ and M_3 has the form

$$M_3 = \begin{pmatrix} a & 0 & \cdots & 0 \\ 0 & c_1^1 & \cdots & c_n^1 \\ & & \cdots & \\ 0 & c_1^n & \cdots & c_n^n \end{pmatrix}.$$

Assume now that an entry b_j^i of M_3 is not divisible by a, say $b_j^i = qa + r$, $\deg r < \deg a$, $r \neq 0$. Then add the first row to the i-th one and subtract from the j-th column the resulting first column multiplied by q. This turns the matrix M_3 into an equivalent one with an entry equal to r and so a strictly lower δ. If this is the case, once again, one takes M_3 in the place of M and repeats the whole procedure from scratch. Since, as already, this cannot occur indefinitely, after finitely many steps all the b_j^i will be multiples of a. Then we take $I_0(\lambda) = a$ and, summarizing, we have proved the existence of a matrix $\bar{M}$, equivalent to M and of the form

$$\bar{M} = \begin{pmatrix} I_0(\lambda) & 0 & \cdots & 0 \\ 0 & c_1^1 & \cdots & c_n^1 \\ & & \cdots & \\ 0 & c_1^n & \cdots & c_n^n \end{pmatrix},$$

where $I_0(\lambda)$ is a monic polynomial that divides all the entries of the matrix.

Now the proof ends using induction on n. Assume $n = 1$: if $c_1^1 = 0$ the goal has been achieved; otherwise, just turn c_1^1 into a monic polynomial $I_2(\lambda)$ by an elementary operation. If $n > 1$, by induction, after suitable elementary operations the n-dimensional matrix $M' = (c_j^i)$ takes the form

$$\begin{pmatrix} I_1(\lambda) & & & & & 0 \\ & \ddots & & & & \\ & & I_\ell(\lambda) & & & \\ & & & 0 & & \\ & & & & \ddots & \\ 0 & & & & & 0 \end{pmatrix},$$

where $n \geq \ell \geq 0$, the $I_j(\lambda)$ are monic polynomials and each $I_{j-1}(\lambda)$ divides $I_j(\lambda)$ for $j = 2, \dots, \ell$. The reader may easily check that the greatest common divisor of the entries of a matrix remains unaltered after an elementary operation, which assures that $I_0(\lambda)|I_1(\lambda)$. Then the same elementary operations used to reduce M', performed on $\bar{M}$ using the rows and columns of same index, turn it into a matrix equivalent to M as claimed. □

Remark 11.1.7. The diagonal matrix of the claim of 11.1.6 is a distinguished representative of the equivalence class of M: it is called the *canonical matrix*, or the *canonical form* of M. Its uniqueness will be proved in the next section.

Remark 11.1.8. By substituting the induction steps by iterations of the procedure in lower dimension, the proof of 11.1.6 provides an effective procedure (quite tedious for human beings) for getting the canonical form and the elementary operations leading to it.

11.2 Classification of polynomial matrices

As above, let M be a square $(n+1)$-dimensional polynomial matrix and assume it has rank $\ell + 1$. We associate to it the polynomials $D_1, \dots, D_{\ell+1}$, D_s being the greatest common divisor of all s-dimensional minors of M. Note that, by the definition of the rank, all the D_s are non-zero polynomials, while all minors of dimension greater than $\ell + 1$ are identically zero. We also take $D_0 = 1$. We have:

Lemma 11.2.1. *The polynomials D_s, $s = 1, \dots, \ell + 1$, are not altered by elementary operations.*

Proof. The claim is obvious for elementary operations of type 1. Perform on M an elementary operation by adding to the i-th column a multiple of the j-th one, and fix our attention on an s-dimensional minor. If the minor does not involve the i-th column, it obviously remains unmodified. The same occurs if the minor involves both the i-th and j-th columns, by the properties of the determinants. Lastly, if the minor involves the i-th column but not the j-th one, after operation it appears as the sum of the original minor plus a multiple of another s-dimensional minor (the one involving the same entries as the original one, but for those on the i-th column, which are replaced with their corresponding entries on the j-th column). All together, by the operation, each s-dimensional minor m either remains unaltered or is turned into $m + hm'$, with m' another s-dimensional minor and $h \in k[\lambda]$. Obviously, this does not affect the greatest common divisor D_s of all s-dimensional minors. Since a similar argument applies to rows, the claim is proved. □

By developing it by a row, any $(s+1)$-dimensional minor appears as a linear combination of s-dimensional minors, with polynomials as coefficients: then it is clear from their definition that $D_s|D_{s+1}$, $s = 0, \dots, \ell$. The s-th *invariant factor* of M above is defined as being D_{s+1}/D_s, for $s = 0, \dots, \ell$. By definition, the invariant factors of a polynomial matrix M come indexed, their indices varying from zero to $\operatorname{rk} M - 1$. In the sequel, saying that two matrices have the *same invariant factors* will mean that they have the same rank $\ell + 1$ and equal invariant factors of index s for $s = 0, \dots, \ell$.

Remark 11.2.2. After the definitions of equivalence of matrices and invariant factors, Lemma 11.2.1 may be equivalently stated by saying that equivalent matrices have the same invariant factors, which justifies the use of the adjective invariant. Incidentally, 11.2.1 proves that equivalent matrices have the same rank, which may also be seen directly.

Proposition 11.2.3. *If a square polynomial matrix M has canonical form*

$$\begin{pmatrix} I_0 & & & & & & 0 \\ & I_1 & & & & & \\ & & \ddots & & & & \\ & & & I_\ell & & & \\ & & & & 0 & & \\ & & & & & \ddots & \\ 0 & & & & & & 0 \end{pmatrix},$$

then the invariant factors of M are

$$D_1/D_0 = I_0, \dots, \; D_{\ell+1}/D_\ell = I_\ell.$$

Proof. By 11.2.2, M and its canonical matrix have the same invariant factors which, once computed from the latter, easily turn out to be as claimed. □

Since we have seen in 11.1.6 that $I_{i-1} | I_i$, $i = 1, \dots, \ell$, it follows that:

Corollary 11.2.4. *For $i = 1, \dots, \ell = \operatorname{rk} M - 1$, the $(i-1)$-th invariant factor of M divides the i-th one.*

The information enclosed in the invariant factors is often presented in an equivalent way we will explain next. Assume as above that the matrix M has invariant factors $I_0, \dots, I_\ell$ and denote by $e_1, \dots, e_s$ the irreducible factors of I_ℓ. By including factors with exponent 0 if needed, by 11.2.4 we may factorize the invariant factors in the form

$$\begin{aligned} I_0 &= e_1^{\mu_{1,0}} \dots e_s^{\mu_{s,0}}, \\ I_1 &= e_1^{\mu_{1,1}} \dots e_s^{\mu_{s,1}}, \\ &\vdots \\ I_\ell &= e_1^{\mu_{1,\ell}} \dots e_s^{\mu_{s,\ell}} \end{aligned}$$

where, by 11.2.4, $0 \leq \mu_{i,0} \leq \cdots \leq \mu_{i,\ell}$, and $\mu_{i,\ell} > 0$ for $i = 1, \ldots, s$. Then list, repeated if this is the case, the factors appearing in the above decompositions,

$$e_1^{\mu_{1,0}}, e_1^{\mu_{1,1}}, \ldots, e_1^{\mu_{1,\ell}}, \ldots, e_s^{\mu_{s,0}}, e_s^{\mu_{s,1}}, \ldots, e_s^{\mu_{s,\ell}},$$

and drop those equal to one. The resulting unordered list is called the *list of elementary divisors* of M, each of its members being called an *elementary divisor* of M. Repetitions do matter, so, in the sequel, we will say *the list of elementary divisors* instead of just *the elementary divisors*, which may be confusing. In particular, saying that two matrices have the *same* (or *equal*) lists of elementary divisors will mean that the elementary divisors of one and another matrix are pairwise coincident, and that coincident elementary divisors appear repeated the same number of times in both lists.

It is clear that the invariant factors determine the list of elementary divisors and, of course, also the rank of the matrix. The converse is true: the invariant factors may be recovered from the elementary factors and the rank of the matrix. For, assume that $e_1, \ldots, e_s$ are the irreducible polynomials some power of which appears in the list of elementary divisors. First, take ℓ to be the rank of the matrix minus one. Then the last invariant factor is I_ℓ and it equals the product of the powers of $e_1, \ldots, e_s$, non-repeated, appearing with maximal exponents in the list of elementary factors. After dropping once each of these powers from the list, $I_{\ell-1}$ is the product of the powers of $e_1, \ldots, e_s$, non-repeated, appearing with maximal exponents in the remaining list, and so on, until getting an empty list, at which point all the invariant factors other than 1 have been obtained.

Remark 11.2.5. Assume that M is a real polynomial square matrix. It may also be seen as a complex polynomial matrix. The procedure described in the proof of 11.1.6, leading from M to its canonical form, uses rational operations only, and therefore it is equally performed and gives rise to the same canonical form, no matter if the matrix is seen as a real or a complex one. In particular the invariant factors of M, which may be read from the canonical form by 11.2.3, are independent of the choice of the base field. A similar claim about the elementary divisors would be false, as the decompositions of the invariant factors in irreducible factors in $\mathbb{R}[\lambda]$ and $\mathbb{C}[\lambda]$ may differ.

The next theorem summarizes what we have obtained regarding the classification of square polynomial matrices modulo equivalence:

Theorem 11.2.6. *If M, M' are square polynomial matrices of the same dimension, then the following conditions are equivalent:*

(i) *M and M' are equivalent.*

(ii) *M and M' have the same invariant factors.*

(iii) *M and M' have the same rank and the same list of elementary divisors.*

(iv) *M and M' have the same canonical form.*

Proof. (i)$\Rightarrow$(ii) has been seen in 11.2.2, (ii)$\Rightarrow$(iv) follows from 11.2.3 and (iv)$\Rightarrow$(i) holds because any matrix is equivalent to its canonical form (11.1.6). To close, the equivalence between (ii) and (iii) has been seen above. □

So far, we have dealt with the equivalence of polynomial matrices from its definition in Section 11.1. An equivalent formulation better suited to our purposes will be presented in the forthcoming Proposition 11.2.9. To obtain it, we need first to see a different way of performing elementary operations.

For any $i = 0, \dots, n$ and $\alpha \in k - \{0\}$, denote by $E_i(\alpha)$ the $(n+1)$-dimensional diagonal matrix all of whose entries on the diagonal are 1 but the i-th one, which is α.

Also, for any $i, j = 0, \dots, n$, $i \neq j$, and $a \in k[\lambda]$, denote by $F^i_j(a)$ the $(n+1)$-dimensional square matrix with all entries on the diagonal equal to 1, the entry on row i and column j equal to a and all the remaining ones equal to 0.

$$E_i(\alpha) = \begin{pmatrix} 1 & & & & \\ & \ddots & & & \\ & & \alpha & & \\ & & & \ddots & \\ & & & & 1 \end{pmatrix}, \quad F^i_j(a) = \begin{pmatrix} 1 & & a \\ & \ddots & \\ & & 1 \end{pmatrix}.$$

We will call the matrices $E_i(\alpha)$, $F^i_j(a)$ *elementary matrices*, note that they are invertible. Multiplying by one them causes an elementary operation; more precisely:

Lemma 11.2.7. *Multiplication of a square polynomial matrix M by $E_i(\alpha)$ on the right (resp. left) causes its i-th column (resp. row) to be multiplied by α, while the remaining ones are not modified.*

Multiplication of M by $F^i_j(a)$ on the right (resp. left) causes the result of multiplying its i-th column (resp. j-th row) by a to be added to its j-th column (resp. i-th row); the columns (resp. rows) other than the j-th (resp. i-th) one are not modified.

Proof. Left to the reader. □

Lemma 11.2.7 assures that any elementary operation on a polynomial matrix M may be produced by multiplying it by an elementary matrix on the suitable side. It follows thus:

Lemma 11.2.8. *Square polynomial matrices of the same dimension M, M' are equivalent if and only if there are elementary matrices $A_1, \dots, A_r, B_1, \dots, B_s$ such that*

$$M' = A_1 \dots A_r M B_1 \dots B_s.$$

An important characterization of the equivalence of polynomial matrices follows:

Proposition 11.2.9. *Square polynomial matrices M, N, of the same dimension, are equivalent if and only if there exist invertible polynomial matrices A, B such that*

$$M = ANB.$$

Proof. The *only if* part is direct from 11.2.8, as elementary matrices are invertible. The *if* part will follow after seeing that both A and B are products of elementary matrices. This results from the next lemma. □

Lemma 11.2.10. *Any invertible polynomial matrix C is a product of elementary matrices.*

Proof. We know from 11.1.1 that $\det C \in k - \{0\}$, hence the n-th invariant factor of C is 1, and therefore so are the remaining invariant factors, because they divide the n-th one (11.2.4). Then, by 11.2.3, the canonical matrix of C is the unit matrix $\mathbf{1}_{n+1}$. Since it is equivalent to C, by 11.2.8,

$$C = A_1 \dots A_r \mathbf{1}_{n+1} B_1 \dots B_s = A_1 \dots A_r B_1 \dots B_s,$$

where $A_1, \dots, A_r, B_1, \dots, B_s$ are elementary matrices, hence the claim. □

The forthcoming results strongly depend on the fact that for certain polynomial matrices of degree one, one may take constant matrices in the places of A and B in 11.2.9, namely:

Proposition 11.2.11. *Degree-one square polynomial matrices of the same dimension*

$$M_0\lambda + M_1, \quad N_0\lambda + N_1,$$

with $\det M_0 \neq 0$, $\det N_0 \neq 0$, *are equivalent if and only if there exist invertible constant matrices $\overline{A}$, $\overline{B}$ such that*

$$N_0\lambda + N_1 = \overline{A}(M_0\lambda + M_1)\overline{B}.$$

Proof. After 11.2.9, the *if* part is obvious and to prove the converse we may assume that there are polynomial invertible matrices A, B for which

$$N_0\lambda + N_1 = A(M_0\lambda + M_1)B, \tag{11.1}$$

or, equivalently,

$$A^{-1}(N_0\lambda + N_1) = (M_0\lambda + M_1)B. \tag{11.2}$$

Divide (11.1.4) A^{-1} on the left by $M_0\lambda + M_1$ and B on the right by $N_0\lambda + N_1$ to get

$$\begin{aligned} A^{-1} &= (M_0\lambda + M_1)Q + R, \\ B &= Q'(N_0\lambda + N_1) + R', \end{aligned} \tag{11.3}$$

where $\deg R < 1$ and $\deg R' < 1$, and therefore R and R' are constant matrices. In (11.2), replace A^{-1} and B using (11.3) to get

$$(M_0\lambda + M_1)(Q - Q')(N_0\lambda + N_1) = (M_0\lambda + M_1)R' - R(N_0\lambda + N_1).$$

If $Q - Q' \neq 0$, then, using that M_0 and N_0 are invertible, the left-hand side has degree at least two, as the reader may easily check. Since the right-hand side has degree at most one, it follows that $Q - Q'$=0 and so

$$R(N_0\lambda + N_1) = (M_0\lambda + M_1)R'. \tag{11.4}$$

Next we will show that R is invertible. To this end, divide A on the left by $N_0\lambda + N_1$:

$$A = (N_0\lambda + N_1)F + G, \quad \deg G < 1. \tag{11.5}$$

Then, using (11.5), (11.2) and the first of the equalities (11.3), it is

$$\begin{aligned} \mathbf{1}_{n+1} &= A^{-1}A = A^{-1}(N_0\lambda + N_1)F + A^{-1}G \\ &= (M_0\lambda + M_1)BF + (M_0\lambda + M_1)QG + RG \\ &= (M_0\lambda + M_1)(BF + QG) + RG. \end{aligned}$$

In the last expression $(BF + QG) = 0$, as otherwise, by 11.1.3, $\mathbf{1}_{n+1}$ would have positive degree. The above equalities give thus $\mathbf{1}_{n+1} = RG$, proving that R is invertible.

Now, the equality (11.4) may be written

$$(N_0\lambda + N_1) = R^{-1}(M_0\lambda + M_1)R'. \tag{11.6}$$

By equating the terms of degree one,

$$N_0 = R^{-1}M_0R',$$

which proves that also R' is invertible, because N_0, M_0 and R are. After this the claim follows from the equality (11.6) by taking $\bar{A} = R^{-1}$ and $\bar{B} = R'$. □

Assume now that M is a square $(n+1)$-dimensional constant matrix; the polynomial matrix

$$C_M(\lambda) = M - \lambda\mathbf{1}_{n+1}$$

is called the *characteristic matrix* of M. Its determinant being the characteristic polynomial of M, it is in particular a non-zero polynomial and therefore the characteristic matrix has rank $n+1$. The invariant factors and the list of elementary divisors of the characteristic matrix of M are called the *invariant factors* and the *list of elementary divisors* of M. The reader may note that these are not the invariant factors and elementary divisors of M if M is taken as a polynomial matrix; the latter being trivial (see Exercise 11.1), this causes no confusion.

Lemma 11.2.12. *The product of the characteristic polynomial of a square* $(n+1)$*-dimensional constant matrix by* $(-1)^{n+1}$ *equals the product of its invariant factors, and also the product of its elementary divisors, repeated as listed.*

Proof. If we take $D_0 = 1$ and D_s to be the greatest common divisor of all s-dimensional minors of the characteristic matrix $M - \lambda\mathbf{1}_{n+1}$, $s = 1, \dots, n+1$, then the characteristic polynomial of M is $(-1)^{n+1}D_{n+1}$. On the other hand, D_{n+1} equals the product of the invariant factors $I_s = D_{s+1}/D_s$, $s = 0, \dots, n$, of M, and also the product of the elementary divisors of M, repeated as listed, by the definition of the list of elementary divisors. □

Remark 11.2.13. By comparing degrees, it follows from 11.2.12 that both the sum of degrees of the invariant factors and the sum of degrees of the elementary divisors, repeated as listed, equal $n+1$.

Square constant matrices of the same dimension M, N are called *similar* if and only if there exists an invertible constant matrix P, of the same dimension, for which

$$N = PMP^{-1}.$$

The similarity of square matrices is, clearly, an equivalence relation. It is easily related to the equivalence of polynomial matrices using 11.2.11:

Lemma 11.2.14. *Square* $(n+1)$*-dimensional constant matrices* M *and* N *are similar if and only if their characteristic matrices are equivalent.*

Proof. If M and N are similar, then from an equality $N = PMP^{-1}$ it follows that

$$N - \lambda\mathbf{1}_{n+1} = PMP^{-1} - \lambda\mathbf{1}_{n+1} = P(M - \lambda\mathbf{1}_{n+1})P^{-1}$$

and the characteristic matrices are equivalent.

Conversely, if the characteristic matrices are equivalent, then by 11.2.11, there are invertible constant matrices A, B such that

$$N - \lambda\mathbf{1}_{n+1} = A(M - \lambda\mathbf{1}_{n+1})B.$$

The equality of the terms of degree one in the above equality gives $AB = \mathbf{1}_{n+1}$ and thus $B = A^{-1}$. The equality of the terms of degree zero gives then $N = AMA^{-1}$ and proves the similarity of M and N. □

We obtain thus the following characterization of the similarity of square constant matrices. It completes the algebraic background we need and its proof is direct from 11.2.14 and 11.2.6, taking into account that all characteristic matrices have rank $n + 1$.

Theorem 11.2.15. *Assume that M, N are square constant matrices of the same dimension. Then the following conditions are equivalent:*

(i) *M and N are similar.*

(ii) *M and N have the same invariant factors.*

(iii) *M and N have the same list of elementary divisors.*

11.3 Projective equivalence of collineations

The action of a projectivity $g\colon \mathbb{P}_n \to \mathbb{P}'_n$ on the collineations $f\colon \mathbb{P}_n \to \mathbb{P}_n$ was defined at the end of Section 1.9 by the rule

$$g(f) = g \circ f \circ g^{-1}$$

and therefore collineations of projective spaces $\mathbb{P}_n$, $\mathbb{P}'_n$,

$$f\colon \mathbb{P}_n \longrightarrow \mathbb{P}_n, \quad f'\colon \mathbb{P}'_n \longrightarrow \mathbb{P}'_n$$

are projectively equivalent if and only if there is a projectivity $g\colon \mathbb{P}_n \to \mathbb{P}'_n$ such that

$$f' = g(f) = g \circ f \circ g^{-1}.$$

In the present section we characterize the projective equivalence of collineations in terms of the elementary divisors and the lists of invariant factors of their matrices.

The next lemma shows that the projective equivalence of collineations is a condition close, but not equivalent, to the similarity of their matrices:

Lemma 11.3.1. *Let f, f' be, respectively, collineations of projective spaces $\mathbb{P}_n$, $\mathbb{P}'_n$ which have matrices M, M'. Then f and f' are projectively equivalent if and only if there are an $(n+1)$-dimensional invertible matrix P and a non-zero $\rho \in k$ for which*

$$M' = \rho P M P^{-1}.$$

Proof. Assume that Δ is a reference of $\mathbb{P}_n$ relative to which f has matrix M and Δ' is a reference of $\mathbb{P}'_n$ relative to which f' has matrix M'. If there is $g : \mathbb{P}_n \to \mathbb{P}'_n$ such that $f' = g \circ f \circ g^{-1}$, let P be a matrix of g relative to Δ, Δ'. Then, by 2.8.9, PMP^{-1} is a matrix of f' relative to Δ' and hence is proportional to M' (2.8.6). Conversely, take $g : \mathbb{P}_n \to \mathbb{P}'_n$ to be the projectivity whose matrix, relative to Δ, Δ', is P (2.8.5), to have $f' = g \circ f \circ g^{-1}$. □

After 11.3.1 it is clear that we have to deal with pairs of matrices, one of which is proportional to a matrix similar to the other, rather than with just pairs of similar matrices. To this end, we introduce a relation between polynomials we will call *homothety:* assume that we have two polynomials $p, p' \in k[\lambda]$; we will say that p' is *homothetic* to p with *dilation ratio* $\rho \in k - \{0\}$, or just ρ-homothetic for short, if and only if

$$p'(\lambda) = \rho^{\deg p} p(\lambda/\rho).$$

Homothetic will mean ρ-homothetic for some $\rho \in k - \{0\}$.

Note that homothetic polynomials have the same leading monomial. The next two lemmas present some elementary properties of homothety. Their proof is left to the reader.

Lemma 11.3.2. *Assume that p' is homothetic to p with dilation ratio ρ. Then,*

(1) *p and ρ determine p'.*

(2) *$\rho = 1$ if and only if $p' = p$.*

(3) *p is homothetic to p' with dilation ratio $1/\rho$.*

(4) *If p'' is homothetic to p' with dilation ratio ρ', then p'' is homothetic to p with dilation ratio $\rho\rho'$.*

In particular the relation *to be homothetic* (with unspecified dilation ratio) is an equivalence relation.

Lemma 11.3.3. (1) *If $p, p', q, q' \in k[\lambda]$ and p' is homothetic to p with dilation ratio ρ, then q' is homothetic to q with dilation ratio ρ if and only if so is $p'q'$ to pq.*

(2) *Polynomials homothetic to an irreducible polynomial are irreducible too.*

(3) *If $p = ap_1^{\alpha_1} \dots p_h^{\alpha_h}$, $a \in k - \{0\}$, is the decomposition of a polynomial in irreducible factors and $q, q_1, \dots, q_h$ are, in this order, homothetic to $p, p_1, \dots, p_h$, all with dilation ratio ρ, then*

$$q = aq_1^{\alpha_1} \dots q_h^{\alpha_h}$$

is the decomposition of q in irreducible factors.

Homothety will be used on invariant factors and elementary divisors. Assume that M_1 and M_2 are square matrices of the same dimension. We will say that the invariant factors of M_2 are *ρ-homothetic* to those of M_2 if and only if M_1 and M_2 have the same number of invariant factors and each invariant factor of M_2 is ρ-homothetic to the invariant factor of M_1 of same index. We will also say that the list of elementary divisors of M_2 is *ρ-homothetic* to that of M_1 if and only if the polynomials ρ-homothetic to the elementary divisors of M_1, repeated as listed, form the list of elementary divisors of M_2. Still, in both cases, *homothetic* will mean ρ-homothetic for some ρ. The next lemma relates the invariant factors and elementary divisors of a constant matrix and those of a proportional one.

Lemma 11.3.4. *Let M be a square constant matrix and $\rho \in k - \{0\}$. Then the invariant factors of ρM are ρ-homothetic to those of M and the list of elementary divisors of ρM is ρ-homothetic to the list of elementary divisors of M.*

Proof. Note first that the characteristic matrix of ρM is

$$C_{\rho M}(\lambda) = \rho M - \lambda \mathbf{1}_{n+1} = \rho(M - (\lambda/\rho)\mathbf{1}_{n+1}) = \rho C_M(\lambda/\rho),$$

proportional to the result of substituting λ/ρ for λ in (all entries of) the characteristic matrix of M. Since this substitution induces the isomorphism of k-algebras

$$\begin{aligned} k[\lambda] &\longrightarrow k[\lambda], \\ p(\lambda) &\longmapsto p(\lambda/\rho), \end{aligned}$$

it is clear that for any minor $m(\lambda)$ of C_M, the minor of $C_{\rho M}$ involving the same rows and columns differs by a constant factor from $m(\lambda/\rho)$. By the same reason, if $D_s(\lambda)$ is the greatest common divisor of all the s-dimensional minors of C_M, then $D_s(\lambda/\rho)$ divides all the s-dimensional minors of $C_{\rho M}$ and is a multiple of any of their common divisors. Then the result of dividing $D_s(\lambda/\rho)$ by its leading coefficient is just the greatest common divisor of all the s-dimensional minors of $C_{\rho M}$. On the other hand, $D_s(\lambda)$ being monic, the result of dividing $D_s(\lambda/\rho)$ by its leading coefficient is the polynomial ρ-homothetic to $D_s(\lambda)$. Then the claim about the invariant factors follows using 11.3.3 (1). In turn, the claim relative to the elementary divisors follows from the one relative to the invariant factors using 11.3.3 (3). □

Characterizations of the projective equivalence of collineations in terms of the principal factors and the lists of elementary divisors of their matrices follow:

Theorem 11.3.5. *Assume that f and f' are, respectively, collineations of projective spaces $\mathbb{P}_n$, $\mathbb{P}'_n$ that have matrices M and M'. The following conditions are equivalent:*

(i) *The collineations f and f' are projectively equivalent.*

(ii) *The invariant factors of M and M' are homothetic.*

(iii) *The lists of invariant factors of M and M' are homothetic.*

Proof. Assume that f and f' are projectively equivalent. By 11.3.1, there is a non-zero $\rho \in k$ such that $M' = \rho N$, where N is a matrix similar to M. Then, by 11.3.4, the invariant factors of M' are homothetic to the invariant factors of N with dilation ratio ρ. Since the latter are the invariant factors of M by 11.2.15, the condition (ii) is satisfied.

Conversely, assume that (ii) is satisfied. If the invariant factors of M' are ρ-homothetic to the invariant factors of M, consider $\rho^{-1}M'$, which is also a matrix of

f'. By 11.3.4, its invariant factors are ρ^{-1}-homothetic to those of M' which in turn are ρ-homothetic to those of M. By 11.3.2, $\rho^{-1}M'$ and M have the same invariant factors and hence are similar matrices by 11.2.15. Using 11.3.1, it follows that f and f' are projectively equivalent, which proves (ii)⇒(i).

Similar arguments prove the equivalence of (i) and (iii). □

Theorem 11.3.5 applies in particular to the case $f = f'$, in which the arbitrariness of the matrices gives:

Corollary 11.3.6. *Any two matrices of the same collineation have homothetic invariant factors and homothetic lists of elementary divisors.*

As a consequence, if the set of invariant factors and the list of elementary divisors are taken up to homothety, then they do not depend on the choice of the matrix of the collineation. It makes thus sense to define the *invariant factors* and the *list of elementary divisors* of a collineation f as being those of any matrix of f, taken up to homothety.

This being done, 11.3.5 may be rewritten:

Corollary 11.3.7. *For any two collineations f, f' of projective spaces of the same dimension, the following conditions are equivalent:*

(i) *f and f' are projectively equivalent.*

(ii) *f and f' have homothetic invariant factors.*

(iii) *f and f' have homothetic lists of elementary factors.*

11.4 Classification of collineations of complex projective spaces

In this section we describe the projective classification of collineations of complex projective spaces and therefore we assume throughout that $k = \mathbb{C}$. Since the only irreducible polynomials in $\mathbb{C}[\lambda]$ are the polynomials of degree one, the list of elementary divisors of (a matrix of) a collineation f of a complex projective space may be written in the form

$$(\lambda - \lambda_1)^{\nu_{1,1}}, \dots, (\lambda - \lambda_1)^{\nu_{1,h_1}}, \dots, (\lambda - \lambda_m)^{\nu_{m,1}}, \dots, (\lambda - \lambda_m)^{\nu_{m,h_m}}, \quad (11.7)$$

where $m > 0$, $\lambda_i \in \mathbb{C}$, $\lambda_i \neq \lambda_j$ for $i \neq j$, $h_i > 0$ and the positive integers $\nu_{i,s}$ have been taken such that $\nu_{i,1} \geq \cdots \geq \nu_{i,h_i}$. The polynomials ρ-homothetic to the above compose the list

$$(\lambda - \rho\lambda_1)^{\nu_{1,1}}, \dots, (\lambda - \rho\lambda_1)^{\nu_{1,h_1}}, \dots, (\lambda - \rho\lambda_m)^{\nu_{m,1}}, \dots, (\lambda - \rho\lambda_m)^{\nu_{m,h_m}}.$$

Thus, as the essential information enclosed in the list of elementary divisors of f we retain the values of $\lambda_1, \dots, \lambda_m$, taken up to a common non-zero factor, and, for each $i = 1, \dots, m$, the degrees $\nu_{i,1}, \dots, \nu_{i,h_i}$ of the powers of $(\lambda - \lambda_i)$ appearing in the list.

We have already dealt with $\lambda_1, \dots, \lambda_m$ in Section 5.7. There, each λ_i was associated to one of the fundamental varieties of f and the ratios λ_i/λ_j were called the invariants of f. Indeed, if M is a matrix of f with list of elementary divisors (11.7), then, by 11.2.12, the characteristic polynomial of M, and hence of a representative φ of f, is

$$c_\varphi = (-1)^{n+1} \prod_{i=1}^{m} \prod_{s=1}^{h_i} (\lambda - \lambda_i)^{\nu_{i,s}}.$$

Hence $\lambda_1, \dots, \lambda_m$ are the eigenvalues of φ, which is the way they were defined in 5.7.2.

Sometimes c_φ is called a *characteristic polynomial* of f. In the sequel we will refer to the unordered set $\{\lambda_1, \dots, \lambda_m\}$, of the eigenvalues of a representative (or matrix) of a collineation f, as a *system of eigenvalues* of f. An arbitrary system of eigenvalues of f is then $\{\rho\lambda_1, \dots, \rho\lambda_m\}$, $\rho \in \mathbb{C} - \{0\}$.

The ordered sequence $(\nu_{i,1}, \dots, \nu_{i,h_i})$ of the exponents appearing in the elementary divisors which are powers of $\lambda - \lambda_i$ is called the *Segre symbol*, and also the *characteristic*, of the eigenvalue λ_i. (Slightly different equivalent definitions do appear in the literature.)

In order to collect all the relevant information in a single object, we consider the class of all sets

$$\{(\rho\lambda_1; \nu_{1,1}, \dots, \nu_{1,h_1}), \dots, (\rho\lambda_m; \nu_{m,1}, \dots, \nu_{m,h_m})\},$$

for $\rho \in \mathbb{C} - \{0\}$, denoted by

$$\kappa(f) = \big\langle(\lambda_1; \nu_{1,1}, \dots, \nu_{1,h_1}), \dots, (\lambda_m; \nu_{m,1}, \dots, \nu_{m,h_m})\big\rangle,$$

and called the *characteristic* of f. Note that, according to its definition, the characteristic of f remains the same if $\lambda_1, \dots, \lambda_m$ is replaced with another system of eigenvalues $\rho\lambda_1, \dots, \rho\lambda_m$, or the pairs $(\lambda_i; \nu_{i,1}, \dots, \nu_{i,h_i})$ are reordered. The reader wanting to avoid formalisms may think of the characteristic $\kappa(f)$ as just a symbol which is agreed not to change if its components are modified in either of the ways described above.

Next is a criterion for the projective equivalence of collineations that follows directly from 11.3.7 and the above definitions.

Corollary 11.4.1 (of 11.3.5). *Two collineations of complex projective spaces of the same dimension are projectively equivalent if and only if they have the same characteristic.*

Remark 11.4.2. Notations being as above, the elements appearing in the characteristic of a collineation of $\mathbb{P}_n$ satisfy:

(1) $\lambda_i \neq 0$, as noted in 5.7.8, and $\lambda_i \neq \lambda_j$, $i, j = 1, \dots, m$, $i \neq j$. Also, $1 \leq m \leq n+1$.

(2) $\nu_{i,1} \geq \dots \geq \nu_{i,h_i} > 0$, $h_i \geq 1$, $i = 1, \dots, m$, by definition.

(3)
$$\sum_{i=1}^{m} \sum_{s=1}^{h_i} \nu_{i,s} = n+1,$$
due to 11.2.13.

Next we will present a complete list of representatives of the projective types of collineations. To this end, fix $a \in \mathbb{C}$ and a positive integer ν, and consider the ν-dimensional square matrix

$$J(a,\nu) = \begin{pmatrix} a & & & \\ 1 & a & & \\ & \ddots & \ddots & \\ & & 1 & a \end{pmatrix},$$

with all entries on the diagonal equal to a, those immediately below equal to one, and the remaining ones equal to zero. $J(a,\nu)$ is called an *elementary Jordan matrix* or a *Jordan block*.

Assume we are given different $\lambda_1, \dots, \lambda_m \in \mathbb{C}$, all non-zero, $1 \leq m \leq n+1$, and, for each $i = 1, \dots, m$, positive integers $\nu_{i,s}$, $s = 1, \dots, h_i$, $h_i > 0$ satisfying $\nu_{i,1} \geq \dots \geq \nu_{i,h_i}$ and $\sum_{i=1}^{m} \sum_{s=1}^{h_i} \nu_{i,s} = n+1$. (Compare with 11.4.2.) Then we consider the square $(n+1)$-dimensional matrix, called a *Jordan matrix*,

$$J = \begin{pmatrix} J(\lambda_1,\nu_{1,1}) & & & & & \\ & \ddots & & & & \\ & & J(\lambda_1,\nu_{1,h_1}) & & & \\ & & & \ddots & & \\ & & & & J(\lambda_m,\nu_{m,1}) & & \\ & & & & & \ddots & \\ & & & & & & J(\lambda_m,\nu_{m,h_m}) \end{pmatrix}, \tag{11.8}$$

composed of $h_1 + \dots + h_m$ Jordan blocks centred on the diagonal, all entries outside them being equal to zero. We have

Proposition 11.4.3. *The list of elementary divisors of the matrix J defined by* (11.8) *above is:*

$$(\lambda - \lambda_1)^{\nu_{1,1}}, \dots, (\lambda - \lambda_1)^{\nu_{1,h_1}}, \dots, (\lambda - \lambda_m)^{\nu_{m,1}}, \dots, (\lambda - \lambda_m)^{\nu_{m,h_m}}.$$

Proof. Let us consider first the characteristic matrix of a single Jordan block $J(\lambda_i, \nu_{i,s})$:

$$C(\lambda_i, \nu_{i,s}) = \begin{pmatrix} \lambda_i - \lambda & & & \\ 1 & \lambda_i - \lambda & & \\ & \ddots & \ddots & \\ & & 1 & \lambda_i - \lambda \end{pmatrix}.$$

Its determinant is $(\lambda_i - \lambda)^{\nu_{i,s}}$, while dropping the first row and the last column gives rise to a $(\nu_{i,s} - 1)$-dimensional minor equal to one. This proves that its invariant factors are $1, \dots, 1, (\lambda - \lambda_i)^{\nu_{i,s}}$, and therefore that by means of elementary operations it may be turned into its canonical form

$$T(\lambda_i, \nu_{i,s}) = \begin{pmatrix} 1 & & & \\ & \ddots & & \\ & & 1 & \\ & & & (\lambda - \lambda_i)^{\nu_{i,s}} \end{pmatrix}.$$

The characteristic matrix of J is composed of the blocks $C(\lambda_i, \nu_{i,s})$. Since these blocks do not share any rows or columns, the elementary operations used to turn each block $C(\lambda_i, \nu_{i,s})$ into $T(\lambda_i, \nu_{i,s})$ may be made on the characteristic matrix of J using the corresponding rows and columns, without affecting rows or columns corresponding to the other blocks. This will turn the characteristic matrix of J into the equivalent diagonal matrix

$$T = \begin{pmatrix} T(\lambda_1, \nu_{1,1}) & & & & & \\ & \ddots & & & & \\ & & T(\lambda_1, \nu_{1,h_1}) & & & \\ & & & \ddots & & \\ & & & & T(\lambda_m, \nu_{m,1}) & & \\ & & & & & \ddots & \\ & & & & & & T(\lambda_m, \nu_{m,h_m}) \end{pmatrix}.$$

Now a straightforward computation shows that the invariant factors of T, which equal those of J, are

$$\begin{array}{c} 1 \\ \vdots \\ 1 \\ (\lambda-\lambda_1)^{\nu_{1,h}}\dots(\lambda-\lambda_m)^{\nu_{m,h}} \\ \vdots \\ (\lambda-\lambda_1)^{\nu_{1,1}}\dots(\lambda-\lambda_m)^{\nu_{m,1}} \end{array}$$

where $h = \max(h_1,\dots,h_m)$ and $\nu_{i,s} = 0$ for $s > h_i$. From this, the claim follows. □

Corollary 11.4.1 and the following proposition completely describe the projective classification of collineations of complex projective spaces:

Proposition 11.4.4. *For any $\lambda_1,\dots,\lambda_m \in \mathbb{C}$ and positive integers $\nu_{i,s}$, $i = 1,\dots,m$, $s = 1,\dots,h_i$, satisfying the conditions of* 11.4.2, *there is a unique projective type of collineations of complex n-dimensional projective spaces whose representatives have characteristic*

$$\langle(\lambda_1;\nu_{1,1},\dots,\nu_{1,h_1}),\dots,(\lambda_m;\nu_{m,1},\dots,\nu_{m,h_m})\rangle.$$

Any collineation having matrix J, as defined by (11.8), *is a representative of that class.*

Proof. Choose any projective space $\mathbb{P}_n$ and an arbitrary reference Δ in it. The matrix J is regular because no λ_i is zero; therefore we are allowed to consider the collineation f of $\mathbb{P}_n$ which has matrix J relative to Δ (2.8.5). Then by 11.4.3, the characteristic of f is as claimed, and so is the characteristic of any collineation projectively equivalent to f, by 11.4.1. The uniqueness follows from 11.4.1 and the last claim is direct from 11.4.3 and 11.4.1. □

The Jordan matrices have provided representatives of all the projective types of collineations in 11.4.4 above. They have a different and interesting use, as they appear as particularly simple matrices of collineations. The next proposition assures that any collineation has a Jordan matrix:

Proposition 11.4.5. *Assume that f is a collineation of a complex n-dimensional projective space $\mathbb{P}_n$. If f has characteristic*

$$\langle(\lambda_1;\nu_{1,1},\dots,\nu_{1,h_1}),\dots,(\lambda_m;\nu_{m,1},\dots,\nu_{m,h_m})\rangle,$$

then there is a reference of $\mathbb{P}_n$ relative to which a matrix of f equals the Jordan matrix J defined by the equality (11.8), *and conversely.*

Proof. Fix a matrix M of f, relative to a certain reference Δ. In view of the definition of the characteristic and 11.3.4, up to replacing M by a scalar multiple, one may assume its list of elementary divisors to be

$$(\lambda - \lambda_1)^{\nu_{1,1}}, \dots, (\lambda - \lambda_1)^{\nu_{1,h_1}}, \dots, (\lambda - \lambda_m)^{\nu_{m,1}}, \dots, (\lambda - \lambda_m)^{\nu_{m,h_m}},$$

which, by 11.4.3, is the list of elementary divisors of J. By 11.2.15, M and J are similar, which means that there exists a regular constant matrix P for which $J = PMP^{-1}$. By 2.2.3, P is a matrix changing coordinates relative to Δ into coordinates relative to another reference Ω. Then, by 2.8.10, J is a matrix of f relative to Ω. The converse is clear after 11.4.3 and the definition of the characteristic. □

The Jordan matrix of 11.4.5 is called a *Jordan matrix* of f. It is determined by the projective type of f (and hence also by f itself), up to a reordering of the groups of blocks corresponding to each λ_i and multiplication of $\lambda_1, \dots, \lambda_m$ by a common non-zero factor, as it may be written after a list of elementary divisors of f. Implicitly taken up to these modifications, it is often mentioned as *the* Jordan matrix of f.

The next proposition provides an easy way to compute the characteristic (or the Jordan matrix) of a collineation from one of its matrices once the corresponding system of eigenvalues is known.

Proposition 11.4.6. *Assume that a collineation f of a complex n-dimensional projective space has matrix M and characteristic*

$$\langle(\lambda_1; \nu_{1,1}, \dots, \nu_{1,h_1}), \dots, (\lambda_m; \nu_{m,1}, \dots, \nu_{m,h_m})\rangle,$$

$\lambda_1, \dots, \lambda_m$ being the eigenvalues of M. For each $i = 1, \dots, m$, we have

$$n + 1 - \operatorname{rk}(M - \lambda_i \mathbf{1}_{n+1}) = h_i$$

and, for $j \geq 2$,

$$n + 1 - \operatorname{rk}(M - \lambda_i \mathbf{1}_{n+1})^j = j h_i - (j-1)\delta_{i,1} - \dots - 2\delta_{i,j-2} - \delta_{i,j-1},$$

where $\delta_{i,\nu}$ is the number of integers in the Segre symbol of λ_i equal to ν.

Proof. The Jordan matrix J in (11.8) has the same elementary divisors as M, and hence is similar to it. Then there is an equality $J = PMP^{-1}$, from which we get

$$(J - \lambda_i \mathbf{1}_{n+1})^j = (PMP^{-1} - \lambda_i \mathbf{1}_{n+1})^j = P(M - \lambda_i \mathbf{1}_{n+1})^j P^{-1}.$$

It follows that $(M - \lambda_i \mathbf{1}_{n+1})^j$ and $(J - \lambda_i \mathbf{1}_{n+1})^j$ are similar and so they have the same rank. We are thus allowed to compute the ranks appearing in the claim using J instead of M. A matrix $(J - \lambda_i \mathbf{1}_{n+1})^j$ is composed of blocks which are the

j-th powers of those of $(J - \lambda_i \mathbf{1}_{n+1})$. The blocks corresponding to an eigenvalue $\lambda_t \neq \lambda_i$ have the form

$$\begin{pmatrix} \lambda_t - \lambda_i & & & \\ 1 & \lambda_t - \lambda_i & & \\ & \ddots & \ddots & \\ & & 1 & \lambda_t - \lambda_i \end{pmatrix}^j,$$

and therefore have maximal rank. The others are

$$\begin{pmatrix} 0 & & & \\ 1 & 0 & & \\ & \ddots & \ddots & \\ & & 1 & 0 \end{pmatrix}^j$$

and an easy computation shows them to have rank $\nu - j$ if $j \leq \nu$, and zero otherwise, ν being the dimension of the block. Then the first equality of the claim directly follows and an easy inductive computation, left to the reader, proves the remaining ones. □

To compute the characteristic of f, just note that, for each eigenvalue λ_i, the first equality of 11.4.6 determines h_i, while the remaining ones allow us to inductively determine the $\delta_{i,\nu}$, for $\nu = 1, 2, \dots$, until getting their sum equal to h_i. From these $\delta_{i,j}$, the Segre symbol of λ_i may be easily computed.

Example 11.4.7. As the reader may easily check, the characteristics of the collineations already seen in Section 5.7 are as follows:

– General plane homologies:

$$\langle (\lambda_1; 1, 1), (\lambda_2; 1) \rangle.$$

– Special plane homologies:

$$\langle (\lambda_1; 2, 1) \rangle.$$

– General three-space homologies:

$$\langle (\lambda_1; 1, 1, 1), (\lambda_2; 1) \rangle.$$

– Special three-space homologies:

$$\langle (\lambda_1; 2, 1, 1) \rangle.$$

– Biaxial collineations (5.7.28):

$$\langle (\lambda_1; 1, 1), (\lambda_2; 1, 1) \rangle.$$

– Collineations of $\mathbb{P}_n$ with $n+1$ isolated fixed points (5.7.10):

$$\big((\lambda_1; 1), \dots, (\lambda_{n+1}; 1)\big).$$

The reader may also note that the plane special homologies have all the same projective type; the same is true of the three-space special homologies. The other items listed above refer to families of projective types, each projective type being determined by the ratios between eigenvalues.

11.5 Classification of collineations of real projective spaces

In this section we will explain the projective classification of collineations of real projective spaces following essentially the same lines as in the complex case. Thus we will assume that $k = \mathbb{R}$ until the end of this section. The differences with the complex case come from the fact that a polynomial is irreducible in $\mathbb{R}[\lambda]$ if and only if it either has degree one or has degree two and no real root. A polynomial $P(\lambda)$ of the latter type has a pair of distinct and conjugate complex roots $z, \bar{z}$. If, in addition, it is monic, an elementary computation shows it to have the form

$$P(\lambda) = (\lambda - a)^2 + b^2,$$

for a uniquely determined pair of real numbers a, b with $b > 0$. Conversely, any polynomial as P above is irreducible in $\mathbb{R}[\lambda]$. Actually, up to replacing z with $\bar{z}$, it is $z = a + bi$.

Fix a matrix M of a collineation f of a real projective space $\mathbb{P}_n$. According to the above, the list of elementary divisors of M is composed of powers of $m \geq 0$ irreducible polynomials of the first degree,

$$(\lambda - \lambda_1)^{\nu_{1,1}}, \dots, (\lambda - \lambda_1)^{\nu_{1,h_1}}, \dots, (\lambda - \lambda_m)^{\nu_{m,1}}, \dots, (\lambda - \lambda_m)^{\nu_{m,h_m}}$$

and powers of $s \geq 0$ irreducible polynomials of the second degree

$$\begin{aligned} &\big((\lambda - a_1)^2 + b_1^2\big)^{\mu_{1,1}}, \dots, \big((\lambda - a_1)^2 + b_1^2\big)^{\mu_{1,r_1}}, \dots, \\ &\qquad \big((\lambda - a_s)^2 + b_s^2\big)^{\mu_{s,1}}, \dots, \big((\lambda - a_s)^2 + b_s^2\big)^{\mu_{s,r_s}}, \end{aligned}$$

where $m + s > 0$ and for $i = 1, \dots, m$ and $j = 1, \dots, s$, the λ_i are distinct real numbers, the $(a_j, b_j) \in \mathbb{R} \times \mathbb{R}^+$ are distinct pairs, $h_i, s_j > 0$ and $\nu_{i,1} \geq \dots \geq \nu_{i,h_i} > 0$, $\mu_{j,1} \geq \dots \geq \mu_{j,r_j} > 0$.

By 11.2.12, the λ_i, $i = 1, \dots, m$, are the (real) eigenvalues of M, while the $(\lambda - a_j)^2 + b_j^2$, $j = 1, \dots, s$, are the second-degree irreducible factors of the characteristic polynomial of M.

For any $\rho \in \mathbb{R} - \{0\}$ the polynomial ρ-homothetic to $(\lambda - a)^2 + b^2$ is

$$\rho^2\big((\lambda/\rho - a)^2 + b^2\big) = (\lambda - \rho a)^2 + (|\rho| b)^2.$$

Therefore, the list of the polynomials ρ-homothetic to the elementary divisors above is obtained by replacing, in each elementary divisor, λ_i with $\rho\lambda_i$, a_j with ρa_j and b_j with $|\rho|b_j$. It is (by 11.3.4) the list of elementary divisors of ρM.

Until the end of the section, when needed, we will shorten some expressions by writing $\bar{\nu}_i$ for the whole sequence $\nu_{i,1}, \dots, \nu_{i,h_i}$ and $\bar{\mu}_j$ for $\mu_{j,1}, \dots, \mu_{j,r_j}$. As in the complex case, $\bar{\nu}_i$ is called the *Segre symbol* of λ_i, while we will call $\bar{\mu}_j$ the *Segre symbol* of $(\lambda - a_j)^2 + b_j^2$.

We collect all the relevant information contained in the list of elementary factors by taking the class of all sets

$$\{(\rho\lambda_1; \bar{\nu}_1), \dots, (\rho\lambda_m; \bar{\nu}_m), ((\rho a_1, |\rho|b_1); \bar{\mu}_1), \dots, ((\rho a_s, |\rho|b_s); \bar{\mu}_s)\},$$

for $\rho \in \mathbb{R} - \{0\}$. This class will be called the *characteristic* of f, written

$$\kappa(f) = \langle(\lambda_1; \bar{\nu}_1), \dots, (\lambda_m; \bar{\nu}_m), ((a_1, b_1); \bar{\mu}_1), \dots, ((a_s, b_s); \bar{\mu}_s)\rangle.$$

By its definition, the characteristic $\kappa(f)$ remains the same after reordering or substituting in it $\rho\lambda_i$ for λ_i, ρa_j for a_j and $|\rho|b_j$ for b_j, $\rho \in \mathbb{R} - \{0\}$. Corollary 11.3.7 and the above definitions directly give:

Corollary 11.5.1 (of 11.3.5). *Two collineations of real projective spaces of the same dimension are projectively equivalent if and only if they have the same characteristic.*

Remark 11.5.2. The notations being as above, the elements appearing in the characteristic $\kappa(f)$ of a collineation of $\mathbb{P}_n$ satisfy:

(1) $m, s \geq 0$, $m + s > 0$, clearly.

(2) $\lambda_i \neq 0$, as noted in 5.7.8, and, by definition, $\lambda_i \neq \lambda_j$, for $i, j = 1, \dots, m$, $i \neq j$.

(3) $b_j > 0$ and $(a_i, b_i) \neq (a_j, b_j)$ for $i, j = 1, \dots, s$, $i \neq j$.

(4) $\nu_{i,1} \geq \dots \geq \nu_{i,h_i} > 0$, $h_i \geq 1$, for $i = 1, \dots, m$, and, similarly, $\mu_{j,1} \geq \dots \geq \mu_{j,r_i} > 0$, $r_j \geq 1$, for $j = 1, \dots, s$, by their own definitions.

(5)
$$\sum_{i=1}^{m} \sum_{\ell=1}^{h_i} \nu_{i,\ell} + 2 \sum_{j=1}^{s} \sum_{\ell=1}^{r_j} \mu_{j,\ell} = n + 1,$$
due to 11.2.13.

In order to show the existence of collineations having the characteristics allowed by the above conditions, we need to introduce a new type of matrices, other than the Jordan blocks. For any positive integer μ and any pair of real numbers a, b, with $b > 0$, we take $K(a,b,\mu)$ to be the 2μ-dimensional square matrix composed of four μ-dimensional square blocks

$$K(a,b,\mu) = \begin{pmatrix} a & & & & -b & & & \\ 1 & a & & & & -b & & \\ & \ddots & \ddots & & & & \ddots & \\ & & 1 & a & & & & -b \\ b & & & 0 & a & & & \\ & b & & & 1 & a & & \\ & & \ddots & & & \ddots & \ddots & \\ & & & b & & & 1 & a \end{pmatrix},$$

or equivalently, using blocks,

$$K(a,b,\mu) = \begin{pmatrix} J(a,\mu) & -b\mathbf{1}_\mu \\ b\mathbf{1}_\mu & J(a,\mu) \end{pmatrix}.$$

In the sequel, the matrices $K(a,b,\mu)$ will be called *K-blocks*. First of all we need:

Lemma 11.5.3. *The matrix $K(a,b,\mu)$ is regular and has a list of invariant factors*

$$1,\dots,1,\big((\lambda-a)^2+b^2\big)^\mu.$$

Proof. Take $z = a + b\boldsymbol{i}$, its conjugate being thus $\bar{z} = a - b\boldsymbol{i}$. It is straightforward to check that the matrix, in blocks,

$$\frac{1}{2}\begin{pmatrix} \mathbf{1}_\mu & \boldsymbol{i}\,\mathbf{1}_\mu \\ \mathbf{1}_\mu & -\boldsymbol{i}\,\mathbf{1}_\mu \end{pmatrix}$$

has inverse

$$\begin{pmatrix} \mathbf{1}_\mu & \mathbf{1}_\mu \\ -\boldsymbol{i}\,\mathbf{1}_\mu & \boldsymbol{i}\,\mathbf{1}_\mu \end{pmatrix}$$

and it also holds that

$$\frac{1}{2}\begin{pmatrix} \mathbf{1}_\mu & \boldsymbol{i}\,\mathbf{1}_\mu \\ \mathbf{1}_\mu & -\boldsymbol{i}\,\mathbf{1}_\mu \end{pmatrix}\begin{pmatrix} J(a,m) & -b\mathbf{1}_\mu \\ b\mathbf{1}_\mu & J(a,m) \end{pmatrix}\begin{pmatrix} \mathbf{1}_\mu & \mathbf{1}_\mu \\ -\boldsymbol{i}\,\mathbf{1}_\mu & \boldsymbol{i}\,\mathbf{1}_\mu \end{pmatrix} = \begin{pmatrix} J(z,\mu) & 0 \\ 0 & J(\bar{z},\mu) \end{pmatrix}.$$

The latter is a regular (because $z, \bar{z} \neq 0$) complex Jordan matrix whose invariant factors, computed using 11.4.3, are those listed in the claim. Furthermore, the above equality shows that it is similar (over $\mathbb{C}$) to $K(a,b,\mu)$, which is thus regular too. Since similar matrices have the same invariant factors (11.2.15) and the invariant factors does not depend on the base field (11.2.5), the second claim follows. □

11.5.4. Assume we have fixed real numbers λ_i, pairs $(a_j,b_j) \in \mathbb{R}\times\mathbb{R}^+$, and finite sequences of positive integers $\nu_{i,1},\dots,\nu_{i,h_i}$ and $\mu_{j,1},\dots,\mu_{j,r_j}$, for $i = 1,\dots,m$

and $j = 1, \dots, s$, satisfying the conditions of 11.5.2. Take K to be the square matrix composed of the blocks

$$J(\lambda_1, \nu_{1,1}), \dots, J(\lambda_1, \nu_{1,h_1}), \dots, J(\lambda_m, \nu_{m,1}), \dots, J(\lambda_m, \nu_{m,h_m}), \dots,$$
$$K(a_1, b_1, \mu_{1,1}), \dots, K(a_1, b_1, \mu_{1,r_1}), \dots, K(a_s, b_s, \mu_{s,1}), \dots, K(a_s, b_s, \mu_{s,r_s})$$

centred on the diagonal, all entries outside the blocks being zero.

The Jordan blocks in K are regular because no λ_i is zero, while the K-blocks are due to 11.5.3; hence K is regular, and, due to the condition (5) in 11.5.2 has dimension $n + 1$. Furthermore, the same arguments used in the proof of 11.4.3, this time using also 11.5.3, allow us to prove:

Proposition 11.5.5. *The list of elementary divisors of the matrix K defined above is*

$$\left\{(\lambda - \lambda_i)^{\nu_{i,\ell_i}}, ((\lambda - a_j)^2 + b_j^2)^{\mu_{j,t_j}}\right\}_{\substack{i=1,\dots,m \\ \ell_i=1,\dots,h_i \\ j=1,\dots,s \\ t_j=1,\dots,r_j}}.$$

We have seen that K is a regular $(n + 1) \times (n + 1)$-matrix. Therefore, after fixing a real projective space $\mathbb{P}_n$ and a reference Δ of it, there is a collineation f of $\mathbb{P}_n$ with matrix K relative to Δ. Then, arguing as in the complex case, we have:

Proposition 11.5.6. *If the real numbers λ_i, the pairs $(a_j, b_j) \in \mathbb{R} \times \mathbb{R}^+$, and the finite sequences of positive integers $\nu_{i,1}, \dots, \nu_{i,h_i}$ and $\mu_{j,1}, \dots, \mu_{j,r_j}$, for $i = 1, \dots, m$ and $j = 1, \dots, s$, satisfy the conditions of* 11.5.2, *then there is a unique projective type of collineations of real n-dimensional projective spaces whose representatives have characteristic*

$$\langle(\lambda_1; \bar{\nu}_1), \dots, (\lambda_m; \bar{\nu}_m), ((a_1, b_1); \bar{\mu}_1), \dots, ((a_s, b_s); \bar{\mu}_s)\rangle.$$

Any collineation with matrix K as above belongs to that class.

The same arguments used in the proof of 11.4.5 prove that in the real case any collineation has a matrix composed of Jordan blocks and K-blocks centred on the diagonal. More precisely

Proposition 11.5.7. *Let f be a collineation of a real projective space $\mathbb{P}_n$. If f has characteristic*

$$\langle(\lambda_1; \bar{\nu}_1), \dots, (\lambda_m; \bar{\nu}_m), ((a_1, b_1); \bar{\mu}_1), \dots, ((a_s, b_s); \bar{\mu}_s)\rangle,$$

then there is a reference of $\mathbb{P}_n$ relative to which a matrix of f equals the matrix K defined in 11.5.4, *and conversely.*

We will call the matrix of 11.5.7 the *reduced matrix* of f. As the reader may check, it is uniquely determined by f up to the ordering of the groups of blocks corresponding to each elementary divisor and the real parameter ρ up to which are determined their coefficients λ_i, a_j, b_j. Other reduced forms of the matrix of f do appear in the literature, see for instance [18], VIII.2, Theorem II, or [23], V.26, both dealing with linear maps rather than with collineations.

Once the irreducible factors of the characteristic polynomial of a matrix of a collineation f are determined, their Segre symbols, and hence the reduced matrix of f, may be computed via 11.4.6 applied to the complex extension of f, see Exercise 11.11.

We close this section with some examples, all relative to collineations with some elementary factor with no real roots (or, equivalently, the notations being as above, with $s > 0$), as these factors are the novelty of the real case. The details are left to the reader.

Example 11.5.8. An elliptic collineation of $\mathbb{P}_1$ has characteristic $\langle (a_1, b_1); 1 \rangle$ and its reduced matrix consists of a single K-block. The reader may recognize this matrix to be proportional to the matrix (5.23) obtained in Section 5.5.

Example 11.5.9. Due to the condition (5) of 11.5.2, the only characteristics of collineations of $\mathbb{P}_2$ with $s > 0$ have the form $\langle (\lambda_1; 1), ((a_1, b_1); 1) \rangle$. Each of the corresponding collineations has an isolated fixed point as its only fundamental variety, while its complex extension has two further isolated fixed points, which are imaginary and complex-conjugate.

Example 11.5.10. In case $n = 3$ the characteristics with $s > 0$ have one of the following forms:

(1) $\langle (\lambda_1; 1), (\lambda_2; 1), ((a_1, b_1); 1) \rangle$,

(2) $\langle (\lambda_1; 1, 1), ((a_1, b_1); 1) \rangle$,

(3) $\langle (\lambda_1; 2), ((a_1, b_1); 1) \rangle$,

(4) $\langle ((a_1, b_1); 1), ((a_2, b_2); 1) \rangle$,

(5) $\langle ((a_1, b_1); 1, 1) \rangle$,

(6) $\langle ((a_1, b_1); 2) \rangle$.

The forms (1) and (4) correspond to collineations whose complex extensions have four isolated fixed points, all imaginary in case (4), two real and two imaginary in case (1). The form (5) corresponds to collineations whose complex extensions are biaxial, with imaginary axes.

11.6 Projective classification of pencils of quadrics

As already defined in Section 9.3 for linear systems of quadrics, the action of projectivities on pencils of quadrics is given by the rule $f(\mathcal{P}) = \{f(Q) \mid Q \in \mathcal{P}\}$, for any pencil of quadrics $\mathcal{P}$ of $\mathbb{P}_n$ and any projectivity $f\colon \mathbb{P}_n \to \widetilde{\mathbb{P}}_n$. Obviously, the image of a regular pencil of quadrics by a projectivity is regular too, so regular pencils of quadrics are projective classified by this action (see Section 1.10). In this section we will describe the projective classification of regular pencils of quadrics. We will not consider the classification of non-regular pencils of quadrics, as it is less interesting and would take us too far. For it, the reader is referred to [18], XII, 10, its algebraic background is in [11], II, XII. Throughout this section, the base field is assumed to be the complex one. Unless otherwise said, all matrices will be assumed to be square and to have dimension $n + 1$.

We begin by proving two algebraic results. The first one is an easy characterization of the degrees of the elementary divisors; it will be useful later on. The second one is a deep result on the equivalence of degree-one symmetric polynomial matrices; it is the key to describe the classification of regular pencils of quadrics.

Proposition 11.6.1. *Assume that the invariant factors $I_0, \dots, I_\ell$, of a complex polynomial matrix $\mathcal{M}$, are written as products of linear factors in the form*

$$\begin{aligned} I_0 &= (\lambda - \lambda_1)^{\mu_{1,0}} \dots (\lambda - \lambda_s)^{\mu_{s,0}}, \\ I_1 &= (\lambda - \lambda_1)^{\mu_{1,1}} \dots (\lambda - \lambda_s)^{\mu_{s,1}}, \\ &\vdots \\ I_\ell &= (\lambda - \lambda_1)^{\mu_{1,\ell}} \dots (\lambda - \lambda_s)^{\mu_{s,\ell}}, \end{aligned} \tag{11.9}$$

where for each factor $\lambda - \lambda_i$ some but not all of the exponents $\mu_{i,j}$ may be zero. Then, for each $i = 1, \dots, s$ and $j = 0, \dots, \ell$, the least of the multiplicities of λ_i as a root of the $(j+1)$-dimensional subdeterminants of $\mathcal{M}$ is

$$\tilde{\mu}_{i,j} = \mu_{i,j} + \dots + \mu_{i,0}.$$

If $\mathcal{M}$ is regular (and hence $\ell = n$), then $\lambda_1, \dots, \lambda_m$ are the roots of $\det \mathcal{M}$ *and the multiplicity of each λ_i, as such, equals $\tilde{\mu}_{i,n}$. In particular,*

$$\sum_{i=1}^{s} \sum_{j=0}^{n} \mu_{i,j} = \sum_{i=1}^{s} \tilde{\mu}_{i,n} = \deg(\det \mathcal{M}).$$

Proof. The first claim follows directly from the definition of the invariant factors in Section 11.2. The second claim follows from the first one after recalling that, again by the definition of the invariant factors, $\det \mathcal{M} = I_0 \dots I_n$. □

Remark 11.6.2. The (possibly repeated) factors $(\lambda - \lambda_j)^{\mu_{i,j}}$, $\mu_{i,j} > 0$, appearing in (11.9) compose the list of elementary divisors of M. According to 11.6.1, their degrees are $\mu_{i,j} = \tilde{\mu}_{i,j} - \tilde{\mu}_{i-1,j}$ (taking $\tilde{\mu}_{-1,j} = 0$); they are thus determined by the multiplicities of the λ_j as roots of the subdeterminants of M.

Proposition 11.6.3. *Assume we have complex polynomial matrices of the form*

$$\mathcal{M} = M + \lambda N \quad \textit{and} \quad \mathcal{M}' = M' + \lambda N'$$

where M, N, M', N' are constant symmetric matrices, $\det N \neq 0$ *and* $\det N' \neq 0$. *Then $\mathcal{M}$ and $\mathcal{M}'$ are equivalent if and only if there is an invertible matrix P such that*

$$P^t \mathcal{M} P = \mathcal{M}'.$$

Proof. The *if* part is obvious from 11.2.9. Conversely, by 11.2.11, we may assume that there are invertible constant matrices A, B for which $A\mathcal{M}B = \mathcal{M}'$ or, equivalently,

$$AMB = M', \quad ANB = N'.$$

The matrices M, N, M', N' being symmetric we have

$$B^t MA^t = M', \quad B^t NA^t = N',$$

which, together with the preceding equalities, give

$$AMB = B^t MA^t$$

and

$$ANB = B^t NA^t,$$

or, equivalently,

$$(B^t)^{-1} AM = MA^t B^{-1} = M\big((B^t)^{-1} A\big)^t$$

and

$$(B^t)^{-1} AN = NA^t B^{-1} = N\big((B^t)^{-1} A\big)^t.$$

By taking $C = (B^t)^{-1} A$, it is

$$CM = MC^t$$

and

$$CN = NC^t.$$

Furthermore the matrix C is invertible because both A and B are. Using induction, the above equalities easily give

$$C^i M = M(C^i)^t$$

and

$$C^i N = N(C^i)^t,$$

for any positive integer i, and hence,

$$\Big(\sum_{i=0}^{r} a_i C^i\Big) M = M \Big(\sum_{i=0}^{r} a_i C^i\Big)^t$$

and

$$\Big(\sum_{i=0}^{r} a_i C^i\Big) N = N \Big(\sum_{i=0}^{r} a_i C^i\Big)^t,$$

for any complex numbers $a_0, \dots, a_r$.

There is an algebraic theorem assuring the existence of a matrix D which has $D^2 = C$, and, furthermore, may be written as a polynomial expression in C, namely $D = \sum_{i=0}^{r} a_i C^i$ for some complex numbers $a_0, \dots, a_r$. For the convenience of the reader we have included a proof of it in a separate section at the end of this chapter (Section 11.8). Obviously such a matrix D is invertible because C is, and by the above it satisfies

$$DM = MD^t$$

and

$$DN = ND^t.$$

Then it is

$$\begin{aligned} M' = AMB &= B^t CMB \\ &= B^t D^2 MB \\ &= B^t DMD^t B, \end{aligned}$$

and, similarly,

$$N' = B^t DND^t B,$$

after which it is enough to take $P = D^t B$. □

Once the required algebraic background has been set, we turn our attention to pencils of quadrics. Assume that $\mathcal{P}$ is a regular pencil of quadrics of a complex projective space $\mathbb{P}_n$ and fix a reference Δ of $\mathbb{P}_n$. Choose matrices M, N of two distinct quadrics $Q, Q' \in \mathcal{P}$, the second one Q' being assumed to be non-degenerate. Then $\det N \neq 0$ and each quadric of $\mathcal{P}$ other than Q' has a unique matrix of the form

$$M + \theta N, \quad \theta \in \mathbb{C}.$$

We will denote by $Q(\theta)$ the quadric of $\mathcal{P}$ which has matrix $M + \theta N$, for $\theta \in \mathbb{C}$, and write $Q(\infty) = Q'$. Note that in this way the parameter $\theta \in \mathbb{C} \cup \{\infty\}$ is the absolute coordinate of $Q(\theta)$ relative to the reference $(Q', Q, Q(1))$ of $\mathcal{P}$ (see 2.1.3). In the sequel we will refer to θ as the absolute coordinate of $Q(\theta)$ relative to the choice of M, N.

Still keep fixed the reference Δ of $\mathbb{P}_n$. We take as *characteristic matrix* of $\mathcal{P}$, relative to Δ, any polynomial matrix of the form

$$C(\lambda) = M + \lambda N$$

where M, N are matrices of two quadrics spanning $\mathcal{P}$, the second one non-degenerate (that is, with $\det N \neq 0$). Note that a characteristic matrix is not determined by $\mathcal{P}$, but depends on the choice of M and N. In turn, a characteristic matrix $C(\lambda)$ of $\mathcal{P}$ determines $\mathcal{P}$, as the matrices of two quadrics spanning $\mathcal{P}$ are recovered from $C(\lambda)$ by the rules

$$M = C(0) \quad \text{and} \quad N = \frac{1}{\lambda}(C(\lambda) - C(0)).$$

In fact one may think of the characteristic matrix as a way of representing the pencil, since, as explained above, giving complex values to the variable λ, $C(\lambda)$ provides matrices of all quadrics in the pencil but the one with matrix N.

Assume that the elementary divisors of the characteristic matrix $M + \lambda N$ are

$$(\lambda - \lambda_1)^{\nu_{1,1}}, \dots, (\lambda - \lambda_1)^{\nu_{1,h_1}}, \dots, (\lambda - \lambda_m)^{\nu_{m,1}}, \dots, (\lambda - \lambda_m)^{\nu_{m,h_m}},$$

where $\lambda_i \neq \lambda_j$ for $i \neq j$, $h_i > 0$ and the positive integers $\nu_{i,s}$ have been taken such that $\nu_{i,1} \geq \dots \geq \nu_{i,h_i}$. By 11.6.1, $\lambda_1, \dots, \lambda_m$ are the roots of the polynomial

$$\det(M + \lambda N)$$

which is called a *characteristic polynomial* of $\mathcal{P}$. Characteristic polynomials of pencils of quadrics already appeared in Section 9.5, see 9.5.7; from now on they will be taken with the added condition $\det N \neq 0$. This assures that the degree of any characteristic polynomial is $n+1$, as an easy computation shows the monomial of higher degree of $\det(M + \lambda N)$ to be $\det N \lambda^{n+1}$. Since $Q(\infty)$ is non-degenerate and $Q(\theta)$, $\theta \in \mathbb{C}$ is degenerate if and only $\det(M + \theta N) = 0$, $\lambda_1, \dots, \lambda_m$ are the absolute coordinates of the degenerate quadrics of $\mathcal{P}$. In the sequel, the degenerate quadrics of $\mathcal{P}$ will be denoted by $Q_i = Q(\lambda_i)$, $i = 1, \dots, m$.

As already done for collineations, we may associate to each λ_i the ordered sequence $(\nu_{i,1}, \dots, \nu_{i,h_i})$ and call it the *Segre symbol* of λ_i. A main point is that this Segre symbol is intrinsically associated to the degenerate quadric Q_i, namely:

Proposition 11.6.4. *Associating to each degenerate quadric Q_i of $\mathcal{P}$ the Segre symbol of λ_i does not depend on the choice of the reference Δ or the matrices M, N.*

Proof. If Ω is another reference of $\mathbb{P}_n$ and T the matrix changing coordinates relative to Ω into coordinates relative to Δ, then T is invertible and we may take $T^t MT$ and $T^t NT$ as matrices of Q and Q' relative to Ω. Then, relative to Ω each Q_i has matrix

$$T^t(M + \lambda_i N)T = T^t MT + \lambda_i T^t NT$$

and so still absolute coordinate λ_i. On the other hand the new characteristic matrix is

$$T^t MT + \lambda T^t NT = T^t(M + \lambda N)T,$$

clearly equivalent to $(M + \lambda N)$ and therefore with the same list of elementary divisors. This proves that the choice of the reference of $\mathbb{P}_n$ does not matter.

Assume now to have chosen matrices

$$\bar{M} = aM + bN \quad \text{and} \quad \bar{N} = cM + dN$$

of another pair of quadrics of $\mathcal{P}$, these quadrics being different (hence $ad - bc \neq 0$) and the second one non-degenerate (hence different from any Q_i, $i = 1, \dots, m$). The latter condition assures that Q_i has a matrix of the form $\bar{M} + \bar{\lambda}_i \bar{N}$. Then, it is

$$\bar{M} + \bar{\lambda}_i \bar{N} = (a + c\bar{\lambda}_i)M + (b + d\bar{\lambda}_i)N,$$

where $a + c\bar{\lambda}_i \neq 0$ because $Q_i \neq Q'$, and so

$$\lambda_i = \frac{b + d\bar{\lambda}_i}{a + c\bar{\lambda}_i}. \tag{11.10}$$

Since

$$\bar{M} + \lambda\bar{N} = (a + c\lambda)M + (b + d\lambda)N = (a + c\lambda)\Big(M + \frac{b + d\lambda}{a + c\lambda}N\Big),$$

for each entry $\bar{F}(\lambda)$ of $\bar{M} + \lambda\bar{N}$ it is

$$\bar{F}(\lambda) = (a + c\lambda)F\Big(\frac{b + d\lambda}{a + c\lambda}\Big),$$

$F(\lambda)$ the entry of $M + \lambda N$ on the same row and column as $\bar{F}$. Take any s-dimensional subdeterminant $\bar{D}(\lambda)$ of $\bar{M} + \lambda\bar{N}$. $\bar{D}(\lambda)$ is a homogeneous polynomial of degree s in the entries of $\bar{M} + \lambda\bar{N}$, hence

$$\bar{D}(\lambda) = (a + c\lambda)^s D\Big(\frac{b + d\lambda}{a + c\lambda}\Big), \tag{11.11}$$

where $D(\lambda)$ is the subdeterminant of $M + \lambda N$ involving the same rows and columns as $\bar{D}(\lambda)$. Assume now that λ_i is a root of multiplicity $r \geq 0$ of D, namely that

$$D(\lambda) = (\lambda - \lambda_i)^r D'(\lambda)$$

and $D'(\lambda_i) \neq 0$. Note that $r \leq \deg D(\lambda) \leq s$. Then, using (11.11) and (11.10),

$$\begin{aligned}\bar{D}(\lambda) &= (a+c\lambda)^s \Big(\frac{b+d\lambda}{a+c\lambda} - \frac{b+d\bar{\lambda}_i}{a+c\bar{\lambda}_i}\Big)^r D'\Big(\frac{b+d\lambda}{a+c\lambda}\Big)\\ &= (a+c\lambda)^{s-r}\frac{(ad-bc)^r}{(a+c\bar{\lambda}_i)^r}(\lambda-\bar{\lambda}_i)^r D'\Big(\frac{b+d\lambda}{a+c\lambda}\Big).\end{aligned}$$

Since $a + c\bar{\lambda}_i \neq 0$ and

$$D'\Big(\frac{b+d\bar{\lambda}_i}{a+c\bar{\lambda}_i}\Big) = D'(\lambda_i) \neq 0,$$

it follows that $\bar{\lambda}_i$ is a root of $\bar{D}(\lambda)$ of multiplicity r.

Now, the subdeterminant $\bar{D}(\lambda)$ being arbitrary, the above equality of multiplicities and 11.6.2 show that the Segre symbol of λ_i relative to the characteristic matrix $M + \lambda N$ equals the Segre symbol of $\bar{\lambda}_i$ relative to the characteristic matrix $\bar{M} + \lambda\bar{N}$, as claimed. □

Notations being as above, after 11.6.4, we may associate to each degenerate quadric Q_i of $\mathcal{P}$ the Segre symbol $(\nu_{i,1}, \dots, \nu_{i,h_i})$ of λ_i and call it the *Segre symbol* of Q_i. We will fix our attention in the set of all ordered pairs consisting of a degenerate quadric of $\mathcal{P}$ and its Segre symbol, namely, in our case,

$$\{(Q_1, (\nu_{1,1}, \dots, \nu_{1,h_1})), \dots, (Q_m, (\nu_{m,1}, \dots, \nu_{m,h_m}))\}.$$

We will call it the *characteristic system* of the pencil. The reader may compare the characteristic system with the characteristic divisor already introduced in Section 9.5: in the characteristic divisor each degenerate quadric has associated a single number, while in the characteristic system it has associated its Segre symbol, which is a finite sequence of positive integers. The characteristic system determines the characteristic divisor, as shown next:

Proposition 11.6.5. *Notations being as above, the multiplicity of the degenerate quadric Q_i in the characteristic divisor is $\nu_{i,1} + \dots + \nu_{i,h_i}$. The variety of double points of Q_i has*

$$\dim \mathcal{D}(Q_i) \leq \nu_1 + \dots + \nu_{h_i} - 1.$$

Proof. The first claim directly follows from 11.6.1, while the second one follows from it and 9.5.10. □

As said above, the characteristic system is a finite set of points of a projective line (the pencil $\mathcal{P}$) with a finite sequence of positive integers associated to each point. Seen in this way, its projective type determines the projective type of the pencil and conversely, namely:

Theorem 11.6.6. *Assume that $\mathcal{P}$ and $\widetilde{\mathcal{P}}$ are regular pencils of quadrics of complex projective spaces of the same dimension, $\mathbb{P}_n$ and $\widetilde{\mathbb{P}}_n$. Then there is a projectivity $f\colon \mathbb{P}_n \to \widetilde{\mathbb{P}}_n$ mapping $\mathcal{P}$ to $\widetilde{\mathcal{P}}$ if and only if there is a projectivity (of one-dimensional projective spaces) $g\colon \mathcal{P} \to \widetilde{\mathcal{P}}$ mapping the degenerate quadrics of $\mathcal{P}$ onto the degenerate quadrics of $\widetilde{\mathcal{P}}$ in such a way that corresponding degenerate quadrics have equal Segre symbols.*

Proof. Assume that there is a projectivity $f\colon \mathbb{P}_n \to \widetilde{\mathbb{P}}_n$ satisfying the conditions of the claim. Choose a reference Δ of $\mathbb{P}_n$ and let M, N be matrices relative to Δ of two different quadrics $Q, Q' \in \mathcal{P}$, the second one non-degenerate; then the pencil $\mathcal{P}$ has characteristic matrix $M + \lambda N$. Take $f(\Delta)$ as reference in $\widetilde{\mathbb{P}}_n$, after which the matrix of f is the unit one and so any quadric of $\mathbb{P}_n$ and its image by f have equal matrices. In particular we may take M and N as matrices of $f(Q)$ and $f(Q')$, which are two different quadrics of $\widetilde{\mathcal{P}}$, the second one non-degenerate. It follows then that $M + \lambda N$ is also a characteristic matrix of $\widetilde{\mathcal{P}}$.

Since $f(\mathcal{P}) = \widetilde{\mathcal{P}}$, the projectivity $\mathbb{Q}(f)$, induced by f on quadrics, see 9.3.4, restricts to a projectivity

$$\begin{aligned} \mathcal{P} &\longrightarrow \widetilde{\mathcal{P}}, \\ Q &\longmapsto f(Q), \end{aligned}$$

which we take as g. Clearly (by 6.4.14), it maps the set of degenerate quadrics of $\mathcal{P}$ onto the set of the degenerate quadrics of $\widetilde{\mathcal{P}}$. If, as above, Q_i is a degenerate quadric of $\mathcal{P}$ with matrix $M + \lambda_i N$, then $g(Q_i) = f(Q_i)$ has the same matrix as Q_i, and hence the same absolute coordinate λ_i. The characteristic matrix being the same for both pencils, it follows that Q_i and $g(Q_i)$ have the same Segre symbol, as claimed.

Conversely, take Δ, M and N as above and, for $\theta \in \mathbb{C}$, write $Q(\theta)$ the quadric of $\mathcal{P}$ with matrix $M + \theta N$ and still $Q_i = Q(\lambda_i)$, $i = 1, \dots, m$, the degenerate quadrics of $\mathcal{P}$. Assume that the Segre symbol of Q_i is $(\nu_{i,1}, \dots, \nu_{i,h_i})$. Then the list of elementary divisors of the characteristic matrix $M + \lambda N$ of $\mathcal{P}$, is

$$(\lambda - \lambda_i)^{\nu_{i,j}}, \quad i = 1, \dots, m, \ j = 1, \dots, h_i.$$

Fix any reference in $\widetilde{\mathbb{P}}_n$ and take matrices $\widetilde{M}$ of $g(Q)$ and $\widetilde{N}$ of $g(Q')$ chosen such that $\widetilde{M} + \widetilde{N}$ is a matrix of $g(Q(1))$. In this way both $Q(\theta)$ and $g(Q(\theta))$ have the same absolute coordinate θ in their pencils. By the hypothesis, the degenerate quadrics of $\widetilde{\mathcal{P}}$ are $g(Q_i)$, $i = 1, \dots, m$. By the above choices, each $g(Q_i)$ has matrix $\widetilde{M} + \lambda_i \widetilde{N}$ and so absolute coordinate λ_i. Also by the hypothesis, each $g(Q_i)$ has Segre symbol $(\nu_{i,1}, \dots, \nu_{i,h_i})$. Then, the absolute coordinates and their corresponding Segre symbols being the same as above, the list of elementary divisors of $\widetilde{M} + \lambda \widetilde{N}$ agrees with the one already found for $M + \lambda N$. It follows thus, by 11.2.6, that $\widetilde{M} + \lambda \widetilde{N}$ and $M + \lambda N$ are equivalent matrices. Therefore, by 11.6.3, there is an invertible constant matrix P for which

$$P^t(\widetilde{M} + \lambda\widetilde{N})P = M + \lambda N$$

and so

$$P^t\widetilde{M}P = M \quad \text{and} \quad P^t\widetilde{N}P = N.$$

Take f to be the projectivity from $\mathbb{P}_n$ to $\widetilde{\mathbb{P}}_n$ which, relative to the already chosen references, has matrix P. The equalities displayed above show that $f(Q) = g(Q) \in \widetilde{\mathcal{P}}$ and $f(Q') = g(Q') \in \widetilde{\mathcal{P}}$, after which $f(\mathcal{P}) = \widetilde{\mathcal{P}}$. □

Theorem 11.6.6 may be rewritten using the absolute coordinates of the degenerate quadrics instead of the degenerate quadrics themselves:

Corollary 11.6.7. *Let $\mathcal{P}$ and $\widetilde{\mathcal{P}}$ be regular pencils of quadrics of complex projective spaces $\mathbb{P}_n$ and $\widetilde{\mathbb{P}}_n$. After fixing references, assume that M, N are matrices of two quadrics in $\mathcal{P}$, the second one non-degenerate, and, similarly, that $\widetilde{M}$, $\widetilde{N}$ are matrices of two quadrics in $\widetilde{\mathcal{P}}$, the second one being also non-degenerate. Assume that the degenerate quadrics of $\mathcal{P}$ are those with matrices $M + \lambda_i N$, $i = 1, \dots, m$, while the degenerate quadrics of $\widetilde{\mathcal{P}}$ are those with matrices $\widetilde{M} + \mu_i\widetilde{N}$, $i = 1, \dots, m'$. Then there is a projectivity $f : \mathbb{P}_n \to \widetilde{\mathbb{P}}_n$ mapping $\mathcal{P}$ to $\widetilde{\mathcal{P}}$ if and only if $m = m'$ and there is a homographic function*

$$\sigma(t) = \frac{at + b}{ct + d}, \quad ad - bc \neq 0,$$

such that

(a) *$c\lambda_i + d \neq 0$ for $i = 1, \dots, m$,*

(b) *$\sigma(\{\lambda_1, \dots, \lambda_m\}) = \{\mu_1, \dots, \mu_m\}$, and*

(c) *the Segre symbol of λ_i relative to $M + \lambda N$ equals the Segre symbol of $\sigma(\lambda_i)$ relative to $\widetilde{M} + \lambda\widetilde{N}$.*

Proof. Call θ (resp. θ') the absolute coordinate on $\mathcal{P}$ (resp. $\mathcal{P}'$) determined by the choice of M, N (resp. $\widetilde{M}, \widetilde{N}$). If the function σ is assumed to exist, just take g of 11.6.6 to be the projectivity which has equation $\theta' = \sigma(\theta)$, and for the converse define σ from the equation of g. □

Theorem 11.6.6 characterizes the projective equivalence of regular pencils of quadrics in terms of a geometric object – the characteristic system – intrinsically related to each pencil. By contrast, the characterization of 11.6.7 is given in numerical terms, using the absolute coordinates and the Segre symbols of the degenerate quadrics. The former are not determined by the pencil; their indeterminacy is described next:

Proposition 11.6.8. *Assume that $\lambda_1, \dots, \lambda_m$ are the absolute coordinates of the degenerate quadrics of a pencil $\mathcal{P}$ relative to a choice of matrices M, N of quadrics spanning P, the second one non-degenerate. Then complex numbers $\bar{\lambda}_1, \dots, \bar{\lambda}_m$ are absolute coordinates of the same degenerate quadrics relative to a second choice of matrices $\bar{M}$, $\bar{N}$ of quadrics spanning $\mathcal{P}$, the second one being also non-degenerate, if and only if there is a homographic function*

$$\sigma(t) = \frac{at + b}{ct + d}, \quad ad - bc \neq 0,$$

such that $c\lambda_i + d \neq 0$ and $\bar{\lambda}_i = \sigma(\lambda_i)$ for $i = 1, \dots, m$.

Proof. If σ is given as in the claim, take $\bar{M} = aM - bN$ and $\bar{N} = -cM + dN$. The corresponding quadrics span the pencil because $ad - bc \neq 0$, while the inequalities $c\lambda_i + d \neq 0$ assure that $\bar{N} = -cM + dN$ is not proportional to any of the $M + \lambda_i N$, and so the quadric defined by $\bar{N}$ is different from any of the degenerate quadrics of $\mathcal{P}$. Then a straightforward computation shows that the matrices $M + \lambda_i N$ and $\bar{M} + \sigma(\lambda_i)\bar{N}$ are proportional, and therefore the $\sigma(\lambda_i)$, for $i = 1, \dots, m$, are the absolute coordinates of the degenerate quadrics of $\mathcal{P}$ relative to the choice of $\bar{M}$ and $\bar{N}$.

Conversely, if the second choice of matrices is $\bar{M} = \alpha M + \beta N$, $\bar{N} = \gamma M + \delta N$, take σ as above, with

$$a = \alpha, \quad b = -\beta, \quad c = -\gamma \quad \text{and} \quad d = \delta.$$

Then $ad - bc = \alpha\delta - \beta\gamma \neq 0$ because $\bar{M}$ and $\bar{N}$ are assumed to be linearly independent. The quadric with matrix $\bar{N}$ being non-degenerate, no one of the matrices $M + \lambda_i N$ is proportional to $\bar{N} = -cM + dN$, which assures $c\lambda_i + d \neq 0$ as required. Again a direct checking shows that the matrices $M + \lambda_i N$ and $\bar{M} + \sigma(\lambda_i)\bar{N}$ are proportional for $i = 1, \dots, m$, after which the proof is complete. □

In the sequel, we call ordered sequences of m complex numbers $\lambda_1, \dots, \lambda_m$ and $\bar{\lambda}_1, \dots, \bar{\lambda}_m$ *homographic* if and only if they satisfy the condition of 11.6.8, namely if and only if there is a homographic function σ as in 11.6.8 whose values at $\lambda_1, \dots, \lambda_m$ are finite and equal to $\bar{\lambda}_1, \dots, \bar{\lambda}_m$, in this order. It is straightforward to check that being homographic is an equivalence relation on $\mathbb{C}^m$.

Proceeding as for collineations, we may associate to each regular pencil of quadrics $\mathcal{P}$ the class of all sets

$$\{(\lambda_1; \nu_{1,1}, \dots, \nu_{1,h_1}), \dots, (\lambda_m; \nu_{m,1}, \dots, \nu_{m,h_m})\},$$

where $\lambda_1, \dots, \lambda_m$ are finite absolute coordinates of the degenerate quadrics of $\mathcal{P}$ relative to a reference of $\mathcal{P}$ (or to a choice of matrices M, N as above), and

$(\nu_{i,1},\dots,\nu_{i,h_i})$ is the Segre symbol of the degenerate quadric with coordinate λ_i. We will represent this class by

$$\left|(\lambda_1;\nu_{1,1},\dots,\nu_{1,h_1}),\dots,(\lambda_m;\nu_{m,1},\dots,\nu_{m,h_m})\right| \tag{11.12}$$

and call it the *characteristic* of $\mathcal{P}$.

Remark 11.6.9. By the definition of characteristic and 11.6.8, the complex numbers $\lambda_1,\dots,\lambda_m$ appearing in (11.12) are determined up to replacement with homographic $\bar{\lambda}_1,\dots,\bar{\lambda}_m$. Therefore, up to three of them may be given already fixed values, usually taken among 0, 1, -1.

After the definition of characteristic, Corollary 11.6.7 obviously translates into:

Corollary 11.6.10. *Regular pencils of quadrics of complex projective spaces of the same dimension are projectively equivalent if and only if they have the same characteristic.*

Proposition 11.6.11. *If the characteristic of an arbitrary regular pencil of quadrics of $\mathbb{P}_n$ is written in the form of* (11.12)*, then it holds that*

(1) $1\le m\le n+1$,

(2) $\lambda_i\ne\lambda_j$ *for* $i\ne j$, $i,j=1,\dots,m$,

(3) $\nu_{i,1}\ge\dots\ge\nu_{i,h_i}>0$ *and* $h_i\ge 1$ *for* $i=1,\dots,m$,

(4) $\sum_{i=1}^{m}\sum_{s=1}^{h_i}\nu_{i,s}=n+1$, *and*

(5) *either* $m>1$ *or* $h_1<n+1$.

Proof. Conditions (1) to (3) directly follow from the definition of the characteristic. Condition (4) results from 11.6.1, as the degree of any characteristic polynomial is $n+1$. Regarding condition (5), assume otherwise that $m=1$ and $h_1\ge n+1$. Then $h_1=n+1$ because $h_1>n+1$ is not allowed by condition (4). The list of elementary divisors consists thus of $\lambda-\lambda_1$ repeated $n+1$ times. This forces the 0-th invariant factor to be $I_0=\lambda-\lambda_1$ and therefore all entries of the characteristic matrix to be multiples of $\lambda-\lambda_1$. It is thus

$$M+\lambda N=(\lambda-\lambda_1)A$$

for a certain matrix A. It turns out that $M=\lambda_1 A=\lambda_1 N$ against the hypothesis of being M,N matrices of two different quadrics. □

Next we will see that any set

$$\{(\lambda_1;\nu_{1,1},\dots,\nu_{1,h_1}),\dots,(\lambda_m;\nu_{m,1},\dots,\nu_{m,h_m})\},$$

whose entries satisfy the conditions of 11.6.11, actually represents the characteristic of a regular pencil of quadrics of a complex $\mathbb{P}_n$. To this end let us consider, for any complex number α and any positive integer ν, the ν-dimensional symmetric matrix

$$S(a,\nu) = \begin{pmatrix} & & & a \\ & & a & 1 \\ & \cdot^{\cdot^{\cdot}} & \cdot^{\cdot^{\cdot}} & \\ a & 1 & & \end{pmatrix},$$

with all entries on the secondary diagonal equal to a, those immediately below equal to one, and the remaining ones equal to zero. Take also

$$T(\nu) = \begin{pmatrix} & & -1 \\ & -1 & \\ \cdot^{\cdot^{\cdot}} & & \\ -1 & & \end{pmatrix},$$

still ν-dimensional and whose only non-zero entries are those on the secondary diagonal, all equal to -1.

Assume we are given complex numbers λ_i, $i = 1,\dots,m$, and, for each i, positive integers $\nu_{i,s}$, $s = 1,\dots,h_i$, satisfying the conditions of 11.6.11 above. Consider the square matrix

$$S = \begin{pmatrix} S(\lambda_1,\nu_{1,1}) & & & & & \\ & \ddots & & & & \\ & & S(\lambda_1,\nu_{1,h_1}) & & & \\ & & & \ddots & & \\ & & & & S(\lambda_m,\nu_{m,1}) & & \\ & & & & & \ddots & \\ & & & & & & S(\lambda_m,\nu_{m,h_m}) \end{pmatrix}, \tag{11.13}$$

composed of $h_1 + \dots + h_m$ blocks $S(\lambda_i,\nu_{i,j})$ centred on the diagonal, all entries outside them being equal to zero. It is $(n+1)$-dimensional due to condition (4) of 11.6.11. Similarly, take

$$T = \begin{pmatrix} T(\nu_{1,1}) & & & & & \\ & \ddots & & & & \\ & & T(\nu_{1,h_1}) & & & \\ & & & \ddots & & \\ & & & & T(\nu_{m,1}) & & \\ & & & & & \ddots & \\ & & & & & & T(\nu_{m,h_m}) \end{pmatrix}, \tag{11.14}$$

also square and $(n+1)$-dimensional. Note that both S and T are symmetric and T is regular. We have:

Lemma 11.6.12. *The list of elementary divisors of the matrix $S+\lambda T$ is:*

$$(\lambda-\lambda_1)^{\nu_{1,1}},\dots,(\lambda-\lambda_1)^{\nu_{1,h_1}},\dots,(\lambda-\lambda_m)^{\nu_{m,1}},\dots,(\lambda-\lambda_m)^{\nu_{m,h_m}}.$$

Proof. The matrix $-T$ is invertible and it is straightforward to check that the product $(S+\lambda T)(-T)$ equals the characteristic matrix of the Jordan matrix defined by the equality (11.8). These polynomial matrices being thus equivalent by 11.2.9, they have the same list of elementary divisors by 11.2.6. Since this list has been already computed for the latter in 11.4.3 and is as claimed, the proof is complete. □

We are now able to complete the description of all projective types of regular pencils of quadrics:

Theorem 11.6.13. *For any complex numbers λ_i, $i=1,\dots,m$, and positive integers $\nu_{i,s}$, $i=1,\dots,m$, $s=1,\dots,h_i$, satisfying the conditions of* 11.6.11*, there is a projective type of regular pencils of quadrics of n-dimensional complex projective spaces whose representatives have characteristic*

$$\big|(\lambda_1;\nu_{1,1},\dots,\nu_{1,h_1}),\dots,(\lambda_m;\nu_{m,1},\dots,\nu_{m,h_m})\big|.$$

Proof. Note first that the matrices S,T are not proportional because the case $m=1$, $h_1=n+1$ has been excluded. As noted above, T is regular; we may thus take any complex projective space $\mathbb{P}_n$, a reference on it and consider the regular pencil $\mathcal{P}$ spanned by the quadrics with matrices S and T, defined above. Then the characteristic of $\mathcal{P}$ is as claimed due to 11.6.12 and by 11.6.10 all pencils projectively equivalent to $\mathcal{P}$ have the same characteristic, which ends the proof. □

Remark 11.6.14. The projective type of 11.6.13 is unique by 11.6.6. A representative of it is explicitly given in the proof of 11.6.13.

Again in close similarity with the case of collineations, the above matrices provide a particularly simple presentation of any regular pencil of quadrics.

Theorem 11.6.15. *If a regular pencil of quadrics $\mathcal{P}$, of a complex projective space $\mathbb{P}_n$, has characteristic*

$$\big|(\lambda_1;\nu_{1,1},\dots,\nu_{1,h_1}),\dots,(\lambda_m;\nu_{m,1},\dots,\nu_{m,h_m})\big|,$$

then there is a reference Ω of $\mathbb{P}_n$ relative to which the matrix $S+\lambda T$, S and T defined by (11.13) *and* (11.14)*, is a characteristic matrix of $\mathcal{P}$.*

Proof. According to the definition of characteristic, there is a reference Δ of $\mathbb{P}_n$ and, relative to Δ, matrices M, N of quadrics spanning $\mathcal{P}$, the second one non-degenerate, such that the characteristic matrix $M + \lambda N$ has list of elementary divisors

$$(\lambda - \lambda_1)^{\nu_{1,1}}, \dots, (\lambda - \lambda_1)^{\nu_{1,h_1}}, \dots, (\lambda - \lambda_m)^{\nu_{m,1}}, \dots, (\lambda - \lambda_m)^{\nu_{m,h_m}}.$$

The list of elementary divisors of $S + \lambda T$ being the same, by 11.6.3 there is an invertible constant matrix P for which

$$P^t M P = S, \quad P^t N P = T.$$

The matrix P may be taken as the matrix changing coordinates relative to a second reference Δ' into coordinates relative to Δ (2.3.3). Then it is enough to take $\Omega = \Delta'$ and apply the rule for changing coordinates 6.1.12. □

The matrix $S + \lambda T$ of 11.6.15 is called a *reduced* characteristic matrix (or a *reduced form*) of the pencil. Note that it is determined up to reordering of the groups of blocks corresponding to each λ_i and replacement of the sequence $\lambda_1, \dots, \lambda_m$ with a homographic one. Taking it up to these undeterminacies, $S + \lambda T$ is often called *the* reduced characteristic matrix of the pencil. Its corresponding equation

$$(x)^t (S + \lambda T)(x) = 0, \quad (x) = (x_0, \dots, x_n),$$

intended to give an equation of each quadric in the pencil, while λ takes values in $\mathbb{C} \cup \{\infty\}$, is called the *reduced form* or the *reduced representation* of the pencil. It is often written

$$(x)^t (\alpha S + \beta T)(x) = 0,$$

using homogeneous parameters α, β instead of the absolute parameter λ.

If Q and Q' are two non-degenerate quadrics of $\mathbb{P}_n$, then composing the polarity relative to Q with the inverse of the polarity relative to Q' gives rise to a collineation $g_{Q,Q'}$ of $\mathbb{P}_n$ which is said to be *associated* to the pair Q, Q'. If Q, Q' have matrices M, N relative to a certain reference Δ of $\mathbb{P}_n$, then a matrix of $g_{Q,Q'}$ relative to Δ is $N^{-1}M$ (see 6.5.1). The next lemma, together with 11.4.6, may be used for computing the characteristics of regular pencils of quadrics.

Lemma 11.6.16. *Assume that $\mathcal{P}$ is a regular pencil of quadrics spanned by two non-degenerate quadrics Q, Q'. If the collineation associated to Q, Q' has characteristic*

$$\big\langle (\lambda_1; \nu_{1,1}, \dots, \nu_{1,h_1}), \dots, (\lambda_m; \nu_{m,1}, \dots, \nu_{m,h_m}) \big\rangle,$$

then $\mathcal{P}$ has characteristic

$$\big| (\lambda_1; \nu_{1,1}, \dots, \nu_{1,h_1}), \dots, (\lambda_m; \nu_{m,1}, \dots, \nu_{m,h_m}) \big|.$$

Proof. Fix a reference and take matrices M of Q and N of Q', after which $-N^{-1}M$ is a matrix of the associated collineation $g_{Q,Q'}$. Then $M + \lambda N$ is a characteristic matrix of $\mathcal{P}$ and $-N^{-1}M - \lambda \mathbf{1}_{n+1}$ is a characteristic matrix of $g_{Q,Q'}$. They are equivalent polynomial matrices because

$$-N(-N^{-1}M - \lambda \mathbf{1}_{n+1}) = M + \lambda N,$$

and therefore have the same list of elementary divisors. By 11.3.4 the list of elementary divisors of $-N^{-1}M - \lambda \mathbf{1}_{n+1}$ has the form

$$(\lambda - \rho\lambda_1)^{\nu_{1,1}}, \dots, (\lambda - \rho\lambda_1)^{\nu_{m,h_1}}, \dots, (\lambda - \rho\lambda_m)^{\nu_{m,1}}, \dots, (\lambda - \rho\lambda_m)^{\nu_{m,h_m}}.$$

It follows that

$$\big\{(\rho\lambda_1; \nu_{1,1}, \dots, \nu_{1,h_1}), \dots, (\rho\lambda_m; \nu_{m,1}, \dots, \nu_{m,h_m})\big\},$$

is a representative of the characteristic of $\mathcal{P}$ and therefore so is

$$\big\{(\lambda_1; \nu_{1,1}, \dots, \nu_{1,h_1}), \dots, (\lambda_m; \nu_{m,1}, \dots, \nu_{m,h_m})\big\},$$

by 11.6.8. □

The projective types of regular pencils of conics have been already described in Section 9.6 by elementary means. They are relisted in the next example according to their characteristics.

Example 11.6.17 (Regular pencils of conics). The characteristic and the reduced form of a representative of each projective type of regular pencils of conics are listed below. Since there are at most three absolute coordinates λ_i in each characteristic, as allowed by 11.6.9, they are taken among 0, 1, -1.

(i) General pencil:

$$|(0; 1), (1; 1), (-1; 1)|, \quad x_1^2 - x_2^2 - \lambda(x_0^2 + x_1^2 + x_2^2) = 0.$$

(ii) Bitangent pencil:

$$|(0; 1, 1), (1; 1)|, \quad x_2^2 - \lambda(x_0^2 + x_1^2 + x_2^2) = 0.$$

(iii) Simple contact pencil:

$$|(0; 2), (1; 1)|, \quad x_1^2 + x_2^2 - \lambda(2x_0x_1 + x_2^2) = 0.$$

(iv) Double contact pencil:

$$|(0; 3)|, \quad 2x_1x_2 - \lambda(2x_0x_2 + x_1^2) = 0.$$

(v) Triple contact pencil:

$$|(0;2,1)|, \quad x_1^2 - \lambda(2x_0x_1 + x_2^2) = 0.$$

Before listing the projective types of regular pencils of quadrics of complex three-spaces in Example 11.6.18 below, it will be useful to make some comments on the base loci of these pencils. By 9.5.1 and 9.5.5, the base points of a regular pencil $\mathcal{P}$, of quadrics of $\mathbb{P}_3$, may always be presented as the intersection points of two non-degenerate quadrics $Q_1, Q_2 \in \mathcal{P}$. Seen in this way, the base points form what is called an *algebraic space curve of the fourth degree*, or *space quartic*: the adjective algebraic comes from the fact that the curve may be defined by polynomial equations, namely those of Q_1 and Q_2; the degree is said to be four because the number of intersection points of such a curve and a plane, if finite, is four (the reader may check this fact using 9.5.4 and the descriptions of Section 9.6; intersection points need to be counted with suitable multiplicities to cover all cases). However, the intersections of two non-degenerate quadrics are not the only space quartics: there are curves of the fourth degree which are contained in a unique quadric, and which, therefore, need to be obtained as a part (not the whole) of the intersection of this quadric and a surface of higher degree (the curve is then called a *partial intersection*). In order to distinguish them from the latter, the quartics which occur as base loci of pencils of quadrics are called *complete intersection quartics*. The complete intersection quartics composed of lines, conics or skew cubics are fairly easy to describe. The other are called *irreducible*: their study belongs to algebraic geometry and is beyond the scope of this book. In the example below we will limit ourselves to mention their appearance as base loci, calling them just quartics for simplicity. In all three cases in which they appear, namely (i) (iii) and (ix), it may be proved that the the pencil is composed of the quadrics containing the base locus; proving this here is, however, beyond our means.

Example 11.6.18 (Regular pencils of three-space quadrics). Below are listed the characteristic and the reduced form of a representative of each of the projective types of regular pencils of quadrics of complex three-dimensional projective spaces. Descriptions of the degenerate quadrics and base points of the corresponding pencils are also included, as well as an explicit mention of the cases in which the pencil is composed of the quadrics containing the base locus. Checking the details on the explicitly given representatives is left to the reader, which may make use of 11.6.5 and note that in no case may two different quadrics of a regular pencil share a double point, as then, by 9.5.1, all quadrics in the pencil would have that point as a double one. Item (i) below describes infinitely many projective types, depending on the values of δ. The other items describe a single projective type each. Again, up to three of the absolute coordinates appearing in the characteristics are taken among $0, 1, -1$, as allowed by 11.6.9. The heading of each item is intended just to identify it, not to give a complete description of the corresponding pencils.

(i) Quadrics through a non-singular quartic:

$$|(0;1),(1;1),(-1;1),(\delta,1)|, \quad \delta \in \mathbb{C}-\{0,1,-1\},$$
$$x_1^2 - x_2^2 + \delta x_3^2 - \lambda(x_0^2 + x_1^2 + x_2^2 + x_3^2) = 0.$$

The degenerate quadrics are four ordinary cones, no vertex of which is a base point. The base locus is a quartic curve.

(ii) Quadrics through a pair of non-coplanar conics sharing two points:

$$|(0;1,1),(1;1),(-1;1)|, \quad x_2^2 - x_3^2 - \lambda(x_0^2 + x_1^2 + x_2^2 + x_3^2) = 0.$$

The degenerate quadrics are a plane-pair $\pi_1 + \pi_2$ and two ordinary cones Q_1, Q_2. The intersection of the plane-pair and either of the cones, say Q_1, is composed of the conics $C_1 = \pi_1 \cap Q_1$ and $C_2 = \pi_2 \cap Q_1$. By 9.5.1, the set of base points is $(\pi_1 + \pi_2) \cap Q_1 = C_1 \cup C_2$. Both C_1 and C_2 are non-degenerate and taking $r = \pi_1 \cap \pi_2$, $C_1 \cap r = Q_1 \cap r = C_2 \cap r$ is a pair of distinct points. These pencils already appeared in Example 9.5.21 and, as seen there, they are composed of the quadrics containing $C_1 \cup C_2$.

(iii) Quadrics through a nodal quartic:

$$|(0;2),(1;1),(-1;1)|, \quad x_1^2 + x_2^2 - x_3^2 - \lambda(2x_0x_1 + x_2^2 + x_3^2) = 0.$$

The degenerate quadrics are three ordinary cones. The vertex p of one of them, say Q_1, is a base point. Therefore, by 9.10.1, all quadrics of the pencil but Q_1 have the same tangent plane at p. The base locus is again a quartic curve, which, unlike the quartic of case (i), has a singular point in which the curve transversally intersects itself (these singular points are called *nodes*).

(iv) Quadrics through a skew quadrilateral:

$$|(0;1,1),(1;1,1)|, \quad x_2^2 + x_3^2 - \lambda(x_0^2 + x_1^2 + x_2^2 + x_3^2) = 0.$$

The degenerate quadrics are two plane-pairs; their intersection is composed of four lines $\ell_1, \ell_2, \ell_3, \ell_4$ that are the sides of a skew quadrilateral. The set of base points is $\ell_1 \cup \ell_2 \cup \ell_3 \cup \ell_4$ and, as already seen in Example 9.5.20, the quadrics containing it compose the pencil.

(v) Quadrics through a conic and two coplanar lines through different points of the conic:

$$|(0;2),(1;1,1)|, \quad x_1^2 + x_2^2 + x_3^2 - \lambda(2x_0x_1 + x_2^2 + x_3^2) = 0.$$

The degenerate quadrics are one plane-pair and one ordinary cone, the vertex of the cone lying on just one of the planes of the pair. The situation is as in case (ii) but for being one of the conics C_1, C_2 a line pair, the other still being non-degenerate. Also this case appeared in Example 9.5.21. Still the set of base points is $C_1 \cup C_2$ and the pencil is the set of quadrics going through $C_1 \cup C_2$.

(vi) Quadrics through a twisted cubic and one of its chords:

$$|(0;2),(1;2)|, \quad x_1^2 + 2x_2x_3 + x_3^2 - \lambda(2x_0x_1 + 2x_2x_3) = 0.$$

The degenerate quadrics are two ordinary cones sharing one generator ℓ. The set of base points is composed of ℓ and a twisted cubic Γ that intersects ℓ at the vertices of the cones. As already proved in 9.12.11, the pencil is the set of quadrics going through $\Gamma \cup \ell$.

(vii) Quadrics tangent along a non-degenerate conic:

$$|(0;1,1,1),(1;1)|, \quad x_3^2 - \lambda(x_0^2 + x_1^2 + x_2^2 + x_3^2) = 0.$$

The degenerate quadrics are a double plane 2π and an ordinary cone Q_1. The base points are those of the conic $C = Q_1 \cap \pi$, which is non-degenerate because the vertex of Q_1 cannot lie on π. As already seen in Example 9.5.22 the members of the pencil are the quadrics tangent to Q_1 along C.

(viii) Quadrics through two non-coplanar conics touching at a point:

$$|(0;2,1),(1;1)|, \quad x_1^2 + x_3^2 - \lambda(2x_0x_1 + x_2^2 + x_3^2) = 0.$$

The degenerate quadrics are a plane-pair $\pi_1 + \pi_2$ and an ordinary cone Q_1, the line $\ell = \pi_1 \cap \pi_2$ being properly tangent to Q_1. As in case (ii), both the conics $C_1 = Q_1 \cap \pi_1$ and $C_2 = Q_1 \cap \pi_2$ are non-degenerate, but in this case they are tangent to ℓ at the same point. Also these pencils appeared in Example 9.5.21 and, as seen there, their members are the quadrics containing $C_1 \cup C_2$.

(ix) Quadrics through a cuspidal quartic:

$$|(0;3),(1;1)|, \quad 2x_1x_2 + x_3^2 - \lambda(2x_0x_2 + x_1^2 + x_3^2) = 0.$$

The degenerate quadrics are two ordinary cones. The base locus is a quartic curve which, like the one in case (iii), has a singular point. This time, however, the singular point belongs to a different type of singular points which are called *cusps*.

(x) Quadrics tangent along a pair of lines:

$$|(0;2,1,1)|, \quad x_1^2 - \lambda(2x_0x_1 + x_2^2 + x_3^2) = 0.$$

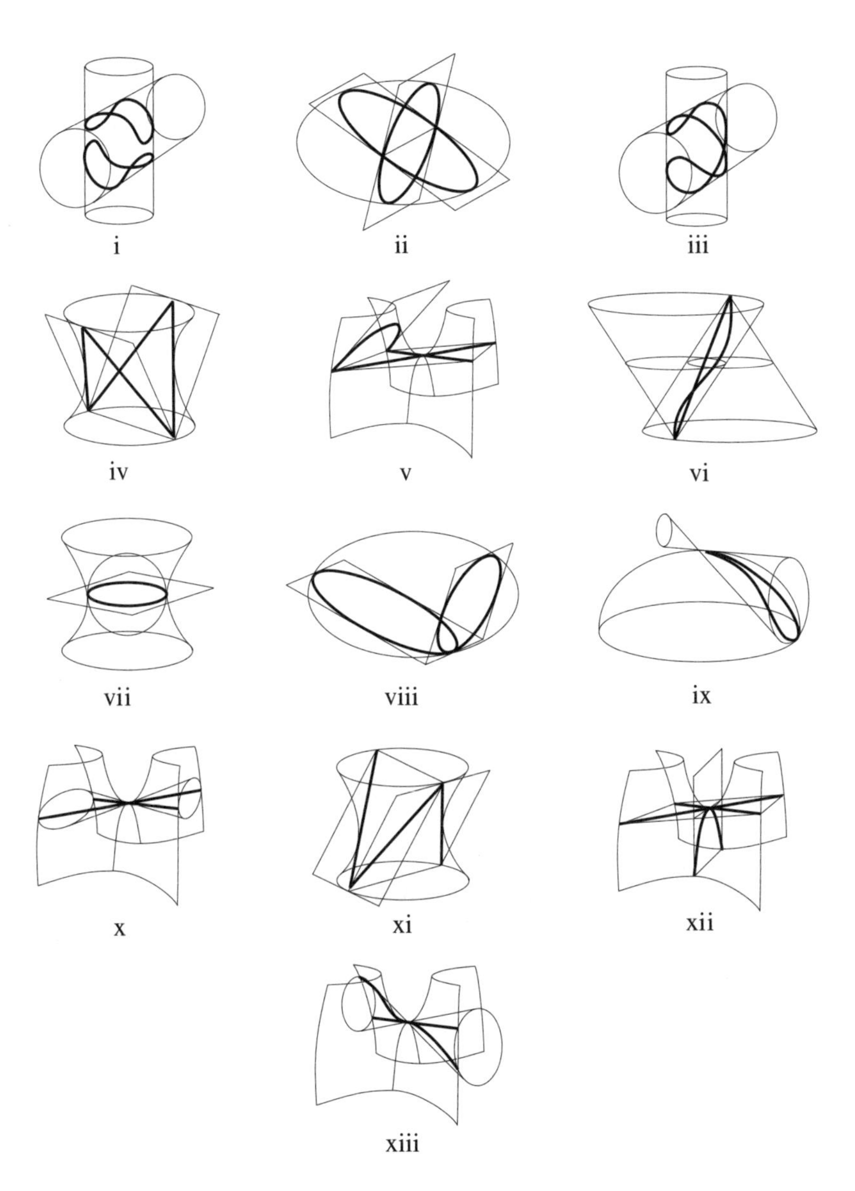

Figure 11.1. Sketches of regular pencils of quadrics of $\mathbb{P}_3$, in the situations described in the corresponding items of Example 11.6.18.

The only degenerate is a double plane 2π with π tangent at a point to one (and hence to all) of the other quadrics in the pencil. The base points are thus those of the line-pair intersection of π and any of the non-degenerate quadrics of the pencil. By 9.10.1, these quadrics have the same tangent plane at any of the points of the pair of lines.

(xi) Quadrics through three lines, tangent along one of them:

$$|(0; 2, 2)|, \quad x_1^2 + x_3^2 - \lambda(2x_0x_1 + 2x_2x_3) = 0.$$

The only degenerate is a plane-pair $\pi_1 + \pi_2$. The line $\ell = \pi_1 \cap \pi_2$ is contained in one, and hence in all of the other quadrics of the pencil. If Q is anyone of them, then $Q \cap \pi_i = \ell + \ell_i$, $i = 1, 2$, the lines ℓ_1, ℓ_2 are skew and meet ℓ (see Section 8.3), and the set of base points is $\ell \cup \ell_1 \cup \ell_2$. Any $p \in \ell$ being double for $\pi_1 + \pi_2$, by 9.10.1, all the other quadrics in the pencil have the same tangent plane at p.

(xii) Quadrics through a conic and two lines meeting on the conic:

$$|(0; 3, 1)|, \quad 2x_1x_2 - \lambda(2x_0x_2 + x_1^2 + x_3^2) = 0.$$

The only degenerate is a plane-pair $\pi_1 + \pi_2$. The situation is as in (viii) but for one of the conics, C_1, being a pair of lines whose double point is the contact point of the other, C_2, with $\ell = \pi_1 \cap \pi_2$. Again this case is part of Example 9.5.21 and, as seen there, the pencil is composed of the quadrics containing $C_1 \cup C_2$.

(xiii) Quadrics through a twisted cubic and one of its tangents:

$$|(0; 4)|, \quad 2x_1x_3 + x_2^2 - \lambda(2x_0x_3 + 2x_1x_2) = 0.$$

The only degenerate is an ordinary cone, one of whose generators ℓ is contained in all the quadrics of the pencil. The set of base points is $\Gamma \cup \ell$ where Γ is a twisted cubic tangent to ℓ at the vertex of the cone. Corollary 9.12.11 still applies and shows the pencil is composed of the quadrics going through $\Gamma \cup \ell$.

11.7 Projective classification of correlations

In this section we study the projective equivalence of correlations of complex projective spaces, and partially describe their projective classification. Again, the base field is assumed to be the complex one throughout.

Correlations of projective spaces were defined in Section5.8 as being projectivities mapping points to hyperplanes of the same projective space; in other words, they are projectivities

$$g\colon \mathbb{P}_n \longrightarrow \mathbb{P}_n^\vee$$

between a projective space $\mathbb{P}_n$ and its dual $\mathbb{P}_n^\vee$. The image of the above g by a projectivity

$$f : \mathbb{P}_n \longrightarrow \mathbb{P}'_n$$

was defined to be

$$f(g) = (f^\vee)^{-1} \circ g \circ f^{-1}$$

and therefore correlations g, g', of projective spaces $\mathbb{P}_n$ and $\mathbb{P}'_n$ respectively, are projectively equivalent if and only if there is a projectivity $f : \mathbb{P}_n \to \mathbb{P}'_n$ for which

$$g' = (f^\vee)^{-1} \circ g \circ f^{-1}.$$

Projective equivalence of correlations may be easily characterized in terms of their matrices:

Lemma 11.7.1. *Assume that correlations g and g' have matrices M and M' relative to references Δ and Δ' of their corresponding spaces $\mathbb{P}_n$ and $\mathbb{P}'_n$. Then g and g' are projectively equivalent if and only if there is a regular $(n+1) \times (n+1)$-matrix P for which*

$$M' = P^t M P.$$

Proof. The sufficiency of the condition is clear by just taking the projectivity $f : \mathbb{P}_n \to \mathbb{P}'_n$ which has matrix P^{-1} relative to the references Δ, Δ'. Conversely, if $f(g) = g'$ for a projectivity $f : \mathbb{P}_n \to \mathbb{P}'_n$ and T is a matrix of f relative to Δ, Δ', then it is

$$M' = \rho (T^{-1})^t M T^{-1},$$

for a certain non-zero complex number ρ. It suffices to take $P = \sqrt{\rho} T^{-1}$. □

For any matrix M of a correlation g, the degree-one polynomial matrix

$$\mathcal{M} = M - \lambda M^t$$

will be called a *characteristic matrix* of g. It clearly is a regular polynomial matrix. Our main point is:

Proposition 11.7.2. *Correlations g, g', of complex projective spaces of the same dimension, with characteristic matrices $\mathcal{M}$, $\mathcal{M}'$, are projectively equivalent if and only if $\mathcal{M}$ and $\mathcal{M}'$ are equivalent polynomial matrices.*

Proof. Assume that $\mathcal{M} = M - \lambda M^t$ and $\mathcal{M}' = M' - \lambda (M')^t$, with M and M' matrices of g and g', respectively. If g and g' are projectively equivalent, then, by 11.7.1 above, there is a regular matrix P for which $M' = P^t M P$ and so, also $(M')^t = P^t M^t P$. It follows that

$$\mathcal{M}' = M' - \lambda (M')^t = P^t (M - \lambda M^t) P = P^t \mathcal{M} P$$

and $\mathcal{M}$ and $\mathcal{M}'$ are equivalent by 11.2.9.

Conversely, assume that $\mathcal{M}$ and $\mathcal{M}'$ are equivalent and so, by 11.2.11, that there are regular $(n+1)$-dimensional constant matrices A and B for which

$$\mathcal{M}' = A\mathcal{M}B,$$

that is,

$$M' = AMB \quad \text{and} \quad (M')^t = AM^tB,$$

or, equivalently,

$$M' = AMB \quad \text{and} \quad M' = B^tMA^t,$$

from which it follows that

$$AMB = B^tMA^t.$$

The same arguments used in the proof of 11.6.3 apply here to show the existence of a regular matrix D such that $DM = MD^t$ and $D^2 = (B^t)^{-1}A$. Then, still arguing as in the proof of 11.6.3, it holds that

$$M' = B^tDMD^tB$$

and therefore it is enough to use 11.7.1 with $P = D^tB$. □

Since any correlation is projectively equivalent to itself, Proposition 11.7.2 assures in particular that any two characteristic matrices of the same correlation are equivalent and therefore have the same list of elementary divisors. We may thus associate to each correlation g the list of elementary divisors of any of its characteristic matrices and call it the *list of elementary divisors of* g. Then, 11.7.2 directly gives:

Theorem 11.7.3. *Correlations g, g', of complex projective spaces of the same dimension, are projectively equivalent if and only if they have the same list of elementary divisors.*

Assume that the list of elementary divisors of a correlation g is written

$$(\lambda-\lambda_1)^{\nu_{1,1}},\dots,(\lambda-\lambda_1)^{\nu_{1,h_1}},\dots,(\lambda-\lambda_m)^{\nu_{m,1}},\dots,(\lambda-\lambda_m)^{\nu_{m,h_m}},$$

where, as in former cases, $m > 0$, $\lambda_i \in \mathbb{C}$, $\lambda_i \neq \lambda_j$ for $i \neq j$, $h_i > 0$ and the positive integers $\nu_{i,s}$ are taken such that $\nu_{i,1} \geq \cdots \geq \nu_{i,h_i}$.

The complex numbers $\lambda_1,\dots,\lambda_m$, are projective invariants of g by 11.7.3. They will be called the *characteristic roots* of g. If M is any matrix of g, by 11.6.1, the characteristic roots $\lambda_1,\dots,\lambda_m$ are the roots of the polynomial

$$\det\mathcal{M} = \det(M-\lambda M^t).$$

In particular, $\lambda_i \neq 0$ for $i = 1,\dots,n$, because the matrix M is regular. Furthermore, also by 11.6.1, each λ_i has multiplicity $\nu_{i,1}+\cdots+\nu_{i,h_i}$ as a root of $\det\mathcal{M}$.

It follows that det $\mathcal{M}$ is determined by g up to a constant factor, which may be also checked by direct computation. In the sequel det $\mathcal{M}$ will be called a *characteristic polynomial* of g. As for pencils of quadrics, any characteristic polynomial of a correlation has degree $n+1$ and so, again by 11.6.1, $\sum_{i=1}^{m}\sum_{s=1}^{h_i}\nu_{i,s}=n+1$.

The decreasing sequence of exponents of $\lambda-\lambda_i$ in the list of elementary divisors will be called, as in former cases, the *Segre symbol* of λ_i. Collecting all pairs composed of a characteristic root and its Segre symbol gives rise to the set

$$\kappa(g)=\{(\lambda_1;\nu_{1,1},\dots,\nu_{1,h_1}),\dots,(\lambda_m;\nu_{m,1},\dots,\nu_{m,h_m})\}, \tag{11.15}$$

which encloses all the information provided by the list of elementary divisors. We will call $\kappa(g)$ the *characteristic* of the correlation g. A direct consequence of this definition and 11.7.3 is:

Corollary 11.7.4. *Correlations g, g', of complex projective spaces of the same dimension, are projectively equivalent if and only if they have equal characteristic.*

The reader may note that, unlike the cases of collineations and pencils of quadrics, the characteristic roots $\lambda_1,\dots,\lambda_m$ of a correlation g are fully determined by g. This is the reason why the characteristic of g has been taken to be just the above set, and not its class modulo an equivalence originated by the indeterminacy of $\lambda_1,\dots,\lambda_m$, as in former cases.

Example 11.7.5. Assume that h is a null-system of $\mathbb{P}_n$. Any matrix M of h being skew-symmetric, h has characteristic matrix

$$M-\lambda M^t=(1+\lambda)M$$

and so its only characteristic root is -1. An easy computation from the characteristic matrix shows the list of elementary divisors of h to be composed of $\lambda+1$ repeated $n+1$ times. This in particular re-proves that all the null-systems of complex projective spaces of a fixed dimension have the same projective type (5.8.14). The same holds for polarities, but for $\lambda-1$ being their only elementary divisor.

Assume that g is a correlation of $\mathbb{P}_n$ which has matrix M relative to a reference Δ. As introduced in Section 5.8, its dual $g^\bullet$ is also a correlation of $\mathbb{P}_n$ and it has matrix M^t relative to Δ. The composition $\hat{g}=(g^\bullet)^{-1}g$ is then a collineation of $\mathbb{P}_n$ with matrix $(M^t)^{-1}M$ relative to Δ. We will call $\hat{g}$ the collineation *associated* to g. By 5.8.11, the correlations whose associated collineation is the identical map are the polarities and the null-systems. We have:

Proposition 11.7.6. *If $f:\mathbb{P}_n\to\mathbb{P}'_n$ is any projectivity, then $f(\hat{g})=\widehat{f(g)}$. In particular projectively equivalent correlations have projectively equivalent associated collineations.*

Proof. We have

$$\begin{aligned} f(\hat{g}) &= f \circ \hat{g} \circ f^{-1} \\ &= f \circ (g^{\bullet})^{-1} \circ g \circ f^{-1} \\ &= f \circ (g^{\bullet})^{-1} \circ f^{\vee} \circ (f^{\vee})^{-1} \circ g \circ f^{-1} \\ &= ((f^{\vee})^{-1} \circ g^{\bullet} \circ f^{-1})^{-1} \circ (f^{\vee})^{-1} \circ g \circ f^{-1} \\ &= \big(f(g^{\bullet})\big)^{-1} \circ f(g). \end{aligned}$$

Then, by 5.8.9, $f(g^{\bullet}) = f(g)^{\bullet}$ and so

$$f(\hat{g}) = \big(f(g)^{\bullet}\big)^{-1} \circ f(g) = \widehat{f(g)},$$

as claimed. The reader may provide a similar proof using matrices. □

It is easy to obtain the characteristic of $\hat{g}$ from the characteristic of g:

Corollary 11.7.7. *The characteristic of a correlation is a representative of the characteristic of its associated collineation.*

Proof. If a correlation g has matrix M, then $(M^t)^{-1}M$ is a matrix of the associated collineation $\hat{g}$ and a characteristic matrix of $\hat{g}$ is

$$(M^t)^{-1}M - \lambda \mathbf{1}_{n+1} = (M^t)^{-1}(M - \lambda M^t),$$

clearly equivalent to the characteristic matrix $M - \lambda M^t$ of g. These characteristic matrices have thus the same list of elementary divisors and the claim follows. □

Remark 11.7.8. We have seen in 11.7.6 that projectively equivalent correlations have projectively equivalent associated collineations. In spite of some wrong claims appearing in old and not so old literature, the converse is not true, essentially because the characteristic of $\hat{g}$ does not determine the characteristic of g. For an easy counterexample, consider a polarity and a null-system of a projective space $\mathbb{P}_n$: they are not projectively equivalent due to 5.8.4, while, as recalled above, they both have the identity map as associated collineation, by 5.8.11. A different counterexample may be found in Exercise 11.16.

Remark 11.7.9. The fact that a characteristic matrix $M - \lambda M^t$ of a correlation g is equivalent to the characteristic matrix $(M^t)^{-1}M - \lambda \mathbf{1}_{n+1}$ of its associated collineation $\hat{g}$, seen in the proof of 11.7.7, allows us to compute the list of the elementary divisors (or the characteristic) of g through 11.4.6. However, in doing so, one has to use just the matrix $(M^t)^{-1}M$ of $\hat{g}$, as using a proportional one would give rise to eigenvalues of the collineation which are proportional, but not necessarily equal, to the characteristic roots of the correlation.

The points $p \in \mathbb{P}_n$ for which $g(p) = g^\bullet(p)$ are the fixed points of the associated collineation $\hat{g}$. They are thus the points of the fundamental varieties of $\hat{g}$, called the *fundamental varieties* of g in the sequel. Since the characteristic roots of g compose a system of eigenvalues of $\hat{g}$, each fundamental variety V_i comes associated to one and only one characteristic root λ_i of g. After fixing a projective reference and a matrix M of g, the points of V_i are those $[x_0, \dots, x_n]$ for which

$$(M - \lambda_i M^t) \begin{pmatrix} x_0 \\ \vdots \\ x_n \end{pmatrix} = \begin{pmatrix} 0 \\ \vdots \\ 0 \end{pmatrix}.$$

The next proposition provides in particular a geometric interpretation of the characteristic roots of a correlation other than a null system or a polarity. The reader may use it to re-prove the projective invariance of the characteristic roots, and may also compare it with 5.7.18.

Proposition 11.7.10. *Assume that g is a correlation of a complex projective space, not a null-system or a polarity, whose characteristic roots and corresponding fundamental varieties are λ_i, V_i, $i = 1, \dots, m$. Denote by I the incidence set of g and take*

$$U = \mathbb{P}_n - V_1 \cup \dots \cup V_m \cup g^{-1}(V_1^*) \cup \dots \cup g^{-1}(V_m^*) \cup I.$$

Then $U \neq \emptyset$ and for any point $p \in U$:

(1) *$g(p)$ and $g^\bullet(p)$ span a pencil $\mathcal{H}_p$ of hyperplanes of $\mathbb{P}_n$.*

(2) *There is a unique hyperplane $H \in \mathcal{H}_p$ containing p. Furthermore $H \neq g(p), g^\bullet(p)$.*

(3) *For each $i = 1, \dots, m$, there is a unique hyperplane $H_i \in \mathcal{H}_p$ containing V_i.*

(4) *$(g(p), g^\bullet(p), H, H_i) = \lambda_i$.*

Proof. Since g is not a null-system or a polarity, $\hat{g}$ is not the identical map and therefore no V_i has dimension n. No V_i being empty either, if

$$T = V_1 \cup \dots \cup V_m \cup g^{-1}(V_1^*) \cup \dots g^{-1}(V_m^*),$$

then, by 2.4.4, we may take $p_1 \in \mathbb{P}_n - T$. Again because g is not a null-system, I is the set of points of a quadric and we may take $p_2 \in \mathbb{P}_n - I$ (by 6.1.5). If $p_1 = p_2$, then $p_1 \in U$ and (1) is proved. Otherwise $\ell = p_1 \vee p_2$ is a line which is not contained in T (because $p_1 \in \ell$) and so has $\ell \cap T$ finite. Similarly ℓ is not contained in I (because $p_2 \in \ell$) and therefore $\ell \cap I$ contains at most two points. It follows that $\ell \cap (T \cup I)$ is a finite set and hence not the whole of ℓ, which proves $U \neq \emptyset$.

Take $p \in U$. Then p does not belong to any fundamental variety, which assures $g(p) \neq g^{\bullet}(p)$ and so $g(p)$ and $g^{\bullet}(p)$ span a pencil of hyperplanes $\mathcal{H}_p$ as claimed in (1).

The existence of a hyperplane $H \in \mathcal{H}_p$ containing p is clear. Since $p \notin I$ and I is also the incidence set of $g^{\bullet}$ (5.8.8), $p \notin g(p)$ and $p \notin g^{\bullet}(p)$. This shows that $H \neq g(p)$ and $H \neq g^{\bullet}(p)$, and also that p is not a base point of $\mathcal{H}_p$, from which the uniqueness of H. Claim (2) is thus proved.

Take coordinates in $\mathbb{P}_n$ and assume that g has matrix M and p is $p = [\alpha_0, \dots, \alpha_n]$. Write $\alpha = (\alpha_0, \dots, \alpha_n)^t$, which is a column coordinate vector of p. Then $M\alpha$ and $M^t\alpha$ are column coordinate vectors of the hyperplanes $g(p)$ and $g^{\bullet}(p)$ respectively. The obvious equality

$$\alpha^t M\alpha = (\alpha^t M\alpha)^t = \alpha^t M^t \alpha$$

shows that $\alpha^t(M\alpha - M^t\alpha) = 0$ and therefore that the hyperplane $H \in \mathcal{H}_p$ containing p has column coordinate vector $M\alpha - M^t\alpha$.

Take an arbitrary point $q = [\beta_0, \dots, \beta_n] \in V_i$ and $\beta = (\beta_0, \dots, \beta_n)^t$. Then we get

$$(M - \lambda_i M^t)\beta = 0,$$

or, equivalently,

$$\beta^t(\lambda_i M - M^t) = -\beta^t(M - \lambda_i M^t)^t = 0.$$

In particular we get

$$\beta^t(\lambda_i M\alpha - M^t\alpha) = \beta^t(\lambda_i M - M^t)\alpha = 0$$

and so q belongs to the hyperplane H_i of $\mathcal{H}_p$ with column coordinate vector $\lambda_i M\alpha - M^t\alpha$. By the arbitrariness of q, $V_i \subset H_i$. If another hyperplane of $\mathcal{H}_p$ contains V_i, then all the hyperplanes of $\mathcal{H}_p$ do and, in particular, $g(p) \supset V_i$ against being $p \notin g^{-1}(V_i^*)$. This proves (3).

To close, regarding (4), it is clear from their already computed coordinate vectors that the hyperplanes $g(p)$, $g^{\bullet}(p)$, H, H_i have cross ratio λ_i. □

Proposition 11.7.11. *If the characteristic of a correlation g of $\mathbb{P}_n$ is written in the form of* (11.15)*, then the following conditions are satisfied:*

(1) $1 \leq m \leq n + 1$,

(2) $\lambda_i \neq 0$ *and* $\lambda_i \neq \lambda_j$ *for* $i \neq j$, $i, j = 1, \dots, m$,

(3) $\nu_{i,1} \geq \dots \geq \nu_{i,h_i} > 0$ *and* $h_i \geq 1$ *for* $i = 1, \dots, m$,

(4) $\sum_{i=1}^{m} \sum_{s=1}^{h_i} \nu_{i,s} = n + 1$, *and*

(5) *The reciprocal* λ_j^{-1} *of each characteristic root* λ_j *of* g *is also a characteristic root of* g. *Furthermore,* λ_j *and* λ_j^{-1} *have the same Segre symbol.*

Proof. Conditions (1) to (3) have been seen to be satisfied while defining the characteristic roots and the characteristic of a correlation. Since the degree of any characteristic polynomial is $n + 1$, as noted after its definition, (4) follows from 11.6.1.

For condition (5) let us write a_i^j the entry on column i and row j of M. Fix an integer s, $1 \leq s \leq n + 1$ and take a subdeterminant of the characteristic matrix $M - \lambda M^t$ of order s,

$$\delta(\lambda) = \det\bigl(a_i^j - \lambda a_j^i\bigr)_{(i,j)\in I\times J}$$

where I, J are two subsets of s elements of $\{1, \dots, n + 1\}$ taken with the induced orderings. It is a polynomial in λ of degree at most s. Consider the symmetrically placed subdeterminant

$$\delta^*(\lambda) = \det\bigl(a_j^i - \lambda a_i^j\bigr)_{(i,j)\in I\times J}$$

and note that

$$\begin{aligned}(-\lambda)^s\delta(1/\lambda) &= (-\lambda)^s \det\bigl(a_i^j - \lambda^{-1}a_j^i\bigr)_{(i,j)\in I\times J}\\ &= \det\bigl(-\lambda a_i^j + a_j^i\bigr)_{(i,j)\in I\times J}\\ &= \delta^*(\lambda).\end{aligned}$$

If for an arbitrary non-zero complex number α, $(\lambda - \alpha)^r$ divides $\delta(\lambda)$, then $r \leq s$ and

$$\delta(\lambda) = (\lambda - \alpha)^r h(\lambda),$$

where $h(\lambda)$ is a polynomial of degree at most $s - r$. It results that

$$\begin{aligned}\delta^*(\lambda) &= (-\lambda)^s\delta(1/\lambda)\\ &= (-\lambda)^r(\lambda^{-1} - \alpha)^r(-\lambda)^{s-r}h(\lambda^{-1})\\ &= (-1 + \alpha\lambda)^r(-\lambda)^{s-r}h(\lambda^{-1})\\ &= (\lambda - \alpha^{-1})^r\alpha^r(-\lambda)^{s-r}h(\lambda^{-1})\end{aligned}$$

and so $(\lambda - \alpha^{-1})^r$ divides $\delta^*(\lambda)$, because, clearly, $\alpha^r(-\lambda)^{s-r}h(\lambda^{-1}) \in \mathbb{C}[\lambda]$. The converse is also true because the roles of $\delta(\lambda)$ and $\delta^*(\lambda)$ may be swapped over. While $\delta(\lambda)$ describes the whole of the subdeterminants of order s of $M - \lambda M^t$, $\delta^*(\lambda)$ does the same, after which it follows that if α is a root of multiplicity r of the higher common divisor D_s of all subdeterminants of order s of $M - \lambda M^t$, then so is α^{-1}. This being true for $s = 1, \dots, n + 1$, the same holds for the invariant factors $I_s = D_{s+1}/D_s$, $s = 0, \dots, n$. All characteristic roots λ_i being non-zero, the above holds for $\alpha = \lambda_i$, $i = 1, \dots, m$, and proves that condition (5) is satisfied. □

Condition (5) of 11.7.11 assures that the characteristic roots other than ± 1 come in pairs of distinct reciprocal roots with equal Segre symbols. This in particular imposes a strong restriction on the characteristics of the collineations associated to correlations. Note however that nothing is said about the characteristic roots equal to ± 1.

In order to show the existence of a correlation with a given characteristic satisfying the conditions of 11.7.11, we will restrict ourselves to the case in which the correlation has no characteristic root equal to ± 1. This case is usually called the *simple case*, the correlations with no characteristic root equal to ± 1 being called *simple* correlations. The general case is more complicated and requires further conditions on the Segre symbols of the characteristic roots 1 and -1; for it the reader is referred to [18], XI.6. Assuming a correlation to be simple is a strong restriction: as explained above, the characteristic roots of a simple correlation are arranged in pairs of different and reciprocal characteristic roots with equal Segre symbols; then condition (4) forces $n + 1$ to be even, and so spaces of even dimension have no simple correlations.

Let us consider, for any complex number $\alpha \neq \pm 1$ and any positive integer ν, the 2ν-dimensional square matrix

$$L(\alpha, \nu) = \begin{pmatrix} & & & & & & & \alpha \\ & & & & & & \alpha & 1 \\ & & & & & \iddots & \iddots & \\ & & & & \alpha & 1 & & \\ & & & 1 & 0 & & & \\ & & \iddots & 0 & & & & \\ & 1 & \iddots & & & & & \\ 1 & 0 & & & & & & \end{pmatrix},$$

all of whose entries other than those on the secondary diagonal or immediately below it are zero. Then

$$L(\alpha, \nu) - \lambda L(\alpha, \nu)^t$$
$$= \begin{pmatrix} & & & & & & & \alpha-\lambda \\ & & & & & & \alpha-\lambda & 1 \\ & & & & & \iddots & \iddots & \\ & & & & \alpha-\lambda & 1 & & \\ & & & 1-\alpha\lambda & 0 & & & \\ & & \iddots & \lambda & & & & \\ & 1-\alpha\lambda & \iddots & & & & & \\ 1-\alpha\lambda & \lambda & & & & & & \end{pmatrix}.$$

has determinant equal to $(\alpha\lambda - 1)^\nu (\lambda - \alpha)^\nu$ but for maybe the sign. Dropping the first column and the $(\nu + 1)$-th row gives rise to a subdeterminant that, up to sign, equals

$$(\lambda - \alpha)^{\nu}\lambda^{\nu-1},$$

while dropping the first row and the $(\nu + 1)$-th column gives the subdeterminant

$$(\alpha\lambda - 1)^{\nu},$$

also up to sign. Since $\alpha \neq \pm 1$, the last two subdeterminants share no factor, which assures that the invariant factors of the above matrix are

$$1, \ldots, 1, (\lambda - \alpha)^{\nu}(\lambda - \alpha^{-1})^{\nu}.$$

Assume we are given complex numbers λ_i, $i = 1, \ldots, m$, no one equal to ± 1, and, for each i, positive integers $\nu_{i,s}$, $s = 1, \ldots, h_i$, in such a way that the conditions of 11.7.11 are satisfied. Due to condition (5), m is even and we are allowed to take $\ell = m/2$ and assume the λ_i numbered such that $\lambda_{\ell+i} = \lambda_i^{-1}$, $i = 1, \ldots, \ell$. Consider the square $(n + 1)$-dimensional matrix

$$L = \begin{pmatrix} L(\lambda_1, \nu_{1,1}) & & & & & \\ & \ddots & & & & \\ & & L(\lambda_1, \nu_{1,h_1}) & & & \\ & & & \ddots & & \\ & & & & L(\lambda_\ell, \nu_{\ell,1}) & & \\ & & & & & \ddots & \\ & & & & & & L(\lambda_\ell, \nu_{\ell,h_\ell}) \end{pmatrix}, \tag{11.16}$$

composed of $h_1 + \cdots + h_\ell$ blocks $L(\lambda_i, \nu_{i,j})$ centred on the diagonal, all entries outside them being equal to zero. Since each $L(\lambda_i, \nu_{i,j})$ is $2\nu_{i,j}$-dimensional, L is $(n + 1)$-dimensional due to the condition (5) of 11.7.11.

Lemma 11.7.12. *The list of elementary divisors of the matrix $L + \lambda L^t$ is:*

$$(\lambda - \lambda_1)^{\nu_{1,1}}, \ldots, (\lambda - \lambda_1)^{\nu_{1,h_1}}, \ldots, (\lambda - \lambda_\ell)^{\nu_{\ell,1}}, \ldots, (\lambda - \lambda_\ell)^{\nu_{1,h_\ell}},$$
$$(\lambda - \lambda_1^{-1})^{\nu_{1,1}}, \ldots, (\lambda - \lambda_1^{-1})^{\nu_{1,h_1}}, \ldots, (\lambda - \lambda_\ell^{-1})^{\nu_{\ell,1}}, \ldots, (\lambda - \lambda_\ell^{-1})^{\nu_{1,h_\ell}}.$$

Proof. Proceed as in the proof of 11.4.3. □

The existence of simple correlations with a prescribed characteristic follows:

Theorem 11.7.13. *For any complex numbers $\lambda_i \neq \pm 1$, $i = 1, \ldots, m$, and positive integers $\nu_{i,s}$, $i = 1, \ldots, m$, $s = 1, \ldots, h_i$, satisfying the conditions of* 11.7.11, *there is a projective type of correlations of n-dimensional complex projective spaces whose representatives have characteristic*

$$\{(\lambda_1; \nu_{1,1}, \ldots, \nu_{1,h_1}), \ldots, (\lambda_m; \nu_{m,1}, \ldots, \nu_{m,h_m})\}.$$

Proof. Just take the projective type of any correlation with matrix L, as defined by (11.16), after which the claim follows from 11.7.12 and 11.7.4. □

Remark 11.7.14. As in former cases, the projective type of 11.7.13 is unique, this time by 11.7.4.

The next theorem allows us to take matrices of the form of L above as *reduced matrices* of simple correlations. Its proof uses the same arguments as in the cases of collineations and pencils of quadrics and is left to the reader.

Theorem 11.7.15. *If a simple correlation g, of a complex projective space $\mathbb{P}_n$, has characteristic*

$$\{(\lambda_1; \nu_{1,1}, \dots, \nu_{1,h_1}), \dots, (\lambda_m; \nu_{m,1}, \dots, \nu_{m,h_m})\},$$

then there is a reference Ω of $\mathbb{P}_n$ relative to which g has matrix L, as defined by (11.16).

11.8 Square roots of regular matrices

This section is devoted to proving an algebraic result that has been already used in Sections 11.6 and 11.7, namely that any regular complex matrix M has a square root which may be written as a polynomial expression in M.

Fix M to be a regular complex $(n+1) \times (n+1)$ matrix. For any polynomial

$$P = a_m X^m + \dots + a_1 X + a_0$$

define the matrix $P(M)$ to be

$$P(M) = a_m M^m + \dots + a_1 M + a_0 \mathbf{1}_{n+1}.$$

Denote by $\mathbb{M}$ the $\mathbb{C}$-algebra of the $(n+1)$-dimensional square complex matrices. We will make use of the next two lemmas; the proof of the first one follows from easy computations and is left to the reader.

Lemma 11.8.1. *The map*

$$\begin{aligned} \mathbb{C}[X] &\longrightarrow \mathbb{M}, \\ P &\longmapsto P(M), \end{aligned}$$

is a homomorphism of $\mathbb{C}$-algebras.

Lemma 11.8.2. *There is a non-zero polynomial P with no factor X and such that* $P(M) = 0$.

Proof. Since $\mathbb{M}$ has dimension $d = (n+1)^2$, there is a non-trivial linear dependence relation

$$a_d M^d + \cdots + a_1 M + a_0 \mathbf{1}_{n+1} = 0.$$

Then the polynomial

$$P_1 = a_d X^d + \cdots + a_1 X + a_0$$

is non-zero and $P_1(M) = 0$. Write it $P_1 = X^r P$, where $r \geq 0$ and P has no factor X. Using 11.8.1, it is

$$M^r P(M) = P_1(M) = 0,$$

from which it follows that $P(M) = 0$, by the regularity of M. □

Theorem 11.8.3. *If M is a regular complex matrix, there is $Q \in \mathbb{C}[X]$ such that $Q(M)^2 = M$.*

Proof. Take P as in Lemma 11.8.2; clearly, we may choose P monic and then write it as the product of linear factors

$$P = \prod_{i=1}^{m} (X - \alpha_i)^{\mu_i}$$

where the α_i are non-zero complex numbers and $\alpha_i \neq \alpha_j$ if $i \neq j$.

It will be enough to prove the existence of a polynomial Q for which $Q^2 - X$ is a multiple of P, say

$$Q^2 - X = FP,$$

as then, by 11.8.1,

$$Q(M)^2 - M = F(M)P(M) = 0.$$

Equivalently, it will be enough to prove the existence of a polynomial Q for which $Q^2 - X$ has roots $\alpha_1, \ldots, \alpha_m$ with multiplicities not less than $\mu_1, \ldots, \mu_m$, respectively. This in turn is equivalent to having

$$\begin{aligned} Q(\alpha_i)^2 - \alpha_i &= 0, \\ \left(\frac{d(Q^2 - X)}{dX}\right)_{X=\alpha_i} &= 0, \\ &\vdots \\ \left(\frac{d^{\mu_i - 1}(Q^2 - X)}{dX^{\mu_i - 1}}\right)_{X=\alpha_i} &= 0, \end{aligned} \tag{11.17}$$

for $i = 1, \ldots, m$.

A straightforward inductive computation of the derivatives shows that the above equalities (11.17) may be rewritten

$$\begin{aligned} Q(\alpha_i)^2 - \alpha_i &= 0, \\ Q(\alpha_i)\left(\frac{dQ}{dX}\right)_{X=\alpha_i} &= (H_1)_{X=\alpha_i}, \\ &\vdots \\ Q(\alpha_i)\left(\frac{d^{\mu_i-1}Q}{dX^{\mu_i-1}}\right)_{X=\alpha_i} &= (H_{\mu_i-1})_{X=\alpha_i}, \end{aligned} \tag{11.18}$$

each

$$H_j = H_j(Q, dQ/dX, \dots, d^{j-1}Q/dX^{j-1})$$

being a polynomial expression in Q and its first $j-1$ derivatives.

Now, the first of the equations (11.18) is satisfied if $Q(\alpha_i)$ equals one of the square roots of α_i. This being true, in particular it is $Q(\alpha_i) \neq 0$. Then, assuming by induction that all the equations (11.18) up to the $(j-1)$-th one are satisfied if Q and its first $j-2$ derivatives take certain values $\beta_{i,0}, \dots, \beta_{i,j-2}$ for $X = \alpha_i$, it is clear that the j-th equation is satisfied if

$$\left(\frac{d^{j-1}Q}{dX^{j-1}}\right)_{X=\alpha_i} = Q(\alpha_i)^{-1} H_j(\beta_{i,0}, \dots, \beta_{i,j-2}) = \beta_{i,j-1}.$$

Thus, for $i = 1, \dots, m$, the equations (11.18) are satisfied if Q and its first $\mu_i - 1$ derivatives take values $\beta_{i,0}, \dots, \beta_{i,\mu_i-1}$, respectively, for $X = \alpha_i$. Since Hermite's interpolation (see for instance [10], Proposition 5.5.1) guarantees the existence of a polynomial satisfying these conditions, the proof is complete. □

11.9 Exercises

Exercise 11.1. Compute the invariant factors and the list of elementary divisors of a degree-zero polynomial matrix and show that they are determined by the rank of the matrix.

Exercise 11.2. Show that if the characteristic polynomial $c_M(\lambda)$ of a square $(n+1)$-dimensional constant matrix M decomposes in a product of different irreducible factors, then the invariant factors of M are $1, \dots, 1, (-1)^{n+1} c_M(\lambda)$.

Exercise 11.3. Prove that a collineation f of a projective space $\mathbb{P}_n$ has a matrix composed of square blocks of dimensions $d_1 + 1, \dots, d_\ell + 1$ centred on the diagonal, all entries outside these blocks being zero, if and only if $\mathbb{P}_n$ is spanned by ℓ independent linear varieties of dimensions $d_1, \dots, d_\ell$, each invariant by f.

Exercise 11.4. Prove that if a collineation of a complex projective space has characteristic

$$\langle(\lambda_1; \nu_{1,1}, \dots, \nu_{1,h_1}), \dots, (\lambda_m; \nu_{m,1}, \dots, \nu_{m,h_m})\rangle,$$

then, for each $i = 1 \dots, m$, the fundamental variety associated to λ_i has dimension $h_i - 1$.

Exercise 11.5. Prove that any collineation and its dual are projectively equivalent. Re-prove Theorem 5.7.9 from this fact.

Exercise 11.6. Assume that a collineation f of a complex projective space $\mathbb{P}_n$ has a list of elementary divisors with a single member (or, equivalently, a Jordan matrix with a single Jordan block). Prove that there is a hyperplane H of $\mathbb{P}_n$ such that any point $p \notin H$ and its successive images $f^j(p)$, $1 \le j \le n$, span $\mathbb{P}_n$. Deduce that in $\mathbb{P}_n$ there is no pair of non-empty supplementary linear varieties both invariant by f, and therefore no matrix of f with a non-trivial decomposition in blocks as the one of Exercise 11.3.

Exercise 11.7. Prove that for a collineation f of $\mathbb{P}_n$, the following conditions are equivalent.

(i) There is a reference relative to which f has diagonal matrix.

(ii) $\mathbb{P}_n$ is spanned by the fundamental varieties of f.

(iii) All the elementary divisors of f have degree one.

Exercise 11.8. The converse being obvious, prove that two square real constant matrices M, N that are similar as complex matrices (that is, there is a regular complex matrix P for which $N = P^{-1}MP$), are also similar as real matrices (the above P may be taken real). *Hint*: Use 11.2.5.

Exercise 11.9. Prove that if a collineation f of a complex projective space $\mathbb{P}_n$ has an elementary divisor of degree higher than one, then there is a line $L \subset \mathbb{P}_n$ invariant by f and with the restriction $f_{|L}$ parabolic.

Exercise 11.10. Prove that two involutions of $\mathbb{P}_{1,\mathbb{R}}$, one hyperbolic and the other elliptic, are not projectively equivalent, while their complex extensions are. Does this contradict Exercise 11.8?

Exercise 11.11. Let f be a collineation of a real projective space and assume that the characteristic polynomial of a matrix of f has irreducible factors $\lambda - \lambda_i$, $i = 1, \dots, m$, and $(\lambda - a_j)^2 + b_j^2$, $j = 1, \dots, s$. Take $z_j = a_j + b_j\boldsymbol{i}$. Prove that

$$\lambda_1, \dots, \lambda_m, z_1, \dots, z_s, \bar{z}_1, \dots, \bar{z}_s$$

is a system of eigenvalues of the complex extension $f_{\mathbb{C}}$ of f. Prove also that the characteristics of each λ_i as eigenvalue of f and $f_{\mathbb{C}}$ are the same, while the characteristic of each $(\lambda - a_j)^2 + b_j^2$ does agree with the characteristics of both z_j and $\bar{z}_j$.

Exercise 11.12. Use Exercises 5.6 and 11.9 to prove that if f is a cyclic collineation of a complex projective space, then all its elementary divisors have degree one. Use this and Exercise 11.11 to show that a cyclic collineation of a real projective space has all its elementary divisors irreducible.

Exercise 11.13. Let f be a collineation of a projective space $\mathbb{P}_n$, $\lambda_1, \dots, \lambda_m$ a system of eigenvalues of f and V_i, W_i^* the fundamental variety and the fundamental bundle corresponding to λ_i. Prove that the number of integers strictly greater than one in the characteristic of λ_i equals $\dim V_i \cap W_i + 1$.

Exercise 11.14. Let $\mathcal{P}$ be a regular pencil of quadrics of a complex projective space $\mathbb{P}_n$. Assume to have fixed a reference of $\mathbb{P}_n$ and let M, N be matrices of quadrics spanning $\mathcal{P}$, the second one degenerate. Prove that using the polynomial matrix $M + \lambda N$ instead of a characteristic matrix of $\mathcal{P}$ would lead to miss some of the entries of the characteristic of $\mathcal{P}$.

Exercise 11.15. Prove that two ordered pairs of non-degenerate quadrics of a complex projective space are projectively equivalent if and only if their associated collineations are. Describe the projective classification of pairs of non-degenerate complex quadrics.

Exercise 11.16. Prove that the correlations of $\mathbb{P}_{3,\mathbb{C}}$ that, relative to a certain reference, have matrices

$$\begin{pmatrix} 0 & 0 & 0 & 2 \\ 0 & 0 & 2 & 1 \\ 0 & 1 & 0 & 0 \\ 1 & 0 & 0 & 0 \end{pmatrix} \quad \text{and} \quad \begin{pmatrix} 0 & 0 & 0 & -2 \\ 0 & 0 & -2 & 1 \\ 0 & 1 & 0 & 0 \\ 1 & 0 & 0 & 0 \end{pmatrix},$$

are not projectively equivalent, but have projectively equivalent associated collineations. Note that neither of the correlations is a polarity or a null-system.

Appendix A
Perspective (for artists)

Perspective sets the rules allowing us to draw plane representations of three-dimensional objects, producing on a correctly placed human eye the same image as the objects themselves. The research on perspective related to the work of the Renaissance painters was very influential in the beginnings of projective geometry. In this appendix we will somewhat close the circle by obtaining the basic rules of perspective from projective geometry. We will pay more attention to foundations than to practice; many of the rules explained below have been taken from [19] and [27], to which the reader is referred for the practical aspects. We will make use of elementary results of Euclidean geometry that should be familiar to the reader, some of them having already appeared as exercises to the preceding chapters.

Related to perspective by the fact that they also deal with plane representations of three-dimensional objects are *descriptive geometry* and *computer vision*. Both have a far more ambitious goal than perspective, namely to provide plane representations in which no information on the original object is lost, and methods for recovering all three-dimensional information from the plane representation. It is maybe worth including a few comments on them, before concentrating on perspective.

Descriptive geometry is rather old: it was initiated by G. Monge at the end of the 18th century and its development was one of the causes of the rebirth of projective geometry at the beginning of the 19th century. Descriptive geometry provides a series of clever techniques to graphically solve, on the drawing board, geometrical problems, as for instance drawing the representation of a line from the representation of two of its points. At the basis of all technical drawing, since its beginnings descriptive geometry has been very important in architecture and industrial design, until eventually sidelined by computer-aided design and numerically controlled machines. Two classical titles on descriptive geometry are [6] and [28].

Computer vision (or geometric computer vision, to be more precise) has had an impressive and very successful development over the last two decades. It provides methods to recover all three-dimensional information on an object (scene) from plane representations (views) of it, obtained by multiple or moving cameras. Computer vision makes intensive use of notions and results of projective geometry, such as cross-ratio, projectivities, correlations, quadrics, Plücker coordinates of lines and many others. To give just an easy example, different views of the same scene are related by projectivities, and the effective determination of (matrices of) these projectivities is a crucial point in the recovering of the scene from its views. Treatment is always analytic, strongly based in computer calculation, in contrast with the exclusively synthetic methods of descriptive geometry. The interested reader may see [7], [16] and [8].

A.1 Basic setting and affine matter

Assume we have fixed a Euclidean three-dimensional space $\mathbb{A}_3$ which will be understood as our ambient space. We will make constant use of the projective closure $\overline{\mathbb{A}}_3$ of $\mathbb{A}_3$, so points may be proper or improper, and lines and planes are assumed to include their improper points. As already, the terms *improper point* and *direction* are taken as synonymous. We fix a proper plane Π, the *picture plane*, and a proper point $O \notin \Pi$, called the *centre of perspective* or *point of sight*. Projecting from O and taking section by Π defines the map

$$\psi : \overline{\mathbb{A}}_3 - \{O\} \longrightarrow \Pi,$$
$$q \longmapsto (p \vee O) \cap \Pi,$$

called the *perspective map* or just the *perspective*; it is a singular projectivity with only singular point O. The image by ψ of any subset Z of $\overline{\mathbb{A}}_3$ is called the *perspective view* or the *representation* of Z; in the sequel it will be denoted by $\tilde{Z}$. Each point q of Z and its representation $\tilde{q}$ being aligned with O, Z and $\tilde{Z}$ have the same appearance if the observer places one eye at O and keeps the other closed, to avoid any stereoscopic effect. A picture camera automatically produces such a representation. Perspective provides rules allowing us to draw on the picture plane the complete representation of a figure from the representation of part of it; for instance one may draw the representation of a series of evenly spaced points on a line from the representation of the first two of them, or the representation of a square on a given plane from the representation of one of its sides.

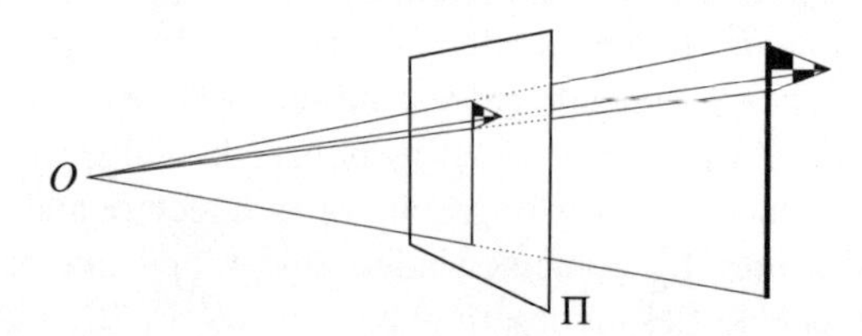

Figure A.1. Perspective map.

Obviously, the perspective map is far from being injective: any point of $q \in \Pi$ is the representation of all points on the line Oq other than O. However, once a plane π not going through O is fixed, ψ restricts to the perspectivity

$$\psi_\pi = \psi_{|\pi} : \pi \longrightarrow \Pi$$

with centre O. In particular, ψ_π is a projectivity, which guarantees that projective relations between parts of a figure of π do still hold between their representations. For instance, aligned points are represented as aligned points and concurrent lines as concurrent lines. Nevertheless, this may not be the case if the relations considered are just affine or metric. Let us examine first in which cases ψ_π is an affinity or a similarity:

Proposition A.1.1. *If the plane π is parallel to the picture plane, then ψ_π is a similarity. Otherwise ψ_π is not an affinity.*

Proof. If π is parallel to Π, then $\pi \cap \Pi$ is the improper line of both planes. Since its points are all invariant by ψ_π, ψ_π is a similarity (6.9.28). Conversely, if π is not parallel to Π, then the improper lines of π and Π are different and so they span the improper plane. Since O is a proper point, ψ_π does not map the improper line of π to the improper line of Π, and therefore it is not an affinity. □

Thus, if the plane π is parallel to the picture plane, then the representations of figures lying on π keep the angles and ratios of distances of the originals, and so drawing them poses no major problem. The case of planes π non-parallel to the picture plane (*receding planes*) is more interesting. Then affine or metric relations are not, in general, preserved by ψ_π, and the main question is how are they modified. For instance, one may ask about the actual angle between the representations of two orthogonal lines on a given receding plane. Our general strategy will be to express the affine and metric relations as projective relations with the improper line or the absolute, and then use projective invariance to identify their corresponding relations after representation. Next we explain how to represent parallelism, which will be the first example of such a procedure.

In the sequel we denote by π_∞ the improper plane of $\bar{\mathbb{A}}_3$. The point of sight O being proper, we have the perspectivity

$$\psi_\infty = \psi_{\pi_\infty} : \pi_\infty \longrightarrow \Pi.$$

Since ψ_∞ is in particular a projectivity, each point (resp. line) of Π is the representation of a uniquely determined improper point (resp. improper line), and incidence relations between improper points and lines are preserved by ψ_∞. The representation of the improper point of a proper line ℓ of $\mathbb{A}_3$ is called the *vanishing point* of ℓ: it belongs to the representation $\tilde{\ell}$ of ℓ. Similarly, the representation of the improper line of a proper plane π is called the *vanishing line* of π. We know from 3.4.9 that two linear varieties are parallel if and only if their improper parts are included; this and the above definitions directly give:

Proposition A.1.2. (a) *Lines of $\mathbb{A}_3$ are parallel if and only if they have the same vanishing point.*

(b) *Planes of $\mathbb{A}_3$ are parallel if and only if they have the same vanishing line.*

(c) *A line and a plane of $\mathbb{A}_3$ are parallel if and only if the vanishing point of the line belongs to the vanishing line of the plane.*

In particular, lines lying on a plane have their vanishing points on the vanishing line of the plane. Non-parallel lines have different vanishing points and therefore:

Corollary A.1.3. *The vanishing line of a plane is spanned by the vanishing points of any pair of non-parallel lines lying on (or just being parallel to) the plane.*

When drawing, the artist decides the placement of some vanishing points and lines on the picture plane, the other being then placed according to the rules of A.1.2. Often a plane of $\mathbb{A}_3$ is distinguished as being the ground plane: the planes parallel to it are then called *horizontal* or *level* and their common vanishing line is called the *horizon* of the picture. Usually drawing is started by deciding which line on the picture plane will be the horizon and then the vanishing points of all lines parallel to the ground (*horizontal* or *level* lines) are placed on it.

Remark A.1.4. The intersection of the level plane through O and the picture plane is the horizon of the picture. Thus all points level with O are represented on the horizon. In many cases, this allows us to recover from the picture the height from the ground of the point of sight O, and so the theoretical position of the artist while painting, see Exercise A.1.

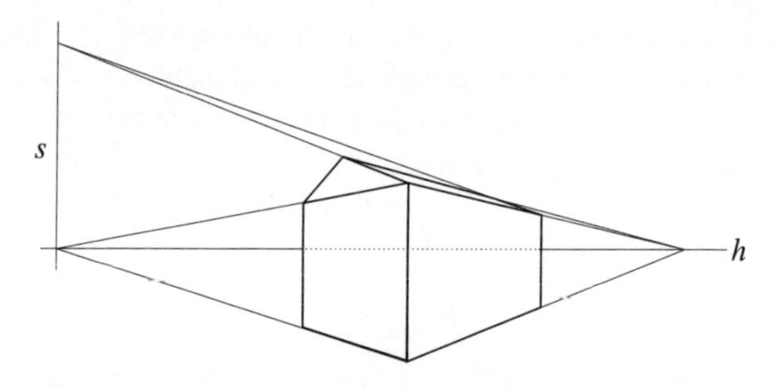

Figure A.2. h is the horizon and s the vanishing line of the plane of the façade. Quoting from [27]: RULE I: *All Receding Level Lines in Nature parallel to one another appear to meet on the Same Spot on the Horizon.* RULE II: *All Receding Lines that are in Nature parallel to one another if inclined upwards appear eventually to meet at a spot that is immediately above that spot where they would have met if they had been Level Lines.*

Proposition A.1.5. *A plane π of $\mathbb{A}_3$ is parallel to the picture plane Π if and only if its vanishing line is the improper line of Π. A line of $\mathbb{A}_3$ is parallel to Π if and only if its vanishing point is an improper point of Π.*

Proof. The plane π is parallel to Π if and only if its improper line is the improper line of Π. Since the latter is obviously invariant by ψ_∞, the first claim follows. A similar argument proves the second. □

Lines non-parallel to the picture plane are called *receding lines*.

The two corollaries below directly follow from A.1.5. The second one follows also from A.1.1:

Corollary A.1.6. *Two different parallel lines are represented as parallel lines if and only if they are parallel to the picture plane.*

Corollary A.1.7. *If a plane π of $\mathbb{A}_3$, $O \notin \pi$, is parallel to the picture plane, then any pair of parallel lines of π are represented as parallel lines. Conversely, if two pairs of parallel lines of π in different directions are represented as pairs of parallel lines, then π is parallel to the picture plane.*

Since $\mathbb{A}_3$ is a Euclidean space, in the sequel the reader may understand the affine ratios as ratios of distances with a sign, namely

$$(p_1, p_2, p_3) = \pm d(p_1, p_3)/d(p_2, p_3),$$

the negative value being taken when p_3 lies between p_1, p_2 and the positive one otherwise. The next proposition provides a rule for representing aligned points with a given affine ratio, which in perspective books is called *dividing a line into equal or unequal parts*:

Proposition A.1.8. *Let ℓ be a line of $\mathbb{A}_3$ not going through O, $\tilde{\ell}$ the line of Π representing it and $p \in \tilde{\ell}$ the vanishing point of ℓ. Choose any proper point $q \in \Pi$, $q \notin \tilde{\ell}$, and any proper line s on Π, parallel to pq and different from it. Clearly $q \notin s$ and we may consider the perspective $\varphi\colon \tilde{\ell} \to s$ with centre q. Then the composition $\varphi \circ \psi_{|\ell}$ is an affinity.*

Proof. Since ℓ is assumed not to contain O, its representation $\tilde{\ell}$ is a line on Π and $\psi_{|\ell}\colon \ell \to \tilde{\ell}$ is the perspectivity with centre O, in particular a projectivity. Since φ is also a projectivity, so is the composition $\varphi \circ \psi_{|\ell}$. By the definition of vanishing point, $\psi_{|\ell}$ maps the improper point of ℓ to p, which in turn, s being parallel to pq, is mapped by φ to the improper point of s. Thus $\varphi \circ \psi_{|\ell}$ maps the improper point of ℓ to the improper point of s and therefore is an affinity, as claimed. □

A direct consequence of the above is:

Corollary A.1.9. *Hypothesis and notations being as in* A.1.8, *let p_1, p_2, p_3 be three different points of ℓ. If $\tilde{p}_i \in \tilde{\ell}$ is the representation of p_i, $i = 1, 2, 3$, then it holds that*

$$(p_1, p_2, p_3) = (\varphi(\tilde{p}_1), \varphi(\tilde{p}_2), \varphi(\tilde{p}_3)).$$

Corollary A.1.9 allows us to represent aligned points having an already prescribed affine ratio, and in particular series of evenly separated aligned points, see Figure A.3. Also the affine ratio of three aligned points p_1, p_2, p_3, non-aligned with O, may be obtained from their representations, by just measuring the affine ratio $(\varphi(\tilde{p}_1), \varphi(\tilde{p}_2), \varphi(\tilde{p}_3))$.

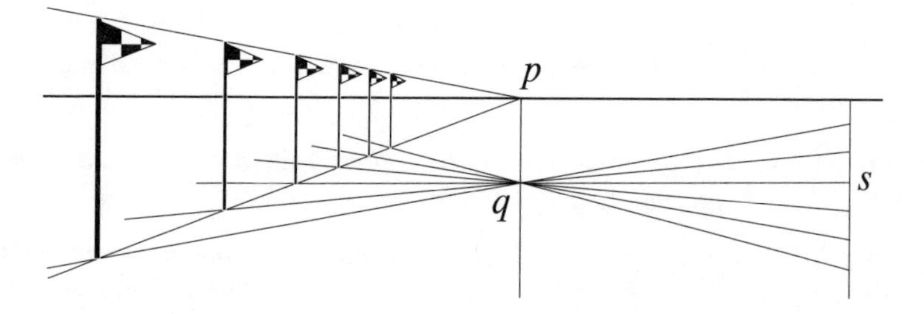

Figure A.3. A perspective view of evenly spaced poles of same height.

A.2 Orthogonality and angles

Representing metric relations would in principle require considering on the picture plane the representation $\psi_\infty(\mathbf{K})$ of the absolute conic $\mathbf{K}$ of $\mathbb{A}_3$. This conic having no points, it is not useful and needs to be replaced with some other objects carrying the same information and better suited to drawing. To this end, we begin by proving an auxiliary lemma which in fact could have been presented in a purely projective context. In the sequel we will denote by L_∞ the improper line of the plane of the picture Π, by F' the direction orthogonal to Π, and by I, J the pair of cyclic points of Π. Then $\mathbf{K} \cap L_\infty = I + J$ and F' is the pole of L_∞ relative to $\mathbf{K}$. Since $F' \notin L_\infty$, we are allowed to consider the harmonic homology h with centre F' and axis L_∞. Then we have:

Lemma A.2.1. *There is a unique non-degenerate conic C of π_∞ for which:*

(a) $\mathcal{P}_{\mathbf{K}} = \mathcal{P}_C \circ h = h^\vee \circ \mathcal{P}_C$, *$\mathcal{P}_{\mathbf{K}}$ and $\mathcal{P}_C$ being the polarities relative to* $\mathbf{K}$ *and C, respectively.*

Furthermore,

(b) *C has real points, and*

(c) $\mathbf{K}$ *and C are bitangent at I, J.*

Proof. The condition (a) obviously determines $\mathcal{P}_C$ and hence C (6.5.2). For the existence, take on π_∞ a reference (p_0, p_1, p_2, U) self-polar with respect to $\mathbf{K}$, with $p_0 = F'$ (and hence $p_1, p_2 \in L_\infty$) (7.1.5) and the unit point U chosen such that $\mathbf{K}$ has reduced equation $x_0^2 + x_1^2 + x_2^2 = 0$ (7.1.8). Then h has matrix

$$\begin{pmatrix} -1 & & \\ & 1 & \\ & & 1 \end{pmatrix}$$

and we take C to be the conic with equation $-x_0^2 + x_1^2 + x_2^2 = 0$: it clearly is non-degenerate with points. Condition (a) is satisfied due to the obvious matricial equalities

$$\begin{pmatrix} 1 & & \\ & 1 & \\ & & 1 \end{pmatrix} = \begin{pmatrix} -1 & & \\ & 1 & \\ & & 1 \end{pmatrix}\begin{pmatrix} -1 & & \\ & 1 & \\ & & 1 \end{pmatrix} = \begin{pmatrix} -1 & & \\ & 1 & \\ & & 1 \end{pmatrix}^t \begin{pmatrix} -1 & & \\ & 1 & \\ & & 1 \end{pmatrix}.$$

For condition (c) just note that $2L_\infty : x_0^2 = 0$ belongs to the pencil spanned by $\mathbf{K}$ and C. □

The representations of C and h are very easy to handle and provide a complete control of orthogonality in terms of vanishing points:

Proposition A.2.2. (a) *The representation $F = \psi_\infty(F')$ of F' is the foot of the perpendicular dropped from O to Π.*

(b) *The representation $D = \psi_\infty(C)$ of C is the circle with centre F and radius the distance $d(O, F)$ from O to Π.*

(c) *The representation $S_F = \psi_\infty \circ h \circ \psi_\infty^{-1}$ of h is the reflection in F.*

(d) *Points p, p' on the picture plane are vanishing points of orthogonal lines if and only if either of them is conjugate with respect to D to the image of the other by S_F.*

(e) *The common vanishing line of the planes orthogonal to a line with vanishing point p is the image by S_F of the polar of p relative to D.*

Proof. Regarding claim (a), just note that, by the definition of ψ_∞, $F = OF' \cap \Pi$, while the choice of F' assures that OF' is orthogonal to Π. The conic D is non-degenerate with points because so is C. Furthermore, since $C \cap L_\infty = I + J$ and ψ_∞ restricts to the identity of L_∞, $D \cap L_\infty = I + J$ and D is a circle. By A.2.1 (c), F' is the pole of L_∞ with respect to C; hence $F = \psi_\infty(F')$ is the pole of $L_\infty = \psi_\infty(L_\infty)$ with respect to $D = \psi_\infty(C)$, that is, the centre of D. We delay proving that the radius of D is $d(OF)$ and turn our attention to claim (c). It is clear that S_F leaves F and all improper points invariant. Choose a proper point $q \in \Pi$, $q \neq F$ and take $q' = Fq \cap L_\infty$; then $q' = \psi_\infty^{-1}(q') = F'\psi_\infty^{-1}(q) \cap L_\infty$, after which

$$(q, S_F(q), F, q') = (\psi_\infty^{-1}(q), h(\psi_\infty^{-1}(q)), F', q') = -1$$

because h is harmonic. All together S_F is the harmonic homology with centre F and axis L_∞, that is (5.7.26), the reflection in F.

Now, for claim (d), by the invariance of conjugation (6.4.4), p' is conjugate to $S_F(p)$ with respect to D if and only if $\psi_\infty^{-1}(p')$ is conjugate to $\psi_\infty^{-1}(S_F(p)) = h(\psi_\infty^{-1}(p))$ with respect to C, which, by A.2.1, is in turn equivalent to being $\psi_\infty^{-1}(p')$ and $\psi_\infty^{-1}(p)$ conjugate with respect to $\mathbf{K}$, that is, orthogonal.

Claim (e) is a direct consequence of claim (d), so it just remains to complete the proof of claim (b). To this end take p, p' to be the ends of a diameter of D and note first that the lines Op, Op' have vanishing points p, p' and so, by the already proved claim (d), are orthogonal. Since on the other hand the triangles OFp and OFp' obviously have a right angle at F and are congruent, by the orthogonality of Op and Op', $\widehat{pOF} = 45°$. After this, the triangle pOF is isosceles and $d(O, F) = d(F, p)$, as wanted. □

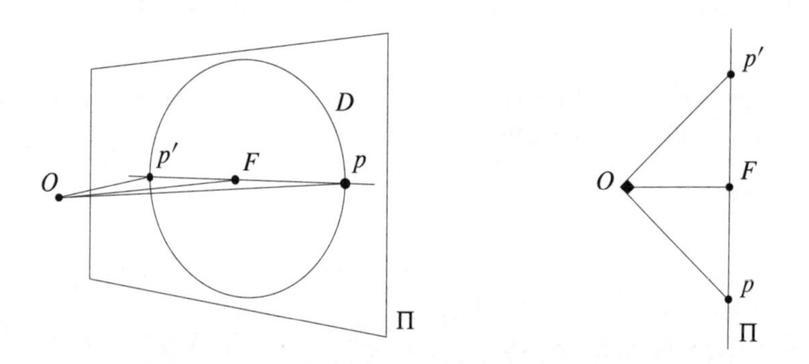

Figure A.4. Perspective and bird's-eye views of end of the proof of Proposition A.2.2: Π is the picture plane, assumed to be orthogonal to the ground, O is the point of sight, F the principal vanishing point and p, p' are taken on the horizon so that they are the diagonal vanishing points.

The point F and the circle D defined in A.2.2 are called the *principal vanishing point* and the *circle of distance*, respectively. The line OF, joining the centre of perspective to the principal vanishing point is called the *line of sight*. The distance $d(OF)$ from the centre of perspective to the picture plane is called the *principal* or *focal distance* and will be denoted by f in the sequel: we have seen in A.2.2 (b) that f is the radius of the circle of distance.

Remark A.2.3. By its own definition (and also by A.2.2 (e)), the principal vanishing point is the vanishing point of all lines orthogonal to the picture plane. If, as rather usual, the picture plane is orthogonal to the ground plane, then the principal vanishing point lies on the horizon of the picture, and conversely.

Remark A.2.4. In the proof of A.2.2 we have seen that $\widehat{pOF} = 45°$, where p is an end of an arbitrary diameter of D, hence an arbitrary point of D. It follows that the points of D are the vanishing points of the lines making with the line of sight OF an angle of 45°. In particular the intersection points of the circle of distance and the horizon are the vanishing points of the level lines making an angle of 45° with the line of sight: they are called the *diagonal vanishing points*.

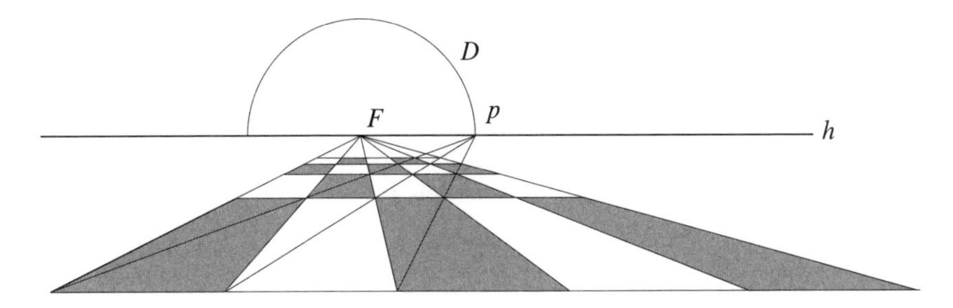

Figure A.5. Perspective view of square tiles on the floor, with one side parallel to the picture plane. h is the horizon, F the principal vanishing point, D the circle of distance and p a diagonal vanishing point. The concurrence of the diagonals at p assures that the tiles are squares, and not just rectangles, by A.2.4. Note that the choice of the focal distance $f = d(F, p)$ is equivalent to the choice of the apparent depth of one square tile. The picture looks quite unrealistic due to the very short focal distance chosen.

After A.2.2, the vanishing line of the planes orthogonal to a line with vanishing point p is $S_F(H_{p,D})$, the line symmetric with respect to F to the polar of p relative to the circle of distance. It is often referred to as the *antipolar* of p. In order to describe an effective construction of the antipolar, we first need:

Proposition A.2.5. *If $p \in \Pi$, $p \neq F$, the antipolar of p is orthogonal to Fp and intersects it at a point q which for p improper is $q = F$, while for p proper satisfies $d(F, p)d(F, q) = f^2$ and leaves F between q itself and p (or, equivalently, $(p, q, F) < 0$).*

Proof. After taking an orthonormal reference on Π with origin F and the first axis through p, D has equation $-f^2 + X^2 + Y^2 = 0$ and p has homogeneous coordinates $(\alpha, 1, 0)$. Then the antipolar of p has equation $-\alpha f^2 + X = 0$ and the claims easily follow. □

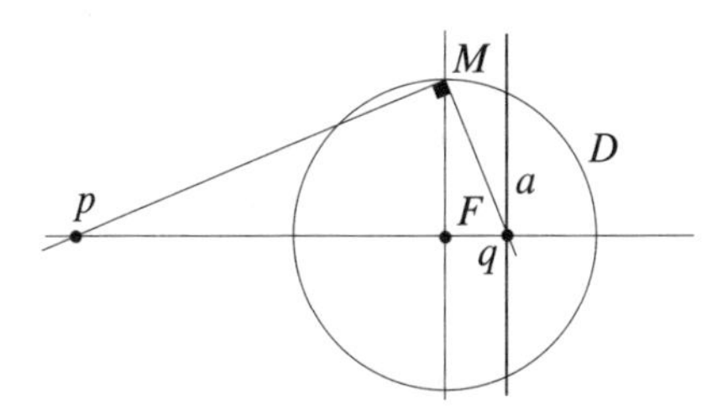

Figure A.6. Construction of the antipolar a of the point p, according to Corollary A.2.6.

We know the antipolar of F to be the improper line L_∞. The next proposition describes how to draw the antipolars of points other than F:

Corollary A.2.6. *Given $p \in \Pi$, $p \neq F$, take M to be one end of the diameter of D orthogonal to Fp. Take s to be the line through M orthogonal to Mp and $q = s \cap Fp$. Then the antipolar of p is the line perpendicular to Fp through q.*

Proof. Follows from A.2.5 and an elementary argument which is left to the reader. □

Remark A.2.7. Clearly, the construction of A.2.6 may be reversed to provide an arbitrary $p \neq F$ from its antipolar, that is, the vanishing point of the lines orthogonal to a receding plane from the vanishing line of the plane.

Corollary A.2.8. *The principal vanishing point is the orthocentre of any triangle whose vertices are vanishing points of three pairwise orthogonal receding lines.*

Proof. Call p_1, p_2, p_3 the vertices of the triangle. Since receding lines have proper vanishing points by A.1.5, they all are proper points and the claim makes sense. By the hypothesis, the antipolar of p_1 is its opposite side $p_2 p_3$. Then, by A.2.5, $p_1 F$ is orthogonal to $p_2 p_3$ and hence it is the altitude through p_1. The same argument applies to p_2 showing that F belongs to the altitude through it, hence the claim. □

Remark A.2.9. Once the principal vanishing point has been determined from the vanishing points of three pairwise orthogonal receding lines using A.2.8, a construction similar to A.2.6 allows us to determine the circle of distance.

Remark A.2.10. There are triangles with orthocentre F whose vertices are not vanishing points of three pairwise orthogonal lines, see Exercise A.3.

After dealing with the representation of orthogonality, we will pay some attention to the representation of lines making a given angle. If the lines lie on a plane parallel to the picture plane, the solution is obvious, by A.1.1. We fix thus a receding plane π of $\mathbb{A}_3$, with improper line ℓ_∞, and assume to have its vanishing line $v = \psi_\infty(\ell_\infty)$ already drawn on Π: v is then a proper line. Let s be the perpendicular to v through F and put $q = v \cap s$ (see Figure A.7). Note first that q and the improper point q_∞ of v are vanishing points of orthogonal lines because, by A.2.5, the antipolar line of q_∞ is just s. Choose any proper point $q_1 \in v$, $q_1 \neq p$ and take q', and q_2 to be, respectively, the intersections of the antipolar of q_1 with $q_1 F$ and v: in this way q_1, q_2 also are vanishing points of orthogonal lines. By A.2.5, q_1, q_2, q' are the vertices of a right triangle with hypotenuse $q_1 q_2$ and F lies on the leg $q_1 q'$, between q_1 and q'. After this, the reader may easily see, by an elementary argument, that q lies between $q_1 q_2$ (for a more sophisticated argument see Exercise A.4). If T is the circle which has a diameter with ends q_1, q_2, this assures that q is interior to T and therefore $T \cap s$ is a pair of real and distinct points $\{X_1, X_2\}$. We have:

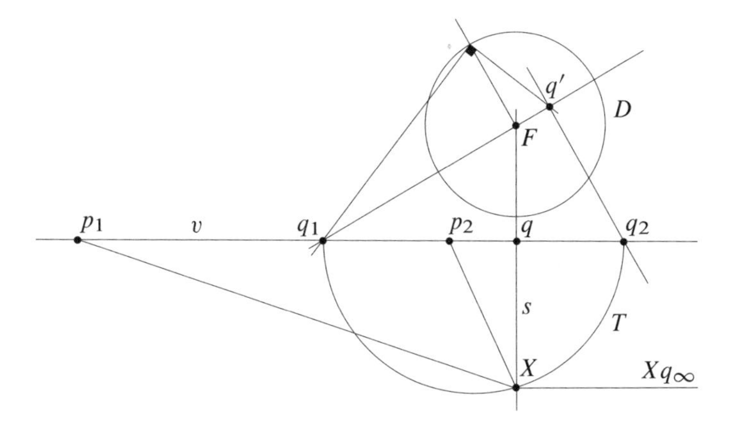

Figure A.7. Construction of one of the conformal centres X of a plane π with vanishing line v. According to Proposition A.2.11, the angle between any two lines of π with vanishing points p_1, p_2 is $\widehat{p_1 X p_2}$.

Proposition A.2.11. (a) *If X is either of the points X_1, X_2, then for any two non-parallel lines ℓ_1, ℓ_2 on π, with vanishing points p_1, p_2,*

$$\widehat{\ell_1 \ell_2} = \widehat{p_1 X p_2}.$$

(b) *If $X \in \Pi$ satisfies the property* (a) above, then it is one of the points X_1, X_2.

Proof. The composition of the restriction of ψ_∞,

$$\psi_{\infty|\ell_\infty} : \ell_\infty \longrightarrow v$$

and the perspective with centre X,

$$\begin{aligned} v &\longrightarrow L_\infty, \\ p &\longmapsto (pX) \cap L_\infty, \end{aligned}$$

is a projectivity $g : \ell_\infty \to L_\infty$ that maps the improper point of any line with vanishing point p to the improper point of pX. Then we will be done after proving that g (its complex extension in fact) maps the cyclic points I', J' of ℓ_∞ to the cyclic points I, J of L_∞, as then the claim will follow from Laguerre's formula and the invariance of the cross ratio.

Denote by τ the involution of orthogonality on ℓ_∞; it has fixed points I', J'. We have noticed above that q, q_∞ are vanishing points of orthogonal lines, and so are q_1, q_2 by its definition. Thus $\{\psi_\infty^{-1}(q), \psi_\infty^{-1}(q_\infty)\}$ and $\{\psi_\infty^{-1}(q_1), \psi_\infty^{-1}(q_2)\}$ are two different pairs of points corresponding by τ. Then the improper points of Xq, Xq_∞ and those of Xq_1, Xq_2 compose two pairs of points corresponding by the involution $g(\tau) = g \circ \tau \circ g^{-1}$ of L_∞. On the other hand, these are pairs of orthogonal directions: the first one because $Xq = s$ and Xq_∞ is parallel to v, and the other because $X \in T$ and q_1, q_2 are the ends of a diameter of T. It follows that $g(\tau)$ and the involution of orthogonality of L_∞ share two different pairs of corresponding points and therefore, by 5.6.10, they agree. Since $g(\tau)$ obviously has fixed points $g(I')$, $g(J')$, this proves that $\{g(I'), g(J')\} = \{I, J\}$ as wanted.

For claim (b) just note that if a point X satisfies claim (a), then on one hand Xq and Xq_∞ are orthogonal, which forces $X \in s$, and on the other also Xq_1 and Xq_2 are orthogonal, after which $X \in T$. Thus $X \in s \cap T = \{X_1, X_2\}$. □

Claim (b) shows in particular that the pair of points X_1, X_2 does not depend on the choice of the point q_1 used to define it. In the sequel we will call the points X_1, X_2 the *conformal centres* of the plane π (due to their role in Proposition A.2.12 below). Obviously, they are the same for all planes parallel to π. Having a conformal centre X determined on the picture plane allows us to recover the angle between two lines on (or parallel to) π as the angle between the lines projecting their vanishing points from X. Conversely, the vanishing point of the lines on (or parallel to) π making a given angle α with a line on π with vanishing point p is determined as the intersection point of v and the line through X making angle α with pX.

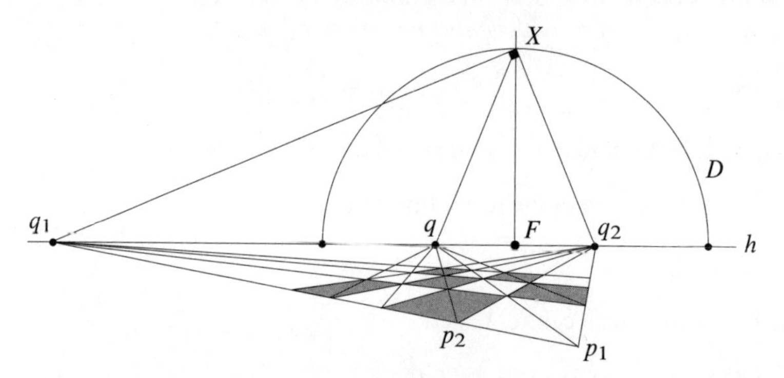

Figure A.8. Another perspective view of square tiles on the floor, this time with no side parallel to the picture plane. Again, h is the horizon, F the principal vanishing point and D the circle of distance. The end X of the diameter of D orthogonal to h is a conformal centre of the horizontal planes (see Exercise A.5). The representations p_1, p_2 of two adjacent vertices being given, q_1 is the common vanishing point of the tile side spanned by these vertices and all tile sides parallel to it. The point $q_2 \in h$, constructed according to A.2.6, is the vanishing point of the level lines orthogonal to $p_1 p_2$, and hence of the other tile sides. The point $q \in h$ has been taken such that qX is a bisector of q_1X and q_2X, after which, by A.2.11, it is the vanishing point of one diagonal of each tile. This allows us to complete the representation of the first tile, and then representing further tiles is straightforward.

Still assume to have selected a receding plane π, assume $O \notin \pi$ and, as above, denote by v the vanishing line of π, by s the perpendicular dropped from the principal vanishing point F to v and put $q = s \cap v$. Take X to be one of the conformal centres of π, choose any proper line e of Π not going through X, parallel to v and different from it; let φ be the general homology of Π which has centre X, axis e, and maps q to the improper point of s. The next proposition asserts that the representation of any figure of π is turned into the shape and proportions of the original figure by the action of φ:

Proposition A.2.12. *The composition of the perspectivity ψ_π and the homology φ is a similarity $\varphi \circ \psi_\pi : \pi \to \Pi$.*

Proof. The homology φ leaves invariant the improper point of v, because it belongs to the axis, and maps q to an improper point; it maps thus v to the improper line of Π and therefore $\varphi \circ \psi_\pi$ is an affinity. Furthermore, the lines through the conformal centre X being invariant, each $p \in v$ is mapped by φ to the improper point of the line pX. Then, the restriction of $\varphi \circ \psi_\pi$ to the improper line of π is just the projectivity g used in the proof of A.2.11. Since we have seen there that g maps the cyclic points of π to the cyclic points of Π, so does $\varphi \circ \psi_\pi$ and we are done. □

Since there is an easy construction of the images of points by a general homology (see Figure 5.6), Proposition A.2.12 provides a method that solves at once all problems that arise when representing a given figure $\mathcal{F}$ that lies on a receding plane π: first a scaled down copy $\mathcal{F}'$ of $\mathcal{F}$ is sketched on the picture plane, and then images by φ^{-1} of suitably chosen points of $\mathcal{F}'$ allow us to draw the wanted representation $\psi_\pi(\mathcal{F})$. Figure A.9 provides an example of the procedure.

A.3 Exercises

Exercise A.1. Assume that the height from the ground of the building pictured in Figure A.2 is about 10 m. and determine the height from the ground at which the point of sight was placed. *Answer:* at about 5 m.

Exercise A.2. Show that the antipolar of any point $p \in \Pi$ is the polar of p relative to the representation $\psi_\infty(\mathbf{K})$ of the absolute $\mathbf{K}$.

Exercise A.3. Let T be a triangle of Π with orthocentre the principal vanishing point F. Prove that the vertices of T are vanishing points of three pairwise orthogonal lines if and only if two of the products of the distances from F to a side of T and its opposite vertex are coincident (in which case they equal the third product too).

Exercise A.4. Assumptions and notations being as in the proof of A.2.11, prove that q lies between q_1, q_2 by showing first that q_1, q_2 and q, q_∞ are pairs of points conjugate with respect to $\psi_\infty(\mathbf{K})$, and then using Proposition 5.6.11 and Exercise 3.8.

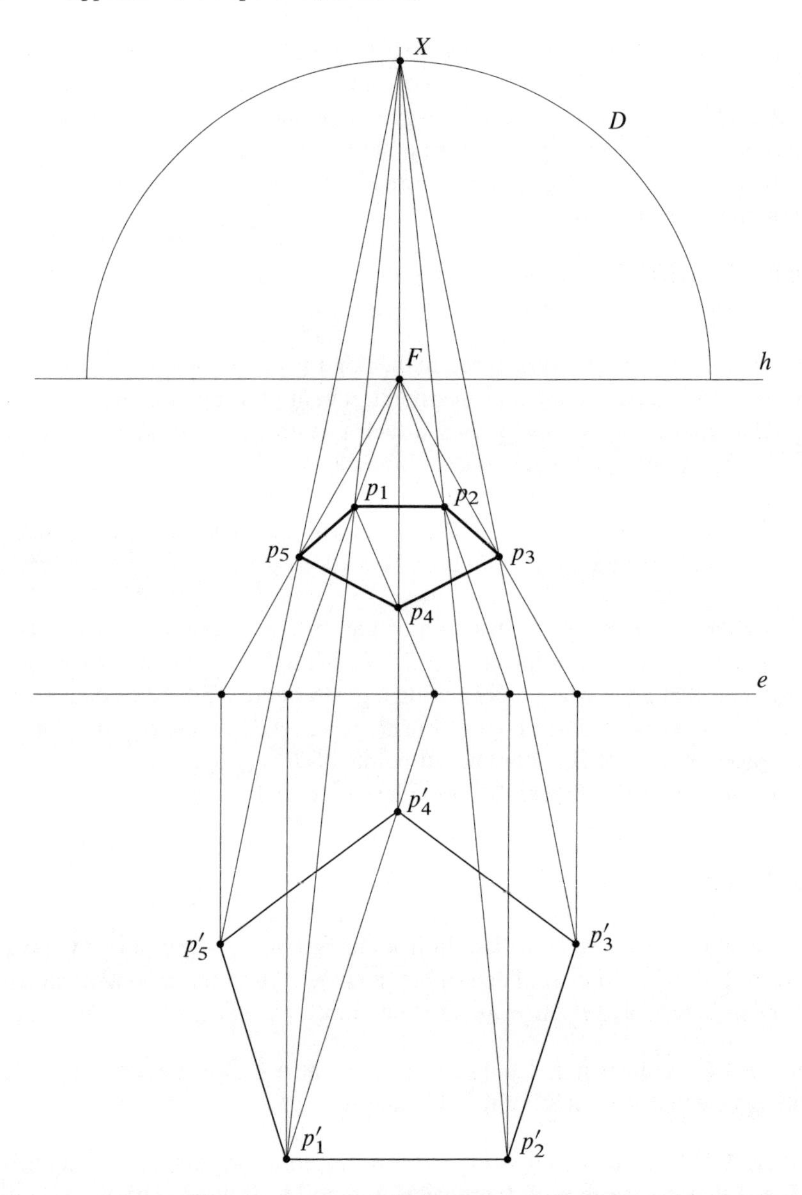

Figure A.9. A perspective view of a regular pentagon on a level plane, obtained using one of the homologies of Proposition A.2.12. h is the horizon, F the principal vanishing point, D the circle of distance and X one of the conformal centres of the horizontal planes. The homology φ has centre X, axis e and maps F to the improper point of the vertical lines. Images of points by φ or φ^{-1} have been determined as described in Figure 5.6. Assuming that the representations p_1, p_2 of two adjacent vertices are already given, one takes p'_3, p'_4, p'_5 to be the remaining vertices of a regular pentagon with adjacent vertices $p'_1 = \varphi(p_1)$, $p'_2 = \varphi(p_2)$. Then $p_i = \varphi^{-1}(p'_i)$, $i = 1, \dots, 5$, are the vertices of a foreshortened pentagon as wanted.

Exercise A.5. Prove that if a plane π is orthogonal to the picture plane, then its vanishing line v is a diameter of the circle of distance D and its conformal centres are the ends of the diameter of D orthogonal to v.

Exercise A.6. Prove that the homologies of Proposition A.2.12 are the only ones whose composition with ψ_π is a similarity.

Exercise A.7. Draw the representation of a (clear) cube from the representation of two adjacent vertices p, p' and the vanishing line of a face π containing them, in the following cases, which in perspective books are referred to as *one point*, *two points* and *three points perspective*, respectively:

(1) π is parallel to the picture plane.

(2) pp' is parallel to the picture plane, but π is neither parallel nor orthogonal to it.

(3) pp' is neither parallel nor orthogonal to the picture plane.

Appendix B
Models of non-Euclidean geometries

In this appendix we will make use of projective techniques to build *models* of plane Euclidean and non-Euclidean geometries. Roughly speaking, such a model is a geometric setting, consisting of points, lines and relations between them, that satisfies the first four Euclid's postulates together with the postulate of existence of either a unique (Euclidean geometry), or many (hyperbolic geometry), or no (elliptic geometry), parallel to a given line ℓ through a given point $p \notin \ell$. Actually, we will present a unified model, which may be taken depending on a parameter $\varepsilon \in [-1, 1]$ in such a way that it is a model of hyperbolic geometry if $\varepsilon < 0$, of Euclidean geometry if $\varepsilon = 0$ and of elliptic geometry if $\varepsilon > 0$ (*parametric model*).

Thanks to the use of a projective background, the models presented here provide a quick understanding and easy handling of the main points of the non-Euclidean geometries, including of course those in which they differ from Euclidean geometry.

B.1 Euclidean and non-Euclidean geometries

This section contains a quick and informal overview of the fundamentals of Euclidean and non-Euclidean geometries. Although not needed in the sequel from a strictly logical viewpoint, it may provide a better understanding of the role and meaning of the models presented next. For a detailed and very appealing presentation of both the contents and the history of Euclidean and non-Euclidean geometries, the reader is referred to M. J. Greenberg's book [13].

Euclid's *Elements of geometry* (written around 300 B.C.) is a thirteen-volume compendium of the geometry known by the Greeks at the time. It is at no doubt the most influential mathematical text of all times: until the discovery of the non-Euclidean geometries in the first half of the nineteenth century, it was the undisputed basis of all geometry and, furthermore, until the mid-nineteen hundreds, all teaching of elementary geometry was modeled on the *Elements*. Most important to us is that Euclid's *Elements* is structured in a rigorously deductive way, as we today think all mathematical texts must be: each new notion is defined using already defined terms and each new theorem is proved from already proved theorems by the rules of logic. Since an infinite regressing deductive chain is not admissible, the *Elements* starts from a short number of undefined terms, called *primitive terms*, and unproved facts, called *postulates* (or *axioms*). From them Euclid deduced 465 theorems, almost the whole of what is today understood as *elementary geometry*. Euclid's deductive chain has amazingly few flaws (many successors added their own when editing supposedly improved versions of the *Elements*). A major and quite accurate revision of Euclid's treatise was made by D. Hilbert around 1900 in his *Grundlagen der Geometrie* [17], at the price of considerably enlarging the number of axioms. For simplicity we will continue our discussion with reference to Euclid's setting, the reader may see Hilbert's axioms in [13], Chapter 3.

Using today's usual axiomatic style, Euclid's primitive terms may be introduced by saying (primitive terms are in italic) that Euclidean geometry starts from two sets of objects, respectively named *points* and *lines*, a binary relation between them named *to lie on* (a point on a line), a ternary relation between points, read *A lies between B and C* and, finally, once segments and angles have been defined, two equivalence relations, one between segments and the other between angles, both named *congruence*. The phrasings *ℓ goes through*, or *contains*, p are taken as synonymous of *p lies on ℓ*.

Euclid's postulates set the rules according to which the above objects behave. The first one assures there is a unique line through two given different points. The second postulate assures the existence of a unique segment congruent to a given one, lying on a given half-line and with one end equal to the end of the half-line. The third Euclid's postulate assures the existence and uniqueness of a circle with given centre and radius, while the fourth one assures that all right angles are congruent (right angles being already defined as those congruent to their supplementaries).

Before presenting the fifth postulate we set the definition of parallel lines: *two lines are taken as parallel if and only if no point lies on both*, that is, in our usual language, parallel lines are those which do not meet. It is important to note that we are not assuming that two parallel lines are equidistant, or make equal angles with a third, non-parallel, line: proving that these properties are satisfied requires the use of the fifth postulate and therefore, when discussing it, they cannot be taken as guaranteed.

In its most common form (Euclid's original one is a bit more complicated) the fifth postulate assures that given a line ℓ and a point p lying not on ℓ, there is one and only one line through p parallel to ℓ.

Since the ancient Greeks, there was a recurrent feeling that the fifth postulate should be proved from the other four. A lot of wrong proofs were proposed, each making a circular reasoning by using a fact that, no matter how "evident" it may seem, requires in turn the use of the fifth postulate to be proved. Examples of these facts are:

- parallel lines are equidistant,
- there exists a rectangle,
- the sum of angles of a triangle is two times a right angle,
- there exist similar non-congruent triangles.

The latter was correctly proposed as an alternative to the fifth postulate by J. Wallis in the seventeenth century.

Attempts to prove the fifth postulate were very popular by the end of the eighteenth century, making what was called at the time the *theory of parallels*. Eventually, by the independent work of J. Bolyai, C. F. Gauss, and N. I. Lobachevsky around 1830, and the subsequent one of E. Beltrami and F. Klein a few years later, it was realized that the first four Euclid's postulates, together with the assumption that, for any line ℓ and any point p not on ℓ, there are at least two different parallels to ℓ through p (*hyperbolic axiom*), lead to a new, somewhat strange, geometry (*hyperbolic geometry*) with no apparent contradiction. Hyperbolic geometry is said to be a *non-Euclidean geometry* because its postulates are incompatible with Euclid's postulates.

Let us pay some attention to the absence of contradiction in hyperbolic geometry, as models play an essential role there. What Beltrami, and later Klein, proved was that, would the first four Euclidean postulates plus the hyperbolic axiom lead to a contradiction, then the five Euclidean postulates would lead to a contradiction too. In other words, would hyperbolic geometry be contradictory, so would be Euclidean geometry. This is called the *relative consistency* of hyperbolic geometry. The proof was achieved by constructing, in the frame of Euclidean geometry, two sets and four relations, as required by Euclid's primitive terms, satisfying the first four Euclidean postulates plus the hyperbolic axiom. This is called a *model* of hyperbolic geometry. Its existence proves the relative consistency of hyperbolic geometry. For, if the first four Euclidean postulates plus the hyperbolic axiom lead to a contradiction, then this contradiction would appear in the model, which was constructed from Euclid's postulates; thus the latter would lead to a contradiction too. Beltrami's model was constructed using differential geometry, while Klein's model, the one we will explain here, uses projective geometry.

Adding the assumption of non-existence of parallel lines (*elliptic axiom*) to the first four Euclidean postulates easily leads to a contradiction. Nevertheless the *lie between* relation may be substituted with a weaker one and the second Euclidean postulate (or Hilbert's betweenness axioms) suitably modified, in order to make a system of axioms (*elliptic axioms*) including the elliptic axiom and giving rise to a second non-Euclidean geometry, also relatively consistent, called *elliptic geometry*.

There is a last fact to point out here, which greatly enhances the importance of the models: it may be proved that the model of each of the three geometries is essentially unique; namely, two different models of the same geometry may be made isomorphic by means of a couple of bijections between the sets of points and the sets of lines of one and another model preserving the relations *to lie on*, *to lie between* and *to be congruent to*. In this situation one says that the axioms are *categorical*. The categoricity of the axioms implies that once a theorem has been proved in a model of one geometry, then the theorem is true, in all its generality, in that geometry. Once the models have been constructed, we will proceed in this way to prove some selected theorems of the hyperbolic and elliptic geometries.

B.2 The models

As far as possible, we will give a unified presentation of the three models, in order to show their close relationship and the possibility of passing from one to another by the variation of a single real parameter. However, we will pay due attention to the specifics of each case, allowing the reader interested in a single model to make a reading avoiding the other two.

Assume we have chosen for once a real projective plane $\mathbb{P}_2$, say with associated vector space E, and a conic envelope $\mathcal{K} = [\tilde{\kappa}]$ of it, which either is non-degenerate (with points or imaginary), or has rank two and index one. $\mathcal{K}$ will be called the *absolute envelope*. To fix the ideas, the reader may assume to have chosen a system of projective coordinates in $\mathbb{P}_2$ and take $\mathcal{K}$ to be the conic envelope with equation $\varepsilon u_0^2 + u_1^2 + u_2^2 = 0$, $\varepsilon \in [-1, 1]$, which is degenerate of index one if $\varepsilon = 0$, non-degenerate with points if $\varepsilon < 0$ and imaginary if $\varepsilon > 0$.

We obtain a model of each of the three geometries according to the projective type of $\mathcal{K}$, namely:

Euclidean model. $\mathcal{K}$ is degenerate, with rank 2 and index 1. Then $\mathcal{K}$ envelops a pair of imaginary points $I + J$ of a well-determined line ℓ_∞ of $\mathbb{P}_2$ (the singular line of $\mathcal{K}$). $\mathcal{K}$ may be recovered as the envelope of the pair $I + J$ of ℓ_∞(6.8.2): hence $\mathcal{K}$ and the couple ℓ_∞, $I + J$ determine each other and are thus equivalent data. The reader may note that the situation is as the one described in Section 6.9: I, J are called the cyclic points, and ℓ_∞ the improper line, of the model.

Hyperbolic model. $\mathcal{K}$ is real non-degenerate. It envelops a real non-degenerate conic $\mathbf{K}$ of $\mathbb{P}_2$ called the *absolute conic* of the model. $\mathbf{K}$ determines and is in turn determined by $\mathcal{K}$ and hence the model may be also defined by giving $\mathbf{K}$ instead of $\mathcal{K}$.

Elliptic model. $\mathcal{K}$ is imaginary. The situation is as in the hyperbolic model but for the enveloped absolute conic $\mathbf{K}$ being imaginary.

The reader may note that in each case the model is essentially unique, because the projective type of the absolute envelope is determined, and therefore between any two models of the same geometry there is a projectivity mapping one absolute envelope to the other. Poles and polars relative to $\mathbf{K}$ (or $\mathcal{K}$) are called *absolute poles* and *absolute polars*.

Definition B.2.1. The *points* of the model are the points of $\mathbb{P}_2$ lying on no (real) line of $\mathcal{K}$.

Euclidean model. The improper line ℓ_∞ is the only real line of $\mathcal{K}$. Therefore the points of the model, also called *proper points*, are all the points of $\mathbb{P}_2$ but those lying on ℓ_∞, which in turn are called *improper*. The situation is as usual in plane metric geometry.

Hyperbolic model. The lines of $\mathcal{K}$ are the lines tangent to $\mathbf{K}$. It follows from the definitions of interior and exterior points in Section 7.4 that the points of the model are the points interior to $\mathbf{K}$. The remaining points of $\mathbb{P}_2$ will be used in our arguments, but they are not points of the model. From them, those belonging to $\mathbf{K}$ are called *ideal points*, while those exterior to $\mathbf{K}$ are called *ultra-ideal*. The points of the model will be called *actual points* if some confusion with the other points of $\mathbb{P}_2$ may occur.

Elliptic model. All points of $\mathbb{P}_2$ are points of the model.

Definition B.2.2. The lines of the model are the lines of $\mathbb{P}_2$ which contain at least one point of the model.

Euclidean model. The lines of the model are all the lines of $\mathbb{P}_2$ but the improper one.

Hyperbolic model. By 7.4.8 and 7.4.11, the lines of the model are the chords of $\mathbf{K}$ with real ends; they are called *actual lines*. Any actual line contains thus two ideal points. Neither the tangent lines to $\mathbf{K}$, nor the chords of $\mathbf{K}$ with imaginary ends are lines of the model; they are called *ideal lines* and *ultra-ideal lines*, respectively.

Elliptic model. The lines of the model are the lines of $\mathbb{P}_2$.

Definition B.2.3. If p and ℓ are a point and a line of the model, it is said that p *lies on* ℓ if and only if $p \in \ell$. In other words, the relation *to lie on* is defined as the restriction, to the points and lines of the model, of the usual relation *to lie on* between points and lines of $\mathbb{P}_2$.

Since $p_1 \vee p_2$ is the only line through any two distinct points $p_1, p_2 \in \mathbb{P}_2$, in all three cases Euclid's first postulate is obviously satisfied, as, in the Euclidean and hyperbolic cases, $p_1 \vee p_2$ is a line of the model because p_1, p_2 are assumed to be points of the model.

We may now discuss parallelism in the different models. In all geometries two lines are taken as *parallel* if and only if there is no point lying on both. This is Euclid's original definition and, for convenience, we will use it throughout this appendix. Note that, unlike the more usual modern definition, Euclid's definition does not take the lines as parallel to themselves and so, in modern terms, Euclid's parallels are parallel and different lines. The situation in the different models is as follows:

Euclidean model. Two different lines are parallel if and only if they have the same improper point, as in the projective closure of an affine plane.

Hyperbolic model. Two different lines are parallel if and only if their intersection point is either ideal or ultra-ideal. In the first case we will call them *asymptotically parallel lines*, while in the second they will be called *divergently parallel lines*[1].

Elliptic model. There are no parallel lines.

[1]Names for these lines differ in the literature, we follow [13].

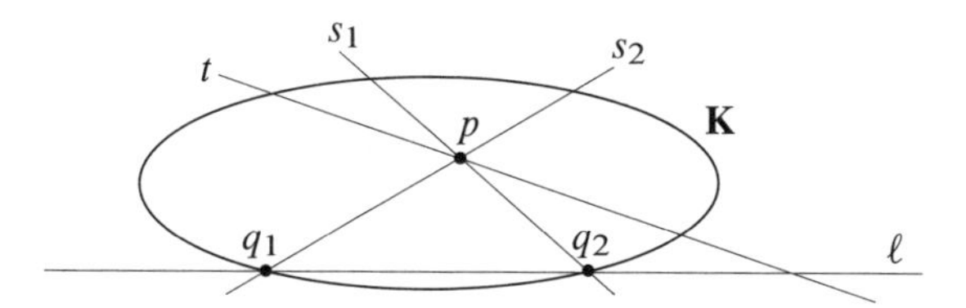

Figure B.1. Hyperbolic parallelism: If **K** is the absolute conic of the hyperbolic model, then p and ℓ are a point and a line of the model, q_1 and q_2 are the ideal points of ℓ, s_1 and s_2 are the two asymptotic parallels to ℓ through p and t is one of the divergently parallel lines to ℓ through p.

To check the validity of the fifth postulate, assume that p and ℓ are, respectively, a point and a line of the model, p lying not on ℓ:

Euclidean model. If q is the improper point of ℓ, then the lines parallel to ℓ are the lines going through q other than the improper line and ℓ itself. It follows that $p \vee q$ is the only parallel to ℓ through p and the fifth postulate is, indeed, satisfied.

Hyperbolic model. Call q_1, q_2 the ideal points of ℓ (its ends as a chord of the absolute **K**): clearly there are two lines asymptotically parallel to ℓ through p, namely $p \vee q_1$ and $p \vee q_2$ and furthermore all lines joining p to an ultra-ideal point of ℓ are divergently parallel to ℓ. In particular, the hyperbolic axiom is satisfied.

Elliptic model. Obviously there is no parallel to ℓ through p and the elliptic axiom is satisfied.

Two lines of the model are called *perpendicular* (or *orthogonal*) if and only if they are conjugate with respect to the absolute envelope $\mathcal{K}$.

Euclidean model. Two lines are perpendicular if and only if they are conjugate with respect to the envelope of the cyclic points. By 6.8.2, this occurs if and only if the improper points of the lines harmonically divide the cyclic ones, as in the projective closure of a Euclidean plane (see 6.9.8). In particular the (proper) lines perpendicular to a given ℓ are those going through the fourth harmonic of the cyclic points and the improper point of ℓ; it follows that any two lines perpendicular to ℓ are parallel and, conversely, any line parallel to a line perpendicular to ℓ is perpendicular to ℓ too.

Hyperbolic model. According to 6.6.1, two lines are perpendicular if and only if either of them contains the absolute pole of the other. A line ℓ of the model being a chord of **K** with real ends, its pole q is an ultra-ideal point and the lines of the model through q are those perpendicular to ℓ. Still any two lines perpendicular to ℓ are parallel (divergently parallel in fact), but, unlike the Euclidean case, a line parallel to a line perpendicular to ℓ need not be perpendicular to ℓ. Since q is not a point of the model, for any point p of the model, $p \neq q$ and the line $p \vee q$ is the only perpendicular to ℓ through p. The intersection point of two perpendicular lines is the pole of the line joining their poles, and so it is an actual point due to 7.4.6. Hence perpendicular lines are never parallel.

Elliptic model. The lines perpendicular to ℓ are those going through the absolute pole q of ℓ. For any $p \neq q$ the line $p \vee q$ is the only line through p perpendicular to ℓ.

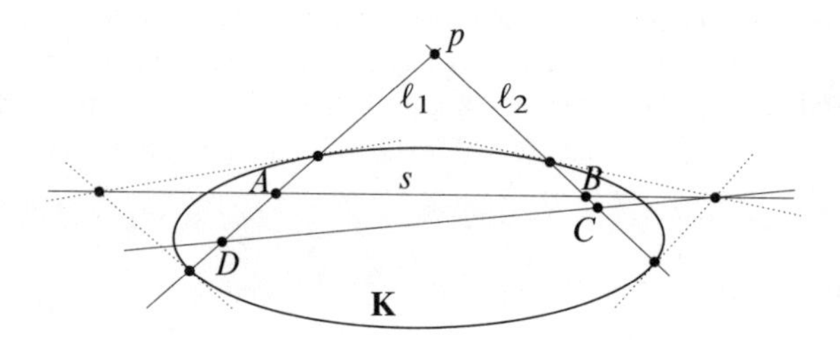

Figure B.2. Hyperbolic perpendicularity: ℓ_1 and ℓ_2 are divergently parallel lines and s is their only common perpendicular. s is an actual line because it is the absolute polar of the ultra-ideal point $p = \ell_1 \cap \ell_2$. The quadrilateral $ABCD$ has right angles at A, B and C, but not at D.

To close this section we examine in which cases two given distinct lines ℓ, ℓ' have a common perpendicular.

Euclidean model. If ℓ and ℓ' are not parallel, then they have distinct improper points, which in turn have distinct fourth harmonics with respect to I, J; by the above, they have no common perpendicular. If ℓ and ℓ' are parallel, they have the same improper point and so any line perpendicular to one of them is also perpendicular to the other.

Hyperbolic model. Call q the intersection point of ℓ and ℓ'. If a line s is perpendicular to ℓ and ℓ', then s goes through the absolute poles of ℓ and ℓ' and therefore is the absolute polar of q. The polar of q is an actual line if and only if q is ultra-ideal, hence if ℓ and ℓ' are divergently parallel lines, then they have a unique common perpendicular, which is the absolute polar of their common ultra-ideal point; otherwise they have no common perpendicular. There are no rectangles, due to the uniqueness of the common perpendicular.

Elliptic model. Arguing as in the hyperbolic case, the lines ℓ and ℓ' have as their only common perpendicular the absolute polar of their intersection point. Again, rectangles do not exist.

From this point onwards we will not continue the development of the Euclidean model, which should be familiar enough to the reader, and concentrate ourselves on the hyperbolic and elliptic models. In fact, Exercise 5.29 may be used to show that the Euclidean model introduced here is just a Euclidean plane (in the sense of

Section 6.9) presented within its projective closure, its scalar product being a suitable representative of $I + J$ (see Exercise 6.6). Then, all the contents of Section 6.9 for $n = 2$ applies to the Euclidean model and its development may be continued in the usual way, by defining *betweenness* from affine ratio (see Exercise 3.8), congruence using motions, and so on.

B.3 Hyperbolic distance

First of all we prove a purely projective lemma that will help us to compute non-Euclidean distances and angles:

Lemma B.3.1. *Let $C = [\eta]$ be a conic of a real projective plane $\mathbb{P}_2$. Assume that $p_1 = [v_1]$ and $p_2 = [v_2]$ are distinct points of $\mathbb{P}_2$ that do not belong to C and span a chord of C with (real or imaginary) ends q_1, q_2. If ρ is the cross ratio $\rho = (p_1, p_2, q_1, q_2)$, then*

$$\rho + \frac{1}{\rho} = \frac{4\eta(v_1, v_2)^2}{\eta(v_1, v_1)\eta(v_2, v_2)} - 2.$$

Proof. If $q_j = [\alpha_j v_1 + v_2]$, $j = 1, 2$, then $\alpha_1 \neq \alpha_2$, $\rho = \alpha_2/\alpha_1$ and (6.3.3)

$$\alpha_j^2 \eta(v_1, v_1) + 2\alpha_j \eta(v_1, v_2) + \eta(v_2, v_2) = 0$$

for $j = 1, 2$. It follows that

$$\alpha_1 + \alpha_2 = -\frac{2\eta(v_1, v_2)}{\eta(v_1, v_1)} \quad \text{and} \quad \alpha_1 \alpha_2 = \frac{\eta(v_2, v_2)}{\eta(v_1, v_1)}.$$

Then,

$$\rho + \frac{1}{\rho} = \frac{\alpha_1^2 + \alpha_2^2}{\alpha_1 \alpha_2} = \frac{(\alpha_1 + \alpha_2)^2 - 2\alpha_1\alpha_2}{\alpha_1 \alpha_2} = \frac{4\eta(v_1, v_2)^2}{\eta(v_1, v_1)\eta(v_2, v_2)} - 2,$$

as claimed. □

From now on and till the end of this section, we assume to be in the hyperbolic case and denote by κ a representative of the absolute $\mathbf{K}$. Take distinct actual points $p_1 = [v_1]$ and $p_2 = [v_2]$ and let q_1, q_2 be the ideal points of the line $p_1 p_2$. First of all we need:

Lemma B.3.2. *It holds that $(p_1, p_2, q_1, q_2) > 0$.*

Proof. Since $p_1 p_2$ is a chord of $\mathbf{K}$ with real ends, by 6.3.5,

$$\kappa(v_1, v_2)^2 > \kappa(v_1, v_1)\kappa(v_2, v_2).$$

On the other hand, p_1 and p_2 being both interior to $\mathbf{K}$, by 7.4.10 they have the same sign, that is,

$$\kappa(v_1, v_1)\kappa(v_2, v_2) > 0.$$

It follows that

$$\frac{\kappa(v_1, v_2)^2}{\kappa(v_1, v_1)\kappa(v_2, v_2)} > 1,$$

which, using B.3.1, gives $\rho + \rho^{-1} > 0$, and so $\rho = (p_1, p_2, q_1, q_2) > 0$, as claimed. □

Lemma B.3.2 above allows us to define the distance between p_1 and p_2 as:

$$d(p_1, p_2) = \frac{1}{2} |\log(p_1, p_2, q_1, q_2)|, \tag{B.1}$$

where q_1, q_2 are the ideal points of the line $p_1 p_2$ and log stands for natural logarithm.

The above is called the *hyperbolic distance* and also the *Lobachevsky distance*. Of course, taking q_1, q_2 in the opposite order turns the cross ratio into its inverse and therefore does not affect the final result. By the same argument applied to p_1, p_2, $d(p_1, p_2) = d(p_2, p_1)$.

The definition is extended to the case $p_1 = p_2$ by taking $d(p_1, p_1) = 0$. The reader may note that then the equality (B.1) above still makes sense if q_1, q_2 are taken to be the ideal points of an arbitrary line through p_1, and correctly gives $d(p_1, p_1) = 0$.

The following easy calculation will serve as an example, and will also show that the distance $d(p, p')$ behaves as one may expect when one of the points p, p' moves on the line they determine.

Example B.3.3. Assume that p, p' are actual points of the hyperbolic model belonging to a line ℓ. Conjugation and polarity being those relative to the absolute conic $\mathbf{K}$, let p_1 be the point of ℓ conjugate to p, and p_2 be the pole of ℓ. Then p, p_1, p_2 are the vertices of a self-polar triangle. By 7.1.8, they may be taken as the vertices of a projective reference of $\mathbb{P}_2$ relative to which the absolute conic has reduced equation. Since p is interior, the other two vertices are exterior (7.4.6), and so the equation of the absolute may be taken to be

$$-x_0^2 + x_1^2 + x_2^2 = 0.$$

Clearly $p = [1, 0, 0]$ and the ideal points of $\ell : x_2 = 0$ are $q_1 = [1, 1, 0]$ and $q_2 = [1, -1, 0]$. Furthermore, p' is an actual point of ℓ if and only if $p' = [1, x, 0]$, $-1 < x < 1$. Computing directly from the definition,

$$d(p, p') = \frac{1}{2} |\log(p, p', q_1, q_2)| = \frac{1}{2} \left| \log \left(\frac{1 + x}{1 - x} \right) \right|,$$

after which it is easy to see that $d(p, p')$ remains the same if the coordinate x is turned into its opposite, and also that for $0 \leq x < 1$, $d(p, p')$ is a strictly increasing continuous function of x with range $[0, \infty)$.

Next is a useful formula for the hyperbolic distance:

Proposition B.3.4. *If $p_1 = [v_1]$ and $p_2 = [v_2]$ are points of the hyperbolic model and κ is a representative of the absolute conic* **K**, *then*

$$d(p_1, p_2) = \arg\cosh\left(\frac{|\kappa(v_1, v_2)|}{\sqrt{\kappa(v_1, v_1)\kappa(v_2, v_2)}}\right),$$

where the positive determination of arg cosh *is being taken.*

Proof. Put $d = d(p_1, p_2)$. If q_1, q_2 are the ideal points of the line $p_1 p_2$, or those of any line through p_1 in case $p_1 = p_2$, take $\rho = (p_1, p_2, q_1, q_2)$. Then according to the definition of distance,

$$\rho + \frac{1}{\rho} = e^{2d} + e^{-2d} = 2\cosh 2d.$$

Then, Lemma B.3.1 gives

$$\cosh 2d = \frac{2\kappa(v_1, v_2)^2}{\kappa(v_1, v_1)\kappa(v_2, v_2)} - 1$$

and the claimed equality follows using the identity $\cosh(2\alpha) = 2\cosh^2(\alpha) - 1$. □

Remark B.3.5. The formula of B.3.4 becomes more explicit using coordinates: if p_1 and p_2 have column coordinate vectors $(x) = (x_0, x_1, x_2)^t$ and $(y) = (y_0, y_1, y_2)^t$, and **K** has matrix M, then

$$d(p_1, p_2) = \arg\cosh\left(\frac{|(x)^t M(y)|}{\sqrt{(x)^t M(x) \cdot (y)^t M(y)}}\right),$$

still using the positive determination of arg cosh.

The distance $d(p, \ell)$, from a point p to a line ℓ, is defined as in Euclidean geometry: if t is the (only) perpendicular to ℓ through p, then $q = t \cap \ell$, the foot of the perpendicular to ℓ from p, is a proper point (see Section B.2) and one takes $d(p, \ell) = d(p, q)$. Next is a formula for it:

Proposition B.3.6. *If p and ℓ are a point and a line, both actual, then*

$$d(p, \ell) = \arg\sinh\left(\frac{|\kappa(v, w)|}{\sqrt{-\kappa(v, v)\kappa(w, w)}}\right),$$

where still κ is a representative of **K**, *v is a representative of p and w a representative of the absolute pole of ℓ.*

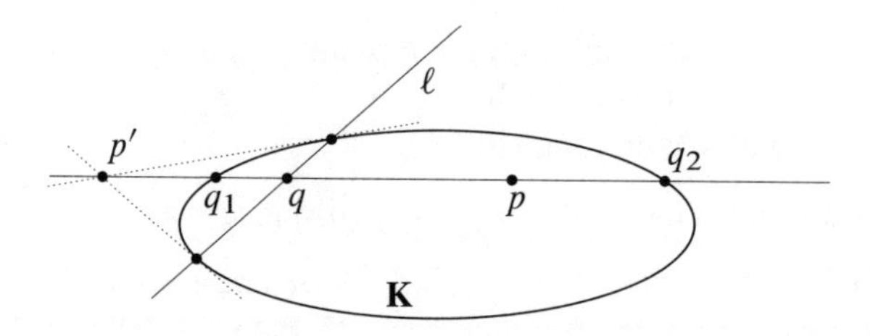

Figure B.3. Points and lines in the proof of B.3.6.

Proof. Let $p' = [w]$ be the pole of ℓ, $q = \ell \cap pp'$ the foot of the perpendicular to ℓ from p and q_1, q_2 the ends of pp'. By 2.9.6 and 6.4.11,

$$(q, p, q_1, q_2)(p, p', q_1, q_2) = (q, p', q_1, q_2) = -1,$$

from which,

$$d = d(p, \ell) = d(q, p) = \frac{1}{2}|\log(q, p, q_1, q_2)| = \frac{1}{2}|\log(-(p, p', q_1, q_2))|.$$

It follows that

$$2\cosh(2d) = -(p, p', q_1, q_2) - \frac{1}{-(p, p', q_1, q_2)}$$

and so, by B.3.1,

$$\cosh 2d = \frac{2\kappa(v, w)^2}{-\kappa(v, v)\kappa(w, w)} + 1.$$

Then the claim follows using the identity $\cosh(2z) = 2\sinh^2(z) + 1$. □

Also in this case, there is a more explicit formula, namely:

Corollary B.3.7. *The distance from a point p with column coordinate vector $(x) = (x_0, x_1, x_2)^t$ to the line ℓ with row coordinate vector $(u) = (u_0, u_1, u_2)$ is*

$$d(p, \ell) = \arg\sinh\left(\frac{|(u)(x)|}{\sqrt{-(x)^t M(x) \cdot (u) M^{-1}(u)^t}}\right),$$

where M is a matrix of **K**.

Proof. A column coordinate vector of p' is $M^{-1}(u)^t$, and so, taking κ the representative of **K** with matrix M, and v, w the representatives of p, p' with coordinate vectors (x) and $M^{-1}(u)^t$, respectively,

$$\kappa(v, w) = (x)^t M M^{-1}(u)^t = (x)^t(u)^t = (u)(x),$$
$$\kappa(w, w) = (u)M^{-1} M M^{-1}(u)^t = (u)M^{-1}(u)^t.$$

Since, obviously, $\kappa(v, v) = (x)^t M(x)$, the claim follows. □

B.4 Elliptic distance

Throughout this section we assume to be in the elliptic case. We will make use of the following projective fact:

Lemma B.4.1. *Let C be a conic of a real projective plane $\mathbb{P}_2$ and p_1, p_2 distinct points of $\mathbb{P}_2$ spanning a chord of C with imaginary ends q_1, q_2. Then*

$$|(p_1, p_2, q_1, q_2)| = 1.$$

Proof. The ends q_1, q_2 being imaginary, they are complex-conjugate. Then, the bar meaning complex conjugation,

$$\overline{(p_1, p_2, q_1, q_2)} = (p_1, p_2, q_2, q_1) = (p_1, p_2, q_1, q_2)^{-1},$$

from which the claim. □

Take distinct points $p_1 = [v_1]$, $p_2 = [v_2]$ and let q_1, q_2 be the intersection points of the line $p_1 p_2$ and the absolute conic $\mathbf{K} = [\kappa]$. Since $\mathbf{K}$ has no real points, q_1, q_2 are imaginary. By Lemma B.4.1 above, we may write $(p_1, p_2, q_1, q_2) = e^{2\varphi i}$, where $\varphi \in [-\pi/2, \pi/2]$. We define the *elliptic distance* between p_1 and p_2 as

$$d(p_1, p_2) = |\varphi|.$$

As for the hyperbolic distance, swapping over q_1, q_2 or p_1, p_2 does not affect the final result.

Equivalently, we take

$$d(p_1, p_2) = \left|\frac{1}{2i} \log(p_1, p_2, q_1, q_2)\right|, \tag{B.2}$$

where $\log(p_1, p_2, q_1, q_2)/2i$ is real by Lemma B.4.1, and belongs to $[-\pi/2, \pi/2]$ after a suitable choice of the determination of log. Thus $d(p_1, p_2) \in [0, \pi/2]$.

As in the hyperbolic case, the definition is extended to the case $p_1 = p_2$ by taking $d(p_1, p_1) = 0$. The equality (B.2) still holds in this case provided q_1, q_2 are taken to be the intersection points of the absolute conic and an arbitrary line through $p_1 = p_2$, as then $(p_1, p_2, q_1, q_2) = 1$.

Proposition B.4.2. *If $p_1 = [v_1]$ and $p_2 = [v_2]$ are points of the elliptic model and κ is a representative of the absolute conic $\mathbf{K}$, then*

$$d(p_1, p_2) = \arccos\left(\frac{|\kappa(v_1, v_2)|}{\sqrt{\kappa(v_1, v_1)\kappa(v_2, v_2)}}\right),$$

the determination of arccos *being taken such that* $0 \leq d(p_1, p_2) \leq \pi/2$.

Proof. Put $d = d(p_1, p_2)$. If q_1, q_2 are as above, take $\rho = (p_1, p_2, q_1, q_2)$. Then according to the definition of distance,

$$\rho + \frac{1}{\rho} = e^{2di} + e^{-2di} = 2\cos 2d.$$

After this the claimed equality follows from Lemma B.3.1, as in the proof of B.3.4. □

Remark B.4.3. Again, there is a more explicit formula using coordinates:

$$d(p_1, p_2) = \arccos\left(\frac{|(x)^t M(y)|}{\sqrt{(x)^t M(x) \cdot (y)^t M(y)}}\right),$$

where p_1 and p_2 have column coordinate vectors $(x) = (x_0, x_1, x_2)^t$ and $(y) = (y_0, y_1, y_2)^t$, $\mathbf{K}$ has matrix M and still the determination of arccos is chosen such that $0 \leq d(p_1, p_2) \leq \pi/2$.

Example B.4.4. Assume we have fixed a line ℓ and points p, p' on it. Arguing as in Example B.3.3, we choose a projective reference of $\mathbb{P}_2$ relative to which ℓ is $x_1 = 0$, $p = [1, 0, 0]$, $p' = [1, x, 0]$, this time with just $x \in \mathbb{R}$, and $\mathbf{K}$ has equation $x_0^2 + x_1^2 + x_3^2 = 0$. Then, the matrix of κ is the unit one and computing from B.4.3 it results that

$$d(p, p') = \arccos\left(\frac{1}{\sqrt{1 + x^2}}\right),$$

with the same conventions as in B.4.2. After an easy calculation, this equality may be rewritten as

$$d(p, p') = \arctan(|x|),$$

the determination of arctan being the one with values in $(-\pi/2, \pi/2)$. It is thus clear that while x takes positive values from 0 to ∞, $d(p, p')$ uniformly increases from 0 to $\pi/2$, and it uniformly decreases from $\pi/2$ to 0 when x takes negative values from ∞ to 0. The behaviour of the elliptic distance is thus quite different from the hyperbolic or Euclidean distance; it is like measuring distances on a closed loop rather than on an unlimited line. This will be explained further below.

There is a close relationship between elliptic geometry and the geometry on a sphere of the ordinary Euclidean three-space. The sphere, with its great circles as lines, would make a model of elliptic geometry but for the fact that two different great circles intersect at two (diametrally opposite) points and therefore the first Euclid's postulate is not satisfied. If to overcome this, one identifies the diametrally opposite points of the sphere, then it results in just the model we are dealing with. Indeed, start from a three-dimensional Euclidean space $\mathbb{A}_3$, with associated vector space E and scalar product κ. The projective plane $\mathbb{P}_2 = \mathbb{P}(E)$, together with

$\mathbf{K} = [\kappa]$ as absolute conic, is a model of elliptic geometry. Fix a point $O \in \mathbb{A}_3$ and take S to be the sphere with centre O and radius one. Consider the map

$$\begin{aligned} \tau \colon S &\longrightarrow \mathbb{P}_2, \\ p &\longmapsto [p - O]. \end{aligned}$$

For any $p = [v] \in \mathbb{P}_2$, $\tau^{-1}(p) = \{A_1, A_2\}$, where A_1, A_2 are the ends of the diameter $O + \langle v \rangle$ of S. Thus τ is exhaustive and maps two different points to the same image if and only if they are diametrally opposite. If $\ell = [F]$ is a line of $\mathbb{P}_2$, then $U = O + F$ is a diametral plane of S and $\tau^{-1}(\ell) = U \cap S$ is a great circle. Conversely, any great circle of S is $S \cap (O + F)$, with $F \subset E$ a subspace of dimension two, and therefore has the line $[F]$ of $\mathbb{P}_2$ as image. Regarding distances, if, as above, $p = [v] \in \mathbb{P}_2$ has $\tau^{-1}(p) = \{A_1, A_2\}$ and $q = [w] \in \mathbb{P}_2$ has $\tau^{-1}(p) = \{B_1, B_2\}$, then

$$A_i = O \pm \frac{1}{\sqrt{\kappa(v, v)}} v, \quad B_i = O \pm \frac{1}{\sqrt{\kappa(w, w)}} w$$

and so,

$$\cos(\widehat{A_1 A_2\, B_1 B_2}) = \frac{\kappa(v, w)}{\pm\sqrt{\kappa(v, v)\kappa(w, w)}}.$$

Using B.4.2, it follows that the elliptic distance $d(p, q)$ is the shortest of the lengths of the arches of great circle joining A_i to B_j, $i, j = 1, 2$, on S. The elliptic distance may thus be viewed as the length of an arch of circle (of a half-circle with identified ends, to be precise), which explains why it behaves like an angle.

Still in the elliptic case, assume that ℓ is a line and p a point other than the pole of ℓ. Then there is a unique perpendicular to ℓ through p and the distance from p to ℓ is defined, as in the Euclidean and hyperbolic cases, as the distance from p to the foot of the perpendicular to ℓ through p. If p is the absolute pole of ℓ, then for any $q \in \ell$ the line pq is perpendicular to ℓ. Let q_1, q_2 be the (imaginary) intersection points of ℓ and $\mathbf{K}$. By 6.4.23 it is $(p, q, q_1, q_2) = -1$ and so $d(p, q) = \pi/2$ for all $q \in \ell$. After this, the distance to the line ℓ from its absolute pole p is taken $d(p, \ell) = d(p, q) = \pi/2$ for any $q \in \ell$. Next is a nice relation between the distance to a line and the distance to its pole. It has no hyperbolic analogue because in the hyperbolic case the absolute polars of the actual points are ultra-ideal lines.

Proposition B.4.5. *In the elliptic case, if p' is the absolute pole of a line ℓ, then, for any point p,*

$$d(p, \ell) + d(p, p') = \pi/2.$$

Proof. In case $p = p'$ the claimed equality is clear because $d(p, p') = 0$ and $d(p, \ell) = \pi/2$. Otherwise take $q = \ell \cap pp'$; then $d(p, \ell) = d(p, q)$. As in the proof of B.3.6, if q_1, q_2 are the ends of the chord pp' of $\mathbf{K}$, by 2.9.6 and 6.4.11,

$$(q, p, q_1, q_2)(p, p', q_1, q_2) = (q, p', q_1, q_2) = -1,$$

from which,

$$d(p,\ell) + d(p,p') = \frac{1}{2i}\log(q,p,q_1,q_2) + \frac{1}{2i}\log(p,p',q_1,q_2) = \pi/2. \qquad \square$$

B.5 Betweenness

As explained in Section B.1, there is no betweenness relation in elliptic geometry; the weaker projective relation *to separate* of the end of Section 2.10 is used instead. Thus, throughout this section we will deal with the hyperbolic case only. After introducing hyperbolic distance we are able to define Euclid's fourth primitive term in the hyperbolic model:

Definition B.5.1. If p', p'' are distinct points of the hyperbolic model, then a third point p *lies between* p' and p'' if and only if it is aligned with p' and p'' and

$$d(p',p'') = d(p',p) + d(p,p'').$$

Call q_1, q_2 the ideal points of the line containing p', p''. The equality in the above definition may be equivalently written:

$$|\log(p',p'',q_1,q_2)| = |\log(p',p,q_1,q_2)| + |\log(p,p'',q_1,q_2)|. \tag{B.3}$$

On the other hand, by the multiplicativity of the cross ratio 2.9.6,

$$\log(p',p'',q_1,q_2) = \log(p',p,q_1,q_2) + \log(p,p'',q_1,q_2).$$

Since $\log(p',p,q_1,q_2) \neq 0$ and $\log(p,p'',q_1,q_2) \neq 0$, the equality (B.3) is satisfied if and only if $\log(p',p,q_1,q_2)$ and $\log(p,p'',q_1,q_2)$ have the same sign, or, equivalently, if and only if (p',p,q_1,q_2) and (p,p'',q_1,q_2) are either both lesser than one or both greater than one. Since

$$1 - (p',p,q_1,q_2) = (p',q_1,p,q_2)$$

and similarly for the other cross ratio, we have proved:

Proposition B.5.2. *Assume that p', p'' are distinct points of the hyperbolic model, and that the line $p'p''$ has ideal points q_1, q_2. Then p lies between p' and p'' if and only if p is aligned with p', p'' and*

$$(p',q_1,p,q_2)(p,q_1,p'',q_2) > 0.$$

If p_1, p_2 are distinct points of the hyperbolic model, the *segment* with ends p_1, p_2 is defined as the set composed of p_1, p_2 and all points lying between them.

Presenting the notions (half-lines, half-planes, angles, etc.) and properties (*separation properties*) the betweenness relation gives rise to, would take us too far. The interested reader may see [13], Chapter 3, and also Exercise B.3.

B.6 Angles between lines

In this section we will associate to each pair of non-parallel lines ℓ_1, ℓ_2 a real number $\widehat{\ell_1\ell_2} \in [0, \pi/2]$, which will be taken as the measure of the angle between the lines. We will not enter considering angles (defined as pairs of half-lines with the same end) and their measure, as this requires some extra work involving betweenness. If measure of angles is considered, then two of the four angles determined by two lines ℓ_1, ℓ_2 measure $\widehat{\ell_1\ell_2}$, while the other two measure $\pi - \widehat{\ell_1\ell_2}$.

The definition being the same in all cases, assume to be in either of the three models and to have fixed a pair of non-parallel lines $\ell_1 = [\omega_1], \ell_2 = [\omega_2], \omega_1, \omega_2 \in E^\vee$. Let p be the point of the model shared by ℓ_1 and ℓ_2. Call s_1, s_2 the lines of the absolute envelope $\mathcal{K}$ through p. By the definition of points of the model, in all cases, s_1, s_2, are imaginary. Therefore, Lemma B.4.1 applies to p^* and $\mathcal{K}$ in $\mathbb{P}_2^\vee$ and allows us to write

$$(\ell_1, \ell_2, s_1, s_2) = e^{2i\alpha},$$

with $\alpha \in [-\pi/2, \pi/2]$. We define $\widehat{\ell_1\ell_2} = |\alpha|$. As for the elliptic distance, swapping over ℓ_1, ℓ_2 or s_1, s_2 in the above equality turns α into $-\alpha$ and so leaves $\widehat{\ell_1\ell_2}$ unchanged. Equivalently,

$$\widehat{\ell_1\ell_2} = \left|\frac{1}{2i}\log(\ell_1, \ell_2, s_1, s_2)\right|, \tag{B.4}$$

the determination of log being taken such that $\widehat{\ell_1\ell_2} \in [0, \pi/2]$.

The real number $\widehat{\ell_1\ell_2} = |\alpha|$ will be called the *measure of the angle* between ℓ_1 and ℓ_2 or just the *angle* between ℓ_1 and ℓ_2 if no confusion may result. If the lines are orthogonal, then, by 6.4.2,

$$(\ell_1, \ell_2, s_1, s_2) = -1$$

and so $\alpha = \pi/2$, as one could expect.

Again, there is nothing new in the Euclidean case, as then $\{s_1, s_2\} = \{pI, pJ\}$ and so the equality (B.4) above is just an unoriented version of Laguerre's formula 6.9.11.

Proposition B.6.1. *If $\ell_1 = [\omega_1]$ and $\ell_2 = [\omega_2]$ are non-parallel lines of either model and $\tilde{\kappa}$ is a representative of the absolute envelope $\mathcal{K}$, then*

$$\widehat{\ell_1\ell_2} = \arccos\left(\frac{|\tilde{\kappa}(\omega_1, \omega_2)|}{\sqrt{\tilde{\kappa}(\omega_1, \omega_1)\tilde{\kappa}(\omega_2, \omega_2)}}\right),$$

the determination of arccos *being taken such that* $0 \leq \widehat{\ell_1\ell_2} \leq \pi/2$.

Proof. Same as for B.4.2. □

The remainder of this section is devoted to showing some relevant facts that are characteristic of the non-Euclidean geometries. Assume first to be in the hyperbolic case and to have fixed a line ℓ and a point p lying not on ℓ. Let t be the perpendicular to ℓ through p and take $q = t \cap \ell$, the foot of the perpendicular t. The angle α between t and one of the asymptotic parallels to ℓ through p (which is the same for both asymptotic parallels, as we will see in the proof of Theorem B.6.2 below) is called the *angle of parallelism* of ℓ and p; $\pi - 2\alpha$ measures the part of the pencil p^* composed by lines parallel to ℓ. There is a very relevant formula relating the angle of parallelism to the distance from p to ℓ:

Theorem B.6.2 (Bolyai–Lobachevsky). *If α is the angle of parallelism of ℓ and p and d the distance from p to ℓ, then*

$$e^{-d} = \tan \frac{\alpha}{2}. \tag{B.5}$$

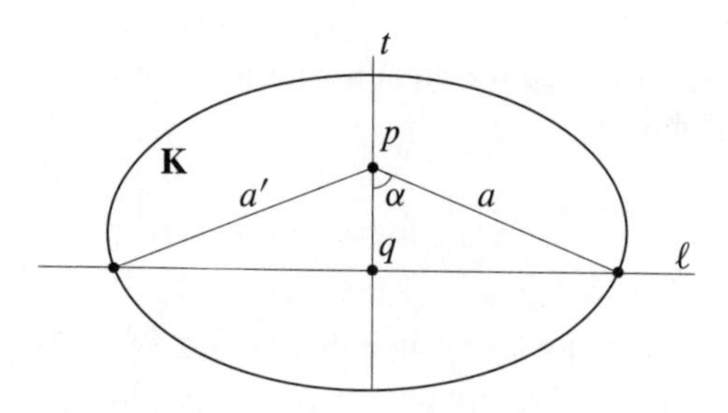

Figure B.4. The asymptotic parallels a, a' to ℓ through p and the angle of parallelism α.

Proof. As in Example B.3.3, take a reference with first vertex q, the second vertex on ℓ, self-polar with respect to $\mathbf{K}$ and relative to which $\mathbf{K}$ has equation $-x_0^2+x_1^2+x^2 = 0$. Then $q = [1,0,0]$, ℓ has equation $x_2 = 0$ and, the reference being self-polar, t has equation $x_1 = 0$. Assume $p = [1,0,b]$, $0 < |b| < 1$: since the ideal points on ℓ are $[1,0,\pm 1]$,

$$\begin{aligned} d = d(p,q) &= \frac{1}{2}|\log([1,0,0],[1,0,b],[1,0,1],[1,0,-1])| \\ &= \frac{1}{2}\left|\log\left(\frac{1+b}{1-b}\right)\right| = \frac{1}{2}\log\left(\frac{1+|b|}{1-|b|}\right). \end{aligned} \tag{B.6}$$

On the other hand, the asymptotic parallels to ℓ through p are the joins of p and each of the ideal points on ℓ, which are $[1,\pm 1,0]$. They have thus equations $bx_0 \pm bx_1 - x_2 = 0$ and therefore are represented by the forms with components

$(b, \pm b, -1)$. The line $t : x_1 = 0$ is in turn represented by the form with components $(0, 1, 0)$, while a matrix of the absolute envelope is

$$\begin{pmatrix} -1 & 0 & 0 \\ 0 & 1 & 0 \\ 0 & 0 & 1 \end{pmatrix}^{-1} = \begin{pmatrix} -1 & 0 & 0 \\ 0 & 1 & 0 \\ 0 & 0 & 1 \end{pmatrix}.$$

Applying B.6.1, we get $\cos\alpha = |b|$,no matter which asymptotic parallel is used. After substitution in (B.6), it results that

$$d = \frac{1}{2}\log\left(\frac{1+\cos\alpha}{1-\cos\alpha}\right) = \frac{1}{2}\log\left(\tan\frac{\alpha}{2}\right)^{-2} = -\log\left(\tan\frac{\alpha}{2}\right),$$

and so the claimed formula. □

It is worth noting that the equality (B.5) relates an angle to a distance, not to a ratio of distances. A similar equality cannot hold in Euclidean geometry, as the similarities (or just the homotheties) of a Euclidean plane preserve angles and modify distances. In this sense, the equality (B.5) suggests that, in hyperbolic geometry, angles and distances (shape and size) of figures are related in such a way that the latter cannot be modified without altering the former. A similar phenomenon does occur in the elliptic case, see Exercise B.7. The next example further illustrates this fact:

Example B.6.3. We will see that, in the hyperbolic case, increasing the size of a right triangle, leaving constant one of the acute angles, causes the other acute angle to decrease approaching zero. As the reader may easily check, there is a reference relative to which the absolute conic has equation $-x_0^2 + x_1^2 + x_2^2 = 0$ and such that the triangle has one vertex at $B = [1, 0, 0]$, hypotenuse $x_1 - bx_2 = 0$, $b > 0$ and the right angle at $A = [1, a, 0]$, $0 < a < 1$ (see Figure B.5). One leg is then $x_2 = 0$, and the other is $ax_0 - x_1 = 0$ (because the pole of $x_2 = 0$ is $[0, 0, 1]$). Then the third vertex will be $C = [b, ab, a]$ once we have guaranteed it to be an actual point. The actual points are the points with same sign as $[1, 0, 0]$, that is, after the choice of the equation of the absolute, the points with $-x_0^2 + x_1^2 + x_2^2 < 0$: C is thus an actual point if and only if

$$a < \frac{b}{\sqrt{1+b^2}},$$

and from now on we assume this inequality to be satisfied.

If the value α of the angle at C is computed using B.6.1, it results in the equality

$$\cos\alpha = \frac{1}{\sqrt{1-a^2}\sqrt{1+b^2}},$$

whose right-hand side is a continuous and uniformly increasing function of a for $0 < a < b/\sqrt{1+b^2}$ with limit 1 for $a \to b/\sqrt{1+b^2}$. We have seen in Example B.3.3

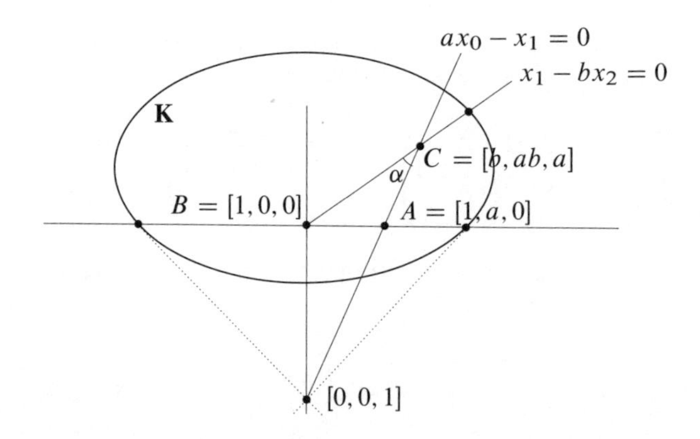

Figure B.5. The hyperbolic right triangle of Example B.6.3. The apparent measures of the pictured angles are not their hyperbolic measures.

that while a increases, so does $d(B, A)$, and we see now that at the same time the angle α uniformly decreases and has limit 0 for $a \to b/\sqrt{1+b^2}$. The reader may note that, at the limit, C becomes an ideal point and therefore the hypotenuse BC and the leg AC become asymptotically parallel.

Example B.6.4. Assume now to have a right triangle, with vertices A, B, C and right angle at A, in either of the models. In the elliptic and hyperbolic cases, each leg contains the pole of the other and therefore, as in former cases, we take A and the poles of the legs as the vertices of a self-polar projective reference and choose the unit point such that the absolute conic has equation $\varepsilon x_0^2 + x_1^2 + x_2^2 = 0$, with $\varepsilon = 1$ in the elliptic case and $\varepsilon = -1$ in the hyperbolic one. In the Euclidean case we take the projective reference associated to an orthonormal reference with origin A and vectors on the legs AB, AC. Then an equation of $\mathcal{K}$ is $u_1^2 + u_2^2 = 0$.

In all cases we have $A = [1, 0, 0]$, $B = [1, b, 0]$ and $C = [1, 0, c]$, with $|b| < 1$ and $|c| < 1$ in the hyperbolic case, and $\mathcal{K} : \varepsilon u_0^2 + u_1^2 + u_2^2 = 0$ with $\varepsilon = 0, 1, -1$ in the Euclidean, elliptic and hyperbolic cases, respectively. Forms representing the legs and the hypothenuse have components $(0, 1, 0)$, $(0, 0, 1)$ and $(-bc, c, b)$. If β and γ are the measures of the angles at B and C, by B.6.1,

$$\cos\beta = \frac{|b|}{\sqrt{\varepsilon b^2c^2 + b^2 + c^2}} \quad \text{and} \quad \cos\gamma = \frac{|c|}{\sqrt{\varepsilon b^2c^2 + b^2 + c^2}}.$$

It follows that

$$\cos(\beta + \gamma) = \frac{|bc|}{\varepsilon b^2c^2 + b^2 + c^2}\left(1 - \sqrt{\varepsilon b^2 + 1}\sqrt{\varepsilon c^2 + 1}\right).$$

In all cases $|bc|/(\varepsilon b^2c^2 + b^2 + c^2) > 0$, in the hyperbolic case due to the conditions $|b| < 1$ and $|c| < 1$, and we have:

Euclidean model: $\varepsilon = 0$ gives $\cos(\beta + \gamma) = 0$ and so $\beta + \gamma = \pi/2$.
Hyperbolic model: $\varepsilon = -1$ gives $\cos(\beta + \gamma) > 0$ and so $\beta + \gamma < \pi/2$.
Elliptic model: $\varepsilon = 1$ gives $\cos(\beta + \gamma) < 0$ and so $\beta + \gamma > \pi/2$.

Since an altitude of an arbitrary triangle may be used to decompose it in two right triangles, the sum of the acute angles of which equals the sum of the angles of the initial triangle, applying the above to both right triangles shows one of the most noteworthy differences between Euclidean, hyperbolic and elliptic geometry, namely:

Theorem B.6.5. *The sum of the angles of a triangle equals $\pi/2$ in the Euclidean case, is strictly less than $\pi/2$ in the hyperbolic case, and strictly greater than $\pi/2$ in the elliptic case.*

The trigonometric relations between the elements of a right triangle follow also from a direct computation. We deal with the hyperbolic case and leave the elliptic one to the reader:

Proposition B.6.6. *In the hyperbolic case, let A, B, C be the vertices of a right triangle with right angle at A. Write β and γ the measures of the angles at B and C, respectively, and $a = d(B, C)$, $b = d(A, C)$, $c = d(A, B)$. Then the following equalities hold:*

$$\sin\beta = \frac{\sinh b}{\sinh a}, \quad \cos\beta = \frac{\tanh c}{\tanh a},$$

$$\sin\gamma = \frac{\sinh c}{\sinh a}, \quad \cos\gamma = \frac{\tanh b}{\tanh a},$$

$$\cosh b = \frac{\cos\gamma}{\sin\beta}, \quad \cosh c = \frac{\cos\beta}{\sin\gamma},$$

$$\cosh a = \cosh b \cosh c = \cot\beta \cot\gamma.$$

Proof. Take the coordinates as in Example B.6.4, so $A = [1, 0, 0]$, $B = [1, x, 0]$, $C = [1, 0, y]$, and K has equation $-x_0^2 + x_1^2 + x_2^2$. Then, using B.3.4 and the identity $\sinh^2(z) = \cosh^2(z) - 1$, it results in

$$\cosh a = \sqrt{\frac{1}{(1 - x^2)(1 - y^2)}}, \quad \sinh a = \sqrt{\frac{x^2 + y^2 - x^2 y^2}{(1 - x^2)(1 - y^2)}},$$

$$\cosh b = \sqrt{\frac{1}{1 - y^2}}, \quad \sinh b = \sqrt{\frac{y^2}{1 - y^2}}$$

and

$$\cosh c = \sqrt{\frac{1}{1 - x^2}}, \quad \sinh c = \sqrt{\frac{x^2}{1 - x^2}}.$$

Furthermore, from former computations we have:

$$\cos\beta = \sqrt{\frac{y^2}{x^2+y^2-x^2y^2}}, \quad \sin\beta = \sqrt{\frac{x^2(1-y^2)}{x^2+y^2-x^2y^2}}$$

and

$$\cos\gamma = \sqrt{\frac{x^2}{x^2+y^2-x^2y^2}}, \quad \sin\gamma = \sqrt{\frac{y^2(1-x^2)}{x^2+y^2-x^2y^2}}.$$

Checking the claimed equalities from the former ones, using the identity $\tanh(z) = \sinh(z)/\cosh(z)$, is direct. □

The reader may compare the equalities in the first two rows to their analogues for a Euclidean right triangle. The first equality in the last row is a sort of Pythagorean theorem. If, in any of these equalities, the hyperbolic functions are substituted with their Taylor series at 0 and the terms of higher infinitesimal order are dropped, then it results in the analogous Euclidean equality. Because of this, it is sometimes said that hyperbolic trigonometry becomes Euclidean for infinitesimal triangles. The remaining three equalities have no Euclidean analogue.

B.7 Circles and similar curves

In this section we will make frequent use of the properties of the bitangent pencils (9.5.19, 9.6.3) without further reference. A conic C of either model is called a *circle* if and only if it is non-degenerate, contains points of the model and its envelope C^* is bitangent to the absolute envelope.

In the Euclidean case the latter condition is equivalent to having $C \cap \ell_\infty = I + J$ and so, once again, the situation is the usual one in Euclidean geometry (sec Section 10.1).

In the hyperbolic and elliptic cases, C^* and $\mathcal{K}$ are bitangent if and only if so are C and **K**. Thus, in these cases, the circles may be equivalently defined as the non-degenerate conics that contain points of the model and are bitangent to the absolute conic.

Still in the hyperbolic or the elliptic case, a circle C and the absolute K span a pencil of bitangent conics $\mathcal{P}_C$. The common tangents to all non-degenerate conics in $\mathcal{P}_C$ at the base points compose a line-pair which is one of the degenerate conics of $\mathcal{P}_C$. Its double point O is called the *centre* of C. The other degenerate conic of $\mathcal{P}_C$ is two times the line δ spanned by the base points. The line δ is the polar of O relative to any of the non-degenerate conics of $\mathcal{P}_C$. So, in particular, δ is the absolute polar of the centre O, and also its polar relative to C: it is called the *directrix* of C.

According to the definitions above, all non-degenerate conics of $\mathcal{P}_C$ containing proper points are circles with centre O and directrix δ; $\mathcal{P}_C$ is often referred to as a *pencil of concentric circles* or the *pencil of circles with centre O*, although not all conics of $\mathcal{P}_C$ are circles.

Similar definitions in the Euclidean case give the usual centre, and the improper line as directrix; being the same for all circles, the latter has little interest and is never mentioned in the Euclidean context.

Clearly, in no case does the centre O belong to $\mathbf{K}$, which in the hyperbolic case leaves two possibilities, namely:

(a) The centre is an actual point: then the directrix, which is its absolute polar, is an ultra-ideal line and $\mathcal{P}_C$ has imaginary base points.

(b) The centre is an ultra-ideal point: then the directrix is an actual line and $\mathcal{P}_C$ has real, necessarily ideal, base points.

In the hyperbolic case, it is usual to name circles only the circles with actual centre, the circles with ultra-ideal centre being then called *equidistant curves*, or just *equidistants*. The name is due to the fact that all points of an equidistant are at the same distance from its directrix, as we will see in B.7.1 below. We will follow this convention in the sequel: from now on, in the hyperbolic case, *circle* stands for *circle with actual centre*.

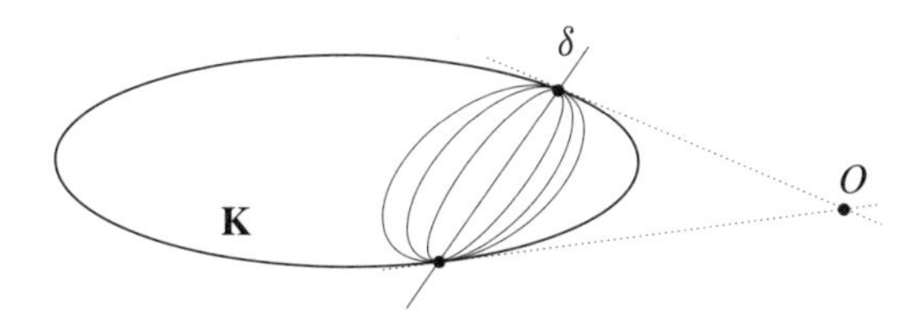

Figure B.6. Circles (equidistants if the elements of the figure are taken as real) with centre O and directrix δ.

In the hyperbolic case, there is a third type of curves that are in many senses similar to the circles and equidistants: they are the non-degenerate conics which contain actual points and span with $\mathbf{K}$ a triple contact pencil: they are called *horocycles*. Since a triple contact pencil has a single, necessarily real, base point (9.6.5), a horocycle has a single ideal point which is called the *ideal point* and also the *centre* of the horocycle. As for the circles, the absolute polar of the centre is called the *directrix* of the horocycle. If t is any tangent to $\mathbf{K}$, then $2t$ and $\mathbf{K}$ span a triple contact pencil whose non-degenerate conics containing actual points are the horocycles with directrix t and ideal point the contact point of t. Horocycles are often presented as limits of equidistants (resp. circles) whose directrices (resp. centres) become ideal.

Next we will do some computations in the non-Euclidean cases. Fix coordinates and assume that p denotes an arbitrary point of $\mathbb{P}_2$, with column coor-

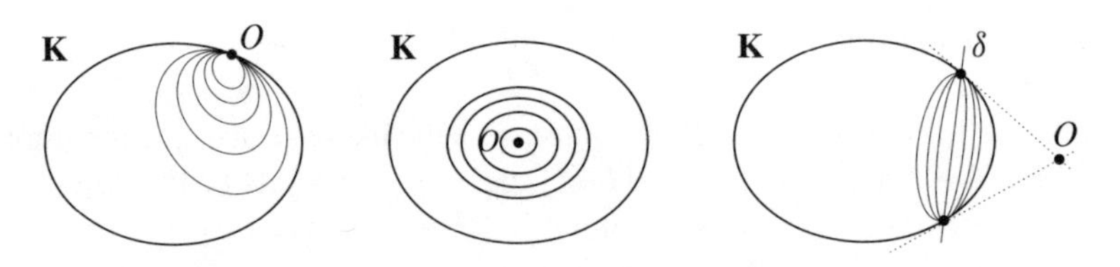

Figure B.7. From left to right, horocycles, circles and equidistants.

dinate vector $(x) = (x_0, x_1 x_2)^t$, and that M is a matrix of **K**. In the hyperbolic case, up to replacing M with $-M$, we assume that the actual points are the points p with $(x)^t M(x) > 0$. Choose a point $O \in \mathbb{P}_2$ with coordinate column vector $(a) = (a_0, a_1, a_2)^t$: its absolute polar δ has then row coordinate vector $(u) = (u_0, u_1, u_2) = (M(a))^t = (a)^t M$. The equation (in the x_i) of δ is thus

$$(u)(x) = (a)^t M(x) = 0$$

and so the conics of the pencil $\mathcal{P}$, spanned by **K** and 2δ, **K** excluded, are

$$C_\lambda : ((a)^t M(x))^2 - \lambda (x)^t M(x) = 0, \tag{B.7}$$

for $\lambda \in \mathbb{R}$.

If $O \in$ **K**, then δ is tangent to **K** and $\mathcal{P}$ is a triple contact pencil with base point O and only degenerate member $2\delta = C_0$.

Otherwise $\rho = (a)^t M(a) \neq 0$, δ is a chord of **K** and the pencil $\mathcal{P}$ is a bitangent pencil. Its degenerate members are then $C_0 = 2\delta$ and

$$C_\rho : \big((a)^t M(x)\big)^2 - (a)^t M(a) \cdot (x)^t M(x) = 0,$$

which is the tangent cone to **K** with vertex O.

Assume first to be in the elliptic case, which is easier. Then, **K** being imaginary, $O \notin$ **K**, $(a)^t M(a) \neq 0$ and $\mathcal{P}$ is a bitangent pencil. Using B.4.3, the equation (B.7) of C_λ may be rewritten

$$\lambda = \frac{((a)^t M(x))^2}{(x)^t M(x)} = (a)^t M(a) \cos^2(d(p, O)),$$

where always $(x)^t M(x) \neq 0$ because **K** is imaginary. Thus a point p belongs to C_λ if and only

$$d(p, O) = \arccos \sqrt{\frac{\lambda}{(a)^t M(a)}}.$$

Since we have seen in Example B.4.4 that the range of the elliptic distance is $[0, \pi/2]$, it follows that C_λ has points if and only if

$$0 \leq \frac{\lambda}{(a)^t M(a)} \leq 1.$$

So, the circles with centre O are the C_λ with

$$0 < \frac{\lambda}{(a)^t M(a)} < 1.$$

Assume now to be in the hyperbolic case. Fix $\lambda \neq 0$. It is clear from the equation (B.7) that for any point $p = [x_0, x_1, x_2] \in C_\lambda$, either $(x)^t M(x) = 0$, and then p belongs to **K**, or $(x)^t M(x)$ has the same sign as λ. Due to the choice of the matrix M, it follows that all the non-ideal points of C_λ are actual if $\lambda > 0$, while they all are ultra-ideal if $\lambda < 0$. Note, however, that still C_λ may have no points.

We will distinguish three cases:

Horocycles. Assume that $O \in \mathbf{K}$: then the pencil $\mathcal{P}$ is a third-order contact one with base point O. No conic in $\mathcal{P}$ is imaginary and therefore those with $\lambda > 0$ are non-degenerate and contain actual points. It follows that the horocycles with ideal point O are the conics C_λ with $\lambda > 0$.

Circles. Assume O to be actual, and hence $(a)^t M(a) > 0$. Using B.3.5, the condition for a non-ideal point p to belong to C_λ reads

$$\lambda = \frac{((a)^t M(x))^2}{(x)^t M(x)} = (a)^t M(a) \cosh^2(d(p, O)), \tag{B.8}$$

where $(x)^t M(x) \neq 0$ because the point p is assumed to be non-ideal. Hence the actual points of C_λ are those and only those whose distance to O is

$$d(p, O) = \arg\cosh \sqrt{\frac{\lambda}{(a)^t M(a)}}.$$

Since we have seen in Example B.3.3 that the range of the hyperbolic distance is $[0, \infty)$, it follows from the equation (B.8) that C_λ has actual points if and only if $\lambda \geq (a)^t M(a)$, and so the circles with centre O are the C_λ with $\lambda > (a)^t M(a)$.

Equidistants. As the only remaining possibility, assume O to be ultra-ideal: then ℓ is an actual line and $(a)^t M(a) < 0$. Using B.3.7, the condition for a non-ideal point p to belong to C_λ is now

$$\lambda = \frac{((u)(x))^2}{(x)^t M(x)} = -(u) M^{-1} (u)^t \sinh^2(d(p, \ell)), \tag{B.9}$$

where, as above, $(x)^t M(x) \neq 0$. This time the actual points of C_λ are those and only those whose distance to ℓ is

$$d(p, \ell) = \arg\sinh \sqrt{\frac{\lambda}{-(u) M^{-1} (u)^t}}.$$

By its definition, the distance to a line has the same range as the distance to a point, which is $[0,\infty)$. So, it the present case, since $-(u)M^{-1}(u)^t = -(a)^t M(a) > 0$, it follows from (B.9) that C_λ has actual points if and only if $\lambda \geq 0$ and so, the equidistants with directrix δ are the C_λ with $\lambda > 0$.

The next two propositions summarize what we have obtained above. The first one guarantees in particular the expected equidistance properties of circles and equidistants.

Proposition B.7.1. (a) *In the elliptic case, all points of any circle C lie at the same distance d from its centre O. Furthermore, any point at distance d from O belongs to C.*

(b) *In the hyperbolic case, all points of any circle C are actual and lie at the same distance d from its centre O. Furthermore, any actual point at distance d from O belongs to C.*

(c) *In the hyperbolic case, any equidistant D contains two ideal points; all its other points are actual and lie at the same distance d from its directrix δ. Furthermore, any actual point at distance d from δ belongs to D.*

(d) *In the hyperbolic case, any horocycle contains a single ideal point and all its other points are actual.*

Proposition B.7.2. *In any non-Euclidean model, after fixing a reference, assume that a point O and a line ℓ have $(a) = (a_0, a_1, a_2)^t$ and $(u) = (u_0, u_1, u_2)$ as column and row coordinate vectors, respectively. Take variables x_0, x_1, x_2 and write $(x) = (x_0, x_1, x_2)^t$. Then,*

(a) *In the elliptic model, the circles with centre O are the conics with equations*

$$((a)^t M(x))^2 - \lambda(x)^t M(x) = 0, \quad 0 < \frac{\lambda}{(a)^t M(a)} < 1.$$

(b) *In the hyperbolic model, up to replacing M with $-M$, assume that the actual points are the points $[x_0, x_1, x_2]$ for which $(x)^t M(x) > 0$. Then:*

(1) *Assuming O ideal, the horocycles with ideal point O are the conics with equations*

$$((a)^t M(x))^2 - \lambda(x)^t M(x) = 0, \quad \lambda > 0.$$

(2) *If O is actual, then the circles with centre O are the conics with equations*

$$((a)^t M(x))^2 - \lambda(x)^t M(x) = 0, \quad \lambda > (a)^t M(a).$$

(3) *If ℓ is actual, then the equidistants with directrix ℓ are the conics with equations*

$$((u)(x))^2 - \lambda(x)^t M(x) = 0, \quad \lambda > 0.$$

B.8 Transformations

By definition, the transformations of a model $(\mathbb{P}_2, \mathcal{K})$ are the projectivities of $\mathbb{P}_2$ that leave invariant the absolute envelope $\mathcal{K}$. They obviously form a subgroup of $\mathrm{PG}(\mathbb{P}_2)$ and so, in particular, a group.

In the Euclidean case, the invariance of $\mathcal{K}$ is clearly equivalent to the invariance of the pair of cyclic points; therefore the transformations of a Euclidean model are just the similarities (6.9.21).

In the hyperbolic and elliptic cases, a projectivity leaves $\mathcal{K}$ invariant if and only if it leaves invariant $\mathbf{K}$; the latter condition is usually taken as the definition of the transformations. In the sequel we will refer to the transformations of the hyperbolic (resp. elliptic) model as the *hyperbolic* (resp. *elliptic*) *transformations*.

All definitions in a model being made in terms of projective relations to the absolute, all notions introduced are kept invariant by the transformations of the model. Thus, for instance, transformations of the hyperbolic model map actual points to actual points, ideal points to ideal points and ultra-ideal points ultra-ideal points, and the same for lines. In all models, both parallelism and orthogonality are preserved, the invariance of the cross ratio assures that the angle between two lines equals the angle between their images, and the same holds for the distance between points in the hyperbolic and elliptic cases.

In Euclidean geometry two figures are called congruent if and only if they are mapped one into another by a motion, which is a transformation preserving distances. Since hyperbolic and elliptic transformations also preserve distances, in the hyperbolic and elliptic models two figures are called *congruent* if and only if a transformation of the model maps one into the other. This applies in particular to segments and angles, and so defines in both the hyperbolic and the elliptic models the two remaining Euclid's primitive terms, namely *congruence of segments* and *congruence of angles*. Being congruent is an equivalence relation just because the transformations of the model form a group. The reader may note, however, that in the Euclidean case figures mapped one into another by a transformation of the model are just similar, but non-necessarily congruent, figures (see Section 6.9).

Before giving a description of the different types of transformations of the hyperbolic and elliptic models, we will show an important example. Assume to be in the hyperbolic case and consider a harmonic homology f of $\mathbb{P}_2$ whose axis s is the absolute polar of its centre O. The homology f is then determined by the choice of O, and also by the choice of s. We will see that f is a transformation of the model. Indeed, the points of $\mathbf{K} \cap s$ are obviously invariant by f, and any other point of $\mathbf{K}$ is mapped by f to the other end of the chord it spans with O, by 6.4.23 and the condition of f being harmonic. Since $\mathbf{K}$ is determined by the set of its points, $f(\mathbf{K}) = \mathbf{K}$. Obviously O is not an ideal point, as in such a case it would belong to its absolute polar s. We have thus two possibilities, namely:

(a) **The centre of homology O is an actual point:** O is of course fixed by f and, all lines through O being invariant, for any actual $p \neq O$, p, $f(p)$ and O are aligned. Furthermore, if q_1, q_2 are the ends of the chord pO, then f maps them to each other and

$$(p, O, q_1, q_2) = (f(p), f(O), f(q_1), f(q_2)) = (f(p), O, q_2, q_1),$$

from which $d(p, O) = d(f(p), O)$. Thus, f leaves O invariant, and for any actual point $p \neq O$, p and $f(p)$ are distinct, aligned with O and at equal distance from it: f is called the *hyperbolic reflection* in O.

(b) **The centre of homology O is an ultra-ideal point:** Then the axis s is an actual line all of whose points are fixed; furthermore, for any actual p, $f(p)$ belongs to the invariant line Op, which is the perpendicular to s through p. If $q = s \cap Op$, as in the former case,

$$(p, q, q_1, q_2) = (f(p), f(q), f(q_1), f(q_2)) = (f(p), q, q_2, q_1)$$

and hence $d(p, s) = d(f(p), s)$. Thus in this case f leaves fixed all points of s and for any actual point p lying not on s, p and $f(p)$ are distinct, span a line perpendicular to s and lie at equal distance form s: f is called the *hyperbolic reflection* in s.

In the elliptic case, harmonic homologies whose axis s is the absolute polar of its centre O still are transformations of the model. The reader may easily check this using coordinates relative to a self-polar reference with one vertex O and hence the other on s. The same arguments used in the hyperbolic case show that for any point p, $p \neq 0$ and $p \notin s$, p and $f(p)$ are distinct points that lie at equal distances from O, lie also at equal distances from s, and span a line that goes through O and is perpendicular to s. In this case f has thus the properties of the two types of hyperbolic reflections: it is called the *elliptic reflection in O*, and also the *elliptic reflection in s*.

Hyperbolic transformations. Back to considering all transformations, assume first to be in the hyperbolic case. The absolute $\mathbf{K}$ is then a non-degenerate conic with points and therefore (8.2.22) mapping each hyperbolic transformation to its restriction to the absolute is an isomorphism between the group of transformations of the model and the projective group of $\mathbf{K}$: for each projectivity g of $\mathbf{K}$ there is one and only one transformation of the model whose restriction to $\mathbf{K}$ is g. We will proceed thus to describe the hyperbolic transformations in terms of their restrictions to the absolute. In the sequel, for any hyperbolic transformation f we will write $f_{|\mathbf{K}} = \hat{f}$.

Note first that if f is a non-identical hyperbolic transformation, then $\hat{f}$ is non-identical too. Therefore $\hat{f}$ has associated its cross-axis (see Section 8.2), which we will call the *axis* of f. The absolute pole of the cross-axis will be called the *centre* of f. We have:

Lemma B.8.1. *Both the axis and the centre of a non-identical hyperbolic transformation f are invariant by f.*

Proof. By 8.2.16, the axis s may be spanned by points $pf(q) \cap qf(p)$, $p, q \in \mathbf{K}$, whose images, by 8.2.13, belong to s. The invariance of the centre follows from the invariance of the axis (by 6.4.4). □

After fixing a non-identical hyperbolic transformation f, we will distinguish three cases, namely:

(a) $\hat{f}$ **has two (real) fixed points.** Then the axis s is an actual line, whose ideal points q_1, q_2 are the fixed points of $\hat{f}$ (8.2.15). Consequently, the centre is an ultra-ideal point.

The distance to s being invariant by f, the set of points of any equidistant D with directrix s is invariant by f due to B.7.1, and hence so is D itself. The restriction $f_{|D}$ obviously has q_1 and q_2 as fixed points. By projecting the ideal points from q_1 and taking section by D it follows that $f_{|D}$ is not the identity and has the same modulus as $\hat{f}$. Let us distinguish two subcases:

(a.1) $\hat{f}$ **has positive modulus.** Then f is called a *hyperbolic translation* with axis s. For each equidistant D with directrix s, the set of actual points of D splits in the two segments of D with ends q_1, q_2 (2.10.4). By 5.5.10 and the above, f leaves invariant each of these segments. Assume that p is an actual point and take $p' = pO \cap s$; then $p'' = f(p)O \cap s = f(p')$ and so p' and $f(p')$ are the orthogonal projections of p and $f(p)$, respectively, on s. By 5.5.7, the cross ratio $(p', f(p'), q_1, q_2)$ is the same for all $p' \in s$ and so $d(p', p''))$ is independent of the choice of p. Summarizing, a hyperbolic translation f leaves invariant its axis

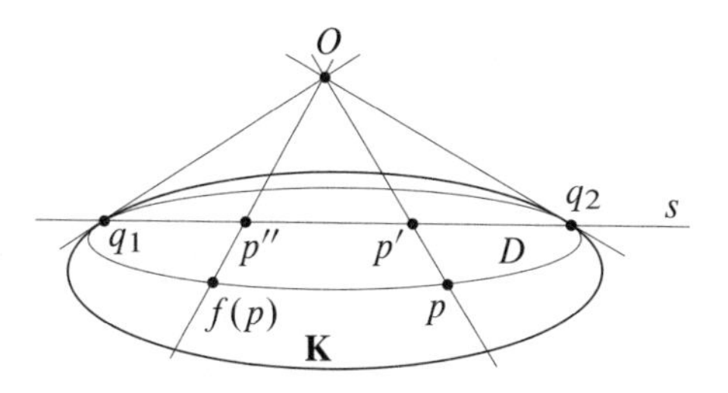

Figure B.8. Hyperbolic translation f with axis s and centre O. D is invariant and $d(p', p'')$ is independent of p.

s and the segments with ideal ends of all equidistants with directrix s; furthermore, the distance between the orthogonal projections on s of any actual p and $f(p)$ is the same for all p. The reader may compare with the Euclidean case, in which parallel lines take the role of the segments of the equidistants.

(a.2) $\hat{f}$ **has negative modulus.** Then f is called a *hyperbolic glide*. Arguing as above, if f is a hyperbolic glide, then, for any equidistant D with directrix s, the segments of D with ends q_1, q_2 are mapped to each other by f. The hyperbolic reflection in s appears as a particular case of hyperbolic glide; its restriction to **K** has modulus -1 and therefore is the involution with fixed points q_1, q_2. Assume that g is the hyperbolic reflection in s and f an arbitrary hyperbolic glide with axis s, $f \neq g$. Then $\hat{f} \circ \hat{g} \neq \mathrm{Id}$, leaves q_1, q_2 fixed and has positive modulus. Therefore $h = f \circ g$, which has $\hat{h} = \hat{f} \circ \hat{g}$, is a hyperbolic translation with axis s. Since then $f = h \circ g$, it follows that any hyperbolic glide with axis s, other than the hyperbolic reflection in s, appears as the composition of the latter and a hyperbolic translation with axis s, just as for Euclidean glides.

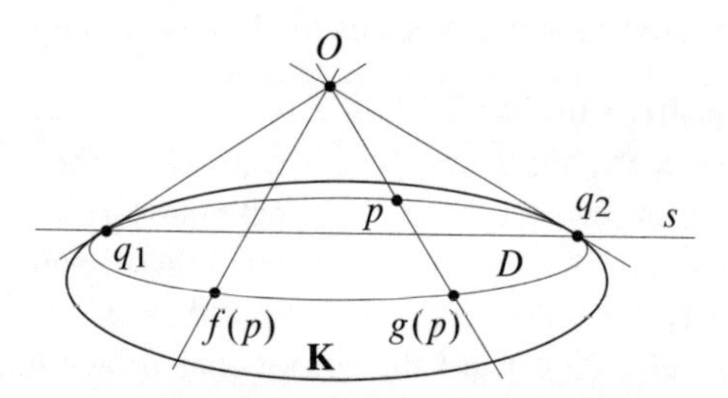

Figure B.9. Hyperbolic glide f with axis s and centre O, as the composition of the hyperbolic reflection g in s and a hyperbolic translation with axis s.

(b) $\hat{f}$ **has no (real) fixed points.** Then f is called a *hyperbolic rotation* with centre O. The centre O is an actual point, while the axis s is an ultra-ideal line. The hyperbolic reflection in a point O appears as a particular case of hyperbolic rotation with centre O, because its restriction to **K** is the involution of **K** with centre O. Again due to the preservation of distances and B.7.1, any hyperbolic rotation with centre O leaves invariant any circle with centre O.

If $h = (f^{\vee})^{-1}_{|O^*}$ is the projectivity of O^* mapping each line ℓ to $f(\ell)$, then h is the identical map if and only f is the hyperbolic reflection in O. Otherwise h has the (imaginary) tangents ℓ_1, ℓ_2, from O to **K**, as fixed lines, because their contact points are the fixed points of $\hat{f}$. By 5.5.16, the cross ratio

$(\ell, f(\ell), \ell_1, \ell_2)$ is the same for all $\ell \in O^*$ or, equivalently, by the definition of angle between lines, the angle $\widehat{\ell f(\ell)}$ is independent of ℓ. All together, if f is a hyperbolic rotation with centre O and p is any actual point other than O, then p and $f(p)$ belong to the same circle with centre O and the angle $\widehat{Op\, Of(p)}$ is the same for all p.

(c) $\hat{f}$ **has a single fixed point.** These transformations are called *parallel displacements* and have no analogue in Euclidean geometry. Both its centre O and its axis s are ideal, s being the tangent to $\mathbf{K}$ at O. We will check next, using coordinates, that the centre O of an arbitrary parallel displacement f is its only (real or imaginary) fixed point, its axis being then its only fixed line, by 5.7.9. We will see also that f leaves invariant any horocycle with ideal point O. By 5.5.14, there is a reference Ω of $\mathbf{K}$, with first vertex O, relative to which $\hat{f}$ has matrix

$$\begin{pmatrix} 1 & 1 \\ 0 & 1 \end{pmatrix}.$$

We take in $\mathbb{P}_2$ the reference Δ associated to Ω; $\mathbf{K}$ has equation $x_1^2 - x_0x_2 = 0$ relative to it, and we may take the matrix (8.6), appearing in the proof of 8.2.22, as a matrix of f; in the present case it is

$$\begin{pmatrix} 1 & 2 & 1 \\ 0 & 1 & 1 \\ 0 & 0 & 1 \end{pmatrix}.$$

Then it is direct to check that O is the only fixed point of f. The tangent to $\mathbf{K}$ at O being $s : x_2 = 0$, the conics of the triple contact pencil spanned by $\mathbf{K}$ and $2s$, other than $\mathbf{K}$, have equations

$$x_2^2 + \lambda(x_0x_2 - x_1^2) = 0, \quad \lambda \in \mathbb{R}.$$

Again a direct checking, using the above matrix, shows that any of these conics, and in particular any horocycle with ideal point O, is invariant by f.

Elliptic transformations. All non-identical elliptic transformations are called elliptic rotations. The next proposition describes their fixed points:

Proposition B.8.2. *An elliptic rotation f has a single isolated fixed point O, which is called its* centre. *The absolute polar of O, called the* axis *of f, is an invariant line which either:*

(a) *contains no fixed point, in which case the centre O is the only fixed point of f, or*

(b) *is a line of fixed points, in which case f is the elliptic reflection in O.*

Proof. Since the characteristic polynomial of a matrix of f has degree three, it has at least one real root and therefore f has at least one fixed point p; the absolute polar ℓ of p is fixed too. We will assume first that $f_{|\ell}$ has no fixed points. Then f has no fixed point other than p, as otherwise the absolute polar ℓ' of a fixed point $p' \neq p$ would be fixed causing $\ell \cap \ell'$ to be a fixed point. Thus, in this case, $O = p$ is the only isolated fixed point and the situation is as described in (a).

Assume now that there is a fixed point $p' \in \ell$. Then its absolute polar ℓ' is fixed and $p'' = \ell \cap \ell'$ is fixed too. Since **K** has no point, it is direct to check that p, p', p'' is a self-polar triangle. We complete it to a reference of $\mathbb{P}_2$ relative to which **K** has equation $x_0^2 + x_1^2 + x_2^2 = 0$ (7.1.8). Then any matrix of f is diagonal, say

$$\begin{pmatrix} \lambda_0 & 0 & 0 \\ 0 & \lambda_1 & 0 \\ 0 & 0 & \lambda_2 \end{pmatrix},$$

and the conditions for leaving **K** invariant read $\lambda_0^2 = \lambda_1^2 = \lambda_2^2$. Since $f \neq \mathrm{Id}_{\mathbb{P}_2}$, two eigenvalues are equal and the other is their opposite. The former give rise to a line of fixed points and the latter to an isolated fixed point: there is thus a single isolated fixed point O, its polar is a line of fixed points and, clearly, f is the elliptic reflection in O, as claimed in (b). □

The same arguments used in the hyperbolic case allow us to prove that if f is an elliptic rotation, with centre O and axis s, then all circles with centre O are invariant by f and, for $p \neq O$, the angle of the lines $Op, Of(p)$, as well as the distance between the orthogonal projections of p and $f(p)$ on s, are independent of the choice of p.

To close we will make some heuristic comments about an essential difference that occurs between the Euclidean and the non-Euclidean geometries. Consider a parametric model $(\mathbb{P}_2, \mathcal{K}_\varepsilon)$, $\mathcal{K}_\varepsilon : \varepsilon u_0^2 + u_1^2 + u_2 = 0$, $\varepsilon \in [-1, 1]$: it is as a family of models that are hyperbolic for $\varepsilon < 0$, Euclidean for $\varepsilon = 0$ and elliptic for $\varepsilon > 0$; in it the Euclidean model appears as a degenerate case ($\mathcal{K}_\varepsilon$ is degenerate) making the transition between the hyperbolic and elliptic models. The fact is that the data of the Euclidean model – $\mathbb{P}_2$ plus the cyclic points – are weaker than those of the hyperbolic and elliptic models – $\mathbb{P}_2$ plus a non-degenerate conic. This results in a larger group of transformations and therefore a less rigid structure for the Euclidean geometry, if compared to the non-Euclidean geometries. The higher rigidity of the non-Euclidean geometries is shown by the existence of formulas relating distances to angles, such as those in B.6.2 and B.6.6, and, notably, by the possibility of intrinsically defining a non-trivial distance which, therefore, is invariant by the transformations of the geometry (*absolute distance*), as done in Sections B.3 and B.4. (See also Exercise B.13.) Of course there is no non-trivial absolute distance in Euclidean geometry. Indeed, an invariant distance d has $d(p, p') = d(q, q')$ for arbitrary pairs of actual points p, p' and q, q', because there is always similarity mapping with p to q and p' to q'. By contrast, the reader may note that there is an absolute measure of angles in each of the three geometries.

Having no absolute distance in our supposedly Euclidean physical ambient space is the reason why we need to base the definition of the unit of distance on a physical object (the Earth, or a platinum bar, or certain radiation). The mathematicians J. B. J. Delambre and P. F. A. Méchain, designated by the French Academy of Sciences, measured the arch of meridian from Dunkerque to Barcelona during the years 1792–1798, in order to set the definition of metre. Would our universe be (measurably) non-Euclidean, they could have avoided a lot of trouble (see [1] for the history of the meridian mission). Indeed, in a non-Euclidean geometry, an intrinsic unit of length is set by its own definition of distance. Other unities may be defined by introducing a constant factor in the equalities (B.1) and (B.2) of Section B.3. Equivalently, one may proceed in an indirect way, by taking as unit of length a distance intrinsically related to some already fixed angle; for instance, in the hyperbolic case, one may take as unit of length the distance from a point to a line giving an already chosen angle of parallelism, see B.6.2 and the comment after it.

B.9 Exercises

Exercise B.1. Prove that the actual lines of the hyperbolic model are the lines that, taken as points of $\mathbb{P}_2^\vee$, are exterior to $\mathcal{K}$.

Exercise B.2. Conventions and notations being as at the end of Section B.3, take in A_3 an orthonormal reference with origin the centre O of S and vectors e_1, e_2, e_3, and take in $\mathbb{P}_2$ the projective reference that has adapted basis e_1, e_2, e_3.

(1) Relate the coordinates of p and $\tau(p)$, for $p \in S$.

(2) Compute the equation of the plane of $\mathbb{A}_3$ containing $\tau^{-1}(\ell)$ from the equation of an arbitrary line ℓ of $\mathbb{P}_2$.

(3) Prove that the angle between two lines ℓ_1, ℓ_2 of $\mathbb{P}_2$ equals the angle between the great circles $\tau^{-1}(\ell_1)$ and $\tau^{-1}(\ell_2)$.

Exercise B.3. In the hyperbolic case, assume that p is an actual point and ℓ a line through p. Take coordinates as in Example B.3.3 and p', p'' to be the points $p' = [1, x, 0]$ and $p'' = [1, y, 0]$.

(1) Prove that $p = [1, 0, 0]$ lies between p' and p'' if and only if $xy < 0$.

(2) Assume that $x < y$ and prove that the segment with ends p', p'' is $\{[1, z, 0] \mid x \leq z \leq y\}$.

(3) Prove that taking $p', p'' \in \ell$, $p', p'' \neq p$, as related if and only if p does not lie between p', p'', defines an equivalence relation on the set of the actual points of ℓ other than p, which has two equivalence classes h'_1, h'_2.

(4) The sets $h_i = h'_i \cup \{p\}$ are called the *half-lines* (or *rays*) on ℓ with end p. Prove that, one is $h_i = \{[1, z, 0] \mid 0 \leq z < 1\}$ and the other $h_j = \{[1, z, 0] \mid -1 < z \leq 0\}$, $\{i, j\} = \{1, 2\}$. It follows that $h_1 \cap h_2 = \{p\}$, while $h_1 \cup h_2$ is the set of all the actual points of ℓ.

Exercise B.4. Use B.4.5 to give formulas for the elliptic distance to a line similar to those of B.3.6 and B.3.7.

Exercise B.5. The *length* of a segment being defined as the distance between its ends, prove that, in the hyperbolic model, two segments are congruent if and only if they have the same length.

Exercise B.6. Use Example B.3.3 and Exercise B.5 to prove that Euclid's second postulate is satisfied in the hyperbolic model.

Exercise B.7. Pose an example similar to B.6.3 in the elliptic model.

Exercise B.8. Prove that in both the elliptic and the hyperbolic models all lines through the centre of a circle (still called diameters) are chords with real ends, and that the lines tangent to a circle at the ends of any diameter are perpendicular to the diameter.

Exercise B.9. Show that:

(1) The composition of two hyperbolic reflections in two different non-parallel lines is a hyperbolic rotation, and any hyperbolic rotation may be obtained in this way.

(2) The composition of two hyperbolic reflections in two ultraparallel lines is a hyperbolic translation, and any hyperbolic translation may be obtained in this way.

(3) The composition of two hyperbolic reflections in two asymptotically parallel lines is a parallel displacement, and any parallel displacement may be obtained in this way.

Hint: Deal with just the restrictions to the absolute conic.

Exercise B.10. Prove that if f is a hyperbolic glide with directrix s, then the distance between the orthogonal projections of p and $f(p)$ on s is the same for all actual points p.

Exercise B.11. Let H, H' be two horocycles with the same ideal point O. Prove that the distance between the actual points of H and H' lying on a variable actual line through O, is constant. *Hint*: Use a parallel displacement with centre O.

Exercise B.12 (*Classification of hyperbolic transformations*). Two hyperbolic transformations f, f' of the hyperbolic model are called congruent if and only if there is a third hyperbolic transformation g of the model such that $f' = g \circ f \circ g^{-1}$; check that the congruence of hyperbolic transformations is an equivalence relation. Using 8.2.22 and the results of Section 5.5, list and describe all congruence classes of hyperbolic transformations. Give geometric interpretations of the invariants used.

Exercise B.13 (*Invariance of circles*). Let f be a transformation of the elliptic or the hyperbolic model leaving fixed a point O of the model. If q_1, q_2 are the intersection points of the absolute and the absolute polar of O, prove that the conic envelopes $\mathcal{K}$, $2O^*$ and $q_1^* + q_2^*$ are distinct and invariant by f. Prove also that they belong to the same pencil $\mathcal{P}$. Prove that $\mathcal{P}$ contains the envelopes of all the circles with centre O. Prove that all conic envelopes in $\mathcal{P}$ are invariant by f and so, in particular, all circles with centre O are invariant by f. Find the gap of a similar argument in the Euclidean model, the last claim being obviously non-true there.

Bibliography

[1] K. Alder, *The measure of all things: the seven-year odyssey and hidden error that transformed the world*. The Free Press, New York, 2002. 599

[2] E. Bertini, *Introduzione alla geometria proiettiva degli iperspazi*. Giuseppe Principato, Messina, 1923. xiv, 183

[3] E. Casas-Alvero, *Singularities of plane curves*. London Mathematical Society Lecture Note Series 276, Cambridge University Press, Cambridge, 2000. 388

[4] G. Castelnuovo, *Lezioni di geometria analitica*. S. E. Dante Alighieri, Roma, 1924. xiv

[5] F. Conforto, *Le superficie razionali*. N. Zanichelli, Bologna, 1945. 334

[6] G. Fano, *Lezioni di geometria descrittiva*. G. B. Paravia, Torino, 1932. 553

[7] O. Faugeras, *Three-dimensional computer vision*. The MIT Press, Cambridge, Mass., and London, England, 1993. 553

[8] O. Faugeras, Q. T. Luong, and T. H. Papadopoulo, *The geometry of multiple images*. The MIT Press, Cambridge, Mass., and London, England, 2001. 553

[9] J. V. Field and J. J. Gray, *The geometrical work of Girard Desargues*. Springer-Verlag, New York, Berlin, Heidelberg, 1987. 390

[10] P. A. Fuhrmann, *A polynomial approach to linear algebra*. Universitext, Springer-Verlag, New York, Berlin, Heidelberg, 1996. 550

[11] F. R. Gantmacher, *The theory of matrices*. AMS Chelsea Publishing, Providence, Rhode Island, 1959. 493, 520

[12] J. Gray, *Worlds out of nothing*. Springer Undergraduate Mathematics Series, Springer-Verlag, New York, Heidelberg, Berlin, 2007. 107

[13] M. J. Greenberg, *Euclidean and non-Euclidean geometries*. W. H. Freeman and Company, New York, 1993. 568, 569, 572, 582

[14] W. H. Greub, *Linear algebra*. Die Grundlehren der Mathematischen Wissenschaften 97, Springer-Verlag, Berlin, Göttingen, Heidelberg, 1963. 267

[15] J. Harris, *Algebraic geometry*. Graduate Texts in Mathematics 133, Springer-Verlag, New York, 1992. 339

[16] R. Hartley and A. Zisserman, *Multiple view geometry in computer vision*. Cambridge University Press, Cambridge, 2000. 255, 553

[17] D. Hilbert, *The foundations of geometry*. 10th English edition of the German 2nd edition, The Open Court Publishing Co., La Salle, Ill., 1971. 569

[18] V. W. D. Hodge and D. Pedoe, *Methods of algebraic geometry*. Cambridge University Press, Cambridge, 1994. 339, 493, 519, 520, 546

[19] G. K. Francis, *A topological picturebook*. Springer-Verlag, New York, 1987. 553

[20] F. Klein, *Elementary mathematics from an advanced standpoint: geometry*. Dover, New York, 1939. 94

[21] F. Klein, *Le programme d'Erlangen*. Gauthier-Villars, Paris, 1974. 94

[22] D. Mumford, *Algebraic geometry I. Complex projective varieties*. Grundlehren der Mathematischen Wissenschaften 221, Springer-Verlag, Berlin, Heidelberg, New York, 1970. 412

[23] O. Schreier and E. Sperner, *Modern algebra and matrix theory*. Chelsea Publishing Company, New York, 1959. 16, 493, 519

[24] O. Schreier and E. Sperner, *Projective geometry of n dimensions*. Chelsea Publishing Company, New York, 1961. xiv

[25] J. G. Semple and G. T. Kneebone, *Algebraic projective geometry*. Oxford Classic Texts in the Physical Sciences, Oxford University Press, Oxford, 1963. xiv, 388

[26] O. Veblen and J. W. Young, *Projective geometry*. Ginn and Co., Boston, Mass., 1910. 52

[27] R. Vicat Cole, *Perspective for artists*. Dover Publications Inc., New York, 1976. 553, 556

[28] J. T. Watts and E. F. Rule, *Descriptive geometry*. Prentice-Hall Inc., New York, 1946. 553

Symbols

$Q_{\mathbb{C}}$	complex extension of the quadric Q, 202
$\lvert Q \rvert$	set of points of the quadric Q, 195
$\mathbb{Q}(\mathbb{P}_n)$	space of quadrics of $\mathbb{P}_n$, 365
$\mathbb{Q}(f)$	projectivity induced on quadrics, 366
(q_1, q_2, q_3, q_4)	cross ratio of q_1, q_2, q_3, q_4, 57
(q_1, q_2, q_3)	affine ratio of q_1, q_2, q_3, 74
$\mathbb{R}$	the real field xv
$\mathbb{R}^+$	the set of positive real numbers, xv
$\mathbf{r}(Q)$	rank of the quadric Q, 212
$\mathbf{r}'(Q)$	rank of the improper section of the quadric Q, 284
$S^2(E)$	space of symmetric bilinear forms, 193
$\mathcal{S}(E)$	set of subspaces, 8
$\mathcal{S}_d(E)$	set of subspaces of dimension d, 7
$\mathrm{Sing}(f)$	singular variety of f, 183
$[v]$	point represented by v, 5
$[x_0, \dots, x_n]$	point with homogeneous coordinates $x_0, \dots, x_n$, 37

Index